AF671501

Lecture Notes in Civil Engineering

Volume 754

Lecture Notes in Civil Engineering (LNCE) publishes the latest developments in Civil Engineering—quickly, informally and in top quality. Though original research reported in proceedings and post-proceedings represents the core of LNCE, edited volumes of exceptionally high quality and interest may also be considered for publication. Volumes published in LNCE embrace all aspects and subfields of, as well as new challenges in, Civil Engineering. Topics in the series include:

- Construction and Structural Mechanics
- Building Materials
- Concrete, Steel and Timber Structures
- Geotechnical Engineering
- Earthquake Engineering
- Coastal Engineering
- Ocean and Offshore Engineering; Ships and Floating Structures
- Hydraulics, Hydrology and Water Resources Engineering
- Environmental Engineering and Sustainability
- Structural Health and Monitoring
- Surveying and Geographical Information Systems
- Indoor Environments
- Transportation and Traffic
- Risk Analysis
- Safety and Security

To submit a proposal or request further information, please contact the appropriate Springer Editor:

– Pierpaolo Riva at pierpaolo.riva@springer.com (Europe and Americas);
– Swati Meherishi at swati.meherishi@springer.com (Asia—except China, Australia, and New Zealand);
– Wayne Hu at wayne.hu@springer.com (China).

All books in the series now indexed by Scopus and EI Compendex database!

Baojie He · Tian Chen · Yangli Li ·
Ali Cheshmehzangi · Peng Zeng · Guofang Zhai
Editors

Climate-Resilient, Low-Carbon, and Sustainable Cities

Editors
See next page

ISSN 2366-2557 ISSN 2366-2565 (electronic)
Lecture Notes in Civil Engineering
ISBN 978-981-95-2168-5 ISBN 978-981-95-2169-2 (eBook)
https://doi.org/10.1007/978-981-95-2169-2

This work was supported by National Natural Science Foundation of China.

This Springer imprint is published by the registered company Springer Nature Singapore Pte Ltd.
The registered company address is: 152 Beach Road, #21-01/04 Gateway East, Singapore 189721, Singapore

Editors
Baojie He
School of Architecture and Urban Planning
Chongqing University
Chongqing, China

Yangli Li
School of Civil Engineering
and Architecture
Southwest University of Science
and Technology
Mianyang, China

Peng Zeng
School of Architecture
Tianjin University
Tianjin, China

Tian Chen
School of Architecture
Tianjin University
Tianjin, China

Ali Cheshmehzangi
School of Architecture, Design
and Planning
University of Queensland
Brisbane, QLD, Australia

Guofang Zhai
School of Architecture and Urban Planning
Nanjing University
Nanjing, Jiangsu, China

Baojie He dedicates this book to all his collaborators and colleagues at the Centre for Climate-Resilient and Low-Carbon Cities. This book is also dedicated to all invited speakers, moderators, and participants of the International Conference of Urban Climate and Urban Design.

Tian Chen dedicates this book to the collaborators and researchers in the field of urban design, and a special dedication to all Ph.D. and Masters graduated from his research team.

Yangli Li dedicates this book to the researchers in the field of urban resilience and urban design, and to the partners and friends who have made tremendous efforts for the organization of this conference and the publication of the proceedings.

Ali Cheshmehzangi dedicates this book to researchers around the world who are passionate about making a genuine and impactful difference and contribute effectively and positively to our fields of architecture, design, and planning/urbanism.

Peng Zeng dedicates this book to the researchers in the field of urban and rural planning, and to colleagues working on School of Architecture, Tianjin University and Engineering Technology Innovation Center for Land and Sea Integrated Key Zone Territorial Space Planning and Governance, Ministry of Natural Resources.

Guofang Zhai dedicates this book to researchers in the field of urban resilience, and to colleagues working in the Research Center for Urban Safety Development, Nanjing University.

Scientific Organizing Committee

Conference Chair

Tian Chen, Professor, Ph.D., Tianjin University, China

Conference Consultant

Kang Jian, Professor, Academician, University College London, UK

Academic Chairs

Bao-Jie He, Professor, Ph.D., Chongqing University, China
Yangli Li, Lecturer, Ph.D., Southwest University of Science and Technology, China

Scientific Organizing Committee Members

Kang Jian, Professor, Academician, University College London, UK
Ayyoob Sharifi, Professor, Ph.D., Hiroshima University, Japan
Ali Cheshmehzangi, Professor, Ph.D., The University of Queensland, Australia
Jian Zeng, Professor, Ph.D., Tianjin University, China
Guofang Zhai, Professor, Ph.D., Nanjing University, China
Peng Zeng, Professor, Ph.D., Tianjin University, China
Bin Jia, Professor, Ph.D., Southwest University of Science and Technology, China
Shifu Wang, Professor, Ph.D., South China University of Technology, China
Quanhua Hou, Professor, Ph.D., Chang'an University, China

Jian Xu, Professor, Ph.D., Yunnan University, China
Wanmin Zhao, Professor, Ph.D., Chongqing University, China
Wei Zhao, Professor, Ph.D., Sichuan University, China
Yunyan Li, Assistant Professor, Ph.D., Chongqing University, China
Yupeng Wang, Professor, Ph.D., Xi'an Jiaotong University, China
Jiangbo Wang, Professor, Ph.D., Nanjing Tech University, China
Guiping Ye, Professor, Ph.D., City University of Macau, China
Attia Shady, Professor, Ph.D., University of Liège, Belgium
Samad Sepasgozar, Associate Professor, Ph.D., University of New South Wales, Australia
Amirhosein Ghaffarianhoseini, Associate Professor, Ph.D., Auckland University of Technology, New Zealand
Waqas Ahmed Mahar, Ph.D., BUITEMS, Pakistan
Joni Jupesta, Ph.D., University of United Nations, Japan
Siliang Yang, Ph.D., Leeds Beckett University, UK
Beta Paramita, Ph.D., Universitas Pendidikan Indonesia, Indonesia
Jinda Qi, Ph.D., National University of Singapore, Singapore
Jack Ngarambe, Ph.D., Kyung Hee University, South Korea
Rongrong Cheacharoen, Ph.D., Chulalongkorn University, Thailand
Zhikang Bao, Ph.D., Heriot-Watt University, UK
Yijun Shi, Assistant Professor, Ph.D., Zhejiang A&F University, Hangzhou, China
Meng Meng, Assistant Professor, Ph.D., South China University of Technology, Guangzhou, China

Local Organizing Committee Members

Mingying Zeng, Professor, Ph.D., Southwest University of Science and Technology, China
Zhaoqiang Zhang, Professor, Ph.D., Southwest University of Science and Technology, China
Yan Li, Lecturer, Ph.D., Southwest University of Science and Technology, China
Yisha Liu, Lecturer, Southwest University of Science and Technology, China
Xiaowei Shang, Lecturer, Southwest University of Science and Technology, China
Yu Wang, Assistant Professor, Ph.D., Southwest University of Science and Technology, China

Student Organizing Committee Members

Mingqiang Yin, Chongqing University, China
Lu Che, Southwest University of Science and Technology, China
Erli Zeng, University of Tokyo, Japan
Xingli Wang, Southwest University of Science and Technology, China

Preface I

The entrenched urban–rural dual structure in traditional urbanization has solidified the developmental divide between cities and rural areas, with imbalanced development trends becoming increasingly pronounced. This has resulted in frequent disasters, severe energy consumption, intensified environmental pollution, and other pressing issues. From the perspectives of "resilience, low-carbon, and sustainability," this conference focuses on two primary directions—urban development and the construction of beautiful countryside—to conduct in-depth discussions on how to promote high-quality urban growth and advance the development of picturesque rural areas through integrated urban–rural planning and design.

Based on the Mianyang Science and Technology City (MSTC) Talent Summit, we launched the 2024 Conference on High-Quality Urban Development and Beautiful Countryside Construction at Mianyang, Sichuan, China. It is of great significance for the construction of low-carbon cities and sustainable urban development in China to have globally renowned scholars in the field of urban planning and design contribute their ideas. The MSTC is the only national science and technology city approved for construction by the State Council and the State Council in September 2000. It is an important region for the comprehensive innovation and experiments of the national system, and it shoulders the important mission of exploring the path and accumulating experience for the construction of new urbanization.

We invited academicians, senior experts, and young scholars from universities and research institutions to deliver speeches and talks, exchange the latest research results, and discuss how to alleviate the challenges and problems caused by local and global climate change through urban planning and design. The conference is expected to generate suggestions and strategies for climate change mitigation and adaptation, high-quality urban development, rural revitalization, and promote the development of disciplines related to urban climate and urban design.

Locally, by holding an internationally influential conference, we aim to make high-level talents from around the world know Mianyang, seek paths for the high-quality development of the city, and create excellent urban and rural spatial quality. The efficient and convenient collaboration among academia, government, and enterprise provides a sound environmental foundation for attracting talents, while the excellent urban and rural spatial quality provides a fundamental environment. By actively organizing such a high-level academic conference, we provide talents with a livable, business-friendly, innovative, enjoyable, and sustainable urban and rural environment, which is an important tool for retaining talents and promoting the innovative and sustainable development of Mianyang city.

Mianyang, China Yangli Li

Preface II

Human beings are under the risks and threats of many challenges. Climate change, associated with greenhouse (GHG) emissions and anthropogenic activities, is the most severe one with significantly environmental, economic, and social consequences. Climate-related disasters and impacts are non-linear, where the impacts are unbearable once the warming level exceeds 1.5 °C. However, we are standing on a warming level of 1.1–1.2 °C which is very close to the critical threshold by 2030–2035. There is an urgent to address climate change before our human beings are locked into the irreversible circumstances.

Even though, many other challenges such as urbanization, environmental deterioration, economic growth, future development, and pandemic are also on, bringing prominent threats to human health, safety, comfort, and well-being. Many of those challenges including climate change can be interlinked. For instance, climate-related disasters such as extreme heat and intense precipitation evolve into urban overheating and urban flooding. It is the unsustainable urban planning and design that extensively integrate hard surfaces into cities and reduce climate-proofing capacity. The release of air pollutants during urban production and operation causes airborne pollution, while the increase of building height and density, the reduction of urban greening, and replacement of open spaces reduces the capacity of pollutant dispersion.

Given this, the 2024 Conference on High-Quality Urban Development and Beautiful Countryside Construction was held in November, 2024, Mianyang, China, with the attendance of experts and researchers from academia and industries in the field of urban planning and design, architecture, atmospheric environment, ecology, geography, and economy to share the ideas, thoughts, technologies, and practices on how to address climate change and enhance sustainable development. The conference covers eight themes, including

- Sustainable Development of Urban and Rural Areas
- Low-Carbon and Healthy Cities
- Sustainable Urban and Rural Food System
- Regional Ecological Safety and Urban-Rural Resilient Development
- Resilience, Safety, and Development

- Planning Techniques for Carbon Neutrality
- Urban Renewal for Urban Vitality and Public Health
- Construction of Resilience Network

A total of 174 submissions were submitted to the conference, but we had only accepted 101 papers with new findings and outstanding contributions in conference proceeding books based on critical peer review. This volume of *Climate-Resilient, Low-Carbon and Sustainable Cities* includes five parts:

- Climate Change and Impacts
- Urban Climate Change and Citizens' Wellbeing
- Decarbonization of the Built Environment
- Climate-Responsive and Sustainable Planning
- Renewal for Space Quality Improvement

Overall, this edited conference proceeding is expected to enable more experts and researchers to understand the methods and theories on the climate-induced impact and assessment, climate change mitigation and adaptation, urban environmental sustainability improvement, sustainable urban–rural planning and design, and urban resilience to emerging challenges.

Chongqing, China — Baojie He
Tianjin, China — Tian Chen
Mianyang, China — Yangli Li
Brisbane, Australia — Ali Cheshmehzangi
Tianjin, China — Peng Zeng
Nanjing, China — Guofang Zhai

Acknowledgements

This collection of papers was able to be published successfully thanks to the support and assistance of many organizations and individuals.

Yangli Li and Bao-Jie He appreciate the sponsorship of the Organization Department of Mianyang Government, Southwest University of Science and Technology, for this International Conference on High-Quality Urban Development and Beautiful Countryside Construction. In addition, Bao-Jie He appreciates the National Natural Science Foundation of China (No. W2422003).

We appreciate the support from the School of Civil Engineering and Architecture at Southwest University of Science and Technology, School of Architecture and Urban Planning at Chongqing University, School of Architecture at South China University of Technology, and School of Architecture at Tianjin University for their assistance in organizing the conference.

We appreciate the support from the School of Architecture and Urban Planning at Chongqing University, School of Architecture at Tianjin University, China Association for Disaster Prevention, and Chinese Society for Urban Studies, and Mianyang Land and Space Planning Association for their strong support for conference propaganda and promotion.

We appreciate Prof. Tian Chen at Tianjin University, Prof. Bao-Jie He at Chongqing University, and Dr. Yangli Li at Southwest University of Science and Technology for chairing the conference. Many thanks go to keynote and invited speakers, international scientific organizing committee members, local organizing committee members, and reviewers all over the world.

Special thanks go to Dr. Li Yan, Ms. Lu Che, Ms. Erli Zeng, Mr. Mingqiang Yin, Ms. Yisha Liu, Ms. Xingli Wang, and Mr. Xiaowei Shang for their tremendous contributions to funding application, conference organization, paper collection, and feedback collection.

Keynote and Invited Speakers

Keynote Speakers

Zhiqiang Wu, Professor, Academician
School of Architecture and Urban Planning, Tongji University, China
Frontiers of Urban and Rural Planning Discipline

Kang Jian, Professor, Academician
The Bartlett School of Architecture, University College London, UK
Low-Carbon Urban and Rural Development and Architectural Design

Invited Speakers

Wanmin Zhao, Professor
School of Architecture and Urban Planning, Chongqing University, China
Exploration and Practice of Scientific Theory on Human Settlements in Mountainous Areas

Quanhua Hou, Professor
School of Architecture, Chang'an University, China
Research on the Mechanism of Water and Carbon Cycle and the Management of the Three-Life Space in the Loess Plateau Basin Under the Background of Warming and Humidification

Yunyan Li, Associate Professor
School of Architecture and Urban Planning, Chongqing University, China
Research on Flood Resilience in Waterfront Areas from the Perspective of High-Quality Development

Ali Cheshmehzangi, Professor
School of Architecture, Design and Planning, University of Queensland, Australia
Time to Make Climate-Resilience Central to High-Quality Urbanization and Sustainable Urbanism Agenda

Yupeng Wang, Professor
School of Human Settlements and Civil Engineering, Xi'an Jiaotong University, China
Research on Parametric Design Methods for Sustainable Cities

Jian Xu, Professor
School of Architecture and Planning, Yunnan University, China
The Vulnerability, Adaptability and Resilience of the Traditional Human Settlement Environment in Plateau and Mountainous Areas with Complex Conditions

Wei Zhao, Professor
School of Architecture and Environment, Sichuan University, China
Rural and Urban Areas in the Chengdu Plain from the Perspective of Human Settlements Construction

Jiangbo Wang, Professor
The School of Architecture, Nanjing Tech University, China
The Changes from Sea to Land: The Spatio-temporal Evolution Characteristics of the Ecological Pattern in the Guangdong-Hong Kong-Macao Greater Bay Area

Mingying Zeng, Professor
School of Civil Engineering and Architecture, Southwest University of Science and Technology, China
Research and Practice of Sustainable Landscape Design Based on the SITES Evaluation System

Reviewers

Ali Cheshmehzangi, University of Queensland, Australia
Bao-Jie He, Chongqing University, China
Beta Paramita, Universitas Pendidikan Indonesia, Indonesia
Bin Li, Polytechnic University of Turin, Italy
Chen Liang, Hebei University of Technology, China
Chenxi Li, Xi'an University of Architecture and Technology, China
Chunping Miao, Chang'an University, China
Deo Prasad, University of New South Wales, Australia
Dongrui Han, Shandong Academy of Agricultural Sciences, CAS, China
Donny Koerniawan, Perencanaandan Pengembangan Kebijakan Teknologi Bandung University, Indonesia
Genyu Xu, Yunnan University, China
Gloria Pignatta, University of New South Wales, Australia
Guoen Wei, Nanjing University, China
Guojin Qin, Shanghai Jiao Tong University, China
Haoxiang Zhan, Universiti Malaya, Malaya
Hui Luo, Nanjing Forestry University, China
Huimin Liu, Wuhan University, China
Jack Ngarambe, Kyung Hee University, Korea
Jinda Qi, National University of Singapore, Singapore
Jun Yang, Northeastern University, China
Junhui Hu, Henan University of Sciences and Technology, China
Junqing Tang, Peking University, China
Junsong Wang, South China University of Technology, China
Junxue Zhang, Jiangsu University of Science and Technology, China
Karam Al-Obaidi, Sheffield Hallam University, UK
Ke Xiong, Chongqing University, China
Chun Liang, Hunan University, China
Liang Wang, Southwest University Science and Technology, China
Lidu Shen, Institute of Applied Ecology, Chinese Academy of Sciences, China
Lili Dong, Chongqing Jiaotong University, China

Linchuan Yang, Southwest Jiao Tong University, China
Linghua Duo, East China University of Technology, China
Lingye Yao, The Chinese University of Hong Kong, China
Meizi You, New York University Shanghai, China
Meng Meng, South China University of Technology, China
Mengyuan Jia, Beijing University of Civil Engineering and Architecture, China
Pingan Ni, Xi'an University of Architecture and Technology, China
Rui Zhang, School of Architecture, Tianjin University, China
Shady Attia, University of Liege, Belgium
Shutian Zhou, Nantong University, China
Siliang Yang, Mott MacDonald, Leeds Office, UK
Simei Wu, Xi'an University of Architecture and Technology, China
Tian Chen, Tianjin University, China
Wei Xue, Jianghan University, China
Xiao Liu, South China University of Technology, China
Xin Dong, Chongqing University, China
Yanan Liu, Chongqing Jiaotong University, China
Yangli Li, Southwest University of Science and Technology, China
Yao Mao, Southwest University, China
Yifu Ou, The University of Hong Kong, China
Yihan Wu, Huazhong University of Science and Technology, China
Yijun Shi, Zhejiang A&F University, China
Yu Luo, The University of Kitakyushu, Japan
Yuejing Gao, Xi'an University of Science and Technology, China
YuWen Lu, School of Architecture and Urban Planning, Nanjing University, China
Ziqi Zhao, Institute of Atmospheric Environment, China Meteorological Administration, China

Contents

Editors and Contributors

About the Editors

Baojie He is a Professor of Urban Climate and Sustainable Built Environment at Chongqing University. He is the Leader of Centre for Climate-resilient and Low-Carbon Cities at Chongqing University and an Honorary Research Fellow of NERPS at Hiroshima University. Prior to these, He worked at UNSW Sydney as a Ph.D. candidate, Research Associate, and Postdoc Research Fellow. His research is on heat-resilient urban planning and design and net-zero carbon built environment. He has published more than 200 peer-reviewed papers in high-ranking journals in reputable conferences/seminars. He has been involved in several large research projects on urban climate and built environment in China and Australia. He acts as Editor-in-Chief, Associate Editor, and Editorial Board Member of many international journals. He received the received the titles of the Most Cited Chinese Researchers in 2023 and 2024, Highly Cited Researcher (Clarivate) in 2022, 2023, and 2024. Bao-jie is a World's top 2% Scientist (Stanford University) from 2020 onwards.

Tian Chen is Professor of the School of Architecture, Tianjin University, Doctoral Supervisor, Director of the Institute of Urban Space and Urban Design, and Master of Urban Planning and Design in Tianjin. He is the Standing Director of the 6th Council of the Chinese Society of City Planning, Member of the Teaching Steering Committee for Architecture-related Majors of the Ministry of Education, Deputy Director of the Teaching Steering Committee for Urban and Rural Planning Majors, Director of the Urban Design Branch of the Architectural Society of China, Standing Director of the Resilient City Professional Committee of the Chinese Society for Urban Studies. He has presided over or participated in more than 20 national and provincial-level scientific research projects. He has presided over more than 80 practical projects of architectural planning and design. More than a hundred master's and doctoral students have been trained by him. He has published 12 monographs and translations.

Yangli Li is a Lecturer and Master's Supervisor in the Department of Urban and Rural Planning, School of Civil Engineering and Architecture, Southwest University of Science and Technology. His main research directions are ecological city planning, ecological city design, and urban water ecological space planning. He is a Member of the Standards and Planning Professional Committee of the China Association for Disaster Prevention and Member of the Resilient City Professional Committee of the Chinese Society for Urban Studies. A total of 28 academic papers have been published by him. He has presided over one provincial and ministerial-level research project, participated in 13 national-, provincial and ministerial-level, university-level, and horizontal scientific research projects, contributed to the compilation of one monograph and one group standard, and am currently compiling one monograph.

Ali Cheshmehzangi is the Head and Professor of the School of Architecture, Design and Planning at the University of Queensland, and Editor-in-Chief of Springer's Urban Sustainability (US) book series. Since 2021, he has been ranked among the top 2% of global scientists by Stanford University and is one of the top 25 scholars in the field of urban sustainability research worldwide. Before joining the University of Queensland, Ali held several strategic leadership and senior management positions. For instance, the Vice-President for international engagement and global partnerships, the founding Director and Head of the Center for Education and Research Innovation, Head of the Department of Architecture and the Built Environment, the founding Director of the Urban Innovation Lab, the Director of the University-wide Teaching and Learning Platform, the Director of the International Research Network for Rural and Urban Development, Head of the Sustainable Built Environment Research Group, the co-director of university-wide research Focus areas, and sustainable energy, Director of the Source Technology Center and Interim Director of the Digital Design Lab. Ali has published over 500 journal articles, articles conference papers, book chapters and reports, and has 40 academic books.

Peng Zeng is Vice-Dean and Professor of the School of Architecture, Tianjin University, Deputy Director of the National Steering Committee for Professional Degree Postgraduate Education in Urban Planning Member of the National Evaluation Committee for Urban and Rural Planning Specialties in Higher Education Institutions, Director of Engineering Technology Innovation Center for Land and Sea Integrated Key Zone Territorial Space Planning and Governance, Chief Editor of Urban Journal Director of the Urban Planning Society of China, Member of the Academic Work Committee, Member of the Small Town Planning Branch, and Member of the Urban Renewal Branch, Party Secretary and Vice-Chairman of Tianjin Urban Planning Society, Team Lleader of the Special Project for the Construction of innovative teams in Tianjin. He has been awarded the title of "Outstanding Science and Technology Worker of Tianjin", and has led and participated in numerous national major and key scientific research projects. He has also led over 50 planning and practice projects and received multiple provincial and ministerial-level awards.

Guofang Zhai is the Director of the Research Center for Urban Safety Development, Nanjing University, Professor, Doctoral Supervisor, concurrently serving as an expert in the compilation of the "National Territorial Space Planning Outline (2021–2035)," Member of the National Higher Education Urban and Rural Planning Professional Evaluation Committee of the Ministry of Housing and Urban-Rural Development, Director of the China Association for Disaster Prevention, and Director of the Planning and Standards Professional Committee. He holds positions such as Director of the Urban Planning Society of China, Deputy Director of the Academic Committee of Urban Safety and Disaster Prevention Planning, Member of the China National Committee of the United Nations "Future Earth" Program (CNC-FE), and Vice-Chairman of the China Branch of the International Society for Risk Analysis. He has served as the Vice-Dean of the School of Architecture and Urban Planning at Nanjing University and the Director of the Urban Planning Department, as well as a Member of the National Expert Committee for the 13th Five-Year Plan.

Contributors

Yibin Ao College of Environment and Civil Engineering, Chengdu University of Technology, Chengdu, China

Xiaopu Bi School of Architecture and Urban Planning, Nanjing University, Nanjing, China

Qianwen Cai School of Architecture and Urban Planning, Shenzhen University, Shenzhen, China

Hao Cao Anhui Province Transportation Holdings Group Co., Ltd., Hefei, Anhui, China

Saijun Cao Chong Qing Architectural Design Institute Co., Ltd. (CQADI), Chongqing, China

Guangzhong Chen School of Public Administration, China University of Geosciences Wuhan, Wuhan, China

Xiaohong Chen Build Planet Zero, Northumbria University, Newcastle Upon Tyne, UK

Yifei Chen Faculty of Civil Engineering and Mechanics, Kunming University of Science and Technology, Kunming, China

Yutong Chen School of Architecture and Urban Planning, Shenzhen University, Shenzhen, China

Zheng Chen Guangdong Yi Lv Ecological Technology Development CO., LTD., Guangzhou, China

Bin Cheng School of Civil Engineering and Architecture, Southwest University of Science and Technology, Mianyang, Sichuan, China

Wenyin Cheng College of Horticulture, China Agricultural University, Beijing, China

Xiang Cheng School of Architecture and Urban Planning, Chongqing University, Chongqing, China

Dongjin Cui School of Architecture and Urban Planning, Shenzhen University, Shenzhen, China

Chonghao Dai Faculty of Civil Engineering and Mechanics, Kunming University of Science and Technology, Kunming, China

Hui Deng Wuxi Taihu University, Wuxi, China

Qing Deng College of Forestry and Landscape Architecture, Anhui Agricultural University, Hefei Anhui, China

Shihan Deng Faculty of Civil Engineering and Mechanics, Kunming University of Science and Technology, Kunming, China

Fenghua Dong School of Civil Engineering, Shandong University, Jinan, Shandong, China;
Urban and Rural Human Settlement Environment Research Center, Shandong University, Jinan, Shandong, China

Lili Dong School of Architecture and Urban Planning, Chongqing Jiaotong University, Chongqing, China

Yang Du Henan University of Science and Technology, Luoyang, China

Xinyi Duan School of Civil Engineering and Architecture, Southwest University of Science and Technology, Mianyang, China

Yuli Fan School of Public Administration, China University of Geosciences Wuhan, Wuhan, China

Guangyuan Feng School of Architecture and Urban Planning, Nanjing University, Nanjing, China

Wenyan Feng School of Civil Engineering and Architecture, Southwest University of Science and Technology, Mianyang, China

Xu Feng College of Horticulture, China Agricultural University, Haidian District, Beijing, China

Juanlin Fu School of Civil Engineering and Architecture, Southwest University of Science and Technology, Mianyang, China

Linsheng Gu Sichuan Earthquake Early Warning and Emergency Research Society, Chengdu, China

Zhuixin Guan School of Civil Engineering and Architecture, Southwest University of Science and Technology, Mianyang, Sichuan, China

Jie Guo Sichuan Meteorological Service Center, Chengdu, China

Liuyin Guo School of Civil Engineering and Architecture, Southwest University of Science and Technology, Mianyang, Sichuan, China

Wen Guo School of Civil Engineering and Architecture, Southwest University of Science and Technology, Mianyang, China

Xiaojun Guo Institute of Mountain Hazards and Environment, CAS, Chengdu, China

Bao-Jie He Centre for Climate-Resilient and Low-Carbon Cities, School of Architecture and Urban Planning, Chongqing University, Chongqing, China;
Smart City Research Institute in Liyang, Chongqing University, Liyang, Jiangsu, China

Baojie He Centre for Climate–Resilient and Low–Carbon Cities, School of Architecture and Urban Planning, Ministry of Education, Chongqing University, Chongqing, China;
Key Laboratory of New Technology for Construction of Cities in Mountain Area, Ministry of Education, Chongqing University, Chongqing, China;
School of Architecture, Design and Planning, University of Queensland, Brisbane, Australia

Lu He Chong Qing Architectural Design Institute Co., Ltd. (CQADI), Chongqing, China

Simin He School of Architecture and Urban Planning, Chongqing Jiaotong University, Chongqing, China

Xinheng He School of Architecture and Urban Planning, Guangdong University of Technology, Guangzhou, China

Yanling He School of Civil Engineering and Architecture, Southwest University of Science and Technology, Mianyang, Sichuan, China

Zhu He Faculty of Civil Engineering and Mechanics, Kunming University of Science and Technology, Kunming, China

Tingting Hou School of Civil Engineering and Architecture, Zhejiang University of Science and Technology, Hangzhou, China

Hongyu Hu Anhui Province Transportation Holdings Group Co., Ltd., Hefei, Anhui, China

Jie Huang School of Architecture, Guangzhou City University of Technology, Guangzhou, China

Songyi Huang School of Architecure, South China University of Technology, Guangzhou, China

Yao Huang Sichuan Meteorological Service Center, Chengdu, China

Yuanzhong Huang Shaanxi City Investment and Operation Group Co., Ltd, Xi'an, Shaanxi, China

Huan Hu School of Civil Engineering and Architecture, Southwest University of Science and Technology, Mianyang, China

Junhui Hu Henan University of Science and Technology, Luoyang, China

Jiaxin Jiang Planning and Industrial Research Institute, CCCC Third Harbor Consultants Co., Ltd., Shanghai, China

Hongmin Ji Sichuan Rural Economy Comprehensive Information Center, Chengdu, China

Jing Jin College of Forestry and Landscape Architecture, Anhui Agricultural University, Hefei Anhui, China

Shengkai Lei Sichuan Provincial Key Laboratory of Plateau and Basin Rainstorm Drought and Flood Disaster, Institute of Plateau Meteorological, China Meteorology Administration, Chengdu, China;
Sichuan Provincial Rural Economic Information Center, Chengdu, China;
Provincial Key Laboratory of Software Automatic Generation and Intelligent Services, Chengdu University of Information Technology, Chengdu, China;
Sichuan Vocational and Technical College of Communications, Chengdu, China

Bingbing Leng Changchun Power Supply Company, State Grid (Jilin Province) Electric Power Co., Ltd., Changchun, China

Shuang Liang School of Civil Engineering and Architecture, Southwest University of Science and Technology, Mianyang, China;
College of Civil Engineering and Architecture, Southwest University of Science and Technology, Mianyang, China

Yuying Liang College of Civil Engineering and Architecture, Southwest University of Science and Technology, Mianyang, China

Qipeng Liao School of Public Administration, China University of Geosciences Wuhan, Wuhan, China

Bin Li Department of Architecture and Design, Politecnico di Torino, Torino, Italy;
Energy Saving Technology Research Institute, South China University of Technology, Guangzhou, China;
Architectural Design and Research Institute Co., Ltd., South China University of Technology, Guangzhou, China;
State Key Laboratory of Subtropical Building Science, South China University of Technology, Guangzhou, China

Bing Li Shaanxi City Investment and Operation Group Co., Ltd, Xi'an, Shaanxi, China

Chunyu Li Housing and Urban-Rural Development Bureau of Kaifeng, Kaifeng City, Henan Province, China

Congyu Li School of Civil Engineering and Architecture, Southwest University of Science and Technology, Mianyang, China

Dingran Li School of Architecture and Urban Planning, Chongqing Jiaotong University, Chongqing, China

Fan Li School of Architecture, Southwest Jiaotong University, Chengdu, China

Hongsen Li Institute of Qilu Culture, Shandong Normal University, Jinan, China

Junrong Li China Construction Engineering Design and Research Institute Co., Ltd., Jinan, Shandong, China

Mingyang Li College of Management Science, Chengdu University of Technology, Chengdu, China

Tong Li Faculty of Civil Engineering and Mechanics, Kunming University of Science and Technology, Kunming, China

Xiaolan Li Zhejiang Southeast Architectural Design Group Co. Ltd., Hangzhou, China

Xiaoshuang Li Sichuan Rural Economy Comprehensive Information Center, Chengdu, China

Yangli Li School of Civil Engineering and Architecture, Southwest University of Science and Technology, Sichuan, China

Hankun Lin School of Architecture and Urban Planning, Guangdong University of Technology, Guangzhou, China

Qiaoyuan Lin School of Civil Engineering and Architecture, Zhejiang University of Science and Technology, Hangzhou, China

Danfeng Liu Henan University of Science and Technology, Luoyang, China

Hongyang Liu Sichuan Provincial Key Laboratory of Plateau and Basin Rainstorm Drought and Flood Disaster, Institute of Plateau Meteorological, China Meteorology Administration, Chengdu, China;
Sichuan Provincial Rural Economic Information Center, Chengdu, China;
Provincial Key Laboratory of Software Automatic Generation and Intelligent Services, Chengdu University of Information Technology, Chengdu, China

Ruiling Liu School of Civil Engineering and Architecture, Southwest University of Science and Technology, Mianyang, China

Yuhong Liu School of Arts (School of Architecture), Nantong University, Nantong, China

Yuzhen Liu School of Civil Engineering and Architecture, Southwest University of Science and Technology, Mianyang, China

Zhenxuan Liu School of Architecture and Urban Planning, Guangdong University of Technology, Guangzhou, China

Zhihua Liu China Construction Engineering Design and Research Institute Co., Ltd., Jinan, Shandong, China

Yi Long College of Environment and Civil Engineering, Chengdu University of Technology, Chengdu, China

Yu Long School of Civil Engineering and Architecture, Southwest University of Science and Technology, Mianyang, China

Mengmin Lu College of Forestry and Landscape Architecture, Anhui Agricultural University, Hefei Anhui, China

Yixiang Lu China Northwest Architecture Design and Research Institute Co., Ltd, Xi'an, China

Wenjiang Luo School of Civil Engineering and Architecture, Southwest University of Science and Technology, Mianyang, China

Yu Luo Faculty of Environmental Engineering, Kitakyushu University, Fukuoka, Japan

Dawei Lv Chong Qing Architectural Design Institute Co., Ltd. (CQADI), Chongqing, China

Xiaowen Ma College of Arts, Qingdao University of Science and Technology, Qingdao, China;
Faculty of Innovation and Design, City University of Macau, Macau, China

Yue Ma Chong Qing Architectural Design Institute Co., Ltd. (CQADI), Chongqing, China

Ziyang Ma Faculty of Civil Engineering and Mechanics, Kunming University of Science and Technology, Kunming, China

Elliot Machin School of Built Environment, Engineering and Computing, Leeds Beckett University, Leeds, UK

Mengyue Mei College of Architecture and Urban Planning, Tongji University, Shanghai, China

Qipeng Mu College of Architecture and Urban Planning, Qingdao University of Technology, Qingdao, China

Ming Nie College of Horticulture, China Agricultural University, Haidian District, Beijing, China

Chuanyang Pan College of Horticulture, China Agricultural University, Haidian District, Beijing, China

Panyu Peng College of Environment and Civil Engineering, Chengdu University of Technology, Chengdu, China

Deo Prasad School of Built Environment, University of New South Wales, Sydney, Australia

Zhongjian Shen School of Civil Engineering, Shandong University, Jinan, Shandong, China;
Urban and Rural Human Settlement Environment Research Center, Shandong University, Jinan, Shandong, China

Manjiang Shi School of Civil Engineering and Architecture, Southwest University of Science and Technology, Mianyang, China

Yijun Shi College of Landscape Architecture, Zhejiang Agricultural and Forestry University, Hangzhou, China;
College of Landscape Architecture, Zhejiang Agriculture and Forestry University, Zhejiang, China

Yuhang Shi College of Landscape Architecture, Zhejiang Agricultural and Forestry University, Hangzhou, China;
College of Landscape Architecture, Zhejiang Agriculture and Forestry University, Zhejiang, China

Han Song China Northwest Architecture Design and Research Institute Co., Ltd., Xi'an, China

Qinyu Song College of Horticulture, China Agricultural University, Beijing, China

Taiguan Song School of Municipal and Environmental Engineering, Jilin Jianzhu University, Changchun, Jilin, China

Zhenyu Song Armed Police College, Chengdu, China

Xiaoning Su China Northwest Architecture Design and Research Institute Co., Ltd., Xi'an, China

Tongyu Sun College of Architecture and Urban Planning, Tongji University, Shanghai, China

Yutao Sun Changchun Power Supply Company, State Grid (Jilin Province) Electric Power Co., Ltd., Changchun, China

Zhihao Sun Wuhan Natural Resources Conservation and Utilization Center, Wuhan, China

Junqing Tang School of Urban Planning and Design, Shenzhen Graduate School, Peking University, Shenzhen, China

Zhao Tian School of Civil Engineering and Architecture, Southwest University of Science and Technology, Mianyang, Sichuan, China

Xin Tu Department of Geography, Tianjin Normal University, Tianjin, China

Dachuan Wang School of Civil Engineering and Architecture, Southwest University of Science and Technology, Mianyang, Sichuan, China

Dongning Wang Xi'an University of Architecture and Technology, Xi'an, China

Lin Wang Faculty of Innovation and Design, City University of Macau, Macau, China

Lu Wang Faculty of Civil Engineering and Mechanics, Kunming University of Science and Technology, Kunming, China

Qianwen Wang Department of Geography, Tianjin Normal University, Tianjin, China

Qing Wang Sichuan Rural Economy Comprehensive Information Center, Chengdu, China

Siqi Wang School of Civil Engineering and Architecture, Southwest University of Science and Technology, Mianyang, China

Tianran Wang School of Civil Engineering, Shandong University, Jinan, Shandong, China;
Urban and Rural Human Settlement Environment Research Center, Shandong University, Jinan, Shandong, China

Xingli Wang School of Civil Engineering and Architecture, Southwest University of Science and Technology, Sichuan, China

Yan Wang Department of Engineering Management, Sichuan College of Architectural Technology, Deyang, China

Zeyi Wang Changchun Power Supply Company, State Grid (Jilin Province) Electric Power Co., Ltd., Changchun, China

Zhihao Wang Faculty of Civil Engineering and Mechanics, Kunming University of Science and Technology, Kunming, China

Zhitian Wang Anhui Province Transportation Planning and Design Institute Co., Ltd., Hefei, Anhui, China

Lu Wei School of Civil Engineering and Architecture, Southwest University of Science and Technology, Mianyang, Sichuan, China

Silin Wu School of Civil Engineering and Architecture, Southwest University of Science and Technology, Mianyang, Sichuan, China

Tong Wu Chong Qing Architectural Design Institute Co., Ltd. (CQADI), Chongqing, China

Zhonghua Wu Hubei Provincial Department of Ecology and Environment, Xiangyang Ecological Environment Monitoring Center, Xiangyang, China

Ling Xiang School of Civil Engineering and Architecture, Southwest University of Science and Technology, Mianyang, China

Yiqiang Xiao School of Architecure, South China University of Technology, Guangzhou, China

Chongshi Xie Chong Qing Architectural Design Institute CO., LTD (CQADI), Chongqing, China;
School of Architecture and Urban Planning, Chongqing University, Chongqing, China

Jing Xie School of Civil Engineering and Architecture, Southwest University of Science and Technology, Mianyang, Sichuan, China

Enpeng Xu College of Arts, Qingdao University of Science and Technology, Qingdao, China

Yiwen Xu Xuzhou University of Technology, Xuzhou, Jiangsu, China

Wei Xue School of Digital Construction and Blasting Engineering, Jianghan University, Wuhan, China;
Hubei Key Laboratory of Industrial Fume and Dust Pollution Control, Wuhan, China;
State Key Laboratory of Precision Blasting, Jianghan University, Wuhan, China

Li Yan School of Civil Engineering and Architecture, Southwest University of Science and Technology, Mianyang, China

Dahai Yang Anhui Province Transportation Planning and Design Institute Co., Ltd., Hefei, Anhui, China

Huijun Yang Department of Geography, Tianjin Normal University, Tianjin, China

Jia Yang China Kide Engineering Corporation, Xi'an, China

Ruoxi Yang China Northwest Architecture Design and Research Institute Co., Ltd., Xi'an, China

Siliang Yang Mott MacDonald, Leeds, UK

Yidi Yao School of Design, South China University of Technology, Guangzhou, China

Jiayue Yin School of Civil Engineering and Architecture, Southwest University of Science and Technology, Mianyang, China

Meng Yuan Sichuan Rural Economy Comprehensive Information Center, Chengdu, China;
Sichuan Rural Economic Information Center, Chengdu, China;
Key Laboratory of Sichuan Province for Heavy Rain and Drought Flood Disasters in Plateaus and Basins, Chengdu, China

Minghong Yu School of Civil Engineering and Architecture, Southwest University of Science and Technology, Mianyang, China

Wang Yu School of Civil Engineering and Architecture, Southwest University of Science and Technology, Mianyang, Sichuan, China

Mingying Zeng School of Civil Engineering and Architecture, Southwest University of Science and Technology, Sichuan, China

Guofang Zhai School of Architecture and Urban Planning, Nanjing University, Nanjing, China

Wei Zhai College of Architecture and Planning, The University of Texas at San Antonio, Texas, USA;
College of Architecture, Planning, and Public Affairs, University of Texas at Arlington, Arlington, TX, USA

Wenshuo Zhai China Construction Engineering Design and Research Institute Co., Ltd., Jinan, Shandong, China

Qingming Zhan School of Urban Design, Wuhan University, Wuhan, China

Boheng Zhang Chong Qing Architectural Design Institute CO., LTD (CQADI), Chongqing, China

Gaochuan Zhang School of Civil Engineering and Architecture, Zhejiang University of Science and Technology, Hangzhou, China;
Zhejiang-Singapore Joint Laboratory for Urban Renewal and Future City, Hangzhou, China

Huizi Zhang PowerChina Zhongnan Engineering Co., Ltd., Changsha, China

Jiatong Zhang School of Municipal and Environmental Engineering, Jilin Jianzhu University, Changchun, Jilin, China

Jingxiang Zhang School of Architecture and Urban Planning, Nanjing University, Nanjing, China

Mian Zhang School of Architecture and Urban Planning, Chongqing Jiaotong University, Chongqing, China

Qi Zhang School of Civil Engineering and Architecture, Southwest University of Science and Technology, Mianyang, China

Rui Zhang College of Design and Engineering, National University of Singapore, Singapore, Singapore

Shengyuan Zhang College of Arts, Qingdao University of Science and Technology, Qingdao, China

Tao Zhang Shaanxi City Investment and Operation Group Co., Ltd, Xi'an, Shaanxi, China

Tianyang Zhang Graduate School of Environmental Engineering, The University of Kitakyushu, Fukuoka, Japan

Ye Zhang China Construction Engineering Design and Research Institute Co., Ltd., Jinan, Shandong, China

Yuqing Zhang School of Architecture, South China University of Technology, Guangzhou, China;
Department of Architecture and Design, Politecnico di Torino, Torino, Italy;
State Key Laboratory of Subtropical Building Science, South China University of Technology, Guangzhou, China

Chunrong Zhao School of Civil Engineering and Architecture, Southwest University of Science and Technology, Mianyang, China

Guangyu Zhao Tianjin University Research Institute of Architectural Design and Urban Planning Co., Ltd., Tianjin, China

Jinpeng Zhao Sichuan Rural Economy Comprehensive Information Center, Chengdu, China

Xiang Zhao School of Civil Engineering and Architecture, Southwest University of Science and Technology, Mianyang, China

Siyin Zheng School of Civil Engineering and Architecture, Southwest University of Science and Technology, Mianyang, Sichuan, China

Jiaqi Zhou College of Forestry and Landscape Architecture, Anhui Agricultural University, Hefei Anhui, China

Min Zhou China Northwest Architecture Design and Research Institute Co., Ltd., Xi'an, China

Shutian Zhou School of Arts (School of Architecture), Nantong University, Nantong, China

Xinyu Zhou College of Landscape Architecture, Zhejiang Agricultural and Forestry University, Hangzhou, China;

College of Landscape Architecture, Zhejiang Agriculture and Forestry University, Zhejiang, China

Jialiang Zhu School of Civil Engineering and Architecture, Southwest University of Science and Technology, Mianyang, China

Pengkun Zhu Faculty of Civil Engineering and Mechanics, Kunming University of Science and Technology, Kunming, China

Chang'an Zuo School of Civil Engineering, Shandong University, Jinan, Shandong, China;
China Construction Engineering Design and Research Institute Co., Ltd., Jinan, Shandong, China;
Urban and Rural Human Settlement Environment Research Center, Shandong University, Jinan, Shandong, China

Climate Change and Impacts

Distribution Characteristics of $PM_{2.5}$ in Different Landscape Spaces Within Parks and Their Influencing Factors

Ling Xiang, Chunrong Zhao, Li Yan, Yu Long, and Xinyi Duan

Abstract This paper employed quantitative measurement methods to conduct empirical research on $PM_{2.5}$ concentrations, analyzing the formation and maintenance mechanisms of the microclimate environment in parks and exploring the impact of spatial composition elements on the distribution of $PM_{2.5}$. The study selected six areas with different landscape elements within People's Park in Mianyang City. By monitoring $PM_{2.5}$ concentrations, air temperature, relative humidity, and wind speed across these areas, the research analyzes the relationship between the composition of landscape elements and $PM_{2.5}$ concentrations. The data obtained was processed using statistical analysis and analyzed by plotting using ORIGIN software. It further investigated the correlation between variations in meteorological factors across different areas and $PM_{2.5}$ concentrations, providing theoretical evidence for optimizing the microclimate of parks. The findings of this research are as follows: (1) Significant differences exist in $PM_{2.5}$ concentrations among spaces composed of different landscape elements; (2) Wind speed, air temperature, temperature, and humidity exhibit certain correlations with $PM_{2.5}$ concentrations, with wind speed showing the strongest correlation with fine particulate matter, indicating a significant negative correlation, maximum negative correlation of wind speed can reach 0.74; (3) There was no evident correlation between $PM_{2.5}$ concentrations and the area ratios of water bodies, roads, and pavements; however, the area ratio of buildings demonstrates a moderate positive correlation with $PM_{2.5}$ concentrations. This study provided a theoretical basis and methodology for urban park design based on outdoor environmental assessment.

Keywords Park green space · $PM_{2.5}$ · Landscape elements · Meteorological factors · Mianyang city

L. Xiang · C. Zhao (✉) · L. Yan · Y. Long · X. Duan
School of Civil Engineering and Architecture, Southwest University of Science and Technology, Mianyang, China
e-mail: 406986617@qq.com

B. He et al. (eds.), *Climate-Resilient, Low-Carbon, and Sustainable Cities*, Lecture Notes in Civil Engineering 754, https://doi.org/10.1007/978-981-95-2169-2_1

1 Introduction

As a major outdoor space in urban areas, the air quality of green spaces in parks is crucial to the health of the entire urban population [1]. Since the promulgation of the new 'Environmental Space Quality Standards' in March 2013, $PM_{2.5}$ has become the primary pollutant in air pollution control in China [2]. Controlling the spread of pollutants in pedestrian spaces is one of the effective ways to improve air quality. The reduction of $PM_{2.5}$ can mainly be divided into two aspects: source control and natural sedimentation [3]. Among them, Meteorological factors are the main factors affecting the diffusion and sedimentation of $PM_{2.5}$ in the air. Good natural ventilation can effectively reduce $PM_{2.5}$ concentration, but excessively high wind speeds can make pedestrians uncomfortable, and ground dust can exacerbate pollution [4]. The higher the air humidity, the more easily water vapor interacts with fine particulate matter, allowing particles to adhere to water vapor, thereby reducing $PM_{2.5}$ concentration [5]. The research by Pan et al. pointed out that the $PM_{2.5}$ concentration is negatively correlated with air temperature, positively correlated with relative humidity, and shows a negative correlation with wind speed [6]. Ma et al. showed that wind speed and precipitation play an important role in $PM_{2.5}$ and precipitation play an important role in the horizontal and vertical transport and deposition of $PM_{2.5}$ [7]. Li found a significant negative correlation between temperature and $PM_{2.5}$ concentration when studying the impact of the urban built environment on urban $PM_{2.5}$ concentration [8]. Changes in surface temperature can affect vertical and horizontal convection in the near-surface layer, thereby influencing the distribution of $PM_{2.5}$ in the region. Additionally, as temperature rises, the enhancement of photosynthesis in plants increases their ability to adsorb particulate matter [9]; On cloudy days, the solar radiation reaching the ground is insufficient, which is unfavorable for the settling and diffusion of $PM_{2.5}$. Zhao and Li found that rainfall of 15mm can wash away the dust on plant leaves while studying the characteristics and patterns of atmospheric particulate matter of different particle sizes retained by landscape plants, suggesting a significant negative correlation between precipitation and $PM_{2.5}$ concentration [10]; Liu 's research indicates that frequent washing by rainwater reduces the dust retention capacity, and rainfall causes the main soluble components in atmospheric $PM_{2.5}$ to dissolve, converging into surface run off [11]. Green corridors have a good connection with the prevailing wind direction in the city, and wind speed promotes the dispersion of particulate matter, which can effectively reduce $PM_{2.5}$ concentration.

The different landscape elements in urban parks also influence the distribution of particulate matter concentration in space; even under non-precipitation conditions, plants exhibit a significant dust retention effect [12]. Rough tree bark and dense branches allow plants to reduce air dust content by 18–20% even after shedding leaves in winter; trees have a better dust retention capacity than shrubs and herbaceous plants, and the dust reduction effect of a mixed community structure of trees, shrubs, and herbs is more pronounced [13]. Previous studies have found that the $PM_{2.5}$ concentration levels above the canopy are significantly higher than those at the canopy level, indicating that as height decreases, the dust retention effect of vegetation becomes more pronounced, reaching an optimal level at the canopy layer [14]. Due to the high temperatures in summer, the evaporation of water bodies absorbs heat, leading to a decrease in the surrounding microclimate temperature, which can improve ventilation and the temperature and humidity environment, thereby affecting the diffusion and deposition of $PM_{2.5}$ concentration [15]. Generally speaking, plant communities with a layered structure of trees, shrubs, and grasses have a stronger ability to adsorb $PM_{2.5}$ concentration than communities with a single structure, and the plant coverage, planting density, and number of evergreen plants are positively correlated with $PM_{2.5}$ concentration [16]. Plants have a reduction effect on $PM_{2.5}$ concentration, and meteorological factors can also alter the ability of vegetation to adsorb particulate matter.

Cities are an important source of $PM_{2.5}$, with $PM_{2.5}$ concentration in urban areas significantly higher than in suburban, agricultural, forest, and rural landscape types [17]. The spatial optimization strategies proposed by previous studies focus on the urban scale, but fewer strategies are implemented at the park scale, focusing on the macro layout but not on the micro regulation. Generally speaking, small-scale, moderately dispersed, and polycentric urban built environments are conducive to reducing $PM_{2.5}$ concentrations. Within urban areas, due to population activities and industrial production, the $PM_{2.5}$ concentrations in residential, industrial, and transportation spaces are higher [18].

This study suggests that the variations in landscape elements, local temperature and humidity conditions, and ventilation effects contribute to the uneven spatial distribution of $PM_{2.5}$. Through field tests and analysis, the research examines the relationship between the composition ratio of landscape elements in different spaces and $PM_{2.5}$ concentration. It also compares $PM_{2.5}$ concentration, temperature and humidity, and wind speed across various landscape spaces, while exploring the correlation between meteorological factors and $PM_{2.5}$ concentration. The purpose of this study was to explore the distribution characteristics of $PM_{2.5}$ in different landscape spaces of urban parks, and the goal was to provide a theoretical basis and method for the design of microclimate suitability in different spaces of urban parks.

2 Research Area and Research Methods

2.1 Overview of the Research Area

Mianyang, with a subtropical humid climate, recorded 365 valid monitoring days in 2023, yielding 294 compliant days and an 80.5% compliance rate, indicating significant air pollution (Fig. 1).

The research site is located in People's Park, Fucheng District, Mianyang City, Sichuan Province (31°27′N, 104°44′E), where park green spaces are predominant, and motor vehicle exhaust is the main contributor to $PM_{2.5}$ in this area. This study

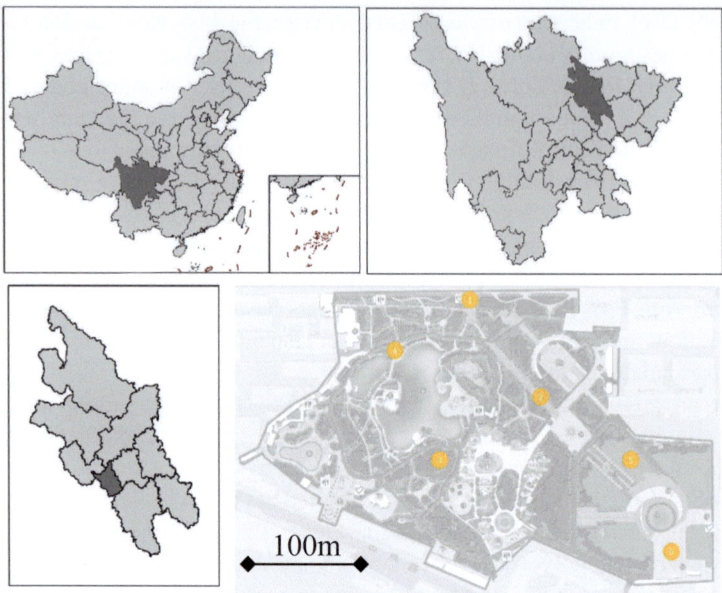

Fig. 1 Location and point distribution map (self-drawn)

is located in the center of the old urban area of Mianyang, surrounded by residential areas, commercial streets and schools. The site contains a small number of buildings, and the outdoor space features a diverse range of vegetation, creating a rich landscape composed of plants, buildings, water bodies, and hard pavements. Six different landscape spaces were selected for study based on different elements, with specific locations (see Fig. 2).

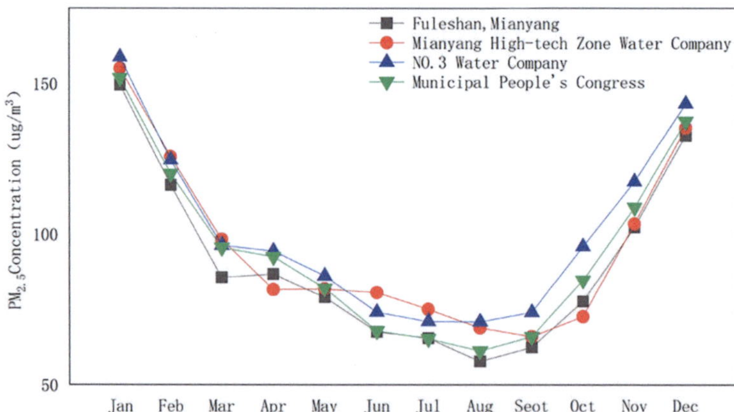

Fig. 2 Mianyang monitoring point data in 2023 (self-drawn)

Table 1 The details of the six points are introduced (self-drawn)

Selected points	Spatial characteristics	Underlying surface properties
1	Traffic node for pure walking near road intersections	Marble
2	Located in the passage of the pedestrian street, with an open space	Marble
3	Situated at the intersection of trees, shrubs, and grass	Lawn, thicket, tree
4	Open square near water	Concrete, tree pit, water
5	Open lawn near streets and squares	Lawn
6	The square is a relatively open, multifunctional space with high ventilation efficiency	Marble

2.2 Research Methods

2.2.1 Selection of Actual Measurement Sample Points

To enhance the practical value of the results, this study focuses on the public space in People's Park, which has high pedestrian activity and frequent resting areas. Six measurement points were selected based on pedestrian flow distribution: north entrance, garden road, joe shrub, lakeside, lawn, and square (see Table 1).

2.2.2 Measurement Method

The testing time is selected from 7:00 to21:00, with fixed-point hourly recordings (data recorded every 20 min), each test lasting 30 s, and 3 sets of instantaneous data recorded. Field measurements are subject to various external factors, and the results may have certain errors, necessitating the correction or screening of the measured data. The measurement steps are as follows: fix the wireless omnidirectional wind speed and temperature recorder and the testo temperature and humidity recorder on a tripod at a height of approximately 1.5 m above the ground, while the research team conducts simultaneous measurements at the six aforementioned measurement points.

2.2.3 Collection and Processing of Measured Data

The experiment involves monitoring $PM_{2.5}$ concentrations within the park's green space, using six evenly distributed fixed points to represent concentrations during the measurement period. Each point is equipped with a wireless anemometer and a Testo 174 humidity and temperature recorder and the BOLANGTONG SMART-128S air quality detector, both of them record data every 20 min, with each instrument obtaining 43 data points at each location. The data obtained was processed using statistical analysis and analyzed by plotting using ORIGIN software. Due to real-world constraints, such as limited instruments and personnel, simultaneous monitoring at various points is not feasible. Transitioning between monitoring points takes 10 min, resulting in mobile cyclic monitoring.

2.2.4 Multiple Linear Regression (MLR)

MLR is a method that combines a dependent variable with multiple independent variables to estimate or predict outcomes, offering a more robust approach than using a single independent variable. In this study, $PM_{2.5}$ is the dependent variable, while meteorological factors serve as independent variables, enabling a quantitative assessment of their impact on $PM_{2.5}$. The regression algorithm's calculation method is as follows:

$$Y = a_0 + a_1X_1 + a_2X_2 + a_3X_3 \ldots + a_kX_k + \varepsilon \tag{1}$$

In the Formula (1): Y is the dependent variable; X_i is the independent variable; a_i is the regression coefficient of the independent variable, used to quantify the impact of the independent variable on the dependent variable; ε is the random error and satisfies $E(\varepsilon) = 0$.

3 Results and Analysis

It can be seen that the $PM_{2.5}$ concentration trends of the surrounding fixed stations are similar and consistent, indicating that most areas in the study zone have $PM_{2.5}$ concentrations at similar regional background levels and are under similar magnitude effects (see Fig. 3). Therefore, in this study, the influencing factors of $PM_{2.5}$ are more dependent on the local urban environment and residents' daily activities rather than background pollution.

3.1 Characteristics of Wind Speed in People's Park

Upon organizing the measured data, it was found that the entire People's Park is almost in a calm wind environment. Based on the average wind speed of six evenly distributed measurement points, the average wind speed is 0.66 m/s, essentially remaining at wind force level 2 or below (see Fig. 3). The wind speeds at the six measurement points indicate that the wind environment in People's Park is relatively

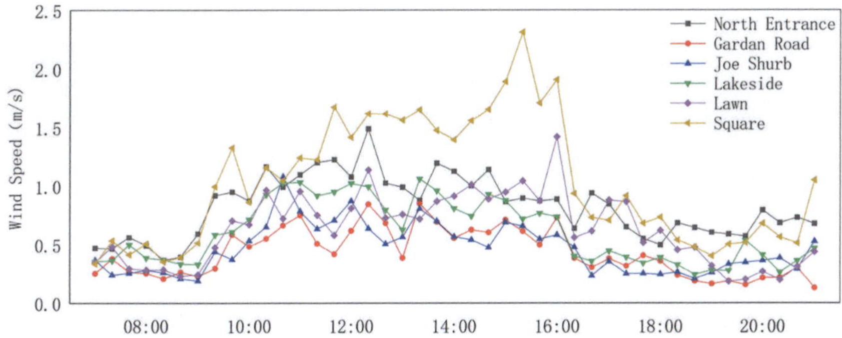

Fig. 3 Wind speed trend at six measurement points (self-drawn)

poor, with a high calm wind rate. The square point is better permeable compared to the other points because there are no tall trees around, so the highest wind speeds occur at this point.

3.2 Spatial Distribution Characteristics of $PM_{2.5}$ in People's Park

Data collected from the six measurement points from 7:00 to 21:00 was statistically analyzed, using the six measurement points as the horizontal axis and their corresponding $PM_{2.5}$ concentrations as the vertical axis, revealing significant differences in $PM_{2.5}$ concentration values among the six measurement points.

The $PM_{2.5}$ concentration at the 6 measurement points has a certain correlation with their spatial differences and the emissions from pollution sources (see Fig. 4). For example, the measurement point located at the main square, close to the main road on the east side, where there is significant crowd activity, so $PM_{2.5}$ concentration is the highest. The measurement point located near the lake has the lowest daily average $PM_{2.5}$ concentration, because the humidity near the lake is high, and $PM_{2.5}$ particles are surrounded by moisture, the mass is aggravated, and the gravity is increased, resulting in a certain degree of deposition, thus reducing the $PM_{2.5}$ concentration.

3.3 $PM_{2.5}$ Concentration Changes in People's Park Over Time

Data collected from 6 measurement points between 7:00 AM and 9:00 PM in September were statistically analyzed. The times of higher $PM_{2.5}$ concentration in People's Park are mainly between 7–8 AM and 7–9 PM (see Fig. 5). This is due to

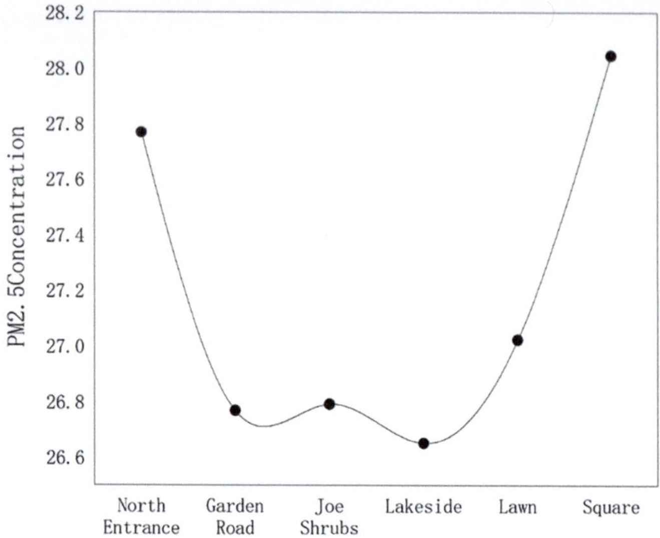

Fig. 4 Daily average concentrations at six points (self-drawn)

increased traffic flow during commuting hours, leading to higher pollution levels; at night, pollution from exhaust emissions along major roads increases and disperses along the roadways. Human activities during the day can last for several hours, generating more pollution emissions, leading to the accumulation of $PM_{2.5}$.

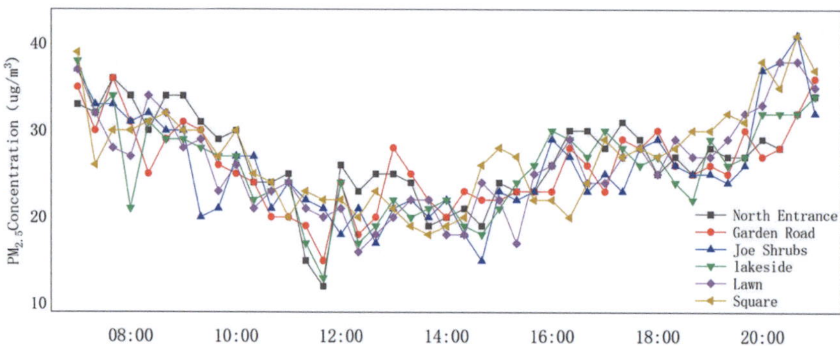

Fig. 5 Temporal trend of $PM_{2.5}$ at six points (self-drawn)

3.4 Correlation Analysis of $PM_{2.5}$ with Different Elements and Measurement Points

According to previous academic research, the minimum distance of landscape impact on microclimate is 10 m; therefore, a circular area with a radius of 10 m around the measurement point is designated as the research area, and the proportions of different landscape elements are statistically analyzed. The area proportions of various landscape elements are calculated from CAD floor plans, where trees, shrubs, herbaceous plants, and ground cover are uniformly classified as vegetation cover [19]. Different landscape elements can affect the spatial distribution of particulate matter concentration, with plants playing a significant role in dust retention (see Fig. 6). Generally, the ranking of $PM_{2.5}$ concentration among different plant community types is as follows: tree-shrub-herbaceous > tree-herbaceous > lawn > shrub-herbaceous > pure coniferous forest [20].

According to the correlation analysis between landscape elements and $PM_{2.5}$ concentration, it can be seen that $PM_{2.5}$ concentration is only positively correlated with buildings, and has no obvious correlation with vegetation, water and pavement, which is different from the previous conclusions (see Fig. 7). The reason may be that the points selected in this study are relatively few, lacking comparable landscape element correlation comparisons; it may also be due to the observation period being in summer, where $PM_{2.5}$ concentration values show little variation, leading to a lack of obvious comparability among landscape elements.

Combining the analysis of various landscape elements with the daily average $PM_{2.5}$ concentration at different points, the gate point and the square point have the highest $PM_{2.5}$ concentrations. Both of these points have a significant proportion of pavement in their landscape elements. The vegetation proportion at the shrub and grass point is the highest, while the water's edge point has the lowest daily average $PM_{2.5}$ concentration. It is also the only point among the six that has a proportion of water bodies. Being close to the lake, the humidification effect gradually

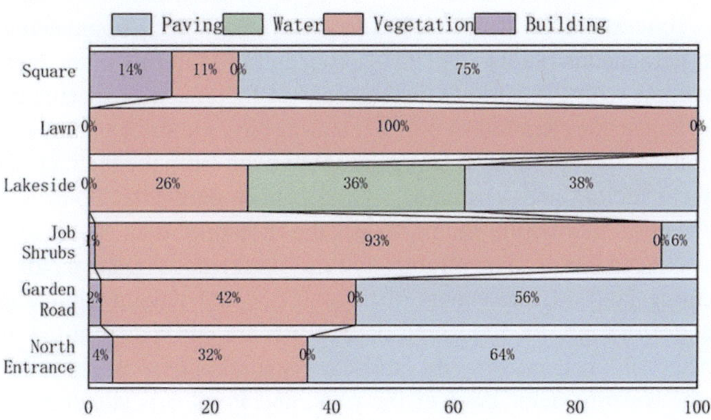

Fig. 6 The spatial proportion of landscape elements (self-drawn)

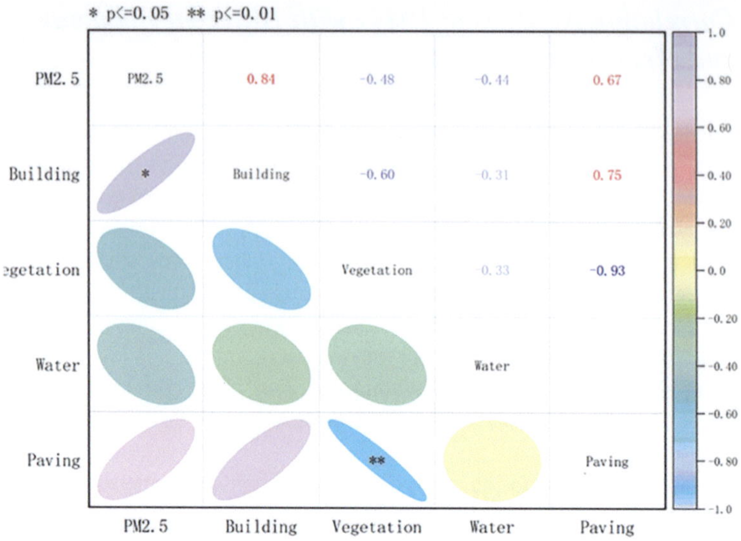

Fig. 7 Correlation analysis of landscape elements (self-drawn)

increases, and the reduction of $PM_{2.5}$ in the air also shows a certain enhancement trend. The interaction of cold and warm air between the lake and the surrounding air is evident, and the dust particles carried in the air tend to spread toward areas with lower humidity, resulting in poorer air quality at the lakeside point.

By conducting a bivariate correlation analysis between the average daily $PM_{2.5}$ concentrations at different points and their corresponding meteorological factors, the following results were obtained (see Fig. 8). The meteorological factors and $PM_{2.5}$ concentrations at the six points all exhibited significant correlations. Point 1 is located at the north entrance of the park, and Point 2 is situated along the main park path. The $PM_{2.5}$ concentrations at these two points show a significant negative correlation with wind speed, indicating that higher wind speeds are associated with lower $PM_{2.5}$ concentrations. Point 3 is located in an area densely populated with trees, shrubs, and herbaceous plants. The $PM_{2.5}$ concentration at this point shows only a weak correlation with the various meteorological factors. It is speculated that the surrounding landscape elements of the selected site have a significant impact on $PM_{2.5}$ concentration, so the measured data results are derived from the interaction between meteorological factors and landscape spatial pattern elements. Point 4 is located next to the central lake within the park, and the measured data shows only a weak correlation between $PM_{2.5}$ concentration and various meteorological factors. Because the surrounding landscape elements of the selected site have a significant impact on $PM_{2.5}$ concentration, the measured data results are derived from the interaction between meteorological factors and landscape spatial pattern elements. Point 5 is located in the western outer lawn of the park, and Point 6 is situated in the hard-paved area of the park's outer zone. The $PM_{2.5}$ concentration at these two points has a

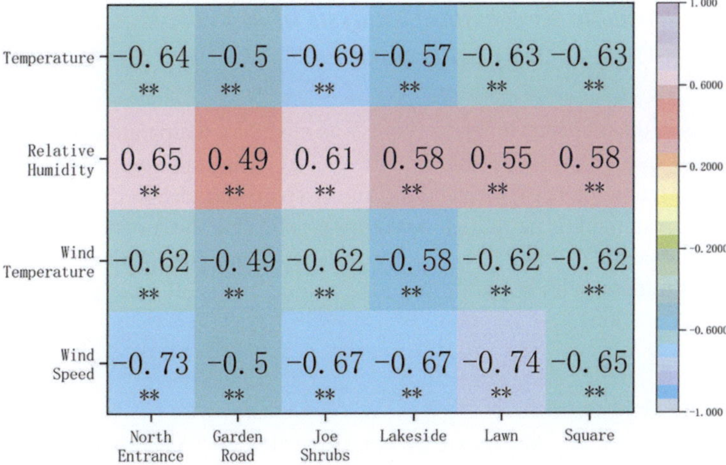

Fig. 8 Correlation analysis of six points with meteorological elements (**indicates $p < 0.01$) (self-drawn)

negative correlation with wind speed, and the correlation is the strongest. Humidity has a significant positive correlation with $PM_{2.5}$ concentration, while temperature has a significant negative correlation with $PM_{2.5}$ concentration.

In summary, the four meteorological factors all have varying degrees of impact on $PM_{2.5}$ concentration. Among them, wind speed has the strongest correlation with $PM_{2.5}$ concentration, which is generally consistent with previous research findings. The correlation between wind speed and $PM_{2.5}$ concentration is particularly pronounced at points 1 and 5, because point 1 is located directly north of the park and the prevailing wind direction that day was from the north, thus having a significant impact on $PM_{2.5}$ concentration. Point 5 is situated in a large area of hard-surfaced square, with no large trees nearby to obstruct the wind, making the area open and well-ventilated, which also significantly affects $PM_{2.5}$ concentration. There is a positive correlation between humidity and $PM_{2.5}$ concentration. In dry environments, fine particulate matter is more likely to remain suspended in the air, while in humid environments, fine particulate matter is more likely to adhere to water vapor, thereby reducing the concentration of fine particulate matter. The average correlation of various meteorological factors with $PM_{2.5}$ concentration is as follows: wind speed (0.67) > temperature (0.61) > wind temperature (0.59) > humidity (0.58). The results indicate that wind speed is the meteorological factor that most affects $PM_{2.5}$ concentration, showing a significant negative correlation. Point 3 is the least affected by meteorological factors among all points, as the surrounding landscape elements at Point 3 are more complex than those at other points, resulting in more factors influencing $PM_{2.5}$ concentration.

3.5 Multivariate Regression Analysis of Meteorological Factors and $PM_{2.5}$ Concentration

The results of the multivariate linear regression model for temperature, precipitation, wind speed, and $PM_{2.5}$ (see Table 2), where $PM_{2.5}$ concentration is the dependent variable and temperature, wind speed and humidity are the independent variables. The calculation results indicate that all monitoring points exhibit a negative correlation with wind speed; Only the $PM_{2.5}$ concentration at monitoring point one shows a positive correlation with humidity and temperature, while all other points exhibit a negative correlation.

Table 2 Multiple linear regression results of $PM_{2.5}$ and meteorological factors (self-drawn)

Pollutants	Region	Multivariate regression model	R^2	Adjusted R^2	P-value
$PM_{2.5}$	North Entrance	$-7.175 + 0.714 \times$ Temperature $- 10.635^{***} \times$ Wind Speed $+ 0.362 \times$ Humidity	0.633	0.604	< 0.001
	Garden Road	$98.679^{*} - 1.577 \times$ Temperature $- 6.491 \times$ Wind Speed $- 0.280 \times$ Humidity	0.566	0.533	< 0.001
	Job Shrubs	$181.023–3.288 \times$ Temperature-$12.466^{***} \times$ Wind Speed $- 0.681 \times$ Humidity	0.407	0.361	< 0.001
	Lakeside	$101.583 – 1.577 \times$ Temperature $- 10.649^{***} \times$ Wind Speed $- 0.275 \times$ Humidity	0.52	0.483	< 0.001
	Lawn	$132.257^{**} - 2.056^{*} \times$ Temperature $- 9.085^{**} \times$ Wind Speed $- 0.533 \times$ Humidity	0.609	0.579	< 0.001
	Square	$202.103^{*} - 3.614^{*} \times$ Temperature $- 2.089 \times$ Wind Speed $- 0.907 \times$ Humidity	0.512	0.474	< 0.001

Note $p > 0.05$ indicates no significance, * indicates $p < 0.05$, ** indicates $p < 0.01$, *** indicates $p < 0.001$

4 Optimization Strategies for Park Green Space Landscape Patterns

The reduction effect of park green space on $PM_{2.5}$ is influenced by various factors. The roads around Mianyang People's Park are all urban trunk roads with heavy traffic and serious exhaust pollution. Internal factors, such as the morphology and quantity of green spaces, play a critical role in determining their effectiveness in reducing $PM_{2.5}$ concentrations. While increasing the number of green spaces can contribute to lower $PM_{2.5}$ levels, the high construction density in built-up areas of Mianyang City limits the feasibility of expanding green space, and the trees in Mianyang People's Park are large trees, which will contribute negatively to the diffusion of $PM_{2.5}$.Therefore, optimizing the landscape morphology of existing green spaces becomes increasingly important for enhancing their $PM_{2.5}$ reduction capabilities.

4.1 Green Space Optimization Strategy

The plants in parks, especially tall trees, hinder the diffusion of $PM_{2.5}$ under high pollution conditions, leading to a weakening of the $PM_{2.5}$ reduction capacity of park green spaces. In planning the surrounding environment, it is essential to minimize the building density outside the park to improve the ventilation conditions of the green spaces. The higher the building irregularity, the more beneficial it is for the reduction of $PM_{2.5}$ in the external environment, indicating that greater vertical variation in buildings is conducive to atmospheric circulation and exchange. Increase the number of trees, combining both trees and grass, appropriately reduce the paved area, and strive to use large-scale, cohesive green spaces while avoiding fragmented green areas.

4.2 Constructing a Suitable Wind Environment

It has been concluded in the previous text that wind speed is the most significant influencing factor for $PM_{2.5}$ concentration, as it controls the transport and distribution of $PM_{2.5}$. Dense plant communities form on the windward boundary of the park, utilizing their barrier function to control spatial enclosure. When the wind carries fine particulate matter, the dense plant community has already formed the first layer of barrier. Leaving certain gaps in the plant community on the leeward side allows $PM_{2.5}$ within the park to disperse into the external space along with the wind's transport and distribution. Controlling the internal plant density and compatibility allows the park to easily achieve the optimal wind speed for plants to reduce particulate matter. Additionally, constructing plant ventilation corridors in certain areas enables the internal space to benefit from multiple air circulation processes over the plant leaf surfaces, thereby reducing $PM_{2.5}$ concentration.

4.3 *Promote Temperature and Humidity Balance*

Light is very important for plant photosynthesis, maintaining the overall light coefficient of the plants. Reduce the plant density in areas with shorter light exposure duration, and increase the plant density in areas with longer light exposure duration. Humidity is difficult to control; higher humidity can affect the distribution of negative ions and weaken the ability of plants to reduce $PM_{2.5}$ [21]. The People's Park contains large water bodies, which significantly reduce the thermal radiation effect through plant shading and seasonal shading nets, thereby diminishing the impact of air vortices generated by radiation convection on the reduction of $PM_{2.5.}$

5 Conclusions and Prospects

This study analyzed the relationship between the combinations of different spatial elements and $PM_{2.5}$ concentration by monitoring $PM_{2.5}$ concentration and internal meteorological factors within the People's Park of Mianyang. It compared the differences in $PM_{2.5}$ concentration, wind speed, and temperature-humidity across different landscape elements, and explored the correlation between various meteorological factors and $PM_{2.5}$ concentration. The following conclusions are drawn.

1. There are significant differences in $PM_{2.5}$ concentration across different landscape spaces, with spatial concentrations ranked from low to high as $4 < 2 < 3 < 5 < 1 < 6$; Open spaces composed of hard-paved squares have the highest $PM_{2.5}$ concentration, while semi-open spaces formed by trees, water bodies, and hard pavements have the lowest concentration.
2. There is no significant correlation between $PM_{2.5}$ concentration and the area ratios of water bodies, roads, and pavements; only the area ratio of buildings showed a moderate positive correlation with $PM_{2.5}$ concentration. The $PM_{2.5}$ concentration at point 3 is least affected by meteorological factors and may also be influenced by the area ratio of vegetation.
3. Different meteorological factors have varying degrees of impact on $PM_{2.5}$ concentration, with the correlation levels ranked from high to low as follows: wind speed > temperature > wind temperature > humidity. Wind speed has the strongest effect on $PM_{2.5}$ concentration, showing a significant negative correlation.

The study elucidated the impact of different landscape elements on $PM_{2.5}$ concentration distribution during summer at a local scale, as well as the relationship between particulate matter in different landscape spaces and temperature, humidity, and wind speed. With the change of seasons, the characteristics of $PM_{2.5}$ pollution and the plant landscape elements exhibit seasonal variations [22]. The study has revealed the influence of various meteorological factors on the distribution of local $PM_{2.5}$ concentration, explaining the distribution patterns of $PM_{2.5}$ within different landscape elements in parks, thereby providing a theoretical basis and scientific guidance for the study of urban park spatial landscape design.

Funding This research was supported by the Humanities and Social Sciences Research Project of the Ministry of Education (Grant No. 24YJCZH22).

References

1. Wang S-Y, He Y-R, Zhao Na et al (2024) Analysis of the temporal and spatial distribution and influencing factors of air pollution in Qingyang. J Longdong Univ 35(05):41–47
2. Zhou SH, Lin RP (2019) Spatial-temporal heterogeneity of air pollution: the relationship between built environment and on-road $PM_{2.5}$ at micro scale. Transp Res Part D—Transp Environ 76:305–322
3. Fang Y, Zhao L, Dou B et al (2024) Reduction effect of green infrastructure on $PM_{2.5}$ in the built-up areas of urban agglomerations in the Yangtze River's middle reaches. Acta Ecologica Sinica 44(01):117–128
4. Abel JGE, Beatriz MMA, Fiorela UAN et al (2024) Temporal variation of the $PM_{2.5}/PM_{10}$ ratio and its association with meteorological factors in a South American megacity: Metropolitan Area of Lima-Callao, Peru. Environ Monitor Assess 196(5):452–452
5. Bian S, Wang Z, Zhang Y et al (2021) Analysis of the relationship between $PM_{2.5}$ pollution characteristics and meteorological factors in Pingdingshan. Papers of the 25th symposium on air pollution prevention and control technology, pp 73–82
6. Pan H (2023) Research on spatial morphology optimization of urban block based on fine particulate matter simulation. Anhui Jianzhu University
7. Ma ZW, Hu XF, Andrew M et al (2016) Satellite-based spatiotemporal trends in $PM_{2.5}$ concentrations China, 2004–2013. Environ Health Perspect 124(2):184–192
8. Li F (2021) Analysis on the influence of the built environment on urban PM2.5 concentration at multiple scales and its governance. Chongqing University
9. Cheng T, Zhu S, Zhang G et al (2020) Seasonal variation of $PM_{2.5}$ in the Beijing-Tianjin-Hebei region in 2018 and its relationship with land surface temperature. Remote Sens Technol Appl 35(6):1457–1466
10. Zhao S, Li X, Li Y et al (2017) Analysis of influence factors and characteristics of $PM_{2.5}$ concentration in a public park and surroundings areas. J Nanjing Forest Univ (Nat Sci Edn) 41(02):187–192
11. Liu X (2006) The studies on bacteriostat and effects of filtering dust of urban forest in Nanjing city. Nanjing Forestry University
12. Lu X (2020) Study on landscape quality evaluation and optimization strategy of recreation space in Beijing country park. North China University Of Technology
13. Qiu L, Liu F, Zhang X et al (2018) Reducing effect of air particulate matter concentration by green spaces with different vegetation structure in city parks. Res Environ Sci 31(10):1685–1694
14. Sun X, Fan Z, Wang B et al (2024) Reducing effect of landscape plant community distribution on urban air pollution. Environ Sci Manag 49(06):157–161
15. Zhu L (2017) The change regularity of particulate matter concentration and dry deposition in different surface. Beijing Forestry University
16. Xu A, Chu B, Xiao X et al (2023) Spatial pattern of plants in streetside green space based on $PM_{2.5}$ reduction: a case study of Suzhou Industrial Park. Chin Landsc Archit 39(02):90–95

17. Shi J, Zhu C, Liu S et al (2024) Insights into source-specific $PM_{2.5}$ mass, light extinction, and health risks in heavily polluted urban and suburban areas, northwestern China. Urban Clim 58:102205
18. Shanyou D, Qian L, Dumei J et al (2021) Exploring the joint impacts of natural and built environments on $PM_{2.5}$ concentrations and their spatial heterogeneity in the context of high-density chinese cities. Sustainability 13(21):11775–11775
19. Yan S, Hong B (2019) $PM_{2.5}$ concentration distribution characteristics in different landscape spaces and influencing factors in urban park. Landsc Plan Des 26(07):101–106
20. Li X, Zhao S, Guo J et al (2014) Effects of different plant communities on the concentration of $PM_{2.5}$ in parks. Mod Landsc Archit 11(11):11–13
21. Fan W, Wu X, Wang M (2022) Differences of dust retention ability and photosynthetic response of five common species of wetland plants in Taiyuan. J Tianjin Normal Univ (Nat Sci Edn) 42(04):52–58
22. Martin EP, Perdrix E, Alleman L et al (2024) Origins of the seasonal variability of $PM_{2.5}$ sources in a rural site in Northern France. Atmos Environ 333120660–120660.

Eggplant Price Forecasting Based on Meteorological Factors

Qing Wang, Zhenyu Song, Xiaoshuang Li, Meng Yuan, Jinpeng Zhao, and Hongmin Ji

Abstract As the foundation of the country and people's livelihood, the price stability of agriculture and food is of great significance to the social and economic development. In recent years, the prices of agricultural products are affected by weather, transportation and policies, and the prices are characterized by non-linear changes, which seriously affects the social stability and people's lives. In order to accurately predict the price of eggplant, this paper proposes a combination model based on CNN_LSTM. Firstly, we use Spearman correlation coefficient (Spearman) to derive the correlation coefficient between meteorological factors and prices to determine the main meteorological factors, secondly, we use CNN convolutional layer to extract the dynamic characteristics of the meteorological factors and price data, and then again, we use LSTM layer to extract the characteristics of the input time-series data, and then we complete the price prediction and output through the output layer. In this paper, the historical meteorological and eggplant price data of Santai County, Sichuan Province in the past 10 years are used, and the CNN, LSTM and CNN-LSTM models are compared and analyzed, and the results show that the RMES of CNN-LSTM model is 0.01, the MAE is 0.06, and the R^2 is 0.99 is better than that of CNN and LSTM, and the predicted value and the actual value have a high degree of fit, and it can be used for the following purposes Agricultural price prediction, agro-meteorological economy, rural revitalization and policy regulation provide mindfulness and new methods.

Keywords CNN · LSTM · CNN_LSTM · Spearman · Eggplant price prediction

Q. Wang · X. Li · M. Yuan · J. Zhao (✉) · H. Ji
Sichuan Rural Economy Comprehensive Information Center, Chengdu, China
e-mail: 421508153@qq.com

Z. Song
Armed Police College, Chengdu, China

B. He et al. (eds.), *Climate-Resilient, Low-Carbon, and Sustainable Cities*, Lecture Notes in Civil Engineering 754, https://doi.org/10.1007/978-981-95-2169-2_2

1 Introduction

Agriculture is the foundation of the state, food is the first of the government, agriculture and food is the foundation of the country, people's livelihood of the foundation of the strong military, agricultural price stability has been the focus of the government's attention, but also the difficulty of regulation [1]. In recent years, with the weather changes, market demand, per capita income, product circulation and policy regulation and other factors, China's agricultural prices are cyclical volatility and non-linear fluctuations [2], "garlic you hard", "ginger a military" and "cilantro assassin". "Cilantro assassin" and other phenomena occur from time to time, seriously affecting the production and circulation of agricultural products, residents' lives and social stability. Therefore, price forecasting for agricultural products can help growers adjust their planting types and planting areas in a timely manner to avoid market risks and ensure market stability [3].

With the wide application of machine learning and deep learning in recent years, scholars at home and abroad have carried out a series of studies on the price prediction of agricultural products using machine learning and achieved remarkable results. Liu Feng [4] successfully predicted the price of cabbage based on ARIMA model and widely applied the model to other fields. Wang Senpei [5] used the GM(1, 1) model to predict the price of kiwifruit based on gray system theory. Junming Fan [6] proposed a soybean futures price prediction method based on LSTM model, and found that LSTM was better than other traditional methods in soybean price prediction. Yao Sheng [7] used an MSIH-VAR model with two moving average terms and one vector autoregressive term to analyze the relationship between garlic prices and various influencing factors and successfully predicted garlic prices.

This paper proposed a CNN-LSTM combination model based on meteorological and eggplant price data. The model effectively combined CNN and LSTM, CNN extracted the dynamic change characteristics of meteorological factors and price data, and LSTM extracted the features of input time series data, effectively improving the prediction accuracy of the model. The effect mechanism of meteorological factors on the price fluctuation of eggplant was better revealed, and new ideas and methods were provided for the price prediction of agricultural products, agricultural meteorological economy, rural revitalization and policy regulation.

2 Data and Processing

2.1 Data Acquisition

The data in this paper were selected from meteorological data and eggplant price data from January 2015 to August 2024 in Santai County, Sichuan Province. Meteorological data were obtained from the Sichuan Tianqin Big Data Cloud Platform, including

Table 1 Data forms (self-drawn)

Data (day)	T (C)	R(Mm)	P (Pa)	RH (%rh)	T0 (°C)	T20 (°C)	T40 (°C)	Wvp (hPa)	Ssh (W/ m^2)	Price (¥)
30-August–24	29.6	0	952.29	6.73	35.48	33.8	31.48	27.6	1.6	3.6
30-August–24	29.6	0	952.29	6.73	35.48	33.8	31.48	27.6	1.6	3.6
29-August–24	31.12	0	951.27	6.34	38.73	34.87	31.62	28.45	5.4	4.6
27-August–24	33.32	0	950.74	5.44	41.63	35.23	31.53	27.31	9.2	4.6
26-August–24	33.05	0	951.81	5.15	40.8	34.88	31.41	25.35	7.6	4.4
24-August–24	33.81	1	952.34	5.61	42.57	35	31.15	28.24	11.4	4.4

daily temperature (T), precipitation (R), evapotranspiration (EVP), barometric pressure (P), humidity (RH), surface temperature (T0), temperature 20 cm below ground (T20), temperature 40 cm below ground (T40), water vapor pressure (WVP), solar radiation (SSH), and the mean dew point temperature (DPT). The price data were obtained from the database of Sichuan Rural Economy Information Center and web crawler, using Sql statement to query the results from the database and export the Csv file, while using Python programming to obtain the price data of eggplants from Huinong.com (www.cnhnb.com) and the First Vegetable.com (www.shucai1.com), and the form of the raw data is shown in Table 1.

2.2 *Data Processing*

The sample data faced problems such as missing, duplicates, redundancy and inconsistency in the scale, in order to solve these problems Excel was used to remove duplicates according to the date. At the same time, linear interpolation [8] and normalization [9] are used to supplement data and eliminate the scale, of which Formula (1) is the maximum-minimum normalization, and Formula (2) is linear interpolation, as follows.

$$x' = \frac{x - x_{min}}{x_{max} - x_{min}} \tag{1}$$

$$\varphi_1(x) = \frac{x - x_1}{x_0 - x_1} y_0 + \frac{x - x_0}{x_1 - x_0} y_1 \tag{2}$$

Through the above linear interpolation and normalization, the missing data were successfully added to complete the data, and the differences in scale were effectively eliminated to normalize the data to between (0, 1) to facilitate the post-model analysis.

3 Models and Methods

3.1 *Spearman's Correlation Coefficient (Spearman)*

Pearson correlation coefficient (Pearson) and Spearman correlation coefficient (Spearman) are commonly used in machine learning to measure the correlation between two variables [10], by plotting the sample data as a scatter plot, it was found that the data has a non-linear relationship and does not follow a normal distribution. Therefore, this paper uses the Spearman correlation coefficient to analyze the correlation between temperature, ground temperature, humidity, and price data. The calculation formula is as follows:

$$\rho = 1 - \frac{6\sum d_i^2}{n^3 - n} \tag{3}$$

Among them $d_i = x_i - y_i$ denotes the difference in ranking between the two variables, n is the capacity of the input sample data, and ρ is the Spearman's coefficient ranging from -1 to 1. $\rho = 1$ for positive correlation, $\rho = -1$ indicates a negative correlation. $\rho = 0$ indicates no correlation.

3.2 *CNN Modeling*

Convolutional neural network (CNN) model is a deep learning model [11], which has a wide range of applications in the fields of image recognition, speech recognition and time series. It consists of input layer, convolutional layer, activation layer, pooling layer and output layer [12], in which the convolutional layer is the core, and the convolutional calculation of input information is used to extract the special data features, and the pooling layer (Pool) adopts the maximum pooling to compress the input features of the convolutional layer to extract the main features, and the data extracted through the features of the convolutional layer and the pooling layer are combined and finally merged with the fully-connected layer (Fully Con) to form a deep learning model [11], which is widely used in the fields of image recognition and time series. The data extracted from the convolutional layer and pooling layer are merged by the Fully Con layer and finally used for data output [13], and the structure of the model is shown in Fig. 1.

3.3 *LSTM Modeling*

Long short-term memory neural network (LSTM) is an improved model based on RNN [14], which effectively overcomes the problem of gradient vanishing or gradient

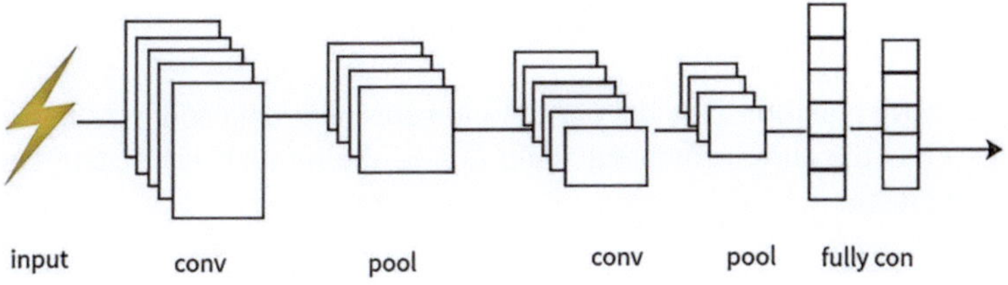

Fig. 1 Structure of the CNN model (self-drawn)

bursting encountered by the traditional RNN when dealing with long sequences [15]. The core of LSTM is the cellular state and the gating mechanism, which consists of the input gates, the output gates and the forgetting gates, and the structure of the model is shown in Fig. 2.

In this case, the forgetting gate is used to choose to forget the data transmitted from the previous unit and to discard or forget some data from the previous moment according to the rule, which is formulated as follows.

$$f_t = \delta(W_f \cdot [h_{t-1}, x_t] + b_f) \quad (4)$$

The output gate determines the final output based on the sigmoid cell state and the tanh layer, which is used for the entire model output with the following equation.

$$O_t = \delta(W_o \cdot [h_{t-1}, x_t] + b_o) \quad (5)$$

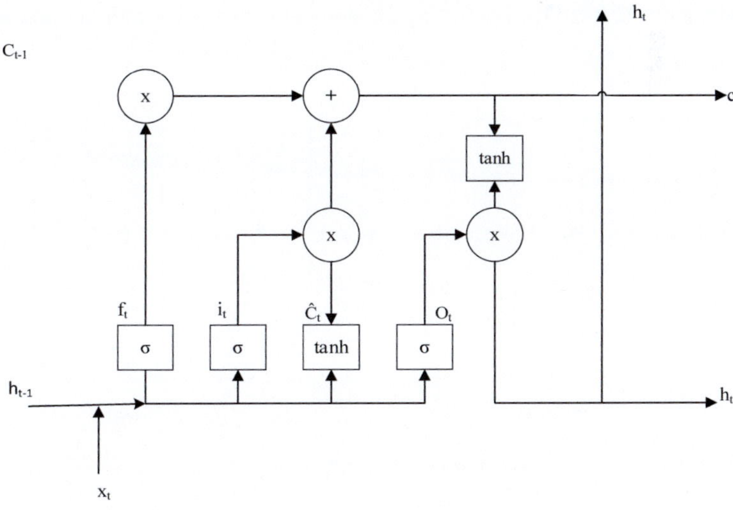

Fig. 2 Structure of the memory unit of the LSTM network (self-drawn)

$$h_t = O_t * tanh(C_t) \tag{6}$$

In the above equation x_t is the current moment input data and σ is the sigmoid function, C_t is the current moment state, h_{t-1} is the implicit layer state from the previous moment.

3.4 CNN_LSTM Model

Meteorological data and price data and so on have high latitude and nonlinear characteristics [16], CNN and LSTM have their own advantages in dealing with time series data, in this paper, we construct a CNN_LSTM combination model by concatenation, which consists of an input layer, a convolutional layer, an LSTM layer, and an output layer, and the structure of the model is shown in Fig. 3.

In the model, the input layer is used to input the main meteorological influences such as temperature, barometric pressure, ground temperature and eggplant price data. The CNN layer performs the convolution operation to extract the features from the input data, which has three convolutional layers (Con1d), three batch normalization layers (Batch), and the activation function is SiLU function, and the data are input into the LSTM layer through the processing. The LSTM layer consists of two LSTM network layers and one Dropout layer [17], in which LSTM selects and calculates the features extracted by CNN, and Dropout randomly discards certain features to avoid the problem of feature selection, and the LSTM layer is composed of two LSTM network layers and one Dropout layer. The LSTM layer consists of two LSTM network layers and one Dropout layer [17], in which the LSTM selects and calculates the features extracted by the CNN, and the Dropout randomly discards certain features to avoid the overfitting phenomenon during the training process.

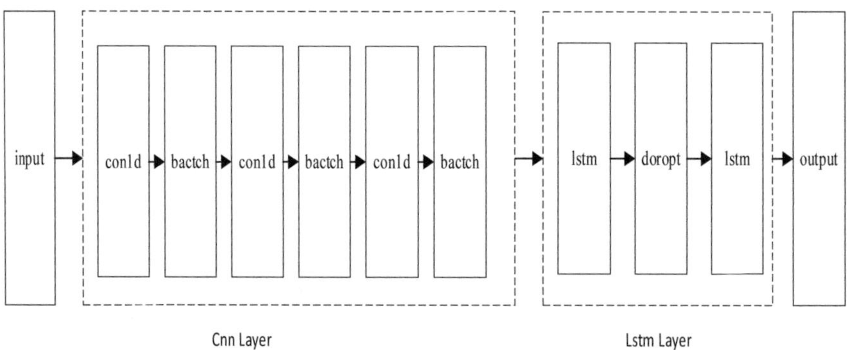

Fig. 3 Structure of the CNN_LSTM model (self-drawn)

3.5 Evaluation Indicators

The price of eggplant has randomness and uncertainty, and there is some error between the predicted value and the real value, in this paper, we chose three indexes of mean square error (RMSE), mean absolute deviation (MAE) and linear regression coefficient of determination R^2 for model evaluation [18] with the following formulas.

$$MSE = \frac{1}{n}\sum_{i=1}^{n}\left(y_i - y_i^{\wedge}\right)^2 \tag{7}$$

$$MAE = \frac{1}{n}\sum_{i=1}^{n}\left|y_i - y_i^{\wedge}\right| \tag{8}$$

$$R^2 = 1 - \frac{\sum_i \left(y_i^{\wedge} - y_i\right)^2}{\sum_i (y^* - y_i)^2} \tag{9}$$

y_i is the actual price. $y_i^{\wedge}$ is the predicted value of the model output, in which the smaller MAE and RMSE indicate that the smaller the error between the predicted value and the actual value, the higher the accuracy of the model, and the value of R^2 is in the range of (0, 1), and the closer the value is to 1, the better the convergence of the model.

4 Experimentation and Analysis

4.1 Parameterization

In this paper, the weather and price numbers from 2015 to 2024 are selected, and the CNN-LSTM model is trained and tested after reweighting, linear interpolation and normalization. The parameters are set as follows: 80% as the training set, 20% as the test set, 3 CNN layers and 2 LSTM layers, and the time step is 14, i.e., the data of the first 14 days are used to predict the price of the next day. The size of the convolutional kernel of the CNN layer is 5 × 5, the convolutional kernel is 32, 64 and 32 respectively, the hidden layer of LSTM is 64 and 128 respectively, the time step is 7, i.e., to predict the price of the next day with the weather and price data of the previous 7 days, the Dropout coefficient is 0.1, the epoch is 300 times, the learning rate is 0.001, and the optimization function is Adam's algorithm.

4.2 *Correlation Analysis of Meteorological Factors and Prices*

Eggplant is a temperature-loving crop with high temperature requirements, with a suitable temperature of 30–35 °C and no <17 °C at night, in addition to being sensitive to changes in light, air pressure and ground temperature [19]. When the weather changes, it affects the growth and yield of eggplant and thus the price. Figure 4 shows the values of correlation coefficients between weather factors and eggplant prices calculated by Spearman and plotted on a heat map.

The graph above visualizes the correlation between temperature, pressure, temperature and price, and the strength of the correlation is determined by the coefficient values. All the correlation coefficients in the graph are not zero, indicating that the selected weather elements and prices are correlated. Among them, barometric pressure (P) and price (Price) are positively correlated, and the rest of air temperature (T), precipitation (R), dew point temperature (DPT), temperature 20 cm below ground (T20), temperature 40 cm below ground (T40) are negatively correlated with price. The absolute value of the correlation coefficient between meteorological factors and prices ranges from 0.09 to 0.61, among which the correlation of solar radiation (SSH) is 0.09 and the correlation is the weakest so that its influence can be ignored, and the absolute value of the correlation coefficients sums up in the average final value of 0.4.

It can be seen from the above figure that the correlation coefficient of average air temperature (T), dew point temperature (DPT), water vapor pressure (WVP), ground temperature (T0), underground 20 cm temperature (T20), underground 40 cm temperature (T40) and air pressure (P) is $\rho > 0.4$. The P values of 2.028e-114, 2.815e-141, 1.458e-145, 1.549e-92, 9.79e-135 and 6.14e-176 were all less than 0.005, indicating that the price of eggplant was significantly correlated with these meteorological

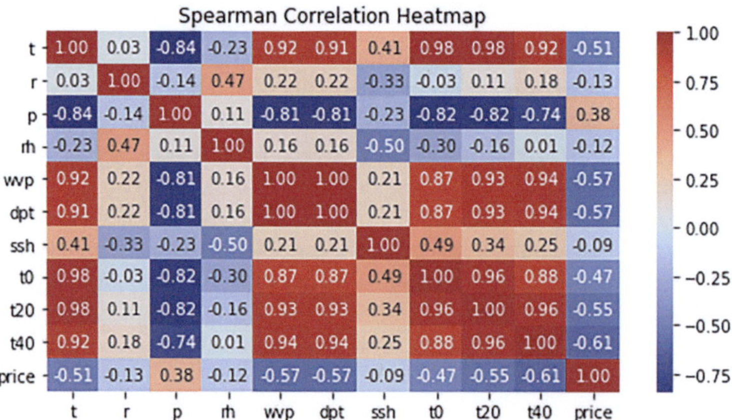

Fig. 4 Spearman heat map (self-drawn)

factors. Therefore, the average air temperature (T), dew point temperature (DPT), water vapor pressure (WVP), ground temperature (T0), underground 20 cm temperature (T20), underground 40 cm temperature (T40) and air pressure (P) were finally selected to input the model.

4.3 Comparative Analysis of Experimental Results

4.3.1 Comparative Analysis of Loss Diagrams

In order to verify the performance of CNN_LSTM model, Pytorch framework was used to call Torch, Sklearn and Numpy libraries to predict the price of eggplant under the same historical data using LSTM, CNN and CNN_LSTM models, respectively, and the comparison of Loss plots is shown below.

The above Loss chart is drawn by using the mean square error (MSE), and the Loss chart can visualize the changes of the function value during the training process of the model, so that it is easy to observe the difference between the predicted price and the actual price and the trend of the change, so as to follow up the tuning of the model. From Fig. 5, it can be seen that the Loss value decreases faster than that of CNN and CNN-LSTM after 300 iterations, and the minimum values are 2.14, 2.04, and 1.37, respectively. Figure 4a shows that the Loss value of LSTM decreases more slowly than that of CNN and CNN-LSTM between 27 and 23, and Fig. 5b shows that the Loss value of CNN decreases more slowly than that of the other two models between 12.5 and 2.8. Figure 5b shows that the Loss value of CNN decreases slower than the other two models between 12.5 and 2.8, and Fig. 5c shows that the Loss value of CNN-LSTM decreases faster than that of CNN and LSTM during the whole training process, the Loss value of the three models shows a slight increase and then decreases during the training process, the CNN shows an increase in the training process up to 35 times, and it decreases when it increases in the 50th time, and the LSTM and CNN-LSTM show an increase in the Loss value up to 15 times, and it decreases when it increases in the 25th time, and it decreases in the 15th time. LSTM and CNN-LSTM training to 15 times began to rise 25 times began to fall back to normal, but CNN-LSTM in the subsequent training process appeared 5 times slight noise fluctuations, but the fluctuation value is relatively small can be ignored, caused by the fluctuation of the above Loss value may exist Learning rate is higher, the Batch Size setting is unreasonable, etc., and need to continue to optimize the follow-up.

4.3.2 Predictive Analysis

In order to visualize the performance of the three models, the predicted and actual values of LSTM, CNN and CNN_LSTM are compared and analyzed, as shown in Fig. 6.

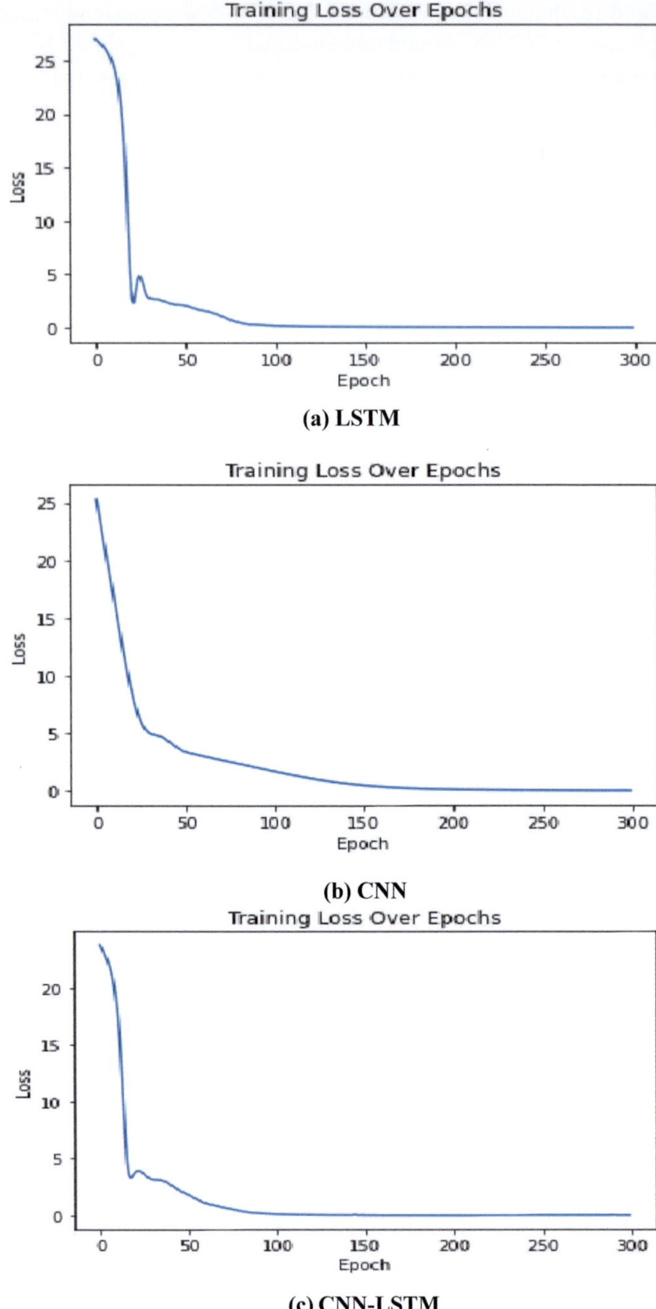

Fig. 5 Loss plots for each model (self-drawn)

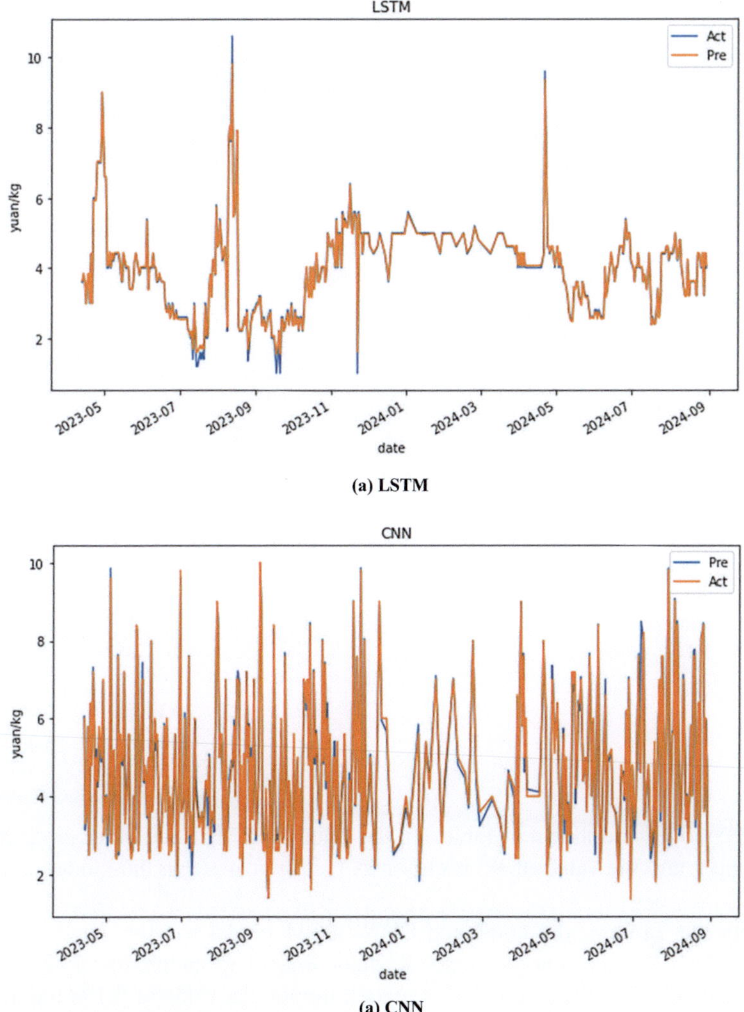

(a) LSTM

(a) CNN

Fig. 6 Predictions of the models (self-drawn)

In Fig. 6, the blue curve represents the actual price of eggplant, and the orange curve represents the predicted price of the model. From the figure, it can be seen that the actual price of eggplant is between 1.6 and 10.5 yuan, with a large range of fluctuation, which is convenient for the subsequent price prediction. From Fig. 6, it can be seen that the difference between the predicted and actual values of LSTM, CNN and LSTM-CNN is relatively small, the trend of the price curve is more or less the same, and the prediction of the mutation point has a small difference, and the predicted value is lower than the actual value. In order to further understand the prediction errors of LSTM, CNN and CNN_LSTM models, the evaluation results of

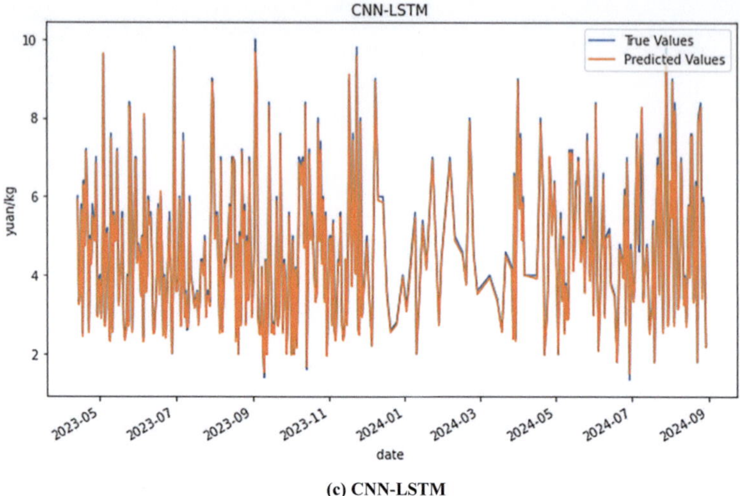

(c) CNN-LSTM

Fig. 6 (continued)

Table 3 Evaluation indicators (self-drawn)

Model	RMES	MAE	R^2
LSTM	0.12	0.07	0.98
CNN	0.2	0.14	0.96
CNN-LSTM	0.1	0.06	0.99

the three models are shown in Table 3, if the values of RMES and MAE of the model are smaller and the value of R^2 is closer to 1, it indicates that the model performs better.

From the Table 3, the combined CNN-LSTM model has the smallest RMES value and MAE value, and the largest R^2 value indicates that the model has the best prediction effect. In the CNN-LSTM model, the RMES value of 0.1 is reduced by 0.02 and 0.1 compared with the single LSTM and CNN models, respectively, the MAE value of 0.06 is reduced by 0.01 and 0.08 compared with the LSTM and CNN, respectively, and the R^2 value of 0.99 increases by 0.01 and 0.03 compared with the LSTM and CNN, respectively, so that this model is superior to the two other model.

5 Conclusions and Future Directions

Eggplant cultivation in Sichuan Province is extensive, mainly distributed in Zigong, Panzhihua, Santai County and Liangshan, etc. The cultivation area is among the highest in the country. As one of the popular vegetables, the price fluctuation of eggplant seriously affects the production flow of agricultural products, residents' lives and social stability [20], while the price of eggplant is affected by many factors such as weather, transportation, holidays and policy regulation [21], so it is difficult to accurately predict the price of eggplant.

Aiming at the nonlinear transformation characteristics of eggplant price, a combined CNN-LSTM model is proposed based on CNN and LSTM. In this paper, the dataset was produced by using the meteorological and price data of Santai County, Sichuan Province for the last 10 years from 2015 to 2024. The correlation value between meteorological factors and prices was calculated by using Spearman correlation coefficient (Spearman), in which the lowest is 0.09, the highest is 0.61, and the mean is 0.4. The main meteorological influences were identified by the coefficient value. The CNN-LSTM model was initially explored to predict the price of eggplant based on the meteorological factors, and the comparative analysis showed that the prediction results of the model and the actual results were more accurately fitted, and the prediction accuracy was higher, and the performance of the model was significantly better than that of LSTM and CNN.

The model predicts and analyzes the price of eggplant based on meteorological factors from a time series perspective, solves the problem of insufficient prediction accuracy of a single model, provides a scientific judgment basis for growers, sellers and managers, and plays a guiding role in stabilizing the market [22]. However, there is still a certain gap between the prediction results and the actual values, which needs to be optimized in the following aspects. Firstly, we should enrich the sample space, collect eggplant price data through multiple channels, increase the sample of eggplant price data, and ensure the authenticity and accuracy of the data. Secondly, natural disasters, transportation, storage, holidays and per capita income are included in the model, so as to analyze from multiple angles and dimensions. The Attention mechanism is added to increase the complexity of the model, and the parameters are optimized. Through the above operations, we continue to optimize the model and improve the accuracy of prediction, so as to provide new ideas and measures for the prediction of agricultural prices, agro-meteorological economy, rural revitalization and policy control.

Acknowledgements This study is supported by the Science and Technology Development Fund for Sichuan Key Laboratory of Plateau and Basin Heavy Rainfall, Drought and Flood Hazards (No. SCQXKJQN202323).

References

1. Hu X (2022) Research on agricultural product price forecasting based on machine learning. Sichuan: Sichuan Agric Univ 1–62
2. Ren W (2011) An analysis of price forecasting methods in agricultural market. China Agron Bull 27(26):209–212
3. Yu S, Gong S (2016) Food price prediction based on Lasso and support vector machine. J Hunan Univ (Social Science, Edition) 30(01):71–75

4. Liu F, Wang R, Li C (2009) The application of ARIMA model in the prediction of agricultural commodity prices. Comput Eng Appl (25):238–239, 248
5. Wang S, Li X, Zhao J, Guo Y (2021) Analysis and prediction of kiwifruit price based on gray prediction model. Agric Sci Technol Manag 40(02):67–71
6. Fan J, Liu H, Hu Y (2021) Soybean futures price prediction based on LSTM deep learning. Price Month12(5):7–15
7. Yao S (2021) Research on price fluctuation characteristics of small agricultural products and their influencing factors—taking garlic storage-resistant agricultural products as an example. Price Theory Pract 2021(08):100–103+186. https://doi.org/10.19851/j.cnki.CN11-1010/F.2021.08.252
8. Duan Q, Zhang L, Wei F et al (2017) Forecasting model and validation of aquatic product price based on time series GA-SVR. J Agric Eng 33(01):308–314
9. Li D, Huang H (2017) The empirical test of the influencing factors of the price fluctuation of agricultural products in China. Stat Policy 2017(20):136–139. https://doi.org/10.13546/j.cnki.tjyjc.2017.20.034
10. Zhang S, Xu K, Xiao H (2021) Gray correlation analysis of meteorological factors with corn failure and yield in hilly areas of central Sichuan. J Sichuan Univ Agric Univ 39(5):666–674
11. Zhang S, Song C. Research on model stock price prediction based on CNN-LSTM. Cooperative Economics and Technology
12. Wang G, Pan D, Liu X (2022) Construction of agricultural product price prediction model based on gated recurrent unit network. J Shenyang Normal Univ (Nat Sci Edn) 40(05):451–456
13. Wang Z, Chen X, Pang T et al (2021) A hog price prediction method base on improved time convolutional network. J China Agric Univ 26(12):137–144
14. Li F, Li Z, Liu S, Shi T (2019) Application of LSTM-RNN in continuous casting lower slag prediction system. Fujian Comput 35(7):181–183
15. Zhang B, Ren W, Wu B (2021) A short-term price prediction method for agricultural products based on LSTM model. Trop Agric Sci 41(04):131–136
16. Xue Z, Li J, Sun Y, Wang Z (2014) Effects of meteorological conditions on the price of green vegetables and its relationship model. Shanghai J Agric 30(5):114–119
17. Chen L, Ting X, Zhang Y, He J (2024) A prediction method of red date futures price based on LSTM. Agric Technol 44(01):162–165
18. Yang H, Zhao X, Wang L (2023) A review of data normalization. Method Comput Eng Appl 59(3):13–22
19. Luo Y (2018) Difficulties and suggestions on the integration of big data and agricultural product prices-taking Guizhou agricultural economy network as an example. Mod Agric Sci Technol 2018(12):293–296
20. Yang X (2018) An empirical study on the impact of drought disaster on local agricultural product price changes in Yunnan Province—based on panel data of 16 prefectures. Chin Agric Bull 34(24):95–100
21. Guo D, Tan Y (2020) Research on price fluctuation of agricultural products and it stochastic factors. Price Theory Pract 2020(03):60–70+177. https://doi.org/10.19851/j.cnki.cn11-1010/f.2020.03.358
22. Leng K, Chen J, Liu L (2017) Research on the impact of agricultural products price fluctuation on residents' food consumption behavior. Res Bus Econ Res 2017(13):103–110

BY NC ND

Urban Resilience Assessment Methods and Enhancement Strategies Under Public Health Emergencies: A Case Study of Northeast China

Xiaopu Bi, Guangyuan Feng, and Jingxiang Zhang

Abstract In recent years, resilient cities have become an important topic in urban development. The global COVID-19 pandemic has raised widespread concerns regarding the prevention and control of public safety incidents. In the post-pandemic era, enhancing cities' capacity to withstand and recover from large-scale public safety events has become an urgent issue. This study concentrates on 35 prefecture-level cities in Northeast China and constructs an indicator system based on five dimensions: social security, economic development, urban environment, intelligence, and resilience. By combining the theory of Complex Adaptive Systems (CAS) with urban resilience, and utilizing techniques such as geographic detectors and the NK model, the study provides strategies for enhancing urban resilience. The study identifies socio-economic factors, social security, and the built environment as the core influences on urban resilience in Northeast China. Socio-economic factors account for the largest contribution, at 58.7%, followed by social security at 16.6%, and the built environment at 10.2%. Although the overall resilience level in the region is relatively high, adaptability remains low: 0% of cities are classified as "Adaptation," 77.1% as "Local Adaptation," and 22.9% as "Non-adaptation." These findings indicate that the high resilience levels are primarily attributable to historically low socio-economic development, which has minimized the relative impact of pandemic shocks. The study categorizes cities based on their adaptive capacity into three types: economic priority, livelihood security, and planning and construction, and proposes corresponding strategies for improving resilience for each type of city.

Keywords Urban resilience · Public safety · Geodetector · CAS

Xiaopu Bi and Guangyuan Feng—These authors contributed equally.

Foundation: Major Program of National Fund of Philosophy and Social Science of China (24&ZD148).

X. Bi · G. Feng · J. Zhang (✉)
School of Architecture and Urban Planning, Nanjing University, Nanjing, China
e-mail: 3593786@163.com

B. He et al. (eds.), *Climate-Resilient, Low-Carbon, and Sustainable Cities*, Lecture Notes in Civil Engineering 754, https://doi.org/10.1007/978-981-95-2169-2_3

1 Introduction

The worldwide outbreak of the novel coronavirus (COVID-19) pandemic has caused significant damage to China's socio-economic development, representing a major public health emergency with wide-ranging social impacts. This event has served as a wake-up tocsin for countries all around the world. Although China's economy has not fully recovered from the pandemic's aftermath, the threat of large-scale public safety events similar to COVID-19 remains long-term, making proactive prevention essential. In this context, how can cities better withstand the impact of such events? How can cities quickly stabilize their socio-economic systems after public safety incidents? These questions have become central topics in academic discussions and pressing issues for society at large.

To address these questions, this study evaluates cities' resistance and recovery capabilities under the impact of sudden public safety events, based on resilience theory. Additionally, using the NK model from Complex Adaptive Systems (CAS) theory, the study regards the resilience level of a city during a public health emergency as its adaptive capacity, aiming to identify the optimal path for enhancing urban resilience.

Resilience research is currently a hot topic, with urban disaster resilience assessment being an important application direction [1]. This shift from a vulnerability perspective to a resilience perspective in disaster prevention and mitigation represents a new approach to urban risk governance, emphasizing improved disaster response and post-disaster recovery capabilities [2]. Current research on urban resilience mainly focuses on climate change, flooding, and other natural disasters, with considerable attention given to cities' resilience against natural hazards. However, the impact of infectious disease outbreaks on urban development is just as serious as other disasters, but it has not received as much focus. In recent years, with the increasing frequency of public health emergencies, this issue has gradually gained widespread attention from scholars worldwide. For example, Jennifer B. Nuzzo and others analyzed healthcare system resilience during the 2014–2016 Ebola outbreak [3]; Glenn McCartney and colleagues examined the resilience mechanisms of Macau, a tourism city, in the course of the COVID-19 pandemic [4]; Ayyoob Sharifi examined the effects of the COVID-19 pandemic on urban systems and its consequences for urban planning [5]; and Tong Yun et al. evaluated the resilience of Chinese cities post-COVID-19 using migration data [6]. These studies have provided valuable insights into the resilience mechanisms of cities during pandemics and have offered methods for future research.

Existing studies on urban resilience exhibit two primary limitations: (i) a predominant focus on specific sectors within urban systems, often neglecting a holistic assessment of overall urban resilience levels, and (ii) an emphasis on analyzing factors influencing resilience, with limited exploration of mechanisms for enhancing resilience. To address these gaps, this study aims to: (i) develop an evaluation framework for assessing overall urban resilience levels and (ii) apply the fitness landscape model (NK model) to identify optimal pathways for enhancing urban resilience, offering practical guidance for future urban planning and development.

2 Literature Review

2.1 Key Concepts of Urban Resilience

Urban resilience means a city's ability to maintain or rapidly restore its original functions and self-improve under limited adaptive capacity when faced with disturbances. It encompasses both the social-ecological and socio-technical networks within the urban system at various spatial and temporal scales. As an emerging topic in urban studies, urban resilience still faces debates regarding its definition, the concept of resilience, system equilibrium, system transformation mechanisms, and the time scale for resilience enhancement [7]. The concept of resilience has progressed from engineering resilience to ecological resilience, and then to evolutionary resilience. Evolutionary resilience moves away from the focus on a stable equilibrium, highlighting that "resilience is a dynamic characteristic of systems, closely linked to their ongoing adaptive capacity," offering a more compelling theoretical foundation [8].

This study focuses on the urban area as the scope of urban research, combining the resilience concept from evolutionary resilience with the concept of fitness in Complex Adaptive Systems (CAS) theory. The aim is to establish a resilience evaluation system that comprehensively reflects a city's resistance and recovery capabilities to external disturbances, while providing an optimized pathway for enhancing urban resilience over long time scales.

2.2 Application of CAS Theory in Urban Resilience Research

As the complexity of cities has garnered increasing attention from scholars, CAS (Complex Adaptive Systems) theory has found growing application in urban research. Proposed by Holland, CAS theory emphasizes that complex systems are the result of continuous evolution by multiple adaptive agents interacting with and adapting to their environment. As a third-generation system theory, it offers new perspectives for studying the behavior of complex systems [9, 10]. In urban resilience research, some scholars have combined this theory with resilience concepts. For instance, Zhou Xia and others integrated the concept of resilience with fitness, constructing an evaluation system for community resilience from three perspectives: adaptability, network structure, and diversity [11]. Similarly, Chen Zhirui and colleagues applied CAS theory to develop a mechanism for enhancing the resilience of Beijing's urban system [12].

These studies demonstrate the broad applicability of CAS theory in urban resilience research and provide valuable guidance for integrating CAS theory with resilience theory. This study, based on CAS theory, applies the NK model to analyze resilience enhancement pathways, offering more scientifically grounded and feasible strategies for improving urban resilience.

3 Study Area and Data Sources

3.1 Study Area

China's Northeast region is made up of three provinces—Heilongjiang, Jilin, and Liaoning—encompassing 36 prefecture-level administrative units and covering an area of about 808,400 km^2. Based on the results of the seventh national census, the resident population is about 97.91 million. In 2021, the regional GDP of the Northeast was approximately 5.57 trillion ¥. This study selects 35 prefecture-level cities (autonomous prefectures) from the three provinces, excluding the Greater Hinggan Mountains area, as the research area.

The Northeast is a key agricultural region in China, playing an important role in national ecological security, food security, industrial security, energy security, and national defense. In recent years, however, the region has faced challenges such as overcapacity and population decline, leading to slow economic development. Moreover, related studies have shown that viruses have a higher transmission risk in cold and dry climates [13, 14]. Therefore, taking the Northeast region as the research scope can not only provide scientific references for the resilience construction of cities in the Northeast region, but also provide theoretical support from a strategic perspective; It can also provide universal reference cases and experiences for other urban areas.

3.2 Data Sources

This study utilizes data from the 2021 statistical yearbooks of Heilongjiang, Jilin, and Liaoning provinces, as well as the seventh national census data, to reflect the socio-economic conditions of each city (autonomous prefecture). The data is sourced from official local government websites, ensuring a high level of reliability. POI data related to medical, educational, shopping facilities, residential areas, hotels, convention centers, and sports stadiums was obtained from the Amap Open Platform (https://lbs.amap.com/), to reflect the state of urban infrastructure. Information on tertiary hospitals and top-tier hospitals was sourced from the National Health Commission's Medical Institutions Query Website (http://zgcx.nhc.gov.cn: 9090/unit/index), to assess the level of healthcare systems in each city (autonomous prefecture). Air quality data for 2021 was obtained from the National Pollution Source Monitoring Information Management and Sharing Platform (https://wryjc.cnemc.cn/), to evaluate the environmental quality of cities. Data from Baidu Migration Big Data Platform (https://huiyan.baidu.com/) for the period between April 15–21, 2022, as well as 2021 migration data, was used to reflect changes in urban population flow.

4 Methodology

4.1 Selection of Urban Resilience Evaluation Indicators

Various resilience evaluation frameworks have been established globally. For example, the system developed by the Rockefeller Foundation includes four dimensions: "Urban Systems and Services," "Health and Well-being" "Leadership and Strategy" and "Society and Economy". Similarly, Japan's Kitakyushu City Center proposed an index based on institutions, hardware, and software, while Hosei University introduced a four-dimensional system focusing on urban, citizen, administrative, and comprehensive indicators. Among these, the Rockefeller Foundation's system is more comprehensive and practical for guiding implementation [15].

This study adapts the Rockefeller Foundation's resilience framework, adding indicators to reflect a city's recovery capacity during public health crises. The resulting system evaluates resilience across five dimensions: social security, economic development, urban environment, intelligence, and recovery.

(1) Social Security: Measures the city's ability to maintain health and livelihoods during emergencies, including indicators of healthcare system robustness and vulnerability, such as the elderly population proportion, healthcare coverage, and unemployment benefits.
(2) Economic Development: Assesses the city's economic capacity to withstand public health shocks, including economic strength indicators and healthcare investment capacity.
(3) Urban Environment: Evaluates whether the city's infrastructure and external environment are conducive to effective epidemic control. This includes indicators of urban infrastructure, air quality (which impacts health) [16], and arable land for food security.

To assess urban planning, this study includes two indicators: facility diversity and living convenience, based on 15-min walking distance criteria. Convenience is measured by the average number of medical and shopping facilities within 1250 m of residential areas. Facility diversity is quantified using the Simpson Diversity Index to assess the variety of facilities (medical, shopping, residential, and business) in a 1250 m * 1250 m grid. The formula is as follows:

$$D = 1 - \sum_{i=1}^{S} P_i^2 \tag{1}$$

where P_i represents the proportion of facility type i in the total number of facilities within the grid, and S is the number of facility types within the grid.

(4) Intelligence: Reflects the city's governance capacity and the self-organizational ability of residents. It includes indicators of urban governance capabilities, such as city size and research capacity, as well as indicators of residents' decision-making ability.
(5) Recovery: It assesses a city's capacity to bounce back following a public health emergency. Research indicates that the Relative Recovery Index (RRI) and the Recovery Gap Index (RGI), derived from Baidu migration data, effectively capture the socio-economic recovery levels of cities after public health crises [5]. This study uses Baidu migration data from the week following April 15 to assess the recovery capacity of cities in Northeast China.

The Relative Recovery Index (RRI) is defined as:

$$RRI_{i,t} = \frac{A_{i,t}}{B_{i,t}} \tag{2}$$

Here, $RRI(i, t)$ denotes the relative recovery index for city i at time t, $A(i, t)$ refers to the migration index for city i at time t, and $B(i, t)$ represents the migration index for city i at the same time t.

The Recovery Gap Index (RGI) is defined as:

$$RGI_{i,t} = \frac{A_{i,t} - B_{i,t}}{T_i} \tag{3}$$

In this case, $RGI(i, t)$ stands for the recovery gap index for city i at time t, $A(i, t)$ represents the migration index for city i at time t, $B(i, t)$ indicates the migration index for city i at time t, and $T(i)$ refers to the number of days in the phase for city i.

4.2 *Comprehensive Evaluation Method of Indicators*

Among the various urban resilience evaluation methods, the composite index approach is commonly used because of its simplicity and ease of implementation. However, this method also encounters the problem of subjectivity in weight assignment [17]. To overcome this, the study employs two objective weighting techniques: the CRITIC method and the TOPSIS method, selecting one based on different scenarios. These methods are utilized to determine the indicator weights, and the Natural Breaks (Natural Jenks) method is used to categorize the overall scores.

(1) CRITIC Method: The CRITIC method is an objective weighting approach that relies on the intrinsic variability of the data. This evaluation method determines indicator weights based on two factors: data variability and indicator conflict. Variability reflects the degree of fluctuation in the sample values of an indicator, represented by the standard deviation; indicator conflict is expressed through the correlation coefficient between indicators. The higher the similarity between indicators, the more information they reflect, and the lower their weight [18]. Since the resilience indicator system used in this study has a large disparity in the number of evaluation indicators across different dimensions, this method helps avoid domination of the evaluation result by dimensions with more indicators.
(2) TOPSIS Method: The TOPSIS method assesses the relative quality of samples by measuring the distance between the evaluation object and the ideal best and worst solutions. It employs the entropy weight method to objectively assign weights to the indicators. By constructing the best and worst solutions and evaluating the distance between each sample and these solutions, the scores for each sample are determined [19]. This method is used in the study for further dimensionality reduction of the evaluation factors derived from the analysis.
(3) Natural Breaks Method: The Natural Breaks method is a classification method based on inherent natural groupings in the data. After classification, the differences between classes are maximized, and the differences within each class are minimized [20]. This method effectively categorizes resilience levels and factor scores.

4.3 Analysis of Influencing Factors and Improvement Pathways

To further identify the key factors influencing urban resilience, and provide valuable insights for improving cities in the Northeast, the study uses factor analysis and the Geodetector method to explore the main factors influencing urban resilience. Additionally, the NK model is applied to further study the resilience improvement pathways.

(1) Factor Analysis: The study simplifies the indicators influencing urban resilience through factor analysis, reducing them to a few key factors for subsequent analysis.
(2) Geodetector Method: Geodetector is a statistical technique employed to examine spatial heterogeneity, assess explanatory factors, and analyze the interactions between variables [21]. In this study, Geodetector is employed to detect the leading factors influencing urban resilience, identify each factor's indicators, and analyze the interaction relationships between the factors, providing a basis for determining resilience improvement pathways.
(3) NK Model: The NK model is a commonly used quantitative analysis method in the application of CAS theory and fitness landscape theory. This method focuses on describing the interaction relationships among various elements in a complex

system and their effects on the overall system's fitness. It can help identify both the optimal fitness subject and the best improvement path for fitness, making it suitable for studying complex systems like cities [22]. In this study, the NK model treats the cities in the research area as species with N genes, and views the resilience enhancement of cities during the pandemic as the species' continuous adaptation to the new environment.

5 Results

5.1 *Urban Resilience Evaluation Results*

The study uses the CRITIC method to perform a comprehensive analysis of various indicators after excluding dimensional differences. The urban resilience evaluation results for Northeast China are shown in Fig. 1. The findings reveal that 7 cities (20%) have high resilience, 7 cities (20%) have relatively high resilience, 7 cities (20%) have average resilience, 5 cities (14.3%) have relatively low resilience, and 9 cities (25.7%) have low resilience. Overall, urban resilience in Northeast China is relatively strong, with a higher proportion of cities falling into the "relatively high" and "high" resilience categories.

The cities with better resilience can be broadly categorized into two types: (1) cities with strong economic development and a high capacity to resist public health emergencies, such as Harbin, Changchun, Shenyang, and Dalian; and (2) cities with weaker economic development but strong self-recovery abilities post-disaster, such as Yichun and Heihe.

In terms of spatial distribution, cities in the southern part of Liaoning Province generally exhibit lower resilience, while cities in the northern provinces of Jilin and Heilongjiang tend to have higher resilience. This could be attributed to the larger population and relatively stronger economic development in Liaoning, where public health emergencies significantly disrupt social and economic operations. In contrast, cities in the northern provinces have weaker economies, making them less susceptible to economic impacts from public health crises. Additionally, the generally better natural environmental conditions in the northern cities may also contribute to their higher resilience levels.

5.2 *Resilience Enhancement Path Analysis*

The study conducted KMO and Bartlett's tests on 31 evaluation indicators across four dimensions—social security, economic level, urban environment, and intelligence—after standardization. The results indicated significant correlations between the indicators, justifying the use of factor analysis. Factors with eigenvalues >1 were selected as the primary influencing factors. The initial factors were rotated using the

Fig. 1 Study area: the Northeast region of China (self-drawn)

Kaiser Normalization Maximum Variance method, and indicators with correlations >0.5 were used to interpret the rotated factors. The factor analysis revealed seven main factors, with a cumulative variance percentage of 82.46% (see Table 1).

Table 1 The conceptual framework of the urban resilience (self-drawn)

Concepts	Understanding of resilience	Theory
Engineering resilience	The ability of a system to return to its original state after being disturbed	Engineering mechanics
Ecological resilience	The ability to form a new equilibrium state after being disturbed, rather than emphasizing the system's capacity for restoration	Ecology
Evolutionary resilience	The ability of a system to continuously learn, adapt, and transform itself to accommodate new environments when external disturbances occur	Evolution theory

The seven factors were input into the Geodetector for spatial differentiation, factor detection, and interaction analysis. The results show that each of the seven factors explains a limited portion of the spatial variation in urban resilience. However, when the factors interact, they significantly improve the explanation of the spatial differentiation of urban resilience (see Tables 2 and 3).

The factors with significant interactions, closely related to urban economic development, namely Factors 1 and 4, were grouped as socio-economic factors to highlight the role of economic development in urban resilience. Factors 3, 5, and 6, which show a high correlation with the medical system and social security indicators, were grouped as social security factors, reflecting the city's ability in public welfare and healthcare. Factors 2 and 7, which are more closely related to urban infrastructure indicators, were classified as built environment factors, representing the quality of the city's internal construction environment.

These three categories of factors were used as adaptive factors in the NK model, with their interaction relationships corresponding to the hierarchical structure of factors in the model. The analysis was then conducted using Geodetector to determine the fator contribution values and interaction relationships (see Tables 4 and 5, Fig. 2).

The socio-economic factor (X1) has the highest fitness contribution value of 0.586868, followed by the social security factor (X2) with a contribution value of 0.165623, and the built environment factor (X3) with a contribution value of 0.102235. The socio-economic factor and social security factor are independent of each other, while non-linear enhancement relationships exist between the socio-economic and built environment factors, as well as between the social security and built environment factors. According to this analysis, the three factors are classified into two states ("0" for low scores and "1" for high scores) using the Natural Breaks method. This results in eight possible fitness combinations in the NK model (as shown in Table 6).

From the fitness contribution values and their interaction relationships, it can be concluded that the "111" combination represents the city with the highest adaptive capacity. Among the combinations with two high factor scores, "101" exhibits the strongest adaptability, followed by "110" and "011." In combinations with one high factor score, "100" shows the strongest adaptability, followed by "010" and "001." The "000" combination represents the weakest adaptability. This analysis leads to the identification of the optimal path for enhancing urban resilience (Table 7).

5.3 *Urban Resilience Enhancement Strategies*

The current adaptation capacity of cities in Northeast China to public health emergencies is generally poor, with no cities fully adapted. Most cities exhibit partial adaptation, while a few cities are non-adaptive. Specifically, 14 cities (40.0%) have adaptation types "101", "110", and "011"; 13 cities (37.1%) are categorized as "100", "010", and "001"; and 8 cities (22.9%) have the adaptation type "000" (Fig. 3).

Table 2 Factor analysis results (self-drawn)

Evaluation indicators	Factor 1	Factor 2	Factor 3	Factor 4	Factor 5	Factor 6	Factor 7
Number of universities and research institutes within the city	0.969						
GDP (in 100 million ¥)	0.963						
Local public fiscal revenue (in 10,000 ¥)	0.938						
Urban population size (number of people)	0.924						
Number of isolation facilities	0.913						
Recovery gap index	0.912						
Number of industrial enterprises above designated size	0.887						
Number of people with a college degree or higher per 100,000 people	0.817						
Number of grade-A hospitals	0.784						
Urban per capita disposable income (¥)	0.636	0.569					
Ratio of permanent urban population to total city population (%)	0.522	0.504					
Per capita urban road area (m^2)		0.866					
Per capita urban drainage pipeline length (m)		0.782					
Per capita GDP (¥)	0.579	0.646					
Ratio of water production capacity to urban permanent population		0.646					
Percentage of population aged 15–59 (%)			−0.806				
Percentage of population aged 65 and above (%)			0.805				
Ratio of urban employees enrolled in basic pension insurance to the population aged 65 and above			0.703				
Per capita grain sown area (m^2)			−0.667				

(continued)

Table 2 (continued)

Evaluation indicators	Factor 1	Factor 2	Factor 3	Factor 4	Factor 5	Factor 6	Factor 7
Urban registered unemployment rate (%)			0.664				
Annual average PM2.5 concentration (micrograms per m^3)				0.832			
Annual average PM10 concentration (micrograms per m^3)				0.798			
Annual average AQI index				0.793			
Per capita healthcare and family planning expenditure (¥)				−0.621			
Number of healthcare technicians per 10,000 people					0.779		
Number of hospital beds per 10,000 people					0.709		
Facility diversity index					0.678		
Number of hospitals per 10,000 people					0.549		
Number of healthcare institutions per 10,000 people						−0.752	
Number of grade-III hospitals per 10,000 people						0.725	
Ratio of urban employees enrolled in basic medical insurance to urban population						0.584	
Per capita park green space area (m^2)						0.855	
Relative recovery index						0.511	

Table 3 Factor detector results (I) (self-drawn)

	Factor 1	Factor 2	Factor 3	Factor 4	Factor 5	Factor 6	Factor 7
q statistic	0.300261	0.028594	0.201385	0.276929	0.179475	0.019628	0.046848
p value	0.526267	0.994125	0.350822	0.106412	0.637594	0.988436	0.999889

Table 4 Interaction detector results (I) (self-drawn)

	Factor 1	Factor 2	Factor 3	Factor 4	Factor 5	Factor 6	Factor 7
Factor 1	0.300261						
Factor 2	0.432202	0.028594					
Factor 3	0.634594	0.420495	0.201385				
Factor 4	0.739317	0.41992	0.681339	0.276929			
Factor 5	0.635188	0.294634	0.441669	0.601995	0.179475		
Factor 6	0.465791	0.26371	0.463073	0.470637	0.286148	0.019628	
Factor 7	0.506642	0.176331	0.384495	0.312865	0.351194	0.203508	0.046848

Table 5 Factor detector results (II) (self-drawn)

	Socio-economic factors (X1)	Social security factors (X2)	Built environment factors (X3)
q statistic	0.586868	0.165623	0.102235
p value	0.014435	0.871568	0.919483

Based on the NK model, cities with different adaptation combinations can follow distinct optimization paths to rapidly enhance their resilience (Fig. 4). The cities are classified into three types based on their optimal improvement directions, with corresponding strategies for each type:

(1) Economic priority type: This type includes cities with adaptation combinations "011", "010", "001", and "000". The main challenge for these cities is their weak economic development, which hinders their fiscal capacity to respond to public health crises. These cities should focus on identifying key industries, optimizing their industrial structure, and prioritizing economic development.
(2) Social welfare priority type: This type includes cities with adaptation combination "101". These cities have strong economic development and relatively well-developed urban infrastructure. They should increase fiscal investment in healthcare, improve per capita healthcare resources, and gradually enhance social welfare systems, such as healthcare and unemployment benefits, to strengthen their resilience during public health emergencies.
(3) Urban infrastructure priority type: This type includes cities with adaptation combinations "110" and "100". These cities have a good economic foundation but relatively underdeveloped urban infrastructure. They should focus on increasing investment in infrastructure, improving urban facilities, and adding green spaces to enhance the quality of the urban living environment, in order to better respond to the impact of public health crises on urban facilities (Fig. 5).

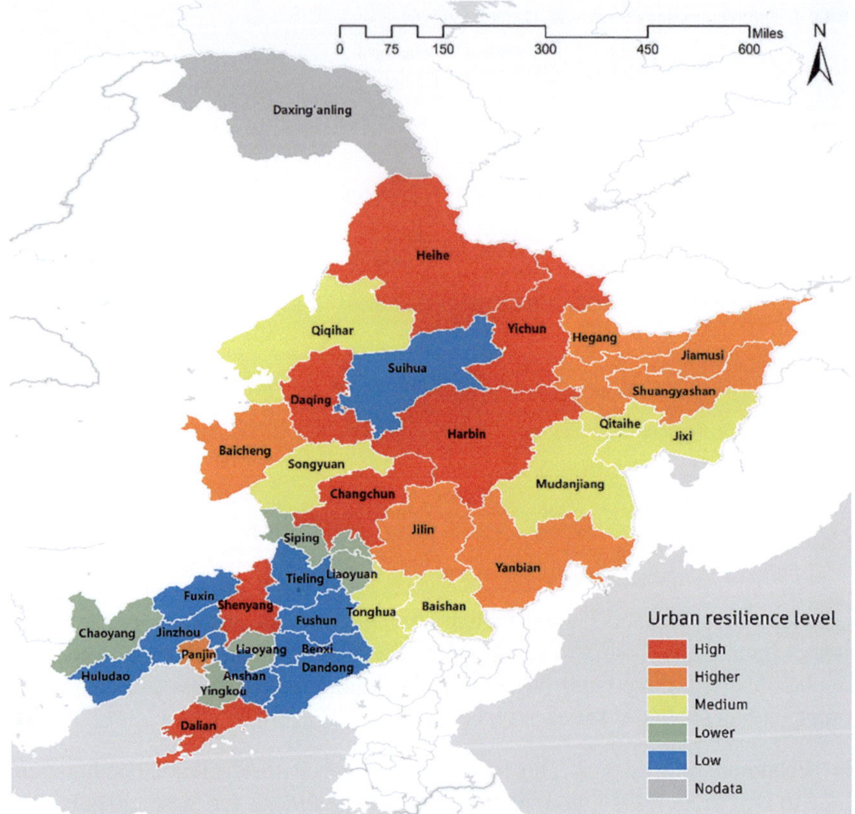

Fig. 2 Evaluation results of urban resilience in northeast China (self-drawn)

Table 6 Interaction detector results (II) (self-drawn)

	Socio-economic factors (X1)	Social security factors (X2)	Built environment factors (X3)
Socio-economic factors (X1)	0.586868		
Social security factors (X2)	0.753668	0.165623	
Built environment factors (X3)	0.767937	0.350645	0.102235

Table 7 Explanation of different types of urban adaptation (self-drawn)

Adaptation type	Fitness combination	Adaptability performance
Adaptation	111	This type of city has a strong economic level, robust social security capabilities, and a good urban construction environment, resulting in a high resilience level when facing the impact of pandemics
Local adaptation	101	This type of city has a good economic development level and a favorable urban construction environment, but lacks sufficient social security capabilities
	110	This type of city has a good economic development level and strong social security capabilities, but its urban construction environment is inadequate
	011	This type of city has strong social security capabilities and a favorable urban construction environment, but its economic development is weak
	100	This type of city has a good economic development level, but lacks social security capabilities and has an inadequate urban construction environment
	010	This type of city has strong social security capabilities, but its economic development level is poor and its urban construction environment is insufficient
	001	This type of city has a good construction environment, but both its economic development and social security capabilities are poor
Non-adaptation	000	This type of city has a low economic development level, insufficient social security capabilities, and a poor urban construction environment, resulting in a low resilience level, indicating an area that urgently needs improvement

6 Conclusion and Discussion

6.1 Conclusion

This study combines the concept of fitness from CAS theory with urban resilience, using analysis methods such as the NK model and Geodetector to evaluate the resilience levels of cities in Northeast China during public health emergencies and analyze their adaptive capacity during the pandemic. The findings are as follows:

(1) The resilience levels of cities in Northeast China show an overestimation. Although the overall resilience level of the region is relatively high, individual cities generally exhibit poor adaptive capacity, with most cities categorized as partially adaptive or non-adaptive. Even cities with higher resilience levels tend to have lower adaptive capacities. This phenomenon may be attributed to the generally lower socio-economic development in Northeast cities, which allows

	Socio-economic factors (X1)	Social security factors (X2)	Built environment factors (X3)
Socio-economic factors (X1)	0.586868		
Social security factors (X2)	0.753668	0.165623	
Built environment factors (X3)	0.767937	0.350645	0.102235

Fig. 3 Interaction of different factors (self-drawn)

for rapid recovery to the original state after a public health emergency, rather than demonstrating strong self-adaptive capacities.

(2) The combination of the NK model and Geodetector offers a better reflection of the interactions between factors, revealing the path of fitness improvement. By integrating urban resilience with CAS theory, a more comprehensive understanding of the underlying causes of urban resilience can be achieved, leading to scientifically sound and reasonable optimization paths.

6.2 *Research Insights and Experiences*

The research results in the Northeast region indicate that for areas with low levels of socio-economic development, the resilience level of cities is not entirely equivalent to their own adaptability, and areas with high resilience levels may not necessarily have high adaptability. Therefore, in the actual assessment of urban resilience level, it is necessary to pay attention to identifying the phenomenon of inflated urban resilience level and conduct a scientific and reasonable comprehensive evaluation based on specific indicators such as adaptability.

In addition, in terms of the comprehensive improvement of urban resilience level, it is necessary to first establish a comprehensive explanatory model for the spatial differentiation characteristics of urban resilience with comprehensive coverage of influencing factors. Based on the analysis results, the adaptability combination type of the city itself should be clarified, and the best improvement direction for different cities (such as social economy, livelihood security, planning and construction, etc.)

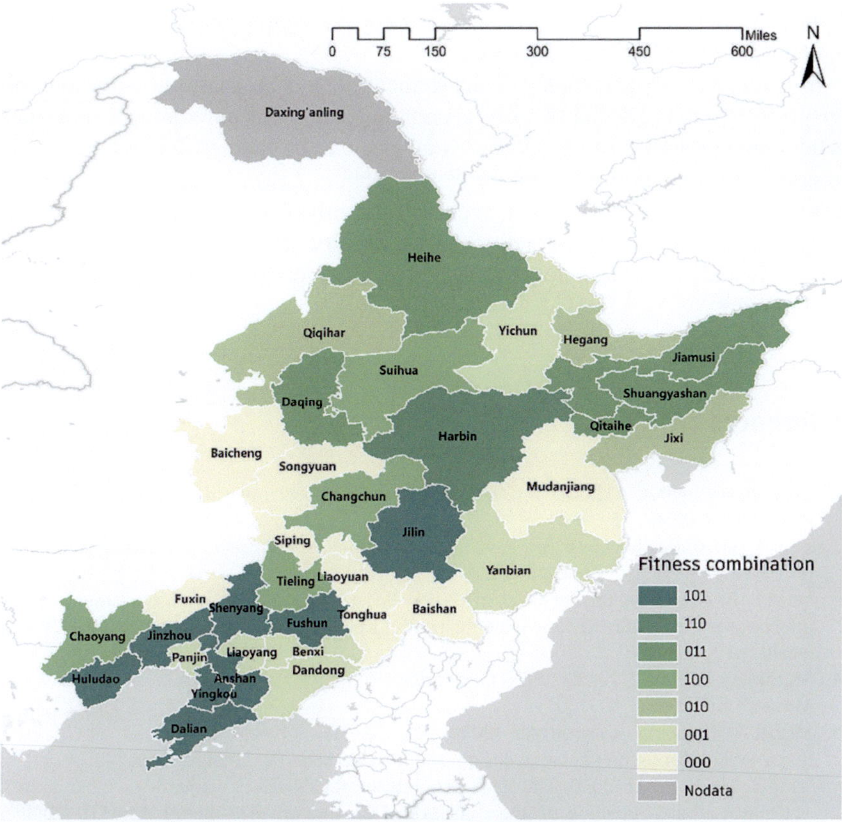

Fig. 4 Types of urban adaptation in northeast China (self-drawn)

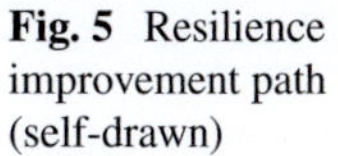

Fig. 5 Resilience improvement path (self-drawn)

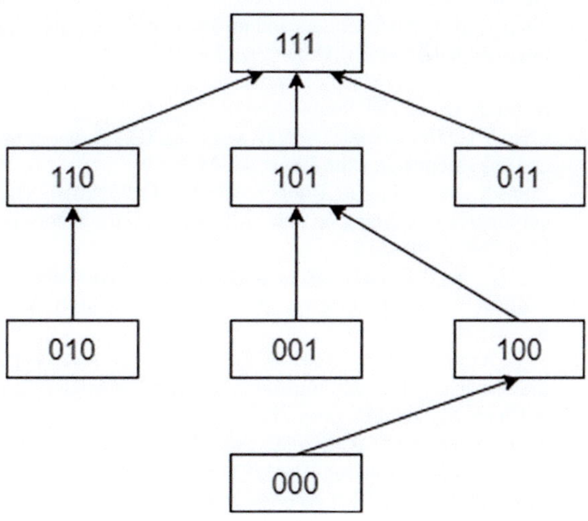

should be determined. Then, corresponding specific optimization strategies should be proposed according to local conditions.

In summary, urban resilience enhancement should be analyzed from multiple perspectives, such as social security, economic level, urban environment, and smart technologies. Based on this, specific strategies can be tailored for different cities according to their optimal improvement direction, categorizing them into types such as economic priority, social security, and urban planning and construction. By focusing on the unique conditions of each city, it becomes possible to more effectively address public safety events and systematically enhance the resilience of modern cities.

References

1. Yang M, Huang B, Cui C, Xiao Z (2016) Review and prospect: urban disaster resilience. Urban Plann Forum 01:48–55
2. Zhou L (2016) Resilient city: risk governance and indicator construction. J Beijing Admin Inst 02:13–20
3. Nuzzo JB, Meyer D, Snyder M et al (2019) What makes health systems resilient against infectious disease outbreaks and natural hazards? Results from a scoping review. BMC Public Health 19
4. Mccartney G, Pinto J, Liu M (2021) City resilience and recovery from COVID-19: the case of Macao. Cities 112(2):103130
5. Sharifi A, Khavarian-Garmsir AR (2020) The COVID-19 pandemic: impacts on cities and major lessons for urban planning, design, and management. Sci Total Environ 749:142391
6. Tong Y, Ma Y, Liu H (2020) The short-term impact of COVID-19 epidemic on the migration of Chinese urban population and the evaluation of Chinese urban resilience. Acta Geographica Sinica 75(11):2505–2520
7. Meerow S, Newell JP, Stults M (2016) Defining urban resilience: a review. Landsc Urban Plan 147:38–49
8. Shao Y, Xu J (2015) Understanding urban resilience: a conceptual analysis based on integrated international literature. Urban Plan Int 30(02):48–54
9. Tian D (2019) A review of studies on the combination of complexity science and urban space research. Urban Dev Stud 26(04):25–30
10. Chen Y (2001) Complex adaptive systems (CAS) theory and its applications: origin, content, and enlightenment. Chin J Syst Sci 04:35–39
11. Zhou X, Bi T, Ding R, Rong Y, Sun L, Ma W (2010) The construction strategy of resilient communities in Xiong'an new area: based on the theory of complex adaptive system. Urban Stud 26(03):108–115
12. Chen Z, Qiu B, Chen H (2021) Research on resiliency level promotion and robustness strategy of resilient city system: from complex adaptive systems (CAS) perspective. Urban Stud 28(08):1–9
13. Cl A, Akhla B, Jchfa C et al (2020) A mechanism-based parameterization scheme to investigate the association between transmission rate of COVID-19 and meteorological factors on plains in China. Sci Total Environ 737
14. Xu H, Yan C, Fu Q et al (2020) Possible environmental effects on the spread of COVID-19 in China. Sci Total Environ 731:139211
15. Li T, Niu P, Gu C (2014) A review on research frameworks of resilient cities. Urban Plan Forum 05:23–31

16. Yao Y, Pan J, Wang W et al (2020) Association of particulate matter pollution and case fatality rate of COVID-19 in 49 Chinese cities. Sci Total Environ 741:140396
17. Zhao R, Fang C, Liu H (2020) Progress and prospect of urban resilience research. Progr Geogr 39(10):1717–1731
18. Milic M, Zupac G (2012) An objective approach to determining criteria weights. Vojnotehniki Glasnik 60(1):39–56
19. Serafim O et al (2004) Compromise solution by MCDM methods: a comparative analysis of VIKOR and TOPSIS. Eur J Oper Res 156(2):445–455
20. Pickle BL (2002) Evaluation of methods for classifying epidemiological data on choropleth maps in series. Ann Assoc Am Geogr 92(4):662–681
21. Wang J, Xu C (2017) Geodetector: principle and prospective. Acta Geographica Sinica 72(01):116–134
22. Hordijk W, Kauffman SA, Stadler PF (2020) average fitness differences on NK landscapes. Theory Biosci 139(1)

Towards Urban Resilience: An Innovative Theoretical Framework for Comprehensive Multi-hazard Risk Assessment of Urban Disasters

Shutian Zhou, Yuhong Liu, and Guofang Zhai

Abstract In the context of building resilient cities, this paper discusses the risk assessment and management of urban natural disasters, aiming at the problem that cities are frequently affected by natural disasters due to dense urban population and economic activities. In view of the lack of a unified and completed risk assessment framework for multi-hazard risk prevention and urban resilience targets, this paper draws on the research achievements in risk management, disaster system theory, property assessment and value of statistical life (VSL), then builds a comprehensive theoretical framework for urban natural disaster risk assessment. The framework not only focuses on the direct economic losses caused by urban natural disasters, but also innovatively considers the additive effects of different types and frequencies of hazards, as well as the resulting compound consequences, thus achieving an integrated and systematic assessment of urban place-based relative composite risk. Through interdisciplinary methods, this study broadens the theoretical horizon of urban multi-disaster risk assessment and provides important theoretical support and framework reference for the practice of resilient city construction. By applying the framework, policy makers and managers can identify, assess and respond to urban natural disaster risks more scientifically, and develop more effective and forward-looking risk management strategies to enhance the resilience of cities.

Keywords Comprehensive assessment · Multi-hazard risk assessment · Urban natural disasters · Theoretical framework · Relative composite risk index

---

S. Zhou (✉) · Y. Liu
School of Arts (School of Architecture), Nantong University, Nantong, China
e-mail: shutian_zhou@ntu.edu.cn

Y. Liu
e-mail: liu_yh1993@163.com

G. Zhai
School of Architecture and Urban Planning, Nanjing University, Nanjing, China
e-mail: Guofang_zhai@nju.edu.cn

B. He et al. (eds.), *Climate-Resilient, Low-Carbon, and Sustainable Cities*, Lecture Notes in Civil Engineering 754, https://doi.org/10.1007/978-981-95-2169-2_4

1 Introduction

The concept of risk has been defined as a combination of event consequences and their likelihood of occurrence (ISO31000:2009, definition 1.1), which has been extensively applied theoretically and practically across various disciplines and technological fields. Specifically, disaster risk encompasses both the impacts of a disaster and its probability of occurrence. Disaster system theory further posits that disaster risk also involves natural hazards coupled with societal exposure and vulnerability to them [1]. When assessing potential disaster impacts, it is crucial to consider the exposure and vulnerability of the affected entities. Currently, quantitative risk assessment of a particular disaster primarily involves analyzing historical data to identify risk types, intensities, and frequencies, as well as employing computer simulations for consequence predictions. Urban multi-hazard risk assessment necessitates integrating multiple disasters with diverse sources and characteristics into a unified spatial system, considering their impacts on personal safety and property [2]. However, accurately representing the intensity and frequency differences among disasters within current risk coupling and superposition processes remains challenging.

Moreover, there is a lack of well-established and universally accepted frameworks for urban risk management, especially for multi-hazard disaster mitigation and resilience goals. Key questions remain unanswered, such as how to comprehensively consider diverse urban disaster consequences and their occurrence probabilities, integrate non-economic and economic losses from multiple disasters into an unified dimension for risk management, balance potential disaster risks with prevention, reduction, and emergency response levels across a city, and utilize risk assessment results to support urban resilience planning.

Existing theories and methods are still limited in providing complete and systematic answers to these questions. This study proposes a comprehensive framework for urban natural disaster risk assessment, leveraging theories and methods from risk management, disaster system theory, property evaluation and value of statistical life. Our hypotheses are: (1) the risk of an urban disaster with a specific intensity or frequency can be evaluated through quantified measurements of expected direct loss; and (2) by using the concept of annual expected direct loss per urban spatial unit, disaster risk can be quantified, accumulated, visualized, and compared across multi-hazard scenarios, locations, and time periods, aiding in the analysis of comprehensive risk distribution patterns.

The following section studies the urban disaster risk literatures, which particularly review relevant theories in risk and disaster loss assessment, and then discusses several key technologies, including key aspects, assessment logics, models and systems. Section 3 attempts to demonstrate how we are going to represent disaster risk, quantify the economic loss value and the value of life loss in disaster scenarios, and understand the comprehensive disaster risk in urban areas, in order to provide the theoretical basis contributing to the comprehensive urban disaster risk assessment. A theoretical assessment framework is then proposed in Sect. 4 including its constituent steps of key methods, models, algorithms and suggested index system. Section 5 addresses the application scope, key features, limitations and potential contribution of the proposed framework.

2 Literature

2.1 Understanding Risk, Disaster Risk and Urban Resilience

The term 'risk' has been assigned numerous meanings across various domains, including insurance, economics, engineering, and catastrophology [3–8]. Regardless of the context, risk is typically characterized by two fundamental aspects: potential consequences or events and their respective probabilities of occurrence. According to definition 1.1 of *ISO31000:2009*, the magnitude of risk is expressed as a combination of these consequences and their associated probabilities. The expected value of disaster risk can be calculated by multiplying each potential outcome by its corresponding probability and summing these values.

Similarly, natural disaster risk pertains to the anticipated level of damage led by a particular natural phenomenon that may manifest in a specific area at a given time [3]. Huang introduced a foundational model for comprehensive natural disaster risk assessment [9]. This model utilizes various functions, i.e. the probability density functions or dose–response functions, to synthesize risk and create a risk matrix depicting disaster consequences and their likelihoods. This matrix facilitates the determination of disaster risk levels.

Drawing from disaster system theory, disaster risk is a composite of hazards, coupled with vulnerability and societal exposure. Research on disaster risk assessment has predominantly focused on hazard factors from a narrow sense [10, 11] or, more holistically, on evaluating three constituent systems, which are environment, hazard, and disaster-bearing body. The risk assessment framework encompassing hazard, exposure, and vulnerability has been extensively employed in numerous studies and endorsed by the United Nations Intergovernmental Panel on Climate Change (e.g., [12–21]).

Over time, the scope of disaster risk assessment research has progressively broadened (as illustrated in Table 1). It now includes considerations of a community's capacity to confront risks [22–24], its anti-disaster capabilities and resilience [25–28], and infrastructure resilience [29]. Zhang et al. established an assessment framework for natural disaster risks and comprehensive response capabilities. Using the gray correlation method, they analyzed trends and disparities in the coordination between natural disaster risks and comprehensive response capabilities for Chongqing, Wuhan, and Shanghai from 2014 to 2020 [30]. The primary paradigm in natural disaster risk assessment remains the establishment of assessment models for disaster-inducing factors, disaster-hosting environments, and disaster-affected entities, grounded in disaster system theory. These models aim to simulate and predict

Table 1 Function expressions of disaster risk calculation (self-drawn)

Calculation formulas	Representative study
$Risk = f(H, V)$	[31, 32]
$Risk = f(H, V, R)$	[33]
$Risk = f(H, V, E)$	[13, 18, 20, 21, 34]
$Risk = f(H, V, E, C)$	[22–24]
$Risk = f(H, V, E, \mathrm{P})$	[25, 26, 29, 30]

Notes H denotes the hazard or possibility of disaster; *V* denotes vulnerability; *R* denotes resilience; *E* denotes exposure; C denotes disaster capacity; *P* denotes anti-disaster capability

the likelihood of disaster occurrence and its potential impact intensity, considering disaster causes, natural conditions, socio-economic factors, and disaster prevention infrastructure.

Resilience in disaster analysis represents the system's capability to absorb interference [35] and reorganize while maintaining its function, identity, structure, and feedback mechanisms [36]. The connotation of resilience often encompasses relevant constituent elements of disaster capacity and anti-disaster capability. These elements can be classified into three main categories: disaster types (e.g., earthquakes, floods, and geological disasters), the disaster development process (e.g., pre-disaster prevention, disaster relief, and recovery stages), and index types (e.g., social, economic, ecological, physical, and managemental).

Research on resilience assessment primarily focuses on dimensions such as ecological resilience, engineering resilience, economic resilience, and social resilience [37, 38]. Specific content includes comprehensive urban resilience assessment [39], economic resilience assessment [40], community resilience assessment [41], and infrastructure resilience assessment [42]. Previous research has continuously improved and refined resilience assessment methodologies, proposing models such as the CDRI index, the TOSE framework, the DROP assessment framework, and the BRIC model [26, 41, 43]. Additionally, several international organizations and research institutions have published comprehensive resilience evaluation systems and operational manuals [36, 44]. Considering the complex urban characteristics, Shi et al. has constructed a theoretical assessment framework for urban system resilience from the dimensions of system environment, system elements, and system structure [45].

However, the imperfections and confidentiality concerns surrounding historical disaster databases pose limitations on the application of sophisticated, large-scale disaster risk and urban resilience models. Advances in computer technology and the proliferation of interdisciplinary research have facilitated continuous improvements and applications of various models and methodologies from other domains within disaster risk assessment research. Consequently, there is a growing emphasis on exploring suitable multidisciplinary approaches for advancing disaster risk assessment.

2.2 Literature on Risk Assessment of Natural Disasters

Risk assessments serve as straightforward tools for comparing the relative economic and life loss risks across different regions, as well as evaluating the varied contributions of diverse factors. Initially, they concentrated on assessing specific risks and have since advanced, shifting from qualitative to quantitative methodologies, from dealing with definiteness to indeterminacy, and finally from probabilistic to ambiguous uncertainty as time progressed [8].

Within the disaster system theory, natural disaster-causing factors study is the staring step in disaster risk assessment. This process typically entails the investigation, examination, and simulation of the timing, scale, intensity, and spatial distribution of disaster occurrences. Notably, factors contributing to different disasters exhibit significant variability. Exposure studies serve as a bridge between research on hazard factors and the vulnerability of disaster-affected entities. Classification and statistical analysis of census data, socio-economic data, land use, and infrastructure information are widely utilized in this context [46–48]. Among these, the acquisition and analysis of spatially refined population and housing construction data have emerged as a prominent trend.

The concept of vulnerability has grown in complexity, encompassing multiple disciplines such as catastrophology, sociology, economics, management science, environmental science, and sustainability. Despite the diverse research priorities, a universally accepted and unified assessment framework remains elusive. Previous studies on disaster vulnerability predominantly relied on simulation experiments based on building vulnerability analysis to determine the vulnerability functions of building failure rates under varying disaster intensities [49–52]. Literature on disaster response or prevention evaluation often focuses on emergency response capability, urban earthquake resistance, disaster resilience, vulnerability and exposure assessment of areas [29, 41, 48, 53, 54].

Publications focusing on disaster risk assessment from the vulnerability perspective have proposed intricate indicator systems and assessment models. Notable examples include the Hazard-of-Place model [49], the Pressure-of-Release framework [55], the Airlie-House Vulnerability approach [53], and the Urban-Earthquake-Vulnerability model [56]. In practical scenarios, like earthquake and flood mitigation strategies, the choice of indicators is often swayed by elements including the scale of the study subject, the presence of historical calamity records, and the ease of obtaining indicator data. As a result, indices are typically streamlined, with particular socio-economic metrics acting as surrogates for the area's socio-economic susceptibility. Within urban disaster risk analysis, the accuracy of detailed, small-scale index data, such as those concerning streets and neighborhoods, greatly influences the robustness of the outcomes.

Traditional risk assessments, typically confined to the analysis of individual hazards or specific targets, may underestimate the risk level. As disaster risk awareness escalates, the outputs of single-disaster risk assessments inadequately address the comprehensive urban disaster prevention and mitigation objectives, necessitating a transition to comprehensive multi-hazard risk assessments. Existing research primarily builds upon disaster system theory, employing two distinct risk synthesis methodologies. Recently, the methods adopted for multi-hazard risk assessment in related studies can be categorized into two perspectives: superposition and coupling. The fundamental distinction between them lies in the consideration of disaster interactions. The superposition perspective measures multi-hazard risk by synthesizing risk results [57–61] or hazard factors [62–64], often without considering spatial comparisons and the effects of diverse hazards. Conversely, the methods with coupling perspective assesses multi-hazard risks through simulations of various coupling relationships, such as adjoint, triggering, coupling relationships and disaster chains, cascade effects and domino effects [65–68].

However, the majority of research endeavors and practical applications continue to concentrate on single-hazard risk assessments, incorporating perspectives on disaster superposition and coupling. The superposition methodologies encounter challenges in precisely quantifying the intensity and frequency variations among diverse disasters, resulting in discrepancies between comprehensive risk assessments and real-world scenarios. The investigation of multiple coupling relationships among disasters remains confined to theoretical research, evolving from trigger relationships and slow variation coupling towards holistic risk quantification. Consequently, most methodologies heavily depend on coupling rules among disasters sharing similar causative factors. Hence, an innovative approach to multi-hazard risk assessment is crucial to surpass the limitations and establish a more intuitive and scientific foundation for further risk management.

2.3 Literature on Loss Assessment of Natural Disasters

Quantitative ex-ante assessment methods for disaster losses primarily rely on the induction of historical empirical data with further computer numerical simulations. Advances in computer technology have significantly enhanced the assessment process and its visualization through the utilization of diverse software applications. Platforms such as ArcGIS are employed for spatial analysis and visualization in frameworks like HAZUS [69–71], and HAZ-China [72] systems. Additionally, MATLAB is used in the Economic Loss Estimation and Recovery (ELER) framework for the Organization for Economic Co-operation and Development (OECD) to integrate input–output methodologies, globally recognized theoretical models, and multi-level evaluation techniques [73].

When examining historical disasters, it is crucial to consider inflation and other pertinent factors. Beyond inflation, local wealth and demographic variables, disaster-induced regional losses are also intricately linked to land use patterns and public risk awareness [74]. Disaster losses are typically categorized into non-economic and economic losses, with economic losses further subdivided into direct and indirect losses, each necessitating distinct assessment methodologies. For instance, direct economic loss evaluation often employs the replacement cost approach, income present value method, current market price method. Conversely, indirect economic losses are frequently assessed using frameworks such as the input–output model and the computable general equilibrium model.

While casualties in disasters are inevitable, there remains a scarcity of models that integrate the impact of human life loss with economic loss within a unified disaster risk assessment framework. The number of fatalities serves as a pivotal index in determining the severity of consequences but is challenging to incorporate into risk assessments due to the mismatched measurement dimensions between life loss and property evaluations. Economic evaluations of life loss in disasters can be derived from the theory and methodologies underpinning the value of life. According to the environmental economic theory, the monetary value of a rational person's life can be quantified through statistical methods [75–77]. The quantitative assessment of the VSL primarily rests on two theoretical approaches: (1) human capital value theory (corresponding to the human capital approach) and (2) risk trading theory (corresponding to methods such as the wage risk method, consumer market method, and contingent valuation method) [78]. The traditional human capital method is often deemed to underestimate the value of human life due to its neglect of welfare preferences and the implication that higher income equates to higher life value [77]. Methods grounded in risk-trading theory and cost–benefit analysis are more extensively applied in labor market analysis, environmental policymaking, traffic safety research, and effectiveness assessments of food and medical treatments [79–83], particularly during the COVID-19 pandemic [84, 85].

In the realm of natural disasters, there is a notable dearth of historical research on VSL losses within disaster risk analysis and reduction, attributed to the unique spatial patterns of such public hazards that resist economic mitigation strategies like market transactions. Zhou et al. employed an enhanced human capital method to quantify VSL losses resulting from earthquakes, offering a framework for the monetary valuation of both direct and average effects of mortality across various age demographics within the affected community [86]. This approach signifies an innovative intersection within the multidisciplinary fields of disaster science, economics, and urban risk management planning. Potentially due to ethical controversies, the process of evaluating life loss remains ambiguous, relying solely on vulnerability indicators of disaster-affected entities such as population density and economic indicators for characterization. The VSL theory in environmental economics offers a pathway to study life loss in disasters for disaster risk assessment. Economic perspectives on disaster risk can effectively identify and quantify life and economic losses within the same context, thereby contributing valuable insights for risk projection, evaluation, and mitigation.

3 Theoretical Base to Understand Comprehensive Urban Natural Disaster Risk

3.1 Representation of Disaster Risk

Risk analysis primarily revolves around the examination of its probability and consequence, which constitute the fundamental components of risk. The occurrence of natural disasters is generally perceived as uncertain; however, within a specific region, local disasters like earthquakes, storms, and floods, adhere to the principle that "greater intensity is associated with a lower probability of occurrence" [87].

As defined by the UNDRO, natural disasters represent the expected loss value in terms of both life and property, as well as economic activities resulting from the consequences of a particular natural phenomenon, specific risks, and risk factors [87]. Consequently, a rational and scientific assessment of disaster risk can be achieved by calculating its expected loss. When disaster risk is analyzed from an economic perspective, utilizing the concept of "expected loss value of disaster," the magnitude of risk associated with different types of disasters or the same disaster with varying intensities can be directly compared (see Fig. 1). Furthermore, within the same spatial unit, the risks posed by different disasters and the risks of the same disaster with varying intensities can be superimposed, thereby accounting for the differences in frequency and consequences among various disasters.

3.2 Quantification of Economic Loss in Disasters

The loss assessment of disaster consequence inevitably involves the valuation of asset damage in disaster. In particular, the essence evaluating the direct economic loss is to quantify the asset value lost where the cost changes, market supply and demand adjustments, price fluctuations, exchange rate fluctuations and other factors all impact the actual monetary cost of purchasing or replacing a property. For disaster risk, and the disaster loss assessment, including economic loss and non-economic loss, mature methods of asset evaluation theory can effectively monetize disaster losses scientifically. Thus, evaluators need to choose, adjust and improve measurement methods for different types and characteristics of asset losses in a certain disaster, including the replacement cost, income present value or current market price methods etc.

3.3 Quantification of Life Loss in Disasters

The VSL theory posits that the value of human life can be monetized through statistical methods. To comprehend the VSL loss incurred by disasters, two points are

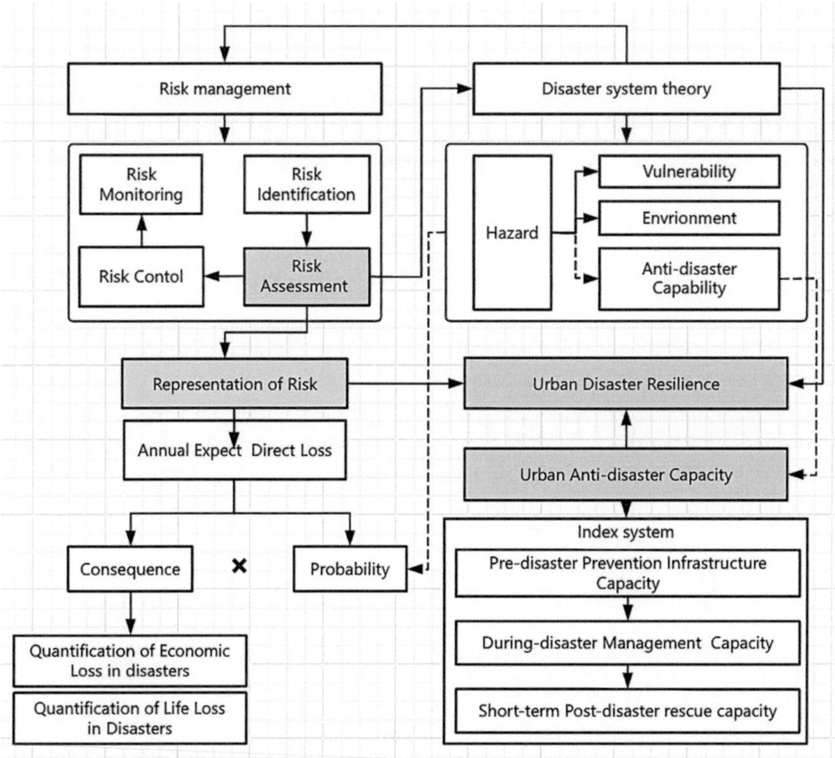

Fig. 1 A flow chart for understanding the proposed theoretical framework for comprehensive multi-hazard risk assessment of urban disasters towards urban resilience (self-drawn)

crucial. Firstly, in disaster economics, the assessment of VSL primarily focuses on the risk and consequences of casualties caused by disasters, without considering the life and death of any specific individual in disasters or the value of any individual's life. Secondly, in natural disaster scenarios such as earthquakes, which are unpredictable in timing, it is often challenging for individuals to purchase specific protective goods to reduce the risk of death or increase the probability of fatality in exchange for economic compensation, rendering risk trading-related methods less rational for evaluating life loss in disasters. Consequently, the human capital method becomes more appropriate, as it emphasizes an individual's contribution to society and their family, implying that the loss of VSL in disasters also entails the loss of that contribution.

3.4 Comprehensive Understanding of Disaster Risk

According to disaster system theory, the probability of disaster occurrence and its associated losses are influenced by subsystems such as disaster-causing factors, environment, and vulnerability. Local anti-disaster capacity, including factors like the disaster prevention and mitigation infrastructure, disaster warning and management levels, and post-disaster emergency response capabilities, which are often linked to vulnerability studies, have practical implications on the consequences of uncertain disaster events. Therefore, these factors cannot be overlooked when assessing the comprehensive disaster risk of a location.

Since the probability of urban natural disasters cannot be altered artificially, various methods and infrastructures for preventing, mitigating, or controlling disaster consequences should be considered when assessing the influencing factors of comprehensive urban disaster risk. Incorporating the assessed comprehensive disaster risk and local resilience into an integrated system is essential for enhancing urban risk management and disaster reduction strategies, thereby streamlining decision-making processes and planning efforts.

4 Construction of the Assessment Framework

4.1 Overall Disaster Risk Assessment

We examine the risks associated with various types of natural disasters within a one-year timeframe in an urban context. Risk is defined as the product of the annual probability of disaster occurrence in a specific location and the direct loss incurred by the disaster, representing the annual expected direct loss across all probabilities. It is important to clarify that the one-year timeframe does not pertain to a specific year but rather serves as a generic reference, implying that the probability of the given disaster intensity remains consistent across years. Consequently, the aggregate risk faced by a city from multiple disasters within this timeframe is the summation of the annual expected direct losses of all these disasters, taking into account all probabilities.

Building upon the definitions of risk and the concept of annual expected direct loss, the hazard risk of a particular disaster intensity (Risk(I)) is computed as the product of the direct loss (Loss(I)) and its annual probability (Probability(I)), representing the annual expected direct loss in this hazard scenario (see Eq. 1). The impact value encompasses not only economic losses ($L_{DE}(I)$) but also the loss of VSL in disasters ($L_{VSL}(I)$) (Eq. 2). This methodology concurrently considers the multifaceted disaster consequences of and their varying occurrence probabilities within a single metric, integrating both economic and non-economic losses into a unified assessment framework. Given the fundamental principle of natural disasters that higher impact events are less likely to occur, the overall risk of local natural disasters can be articulated as the sum of the annual expected impact values of different scenarios, encompassing all potential natural disasters that a region may face within a one-year period (as delineated in Eq. 3). The pertinent formulas are presented as follows:

$$Risk(I) = Loss(I) * Probability(I) \tag{1}$$

$$Loss(I) = L_{VSL}(I) + L_{DE}(I) \tag{2}$$

$$R_{Overall} = \sum_{m=1}^{M} R_m = \sum_{m=1}^{M} \sum_{t_m}^{T_m} L(t_m)_m * P(t_m)_m \tag{3}$$

where $R_{Overall}$ denotes the overall risk of all disaster scenarios; M denotes the total number of natural hazard types that the place faces; m denotes the mth type of disasters among the total number of M; R_m denotes the risk of the mth type of disasters; T_m denotes the total number of disaster intensity of the mth type of disasters; t_m denotes the t_m th type of disaster intensity of mth type of disasters; $L(t_m)_m$ and $P(t_m)_m$ denote the direct loss and the possibility of occurrence of the mth type of disasters of its t_m th type of disaster intensity.

The procedure for assessing the overall risk involves several key steps: first, identifying the types of disasters pertinent to the study site; second, calculating the annual probability for disaster occurrences of varying intensities and types; third, estimating the direct loss values associated with each disaster scenario; and finally, measuring the risks. The overall risk associated with a single disaster is measured as the summation of the annual expected impact values across all intensity scenarios for that disaster. Once the annual expected direct loss in a particular location has been computed to represent the risks posed by different disasters, these values can be aggregated to determine the overall multi-hazard risk for that location. In this place-based disaster risk measurement model, the inclusivity of a broader range of disaster types leads to a higher overall risk value, thereby enhancing the evaluator's precision in learning the local overall risk situation and its spatial distribution pattern. Particularly, when the risk values for individual geographical units are available, the overall risk pattern of the city can be visualized utilizing ArcGIS software.

4.2 *The Urban Anti-disaster Capacity Assessment*

Local capacity of disaster prevention, mitigation and emergency response of a city is vital to resist various disasters, protect the safety of residents' lives and property, or control the loss and impact of disasters within a certain range. To address the distinct stages of before, during and after a disaster, we introduce the concept of urban anti-disaster capacity. It aims to elucidate and evaluate the local level of disaster prevention, mitigation, and emergency response, encompassing infrastructures, management policies, and emergency relief across these three stages.

The overarching objective of this concept is subsequently divided into three components: the pre-disaster prevention infrastructure capacity, during-disaster management capacity and short-term post-disaster rescue capacity. To quantify these subsystems, we conducted a comprehensive review of relevant literature and practical applications [45, 48, 53, 56, 88, 89]. Based on this review, we attempted to represent each of the three subsystems through a series of indicators within an index system. When identifying place-based indicators, our criteria were guided by the principles of structural simplicity, measurability, the ability to support comparative analysis across different settings, and the facilitation of statistical processing.

Consequently, we propose a comprehensive, whole-process multi-factor index system that spans stages of before, during and after a disaster for assessing the urban anti-disaster capacity of a city as a disaster-bearing entity (Table 2). This index framework can independently assess urban disaster resilience or disaster resistance capacity. In practice, based on the accessibility of regional data, certain indicators may be supplemented or substituted to align more closely with local conditions.

4.3 Comprehensive Risk Assessment of Urban Natural Disasters

A comprehensive scientific analysis of the potential ramifications of one or multiple disasters on local economies, societies, and environments, alongside risk prioritization and spatial distribution assessments, serves as a pivotal reference for fostering the sustainability and security of urban spatial development. The investigation into urban overall disaster risk and urban anti-disaster capacity, as detailed in previous sections, represents a pair of diametrically opposed concepts. Practically, a higher urban disaster risk score coupled with a lower resilience value signifies a more acute vulnerability and impact from disasters, and conversely, a lower risk and higher resilience indicate reduced susceptibility.

In the context of natural disasters, which are inherently unpredictable, a city's enhanced disaster resistance equates to improved sustainability and resilience. Nonetheless, the economic and social costs of maintaining support infrastructure and services must be considered, avoiding an overemphasis on extensive protective measures, thorough preparedness, and swift disaster response. It is crucial to strike a balance between the overall disaster risk and the city's disaster resistance capabilities during the planning and decision-making stages.Identifying vulnerabilities in disaster prevention infrastructures, optimizing disaster response capabilities in specific regions, and effectively balancing disaster response resources within urban areas have emerged as critical considerations for risk management and mitigation efforts.

Table 2 Urban anti-disaster capacity assessment index system (self-drawn)

Index	Sub-system	Indicators	Indicator content
A urban anti-disaster capacity	B1 pre-disaster prevention infrastructure capacity	C1 drainage network density	The drainage capacity of local municipal infrastructure
		C2 per capita emergency shelter coverage	The total Amount and distribution of local emergency shelter space
		c3 emergency shelter count	
		C4 underground civil defense facilities	The supply capacity of civil air defense facility
		……	……
	B2 during-disaster management capacity	C5 civil defense alert system locations	The coverage of local alarming services
		C6 meteorological surveillance network	The local meteorological observation and monitoring level
		C7 emergency command centers	The basis of overall coordination and management of emergencies of local governments
		……	……
	B3 short-term post-disaster rescue capacity	C8 hospital bed density	Facility supply and distribution of emergency relief services provided by the local health system
		C9 medical infrastructure count	
		C10 police precinct distribution	The supply capacity of the local policing response
		C11 firefighting station network	The supply capacity of local emergency relief forces
		C12 transportation grid intensity	The transportation capacity and accessibility for emergency evacuation and rescue works
		……	……

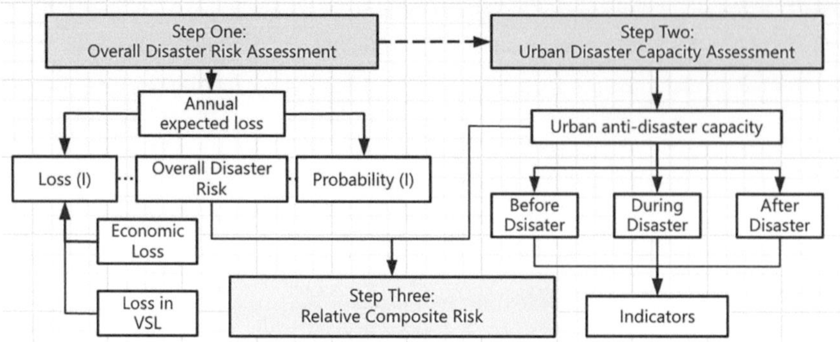

Fig. 2 A flow chart for main 3 steps for comprehensive multi-hazard risk assessment of urban disasters towards urban resilience (self-drawn)

To address this, the concept of a relative composite risk index is introduced, which amalgamates the notions of urban overall disaster risk and urban disaster resilience within a unified framework. This index objectively delineates and evaluates the coordination status between these two parameters, facilitating a more direct representation of mismatches, particularly highlighting urban zones characterized by high risk and low resilience. The value of the relative composite risk index is quantified by the ratio of the standardized values of the overall disaster risk to urban resilience within each evaluation unit (see Fig. 2).

For each evaluation unit, the higher the value of relative composite risk index, the higher the local multi-hazard risk, indicating greater need of improving its urban anti-disaster capacity. This result directly evaluates the equilibrium between potential disaster risks and the corresponding levels of preparedness, alleviation, and crisis management across various urban settings. Additionally, by assessing the subsystems of disaster resistance, it becomes more discernible and efficient to identify areas for enhancement. Utilizing scientific ranking and natural categorization (assisted by ArcGIS software), the outcomes are effectively presented, allowing for a clear visualization of the comparative composite risk index distribution within the research region.

5 Conclusion

5.1 Discussion

The aim of the proposed framework is to comprehensively measure the comprehensive disaster risk of the city. By taking into account the expected overall disaster risks of different kinds, local anti-disaster capacity and finally the urban relative composite risk index, the risk is described, quantified and visualized. Based on risk components, the framework takes into account disaster frequency, local anti-disaster capacity and loss of direct economic value and life value, overcoming the limitations of traditional assessments that ignore differences in disaster frequency and consequences. At the same time, the framework accurately reflects the inverse relationship between the intensity and frequency of natural disasters. The framework has three core features:

(1) Localised risk assessment model: The model conceptualizes urban disaster risk from an economic perspective based on factors such as risk and probability of consequences. By calculating the annual expected direct losses, the model can quantify the level of disaster risk and realize the comparison and summary of different disaster risks. In addition, the model can be applied to assessment units of different scales, such as regional, municipal or community, which enhances the flexibility and applicability of assessment.
(2) Integrated risk assessment across disciplines: The framework extends the perspective of risk assessment by treating non-economic disaster losses (such as loss of VSL) as equally important as property losses. Through integrating theories and methods from the fields of economics, engineering, and disaster science, the framework achieves the integration and application of interdisciplinary knowledge and improves the comprehensiveness and accuracy of assessment.
(3) Guiding local risk management and urban resilience: Through integrated risk assessment, the framework offers a scientific foundation for improving urban disaster prevention and mitigation resources. When the city's expected overall disaster risks and the its local anti-disaster capacity reach a balance, it indicates that the resource allocation is reasonable. The results of the urban relative composite risk index are helpful to identify and monitor the actual risk distribution of cities, and are of great significance for improving urban resilience.

5.2 Limitations

This framework faces limitations: firstly, despite disasters often entailing prolonged impacts, indirect and long-term value losses are excluded due to their complex nature. Secondly, the integrated multi-hazard risk assessment of urban disasters remains imperfect, constrained by data accessibility, albeit adjustable to local contexts.

Application necessitates comprehensive city data, predominantly sourced from open-access platforms. Direct economic loss computations involve extensive data processing through economic methods, grounded in local disaster archives, building information, and socio-economic datasets. The VSL can be approximated using an advanced human capital approach, factoring in expected casualties and injuries across disaster scenarios [86]. Detailed, pre-event damage assessments at community-specific levels enhance overall risk estimation and facilitate the visualization of urban disaster distributions.

Practically, incorporating a limited number of local disasters (two to three) into multi-hazard risk assessments remains viable, offering insights into both individual and aggregate urban risks, thereby informing resilience strategies and infrastructural improvements. Future empirical research can establish a localized disaster prevention database, providing a platform for sustained management and dynamic monitoring to optimize urban disaster risk management strategies.

5.3 Innovations

This paper examines the evolution of urban disaster risk assessment, emphasizing its implications for urban resilience research. Urban resilience, vital for disaster response and socio-economic stability, necessitates precise risk assessments and effective management strategies. Our framework integrates risk management and disaster system theories across disciplines like disaster science, economics, and urban planning, broadening the scope of comprehensive risk assessments.

Guided by economic principles, we propose a location-based multi-hazard risk measurement model for cities, incorporating advancements in risk management, property assessment, value of statistical life, and the disaster system theory. This model defines the scope and consequences of various disasters, integrating their occurrence probabilities and intensities for refined assessments. It also pioneeringly includes the loss of life value in annual expected direct losses, reflecting a people-centered approach.

Furthermore, we introduce a relative comprehensive risk index to balance overall local disaster risk and resilience, providing an alternative for urban disaster risk assessment. Our framework integrates existing loss estimation and risk analysis techniques, offering a novel approach for location-based risk analysis and management. This not only serves as a practical reference for quantitative research on urban multi-hazard risks but also supports the formulation of scientific disaster response and mitigation strategies.

Future research aims to integrate advanced technologies for deeper analysis of expected direct losses from different disasters, particularly those prevalent in urban areas. Through abundant case studies, we will continue to enrich and refine the assessment framework, striving to develop a more comprehensive and accurate resilience indicator system for continuous urban resilience enhancement.

Acknowledgements This study is supported by the Natural Science Foundation of Jiangsu Province (Grants No BK20230618).

References

1. Riddell GA, Delden HV, Maier HR, Zecchin AC (2019) Exploratory scenario analysis for disaster risk reduction: considering alternative pathways in disaster risk assessment. Int J Disaster Risk Reduction 39:101230

2. Kappes MS, Keiler M, Elverfeldt KV, Glade T (2012) Challenges of dealing with multi-hazard risk: a review. Nat Hazards 64(2):1925–1958
3. Frank H (1921) Knight, risk, uncertainty, and profit. Hart Schaffner & Mary, New York
4. Varnes DJ (1984) Landslide hazard zonation: a review of principles and practice. Natural Hazards 3
5. Helm P (1996) Integrated risk management for natural and technological disasters
6. UNEP (2002) Global outlook 3: past, present and future perspectives. Earthscan Publications Ltd., London
7. Hardy CC (2005) Wildland fire hazard and risk: problems, definitions, and context. For Ecol Manage 211(1–2):73–82
8. Li J et al (2023) Construction and empirical testing of comprehensive risk evaluation methods from a multi-dimensional risk matrix perspective: taking specific types of natural disasters risk in China as an example. Nat Hazards
9. Huang CF (2008) A basic model of comprehensive risk assessment. J Appl Basic Sci Eng 16(3):371–381
10. Wu LY, He DJ, Hong W, Ji ZR, You WB, Zhao LL (2014) Research progress on risk assessment and vulnerability of natural disasters. J Catastrophology 29(4):129–135
11. Liu Y, Lu C, Yang X, Wang Z, Liu B (2020) Fine-scale coastal storm surge disaster vulnerability and risk assessment model: a case study of Laizhou bay, China. Remote Sens 12(8):1301
12. Mileti D, Noji E (1999) Disasters by design: a reassessment of natural hazards in the United States
13. Okada N, Tatano H, Hagihara Y (2003) Integrated research on methological development of urban diagnosis for risk and its applications. In: Annual Report of the Disaster Prevention Research Institute, Kyoto University, pp 1–8
14. Merz B, Kreibich H, Thieken A, Schmidtke R (2004) Estimation uncertainty of direct monetary flood damage to buildings. Nat Hazards Earth Syst Sci 4(1):153–163
15. Ge QS, Zou M, Zheng JY (2008) Preliminary study on comprehensive assessment of natural disaster risk in China. Science Press, Beijing
16. Yin ZE, Xu SY, Yin J, Wang J (2010) Small-scale urban rainstorm waterlogging disaster scenario simulation and risk assessment. Acta Geogr Sin 65(5):553–562
17. Kazmierczak A, Cavan G (2011) Surface water flooding risk to urban communities: analysis of vulnerability, hazard and exposure. Landsc Urban Plan 103(2):185–197
18. IPCC (2014) Climate change: impacts, adaptation, and vulnerability. Part A: Global and sectoral aspects. Contribution of working group II to the Fifth Assessment Report
19. Pagliacci, Francesco, Russo, Margherita (2019) Multi-hazard, exposure, and vulnerability in Italian municipalities. https://doi.org/10.4337/9781788970105.00017
20. Díaz-Sarachaga JM, Jato-Espino D (2020) Analysis of vulnerability assessment frameworks and methodologies in urban areas. In: Natural hazards: journal of the international society for the prevention and mitigation of natural hazards, vol 100
21. Anelli D, Tajani F, Ranieri R (2022) Urban resilience against natural disasters: mapping the risk with an innovative indicators-based assessment approach. J Clean Prod 371:133496
22. Davidson RA, Shah HC, Advisor (1997) An urban earthquake disaster risk index report
23. Bollin C, Cárdenas H, Hahn H, Vatsa KS (2003) Disaster risk management by communities and local governments. IDB Publications
24. Ntajal J, Lamptey BL, Mahamadou IB, Nyarko BK (2017) Flood disaster risk mapping in the lower mono river basin in Togo, West Africa
25. Zhang JQ, Okada K, Dona Y (2005) Integrated natural disaster risk management. Urban Disaster Reduction 2:2–5
26. Cutter SL, Ash KD, Emrich CT (2014) The geographies of community disaster resilience. Glob Environ Chang 29(29):65–77
27. Guo E, Zhang J, Ren X, Zhang Q, Sun Z (2014) Integrated risk assessment of flood disaster based on improved set pair analysis and the variable fuzzy set theory in central Liaoning Province, China. Nat Hazards 74(2):947–965

28. Barría P et al (2019) From multi-risk evaluation to resilience planning: the case of central chilean coastal cities. Water 11(3)
29. Arvin M, Beiki P, Hejazi SJ, Sharifi A, Atashafrooz N (2023) Assessment of infrastructure resilience in multi-hazard regions: a case study of Khuzestan Province. Int J Disaster Risk Reduction 88
30. Zhang D, Shi Y, Li Z, Huang Y (2024) Coupling analysis on urban disaster risk and response capacity in Yangtze River economic belt. Yangtze River 3:7–14
31. Maskrey A (1989) Disaster mitigation: a community-based approach. England Oxfam, Oxford
32. UNDHA (1991) Mitigating natural disasters. Phenomena, effects and options. A manual for policy makers and planners
33. UN (2002) Risk awareness and assessment in living with risk. UNISDER, UN, WMO, Asain Disaster Reduction Center
34. Guo G et al (2021) A vulnerability curve method to assess risks of climate-related hazards at county level. Atmosphere 12(8)
35. Boer JD, Wardekker JA, Sluijs JPVD (2010) Frame-based guide to situated decision-making on climate change. Glob Environ Chang 20(3):502–510
36. Walker B et al (2004) Resilience, adaptability and transformability in social-ecological systems. Ecol Soc 9(2)
37. Xu Y, Li G et al (2018) Soil security in peri-urban ecosystems: problems and challenges. Acta Ecologica Sinica 38(15). https://doi.org/10.5846/stxb201801310263
38. Christophe B, Andrew N, Mark D, Martina U, Rachel G (2014) Review article: resilience, poverty and development. J Int Dev 26(5):598–623. https://doi.org/10.1002/jid.2992
39. Li T (2017) New progress in study on resilient cities. Urban Plan Int 32(5):15–25. https://doi.org/10.22217/upi.2015.284
40. Wan B, Zhai G (2017) Research on the post-disaster regional economic growth model based on resilience: Taking the Wenchuan earthquake as an example. Mod City Res 09:41–47. https://doi.org/10.3969/j.issn.1009-6000.2017.09.006
41. Cutter SL et al (2008) A place-based model for understanding community resilience to natural disasters. Glob Environ Chang 18(4):598–606
42. Zhao C, Li N, Fang D (2015) A conceptual framework for modeling critical infrastructure interdependency: using a multilayer directed network model and targeted attack-based resilience analysis. Comput Civil Eng 2015:347–354. https://doi.org/10.1061/9780784479247.043
43. Bruneau M, Reinhorn A (2007) Exploring the concept of seismic resilience for acute care facilities. Earthq Spectra 23(1):41–62. https://doi.org/10.1193/1.2431396
44. Barnett G, Bai X (2007) Urban resilience: a resilience alliance initiative for transitioning urban systems towards sustainable futures. Urban Resilience Res Prospectus
45. Shi Y et al. Assessment methods of urban system resilience: from the perspective of complex adaptive system theory. Cities 112
46. Hanson S, Nicholls R, Ranger N, Hallegatte S, Corfee-Morlot J, Chateau CH (2011) A global ranking of port cities with high exposure to climate extremes. Clim Change 104(1):89–111
47. Smith A, Martin D, Cockings S (2016) Spatio-temporal population modelling for enhanced assessment of urban exposure to flood risk. Appl Spat Anal Policy 9(2):145–163
48. Weis SWM, Agostini VN, Roth LM, Gilmer B, Schill SR, Knowles JE et al (2016) Assessing vulnerability: an integrated approach for mapping adaptive capacity, sensitivity, and exposure. Clim Change 136(3):615–629
49. Cutter SL (1996) Societal vulnerability to environmental hazards. Prog Hum Geogr 20(4):529–539
50. Fan YX, Gao PH, Wang HJ (2003) A theoretical model for fuzzy comprehensive evaluation of regional disaster bearing body vulnerability. J Catastrophology 18(3):20–23
51. Shi Y, Xu SY, Shi C, Sun AL, Wang J (2009) Research progress on flood vulnerability. Prog Geogr 1:41–46
52. Fekete A, Tzavella K, Baumhauer R (2017) Spatial exposure aspects contributing to vulnerability and resilience assessments of urban critical infrastructure in a flood and blackout context. Nat Hazards. https://doi.org/10.1007/s11069-016-2720-3

53. Turner BL, Kasperson RE, Matson PA (2003) A framework for vulnerability analysis in sustainability science. Proc Natl Acad Sci USA 100(14):8074–8079
54. Tang B, Liu XL, Shang ZH (2012) Urban disaster vulnerability and its evaluation index. J Catastrophology 27(4):6–11
55. Blaikie P, Cannon T, Davis I, Wisner B (2004) At risk: natural hazards, people's vulnerability and disasters. De Gruyter
56. Duzgun HSB, Yucemen MS, Kalaycioglu HS, Celik K, Kemec S, Ertugay K et al (2011) An integrated earthquake vulnerability assessment framework for urban areas. Nat Hazards 59(2):917–947
57. Dilley M, Chen RS, Deichmann U, Lerner-Lam AL, Arnold M, Agwe J et al (2005) Natural disaster hotspots: a global risk analysis. International Bank for Reconstruction and Development/The World Bank and Columbia University
58. Gai CC, Weng WG, Yuan HY (2011) Multi-disaster coupled comprehensive risk assessment based on GIS. J Tsinghua Univ: Nat Sci 51(5):627–631
59. Wang HY, Xue H (2013) Compilation of county level comprehensive risk zoning Map of natural disasters: a case study of Luanxian County. J Nat Disasters 3:84–90
60. Bernal GA, Salgado-Gálvez MA, Zuloaga D, Tristancho J, González D, Cardona O (2017) Integration of probabilistic and multi-hazard risk assessment within urban development planning and emergency preparedness and response: application to Manizales, Colombia. Int J Disaster Risk Sci 8(3):270–283
61. Carreño ML, Cardona O, Barbat AH, Suarez DC, Perez MDP, Narvaez L (2017) Holistic disaster risk evaluation for the urban risk management plan of Manizales, Colombia. Int J Disaster Risk Sci 8(3):258–269
62. Xue Y, Chen BZ, Huang CF, Yan JW (2012) Soft hierarchical model for multi-disaster comprehensive risk assessment. Prog Geogr 31(3):353–360
63. Yang J, Wang L, Xu G (2014) Fuzzy comprehensive evaluation of comprehensive disaster risk in Chongqing. Earth Environ 42(2):252–259
64. Shome P, Afroj S (2019) Local Level multi-hazards mapping for natural and manmade hazards: a case study of Chittagong Division
65. Kappes MS, Keiler M, Glade T (2010) From single- to multi-hazard risk analyses: a concept addressing emerging challenges. In: Mountain risks: bridging science to society. Proceedings of the international conference. CERG Editions, Strasbourg, pp 351–356
66. Westen CV, Kappes MS, Luna BQ (2014) Medium-scale multi-hazard risk assessment of gravitational processes. Springer, Netherlands
67. Lu Y (2015) A preliminary study on coupled risk assessment of multiple disasters in coastal cities. J Catastrophology 30(1):211–216
68. Li SS, Yang SN, Liu XF (2017) Multi-disaster spatial-temporal network modeling for non-process: a case study of drought-heatwave coupling in Beijing-Tianjin-Hebei region. Geogr Res 8:1415–1427
69. FEMA, & NIBS (1997) Earthquake loss estimation methodology-HAZUS97, Technical Manual. Federal Emergency Management Agency, Washington, DC
70. Schneider PJ, Schauer BA (2007) HAZUS—its development and its future. Nat Hazard Rev 7:40–44
71. Jaiswal KS, Wald DJ, Earle PS, Porter KA, Hearne M (2011) Earthquake casualty models within the USGS prompt assessment of global earthquakes for response (PAGER) system. Springer, Netherlands
72. Chen HF (2012) HAZ—China earthquake disaster loss evaluation system design and preliminary implementation. Doctoral dissertation, Institute of Engineering Mechanics
73. Hancilar U, Tuzun C, Yenidogan C, Erdik M (2010) ELER software—a new tool for urban earthquake loss assessment. Copernicus GmbH 10:2677–2696
74. Zhai G, Fukuzono T, Ikeda S (2010) An empirical model of efficiency analysis on flood prevention investment in Japan. Rev Urban Reg Dev Stud 15(3):238–254
75. Charles F (1969) The value of life. Harv Law Rev 82(7):1415–1437

76. Schelling TC (1968) The life you save may be your own. In: Chase SB (ed) Problems in public expenditure and analysis. Brookings Institution, Washington, DC
77. Aldy VJE (2003) The value of a statistical life: a critical review of market estimates throughout the world. J Risk Uncertain 27(1):5–76
78. Zhou S, Zhai G, Shi Y, Lu Y (2020) A literature review of urban natural disaster risk assessment. J Catastrophology 35(4):7
79. Blaeij AD, Florax RJGM, Rietveld P, Verhoef ET (2000) The value of statistical life in road safety: a meta-analysis. Tinbergen Inst Discuss Pap 35(6):973–986
80. Viscusi WK (2011) What's to know? Puzzles in the literature on the value of statistical life. J Econ Surv 26(5):763–768
81. Peng XH, Wang CW, Shi QH (2014) Research on the vital statistics value of urban migrant workers: based on the improved characteristic wage model: evidence from Shanghai. Econ Theory Econ Manag 1:52–61
82. Du LJ (2017) Based on wage risk method to estimate of the labor value of life in China. Doctoral dissertation, Zhejiang University of Finance and Economics
83. Zhang GS, Chen Y, Xu C, Chen MM (2018) Value of life, underestimation of occupational injury cost and safety accidents. Econ Res J 53(9):182–198
84. Jonathan C (2020) What is the meaning of (statistical) life? Benefit–cost analysis in the time of Covid-19. Oxford Rev Econ Policy (Supplement_1):56–63
85. Kim YE, Loayza NV (2021) Economic loss from Covid-19 fatalities across countries: a VSL approach. Appl Econ Lett 29(7)
86. Zhou ST, Zhai GF, Lu Y, Shi YJ (2020) Urban seismic risk assessment by integrating direct economic loss and loss of statistical life: an empirical study in Xiamen, China. Int J Environ Res Public Health 17(21)
87. UNDRO (1991) Mitigating natural disaster phenomenal effects and options: a manual for policy makers and planners. UN Publications, New York
88. Zhou B, Zhou JX, Zhou XM, Yang Y (2010) Quantitative analysis of comprehensive ability of urban disaster prevention and mitigation. J Disaster Prev Sci Technol Coll 12(1):104–112
89. Chen W, Tang YM, Zhao J et al (2023) Urban-scale earthquake evacuation simulation based on mobile phone signaling data. J Seismol Res 46(4):611–621. https://doi.org/10.20015/j.cnki.ISSN 1000-0666.2023.0053
90. Walker BH, Abel N, Anderies JM, Ryan P (2009) Resilience, adaptability, and transformability in the Goulburn-Broken Catchment, Australia. Ecol Soc 14(1). https://doi.org/10.1890/08-0048.1

Critical Rainfall of Torrential Floods Induced by Heavy Precipitation Based on the FloodArea Model: A Case of Luba River, China

Meng Yuan, Xiaojun Guo, Yao Huang, and Jie Guo

Abstract Sichuan Province in China is characterized by complex terrain and abundant water resources, with a spatiotemporal distribution of precipitation that is notably non-uniform. Short-term heavy rainfall during the flood season often leads to flash flood disasters, posing significant threats to both lives and property. It's always been difficult to calculate the critical rainfall of mountain flood in mountainous small watershed where water level and discharge data are lacking. In this study, Luba River, the upper reaches of Qujiang River Basin in northeast Sichuan Province in China, is selected as a pilot research area to simulate the flooding of a precipitation process that has caused the most obvious flooding disaster in recent years. Hourly precipitation data in September 2014 were used in this study. The Floodarea Hydrologic model, a two-dimensional unsteady hydrodynamic model developed by German Geomer, was adopted for simulation. In this paper, combined with the information of field investigation, we simulate and reproduce the inundation caused by precipitation in this area. The simulation results achieve a good consistency between the simulated inundation depth of the early warning point and the measured depth of the field survey. The results show that FloodArea model can be applied to the analysis of rainstorm and flood disaster and the study of torrential flood risk rainfall in small watersheds with no hydrological data represented by the Luba River Basin. In addition, we also calculated the critical rainfall threshold of different risk levels of warning points, which can provide some reference for the disaster prevention and mitigation of flash floods in this region.

M. Yuan
Sichuan Rural Economic Information Center, Chengdu, China

Key Laboratory of Sichuan Province for Heavy Rain and Drought Flood Disasters in Plateaus and Basins, Chengdu, China

X. Guo (✉)
Institute of Mountain Hazards and Environment, CAS, Chengdu, China
e-mail: aaronguo@imde.ac.cn

Y. Huang · J. Guo
Sichuan Meteorological Service Center, Chengdu, China

B. He et al. (eds.), *Climate-Resilient, Low-Carbon, and Sustainable Cities*, Lecture Notes in Civil Engineering 754, https://doi.org/10.1007/978-981-95-2169-2_5

Keywords FloodArea model · Critical rainfall · Torrential floods · Disaster prevention and mitigation

1 Introduction

Torrential floods are characterized by the abrupt fluctuations in runoff within small river basins located in mountainous regions. Torrential flood disasters encompass not only flooding but also mud-rock flows and landslides triggered by rainfall in these areas, resulting in significant losses to human life, property, and the national economy [1–3]. In recent years, there has been a notable increase in extreme weather events globally, with precipitation patterns exhibiting increasingly erratic trends; consequently, torrential flood disasters have become more frequent. Given the inherent characteristics of flash floods—namely their sudden onset, intensity, destructiveness, and challenges associated with prevention—the monitoring and early warning systems for such events are of paramount importance and represent critical issues that must be addressed within disaster prevention at and mitigation effort.

In China, particularly in the Sichuan Basin, the landscape is intricate, and summer precipitation is both plentiful and intense, which considerably heightens the risk of torrential flood disasters. Additionally, the population density around the basin is relatively high, exacerbating the effects of such calamities [4, 5]. Research has shown that localized heavy rainfall acts as a key trigger for flash floods, landslides, debris flows, and other associated hazards [6–8]. The critical rainfall threshold is an important factor for predicting and issuing warnings about torrential flood disasters; therefore, examining this threshold is crucial. Given that surface conditions differ across various small watersheds, it becomes essential to simulate precipitation-induced flash flood inundation tailored to each watershed to assist local communities in disaster prevention and mitigation.

In several countries, research on the forecasting and early warning of torrential flood disasters has been conducted for a longer period, resulting in relatively mature methods and technologies [9]. In 1981, Hollingsworth et al. [10] in the United States categorized the risk levels of flash flood occurrences into five distinct tiers and identified the primary influencing factors associated with these events across five classifications. In 1990, Kubota Tetsuya [11] from Japan assessed the likelihood and risks of flash floods by examining two key aspects: effective early rainfall and current rainfall intensity. Since the late 1990s, significant advancements have been made in both theoretical foundations and modeling approaches for flash flood prediction and forecasting. Over the past two decades, related research efforts have gradually increased in China [12–14].

Both domestic and international researchers have delineated two principal methodologies for calculating critical precipitation: 'statistical regression' and 'model simulation.' The 'statistical regression' methodology entails the examination of linear or nonlinear relationships between precipitation levels and water levels in torrential floods ditches, achieved through the aggregation of extensive historical rainfall data from the basin. This process ultimately facilitates the determination of a rainfall threshold that corresponds to the maximum flood discharge capacity of these ditches [15]. However, this approach is not suitable for regions characterized by limited hydrological data. This study concentrates on the Luba River within Qujiang Basin in Sichuan Province. The region boasts a well-established water system and predominantly hilly terrain, leading to a swift transition from rainstorms to flash floods. In this study, we take the Luba River Basin as an example, and use the "Flood-area inundation model simulation" method to simulate and reproduce the rainstorm and inundation process. The advantage of this method is that it can solve the problem of inundation simulation in small river basins where lacking hydrological data. The purpose of this study is to provide scientific insights for flood control and disaster reduction strategies in the region.

2 Data and Methods

2.1 Data

This research employs a variety of data types, which include: (a) Meteorological data: hourly rainfall measurements obtained from regional automatic weather stations, provided by the Sichuan Provincial Meteorological Bureau; (b) Sessential geographic information: including details on hydrological systems and populated areas, sourced from the National Basic Geographic Information Center; (c) Field survey data: information concerning disaster impacts, such as inundation depth and duration, along with recorded geographical coordinates and elevation in the study area were collected through direct measurements conducted by the author; (d) DEM data: known as Digital Elevation Model (Abbreviated as DEM). The DEM utilized in this study consists of global elevation data acquired from NASA's TERRA Earth Observation Satellite. This dataset is supplied by the Institute of Computing Technology at the Chinese Academy of Sciences and features a spatial resolution of 30 m, specifically version ASTER GDEM V2.

2.2 Methods

Calculation of Areal Rainfall

In the analysis of critical rainfall for flood events, "areal rainfall" is typically employed to denote precipitation levels. When data discrepancies arise, some researchers opt to utilize the single station rainfall exhibiting the highest correlation coefficient with areal rainfall within the study area as the initial precipitation condition for statistical or hydrological models. Areal rainfall is defined as the mean precipitation across a specific region or watershed and is calculated by averaging measurements from various points within that area. In hydrology, areal rainfall is generally represented as:

$$\overline{P} = \frac{1}{A}\int_{A} PdA \tag{1}$$

In this study, A represents the spatial extent of the research subject, while P indicates the precipitation within the dA area. Established methodologies for estimating areal rainfall include the arithmetic mean method, Thiessen polygon approach, and inverse distance weighting (IDW) interpolation technique. Due to the sparse distribution of meteorological stations in the Luba River basin, this paper adopts the Thiessen polygon approach for calculating areal rainfall.

FloodArea Hydrodynamic Model

The FloodArea model is a two-dimensional, unsteady hydrodynamic model developed by German Geomer [16]. The FloodArea model is a tool that runs on ArcGIS software. The version of the drawing and simulation software used in this article is ArcGIS 10.1.

The FloodArea model is seamlessly integrated into the ArcGIS platform, where the operational dynamics of each time step are meticulously recorded and represented in raster format. This model leverages DEM for hydrologic-hydrodynamic simulations, wherein hydrodynamic processes are characterized by the unsteady flow evolution within a two-dimensional framework. The primary computational unit of this model is defined as the grid cell, and water movement calculations utilize the Manning–Strickler formula to elucidate flow interactions between each grid cell and its eight adjacent neighbors. In the context of flooding simulations, this formula is employed to accurately compute the inflow velocity of water into neighboring grid cells.

$$V = k_{st} \cdot r_{hy}^{2/3} \cdot I^{1/2} \tag{2}$$

Among them, V represents the flow velocity of water; k_{st} is the hydraulic roughness; r_{hy} is the hydraulic radius, which is the ratio of the fluid cross-sectional area to the wet perimeter; I refers to the slope of the terrain.

The flood submergence depth, also known as the submergence water level elevation, is calculated by subtracting the ground elevation:

$$flow\_depth = water\_level_a - elevation \tag{3}$$

The direction of water flow during inundation is governed by the topographic slope orientation, which indicates the directional aspect of the slopes surrounding the river channel. This slope orientation reflects the direction of maximum elevation change from a given point on the terrain. The calculation formula is as follows (α represents the terrain slope).

$$aspect = 270 - \frac{180}{\pi} \cdot \alpha \tan 2\left[\frac{\partial Z}{\partial Y} + \frac{\partial Z}{\partial X}\right] \tag{4}$$

In this research, the Rainstorm model integrated within the FloodArea framework is utilized to conduct a dynamic simulation of flood evolution processes. This model proficiently visualizes flow direction and velocity at each grid cell in real-time through vector representation, facilitating an evaluation of potential flood risks, inundation dynamics, and water depths [17–19]. A robust correlation between water levels and regional precipitation is established by calibrating input parameters such as elevation, rainfall weight distribution, and surface roughness; this calibration is further corroborated with disaster data acquired from field investigations. Furthermore, a refined relationship between inundation depth and regional precipitation enables the identification of critical thresholds for disasters induced by torrential floods.

3 Simulation of Torrential Flood Dynamics Utilizing the FloodArea Hydrodynamic Model

3.1 Overview of Precipitation Processes of the Luba River

Tongjiang County, located in Sichuan Province and home to the Luba River, experienced a prolonged rainstorm from 20:00 on September 8, 2014, to 08:00 on September 18, 2014. This event brought localized heavy and extremely heavy rainfall that resulted in swollen mountain streams and rivers. According to records from the local flood control office, over 339,800 individuals across 49 townships were affected by varying degrees of disaster due to the rainstorm, leading to economic losses exceeding 1.4 billion yuan.

During this period, out of the 54 regional automatic weather stations within the county, rainfall accumulation at 41 stations surpassed 300 mm. Notably, Nuoshuihe Station reported an unprecedented cumulative rainfall of 717.5 mm.

The Luba River Basin is situated in northern Tongjiang County with altitudes ranging from 502 m to as high as 1895 m and encompasses an area of merely 103 km^2. There are no hydrological observation data available for this basin due to a lack of monitoring infrastructure; currently only two meteorological automatic stations (Tiechang and Nuoshui River) provide precipitation data for this region. Characterized as a typical mountainous basin with limited hydrological information available during the flood event on September 9th, this small basin exemplifies an alpine canyon river where significant surface runoff coefficients facilitate rapid runoff formation under heavy rainfall conditions—thereby increasing susceptibility to flash floods and debris flows which pose substantial risks for local residents.

The hourly precipitation data corresponding with the first flood peak occurrence along the Luba River (from 1:00 to 15:00 on September 9th) was extracted as boundary conditions for simulating torrential flooding scenarios within this river system.

3.2 Conducting Field Assessments and Pinpointing Early Warning Indicators

A calibrated GPS satellite locator was employed to measure and document 5 locations exhibiting inundation traces within and around the Luba River basin, aiming to analyze the occurrence and frequency of flood disasters in previous years. A survey was conducted regarding a significant mountain flood disaster that occurred on September 9, 2014, which had considerable repercussions in recent years. This survey recorded various parameters including inundation depth, river channel characteristics, dam height etc. Additionally, interviews were conducted with local residents to gather insights into the disaster's impact. The distribution of research points and the DEM of the research area are illustrated in Fig. 1, while detailed information about each site is presented in Table 1.

The investigation reveals that the Tang Jia He Ba at Point A is situated near Nuoshuihe Town, within the middle and upper sections of the small watershed under examination, which has a relatively high population density. During the flood season, heavy rainfall and flash floods frequently inundate local farmland; in severe instances,

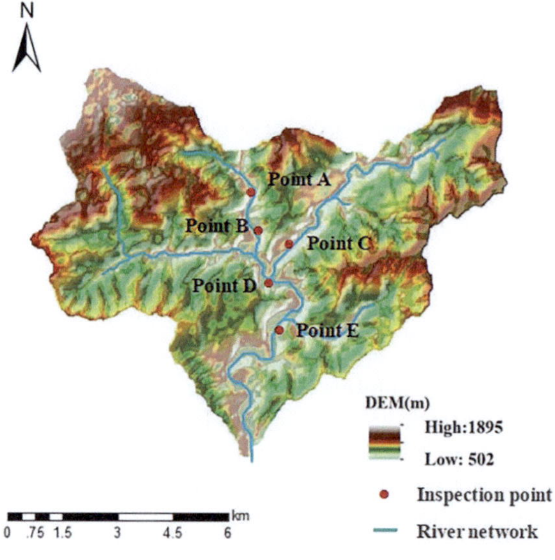

Fig. 1 Geographical representation of the survey locations and DEM of the Luba river (self-drawn)

Table 1 On site survey information of Luba River Basin (self-drawn)

Inspection point number	Inspection point name	Longitude	Latitude
Point A	Tang Jia He Ba	107.182	32.402
Point B	Yu Huang Ba village	107.183	32.390
Point C	Lou Zi vountryside	107.192	32.387
Point D	Nuoshuihe National Geological Park entrance	107.189	32.377
Point E	Jin Jia Ba village	107.187	32.361

flooding can overflow onto highways and submerge residential properties. Consequently, this location has been designated as a flash flood warning point for this small watershed. According to on-site investigation records, the altitude of warning point A was 496 m and the actual submergence depth of warning point A is about 4.05 m. Due to the lack of detailed records from hydrological stations, the time information of the maximum submergence depth was obtained through visits to local villagers, which was approximately after lunch on September 9, 2014 (Fig. 3).

Measurement sites were strategically selected in areas severely impacted by mountain flood disasters. In conducting height difference measurements, priority was given to boundary representative points that experienced inundation at peak water levels during mountain floods, historically flooded areas with frequent inundation events, or regional representative points characterized by elevated terrain that remained unaffected by flooding—these selections were informed by investigations and consultations with local residents.

3.3 Simulation Results of Flash Flood Inundation Process

This study employs the Rainstorm model within the FloodArea framework to simulate and replicate the flood rising dynamics of the Luba River, induced by significant rainfall from 1:00 to 15:00 on September 9, 2014. In alignment with the characteristics of the study area, this chapter applies the Thiessen polygon method for areal rainfall estimation. The Thiessen polygon for the Luba River basin was constructed using ArcGIS (refer to Fig. 2), and a weighted methodology was adopted.

$$AR = \sum_{i=1}^{n} R_i \times A_i / A \tag{5}$$

The formula states that AR represents the area rainfall of the basin, R_i denotes the rainfall at site i, A_i signifies the area associated with site i, A indicates the total size of the basin, and n refers to the number of Thiessen polygons. Below is a diagram illustrating the Thiessen polygons for the study area (Table 2 and Fig. 2).

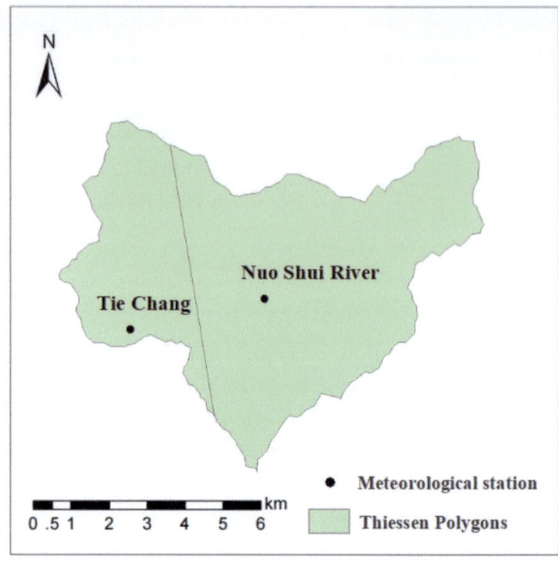

Fig. 2 Weight distribution of Tyson polygon area in the Luba River basin (self-drawn)

Table 2 Area and weight coefficient of Tyson polygon in the Luba River basin (self-drawn)

Station name	Area size (km^2)	Weight coefficient
Tie Chang	28.963499	0.2803
Nuo Shui river	74.356814	0.7197

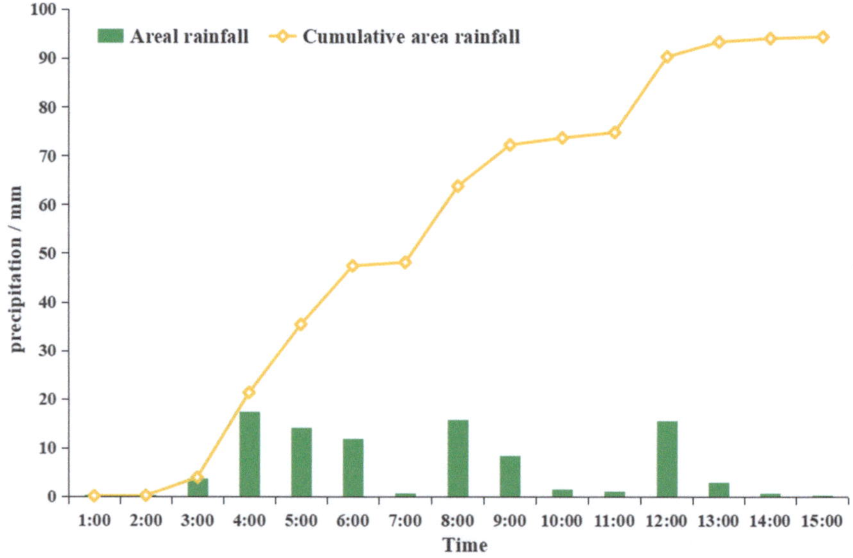

Fig. 3 Hourly area rainfall in Luba River basin (self-drawn)

Input the rainfall weight grid for this process, along with the hourly area rainfall data for this basin from 1 to 15 h on September 9, 2014 (Fig. 3). Utilizing the specified surface roughness parameters, the FloodArea hydrodynamic model is employed to simulate the flooding process induced by heavy precipitation (notably marking the occurrence of the first flood peak). The results of this simulation are presented in Fig. 4.

A comparison of the hourly area rainfall with the extracted hourly inundation results reveals that the hourly area rainfall begins to increase gradually from 3:00 AM, with precipitation intensity escalating between 4:00 AM and 6:00 AM, surpassing 10 mm. At 5:00 AM, water depth rises by more than 1 m; thereafter, it increases rapidly from 6:00 AM onward, resulting in a total rise of 1.79 m in water level. By 8:00 AM, area rainfall reaches 15.69 mm while the water level in the Luba River basin continues to ascend, causing river overflow over the dam and leading to a

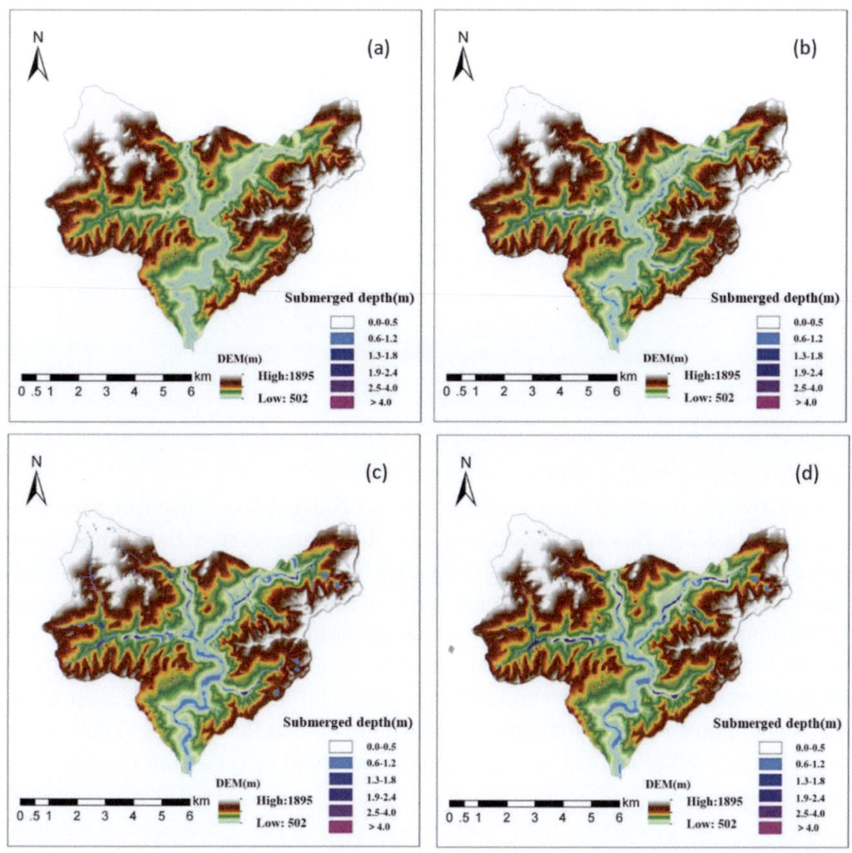

Fig. 4 Flood inundation simulation results of Luba River Basin on September 9, 2014 (self-drawn). **a** 2014–09–09 2:00. **b** 2014–09–09 4:00. **c** 2014–09–09 6:00. **d** 2014–09–09 8:00. **e** 2014–09–09 10:00. **f** 2014–09–09 12:00. **g** 2014–09–9 14:00

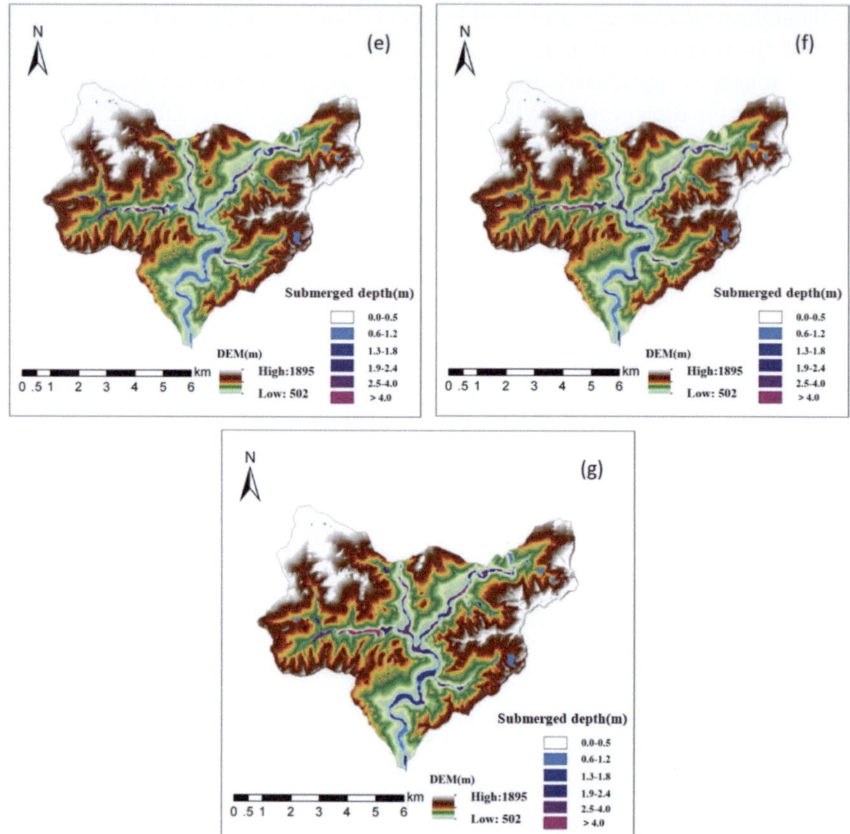

Fig. 4 (continued)

simulated water depth at the warning point exceeding 2.6 m. Given that the dam height is only 2.2 m, this indicates that by 8:00 AM, water levels at the warning point exceed dam height by approximately 0.4 m. By noon (12:00 PM),area rainfall totals reach approximately 15.60 mm as water levels persistently rise and inundation areas expand; consequently, water levels at warning points increase by more than 3.5 m and exceed inundation thresholds by over 1.2 m. The simulation results indicate that both precipitation intensity and duration of heavy rain significantly influence river water level fluctuations; notably, maximum inundation depth at the warning point occurs around noon (13:00 PM),which aligns closely with field investigation findings.

3.4 Assess the Critical Rainfall Threshold for Mountain Flood Disasters

In accordance with the pertinent regulations established by the National Meteorological Administration, the risk of mountain flood rainfall is categorized into four levels: Level Four indicates overtopping of levees; Level Three corresponds to an inundation warning point at 0.6 m; Level Two signifies an inundation warning point at 1.2 m; and Level One denotes an inundation warning point of 1.8 m or higher.

Based on the flood simulation conducted for the Luba River Basin from 1:00 to 15:00 on September 9, 2014, Table 3 presents the hourly simulated water levels at flash flood warning points. In comparison with field investigation records, the maximum actual submerged depth recorded was 4.05 m, while the simulation yielded a result of 3.6719 m—approximately 0.37 m lower than that observed in situ—but both occurrences were temporally aligned: when the simulated value reached thirteen, resident surveys indicated that peak submergence occurred around between 1:00 and 2:00 PM. Referring to the research of Fan [19] and Jiang [20], this error range is acceptable. Thus, these simulation results are deemed reasonable and effective as they align closely with real-world conditions.

The correlation equation is derived from the accumulated hourly rainfall data and the corresponding water levels at the warning point of the Tang Jia He Ba. Simulation results indicate that the maximum inundation depth predicted by the

Table 3 Flash flood warning point (Point A-Tang Jia He Ba) simulated water level by hour (self-drawn)

Time	Hourly area rainfall (mm)	Warning point simulated water depth (m)	Depth of water above dam height (m)
1:00	0.10	0.0000	−2.2000
2:00	0.10	0.0010	−2.1990
3:00	3.74	0.0636	−2.1364
4:00	17.36	0.5762	−1.6238
5:00	14.10	1.2553	−0.9447
6:00	11.98	1.7866	−0.4134
7:00	0.72	2.1540	−0.0460
8:00	15.69	2.6559	0.4559
9:00	8.43	3.1779	0.9779
10:00	1.44	3.2238	1.0238
11:00	1.11	3.4724	1.2724
12:00	15.60	3.5717	1.3717
13:00	3.02	3.6719	1.4719
14:00	0.70	3.6207	1.4207
15:00	0.33	3.5547	1.3547

model is approximately 0.37 m less than the recorded actual inundation height. To enhance both applicability and authenticity of these simulation results, a correction of 0.37 m is applied to the relative inundation level based on river surface measurements, allowing for an accurate calculation of mountain flood risk rainfall using the fitted equation.

A sliding cumulative correlation analysis with a lag time ranging from 1:00 to 15:00 for water levels was conducted (Table 4). It is evident that when rainfall lags exceed 8 h, the correlation coefficient consistently remains above 0.9, indicating stability in trends. Given that timely flash flood risk assessments are crucial for effective defense measures, an assessment interval of 8 h has been selected. Consequently, an accumulated rainfall series over this forward time frame has been constructed as atmospheric precipitation relevant to flood-triggering processes.

The correlation between the simulated water levels at the early warning point and the accumulated rainfall over the initial 1–15 h was examined, revealing a correlation coefficient of 0.9025 for the accumulated area rainfall over an 8-h period. The regression equation representing the relationship between rainfall and water level is established as follows (Fig. 5), which facilitates simultaneous calculations of critical rainfall and corresponding water levels. The vertical axis of this equation represents the cumulative area rainfall in millimeters for a small watershed, while the horizontal axis denotes the inundation depth at Tang Jia He Ba's warning point, measured in meters. Table 5 presents four risk levels associated with cumulative area rainfall at these warning points, serving as a reference for future flash flood warnings in this region (Table 5).

$$y = 16.343x + 1.956 \tag{6}$$

Table 4 Correlation between cumulative N-hour area rainfall and water level (R) (self-drawn)

N	3 h	4 h	5 h	6 h	7 h	8 h	9 h	12 h	15 h
R	0.276	0.477	0.576	0.699	0.818	0.905	0.973	0.974	0.989

Table 5 Different levels of risk rainfall of flash flood warning points (self-drawn)

Risk level	Water level (m)	Cumulative area rainfall (mm)			
		1 h	3 h	6 h	9 h
Level 1	Overtopping	16.615	27.539	35.038	39.609
Level 2	0.6 m	18.869	34.965	47.407	54.720
Level 3	1.2 m	21.123	42.392	59.775	69.830
Level 4	1.8 m	23.377	49.818	72.144	84.941

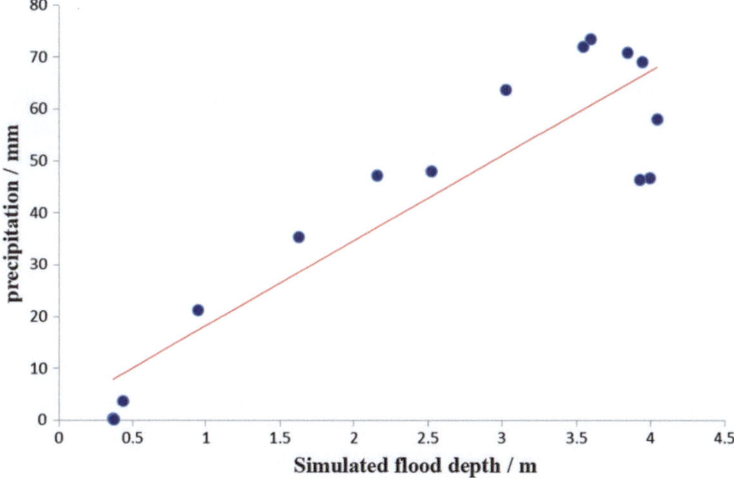

Fig. 5 The correlation between simulated flood depth (m) and the cumulative rainfall over an 8-h period (mm) at the warning point (self-drawn)

4 Conclusion

The FloodArea model is employed to simulate and replicate the flooding scenario triggered by the initial flood peak in the Luba River small watershed of the Qujiang River Basin on September 9, 2014. The simulation results demonstrate a strong consistency between the extracted inundation depth at early warning points and the measured depths from field surveys. These findings indicate that the FloodArea model offers technical support for analyzing storm-induced flash floods and studying rainfall-related flash flood risks in small mountainous basins with limited hydrological data.

From an hour-by-hour simulation perspective, it is evident that both cumulative precipitation and hourly rainfall intensity during rainstorm events significantly influence flood inundation within small watersheds. An analysis of the correlation between simulated water levels at warning points and accumulated rainfall over periods ranging from 1:00 to 15:00 reveals a correlation coefficient exceeding 0.9 for rainfall recorded in the last 8 h, indicating stability in this relationship. To enhance timeliness in flash flood warnings, critical area rainfall thresholds across four levels were calculated based on cumulative area rainfall over an 8-h period, providing insights for disaster prevention and mitigation strategies against flash floods in this region.

5 Discussion

This paper verifies that the Floodarea model can provide reliable technical support for the analysis of rainstorm and flood and the study of risk rainfall of mountain torrents in small watersheds without hydrological data, but there are still some shortcomings, which we will study and improve in the future work:

(1) We need to obtain DEM data with higher precision (e.g. 5 m precision dem), and conduct a more comprehensive and detailed geographic information survey on the disaster warning area to improve the simulation accuracy.

(2) There are few meteorological monitoring stations in small watersheds in mountainous areas, which may lead to low accuracy and insufficient spatial representation in calculating the area rainfall of small watersheds. In addition, in this paper, the time scale of precipitation input by the model is 1 h, which cannot reflect the more detailed spatial and temporal distribution characteristics of precipitation. In the future work, we can consider selecting minute-level precipitation data.

(3) There are not enough cases of torrential rain and flash floods recorded in small watershed in mountainous areas, and the accuracy of the calculation results of risk rainfall in watershed needs to be further improved. We hope to collect more samples to verify the simulation method. In the future, more examples are needed to confirm and make reasonable parameter adjustment to improve the critical rainfall index of mountain torrents in small watersheds.

Acknowledgements This study is supported by the NSFC(42322703), the Sichuan S&T project (2022JDJQ0008) and the Western Light of Young Scholars, CAS.

References

1. Yonglin H, Zhiyi W, Qiang F (2007) Application of GIS in mountain flood disaster warning in Guangxi. Meteorol Res Appl 28(3):30–32
2. Zhuyun Y, Yanyan X, Yingjie X et al (2023) Comparative analysis of precipitation characteristics of geological hazards of different precipitation types in Yunnan Province. J Yunnan Univ: Nat Sci Edn 45(3):634–641
3. Houfu Z, Wenyu W, Shuangwu J et al (2012) Establishment and application of flash flood disaster warning model in Dabie mountain area. J Catastrophology 27(2):49–53
4. Bangxing T (2000) Debris flow in China. The Commercial Press, Beijing, pp 1–115
5. Ronghuan D, Deji L, Long Q (1992) Study on the characteristics and prevention measures of debris flow disaster in mountainous towns in China. Disaster situation analysis and disaster reduction countermeasures of natural disasters in China. Hubei Science and Technology Press, Wuhan, pp 330–336
6. Yuzhong G, Junjiang X, Chunli W et al (2006) Analyzing the causes and forecasting methods of rainstorm and mountain flood disasters. J Nat Disasters 15(4):65–70
7. Boqian H (2006) A critical analysis of rainfall inducing mountain floods in Hebei Province. In: South-to-North water diversion and water science and technology, vol 4, no B06, pp 28–29

8. Dongtao M, Jinshan Z, Zili F et al (2005) Investigating the causes of mud-rock flow and mountain flood disasters during the Yingjiang landslide in Yunnan province: mitigation strategies proposed. J Catastrophology 20(1):67–71
9. Adachi K, Tokushan J, Nakasuji A et al (1977) The risk of earth-rock flow is determined. New Sand Control 30(3):7–16
10. Hollingsworth R, Kovacs GS (1981) Soil slumps and debris flows: prediction and protection. Bull Assoc Eng Geol 38(1):17–28
11. Kubota T, Masaiko, Itagaki A (1990) Prediction of debris flow risk by short-term rainfall at any place in a basin. New Sand Prevent 42(6):ll-17
12. Chuanzheng L, Mingsheng W, Can T (2004) Preliminary study on meteorological warning of geological disasters in China. Geol Bull China 23(4):304–309
13. Li M, Xiangping Z, Bo X (2002) Analysis of precipitation conditions for landslide occurrence in Chongqing. J Mountain Sci 20(2):4
14. Hanzhong F, Linna X, Kangquan Y et al (2011) Analysis of rainfall characteristics of debris flow and landslide based on hourly rainfall. Plateau Mountain Meteorol Res 31(4):18–23
15. Guiya C, Yaming Y (2005) Study on analysis and calculation method of critical rainfall of mountain flood disaster. People Yangtze River 36(12):40–43
16. Geomer (2003) FloodArea—Arcview extension for calculating flooded areas (User manual Version 2.4), Heidelberg
17. Buda S, Tong J, Yeyou G et al (2005) Flood risk dynamic simulation model based on GIS raster data and its application. J Hohai Univ: Nat Sci Edn 33(4):5
18. Mingzhang W, Xin L, Lijun Y et al (2013) Risk rainfall assessment methods of flash floods disaster. Meteorol Month 39(10):1325–1330
19. Jianyong F, Jiusheng S, Min G et al (2014) Research on analysis and calculation method of critical precipitation of mountain torrents in Jiangxi Province. Meteorol Month 38(9):1110–1114
20. Zhihua J, Yizhi Z, Zhe C et al (2014) Simulation of flood disasters based on embedded river raster. Meteorol Month 40(8):1013–1018

Disaster Resilience Assessment of Tourist Places Under the Influence of Multiple Hazards—A Case of Jiuzhaigou Scenic Area, China

Yuzhen Liu, Linsheng Gu, Manjiang Shi, Huan Hu, and Wenjiang Luo

Abstract This study takes the 2017 "8–8" Jiuzhaigou earthquake as the background, based on the disaster system theory, constructs the Jiuzhaigou scenic disaster resilience evaluation index system from the four dimensions of early warning, rescue, disaster preparedness and management, and analyzes the dynamic evolution characteristics of the scenic disaster resilience in the period of 2015–2024 through the entropy weight-TOPSIS method. The dynamic evolution characteristics of disaster toughness in the scenic area. The study found that: (1) the overall toughness presents the trend of "pre-disaster fluctuation—sudden drop in disaster time—continuous rebound in post-disaster period", and the toughness index grows at a slower rate before the earthquake, and there is a sudden drop due to the earthquake in 2017; in the post-disaster recovery period, through the ecological restoration, optimization of the emergency channel and the construction of digital twin platform, the resilience of the scenic area is improved. After the disaster, through ecological restoration, emergency access optimization and digital twin platform construction, the toughness has been significantly improved; (2) at the subsystem level, resilience is the fastest-growing due to vegetation gradient restoration and seismic facilities renovation, early warning benefits from AI monitoring network upgrades, and rescue is limited by mountainous terrain; (3) the research deficiency lies in the dilemma of data isomorphism, lack of theoretical applicability, and the standardization of the evaluation system. It is suggested to strengthen the "ecological-facility" composite resilience network, multi-hazard early warning system and community collaborative governance to provide scientific basis for disaster resilience planning in mountainous scenic areas.

Keywords Multi-hazard · Disaster resilience · Entropy weight-TOPSIS method · Jiuzhaigou scenic area

---

Y. Liu · M. Shi · H. Hu · W. Luo
School of Civil Engineering and Architecture, Southwest University of Science and Technology, Mianyang, China

L. Gu (✉)
Sichuan Earthquake Early Warning and Emergency Research Society, Chengdu, China
e-mail: 1026620864@qq.com

B. He et al. (eds.), *Climate-Resilient, Low-Carbon, and Sustainable Cities*, Lecture Notes in Civil Engineering 754, https://doi.org/10.1007/978-981-95-2169-2_6

1 Introduction

Since the twenty-first century, China's economy has shifted to a period of high-quality development, the consumption level of the population has increased in all aspects, and tourism has gradually become a rigid demand of the people for a better life. The increasing selectivity of the types of tourism activities and destinations has contributed to the overlap between tourism and disaster risks, and tourism destinations, especially mountain tourism destinations, tend to face more disaster risk hazards, as shown in the table. The Wenchuan Earthquake in 2008 damaged several scenic spots and the natural ecological landscapes within them, resulting in serious economic losses. The 7.0 magnitude earthquake in Jiuzhaigou and the 6.8 magnitude earthquake in Luding caused serious damage and casualties to the two 5A scenic spots in Jiuzhaigou and Hailougou [1]. The 7.6 magnitude earthquake in Noto Peninsula in Ishikawa Prefecture, Japan, triggered secondary fires on January 1, 2024, resulting in serious fires around the famous tourist attraction "Wajima Hayashi". The earthquake also caused a serious fire around Wajima Hayashi, which is the most famous tourist attraction in Japan. The earthquake caused secondary fires, which led to severe fires around the famous tourist attraction "Wajima Hayashi" [2]. When a disaster occurs, it generates a huge amount of catastrophic energy, which triggers a series of secondary and derivative disasters, i.e., the disaster chain theory. According to this theory, there is a certain degree of correlation, hierarchy, and complexity among disasters, which may be manifested as a single chain relationship of one cause and one effect, with the disaster being connected to each other, or as a complex characteristic of a tree-like network with multiple causes and multiple effects [3].

In terms of tourism system resilience research, the current research object is mostly centered on tourism economic system, tourism industry, urban tourism system, tourism community resilience research, etc., and the research content mainly includes the definition of conceptual connotation, tourism resilience influencing factors and influencing mechanisms, tourism resilience level evaluation, tourism resilience enhancement strategies, etc. [4]. From the point of view of research scale, most of them are tourism resilience evaluation studies on the national and provincial scale, but there are few quantitative studies on the scale of tourist attractions [5]. In terms of disaster resilience research, the existing studies on the measurement of disaster resilience level tend to focus on some urban and rural areas, and there is less relevant literature on this kind of composite geospatial space of tourist attractions, which, as a tiny unit of the tourism industry, has disaster resilience characteristics consistent with the resilience of the tourism system, which mainly refers to the ability of resisting the disaster risk, restoring the function of the system, and updating the structure of the system in the case of a devastating shock. The ability of system structure.

Based on this, this study combines tourism attractions with the concept of disaster resilience, and based on the disaster system theory, constructs an index system for the evaluation of disaster resilience of tourism attractions from the whole process of disaster prevention, relief, mitigation, and post-disaster recovery, and carries out an

Table 1 Differential analysis of potential disaster risks between tourist attractions and other areas (self-drawn)

Potential disaster risks	Tourist attraction	Other areas
Natural disaster	Mostly in geologically complex areas, vulnerable to geological hazards such as landslides and mudslides	Many face weather disasters such as urban flooding and typhoons
Social environment	Tourist behavior has a certain complexity and is more uncontrollable	Focus more on public safety such as transportation and infrastructure
Ecosystems	The ability of landscape ecosystems to repair themselves directly affects disaster recovery	Ecological resilience of urban green space systems is more of a carbon sink function
Risk perception	Tourists have insufficient knowledge of potential disaster risks in scenic spots, and are prone to leave things to chance	Local residents are relatively more aware of precautionary measures
Protective measures	Specialized plans for specific risks	Universal

empirical analysis, which provides a theoretical basis and reliable data support for the scenic spots as well as the local government to formulate a more scientific and reasonable disaster resilience enhancement policy (Table 1).

2 Analysis of Research Subjects

2.1 Scenic Area Overview

Jiuzhaigou Scenic Area is located in Zhangzha Town, Jiuzhaigou County, Sichuan Province (Fig. 1). The overall elevation of the scenic area is at 1996–4764 m, with the highest point of elevation located at the southernmost Gaerna and the lowest point at the northernmost ditch mouth of Yandong, with the maximum difference in elevation amounting to 2,768 m and the average difference in elevation amounting to more than 1,600 m.

2.2 Disaster Risk Analysis

Jiuzhaigou scenic area is located at the intersection of the Songpan-Ganzi orogenic belt and the West Qinling orogenic belt [6], and the fracture zones that have a great influence on the geologic construction-tectonic environment of the region are

Fig. 1 Jiuzhaigou location map (self-drawn)

mainly the Wenxian-Maqin fracture zone, the Minjiang River fracture zone, and the Huya fracture zone [7]. Therefore, seismic hazard remains one of the major disaster activities in the region. A 7.0-magnitude earthquake ("8–8" earthquake) occurred in Bimang Village, Jiuzhaigou County, on August 8, 2017, and according to the results of the 2021 Jiuzhaigou Scenic Area Geological Hazard Hazard Contingency Investigation, the results show that before the "8–8" According to the results of the 2021 Emergency Investigation of Geological Hazards in Jiuzhaigou Scenic Area, there were 72 geologic hazards in Jiuzhaigou Scenic Area before the earthquake. After the earthquake, the original geological hazards in the scenic area, there are 63 points of disaster hazard intensification, there are 9 points of disaster hazard unchanged. After the "8–8" earthquake, Jiuzhaigou developed a total of 153 geological hazards of collapse, landslide, mudslide and unstable slope [8], with 81 new hazard points. These geologic hazards are mainly distributed in Shuzheng, Changhai, Rize, Zaru, Danzugou and other places.

Therefore, the article takes the "8–8" earthquake as the research background, selects 2015–2024 as the research data sample, constructs the framework of Jiuzhaigou scenic area disaster toughness average system, and studies the evolution characteristics of scenic area disaster toughness level before and after the earthquake in Jiuzhaigou scenic area, and the index parameters in the toughness evaluation system are mainly composed of the relevant departments of Jiuzhaigou Management Bureau, the information records, web data, field research and so on. The index parameters in its toughness evaluation system are mainly composed of the relevant departments of Jiuzhaigou Management Bureau, information records, webpage data, field research and other contents.

3 Constructing a Disaster Resilience Evaluation System

3.1 Construction of Evaluation Index System

Constructing an assessment indicator system is the most commonly used method for toughness evaluation, so the article selects the indicator system method to construct the relevant assessment system. In the process of constructing resilience assessment indicators, the following principles of indicator selection are mainly followed: (1) the principle of scientificity, the construction of the indicator system must be based on the scientific nature to ensure that the indicator system is scientific and reasonable, and the evaluation results are objective and fair; (2) the principle of representativeness, i.e., to use the most direct and concise indicator system to embody the basic situation of disaster resilience of the study area; (3) the principle of operability, The selection of targets needs to take into account external influences and the ease of data acquisition [9]; (4) the principle of comparability, i.e. the selection of quantifiable and operational indicators for comparison and analysis [10].

The article analyzes the complexity of the impact of disasters on the tourism land system, and provides theoretical basis for the construction of the resilience evaluation system for tourism scenic spots from the perspective of multi-factors, multi-dimensions, etc.; and constructs the resilience evaluation system from the aspects of preventive power, rescue power, disaster preparedness and management power, etc., and examines and judges the level of disaster resilience development of the scenic spots. Among them, the target layer is the level of disaster resilience in tourist attractions; the system layer includes resource elements such as early warning, rescue, disaster preparedness and management; the element layer includes 8 indicators such as disaster monitoring and engineering defense; the fourth level is the specific evaluation index layer, and a total of 20 indicators are selected for comprehensive evaluation (Table 2).

3.2 Interpretation of Specific Indicators

From the perspective of mutual coupling of subsystems such as early warning capacity, rescue capacity, disaster preparedness capacity, and management capacity, 20 indicators are selected to measure the state and response level of the system of the tourist places in the face of disaster impacts and to comprehensively reflect the resistance, resilience and updating ability of the tourist places in response to disasters.

Table 2 Jiuzhaigou scenic area disaster resilience evaluation index system (self-drawn)

Target level	System level	Factor layer	Indicator layer	Interpretation of indicators	Type of indicator
Disaster resilience assessment system for tourist sites	Early warning capability	Disaster monitoring	Coverage of disaster early warning monitoring sites (%)	Level of early warning monitoring coverage in disaster-prone areas	+
			Number of daily monitoring patrols (person-times)	Frequency of daily monitoring and patrols carried out in the landscape	+
		Message	Information orientation refinement (%)	Level of communication of disaster information in scenic areas	+
			Timeliness of publicity and reporting	Timeliness of news and publicity coverage of scenic disasters, etc.	+
	Rescue capacity	Team building	Rescue team security (person-times)	Scenic rescue team strength guarantee	+
		Engineered defense	Coverage of emergency shelters (%)	Sites that cater for the temporary sheltering of persons in the event of a disaster	+
			Proportion of buildings meeting defense standards (%)	Percentage of buildings, structures, etc., that meet regional requirements for defense standards	+
	Preparedness capacity	Economic security	Annual number of tourists (thousand)	Annual cumulative number of tourists received at scenic spots	+
			Total tourism income from scenic spots (million)	Annual admission revenue from scenic spots	+

(continued)

Table 2 (continued)

Target level	System level	Factor layer	Indicator layer	Interpretation of indicators	Type of indicator
			Disaster prevention and emergency management costs (million)	Scenic area funding in areas such as disaster prevention and emergency management	+
			Research project support	Promotion of scientific and technological research projects for scenic conservation	+
		Ecological carrying capacity	Green coverage (%)	Forest green coverage in scenic areas	+
			The proportion of days with good ambient air quality (%)	Percentage of good ambient air quality days in a year	+
			Tourist satisfaction (%)	Visitors' satisfaction with the work of scenic spot-related services	+
			Number of related web stories (times)	Number of promotional campaigns in various forms	+
	Management capability	Talent pool	Training coverage for managers (person-times)	Several high-level professional and technical staff introduced	+
			Medical personnel coverage (person-times)	Number of medical aid workers	+
		Information construction	Degree of completeness of contingency plans (%)	Degree of sophistication of disaster contingency plans	+

(continued)

Table 2 (continued)

Target level	System level	Factor layer	Indicator layer	Interpretation of indicators	Type of indicator
			Number of emergency integrated exercises (times)	Integrated exercise on earthquake relief, flood control, and geological disaster emergency response	+
			Number of relevant awareness-raising activities (times)	Number of protection campaigns on emergency response in the landscape	+

(1) Early warning resilience indicator. Early warning force reflects the state of the tourist site before the disaster and the risk-resistant capacity it has. The state indicators include disaster monitoring and information communication ability, and the response indicators include the coverage of disaster early warning monitoring points, the number of daily monitoring patrols, and the degree of perfect information orientation. The coverage of early warning monitoring points and daily monitoring patrols reflect the monitoring capacity of the scenic area before the occurrence of the disaster (under normalized state), and the degree of perfection of information orientation and timeliness of reporting reflect to a certain extent the level of information conveyance and corresponding preparatory work of the scenic area.

(2) Resilience indicator. Rescue capacity reflects the response-ability and resistance ability of tourist places when facing disaster impacts. Status indicators include rescue team construction and engineering defense capability. Response indicators include the construction of rescue teams, the distribution of emergency shelters, and the proportion of buildings (structures) meeting the defense standards. The guarantee of rescue team power reflects the professional rescue ability of the scenic area, and the distribution of refuge places and defense standards of buildings and structures reflect the rescue response ability of the scenic area when facing external disaster impact.

(3) Disaster Preparedness Disaster preparedness reflects the reserve capacity and support capacity of the tourist attractions to cope with disasters in their daily state. The state indicators include economic reserves, ecological environment bearing, and basic guarantee. Response indicators include tourist volume, tourism income, disaster prevention cost investment, scientific research project support, greening rate, network publicity report and other resource elements. The higher tourism income indicates the faster tourism economic growth rate, which indicates the higher tourism development potential of the tourist place,

the disaster prevention expense input and scientific research project support can reflect the economic support that the scenic spot can provide, the vegetation coverage rate reflects the regional ecological environment balance status, representing the ecosystem stability and carbon sink capacity [11]. The satisfaction of tourists and network publicity reports can reflect the basic support and soft power of the scenic area.

(4) Management resilience indicator. Management force reflects the resistance ability of tourist places in the face of disasters at the management level. Status indicators include talent management and information construction. Response indicators include elements such as scientific and technological talents and medical personnel security, the degree of perfection of the emergency plan, the number of comprehensive emergency drills, and the level of information technology services. The introduction of technical talents and medical support can reflect the social security strength of the tourist places in coping with disasters, and the improvement of emergency plans and comprehensive drills side by side reflect the improvement of the construction level of disaster prevention in the tourist places.

3.3 Research Methodology

The number of disaster resilience evaluation indicators constructed in the article is relatively large, and it is difficult to realize the scientific and reasonable assignment of indicator weights by adopting the subjective assignment method. Entropy weight method can make full use of the original information of the evaluation results, reduce the influence of human factors, so that the determination of evaluation index weights is more objective and accurate [12]. TOPSIS method by measuring the gap between a limited number of evaluation objects and the idealized target can be the overall performance of the different evaluation objects for the relative merits of the ranking, a limited number of evaluation objects to achieve multi-objective decision-making analysis [13]. Therefore, the article chooses the objective entropy weight-TOPSIS method to evaluate the disaster resilience level of Jiuzhaigou scenic area, which objectively reflects the dynamics of the development of the regional resilience level and the trend of change. The evaluation steps of method are as follows:

(1) Construction of raw matrices for evaluation indicators and data standardization.

$$X_{\hat{i}j} = \begin{bmatrix} X_{11} & X_{21} & \cdots & X_{m1} \\ X_{12} & X_{22} & \cdots & X_{m2} \\ \vdots & \vdots & \ddots & \vdots \\ X_{1n} & X_{2n} & \cdots & X_{mn} \end{bmatrix} \tag{1}$$

In the above equation, m represents that there are m research subjects in this study; n represents that there are n evaluation indicators. For positive (the bigger the better) and indicators, the treatment is shown in Eq. (2); for negative (the smaller the better) indicators, see Eq. (3).

For positive indicators:

$$r_{ij} = \frac{X_{ij} - X_{\min}}{X_{\max} - X_{\min}} \tag{2}$$

For negative indicators:

$$r_{ij} = \frac{X_{\max} - X_{ij}}{X_{\max} - X_{\min}} \tag{3}$$

In the above equation, the r_{ij} is the standardised value of the jth indicator in the ith system, and $X_{\max}$ represents the maximum value corresponding to the jth indicator in each region, and $X_{\min}$ represents the minimum value corresponding to the jth indicator in each system.

(2) Calculate the information entropy of each indicator and determine the weights of the indicators.

$$E_j = -k\sum_{i=1}^{m} f_{ij}\ln f_{ij} \tag{4}$$

In the above equation, the $f_{ij} = \frac{r_{ij}}{\sum_{i=1}^{m} r_{ij}}$. $k = \frac{1}{\ln n}$, if $f_{ij} = 0$, then define $f_{ij}\ln f_{ij} = 0$. The entropy weight of the jth indicator ω_j is calculated by the formula:

$$\omega_j = \frac{1 - E_j}{\sum_{j=1}^{m}(1 - E_j)} \tag{5}$$

(3) Construction of a weighted decision matrix.

Construct a weighted decision matrix Z based on the standardized matrix and each weighted indicator.

$$Z = \begin{bmatrix} Z_{11} & Z_{21} & \cdots & Z_{m1} \\ Z_{12} & Z_{22} & \cdots & Z_{m2} \\ \vdots & \vdots & \ddots & \vdots \\ Z_{1n} & Z_{2n} & \cdots & Z_{mn} \end{bmatrix} = \begin{bmatrix} r_{11}\omega_1 & r_{21}\omega_2 & \cdots & r_{m1}\omega_n \\ r_{12}\omega_1 & r_{22}\omega_2 & \cdots & r_{m2}\omega_n \\ \vdots & \vdots & \ddots & \vdots \\ r_{1n}\omega_1 & r_{2n}\omega_2 & \cdots & r_{mn}\omega_n \end{bmatrix} \tag{6}$$

Based on the calculation of the weighted decision matrix determine the optimal vector of the matrix Z^+ and the worst vector Z^-.

$$Z^+ = \max\{Z_{1j}, Z_{2j}, Z_{3j}, \ldots Z_{mj}\} \tag{7}$$

$$Z^{-} = \min\{Z_{1j}, Z_{2j}, Z_{3j}, \ldots Z_{mj}\} \tag{8}$$

(4) Calculate the distance of the object from the positive and negative ideal solutions.

$$D_i^{+} = \sqrt{\sum_{i=1}^{n}\left(Z_{ij} - Z_j^{+}\right)^2} \tag{9}$$

$$D_i^{-} = \sqrt{\sum_{i=1}^{n}\left(Z_{ij} - Z_j^{-}\right)^2} \tag{10}$$

(5) Calculate the relative proximity of the evaluation object to positive and negative ideal solutions.

$$C_i = \frac{D_i^{-}}{D_i^{+} - D_i^{-}} \tag{11}$$

C_i The larger the value, the higher the relative proximity of the ith evaluation object to the optimal solution, i.e., the higher the resilience level of the site; conversely, the lower the resilience level of the site.

4 Findings and Analyses

4.1 Evolutionary Characteristics of Disaster Resilience

The entropy weight-TOPSIS method was applied to the raw data to obtain the weights (Table 3) and resilience indices (Fig. 2) of 20 relevant indicators of Jiuzhaigou scenic area in 2015–2024. From the perspective of time scale, the development of the scenic area in 2015–2024 can be roughly divided into three stages, with 2015–early 2017 as the pre-disaster development stage, 2018–2020 as the restoration and reconstruction stage, and 2021–2024 as the upgrading and development stage. According to Fig. 2, in the pre-disaster development stage, the disaster resilience level of the scenic area is relatively low, but the overall trend is upward, and the scenic area was affected by the "8–8" earthquake in 2017, the scenic area came to a brief stagnation, and the overall development level dropped sharply. In the stage of restoration and reconstruction, with the completion of ecology, disaster management, facility construction and other aspects of the work, the scenic area disaster level continues to improve, from the point of view of the toughness index, 2018 has been a large development, in 2019 to catch up with the pre-disaster level of toughness, and 2020 has completely exceeded the pre-disaster level, indicating that three years of restoration and reconstruction has achieved good results, and the scenic area as a whole is in a period of rapid development, and the development of tourism has entered a period of high-speed development. Development has entered a period of high-speed development. In the upgrading development stage, the overall disaster toughness level of the scenic area is gradually rising, but compared to 2023–2024, the upgrading speed is higher than that of 2020–2022, which is mainly due to the impact of the new coronary pneumonia epidemic, the number of its tourists and tourism revenue, etc. decline, and the overall development of the toughness level is slower, to the tourism industry enters the key period of comprehensive development in 2023, and the toughness of its construction The pace of resilience building will accelerate.

4.2 Subsystem Disaster Resilience Evolution Patterns

In addition, the resilience index of each subsystem layer was calculated separately using the above method (Fig. 3), as shown in the figure. The resilience index of each subsystem layer shows a fluctuating upward trend, and the fluctuation of the resilience index of each dimension is obvious in 2015–2017, with a collective trough in 2017 due to the Jiuzhaigou earthquake; in 2018–2024, the early warning power, rescue power, preparedness power, and restoration power are all showing a continuous upward trend, reflecting that the post-disaster resilience building The effectiveness is remarkable. Analyzing from the time dimension, before the earthquake (2015–2017), the resilience index of each dimension is affected by the management of the scenic area and the rhythm of facility construction, showing natural fluctuations. 2017 earthquake, the scenic area's original disaster prevention system suffered damage, the early warning system, rescue channel, disaster preparedness material reserves and other functional failures, resulting in the collective bottom of the four resilience indices. 2018–2024 is the post-disaster reconstruction phase. 2024 for the post-disaster reconstruction stage, early warning power enhancement comes from the scientific deployment of earthquake monitoring network, meteorological early warning facilities; rescue power enhancement benefits from the widening of emergency channels, rescue site optimization and other spatial planning; disaster preparedness power growth relies on the hierarchical construction of the material reserve; resilience enhancement relies on the ecological restoration planning, seismic retrofitting of buildings and other long-term strategies.

Analyzing from the dimension of subsystem layer, the early warning force shows a stable upward trend after 2018, indicating that scenic spots gradually strengthen the front-end sensing ability of disasters by upgrading monitoring equipment (e.g., seismic and meteorological early warning systems) and optimizing the information transfer mechanism, which provides a more adequate time window for emergency response. Rescue force in the late growth rate is significant, reflecting the scenic spot in the rescue team construction, emergency channel dredging, rescue process optimization and other aspects of outstanding results. Disaster-preparedness force shows "stepped" growth characteristics, indicating that the construction of disaster-preparedness material reserve system has a stage. It is possible to improve the layout

Table 3 Jiuzhaigou scenic area disaster resilience evaluation index system and weight (self-drawn)

Target level	System level	Factor layer	Indicator layer	Entropy weight	Combined weights
Disaster resilience assessment system for tourist sites	Early warning capability	Disaster monitoring	Coverage of disaster early warning monitoring sites	0.043	0.21
			Number of daily monitoring patrols	0.089	
		Message	Information orientation refinement	0.041	
			Timeliness of publicity and reporting	0.041	
	Rescue capacity	Team building	Rescue team security	0.041	0.14
		Engineered defense	Coverage of emergency shelters	0.056	
			Proportion of structures meeting defense standard	0.039	
	Preparedness	Economic security	Annual number of tourist	0.050	0.38
			Total tourism income from scenic spots	0.047	
			Disaster prevention and emergency management cost	0.051	
			Research project support	0.056	
		Ecological carrying capacity	Green coverage	0.040	
			The proportion of days with good ambient air quality	0.043	
			Tourist satisfaction	0.047	
			Number of related web stories	0.044	
	Managerial capability	Talent pool	Training coverage for managers	0.086	0.27
			Medical personnel coverage	0.052	
		Information construction	Degree of completeness of contingency plans	0.041	

(continued)

Table 3 (continued)

Target level	System level	Factor layer	Indicator layer	Entropy weight	Combined weights
			Number of emergency integrated exercises	0.043	
			Number of relevant awareness-raising activities	0.047	

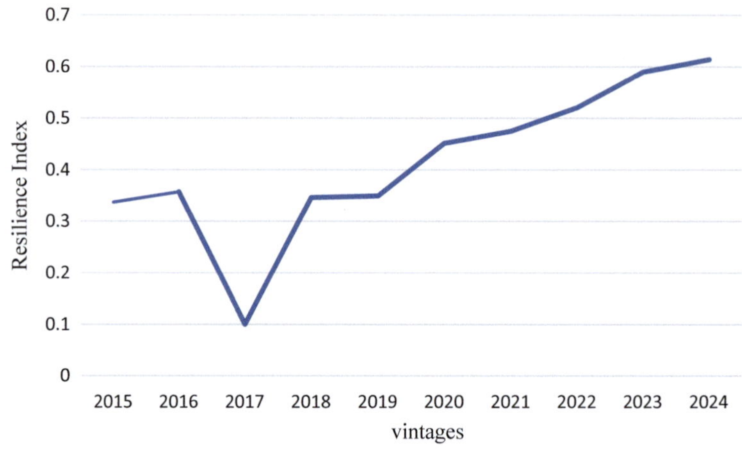

Fig. 2 Evolutionary trend of overall disaster resilience in Jiuzhaigou 2015–2024 (self-drawn)

of material reserve points and optimize the categories and quantities of reserves in phases, so as to gradually enhance the material protection capacity for disaster response. The late growth rate of resilience is leading and the final index is the highest, reflecting the synergistic effect of ecological restoration, facility reconstruction, functional restoration and other comprehensive measures in the scenic area.

5 Challenges and Recommendations

5.1 Challenges

(1) Data acquisition and integration

The study of disaster resilience in scenic areas needs to integrate multi-source data (e.g., ecological environment data, tourist flow data, disaster monitoring data, etc.),

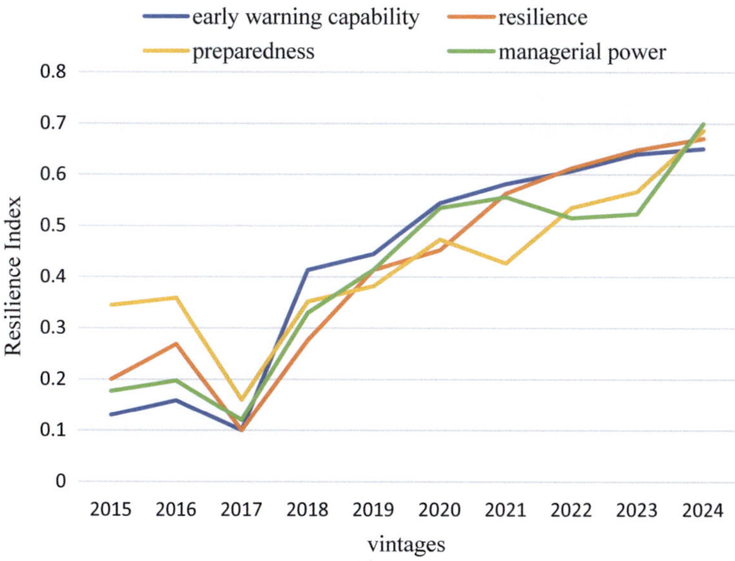

Fig. 3 Evolutionary trends of subsystem layer resilience in Jiuzhaigou, 2015–2024 (self-drawn)

but there is a problem of strong heterogeneity and insufficient real-time data. Part of the scenic area lacks long-term and continuous disaster monitoring system, and the data collection of ecological and tourism activities is fragmented, which leads to large data gaps in the study, making it difficult to construct accurate analysis models and affecting the accuracy of resilience evaluation and prediction.

(2) Inadequate adaptation of theory system

Existing resilience theories mostly originate from urban, community and other fields, and there is insufficient theoretical integration for the unique attributes of scenic spots (e.g., ecological vulnerability, dynamics of tourism activities, and irreplaceability of landscape resources). Interdisciplinary integration (geography, tourism, disaster science, etc.) is still immature, and there is a lack of a systematic theoretical framework adapted to scenic spots, which makes it difficult to guide the practice of resilience enhancement in complex disaster situations.

(3) Dilemma of evaluation system standardization

The geographical environment, disaster types (such as earthquakes, mudslides, rainstorms), and functional positioning of different scenic spots differ significantly, resulting in the lack of uniform standards for the selection of indicators and determination of weights. For example, natural heritage-type scenic spots focus on ecological resilience indicators, while humanistic scenic spots are more concerned about the resilience of facilities and tourism services, so it is difficult to build a universal evaluation system, and the horizontal comparison and popularization and application of the research results are limited.

5.2 Recommendations

Based on the results of the above study, the following recommendations are made for the enhancement of disaster resilience in the Jiuzhaigou Scenic Area.

(1) Strengthening the spatial layout of scenic safety. Firstly, regular disaster risk assessment is conducted to comprehensively analyze and judge disaster risk levels from a multidimensional perspective and formulate appropriate defensive measures. The second is to strengthen the planning of emergency shelters and corresponding evacuation routes to meet the sheltering needs of permanent residents and tourists, rationally allocate emergency rescue forces and emergency supplies, and upgrade the standard of infrastructure and public facilities to enhance disaster resistance.
(2) Building a comprehensive disaster defense system. Firstly, to establish a multi-hazard, all-element disaster comprehensive monitoring, and early warning system with multi-source perception and means integration [14]. The realization of scenic risk management platform and emergency command platform interconnection, the formation of up and down the linkage, and the interconnection of synergistic effect. Secondly, to establish and improve the engineering facilities system that matches the development of the scenic area, to promote the resilience planning of the scenic area disaster prevention engineering facilities, and to enhance the risk-resistant capacity of the disaster prevention engineering.
(3) Improve the comprehensive disaster prevention and management system. One is to improve the scenic disaster prevention and mitigation management system, cohesion of scenic multi-party strength, the establishment of disaster prevention and mitigation of diversified management, and common disaster prevention and mitigation planning for the scenic area to come up with ideas. The second is to actively carry out disaster prevention and mitigation education and publicity activities to improve the ability of residents and tourists to cope with sudden disaster events; actively coordinate multi-departmental participation in emergency drills to improve the practical ability to achieve the purpose of disaster reduction.

6 Conclusions

This study constructs a disaster resilience evaluation model for tourist attractions based on the multi-dimensional aspects of "early warning, rescue, disaster preparedness and management", adopts the research method of determining the weights by combining subjectivity and objectivity, analyzes the evolution of disaster resilience in Jiuzhaigou Tourist Attractions, and draws the conclusions as follows.

(1) From the perspective of overall evolution characteristics, the development of disaster resilience level in Jiuzhaigou scenic area during the study period is generally characterized by "pre-disaster fluctuation—sudden drop in disaster time—continuous rebound after the disaster".
(2) In terms of the evolution pattern of the subsystem layers, the trend of resilience level development among the subsystem layers is basically consistent with the overall disaster resilience trend.

References

1. Junqing H, Han T, Yameng D (2021) A review of tourism disaster management research based on knowledge graph visualization. Disaster Sci 36(01):134–138
2. Guangqi C, Yanqiang W, Mingyao X et al (2024) The 7.6-magnitude earthquake on the Noto Peninsula, Japan, on 1 January 2024: source characteristics, hazard profile and emergency response. Earthquake 44(01):141–152
3. Wei W (2023) Research on the construction of resilient townships in southwestern ethnic areas based on multi-hazard governance. China Emerg Manag Sci 05:50–61
4. Bingwei T, Peijing J, Xudong T et al (2023) Construction of resilience evaluation index system for natural disaster response in tourism communities. China Ecotourism 13(05):890–906
5. Lu W, Ziruo H, Le Y et al (2024) Resilience assessment of tourist attractions under the impact of major public health events. J Resour Ecol 15(03):698–710
6. Jun C, Geng Y, Jie D et al (2021) Study on the landscape changes of the Jiuzhaigou scenic area after the 8–8 earthquake. Sichuan J Geol 41(01):161–166
7. Yusheng L, Chao H, Shujian Y et al (2017) Tectodynamic mechanism of seismic rupture and source rupture of the 7.0 magnitude earthquake in Jiuzhaigou. J Eng Geol 25(04):1141–1150
8. Cao J, Zheng S, Song W et al (2021) Distribution of geological hazards in Jiuzhaigou scenic area after the 8–8 earthquake and their impact on the core landscape of Jiuzhaigou. Sichuan J Geol 41(S1):148–153
9. Zheming S (2021) Research on comprehensive urban disaster prevention planning based on the concept of resilient city. Southwest University of Science and Technology
10. Fucai Y, Yichen Y, Linyu L et al (2023) Resilience level assessment of geological hazards in Beijing. J Beijing Norm Univ (Nat Sci Ed) 59(02):306–312
11. Jingyi W, Jiaolong L, Tianxiang H et al (2023) Spatial and temporal evolution of tourism industry resilience in the ChangZhuTan city cluster. China Ecotourism 13(02):252–267
12. Min W, Shuhao L (2018) Research on measuring the level of high-quality development of China's economy in the new era. Res Quant Econ Tech Econ 35(11):3–20
13. Wu JH (2022) Research on urban resilience evaluation based on integrated disaster prevention perspective. Nanhua University
14. National comprehensive disaster prevention and mitigation plan for the 14th five-year plan. China Prod Saf 17(08):6 (2022)

BY NC ND

Strong Wind Risk Assessments for Urban Trees in an Urban Park in Guangzhou, China

Yiqiang Xiao, Songyi Huang, Zhenxuan Liu, Xinheng He, Zheng Chen, and Hankun Lin

Abstract Typhoons are an extreme weather disaster that strikes coastal cities in south-eastern China every year. This study was based on a tree management project counter measuring to typhoon disasters in a park in Guangzhou, China. Based on the tree risk assessment methods, the research group tempted to explore a much more precise management basing on the wind field characters of the park. Thus, this study conducted some explorations on computational fluid dynamic (CFD) methods on extreme strong wind simulations in the urban environment to find out the field character and support the tree management works in future. Main results were found out in this research: (1) The CFD simulation revealed that the main risk area of the Central Park lied at the southern area, west-northern area, and both paths on the eastern and western sides; (2) The peak wind speed at the height of 5 m and 10 m could reached the wind scale of level 9 to level 11 when the boundary wind speed was set at 24 m/s; (3) With the overlapping of tree locations and wind risk areas, this study could support the tree management in a much more precise way in future. Thus, this study provided a CFD method on urban extreme wind assessment for tree management, which would be valuable for other similar problems in urban environments.

Keywords Tree management · CFD · Strong wind · Urban environment

Y. Xiao · S. Huang
School of Architecure, South China University of Technology, Guangzhou, China

Z. Liu · X. He · H. Lin (✉)
School of Architecture and Urban Planning, Guangdong University of Technology, Guangzhou, China
e-mail: hklin@gdut.edu.cn

Z. Chen
Guangdong Yi Lv Ecological Technology Development CO., LTD., Guangzhou, China

B. He et al. (eds.), *Climate-Resilient, Low-Carbon, and Sustainable Cities*, Lecture Notes in Civil Engineering 754, https://doi.org/10.1007/978-981-95-2169-2_7

1 Introduction

In the context of global climate change, cities are faced with increasing climate risks such as heat island, heat wave, strong wind and rainstorm, and urban resilience has been introduced into practices in many cities [1, 2]. The United Nations Sustainable Development Goal (SDG) "SDG 13: Climate action" target points out that climate change and extreme weather phenomena was increasingly worsening the human living environment. The risk of strong wind related disasters, especially in coastal cities [3], calls for the urgent need to strengthen the resilience capacity of countries to adapt to climate-related disasters. The 26th United Nations Climate Conference (COP26), to be held in 2021, pointed out the urgency of nature-based solutions (NBS) in response to the trend of climate change [4]. Recent researches also highlighted the importance of Green Infrastructure (GI) for climate change mitigation [5].

In recent years, the urban tree management adapting to climatic disasters has developed rapidly in Guangdong-Hong Kong-Macao Greater Bay Area (GBA). The complex spatial structure and extreme wind field changes of high-density cities often cause open space to be affected by gust winds [6], and the threat of damage to urban trees cannot be ignored [7]. The spatial structure and layout of high-density cities are diverse, and the underlying surface is often complicated, resulting in a series of abrupt air distribution [8]. The urban vertical wind gradient is also affected by the high-rise buildings in the dense urban environment, which intensifies the strong wind effect in the open space around the high-rise buildings [9]. Furthermore, cities in the GBA are often affected by typhoons landing in summer and autumn every year, and the risk of strong winds on urban trees in open space is huge.

Typhoons are an extreme weather disaster that strikes coastal cities in southeastern China every year [10]. For the effect of extreme wind, the research focus is mainly on meteorology and structural engineering [11]. Meteorology studies on strong tropical cyclones mainly include the mechanism and wind profile characteristics [12]. In structural engineering, the research object is mainly the wind-induced response and structural safety of building main structure, curtain wall system, and road and bridge structure [13]. However, the description of extreme wind field characteristics in high-density urban open space for tree management is relatively lacking. In practices, the strong wind protection methods for tree management still mainly relied on empirical experiences.

In the previous study, we had conducted some studies on the effects of typhoons on urban trees. Statistical studies had been conducted and highlighted some tree species at high risk of damage from typhoons, including *Ficus microcarpa*, *Ficus virens*, and *Bauhinia variegate*. We also identified some typical urban space with high wind risk including space around high-density urban environment, the boundary between high-density area and open space, and around some transportation infrastructure [14].

Following the above research, this study was based on a tree management project in Guangzhou, China. The tree management group was assigned to evaluate and improve the tree safety for the central park in Guangzhou. Based on the tree risk assessment (TRA) methods including a serious of tree apparent characteristic parameters [15, 16], the research group tempted to explore a much more precise management basing on the wind field characters of the park. Thus, this study conducted some explorations on computational fluid dynamic (CFD) methods on extreme strong wind simulations in the urban environment to find out the field character and support the tree management works in future.

2 Methodology

2.1 Study Area and Trees

The Central Park, located in Yuexiu District, Guangzhou, China, is the first comprehensive park established in Guangzhou, covering an area of 44,600 m^2. The layout of the park is a French pattern garden with a square geometric symmetry form. The main dominant tree species in the park are *Michelia × alba*, *Ficus microcarpa*, *Ficus virens*, *Delonix regia* (Bojer) Raf., *Dracontomelon dao*, *Bombax ceiba* L., *Mangifera indica* L., and *Grevillea robusta* A. Cunn. (see Table 1). As the geometric planning form of the park, the trees mainly planted alongside with the paths on the south-north orientations. In past studies, the research team had found that some tree species were under higher wind risk during typhoons, such as *Ficus microcarpa*, *Ficus virens*, and *Bombax ceiba* L. [14]. These species need more attention in tree management.

As this study tempted to evaluate the wind effects on trees in the park, the tree height related parameters were collected via field survey as well. The Ave. Tree Height was 14.3 m, the Ave. Diameter at Breast Height was 5.0 m, and the Ave.

Table 1 Main dominant tree species in the central park (self-drawn)

Tree species	Quantity	Ave. TH* (m)	Ave. DBH* (m)	Ave. CH* (m)	Ave. CCH* (m)
Michelia × alba	94	12.8	5.4	7.4	9.1
Ficus microcarpa	65	15.8	3.6	12.2	9.7
Ficus virens	47	16.3	3.3	13	9.8
Delonix regia (Bojer) Raf.	22	9.5	3.2	6.3	6.4
Dracontomelon dao	19	12.9	2.7	10.2	7.8
Bombax ceiba L.	11	17.7	8.9	8.8	13.3
Mangifera indica L.	7	13	4.5	8.5	8.8
Grevillea robusta A. Cunn.	7	16	7.8	8.2	11.9
Ave.	N/A	14.3	5.0	9.3	9.6

* TH, Tree Height; DBH, Diameter at Breast Height; CH, Crown Height; CCH, Crown Center Height

Crown Center Height was 9.6 m. Thus, the wind speed analysis surface was defined at heights of 5.0 and 10.0 m in the later simulation studies.

2.2 Strong Wind Parameters

In recent year, there were three main typhoons landed in Guangdong Province and caused serious destroy on facilities, trees, and etc. in Guangzhou. The research team had collected the quantity of damaged trees in past years. The typhoon Nida (No. 1604) in 2016, Hato (No. 1713) and Pakhar (No. 1714) in 2017, and Mangkhut (No. 1804) in 2018 had caused 74, 351, and 144 trees damage separately [10–14].

In order to obtain the near-surface wind speed data of the study area during the typhoon, this study collected the data from the ground meteorological stations within 2 km of the study site, and extracted the 24 h wind speed data during the typhoon landfall. The peak wind speed was found in the typhoon Mangkhut landfall. The average two-minute wind speed at a height of 10 m obtained from the two weather stations was 24.0 m/s and 15.3 m/s respectively, and the maximum wind speed can reach 38.0 m/s. Wind direction data from the two weather stations showed that during the typhoon landfall, the wind direction was mainly concentrated at 0°, 202.5°, 225°, and 337.5° respectively (see Fig. 1). Because the study site is located in the central area of the city, the building density is high, and the distribution of buildings of different heights is complicated, which also has a certain influence on the wind direction distribution of the weather stations.

From the perspective of simulation research, the higher average wind speed was selected as the simulated boundary wind speed (24.0 m/s at 10.0 m height). Considering the differences in the landing and moving paths of different typhoons, this study tested the four angles of 0°, 90°, 180° and 360° for the selection of wind direction, so as to preliminarily analyze the influence of site wind speed distribution under various circumstances. Subsequent studies will further refine the test of the different wind directions.

2.3 GIS Database Platform

In this study, a GIS database was introduced to collected the urban morphology information, such as building floor and footprint, central line of the road, boundary of the urban park. ArcGIS (version 10.2) [17] was used to record and output the building information for the CFD simulation later. The information of urban trees, including tree species and locations, was also collected in the GIS database to support the analysis on urban space and tree risks.

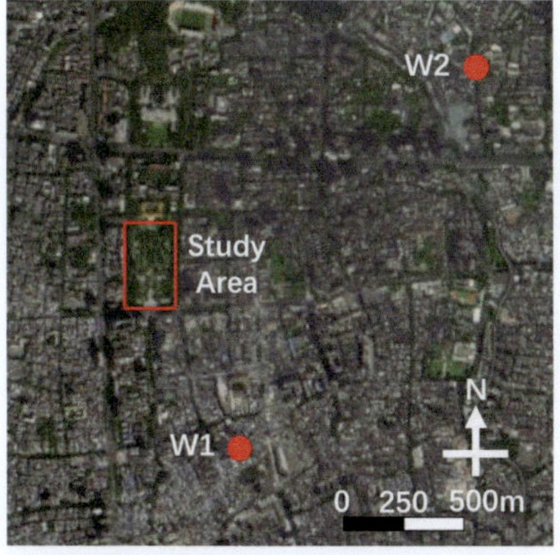

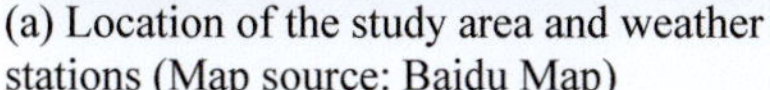
(a) Location of the study area and weather stations (Map source: Baidu Map)

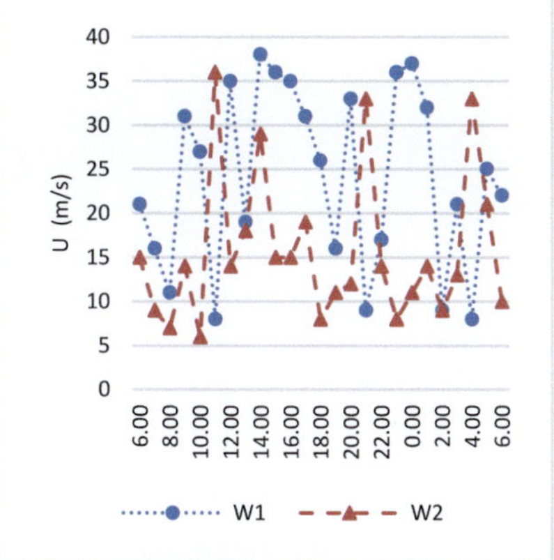

(b) Wind speed recorded by two weather stations during typhoon landfall

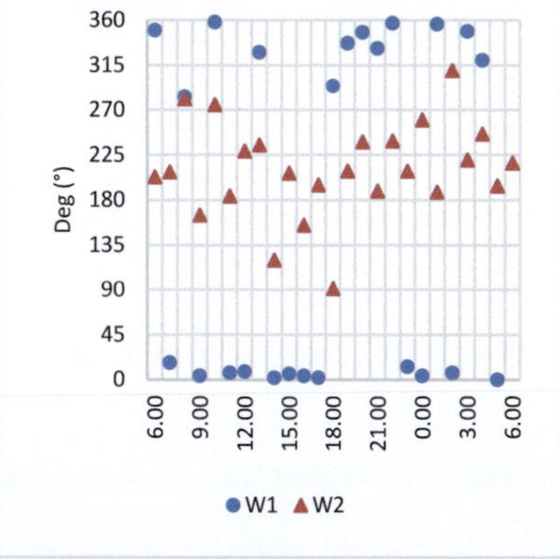

(c) Wind orientations recorded by two weather stations during typhoon landfall

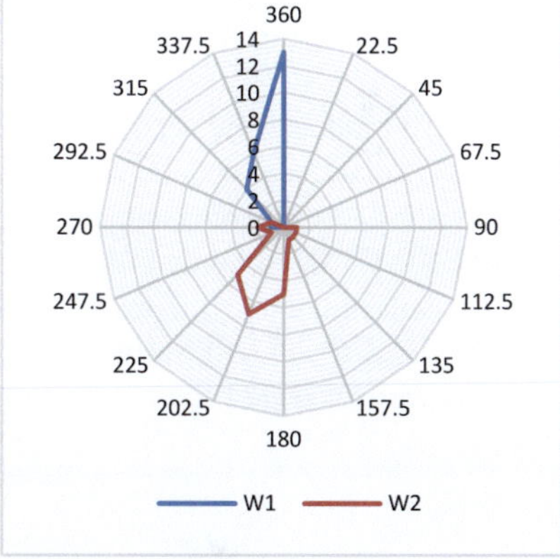

(d) Wind orientations layout

Fig. 1 Wind speed and orientations from two weather stations near the study area (self-drawn). **a** Location of the study area and weather stations. *Map source* Baidu map. **b** Wind speed recorded by two weather stations during typhoon landfall. **c** Wind orientations recorded by two weather stations during typhoon landfall. **d** Wind orientations layout

2.4 *CFD Simulation Settings*

The software PHOENICS (version 2018) [18] was employed to solve the CFD simulation models. The central park scale is 330 m × 200 m ($L \times W$, see Fig. 2). The building volumes in the surrounding simulation area were obtained with GIS data. However, as this study tempted to set up an evaluation platform and provide a fast assessment on wind speed, the tree models were neglected in the CFD model to reduce the computation resource. The computational area of the urban model was 1400 m × 1500 m × 260 m ($W \times L \times H$). The grid of the CFD model was generated

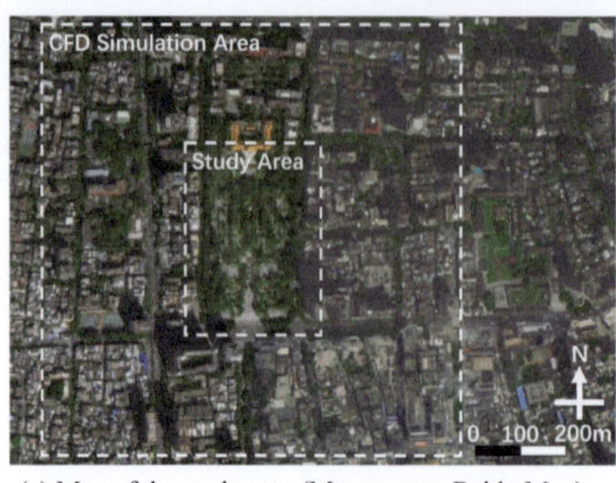

(a) Map of the study area (Map source: Baidu Map)

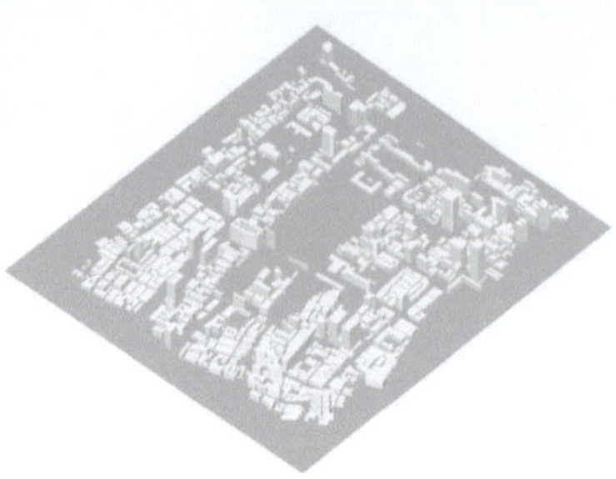

(b) Simulation area with building models

Fig. 2 CFD model of the study area (self-drawn). **a** Map of the study area. *Map source* Baidu map. **b** Simulation area with building models

with the hex grid from the size of 1.0 m to 10.0 m. Total grid quantity was near 7 million.

The solve models were set as Reynolds-averaged Navier–Stokes (RANS) model and the k–ϵ turbulence model. Wind velocity, wind turbulence and energy equations were solved by second order scheme. The pressure–velocity coupling simulation was solved by SIMPLEC scheme [19].

At the setting of the model boundary, the wind inlet boundary information, including wind speed (U), turbulence kinetic energy (K), and dissipation rate (ε), was set as vertical wind profile according to

$$U(z) = \frac{u_*}{k} \ln \frac{z + z_0}{z_0} \tag{1}$$

$$K(z) = \frac{u_*^2}{\sqrt{C_u}} \tag{2}$$

$$\varepsilon(z) = \frac{u_*^3}{k(z + z_0)} \tag{3}$$

where z is the height position, z_0 is the roughness length for an urban area ($z_0 =$ 2.0 m), $C_u = 0.09$, k is the van Karman constant ($k = 0.42$), and u_* is the friction velocity [20, 21]. U (10 m) is set as 24.0 m/s in the model.

After the simulations, results were output and analyzed with CFD postprocessing progress. The U data was obtained from the CFD model with coordinates. Then, the data would be overlapped with the tree locations, which were output from the GIS platform, and be compared in the platform of CAD. Based on the U data and tree information, the tree management group would generate some strategies on the trees with higher wind risk.

3 Results

3.1 Results of Wind Speed Distribution

The results showed the wind field under different wind direction. As the park was an open space in a dense urban area, the difference of wind speed of the park area in different wind directions was not obviously. When the wind direction was set at east, the simulated wind speed was lower than other three directions, because some higher buildings were located on the east side and blocked the wind flow to a certain degree.

The results of maximum U of the tree crown height (5–10 m) is greater than 30.0 m/s, and the average U was 9.7–14.8 m/s (see Table 2). The results of proportions at different wind scales revealed that the U at 10 m and 5 m altitudes mainly distributed at level four to level eight according to Chinese wind scale standard [22]. However, it should be noticed that more than 10% of U was higher than level eight, which would cause tree damage with broken branch or even main trunk lodging in typhoons (see Table 3).

Table 2 Results of wind speed (self-drawn)

Wind direction	Ave. U (m/s)		Max. U (m/s)		Min. U (m/s)	
	$H = 5$ m	$H = 10$ m	$H = 5$ m	$H = 10$ m	$H = 5$ m	$H = 10$ m
E	9.7	10.9	20.6	24.4	1.8	1.9
W	13.9	14.2	30.0	30.0	1.9	2.0
S	11.2	14.2	18.8	30.3	2.0	2.0
N	11.7	14.8	30.0	30.5	2.0	2.1

Table 3 Results of wind scale distribution in the study area (self-drawn)

Wind scale*	U (m/s)	Proportion ($H = 5$ m) (%)	Proportion ($H = 10$ m) (%)
2	1.6–3.3	2.6	2.3
3	3.4–5.4	8.4	7.3
4	5.5–7.9	21.9	15.4
5	8.0–10.7	13.6	13.4
6	10.8–13.8	24.0	20.9
7	13.9–17.1	16.9	18.7
8	17.2–20.7	8.3	10.5
9	20.8–24.4	2.2	3.0
10	24.5–28.4	1.0	6.8
11	28.5–32.6	1.0	1.7

* Wind scale standard source [22]

3.2 Wind Speed Distribution

The U distribution at 10 and 5 m altitudes (see Figs. 3 and 4) showed that when the wind direction was set to the east and west, the open space on the south side of the park was affected by higher U. The U ranged from 16.9 to 30.0 m/s at an altitude of 10 m, and from 13.1 to 24.4 m/s at an altitude of 5 m. When the wind direction was set to north and south, the U was higher on the eastern and western paths of the park. The U ranged from 12.0 to 22.5 m/s at an altitude of 10 m, and from 9.1 to 18.6 m/s at an altitude of 5 m. Figures 5 and 6 also provided the U character on the sections of the study site. The results showed that the U variation was mainly occurred near the surrounding buildings of the park. The trees alongside the paths at the boundary of the park was faced higher wind risk in different scenarios. Thus, the results showed that border trees along the paths on the eastern, southern and western sides of park were under higher risk of U.

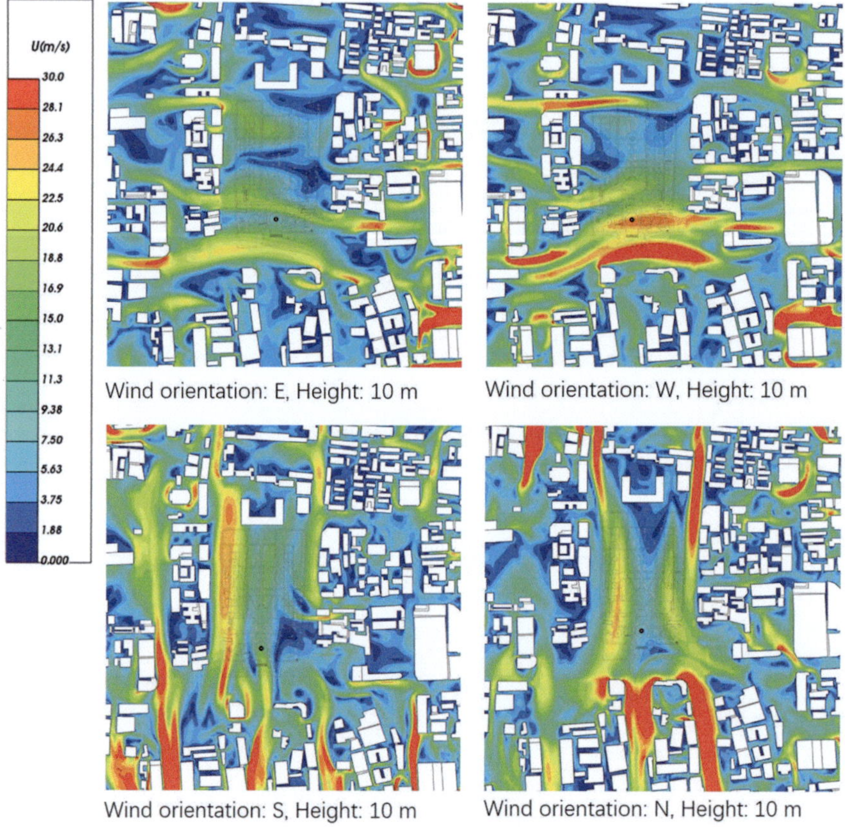

Fig. 3 Results of wind speed distribution at height of 10 m (self-drawn)

U(m/s)
30.0
28.1
26.3
24.4
22.5
20.6
18.8
16.9
15.0
13.1
11.3
9.38
7.50
5.63
3.75
1.88
0.000
Wind orientation: E, Height: 5 m
Wind orientation: W, Height: 5 m
Wind orientation: S, Height: 5 m
Wind orientation: N, Height: 5 m

Fig. 4 Results of wind speed distribution at height of 5 m (self-drawn)

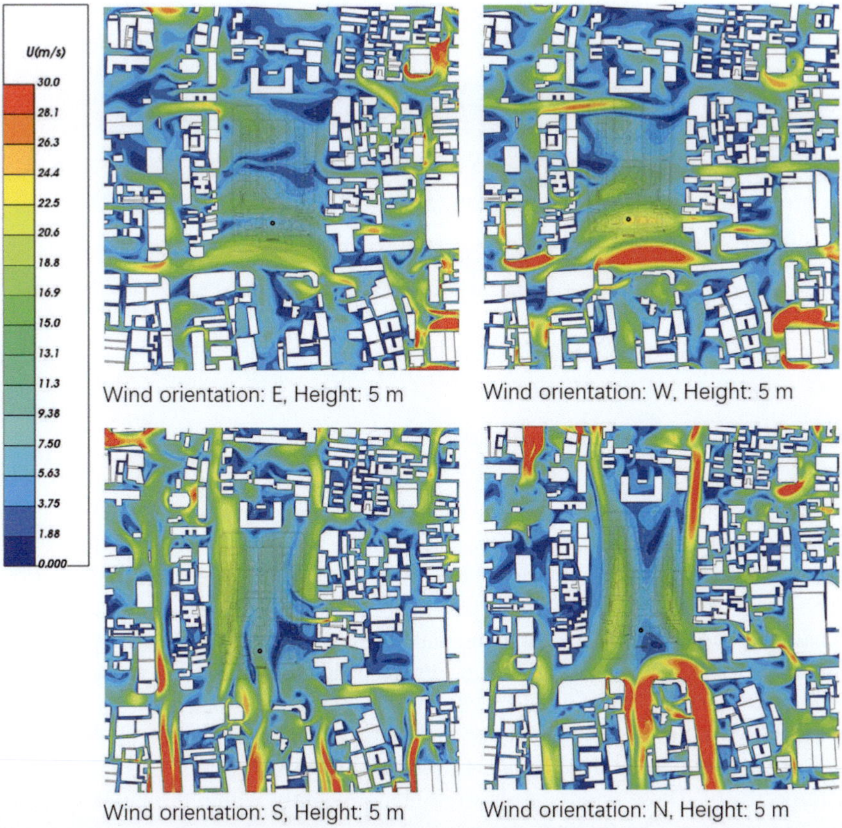

Wind orientation: E, Section: W-E
Wind orientation: W, Section: W-E
Wind orientation: S, Section: W-E
Wind orientation: N, Section: W-E
W E

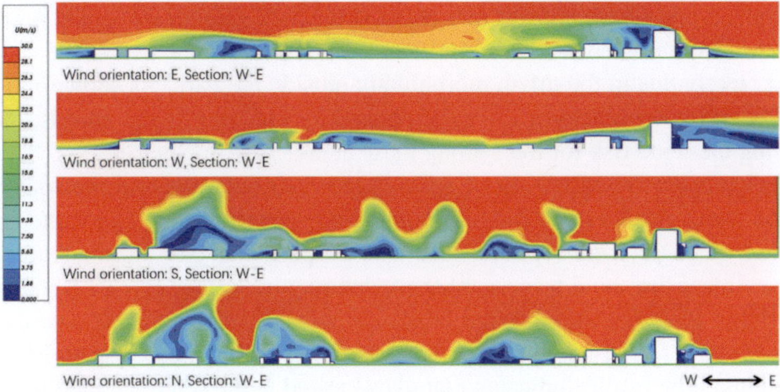

Fig. 5 Results of wind speed distribution at section W–E (self-drawn)

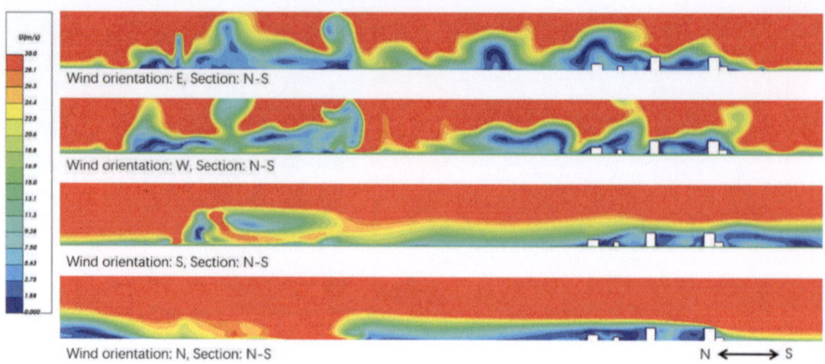

Fig. 6 Results of wind speed distribution at section N–S (self-drawn)

3.3 High Risk Areas Analysis

In order to provide a clearer guidance for tree management in the Central Park, the *U* distribution in the park was further analyzed. As the border trees were mainly planted in four paths in the park, four lines were selected to analyze (see Table 4 and Fig. 7). The average *U* for the four paths is presented in Table 4. The results indicated that the average *U* at a height of 5 m along paths A, B, C, and D was 9.1–16.3 m/s, 9.2–15.2 m/s, 7.5–15.5 m/s, and 9.3–18.6 m/s, respectively. According to the Chinese wind scale standard, these wind speeds were primarily distributed within the range of Level five to Level eight. Similarly, the average *U* at a height of 10 m along paths A, B, C, and D was 12.0–25.2 m/s, 11.2–15.3 m/s, 8.2–15.8 m/s, and 8.1–22.5 m/s, respectively. These wind speeds were predominantly within the range of Level five to Level ten. The findings suggest that the paths in the park were exposed to high wind risks at both tree crown heights, with the risk level at 10 m being higher than that at 5 m. It was also observed that the wind speeds in the central part of the park (paths B and C) were lower than those at the park's boundary (paths A and D). This indicates that wind risk was higher at the boundary compared to the center of the park, likely due to the urban morphology outside the park. As shown in Fig. 4, the roads and open spaces at the park's boundary enhanced wind speeds, thereby increasing the wind risk for trees along these areas.

To enhance the visualization of wind speed distribution across the four paths, the wind speed results from the four simulation scenarios were aggregated into a total wind speed (U_t) and superimposed onto the park map (see Fig. 7). A tree wind risk assessment threshold was established, defining high wind risk as *U* exceeding Level eight (17.2 m/s, strong wind level) in one simulation wind orientation or a total wind speed (U_t) exceeding 51.3 m/s (average *U* higher than Level eight in three orientations). The results indicated that trees located in the southern open space and along both the eastern and western sides of the park were exposed to higher wind risks in the simulations.

Table 4 Results of Ave. U distribution in four paths of the trees in the park (self-drawn)

Positions	Orientations				Ave.
	E	N	W	S	
$H = 5$ m					
Path A	11.0	9.1	11.9	16.3	12.1
Path B	11.2	10.8	15.2	9.2	11.6
Path C	8.3	8.4	15.5	7.5	9.9
Path D	8.3	18.6	13.1	11.6	12.9
$H = 10$ m					
Path A	13.8	13.3	12.0	25.2	16.1
Path B	12.7	13.1	15.3	11.2	13.1
Path C	8.9	10.5	15.8	8.2	10.8
Path D	8.1	22.5	13.8	12.3	14.2

Following the U simulation analysis, the U_t was further integrated with the tree species and their locations on the park's site plan. The tree species were identified and represented with different colored dots. Consequently, the overlapping map clearly delineated high-risk areas for wind damage, offering valuable suggestions and guidelines for future tree management strategies (see Fig. 8).

In this extreme wind simulations study, main tree species in the wind risk areas included *Michelia* × *alba*, *Ficus microcarpa*, *Ficus virens*, *Delonix regia* (Bojer) Raf., *Dracontomelon dao*, and *Grevillea robusta* A. Cunn. Based on the wind risk area assessments, future tree management strategies could be integrated with the Visual Tree Assessment (VTA) method [23] and the Basic Tree Risk Assessment Form (BTRAF) developed by the International Society of Arboriculture (ISA) [24]. By employing these methods, we can identify trees that are at risk due to both poor health conditions (such as tree tilting, hollow trunks, and root rot) and external wind threats, thereby providing targeted maintenance and interventions.

To mitigate wind risks, tree management approaches may include the following measures: installing external support structures or guy wires for tilted trees, selectively pruning branches on the tilted sides, and reducing the windward surface area by thinning branches and leaves. These methods should be carefully evaluated using VTA and BTRAF before implementation to ensure their appropriateness and effectiveness. By combining wind risk and tree risk assessments, we can enhance tree management practices within urban green spaces, ultimately contributing to a safer and more resilient urban forestry system.

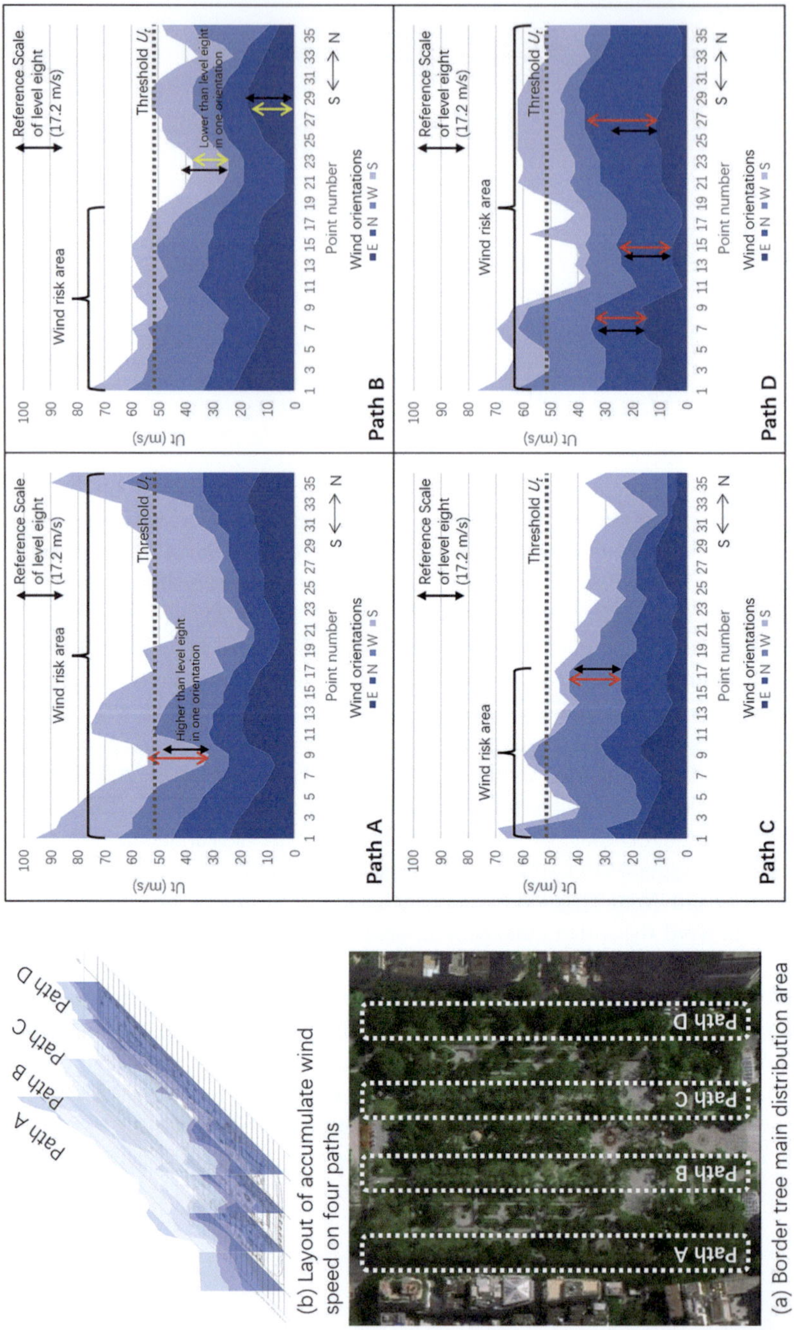

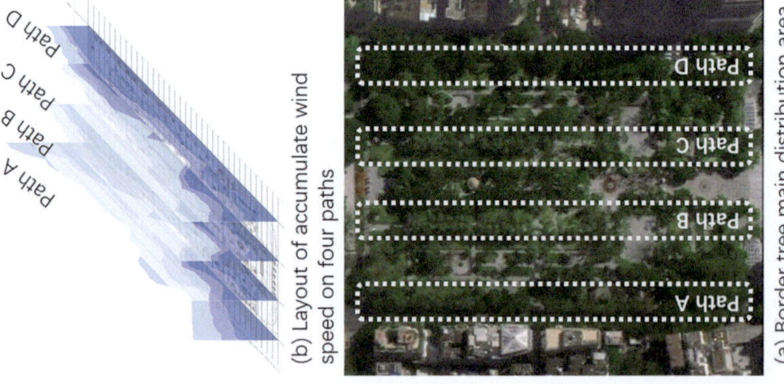

Fig. 7 Results of accumulated wind speed distribution at four main paths in the park (self-drawn)

Main tree species

- *Michelia × alba.*
- *Ficus microcarpa.*
- *Ficus virens.*
- *Delonix regia (Bojer) Raf.*
- *Dracontomelon dao.*
- *Bombax ceiba L.*
- *Mangifera indica L.*
- *Grevillea robusta A. Cunn.*

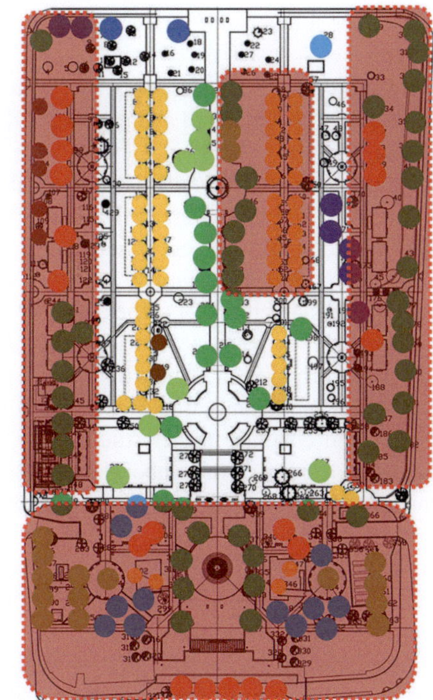

Fig. 8 Overlapping of tree species, locations and wind risk areas (self-drawn)

4 Discussion

This study tested the extreme wind evaluation with CFD simulations to support the tree management in an urban park in Guangzhou. Effective tree management relies on comprehensive tree information monitoring. To this end, various methods have been developed, ranging from top-down approaches such as LiDAR surveys to bottom-up techniques like handheld laser scanning and AI-IoT-based tree tilt monitoring [25, 26]. However, these precise local tree observation technologies often come with high costs, limiting their widespread application. In contrast, computational simulation methods offer a more accessible and cost-effective alternative for assessing wind risks among a large number of trees in urban environments. By integrating these simulations with limited wind monitoring data or combining them with tree information based on a GIS system, we can generate more reliable and actionable results. This approach would significantly enhance future urban tree monitoring and risk alert systems.

This study also included some limitations as follow:

(1) The CFD simulation method was lack of consideration of the variation of the turbulence character and vertical wind profile during the typhoon landing. This study only focused on the fast evaluation of the urban space under the strong

wind effect. A much more precise simulation method should be test and validated in future.

(2) In the simulation scenarios, only four wind directions were tested. More directions should be simulated and supported the tree management in the extended works.

(3) The simulations were conducted with building volumes; however, the trees were not included in the simulation. The drag effects of tree crowns would affect the wind distribution on different heights of the park. In order to provide a better evaluation on tree management, the exist tree characters including the scale of crowns, the diameter of branches, and the root conditions are needed to record and support the tree models in further studies.

(4) This study only evaluated the wind speed with wind scale; however, the threshold wind speed of different tree species was different. Thus, the tree management team would conduct further studies with tree locations, species, and growth characters to support the elaborate tree maintenance.

(5) On the methods of tree management, much more on-site survey should be conducted with the extreme wind evaluation, such as the planting pool and root assessment, insect investigations, tree trunk scanning, and etc. A combination of series methods would support the tree management in a much more scientific and effective way.

(6) In the analysis of the wind field, the relationship between wind speed distribution and urban morphology—such as building height, open space scale, road orientations, and other relevant factors—has not been sufficiently explored. This area requires further investigation and improvement in future research.

5 Conclusion

This study explored the application of CFD methods in simulating extreme strong winds in urban environments, aiming to characterize the wind field and support tree management efforts. The primary findings indicated that the main risk areas in Central Park were located in the southern, west-northern regions, as well as along the eastern and western pathways. The results also revealed that, when the boundary wind speed was set at 24 m/s, the peak wind speeds at heights of 5 and 10 m could reach levels nine to eleven on the Chinese wind scale. This finding highlights the increased risk for tree management in these areas. Using typical typhoon conditions in Guangzhou's central urban area as a case study, this research provided a detailed wind speed distribution and values for Central Park through CFD simulations. The methods developed in this study are expected to support improved tree management not only in Guangzhou but also in other cities and similar scenarios.

References

1. Desa UN (2016) Transforming our world: the 2030 agenda for sustainable development
2. Lei L, Pwc D, Tao DE et al (2021) Review of advances in urban climate study in the Guangdong-Hong Kong-Macau Greater Bay Area, China. Atmos Res
3. Mei W, Xie SP (2016) Intensification of landfalling typhoons over the northwest Pacific since the late 1970s. Nature Geoence
4. https://ukcop26.org/
5. Wong NH, Tan CL, Kolokotsa DD et al (2021) Greenery as a mitigation and adaptation strategy to urban heat. Nat Rev Earth Environ 2(3)
6. Zhang Q, Wu L, Liu Q (2009) Tropical cyclone damages in China 1983–2006. Bull Am Meteor Soc 90(4):489–496
7. Gardiner B, Berry P, Moulia B (2016) Wind impacts on plant growth, mechanics and damage. Plant Sci 245:94–118
8. Huang P, Wang X, Gu M (2013) Study on near-ground wind characteristics of a strong typhoon-wind speed, turbulence intensities, gust factors and peak factors. Disaster Adv 6(5):3–18
9. Liu J, Zhang X, Niu J et al (2019) Pedestrian-level wind and gust around buildings with a 'lift-up' design: assessment of influence from surrounding buildings by adopting LES. Build Simul 12(6):1107–1118
10. Joint Typhoon Warning Center (2015) Annual tropical cyclone report: Western Pacific (Report). United States Navy, United States Air Force
11. He YC, Li QS, Chan PW, Wu JR, Fu JY (2019) Toward modeling the spatial pressure field of tropical cyclones: insights from Typhoon Hato (1713). J Wind Eng Ind Aerodyn 184:378–390
12. Snaiki R, Wu T (2017) Modeling tropical cyclone boundary layer: height-resolving pressure and wind fields. J Wind Eng Ind Aerodyn 170:18–27
13. Wang X, Huang C, Huang P, Yu X (2017) Study on wind characteristics of a strong typhoon in near-ground boundary layer. Struct Design Tall Spec Build 26(5):e1338
14. Hankun L, Songyi H, Yaning G et al (2020) Countermeasures to typhoon risks: perspectives from tree safety and urban morphology. In: Passive and low energy architecture conference 2020 (PLEA 2020): planning post carbon cities
15. Klein RW, Koeser AK, Hauer R et al (2019) Risk assessment and risk perception of trees: a review of literature relating to arboriculture and urban forestry. Arboric Urban For 45(1):26–38
16. Gitta HU (2016) Tree risk assessment methods: a comparison of three common evaluation forms. Environ Hortic
17. http://www.saga-gis.org/en/index.html
18. https://www.cham.co.uk/phoenics.php
19. Shih TH, Liou WW, Shabbir A, Yang Z, Zhu J (1995) A new k–ϵ eddy viscosity model for high Reynolds number turbulent flows. Comput Fluids 24(3):227–238
20. Richards PJ, Hoxey RP (1993) Appropriate boundary conditions for computational wind engineering models using the k-ε turbulence model. Comput Wind Eng 1:145–153
21. Shen C, Shen A, Tian C, Zhou S, Zhu L, Chan P, Li W (2019) Evaluating the impacts of updated aerodynamic roughness length in the WRF/Chem model over Pearl River Delta. Meteorol Atmos Phys 1–14
22. China Meteorological Administration (2012) Wind scale (GB/T 28591-2012)
23. Mattheck C, Breloer H (1994) Field guide for visual tree assessment (VTA). Arboric Assoc J 18(1):1–23
24. International Society of Arboriculture (2013) Basic tree risk assessment form
25. Majid N, Man SW, Xinyu Y, Coco YTK, Qian P, YanShuai D (2024) Urban tree health assessment using multifaceted remote sensing datasets: a case study in Hong Kong. Remote Sens Appl: Soc Environ 36:101347
26. Wai YC, Yu-Hsing W, Siu WC, Pin ST, Mei LL, Hoi LL, Jimmy W, Yun ML (2023) AI-IoT integrated framework for tree tilt monitoring: a case study on tree failure in Hong Kong. Agric For Meteorol 341:109678

BY NC ND

Prediction Model of Geological Disasters Based on Meteorological Features

Hongyang Liu and Shengkai Lei

Abstract This study proposes a geological disaster prediction model based on meteorological characteristics to address the frequent occurrence of geological disaster in rural mountainous areas. This model focuses on geological disaster significantly affected by meteorological conditions, including five types of geological disaster: mudslides, landslides, collapses, flash floods, mudslides, and landslides. Based on geological disaster data and meteorological observation data from 2011 to 2022, a 1:1 negative sample of geological disaster was constructed, and the Spearman correlation coefficient was used to screen out the 15 meteorological characteristic factors with the highest correlation with geological disaster. In terms of model construction, mainstream machine learning algorithms such as XGBoost, LightGBM, CatBoost, and RF were used to compare their accuracy and root mean square error on geological disaster samples. After comprehensive evaluation, the LightGBM algorithm performs the best in geological disaster prediction on this dataset. At the same time, an importance analysis of meteorological features was conducted based on this algorithm, removing meteorological feature factors with high Spearman correlation but little impact on geological disaster prediction, and optimizing the geological disaster prediction model based on LightGBM algorithm. Finally, the ROC curve and confusion matrix were used to evaluate the prediction model, and the AUC value of the model reached 0.99, an increase of 3% compared to before optimization; The accuracy of the model reached 0.98, an increase of 2% compared to before optimization, significantly improving the effectiveness of geological disaster prediction.

H. Liu · S. Lei (✉)
Sichuan Provincial Key Laboratory of Plateau and Basin Rainstorm Drought and Flood Disaster, Institute of Plateau Meteorological, China Meteorology Administration, Chengdu, China
e-mail: 280103626@qq.com

Sichuan Provincial Rural Economic Information Center, Chengdu, China

Provincial Key Laboratory of Software Automatic Generation and Intelligent Services, Chengdu University of Information Technology, Chengdu, China

S. Lei
Sichuan Vocational and Technical College of Communications, Chengdu, China

B. He et al. (eds.), *Climate-Resilient, Low-Carbon, and Sustainable Cities*, Lecture Notes in Civil Engineering 754, https://doi.org/10.1007/978-981-95-2169-2_8

Keywords Geological disaster · Meteorological characteristics · Prediction model · LightGBM

1 Introduction

Geological disasters, such as mudslides, landslides, avalanches, flash floods, landslides and mudslides, often cause huge losses of life and property to human society. The occurrence of these disasters is often closely related to specific meteorological conditions. Therefore, the establishment of a geological disaster prediction model based on meteorological characteristics can detect potential geological disasters in time and take corresponding precautionary measures to enhance the geological disaster prevention ability of mountain villages and reduce the loss of life and property caused by disasters. This will help to protect the lives and properties of rural residents and maintain social stability.

At present, the research on geologic hazards mainly focuses on the mechanism of disaster and disaster risk assessment. Xilin [1] discussed the causes of geohazards such as mudslides in depth, promoted the development of disaster risk assessment methods, and put forward the evaluation model of "Risk degree = Hazard degree × Vulnerability degree". Huang [2] used neural network (ANN) to screen the weighting factors, and drew the geohazard risk level zoning map through the weighting analysis; while Huang [3] used support vector machine (SVM), multilayer perceptron (MLP) and random forest (RF) model to predict the susceptibility of landslides. These studies mainly considered the stable or minimally varying causal factors such as elevation, slope, slope direction, geological structure, vegetation cover and geomorphological type. With regard to the meteorological factors that vary significantly, Yang [4] and others analyzed the relationship between rainfall characteristics and the occurrence of mudslides and landslides with examples. In summary, there are many factors that cause geologic hazards, and when the fixed disaster-causing factors are determined, the dynamic changes of meteorological characteristic factors become the key factors of geologic hazards.

In this paper, the geohazard and meteorological observation data were integrated in order to investigate the correlation between meteorological characterization factors and geohazards. Firstly, Spearman's [5–7] correlation analysis was used for the preliminary screening of meteorological characteristic factors, and then four algorithms, namely, XGBoost [8, 9], LightGBM [10–12], CatBoost [13] and Random Forest [14] (RF) were used to construct the prediction models of geohazards and screen out the optimal prediction models based on the accuracy and root-mean-square error of the models. The optimal prediction model was selected based on the accuracy and root mean square error of the model. Then, based on the results of the characteristic importance analysis of the optimal model, the key meteorological characteristic factors were screened again, and the geohazard prediction model was reconstructed on the basis of these key meteorological characteristic factors, and finally, the predictive efficacy of the model was verified by the confusion matrix and the ROC curves, which was aimed at providing a powerful decision-making support for disaster prevention and mitigation in mountainous villages.

2 Information and Methods

2.1 Data Sources

The data cited in this paper come from many authoritative platforms and official organizations, thus ensuring the authenticity and accuracy of the data.

(1) Geohazard data come from three main sources. First, the Geological Hazards Service Application Platform [15], which has been providing nationwide records of geological hazards from 2011 to 2022, with detailed core information on the address, time and type of hazard. Secondly, the Ganzi Prefecture Natural Resources and Planning Bureau provided detailed data on natural disasters such as mudslides, landslides, and avalanches in Ganzi Prefecture for the years 2022 and 2023. Finally, news reports related to geohazards were further collected through a web search engine, from which key data such as the address, time and type of disasters were successfully extracted.
(2) Meteorological observation data were obtained from the "Meteorological Big Data Cloud Platform—SkyKing", which covers data from January 1, 2011 to the present. These data come from meteorological observation stations near the geohazard sites and include a number of key meteorological characteristics, including: daily precipitation, average and maximum/minimum relative humidity, average and maximum/minimum air temperature, wind direction, wind speed, average and maximum/minimum ground temperature, average and maximum/minimum 5 cm ground temperature, average and maximum/minimum 10 cm ground temperature, average and maximum/minimum 15 cm ground temperature, average and maximum/minimum 20 cm ground temperature, average and maximum/minimum 40 cm ground temperature, average and maximum/minimum barometric pressure, and precipitation data for the previous day, the previous two days, and the previous three days, totaling 33.
(3) Geohazard latitude and longitude information: online batch address query latitude and longitude tool [16], according to the address of geohazard occurrence query corresponding latitude and longitude information.

2.2 Data Pre-processing

Firstly, the collected geohazard data and meteorological data were cleaned, including the removal of duplicates and erroneous data, and the supplementation of incomplete data. Next, the geohazard data and meteorological data were spatially and temporally matched precisely, including the address, time and latitude/longitude of the geohazard, as well as the meteorological data of the nearest meteorological observation station during the corresponding time period.

In terms of time matching, according to the specific date and time of occurrence of geologic hazards, the meteorological data of the corresponding date are obtained. Except for the disaster data provided by Ganzi State Natural Resources and Planning Bureau, the geologic disaster data collected in this paper do not have specific time of occurrence and are only accurate to what kind of geologic disaster occurred in a certain place on a certain day, therefore, the statistical day-scale meteorological data are used as meteorological characteristics.

In the spatial matching section, the latitude and longitude information of a geohazard was first determined based on the address of the geohazard. Then, by calculating the four-dimensional Euclidean distance, which includes longitude, latitude, altitude and land-use type, the meteorological observation site that is closest to the four-dimensional Euclidean distance of the geohazard is selected.

In the data integration stage, the cleaned and matched meteorological data are merged with the geohazard event data to construct a new geohazard meteorological data set. This new dataset contains not only the basic information of geohazard events, but also the matching meteorological data.

In the screening of meteorological characteristic factors, Spearman's correlation coefficient was used to evaluate the correlation between each meteorological characteristic factor and the occurrence of geohazards. According to the ranking of the absolute value of the correlation coefficients, the 15 meteorological characterization factors with the highest absolute value of the correlation coefficients with geohazards were selected (see Table 1 for details), and these meteorological characterization factors will be used in the subsequent model prediction.

2.3 Research Methodology

This study was conducted in the Python 3.11 environment, four algorithms, XGBoost, LightGBM, CatBoost and Random Forest (RF), are used to construct the geohazard prediction model. The meteorological data related to geohazards are selected as positive samples, while the corresponding negative samples are randomly selected in the ratio of 1:1 under the meteorological conditions without geohazards, and the positive and negative samples are integrated to form the total sample of geohazards. In the model construction stage, the total samples were divided into 70% training set and 30% testing set, which were used for model training and performance evaluation, respectively. By comparing the accuracy and root mean square error of the four algorithms on the geohazard samples, the most suitable model for geohazard prediction was selected. Further, the importance of meteorological factors was analyzed, and those meteorological factors with high Spearman's correlation but low impact on geohazard prediction were excluded. Based on the filtered meteorological factors,

Table 1 Variables, descriptions, unit and indicative significance of the 15 meteorological characterization factors (self-drawn)

Variables	Description	Unit	Significance of variables
d_prec	Daily precipitation	mm	Reflecting the impact of daily precipitation rainfall
av_RH	Average relative humidity	%	Reflecting the effect of average daily relative humidity
h_RH	Maximum relative humidity	%	Reflecting the effect of maximum daily relative humidity
l_RH	Minimum relative humidity	%	Reflecting the effect of daily minimum relative humidity
av_temp	Average temperatures	°C	Reflecting the effect of average daily temperatures
h_temp	Maximum temperature	°C	Reflecting the effect of daily maximum temperatures
l_temp	Minimum temperature	°C	Reflecting the effect of daily minimum temperatures
av_gtemp	Average ground temperature	°C	Reflecting the effect of daily mean ground temperature
av_5gtemp	Average 5 cm ground temperature	°C	Reflecting the effect of the daily mean 5 cm ground temperature
ws	Wind speed	m/s	Reflecting the speed of air movement
av_10gtemp	Average 10 cm ground temperature	°C	Reflecting the effect of the daily mean 10 cm ground temperature
b1d_prec	Precipitation on the previous day	mm	Reflecting the impact of daily precipitation rainfall in the preceding period
b2d_prec	Precipitation for the first two days	mm	Reflecting the impact of daily precipitation rainfall in the preceding period
av_atm	Average air pressure	Pa	Reflecting the effect of average daily air pressure
b3d_prec	Precipitation for the first three days	mm	Reflecting the impact of daily precipitation rainfall in the preceding period

the geohazard prediction model was reconstructed and the accuracy of the simulation results was evaluated.

Table 2 Comparison of modeling tests (self-drawn)

Model	Predictive accuracy	Root mean square error
XGBoost	0.94	0.14
LightGBM	0.96	0.13
CatBoost	0.93	0.14
RF	0.95	0.14

3 Results

3.1 Model Evaluation Criteria and Comparison

In this study, the performance of four prediction models was evaluated using two assessment metrics, namely, prediction accuracy and root mean square error. The results are shown in Table 2. The experimental comparison results of each model are shown in detail in the table. In terms of prediction accuracy, all four models show good performance, with their accuracy exceeding 93%. Specifically, the prediction accuracy of LightGBM model is as high as about 96%. Meanwhile, the LightGBM model also performs best in terms of root mean square error. Based on the above results, the LightGBM model was selected for the subsequent feature importance analysis in this study.

3.2 Characterization Importance

In this study, the LightGBM model was constructed using the geohazard samples to screen the meteorological characteristic factors that have a great influence on geohazards, and the specific results are shown in Fig. 1. According to the order of the characteristic importance of the model, the top 12 key meteorological characteristic factors include: daily precipitation, average air pressure, wind speed, average and minimum relative humidity, day-by-day precipitation within the first three days (the previous day, the previous two days, and the previous three days), average and maximum/minimum air temperature, and average 10 cm ground temperature.

According to the results of the analysis of the significance of the characteristics, the meteorological characteristic factors with high Spearman's correlation but less influence on the actual geological hazards were excluded: maximum relative humidity, average ground temperature and average 5 cm ground temperature. After careful screening, the following meteorological factors were identified as having a significant influence on the prediction of geohazards: daily precipitation, mean barometric pressure, wind speed, mean and minimum relative humidity, daily precipitation in the previous three days (the previous day, the previous two days, and the previous three days), mean and maximum/minimum air temperatures, and the mean 10 cm ground temperature. Using these screened key meteorological factors, the geohazard

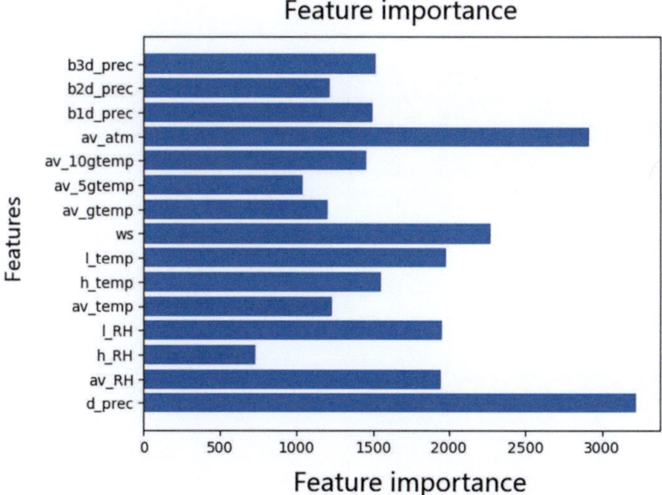

Fig. 1 Importance of LightGBM model features (self-drawn)

prediction model was reconstructed and its simulation results were evaluated for accuracy.

3.3 *Model Evaluation*

The ROC curve provides an effective means of comprehensively assessing the accuracy of geohazard prediction models. The method utilizes the area under the curve (AUC) as the core indicator for assessing the accuracy of the model, with AUC values ranging from 0.5 to 1. Based on the different ranges of AUC values, the following evaluation criteria are set: if the AUC value is between 0.9 and 1 (including 0.9), the prediction effect is regarded as excellent; if the AUC value is between 0.8 and 0.9 (including 0.8), the prediction effect is evaluated to be good; if the AUC value is between 0.7 and 0.8 (including 0.7), the prediction effect is fair; When the AUC value is greater than 0.5 but less than 0.7, the model still shows some predictive ability although the prediction effect is poor. In addition, the confusion matrix also provides another effective method to evaluate the model performance through the accuracy.

In this study, the LightGBM geohazard prediction model was constructed and reconstructed before and after the re-screening of meteorological characteristic factors according to the method described in Sect. 3.2, and the ROC curves and confusion matrices were plotted, which are shown in Figs. 2, 3, 4 and 5. The ROC curve area under the ROC curve of the reconstructed LightGBM model was 0.96, and the accuracy of the model was 0.96. The AUC of the reconstructed model was

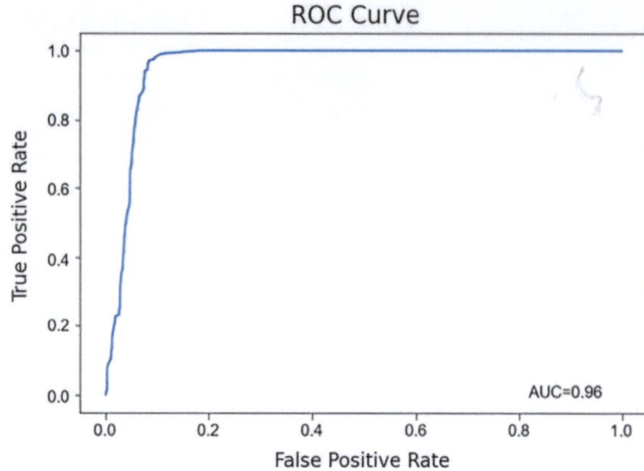

Fig. 2 Constructing the LightGBM model ROC curve (self-drawn)

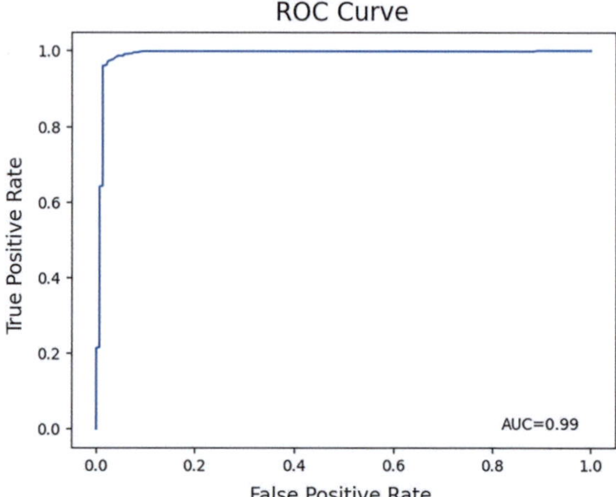

Fig. 3 Reconstructing the LightGBM model ROC curve (self-drawn)

0.99, and the accuracy of the model was 0.98, which indicates that the reconstructed LightGBM model has a very high accuracy in the prediction of geohazards.

4 Discussion

Geohazard data and meteorological observations are collected relatively independently of each other, and therefore a link between the two needs to be established through spatial and temporal matching.

In the process of spatial matching, the meteorological observation stations and the actual addresses of geologic hazards do not exactly coincide, especially in mountainous areas, where the meteorological observation stations and the actual addresses of geologic hazards are often tens of kilometers apart. Therefore, these observations

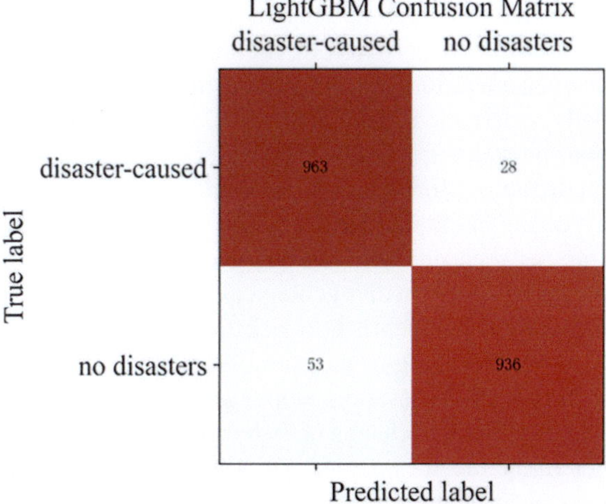

Fig. 4 Constructing the LightGBM model confusion matrix (self-drawn)

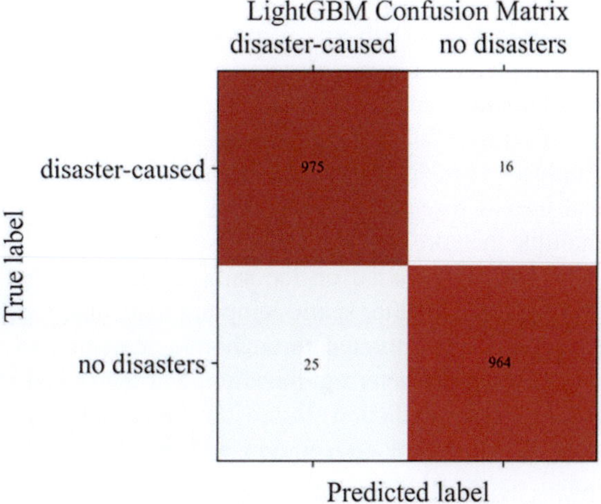

Fig. 5 Reconstructing the LightGBM model confusion matrix (self-drawn)

cannot fully reflect the exact meteorological characteristics at the time of the disaster, but they can still reflect the meteorological conditions at the time of the geohazard to a certain extent. Research on geological hazard prediction [17–19] generally specifies the overall prediction within a designated area, while spatial matching refers to the spatial matching of geological hazard points with meteorological observation stations, which is the point prediction within the area.

In terms of temporal matching, geohazards are often significantly affected by short-term meteorological information. Therefore, the smaller the time scale of the collected geohazard occurrence time, the smaller the time scale of the meteorological characterization factor that can be matched, even to the level of hours or minutes. The time scale of most of the geohazard data obtained in this study is limited to days, and only the time scale of the geohazard data in Ganzi Prefecture reaches the level of minutes. Therefore, the meteorological features studied in this paper have less consideration for short-term meteorological features and more consideration for daily scale meteorological features. In the future, geohazard data with time scale of hour or even minute should be collected as much as possible to further optimize the prediction performance of the model. In related studies, Zhongman [20] used the average precipitation on an annual and monthly time scale to discuss the relationship between precipitation and geological hazards. This study used daily average precipitation to predict geological hazards, which is consistent with the time scale of landslide and debris flow risk zoning conducted by Rui [21].

In the matching of meteorological information, when the fixed disaster-causing factors such as elevation, slope, geological structure and geomorphology are clarified, the dynamic change of meteorological characteristic factors becomes the key to determine the accuracy of geohazard prediction. Therefore, the accuracy and richness of the data provided by meteorological observation stations are particularly important. Especially in the mountainous villages where geological hazards occur frequently, there is a serious lack of meteorological observation capacity, and most of the meteorological observation stations can only observe basic meteorological data such as precipitation and air temperature, and do not have the observation capacity of soil humidity and soil moisture. Improving meteorological observation capacity can significantly enhance the effectiveness of disaster prevention and mitigation services. At present, precipitation is the main meteorological characteristic factor for geological hazard research [22–24]. This study considers the comprehensive impact of multiple meteorological characteristic factors on geological hazards.

In summary, based on the sample sets of geohazard data and meteorological observation data after spatio-temporal matching, a geohazard prediction model was successfully constructed through fine screening of meteorological characteristic factors and parameter optimization, and the model showed good prediction effect in geohazard prediction. Based on this model, the Ganzi Target Release System developed in close cooperation with the Ganzi Meteorological Bureau has been popularized and applied in the whole state of Ganzi, and in the cases of geologic disaster avoidance such as Republican Village of Muxi Town, Luding County, Yorizhi Village of Sun Valley Town, Derong County, and Xuewa Village of Xuewalong Town, Jiulong County, the successful warnings, precise releases, and timely response were made, realizing that there was not a single casualty due to the disaster, and providing strong support for the cause of countryside revitalization.

5 Conclusion

In this study, the Spearman correlation analysis was used to screen out 15 meteorological characteristic factors with the highest absolute value of correlation coefficients with geohazards, and four kinds of geohazard prediction models, namely, XGBoost, LightGBM, CatBoost and Random Forest, were constructed on the basis of these meteorological characteristic factors. Comparing the accuracy and root mean square error of the four prediction models, the results show that the LightGBM prediction model has an accuracy as high as 96% and a root mean square error as low as 13%, which is obviously better than the other three models. Therefore, the LightGBM prediction model was selected for in-depth study.

Then, according to the results of the characteristic importance analysis of the LightGBM model, the meteorological characteristic factors with high Spearman's correlation but low influence on the actual geological hazards were excluded: maximum relative humidity, average ground temperature and average 5 cm ground temperature. The 12 meteorological characteristic factors that have significant influence on geohazard prediction were further screened out, and the geohazard prediction model was reconstructed based on them.

Finally, by reconstructing the LightGBM prediction model, the AUC value of the model reaches 0.99, which is 3% higher than that before optimization, and the accuracy of the model reaches 0.98, which is 2% higher than that before optimization. This result shows that the LightGBM model based on the filtered meteorological characterization factor dataset shows the best effect in geohazard prediction.

Acknowledgements This study is supported by the Open Project of Sichuan Key Laboratory of Software Automatic Generation and Intelligent Service (CUIT-SAG202303); Science and Technology Development Fund Project of Sichuan Key Laboratory of Plateau and Basin Heavy Rainfall, Drought and Flood Disasters (SCQXKJYJXMS202324); Science and Technology Development Fund Project of Sichuan Key Laboratory of Plateau and Basin Heavy Rainfall, Drought and Flood Disasters (SCQXKJQN202323).

References

1. Liu XL, Miao C, Tian CS et al (2016) Comparative analysis of risk assessment of landslides and debris flows of China in 2000 and 2010. Adv Earth Sci 1001–8166
2. Huang RH, Wu Y, Yao HC et al (2024) Research on geohazard risk assessment based on machine learning. Surv Mapp Stand 40(02):53–59
3. Huang FM, Ouyang WP, Jiang SH et al (2024) Predictive modeling of landslide susceptibility considering the principle of spatio-temporal division of training/test sets in machine learning modeling. Earth Sci 49(05):1607–1618
4. Yang ZY, Xu YY, Xu YJ et al (2024) A comparative analysis about precipitation characteristics of geological disasters of different precipitation types in Yunnan Province. J Yunnan Univ: Nat Sci Ed 45(03):634–641
5. He J, Li WP, Zhang DY et al (2024) Analysis of correlation between grounding resistance and soil volumetric water content based on Spearman's rank correlation coefficient. J Mt Meteorol 48(03):86–90
6. Zhang LY, Cui HJ (2023) Local dependence test between random vectors based on the robust conditional Spearman's ρ and Kendall's τ. Acta Math Appl Sin 39(03):491–510
7. Xie WH (2015) Variable screening method of Spearman correlation coefficient. Beijing University of Technology
8. Zhang W, Hu FR, Qi W et al (2023) Evaluation of geologic hazard susceptibility based on XGBoost and cloud model. Chin J Geol Hazards Prev 34(06):136–145

9. Yang H, Ma HB, Sun WN et al (2023) Research on classification and recognition of severe convective weather in Jilin Province based on XGBoost algorithm. Meteorol Disaster Prev 30(02):28–33
10. Liu XW, Huang WB, Jiang YS et al (2021) Study of the classified identification of the strong convective weathers based on the LightGBM algorithm. Plateau Meteorol 40(04):909–918
11. He Z, Shi YH, Li FC et al (2024) Evaluation of landslide susceptibility in Linxia County, Gansu Province based on LightGBM model. J Water Resour Water Eng 35(01):197–205+216
12. Lu HX, Li K, Li L et al (2024) Research on precipitable water vapor prediction method based on LightGBM algorithm. GNSS World China 49(05):110–119
13. Shao J, Wang X, Chang WF et al (2022) Landslide sensitivity modeling based on CatBoost algorithm—taking Dafang County of Bijie City as an example. Comput Digit Eng 50(06):1365–1370
14. Liu L, Yang Y (2023) Landslide susceptibility assessment based on grid cells and random forest. Petrochem Technol 30(06):180–182
15. GHSA Geological Hazards Service Application. http://geohazard.cgl.org.cn/. Accessed 16 July 2024
16. OBAQ Online batch address query latitude and longitude. https://maplocation.sjfkai.com/. Accessed 16 July 2024
17. Shikang X (2023) Research and application of landslide geological disaster prediction model. University of Electronic Science and Technology. https://doi.org/10.27005/d.cnki.gdzku.2023.002024
18. Li Q, Li R (2023) Research on causes of soil and rock landslide deformation and prediction method of geological disasters. China Energy Environ Prot 45(01):149–153
19. Li S (2024) Research on prediction and prevention of environmental geological disasters based on GIS technology. North China Nat Resour 04:109–113
20. Duan ZM, Zhang LH, Xue Y et al (2024) Research on the coupling relationship between the occurrence of geological disasters and precipitation in Hunan Province. Mineral Explor 15(07):1310–1317
21. Sun R, Ding YH, Guo HY et al (2022) GIS-based zoning of rainfall-induced landslide and debris flow risk in Sichuan Province. Plateau Mountain Meteorol Res 42(03):111–116+140
22. Li M, Zhang L, Xiong QH et al (2024) A dynamic assessment method of geological disaster susceptibility and its application based on effective rainfall in Wuhan City. Torrential Rain Disasters 43(05):572–579
23. Xiao RD, Fan LZ, Zhang MD et al (2024) Temporal and spatial variation characteristics of geological disasters and its relationship with precipitation in Nujiang Prefecture of Yunnan. Plateau Mt Meteorol Res 44(02):127–132
24. Li HL, Jiang SY (2024) Risk assessment of geological hazards under multiple rainfall conditions-taking the key survey area of Fuchuan Yao autonomous county in Hezhou city as an example. West-China Explor Eng 36(07):1–4

Application of Stormwater and Sediment Management Based on the ABC Waters Design Guidelines in the Upper Stream Suburban Section of the Mengkibol River, Kluang, Malaysia

Jiaxin Jiang

Abstract In the context of climate change, the increasing frequency of extreme and prolonged rainfall has significantly heightened the risk of urban flooding. Urban waterfront spaces are especially vulnerable, as floods often follow river channels, making adjacent zones primary targets of flood impact. Due to unreasonable zoning or limited urban land availability, these areas frequently lack sufficient space to function as floodable zones that regulate excess stormwater. Additionally, heavy rain-induced runoff can carry pollutants into rivers, threatening local ecosystems and public health. The scouring effect of floods on riverbanks also leads to increased sediment deposition, reducing channel capacity and exacerbating flood risks. Effective flood management must therefore incorporate sediment control strategies. However, current research often treats flood and sediment management as separate issues. This study adopts ABC Waters design guidelines, utilizing runoff routes and sediment transport index to develop an integrated flood retention and sediment management system for the upstream section of Kluang's Mengkibol River, under land constraints. The aim is to enhance the resilience and environmental quality of urban waterfront spaces against future climate uncertainties.

Keywords Resilient waterfront landscape design · Stormwater management · Sediment control · ABC waters · Kluang

1 Introduction

Stormwater management has become critical for most cities' journey toward achieving urban resilience. Under the influence of climate change, unpredictable rainfall patterns and extreme precipitation events have led to an increased frequency

J. Jiang (✉)
Planning and Industrial Research Institute, CCCC Third Harbor Consultants Co., Ltd., Shanghai, China
e-mail: jiangjiaxin@ccccltd.cn

B. He et al. (eds.), *Climate-Resilient, Low-Carbon, and Sustainable Cities*, Lecture Notes in Civil Engineering 754, https://doi.org/10.1007/978-981-95-2169-2_9

of urban flooding [1]. Concurrently, urbanization, through ongoing land development and infrastructure expansion, has significantly increased the area of impervious surfaces within cities [2]. This process has substantially altered the response of urban hydrological and river systems to the built environment [3]. Driven by climate change and urbanization, the stormwater load on urban drainage systems has markedly intensified, presenting significant challenges to city stormwater management [4, 5].

In urban riverfront areas, flooding issues tend to be more frequent and severe. Rivers serve as the natural drainage network within cities. However, the limited capacity of these waterways often makes it challenging to accommodate high-intensity, prolonged rainfall, leading to flooding in riverfront regions. Inadequate planning during the early stages of urban development or the scarcity of urban land has often resulted in insufficient flood buffers along riverbanks. Consequently, built-up areas adjacent to the river become direct targets of flooding, causing significant economic losses, casualties, and burdens of environmental governance. Therefore, implementing targeted stormwater management measures in urban riverfront areas is critical.

Simultaneously, the issue of pollutant diffusion caused by river floods carrying sediments, which threaten public health, should be considered. Sediments are solid particles transported by water flow that settle on riverbeds, banks, or near river mouths. They typically consist of silt, gravel, stones, and organic matter. Sediments are a crucial component of river ecosystems, serving as a nutrient-rich reservoir that influences the exchange of materials within the river [6]. However, riverbed sediments can also act as a repository for accumulated pollutants [7]. Under the influence of water flow, sediments can be transported, leading to changes in riverbed morphology and the movement of solutes within the riverbed. Research indicates that during peak flow events, a significant portion of the biological load may originate from riverbed sediments [8, 9]. Floodwater flow instability significantly affects the movement of sediment particles in the riverbed, thereby influencing the diffusion of pollutants [7]. Then, it can negatively impact downstream river ecosystems and contribute to the spread of diseases.

Moreover, the increased rainfall associated with climate change can lead to a higher discharge of sediments [10]. The accumulation of large amounts of sediment can alter the elevation and cross-sectional geometry of riverbeds [11–15], reducing the water-holding capacity of river channels, weakening their flood storage ability, and ultimately exacerbating flooding [16, 17]. Therefore, sediment management is a critical component of stormwater management. During the early stages of planning and design, practitioners should comprehensively consider both aspects to ensure urban safety, public health, and river ecosystem protection during heavy rainfall events. However, sediment transport and management are often overlooked in flood risk studies and stormwater management practices [18]. Stormwater management and sediment management are frequently treated as two separate issues.

Gray infrastructure represents the traditional approach to stormwater management, primarily focusing on collecting, storing, and conveying stormwater to receiving water bodies [19, 20]. During the last century, urbanization led to significant transformations in river environments. Rigid, artificial channels replaced natural drainage channels to facilitate the development of impervious surface areas [21]. With the continuous improvement and expansion of urban underground drainage networks, structural measures gradually became the dominant means of urban drainage. However, traditional drainage systems have a fixed capacity and tend to deteriorate over time, leading to a decline in performance.

Moreover, the construction and upgrading of drainage facilities require substantial financial investment. Consequently, conventional gray infrastructure struggles to cope with the increasingly severe and unpredictable risks posed by extreme and prolonged rainfall events. As a result, there has been a gradual shift in approaches and strategies for urban stormwater management. Over the past two decades, more sustainable and resilient stormwater management methods have been proposed worldwide, such as implementing blue-green infrastructure. Examples include Stormwater Control Measures (SCM) in the United States, Water Sensitive Urban Design (WSUD) in Australia and New Zealand, Low Impact Development (LID) in China, Sustainable Urban Drainage Systems (SUDS), and Nature-Based Solutions (NBS) in Europe. These blue-green infrastructure approaches often complement existing urban gray infrastructure to achieve more effective, sustainable, and integrated stormwater management [22]. However, these methods focus more on enhancing stormwater retention capacity, with relatively less emphasis on water quality improvement.

Historically, sediment management in river basins and waterways has been a critical issue for water resource managers [23]. Current research on river sediment primarily focuses on sediment transport in large basins, such as the Nile and Yellow River basins. Sediment management efforts are primarily concentrated on reservoirs or open basins, with specific measures typically involving using rigid structures to intercept sediments. In contrast, relatively little research is dedicated to sediment management in urban river channels during flooding.

This study identifies the limitations of existing approaches to stormwater management and sediment control, and explores the application of an integrated methodology within the specific site. Traditionally, stormwater management and river sediment control have been treated as two separate domains in waterfront planning and management, lacking systemic and coordinated solutions. In response, this research proposes an integrated design strategy grounded in Singapore's ABC Waters Design Guidelines, which incorporates both stormwater and sediment management within a unified spatial framework. By emphasizing the interrelationship and synergetic regulation between these two aspects, the study offers a replicable case of practice that contributes to the ecological resilience of urban riverfronts and the advancement of holistic water environment governance.

2 ABC Waters Design Guidelines

Singapore is a typical tropical city with high annual rainfall and limited land resources, posing significant challenges for urban stormwater management systems. Urbanization has dramatically altered the island's water balance [24]. However, given the constraints on land availability, expanding drainage channels to increase the capacity of the stormwater network is not a feasible option. In response, Singapore's national water agency, PUB, launched the Active, Beautiful, Clean (ABC) Waters program in 2006 [25]. The program aims to achieve sustainable stormwater management, quality improvement, and enhanced aesthetic and recreational value of water environments by 2030 [24].

To promote the implementation of this initiative, the government has partnered with developers to establish an integrated stormwater management strategy as outlined in ABC Waters, which comprises source, pathway, and receptor components. The 'source' refers to the origin of stormwater runoff, emphasizing on-site collection and attenuation of runoff through ABC design features. The 'pathway' represents the route through which stormwater is conveyed, focusing on enhancing the capacity of the stormwater transport system, including measures like widening and deepening channels and detention of catchment areas. The 'receptor' denotes locations where stormwater accumulates and may impact infrastructure, emphasizing the arrangement of protective measures. On a micro level, ABC Waters proposes six treatment elements to achieve purification, detention, retention, conveyance, and infiltration. The design features corresponding to these five primary functions are listed in the Table 1.

Although the treatment elements in ABC Waters share similarities with stormwater management methods popular in other countries and regions [26], the design features of ABC Waters place a greater emphasis on water quality improvement [27]. Specifically, it focuses on purifying stormwater runoff and river pollutants, thereby contributing to sediment management by reducing the pollutant content within sediments. While numerous ABC Waters landscape projects have already been implemented within Singapore, this approach has yet to be widely adopted in other tropical cities with similar environmental conditions.

Singapore's northern neighbor, Malaysia, shares similar environmental conditions in certain areas. Flooding is one of the significant environmental challenges facing Malaysia. From November to March each year, the northeast monsoon brings heavy rainfall, which frequently causes widespread flooding. Despite the local government's efforts to implement several initiatives to mitigate flooding, traditional measures have proven insufficient to meet urban and rural areas' long-term flood control needs [28]. Therefore, this study proposes to select Kluang, a city in Johor, Malaysia, that suffers from flood and river pollution, as a pilot city for promoting the ABC Waters design guidelines. Utilizing the ABC Waters design principles as the methodology and a systematic landscape design approach as the outcome, this project aims to provide a reference case for stormwater management and sediment

Table 1 Functions of ABC waters design features (self-drawn)

	Vegetated swale	Bioretention swale	Bioretention basin (rain garden)	Sedimentation basin	Cleansing biotope	Constructed wetland
Design feature	They are natural drainage channels with gentle slopes that slow stormwater flow, promoting sediment settlement and reducing erosion. Often used alongside bioretention systems	They are vegetated channels with integrated bioretention systems. They temporary ponding to enhance infiltration. Runoff is filtered as it seeps through, with cleansed water collected by subsoil pipes	They are vegetated channels with integrated bioretention systems. They temporary ponding to enhance infiltration. Runoff is filtered as it seeps through, with cleansed water collected by subsoil pipes	These ponds temporarily detain stormwater, slowing flow for particle settling, and often serve as pre-treatment for wetlands or as water storage and irrigation ponds	Cleansing biotopes are vertical flow wetlands with recirculation, using nutrient-poor substrates and wetland plants for their water-purifying abilities	They are shallow, vegetated waterbodies used for stormwater management. They consist of an inlet zone for sediment removal, a macrophyte zone for pollutant filtration, and a high-flow bypass channel

Function	● Purification ● Detention ● Conveyance	● Purification ● Detention ● Conveyance ● Infiltration	● Purification ● Detention ● Infiltration	● Purification ● Detention ● Retention	● Purification ● Detention ● Infiltration	● Purification ● Detention ● Retention

● Purification, ● Detention, ● Retention, ● Conveyance, ● Infiltration control in the riverfront areas of Kluang, transforming it into a more resilient and livable city.

3 Study Method and Study Area

3.1 Study Method

In this study, the author employed a series of software tools to visualize the preliminary analysis and design practices. The software used includes Adobe Photoshop 2021, Adobe Illustrator 2021, Rhino 7.0 with Grasshopper, and ArcGIS Pro.

Specifically, Rhino 7.0 with Grasshopper and ArcGIS Pro were integrated with DEM data to generate terrain models and conduct analyses of surface runoff direction and flood extent. Based on the 3D terrain model and satellite imagery, Rhino 7.0 was further utilized for environmental modeling, the drawing of existing and proposed river sections, as well as the development of axonometric diagrams to illustrate design strategies. Finally, the Adobe software suite was used to enhance vector drawings, incorporate raster imagery, and arrange the overall layout, resulting in a clearer and more visually appealing presentation.

3.2 Kluang and the Mengkibol River

Kluang is a city located in the central region of Johor, Malaysia, covering a total area of approximately 28,518.8 km^2, with a council administrative area of about 126.57 km^2 [29]. In 2023, the city had an estimated population of around 328,000 [30]. The primary industries in Kluang include agriculture, manufacturing, and tourism. Kluang experiences a tropical rainforest climate, with an average annual temperature of 25.6 °C and an average annual rainfall of 2,473 mm [31]. Influenced by the northeast and southwest monsoons, Kluang's climate can be categorized into a wet season (from November to March) and a less wet season (from April to October). The city is in a hilly region, with an elevation range between -1 and 941 m above sea level, the highest point being Gunung Lambak in the east. Despite numerous small hills and undulating terrain within Kluang, there are also substantial areas of flat lowlands. While these geographical features provide favorable conditions for agricultural activities, they also pose flooding risks (Fig. 1).

The river system in the central area of Kluang is primarily composed of the Sungai Melantai in the west and the Mengkibol River in the east. Both rivers flow from south to north through the center of Kluang and converge in the northern area to form the Sungai Melantai, which eventually drains into the Sungai Sembrong. The Mengkibol River is the mother river of Kluang, extending approximately 8.5 km from south to north. Based on the functional attributes of the buildings along its banks, the

Fig. 1 Location, topography, and river system of Kluang (self-drawn)

Mengkibol River can be divided into three segments: (1) the upper suburban section, characterized by the presence of villages, industrial parks, and oil palm plantations; (2) the urban core, dominated by commercial and residential buildings; and (3) the lower suburban section, where oil palm plantations are predominant, with a few scattered villages (see Fig. 2). The morphological features of the Mengkibol River include a relatively narrow channel width, ranging from approximately 10 to 25 m, and steep banks with a slope of about 45°. Under climate change's impact, flooding along the banks of the Mengkibol River has become increasingly severe.

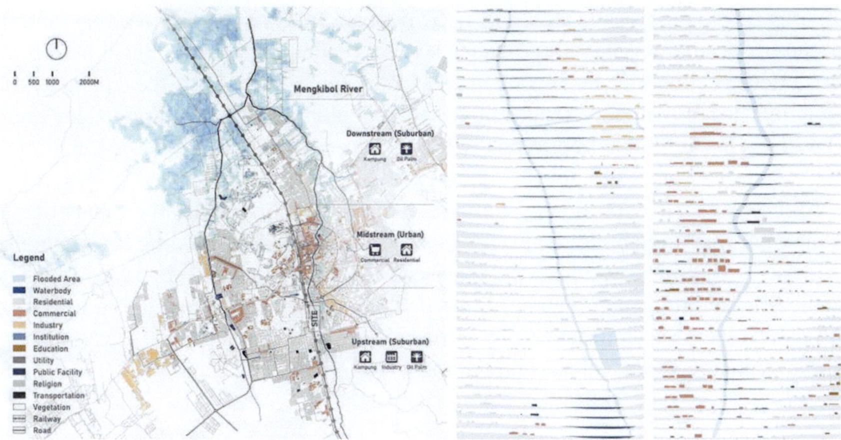

Fig. 2 Flooding extent of Kluang and flooding sections of the Mengkibol River in the urban context (upstream and midstream) (self-drawn)

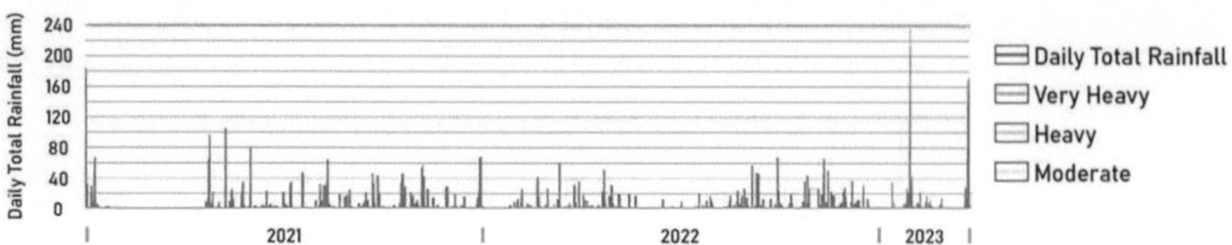

Fig. 3 Rainfall of Kluang (2021–2023) (self-drawn)

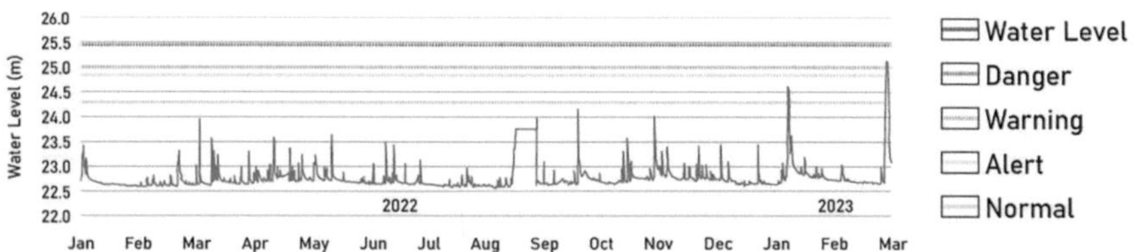

Fig. 4 Water level of the Mengkibol River (2022–2023) (self-drawn)

3.3 Flooding in Kluang

In recent years, the annual rainfall in Kluang has become increasingly unpredictable, with a higher probability of extreme peaks (see Fig. 3). As a result, the maximum water levels of the Mengkibol River during the rainy season have been steadily rising [32], exacerbating the risk of flooding in the surrounding areas (see Fig. 4). In March 2023, Kluang experienced another continuous heavy rainfall, leading to widespread flooding across the city and affecting 1,952 people [33]. The average monthly rainfall reached 209 mm, and the river's water level exceeded the warning level by 0.3 m. The extent of flooding in Kluang increased progressively from the upper to the lower reaches of the river. According to GIS calculations, the total flooded area was 1,847 ha. Of the affected buildings, 77% were residential, 12% were commercial, and 11% served other functions (see Fig. 3).

In recent years, with the development and construction along the banks of the Mengkibol River, the buffer zone area that can accommodate flooding has gradually decreased. Additionally, due to the river's narrow and elongated shape, its capacity for flood retention is significantly limited. These two factors have led to a lack of sufficient blue-green infrastructure in the city to regulate such large volumes of rainfall, causing severe impacts on the local community and economy.

3.4 Sediment Issue Along the Mengkibol River

In addition to flooding, sediment accumulation is another issue for the Mengkibol River. Observations from site investigations indicate that the amount of sediment and

Fig. 5 Sediment and insoluble solid waste along the Mengkibol River (self-drawn)

insoluble solid waste increases progressively from the upstream to the downstream sections of the river (see Fig. 5). Interviews with residents revealed that riverbank development has led to surface runoff carrying various types of pollutants into the river, contributing to increased sediment levels. This process not only results in river pollution but also causes the riverbed to rise, ultimately reducing the river's flood retention capacity and exacerbating flooding along its banks.

This project addresses the issue of sedimentation by utilizing Digital Elevation Model (DEM) data (https://earthexplorer.usgs.gov/). The author employed Grasshopper and GIS to visually represent the surface runoff directions, the Sediment Transport Index (STI), and sediment sources along both sides of the river. Surface runoff illustrates the natural flow paths of rainwater as it carries surface materials across the land. The STI indicates the potential for soil erosion within a specific area; higher STI values suggest a greater ability for sediment transport and a higher likelihood of soil erosion (see Fig. 6). The sources of river sediment can be categorized into four types: construction, agriculture, commercial activities, and erosion (natural or due to flooding). These sources lead to six river pollutants, resulting in a range of negative impacts (see Fig. 7).

4 Application of the Cleansing Sponge Design Framework Based on the ABC Waters Design Guidelines

4.1 Design Area

The site is on the upper west side of the Mengkibol River. On the western side of the river, there are orchards, the rail corridor, and the golf course, while a village occupies the eastern side. The reasons for selecting this site include: (1) It is one of the few open spaces along the riverbank, making it a promising candidate for conversion into blue-green infrastructure with stormwater and flood regulation functions. (2) Being situated upstream, implementing sediment control measures at this site could help reduce sediment accumulation downstream. (3) The land is proposed as residential in the Kluang zoning plan. However, the village across from the site has been severely affected by flooding. If this area is developed into a residential zone in the future, it may also face flood risks, potentially compromising the residents'

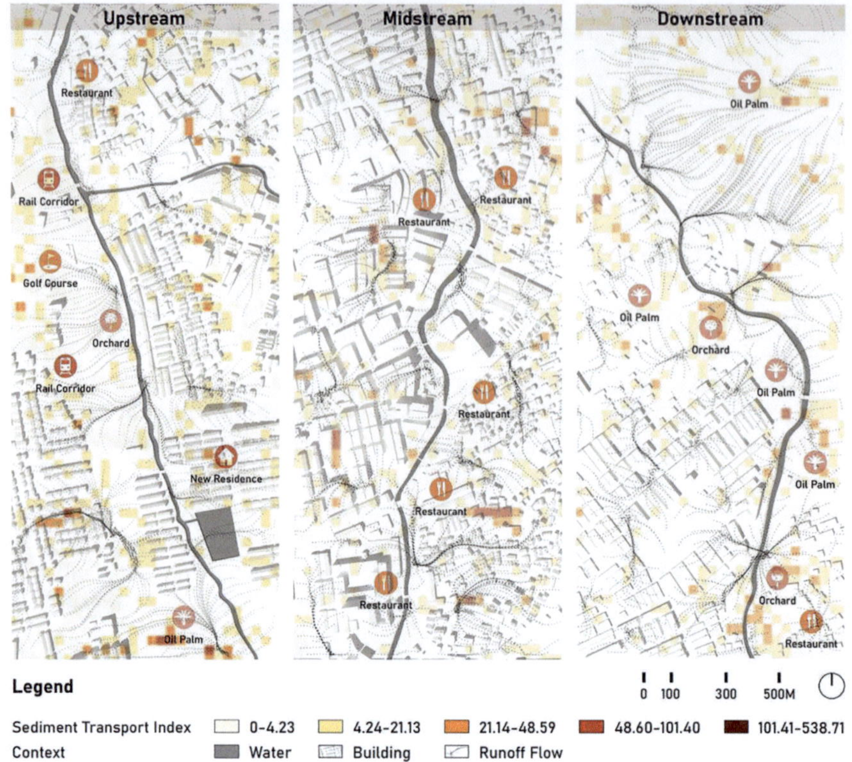

Fig. 6 Sediment transport index along the Mengkibol River, the distribution of possible pollution sources and the surface runoff flow (self-drawn)

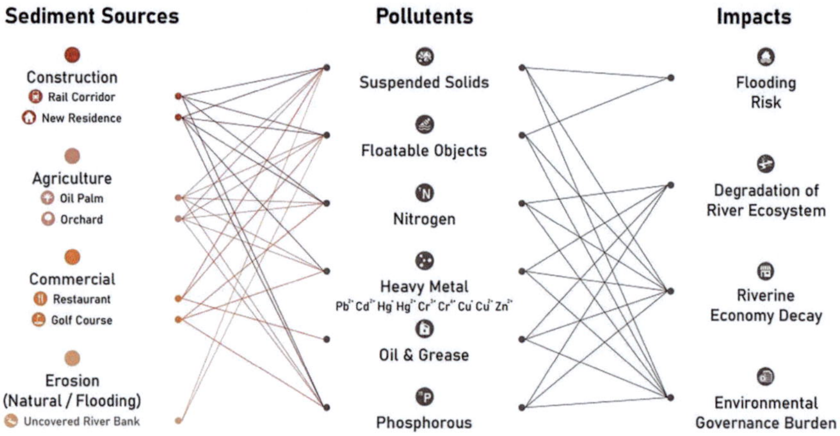

Fig. 7 Sediment sources, pollutants, and their negative impacts on the river (self-drawn)

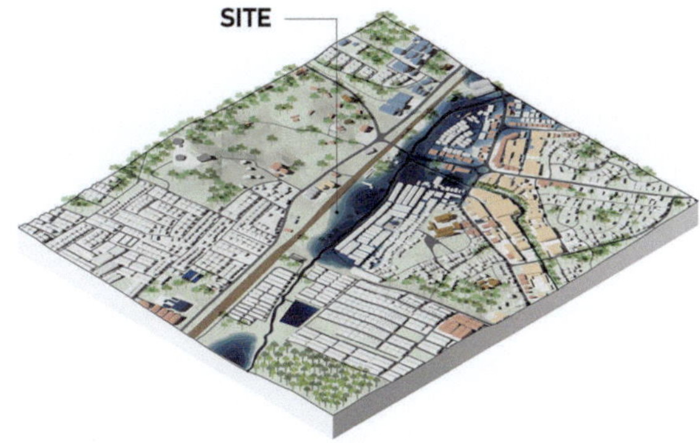

Fig. 8 Flooding scenario simulation of the site (self-drawn)

rights and negatively impacting real estate investment returns (see Fig. 8). Therefore, the author selected this site for intervention, guided by the ABC Waters Design Guidelines principles, to alleviate flooding and sediment issues within the area.

4.2 Three Strategies

Sponge Strategy. This strategy has two primary objectives: first, to increase the water storage capacity of the river channel, and second, to create swales to increase stormwater runoff routes (see Fig. 9). The ultimate goal is to reduce the likelihood of flooding on both sides of the riverbanks. To increase the river channel's water storage capacity, the author proposes three strategies: creating meandering streams along the river channel (S-1-1); widening the river channel (S-1-2); and installing underground water tanks in low-lying areas adjacent to the river while maintaining the above-ground space as a public activity area (S-1-3). When the river exceeds the warning level, it can flow into the underground water tanks. Meanwhile, to increase stormwater runoff routes by creating vegetated swales (S-2-1), rain gardens (S-2-2), and bioretention swales (S-2-3), thereby extending the time it takes for runoff to reach the river channel and delaying the river's peak volume.

Purification Strategy. This strategy aims to reduce sedimentation due to erosion during flooding and purify pollutants in stormwater runoff (see Fig. 10). Riverbanks are the areas most prone to soil erosion, necessitating bank stabilization measures to prevent erosion during flood periods. The erosion can lead to excessive soil entering the river channel, forming sediment, and raising the riverbed. Initially, based on the bank's slope, three stabilization measures will be implemented: vegetation, terraces, and gabions (C-1-1, C-1-2, C-1-3). As for the purification, in the proposed swales along the riverbanks, wetland plants will play a role in adsorbing and purifying pollutants from the runoff, thereby reducing the entry of pollutants into the river and their attachment to sediment. The three main approaches to existing sediment are to capture and remove it. Sediment will be collected using sediment basins (C-2-1),

Fig. 9 Sponge strategy (self-drawn)

topo capturers (C-2-2), and wetland plants (C-2-3). Once the sediment is captured, it will be removed periodically using mechanical means (C-3-1), artificial means (C-3-2), or a combination of both (C-3-3).

Recreation Strategy. This strategy aims to establish a connection between the site and the surrounding residents, meeting their needs for access and recreational activities during both the wet and normal seasons (see Fig. 11). The recreational space can be categorized into point, linear, and surface-based. Point-based spaces include the bridge (R-1-1), water steps (R-1-2), and viewing steps (R-1-3). Linear spaces consist of riverbank trails (R-2-1), slope trails (R-2-2), and water trails (R-2-3). Area-based spaces comprise the amphitheater (R-3-1), event lawn (R-3-2), and sunken playground (R-3-3). The latter two are designed as sunken spaces that serve as open green areas for recreational activities during normal conditions and water storage areas during the wet season.

4.3 Master Plan Deduction

Firstly, the lowest points within the site are selected to serve as the underground water tank or sunken activity spaces, with three low points identified. Since the central and southern points are closer to the village and offer more significant potential for interaction with residents, these two points are designated sunken activity spaces, specifically the event lawn in the central area and the sunken playground in the southern area. The northernmost point is utilized as an underground water tank (Step 1). Additionally, the existing river channel is widened to the west, and a meander is created in the central section to enhance the overall water retention capacity of the river channel (Step 2).

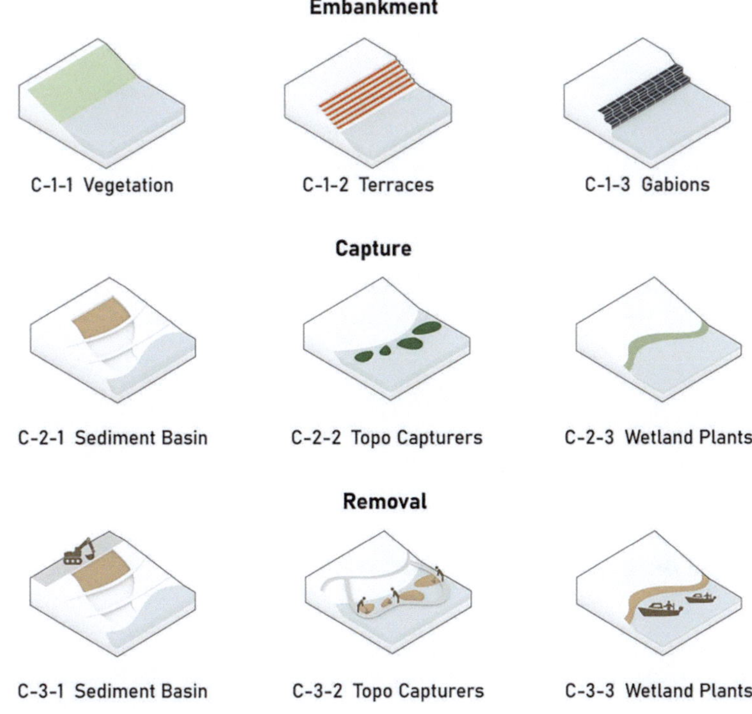

Fig. 10 Purification strategy (self-drawn)

The second strategy involves applying different bank stabilization measures based on the slope gradient to reduce sediment caused by riverbank erosion (Step 3). Specifically, vegetation (10° < slope ≤ 20°), terracing (20° < slope ≤ 30°), and gabion (slope > 30°) stabilize the riverbank. Vegetative cover is primarily arranged on the western side of the river channel, where the slope is relatively gentle and can form a floodplain area during the wet season. Terracing is positioned along the eastern riverbank, bordering the village, where it stabilizes the bank and serves as a recreational area for residents, enhancing the riverside landscape quality. In the northern and southern sections of the eastern bank, gabion are used for stabilization, providing lower interactivity but higher stability on steeper slopes.

The third strategy focuses on sediment capture (Step 4). Areas on the site's western boundary with high STI values, which indicate a higher risk of erosion, are transformed into gently sloped sedimentation ponds to capture external sediment sources, thereby reducing soil erosion in this region. Following this, swales are designed to allow wetland plants along the banks to perform a secondary capture of residual sediment, further minimizing external sediment input into the river. Additionally, based on fluvial dynamics, the flow velocity is slower along the concave bank of the river, making it more conducive to sediment deposition. Therefore, topo capturers are positioned along the concave bank to trap sediment already present in the river,

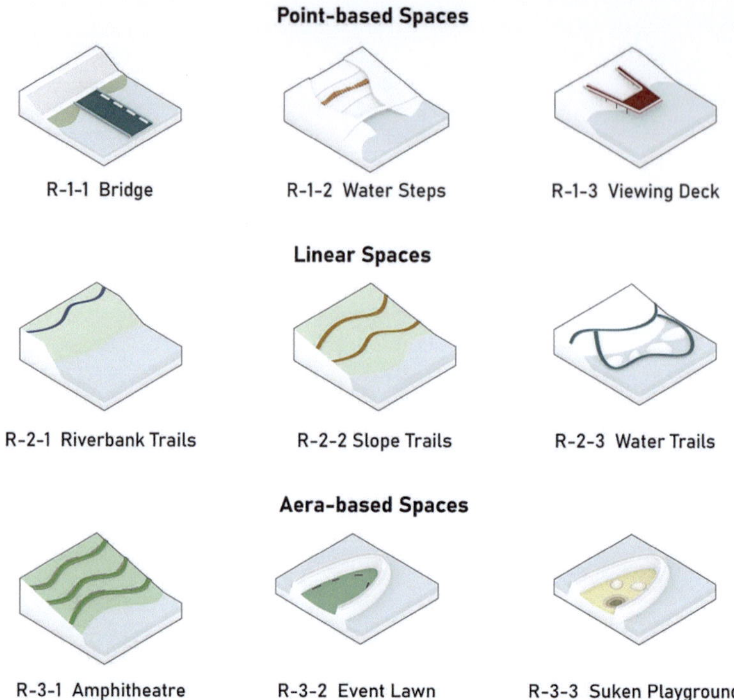

Fig. 11 Recreation strategy (self-drawn)

with wetland plants on the western side also contributing to sediment capture within the river channel (Fig. 12).

The fourth strategy involves arranging a swale system based on surface runoff flow lines to purify the runoff entering the river (Step 5). Since different types of swales have varying capacities for stormwater management, specific swales are positioned according to the concentration of surface runoff. Vegetated swales are placed in

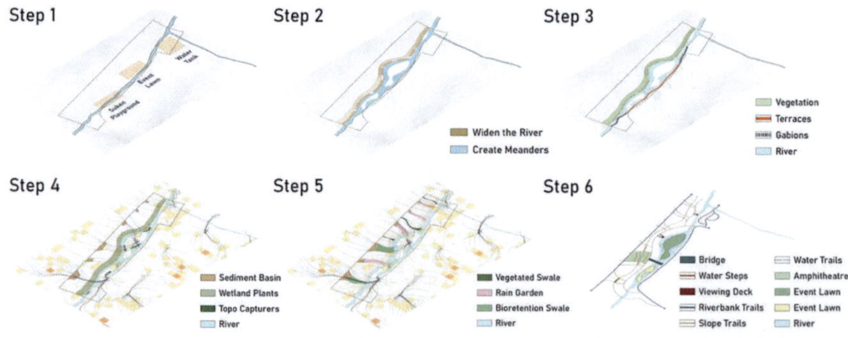

Fig. 12 Master plan deduction (self-drawn)

areas with low runoff concentration, rain gardens in areas with moderate runoff concentration, and bioretention swales in areas with high runoff concentration.

The fifth strategy addresses the placement of various recreational spaces (Step 6). Riverbank, slope, and water trails are arranged sequentially along the contour lines to provide different levels of riverside walking and water interaction experiences. Rain gardens, bioretention swales, and water steps enhance the site's interactivity and spatial interest. A bridge in the central area connects the eastern village with the western amphitheater, strengthening connectivity between the two banks. Water steps link the riverbank to the sunken spaces. The larger sunken space in the central area is designated as an event lawn, providing a venue for residents to host everyday outdoor activities and weekend markets. The smaller sunken space in the southern area is mainly designed as a children's play area (Fig. 13).

4.4 Three Systems

Sponge System. During the wet season or extreme rainfall events, the system efficiently guides runoff and stores water, thereby reducing the risk of flooding along the riverbanks. The functional modules of the sponge system consist mainly of topographical features and wetland vegetation (see Fig. 14), where the former directs water flow, and the latter assists by enhancing soil water infiltration through its root system. When the river reaches a high water level, the vegetated gentle slope area and sunken spaces on the left bank will serve as floodable zones. The vegetation in these floodable areas is designed to be flood-tolerant and has stable root systems to withstand the impacts of inundation during flood events (see Fig. 15).

Purification System. The primary components of this system are wetland plants with water purification capabilities, which function throughout the year to reduce the accumulation of pollutant-laden sediments in the river channel. Plant selection is based on the analysis of pollutant sources mentioned earlier, and their arrangement is determined by their flood tolerance in a vertical layout. During the dry season, runoff from the site's western side flows through vegetated swales, rain gardens, or bioretention swales, allowing the plants to absorb or capture pollutants. In the wet season, the swale system within the floodable areas helps to retain the runoff, with wetland plants purifying the water. The introduction of wetland plants not only improves water quality, enhances landscape aesthetics and increases biodiversity within the site (see Fig. 16).

Recreation System. This system comprises a series of landscape and recreational facilities designed to promote interaction between the surrounding residents and the site. During the dry season, these recreational facilities not only meet the needs of the residents for various daily activities but also serve as venues for festive events. For example, the event lawn can host weekend markets and traditional local performances, while the amphitheater can serve as a space for spectators and everyday relaxation. The landscape features in this area primarily consist of pathways and

Fig. 13 Master plan (self-drawn)

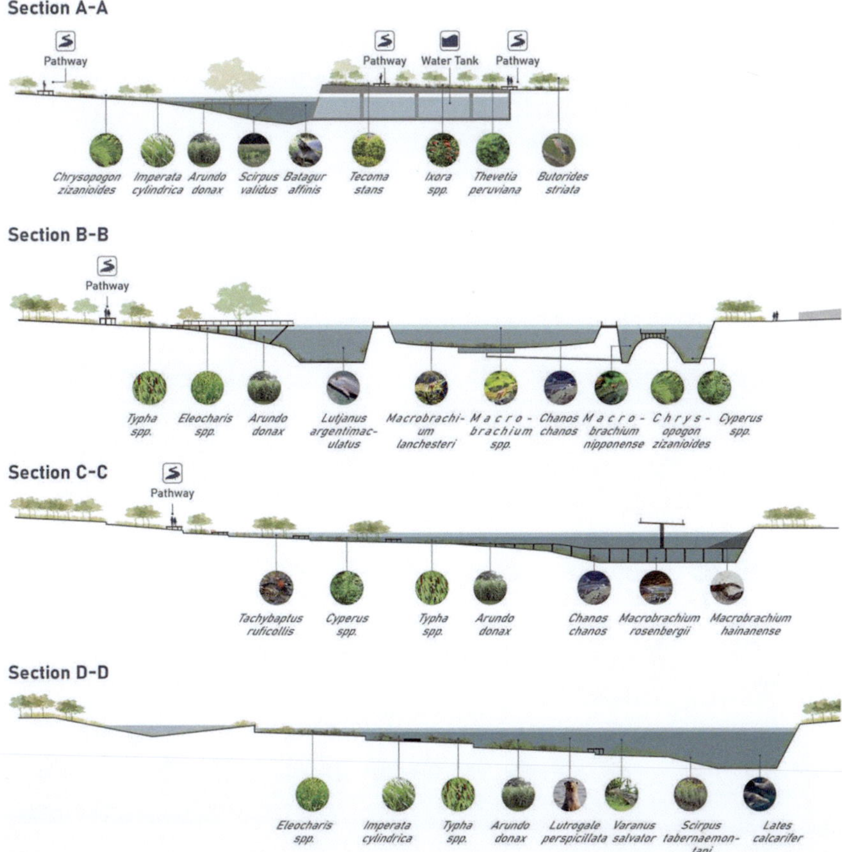

Fig. 14 Rainy season section (self-drawn)

viewing decks to minimize damage from flooding. Most of the pathways within the floodable areas will be submerged during heavy rainfall. However, the pathways near the site's western boundary are situated above the warning water level, ensuring pedestrian accessibility around the site during such events (Fig. 17).

5 Discussion

ABC Waters is an exemplary systemic stormwater management program replicating the natural water cycle within urban environments by integrating engineering techniques and ecological knowledge. As discussed earlier, the significant challenges in urban stormwater management are primarily driven by the intensification of climate change, urbanization, and the constraints posed by traditional grey infrastructure.

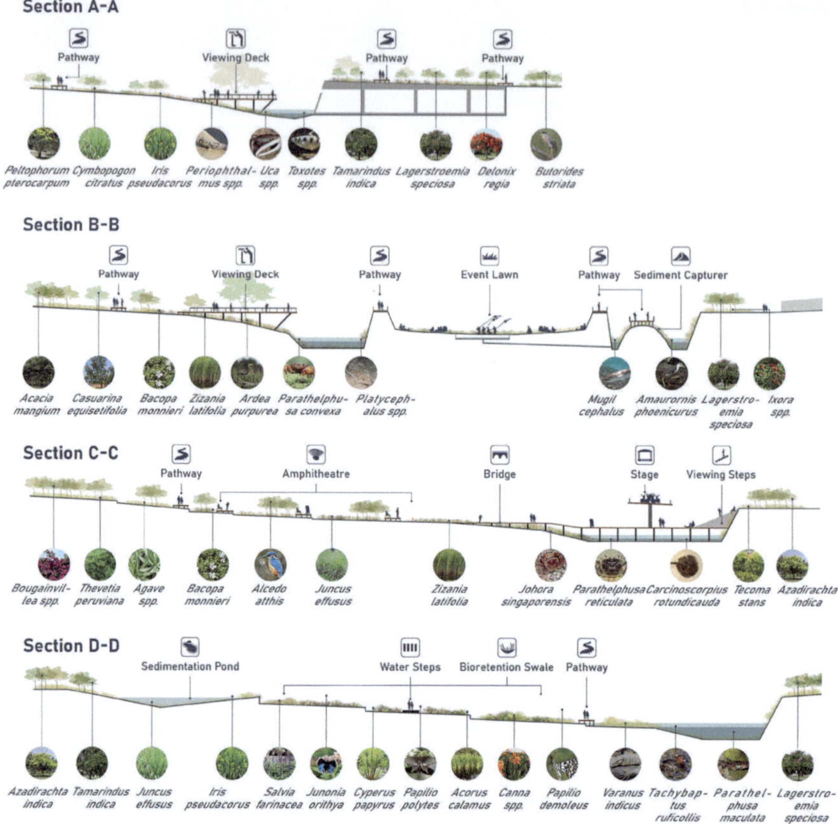

Fig. 15 Less wet season section (self-drawn)

Existing solutions are increasingly inadequate to address these escalating issues at their source. In response, ABC Waters proposes a series of more resilient and efficient blue-green infrastructure solutions, such as bioretention swales, cleansing biotopes, and constructed wetlands, to tackle the unpredictable challenges of urban stormwater. This design approach also emphasizes water quality improvement through wetland vegetation and topographical modifications, reflecting a comprehensive consideration of stormwater and sediment management during heavy rainfall events. Furthermore, ABC Waters underscores the interaction between artificial water systems and surrounding communities and the promotion of environmental education and awareness. These elements highlight its distinction from water management strategies proposed by other countries and regions.

However, ABC Waters is not without controversy. Regarding sediment management, some experts have argued that compared to traditional complex engineering channels, the softscape elements of ABC Waters may lead to increased sediment accumulation in waterways. Indeed, the soft banks, primarily composed of wetland

Vegetation Species	Flood Tolerance	Planting Location	Pollutent Purification Performance										
			Suspended Solids	Nitrogen	Heavy Metal							Oil & Grease	Phosphorous
					Pb^{2+}	Cd^{2+}	Hg^{2+}	Cr^{3+}	Cr^{6+}	Cu^{2+}	Zn^{2+}		
Cyperus spp.	Strong	Below Low Water Level	√	√									√
Scirpus validus	Strong	Low to Middle Water Level	√	√	√					√	√		√
Scirpus tabernaemontani	Strong		√	√						√	√		√
Juncus effusus	Strong	Middle to High Water Level	√	√	√					√	√		√
Cyperus papyrus	Strong		√	√		√	√			√			√
Acorus calamus	Strong		√	√	√	√				√	√	√	√
Canna spp.	General		√	√					√	√		√	√
Zizania latifolia	General		√	√	√	√	√	√	√	√	√		√
Arundo donax	Strong		√	√	√		√						√
Iris pseudacorus	Strong	High to Warning Water Level	√	√						√	√	√	√
Bacopa monnieri	Strong		√	√	√		√			√	√		√
Chrysopogon zizanioides	General		√	√	√					√			√
Imperata cylindrica	General		√	√						√	√		√
Typha spp.	Strong		√	√	√	√	√		√	√	√	√	√
Eleocharis spp.	Strong		√	√	√					√			√
Peltophorum pterocarpum	Weak	Above Warning Water Level	/										
Tamarindus indica	Weak												
Lagerstroemia speciosa	General												
Delonix regia	Weak												
Acacia mangium	Weak												
Casuarina equisetifolia	Weak												
Ixora spp.	Weak												
Bougainvillea spp.	Weak												
Azadirachta indica	General												
Cymbopogon citratus	General												
Thevetia peruviana	Weak												
Agave spp.	Weak												
Tecoma stans	General												
Salvia farinacea	Weak												

Flood Tolerance ● Strong ● General ● Weak

Fig. 16 Planting list and features of the selected plants (self-drawn)

Provision of Habitats for					
Fish	**Crustaceans**	**Amphibians**	**Birds**	**Reptiles**	**Insects**
Chanos chanos	*Carcinoscorpius rotundicauda*	*Lutrogale perspicillata*	*Alcedo atthis*	*Batagur affinis*	*Junonia orithya*
Lates calcarifer	*Johora singaporensis*		*Amaurornis phoenicurus*	*Varanus indicus*	*Papilio demoleus*
Lutjanus argentimaculatus	*Macrobrachium hainanense*		*Ardea purpurea*	*Varanus salvator*	*Papilio polytes*
Mugil cephalus	*Macrobrachium lanchesteri*		*Butorides striata*		
Periophthalmus spp.	*Macrobrachium nipponense*		*Tachybaptus ruficollis*		
Platycephalus spp.	*Macrobrachium rosenbergii*				
Toxotes spp.	*Macrobrachium spp.*				
	Parathelphusa convexa				
	Parathelphusa maculata				
	Parathelphusa reticulata				
	Uca spp.				

Fig. 17 Increased biodiversity in the proposed site (self-drawn)

vegetation, soil, and gravel layers, may contribute to more erosion-related sediment than hardened riverbanks. Nevertheless, these soft banks have the potential to filter runoff from adjacent areas, reducing the likelihood of pollutant-laden sediments entering the waterways. Additionally, appropriate sediment capture measures can be implemented within the water channels. For instance, in the design practices of this study, topo capturers are installed on convex banks to collect sediments, with periodic removal of excess sediment. Furthermore, GIS-based calculations of the site's Sediment Transport Index (STI) can guide the implementation of complex engineering measures, such as gabions and terraces, in areas prone to erosion. By integrating complex engineering solutions with the soft structures of ABC Waters, a more comprehensive strategy for addressing riverbank erosion and flood control can be developed.

Although ABC Waters provides a refined approach to managing urban stormwater runoff, managing riverine flood events and sediment control remains an issue that requires consideration from a watershed-level perspective. Consequently, there are certain limitations within the design guidelines. In other words, if individual sites are not defined at a macro level in terms of their specific roles in stormwater management and sediment control, and if the guidelines are applied indiscriminately, the design features may not achieve their maximum performance. Therefore, the ABC Waters design guidelines should be further developed to include broader watershed-level flood and sediment management considerations to reflect their systematic and intelligent approach better. The primary shortcoming of this study's design is the need for a more macro-level perspective on flood and sediment issues. As a result, the proposed design can only address localized problems to a limited extent and cannot achieve comprehensive watershed management. Flood risks and river pollution will likely persist. Nonetheless, this design offers a practical exploration of applying the ABC Waters guidelines outside Singapore, providing a reference case for stormwater and sediment management in riverfront areas.

Moreover, the limitations of ABC Waters are also evident in that its strategies primarily address design and maintenance without sufficient early intervention and guidance at the urban planning policy level. Proactive measures in policy can better equip cities to face the challenges of climate change. The site in this study exemplifies a case where insufficient consideration was given to flood-prone riverbank areas during the initial planning stages. Consequently, as urban development continues along the riverbanks, the latent risks embedded in the early planning phase have become more apparent, ultimately resulting in flooding. The generation of significant levels of pollutants in the river is also closely linked to management policies. Different types of establishments should be required to treat and monitor the quality of their site-generated runoff before discharging it into the river, ensuring that only runoff meeting established standards is released. Intervention at the policy and urban planning levels is more efficient than attempting to rectify issues through design guidelines alone. Therefore, government agencies and relevant stakeholders must prioritize proactive policy and planning decisions that will help cities respond more effectively and resiliently to the challenges posed by climate change.

From the perspective of urban planning and design, this study argues that localized green infrastructure strategies should be coordinated with large-scale landscape planning to enable a more forward-looking and resilient approach to urban water management. Such integration requires cross-scalar and multi-sectoral collaboration to address the complex dynamics of urban hydrological systems.

First, a data-driven planning approach is essential to enhance understanding and responsiveness to natural hydrological processes. By utilizing multi-source data such as remote sensing, digital elevation models (DEMs), and hydrological simulations, planners can better analyze river dynamics and sediment transport patterns. This scientific basis supports more informed decisions regarding urban spatial layout and future development. Second, water-related spatial planning should be conducted at the watershed scale. Analytical results concerning flood-prone areas and sediment flow pathways can inform the spatial configuration of urban blue-green infrastructure networks. These insights can guide the rational allocation of ecological buffers, flood detention zones, and wetland systems. Such measures serve as a key reference for urban master planning and regional landscape structure design, ensuring that urban development—especially in riparian areas—respects natural hydrological processes and reduces the likelihood of conflict with natural hazards. In this way, the comprehensive management of water environments can move beyond site-specific engineering solutions toward an ecologically integrated planning framework, fostering a more harmonious coexistence between urban growth and natural systems.

6 Conclusion

This study explores the application of ABC Waters design features in riparian spaces. Different types of swales are initially selected and positioned based on runoff routes. Concurrently, the embankment design is optimized by integrating the Slope Terrain Index (STI) with runoff paths to reduce river sedimentation caused by soil erosion. Following this, an analysis of the sources of river pollutants in the surrounding area informs the selection of appropriate vegetation for the swale system, aiming to improve water quality and enhance site biodiversity. Finally, interactive landscape elements are incorporated into the design, leveraging the site's topography to strengthen connections between the area and the community. Unlike conventional studies that treat river flood management and sediment control as separate issues, this research adopts a holistic approach using ABC Waters design guidelines to develop a comprehensive solution. This approach provides a reference case for integrating floodwater and sediment management in riparian areas. It improves urban waterfront spaces' resilience, environmental quality, and community engagement. This study will raise awareness and encourage support for flood management and sediment control efforts, creating more livable and resilient urban environments.

References

1. Batica J (2015) Methodology for flood resilience assessment in urban environments and mitigation strategy development. PhD thesis, Université Nice Sophia Antipolis
2. Tansar H, Li F, Zheng F, Duan H-F (2024) A critical review on optimization and implementation of green-grey infrastructures for sustainable urban stormwater management. AQUA - Water Infrastruct, Ecosyst Soc 73(6):1135–1150
3. Hu J, Li K, Chen Q, Wu J, Li B, Lou Z (2014) Modeling the effects of landscape pattern on water quality in the semi-arid Baiyangdian Basin, China. Water 6(4):976–998
4. Wang M, Liu M, Zhang D, Qi J, Fu W, Zhang Y, Rao Q, Bakhshipour AE, Tan SK (2023) Assessing and optimizing the hydrological performance of grey-green infrastructure systems in response to climate change and non-stationary time series. Water Res 232:119720
5. Zhang X, Jia H (2023) Low impact development planning through a comprehensive optimization framework: current gaps and future perspectives. Resour, Conserv Recycl 190:106861
6. Lei X, Cui B, Zhao H (2010) Study on the simulation of nutrient release from river inner source and its application—a case study of Guangzhou-Foshan river network, China. Procedia Environ Sci 2:1380–1392
7. Karimaei Tabarestani M, Zarrati AR (2015) Sediment transport during flood event: a review. Int J Environ Sci Technol 12(2)
8. Cho KH, Pachepsky YA, Kim JH, Guber AK, Shelton DR, Rowland R (2010) Release of Escherichia coli from the bottom sediment in a first-order creek: experiment and reach-specific modeling. J Hydrol 391:322–332
9. Wilkinson J, McKergow L, Davies-Colley RJ, Ballantine D (2011) Modelling storm-event E. coli pulses from the Motueka and Sherry Rivers in the South Island, New Zealand. N Z J Mar Freshwater Res 45(3):369–393
10. Kido R, Inoue T, Hatono M, Yamanoi K (2023) Assessing the impact of climate change on sediment discharge using a large ensemble rainfall dataset in Pekerebetsu River basin. Hokkaido Prog Earth Planet Sci 10:54
11. Schumm SA, Khan HR (1972) Experimental study of channel patterns. GSA Bull 83(6):1755–1770
12. Parker G (2006) Self-formed straight rivers with equilibrium banks and mobile bed. Part 1. The sand-silt river. J Fluid Mech 89(1)
13. Church M (2006) Bed material transport and the morphology of alluvial river channels. Annu Rev Earth Planet Sci 34:325–354
14. Métivier F, Barrier L (2012) Alluvial landscape evolution: what do we know about metamorphosis of gravel-bed meandering and braided streams? In: Gravel-bed rivers: processes, tools, environments, pp 474–501
15. Dingle EH, Paringit EC, Tolentino PLM, Williams RD, Hoey TB, Barrett B, Long H, Smiley C, Stott E (2019) Decadal-scale morphological adjustment of a lowland tropical river. Geomorphology 333:30–42
16. Slater LJ, Singer MB, Kirchner JW (2015) Hydrologic versus geomorphic drivers of trends in flood hazard. Geophys Res Lett 42(2):370–376
17. Slater LJ, Khouakhi A, Wilby RL (2019) River channel conveyance capacity adjusts to modes of climate variability. Sci Rep 9:12619
18. Vázquez-Tarrío D, Ruiz-Villanueva V, Garrote J, Benito G, Calle M, Lucía A, Díez-Herrero A (2024) Effects of sediment transport on flood hazards: Lessons learned and remaining challenges. Geomorphology 446:108976
19. Duan H-F, Yan H (2016) Multi-objective optimal design of detention tanks in the urban stormwater drainage system: LID implementation and analysis. Water Resour Manag 30(13)
20. Li F, Xufeng Y, Duan H-F (2019) Sustainable design of urban stormwater drainage systems by implementing detention tank and LID measures for flooding risk control and water quality management. Water Resour Manag 33(7)
21. Batica J (2015) Methodology for flood resilience assessment in urban environments and mitigation strategy development. Université Nice Sophia Antipolis

22. Islam A, Hassini S, El-Dakhakhni W (2021) A systematic bibliometric review of optimization and resilience within low impact development stormwater management practices. J Hydrol 599:126457
23. Liu C, Walling DE, He Y (2018) Review: the international sediment initiative case studies of sediment problems in river basins and their management. Int J Sediment Res 33:216–219
24. Goh XP, Radhakrishnan M, Zevenbergen C, Pathirana A (2017) Effectiveness of runoff control legislation and active, beautiful, clean (ABC) waters design features in Singapore. Water 9(8):627
25. PUB Singapore's National Water Agency (2024) Active beautiful clean waters design guidelines. Public Utilities Board, Singapore
26. Fletcher TD, Shuster W, Hunt WF, Ashley R (2014) SUDS, LID, BMPs, WSUD and more—the evolution and application of terminology surrounding urban drainage. Urban Water J 12(7):525–542
27. Yau WK, Radhakrishnan M, Liong S-Y, Zevenbergen C, Pathirana A (2017) Effectiveness of ABC waters design features for runoff quantity control in urban Singapore. Water 9(8):577
28. Chang CK, Ab Ghani A (2024) Quantifying the effects of sediment transport on river channel geometry and flood level—a case study of the December 2014 flood. Int J River Basin Manag 22(3):309–320
29. Kluang Municipal Council (MPK) (2024) Official Portal of Kluang Municipal Council. https://www.mpkluang.gov.my/en. Accessed 13 Oct 2024
30. City Population (2023) Kluang (District, Johor, Malaysia). https://www.citypopulation.de/en/malaysia/admin/johor/0103__kluang/. Accessed 17 Feb 2023
31. Climate-Data.org (2023) Climate Kluang. https://en.climate-data.org/asia/malaysia/johor/kluang-25956/. Accessed 6 April 2023
32. Info Banjir Johor (2023) Flood Information System. https://infobanjirjpsjohor.water.gov.my/. Accessed 03 April 2023
33. The Star (2023) First flood fatality recorded in Kluang. https://www.thestar.com.my/news/nation/2023/03/02/first-flood-fatality-recorded-in-kluang. Accessed 04 April 2023

Urban Climate Change and Citizens' Well-Being

Delineating Street Thermal Environments of Kunming, China: A Study Based on Mobile Measurement

Zhihao Wang, Ziyang Ma, Yifei Chen, Pengkun Zhu, and Lu Wang

Abstract With the acceleration of urbanization, the problem of the urban heat island effect has become increasingly prominent, seriously impacting on the quality of life and health of urban residents. This study used mobile observation method to measure the thermal environment of the street canyon in Kunming's Chenggong District. And select nine street samples with similar lengths to analyze the impact of street canyon orientation and aspect ratio on the thermal environment. Results show that in the morning, north–south oriented streets are about 1.5 °C warmer than east–west oriented streets with the same aspect ratio. In the afternoon high-temperature period, the north–south oriented streets are about 0.9 °C cooler than the east–west oriented ones, indicating better thermal comfort for north–south oriented street canyons compared to east–west oriented ones. Additionally, the relationship between the street aspect ratio and street temperature is associated with solar irradiation, showing a positive correlation in the morning and a negative correlation in the afternoon, with this correlation also being influenced by the street orientation. This study provides a scientific basis for livable urban planning and the optimization of urban microclimate environments.

Keywords Urban heat island · Street thermal environment · Mobile observation · Street orientation · Street aspect ratio

1 Introduction

With the rapid development of the global economy, the urbanization process in China has accelerated as well. China is one of the developing countries with the fastest urbanization processes. According to the 'China Urban Development Report (2022/2023)' released by the Chinese Academy of Social Sciences, the urbanization rate in China reached 66.16% in 2023, an increase of 0.94% from 2022, marking a new

Z. Wang · Z. Ma · Y. Chen · P. Zhu · L. Wang (✉)
Faculty of Civil Engineering and Mechanics, Kunming University of Science and Technology, Kunming, China
e-mail: wanglu_annie@163.com

B. He et al. (eds.), *Climate-Resilient, Low-Carbon, and Sustainable Cities*, Lecture Notes in Civil Engineering 754, https://doi.org/10.1007/978-981-95-2169-2_10

stage in China's urbanization process [1]. The acceleration of urbanization has led to an intensification of the urban heat island effect [2]. Severe urban heat islands not only affect the physical and mental health of residents but also damage the urban environment [3]. High urban temperatures increase the risk of illness among urban residents, raise the incidence of heart and respiratory diseases [4], and significantly increase the mortality and morbidity rates among urban residents [5]. Under the influence of the heat island effect, the form of atmospheric thermal circulation changes, to some extent, hindering the formation of droplets in the air and reducing urban rainfall [6]. In response to the many hazards brought by the urban heat island effect, a large number of researchers have conducted extensive studies on it. The concept of urban heat islands first proposed by Manley in 1958 [7]. In 1987, Oke defined the intensity of heat islands as the temperature difference between the city center and the suburbs [8]. Yang et al. also proposed in the early twenty-first century that the urban heat island phenomenon may be related to factors such as urban morphology, urban surfaces, air pollution, and population density [9]. Some scholars have used meteorological observation, remote sensing monitoring, and numerical simulation to study the formation mechanism, intensity changes, and evolution characteristics of urban heat islands, making significant progress in the study of urban heat island effects.

Street canyon, as the most representative basic urban surface units [10], are the backbone of urban structure and studying their thermal environment can promote understanding of urban heat island effects. Matzarakis et al. constructed the RayMan model to simulate the thermal comfort feedback of pedestrians in different urban block forms, discovering that the morphology of street canyon has great potential to improve pedestrian thermal comfort [11]. Zhao and Liu utilized a street canyon dynamic model for numerical simulation of typical street canyon, and through extensive numerical analysis, compared the impact of various street canyons design factors on their thermal environment. The simulation results quantitatively revealed the influence and extent of each factor on the thermal environment of street canyon [12]. Zhang and Du conducted orthogonal experimental numerical simulation of the thermal environment in Guangzhou's street canyons, and the results indicated that the geometric form and greening of street canyons significantly affect their summer thermal environment. The orientation, aspect ratio, green coverage rate, and ground paving materials of street canyons are all key factors influencing their thermal environment [13]. Elbondira, through field measurement comparisons of microclimate data from 26 street canyons, pointed out that the morphology and greening of street canyon jointly affect the microclimate. In historic low-density residential areas, street trees have a greater impact on the microclimate, while in modern high-density areas, the impact of vegetation on the microclimate is limited, and the aspect ratio and orientation of street canyon become key factors affecting the microclimate [14]. Jin conducted field measurement research on the winter thermal environment of typical residential street canyons in Harbin's old city, and the results showed that street canyons with different orientations have significant temperature differences. The black globe temperature is greatly affected by the continuity of street canyon facades, and the aspect ratio of street canyons can change the temperature and influence the trend of temperature changes [15]. However, studies by various scholars have indicated that the thermal environment varies across different latitudes and longitudes throughout the day. As a low-latitude plateau city in China, Kunming may exhibit distinct characteristics in its thermal environment compared to other cities. Therefore, this paper will focus on Kunming as the target city to investigate the thermal environment changes in street canyons of low-latitude plateau cities.

Research on the thermal environment of urban blocks is beneficial for conducting fundamental analysis of urban heat island effects at the micro-scale. Compared to numerical simulation methods, field measurement methods offer stronger reliability and are suitable for various complex environments. In light of this, this study employs the mobile observation method to record meteorological data for the target streets within the research area, exploring the impact of different factors on the thermal environment of street canyon.

2 Overview of Research Region

Kunming is located in the central part of the Yunnan-Guizhou Plateau in China, between 102°10′–103°40′ East longitude and 24°23′–26°22′ North latitude. It has a subtropical plateau monsoon climate at a low latitude, with a mild climate and an average annual temperature of approximately 15 °C, with an average annual rainfall of about 1000 mm. The study area of this paper is located in Chenggong District, Kunming City. Chenggong District has diverse topography with an elevation of about 1775 m, belonging to a low-latitude plateau region, and the terrain shows a general trend of being higher in the east and lower in the west. The total administrative area of Chenggong District is 510.2 km^2, with a permanent population of about 694,700 people. The development of the economy and the influx of population have led to the continuous expansion of the urban heat island area, and there is a trend of the heat island center continuously moving southward (Fig. 1).

3 Method and Data Processing

Mobile observation involves the use of a mobile platform equipped with meteorological data recording instruments to record meteorological data within the target area. Concurrently, climate data recorded by local small-scale climate stations are utilized. By employing the simultaneous revision method, the data recorded during the mobile observation experiment is revised to the same moment, thereby representing the near-surface air temperature at different locations along the mobile observation route at the same instant [16].

Kunming is a city with distinct wet and dry seasons. The rainy season in Kunming lasts from April to October each year, while the dry season starts in November and continues until March of the following year. In November, the climate in Kunming is

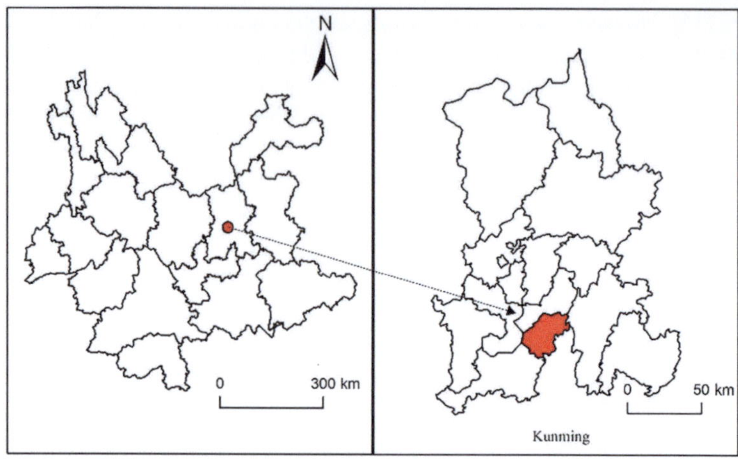

Fig. 1 Overview of research region (self-drawn)

relatively stable, with clear weather and moderate temperatures, which are suitable for conducting outdoor thermal environment measurements. The experiment conducted on November 26th, 2022, under clear skies with an average daily temperature of 18.2 °C. The experiment carried out in five time slots: from 9:00 to 10:00, 11:00 to 12:00, 15:00 to 16:00, 19:00 to 20:00, and 22:00 to 23:00. During the experiment, GPS equipment used to record the observation path, which served as a basis for subsequent data processing and mapping. The mobile observation route started at the intersection of Jingming Road and Lianda Street, proceeded clockwise through Lianda Street, Xingcheng Road, Huanhu Road, Shilong Road, Caiyun Road, Jinxiu Road, and Jingming Road, and stopped at the starting position to form a closed loop, with a total distance of 18.4 km. Based on Baidu Map data, the mobile observation route is illustrated in Fig. 2.

The meteorological data recording instruments and GPS devices have a sampling frequency of 1 s. Due to the non-simultaneous nature of the meteorological data recorded during the mobile observation experiment, and the fact that air temperature tends to increase or decrease linearly over a period of time, it is necessary to normalize the experimental data through the establishment of a mathematical regression equation to ensure they share the same temporal dimension. This paper utilizes the simultaneous revision method employed in the team's previous research [17] to process the subsequent experimental data.

The meteorological data, after simultaneous revision, imported into ArcMap10.8 software. Based on the path data recorded by the GPS device, establish a buffer zone of 60 m. By using the Kriging interpolation method, insert the meteorological data into the buffer zone and the adjust colors to obtain a color band representing the temperature values. Comparing the colors of the color band allows for a more intuitive comparison of temperature changes. The temperature color band maps for the five different time slots of this experiment are shown in Fig. 3.

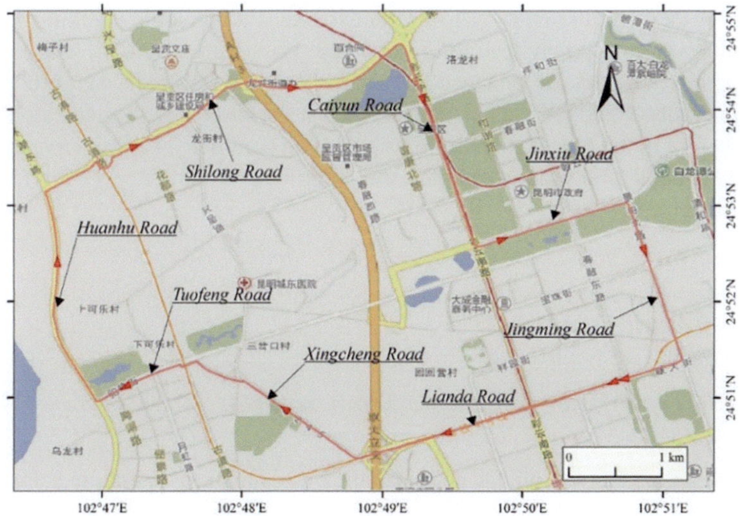

Fig. 2 Roadmap of mobile observation. *Map source* Baidu map

Fig. 3 Temperature stripe chart at five different moments. *Map source* Baidu map

As depicted in Fig. 3, the temperature fluctuations throughout the day across the eight observed urban streets exhibit significant variability. Such variability is potentially attributable to a range of factors, including geographical location, street geometry, and the surrounding environment. The influence of these factors on temperature can alter with changes in time or direction. To elucidate the correlation between specific factors and ambient temperature, it is imperative that subsequent studies conduct both qualitative and quantitative analyses.

Due to the potential existence of various street morphologies within the same street, this variability may lead to errors when conducting quantitative analysis of the experimental data. To enhance the accuracy of the results in data processing, select intervals of 500–600 m in length on each street based on three indicators: road width, height of the blocks on both sides, and road continuity. Each interval set as a sample. Select a total of nine samples, and their locations and street overviews are shown in Table 1.

4 Result and Analysis

4.1 Characteristics of Sample Temperature

Extracted temperatures from the corresponding areas in the temperature color bands of Fig. 3 for five different times based on the sample locations and averaged the temperature data at these specific time points to obtain a comprehensive temperature indicator. Based on the temperature indicators of different samples, construct a temperature box plot to visually display the temperature fluctuations throughout the day. The temperature box plot is shown in Fig. 4, and ΔT means deviation between maximum and minimum temperature in this figure.

As shown in Fig. 4, samples I, II, and III exhibit relatively small temperature variations across the five observation experiments, with daily temperature differences all less than 5 °C. Among them, sample III has the lowest temperature of 15.6 °C and the highest temperature of 18.9 °C, resulting in a temperature variation of 3.3 °C, making it the street with the smallest temperature variation in the study area. In contrast, samples V, VI, VII, VIII, and IX show larger temperature variations, with daily temperature differences all greater than 5 °C. Sample IX has the lowest temperature of 14.9 °C and the highest temperature of 23.0 °C, with a temperature variation of 8.1 °C, which is the largest temperature variation among the streets in the study area. Samples VII and VIII have the lowest temperature of 14.3 °C, which is the lowest among all samples, while sample IX has the highest temperature of 23.0 °C, the highest among all samples. Overall, samples II and III have smaller daily temperature variations and tend to be more stable, whereas samples IX and X have more prominent peaks in their highest temperatures, with more significant temperature variations within the same time frame compared to other sections.

4.2 Analysis of Influencing Factors

Street orientation. The orientation of street canyon is a key component of their geometric form and should not be overlooked in terms of its impact on the thermal environment of street canyons. Street canyons are primarily divided into two types:

Table 1 Sample location and corresponding street overview. *Map source* Baidu map

Samples	Location	Overview
I	Sample I	
II	Sample II	
III	Sample III	
IV	Sample IV	

(continued)

Table 1 (continued)

Samples	Location	Overview
V	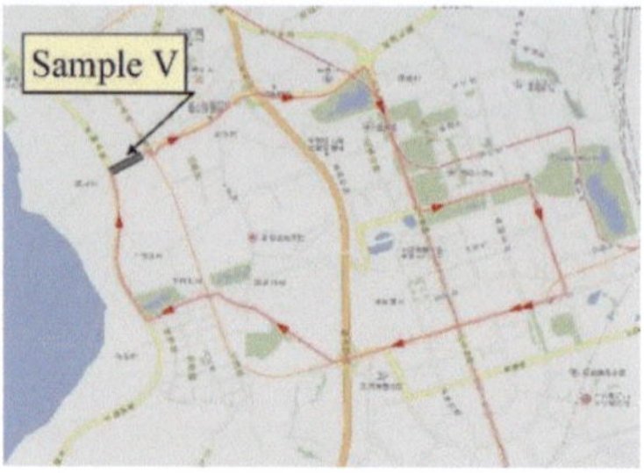 Sample V	
VI	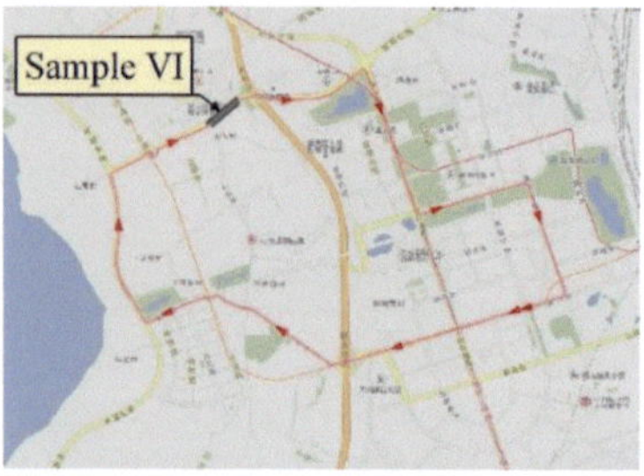 Sample VI	
VII	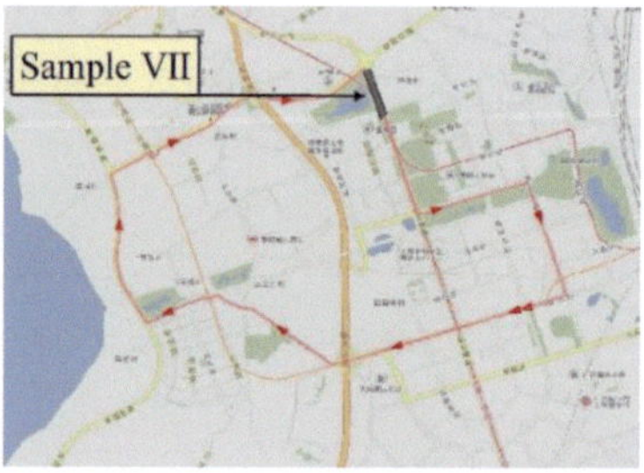 Sample VII	

(continued)

Table 1 (continued)

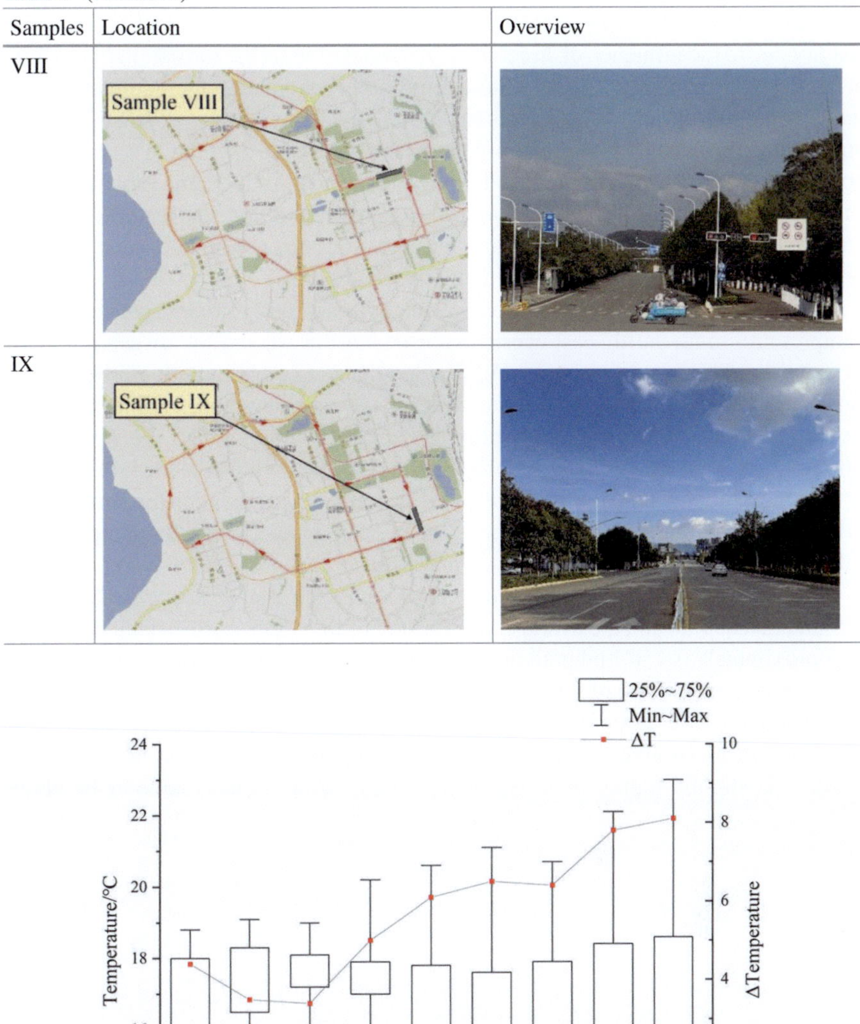

Samples	Location	Overview
VIII		
IX		

Fig. 4 Temperature box plot of nine samples (self-drawn)

south–north and east–west orientations. This difference in orientation can lead to significant variations in near-surface temperature performance and appearance, mainly due to the influence of street canyons orientation on the formation of daytime shadows on the ground [18] Pearlmutter and other scholars, in their study of street canyons climate in the Negev Highlands of Israel, found that east–west oriented street canyons receive more heat for pedestrians during the daytime in summer than south–north oriented street canyons [19]. Similarly, Ali-Toudert and other scholars, in their research on street canyons in arid regions, found that south–north oriented street canyons have a better thermal environment compared to east–west oriented ones, which are more difficult to mitigate thermally [20].

Based on the different orientations of the experimental samples, categorize the samples into south–north, east–west, southeast–northwest, and southwest–northeast orientations. Specifically, samples IV, VII, and IX belong to the south–north orientation, samples I, III, V, and VIII belong to the east–west orientation, sample II belongs to the southeast–northwest orientation, and sample VI belongs to the southwest–northeast orientation. The average temperatures of samples with the same orientation at the same time were used to calculate the mean temperatures for the four groups at different time intervals. The variation of average temperatures for samples with different orientations over time is shown in Fig. 5.

Comparing the south–north oriented samples with the east–west oriented samples, it found that at 9:30 am, the average temperature of the south–north oriented samples is approximately 0.4 °C higher than that of the east–west oriented samples. At 15:30, the average temperature of the south–north oriented samples is about 1.1 °C higher than that of the east–west oriented samples. At 11:30 and 19:30, the average temperatures of both are similar. At 22:30, the average temperature of the east–west oriented samples is slightly higher than that of the south–north oriented samples by about

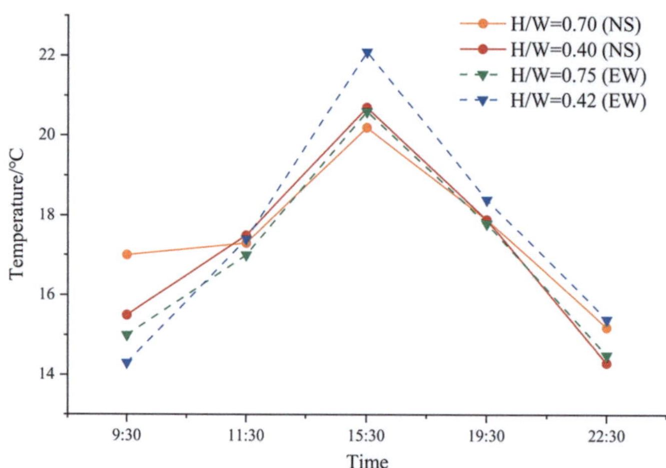

Fig. 5 The variation of average temperatures for samples with different orientations at different times (self-drawn)

0.2 °C. When comparing the southeast–northwest oriented samples with the southwest–northeast oriented samples, it observed that there are significant differences in temperature performance throughout the day: at 9:30, the southeast–northwest oriented samples had temperatures about 1.7 °C higher than the southwest–northeast oriented samples. From 11:30 to 15:30, the temperatures of the southwest–northeast oriented samples rose and exceeded those of the southeast–northwest oriented samples. From 19:30 to 22:30, the temperatures of the southwest–northeast oriented samples decreased and once again became lower than those of the southeast–northwest oriented samples. The temperature variations of the four different oriented samples throughout the day exhibit a pattern of fluctuation, and the rate of temperature change also differs at various times, which is related to the change in the solar azimuth angle throughout the day.

In general, this research found that the heat gain patterns of southeast–northwest and northeast–southwest oriented samples throughout the day align with the conclusions drawn by Ali-Toudert and others in previous studies, suggesting that southeast–northwest oriented street canyons receive more solar radiation in the morning than northeast–southwest oriented street canyons. In the afternoon, the northeast–southwest oriented street canyons receive more solar radiation, resulting in higher ambient temperatures. The south–north oriented samples exhibit higher overall daily temperatures than the east–west oriented samples. Contrary to the aforementioned conclusions, the east–west oriented samples have a better thermal environment compared to the south–north oriented samples. This discrepancy may be attributed to the significant differences in the aspect ratios of the street samples, as found in other scholars' research on geometric form. Further analysis based on the aspect ratio will delve into the impact of street morphology on the thermal environment of street canyons.

Aspect ratio. The street aspect ratio (H/W) is an important metric for describing the geometric form of urban streets, and variations in this ratio are closely related to changes in the thermal environment of street canyons. The aspect ratio is the most relevant urban parameter leading to microclimatic changes within street canyons, affecting the thermal environment by influencing horizontal airflow potential and solar irradiation in street canyons [21]. The aspect ratio significantly impacts the amount of solar energy received by street canyons, with the ground receiving more solar radiation compared to vertical surfaces. Research indicates that street canyons with a low aspect ratio ($H/W \leq 1.5$) have slightly higher near-surface air temperatures during the daytime in summer, while nighttime temperatures are similar, whereas street canyons with a high aspect ratio ($H/W \geq 2.5$) have lower daytime temperatures and higher nighttime temperatures in summer [22, 23].

Among the nine samples selected in this study, there are no large-scale buildings lining the streets continuously along the road, but there is a significant presence of dense, tall crown trees with relatively uniform heights distributed on both sides of the road. This vegetation configuration has a certain impact on the road surface's reception of solar radiation. Therefore, this study defines the aspect ratio H/W as the ratio of the average height of the tree canopy on both sides to the width of the traffic street, which serves as the basis for further research. The street canyons morphological data of the study samples are shown in Table 2.

Based on the different orientations of the streets, this study compares the temperature variations of south–north oriented samples and east–west oriented samples under various aspect ratios.

As shown in Fig. 6, among the three south–north oriented samples, there is a clear phenomenon of temperature change with aspect ratio at 9:30 and 15:30. In the morning, samples with a larger aspect ratio have higher temperatures than those with a smaller aspect ratio, indicating a clear positive correlation between street temperature and aspect ratio at this time. In the afternoon, the temperature increase in samples with a larger aspect ratio is less than that in samples with a smaller aspect ratio, indicating a clear negative correlation between sample temperature and aspect ratio (Fig. 7).

Among the four east–west oriented samples, the trend of temperature change is correlated with the aspect ratio in the same way as the south–north oriented samples, showing a clear relationship at 9:30 and 15:30. The combined results from both the south–north and east–west oriented samples indicate that increasing the street aspect ratio is beneficial for raising the environmental temperature in the morning and for

Table 2 Street canyons morphological data of nine samples. *Date source* field measurement

	I	II	III	IV	V	VI	VII	VIII	IX
H (m)	10.0	7.5	8.0	8.0	6.0	12.0	8.0	8.0	9.0
W (m)	17.5	8.0	8.0	11.5	8.0	17.5	20.0	24.0	35.0
H/W	0.57	0.94	1.00	0.70	0.75	0.69	0.40	0.42	0.26

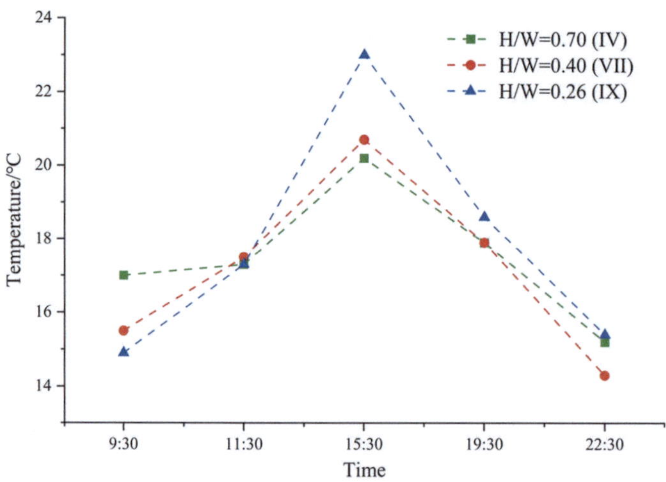

Fig. 6 Temperature variations of samples with different aspect ratios in south–north oriented street canyons (self-drawn)

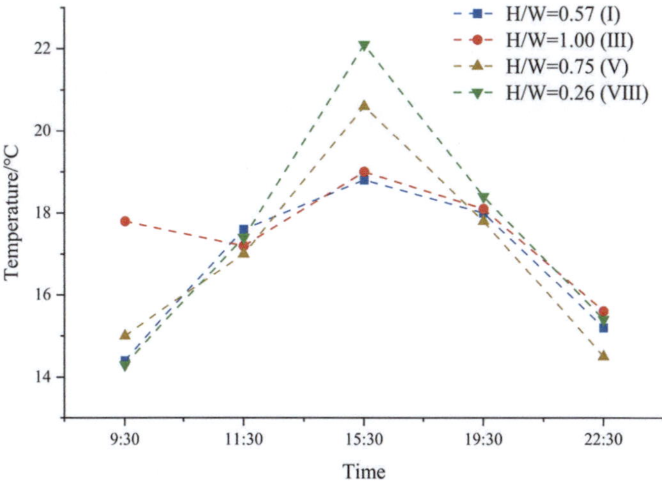

Fig. 7 Temperature variations of samples with different aspect ratios in east–west oriented street canyons (self-drawn)

mitigating high temperatures in the afternoon. Overall, streets with a larger aspect ratio exhibit more stable temperature fluctuations throughout the day. Additionally, comparing streets with similar aspect ratios but different orientations reveal that during the low-temperature period in the morning, the average temperature of north–south oriented streets is approximately 1.5 °C higher than that of east–west oriented streets. Conversely, during the high-temperature period in the afternoon, the average temperature of north–south oriented streets is about 0.9 °C lower than that of east–west oriented streets. These findings indicate that north–south oriented streets exhibit superior thermal comfort throughout the day compared to east–west oriented streets, corroborating the conclusions of other scholars.

In the study by Ruzana et al., it found that increasing the height of street trees in east–west oriented streets yields greater benefits for improving the thermal environment of street canyons than in south–north oriented streets [24]. To explore the benefits of improving the aspect ratio in streets with different orientations in this experiment, sample IV (H/W = 0.70) and sample VII (H/W = 0.40) from the south–north oriented samples, and sample V (H/W = 0.75) and sample VIII (H/W = 0.42) from the east–west oriented samples, were selected for comparison to analyze the temperature changes in different oriented streets when increasing the aspect ratio value.

As shown in Fig. 8, when the street aspect ratio is increased by approximately 75%, at 9:30, the temperature of the south–north oriented samples rises by 1.5 °C, while the temperature of the east–west oriented samples rises by 0.7 °C, with the temperature change in the south–north oriented samples being significantly greater than that in the east–west oriented samples. At 11:30, there is no significant difference between the two groups. From 15:30 to 19:30, the temperature of the south–north oriented

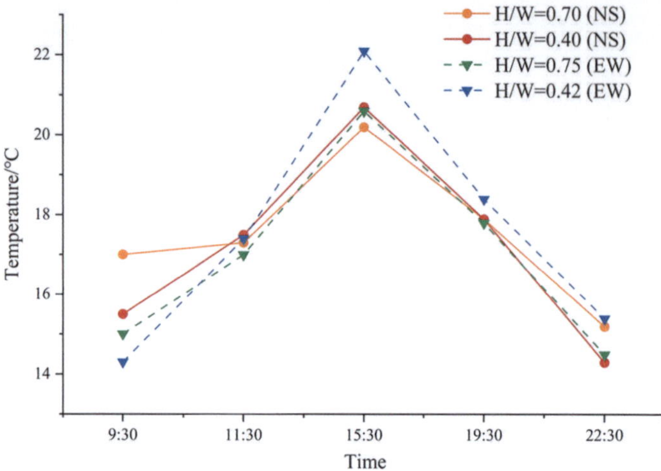

Fig. 8 The temperature change of similar aspect ratio changes on streets with different orientations (self-drawn)

samples decreases by 0.5 °C, while the temperature of the east–west oriented samples decreases by 1.5 °C, with the east–west oriented samples showing a more significant cooling effect than the south–north oriented streets. At 22:30, the temperature of the south–north oriented samples increases by about 0.9 °C, while the temperature of the east–west oriented samples decreases by about 0.9 °C, with the two groups showing opposite temperature changes.

In summary, the impact of street aspect ratio on temperature varies depending on the orientation of the street canyons, and the extent of this impact also differs at different times of the day. At 9:30, there is a positive correlation between street aspect ratio and temperature, with the south–north oriented streets experiencing greater warming benefits from an increase in aspect ratio than the east–west oriented streets. From 15:30 to 19:30, there is a negative correlation between street aspect ratio and temperature, with the east–west oriented streets experiencing greater cooling benefits from an increase in aspect ratio than the south–north oriented streets. Therefore, increasing the height of tree canopies in south–north oriented streets can effectively mitigate the environmental low temperatures caused by heat loss at night, while for east–west oriented streets, increasing the height of tree canopies can effectively alleviate the high temperatures caused by solar radiation in the afternoon, and this cooling benefit from increasing the height on both sides of the street is greater than that for south–north oriented streets.

5 Discussion

When planning and designing cities, the impact of multiple factors such as street canyons orientation and aspect ratio on the thermal environment should not be overlooked. East–west oriented streets benefit more from the cooling effect of increased the height on the sides, so planting tall crown layer trees in east–west oriented streets is of great significance for improving the thermal environment of street canyons. Increasing the street aspect ratio demonstrates benefits for environmental temperature improvements across different times of the day, however, this study's data do not specifically substantiate these findings, indicating a need for further investigation into the underlying causes of these phenomena. Overall, measures such as increasing the height of obstructions on both sides of street canyons and narrowing the width of street canyons can effectively prevent the environmental thermal comfort of street canyons, which has high reference value for urban planning.

6 Conclusion

This study, through mobile observation and data analysis of nine street samples in Chenggong District, Kunming City, has thoroughly investigated the impact of street canyons orientation and street aspect ratio on the thermal environment of street canyons. The findings reveal that:

1. Under the same aspect ratio conditions, south–north oriented street canyons perform better in terms of thermal comfort compared to east–west oriented street canyons, especially during the afternoon when the temperature in south–north oriented streets is 0.9 °C lower than that in east–west oriented streets. This suggests that changes in street canyons orientation can significantly affect the thermal environment, and south–north oriented street canyons may have an advantage in mitigating the urban heat island effect.
2. In the morning, there is a positive correlation between street temperature and aspect ratio, which is more pronounced in south–north oriented streets, with a specific manifestation that for every 75% increase in street aspect ratio, street temperature rises by about 1.5 °C. In the afternoon, there is a negative correlation between street temperature and aspect ratio, which is more pronounced in east–west oriented streets, with a specific manifestation that for every 78% increase in street aspect ratio, street temperature decreases by about 1.5 °C.

Acknowledgements This study is supported by the Science and Technology Projects of Yunnan Province's Universities Serving Key Industries (FWCY-BSPY2024061).

References

1. Shao Y (2022) China urban development report no. 21. China City Press, Beijing
2. Ren Y, Ren G, Zhang A (2010) A review of the impact of urbanization on the change trend of surface temperature. Prog Geogr 29(11):1301–1310
3. Li W, Tang J, Gao X et al (2021) Analysis of heat island characteristics and changes in the main urban area of Shijiazhuang. Sci Technol Eng 21(27):11858–11867
4. Giuseppe M et al (2017) Pattern and determinants of hospitalization during heat waves: an ecologic study. BMC Public Health. https://doi.org/10.1186/1471-2458-7-200
5. Feng L, Li X (2016) Research progress on effects of heat wave on human health. J Environ Health 33(2):182–188
6. Xu X et al (2021) Relationship between urban heat island and climate characteristics change in Kunming city. Bull Sci Technol 37(12):101–107
7. Manley G (1958) On the frequency of snowfall in metropolitan England. Q J R Meteorol Soc 84(359):70–72
8. Oke TR (1987) Boundary layer climate. Methuen, London
9. Yang S (1991) Research on urban ecological environment. Sun Yat-Sen University Press, Guangzhou
10. Nunez M, Oke TR (1977) The energy balance of an urban canyon. J Appl Meteorol 16:11–19
11. Matzarakis A, Rutz F, Mayer H (2007) Modelling radiation fluxes in simple and complex environments—application of the RayMan. Int J Biometeorol 51:323–334
12. Zhao J, Liu J (2007) Numerical simulation of urban street valley thermal environment and planning and design countermeasures. Arch J (03):37–39
13. Du X, Shi Y, Zhang Y (2015) Numerical simulation and design of thermal environment in residential streets and valleys in Guangzhou. Build Sci 31(12):78–87
14. Elbondira TA, Ibrahim M, Fujii M (2019) The impact of urban greening and urban geometry on the microclimate at the neighborhood level in hot arid climates. In: 2019 Advances in science and engineering technology international conferences (ASET)
15. Jin Y (2016) Research on street space design of enclosed residential areas in severe cold areas based on thermo-acoustic environment. Harbin Institute of Technology
16. Li H (2022) Research on urban thermal and humid environment management based on mobile observation—a case study of Kunming city. Kunming University of Science and Technology
17. Zhou Y (2020) Study on thermal and humid environment characteristics of Chenggong new town area, Kunming city. Kunming University of Science and Technology
18. Zhang Y (2016) Energy balance and street valley microclimate. Build Sci 32(10):096
19. Pearlmutter D, Bitan A, Berliner P (1999) Microclimatic analysis of 'compact' urban canyons in an arid zone. Atmos Environ 33(24/25):41434150
20. Ali-Toudert F, Mayer H (2006) Numerical study on the effects of aspect ratio and orientation of an urban street canyon on outdoor thermal comfort in hot and dry climate. Build Environ 41(2):94108
21. Yoshida A, Tominaga K, Watani S (1990) Field measurements on energy balance of an urban canyon in the summer season. Energy Build (15/16):417–423
22. Nakamura Y, Oke TR (1988) Wind, temperature and stability conditions in an east-west oriented urban canyon. Atmos Environ 22(12):26912700
23. Coronel JF, Alvarez S (2001) Experimental work and analysis of confined urban spaces. Sol Energy 70(3):263–273
24. Sanusi R, Johnstone D, May P (2015) Street orientation and side of the street greatly influence the microclimatic benefits street trees can provide in summer. J Environ Qual 01:0039

Thermal Perception and Comfort of Pedestrians in Summer: A Study of Street Canyons in Chongqing, China

Siqi Wang and Shuang Liang

Abstract With global warming, it is particularly important to study pedestrian thermal comfort in cities with high humidity and heat. This research was carried out to assess street thermal conditions and how pedestrians perceive them, with the ultimate aim of improving urban thermal environments. A total of four conditions, i.e., different gender, different age, different SVF and different street orientation conditions, were investigated to assess the contribution of thermodynamic parameters to subjective thermal evaluation. The results showed that the neutral range of thermal comfort for pedestrians in summer was smaller than the range of thermally neutral perceptions. The difference between thermal sensation and subjective comfort of pedestrians of different genders was small, while the difference was larger for pedestrians of different ages. High temperatures made men more uncomfortable compared to women. As age increases, a reduction in pedestrians' thermal perceptual acuity occurs with environmental variation. The effect of air velocity on pedestrian heat perception was strongest on EN-WS oriented streets and weakest on E-W oriented streets. In summary, this study aims to reveal how thermal equivalent temperature elements act on pedestrian subjective thermal sensation and thermal comfort evaluation under different conditions, which can help to intervention methods to ameliorate human thermal experience in unenclosed metropolitan zones.

Keywords Thermal comfort · Thermal sensation · Thermal equivalent temperature element · Urban geometry

S. Wang · S. Liang (✉)
School of Civil Engineering and Architecture, Southwest University of Science and Technology, Mianyang, China
e-mail: 1793230049@qq.com

S. Wang
e-mail: wsq@mails.swust.edu.cn

B. He et al. (eds.), *Climate-Resilient, Low-Carbon, and Sustainable Cities*, Lecture Notes in Civil Engineering 754, https://doi.org/10.1007/978-981-95-2169-2_11

1 Introduction

Under the dual the increasing stress induced by global climate change and population explosion, the urban heat island effect (UHI) is becoming increasingly significant [21, 26], and thermal extremes act as catalysts for both heat-induced pathophysiology and cardiocirculatory disorders [30]. With growing urbanization pressures, combating heat island effects and optimizing outdoor thermal comfort have become imperative for sustainable community development.

As a psychological construct, outdoor thermal comfort (OTC) measures people's thermal sensation and satisfaction in open-air spaces [22]. Lai et al. [16] proposed that the microclimatic characteristics as well as physiological and psychological factors affect OTC [3, 23]. According to statistics, there are about 165 metrics for assessing thermal comfort, and OTC evaluation predominantly utilizes the physiological equivalent temperature (PET) as its principal assessment framework [23]. In addition, thermal equivalent temperature is an index that integrates several factors to judge thermal comfort, and its parameters mainly include air temperature (T_a), globe temperature (T_g), mean radiant temperature (T_{mrt}), air velocity (V_a), Relative Humidity (RH), human activity level, and clothing thermal resistance.

The role of Ta is paramount in governing pedestrian thermal comfort in exterior settings [2]. Lai et al. [15] found the highest correlation the association of atmospheric thermal conditions with occupant comfort levels in environmental ergonomics studies in a Tianjin park. Secondly, urban spatial morphology affects airflow and wind speed to some extent, which in turn affects thermal comfort [7, 16]. Roshan et al. [25] simulated the ventilation cooling potential across four distinct climatic regions, and found that as a natural cooling agent, wind flow may ameliorate thermal discomfort in outdoor settings. Chow et al. [10] found that pedestrians in forested parks in the highly humid and hot urban city of Singapore, in environments of high humidity, usually possessed a less than comfortable experience.

Pedestrians tend to actively adapt to the thermal environment. Pedestrians are willing to choose to sit and stay when the street thermal environment is comfortable for them, on the contrary, when pedestrians feel uncomfortable they choose to walk faster and head to sheltered areas. Elnabawi et al. [13] found that increasing or decreasing clothing or changing clothing materials were used as adaptive behaviors to enhance pedestrians' thermal comfort. However the role of clothing in influencing OTC was limited, this study showed that almost 15% of males chose to wear less clothing while less than 2.0% of females chose to wear less clothing.

Urban geometry is significant in the study of OTC. One of the measures of geometric morphology is the sky view factor (SVF), which was first proposed by Oke [8]. Chiang et al. [9] found that in the summer season, through deep learning-based analysis, SVF was found to exert a substantial positive influence on PET values and subjective thermal perception. Aboelata [1] found a significant correlation between SVF and thermal microenvironment such as T_a and RH on ES-WN streets. Arias-Rosales and LeDuc [5] noted that pedestrians are exposed to more intense solar radiation in street canyons with higher SVF values. Furthermore, Sharmin et al. [27]

argued that street canyon orientation can regulate thermal environments by modifying wind speed and solar radiation exposure, thereby influencing pedestrian comfort.

In summary, T_g, T_a, V_a, RH, activity level, and clothing heat resistance are factors that affect thermal sensation vote (TSV) and thermal comfort vote (TCV) of pedestrians, and the strength of their effects are highly related to individual characteristics and spatial patterns. Most previous studies have focused on how thermal environment parameters in a single environment act on thermal comfort, with little consideration of differences in pedestrian thermal perception and thermal comfort due to individual differences and changes in spatial patterns. Therefore, the present research is designed to: (1) To investigate the differences between pedestrian thermal perception voting and thermal comfort voting; (2) Assess age- and gender-based differences in pedestrians' thermal responses (TSV/TCV) to thermal equivalent temperature factors; (3) Examine the moderating effect of street morphology indicators (SVF and orientation) on thermal equivalent temperature—TSV/TCV associations.

2 Methodology

2.1 Study Area

Chongqing municipality spans latitudes 28°10′N to 32°13′N and longitudes 105°11′E to 110°11′E (Fig. 1). It is widely recognized as one of China's "Four Furnace Cities", distinguished by the greater intensity and prolonged duration of its summer heat compared to the other three. According to the information statistics of the Meteorological Bureau, the occurrence of elevated temperature days above 37 °C in the center of the city is about 15–18 days during the whole year. Temperature mean value in Chongqing Municipality exceeds 20 °C from June to August (Fig. 2).

2.2 Site Description

The scholarly focus in thermal comfort has traditionally been on building interiors [29, 32], campuses [33, 35], and parks [18, 31, 37], the commercial street has received disproportionately limited attention. This study selected three representative commercial streets as research subjects: Shapingba (SR), Guanyinqiao (GT), and Jiefangbei (JR) (Fig. 3). These streets exhibit diversified architectural configurations, well-developed amenities, and comprehensive greening systems. Field investigations revealed their characteristic high-frequency usage patterns and stable pedestrian flows. As vital urban activity hubs, their composite microclimatic mechanisms—incorporating street orientation, building density, vegetation coverage and other multifactorial elements—provide exemplary cases for thermal environment research, thus constituting the empirical study area.

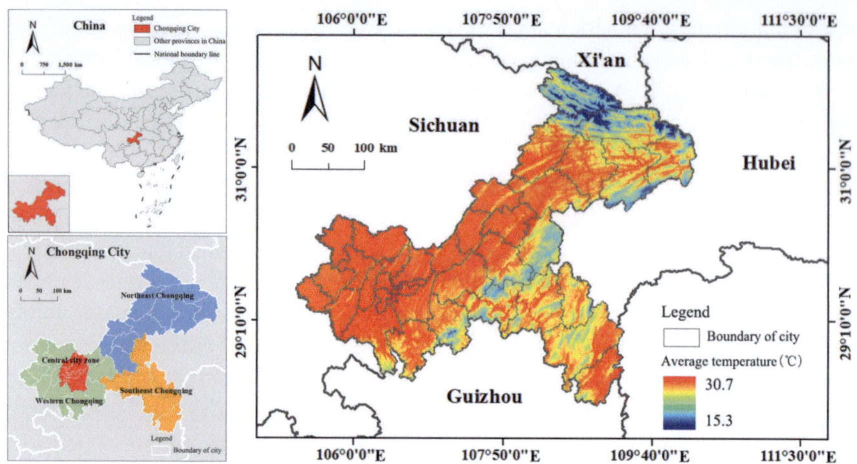

Fig. 1 Location of the study area (self-drawn)

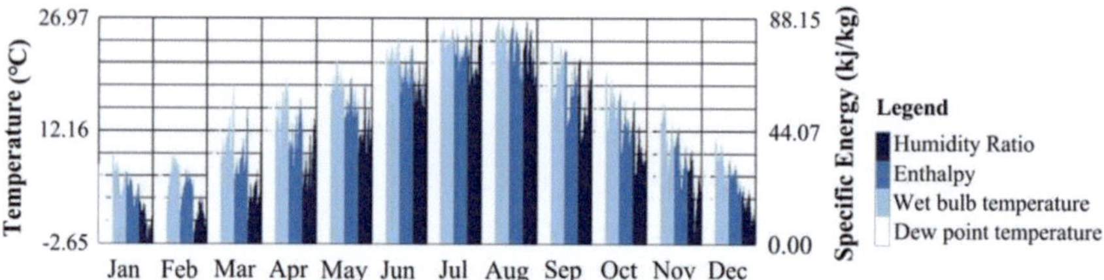

Fig. 2 Temperature and precipitation monthly means in Chongqing (2023) (self-drawn)

2.3 *Field Measurements*

This study measured thermal environment parameters at 1.5 m above ground level using a Testo 400 instrument (Fig. 4, Table. 1), including T_a, T_g, V_a and RH [11, 24]. Measurements were conducted daily from 8:00 to 12:00 and 14:00 to 20:00, with recordings taken at 6-min intervals. In order to minimize the impact of the experiment due to climate change, all measurements were conducted exclusively on clear, cloudless days with similar meteorological conditions during July, spanning a total measurement period of nine days. The study used Rayman pro v3.0 to calculate pedestrian PET, which requires T_a, T_{mrt}, V_a, RH, and time zone, date, and standard parameters of the human body to be entered into it, and the following is the formula for calculating T_{mrt}:

$$MRT = \left[\left(T_g + 275.15\right)^4 + \frac{1.1 \times 10^8 \cdot v^{0.6}}{\varepsilon \cdot D^{0.4}} \left(T_g - T_a\right) \right]^{1/4} - 273.1 \quad (1)$$

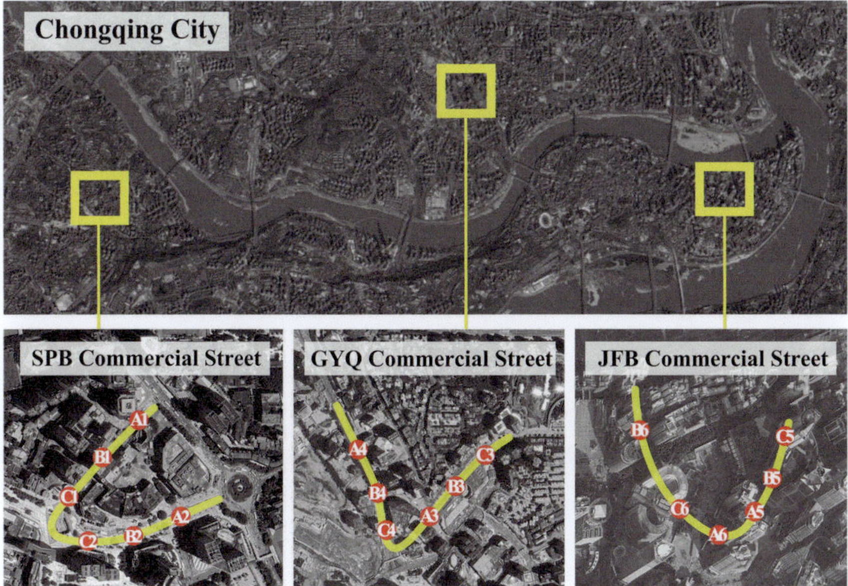

Fig. 3 Location of streets and photos of measurement points (self-drawn)

where T_a is the air temperature (°C); T_g is the black sphere temperature (°C); V_a is the wind speed (m/s); D is the diameter of the sphere (0.15 m); ε is the absorption rate of the earth (0.95).

2.4 *Physical Environmental Parameters*

Pedestrian comfort is significantly influenced by street spatial morphology. The SVF, a quantitative indicator of spatial geometry, is mathematically expressed as the hemispherical projection ratio between visible sky dome area and the total celestial sphere [16], with values ranging from 0 to 1 [17]. A lower SVF value may reduce exposure to solar thermal radiation. Fisheye photography was used to obtain the images, which were processed by Photoshop and finally entered into winScanopy to calculate SVF (Fig. 5). Additionally, street orientation critically determines solar radiation and wind speed within urban canyons [18]. Therefore, this study selects SVF and orientation as key factors for in-depth analysis.

The study identified 18 measurement points with significant SVF variations across three commercial streets, which were classified into three categories: semi-enclosed [0, 0.3) (A1–A6), semi-open [0.3, 0.5) (B1–B6), and open [0.5, 1] (C1–C6). Street orientation analysis revealed that Shapingba (SR) features EN-WS (northeast-southwest) and E-W (east–west) orientations, while both Guanyinqiao (GR) and

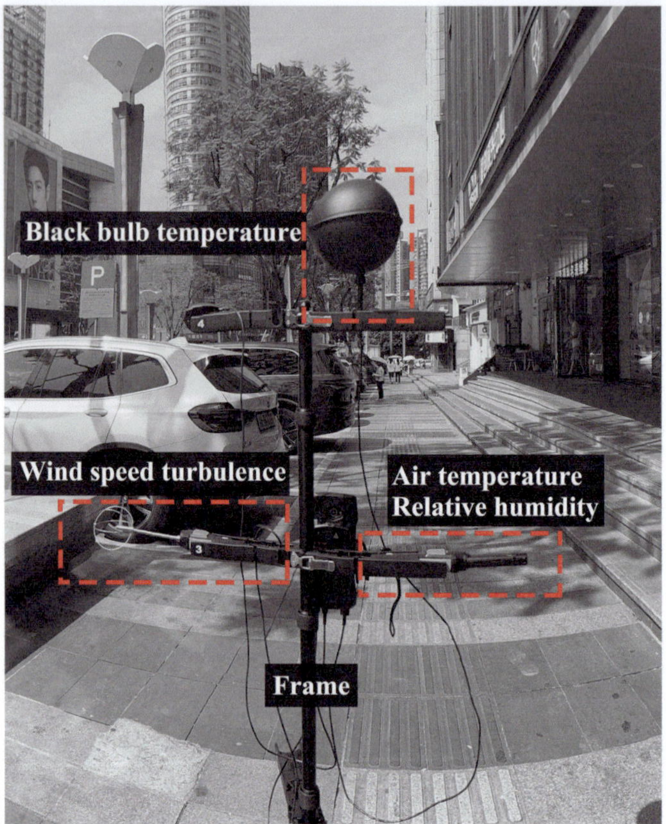

Fig. 4 Schematic diagram of the testo 400 instrument (self-drawn)

Jiefangbei (JR) exhibit EN-WS and WN-ES (northwest-southeast) orientations (Table 2).

2.5 Questionnaire Survey

The questionnaire research is shown in Fig. 6. The interview subjects are local residents of Chongqing. Initial data collection focused on capturing pedestrian attributes: biological sex, age range, physical activity levels, and apparel thermal properties; the second is the real perception vote of the interviewed pedestrians on the street heat environment, including TSV, TCV. all the perception votes are accurate to 0.1. 1,716 questionnaires were collected in this research.

Table 1 Introduction of instrument parameters (self-drawn)

Instrument	Variables	Coverage	Resolution	Response time (min)
Testo 400	Air temperature	0–50 °C	±5 °C	1
	Globe temperature	0–120 °C	±(0.3°C + 0.1%)	1
	Air velocity	0–5 m/s	±(0.03 m/s + 4%)	1
	Relative humidity	5–95%	±3%,RH(10–35%RH); ±2%,RH(35–65%RH); ±3%,RH (65–90%RH)	1

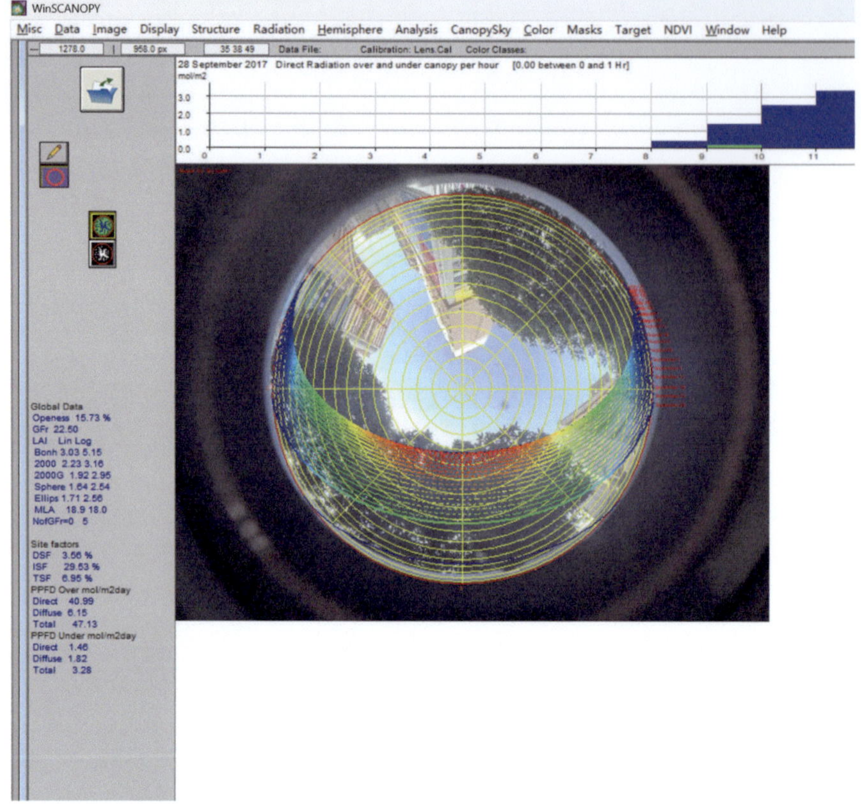

Fig. 5 Calculation of SVF using WinScanopy 2017a (self-drawn)

Table 2 Description of street orientation and points' SVF (self-drawn)

Street	Shapingba commercial street					
Sample points	A1	B1	C1	A2	B2	C2
SVF	0.226	0.316	0.559	0.257	0.363	0.673
Orientation	EN-WS			E-W		
Street	Guanyinqiao commercial street					
Sample points	A3	B3	C3	A4	B4	C4
SVF	0.243	0.418	0.535	0.172	0.385	0.641
Orientation	EN-WS			WN-ES		
Street	Jiefangbei commercial street					
Sample points	A5	B5	C5	A6	B6	C6
SVF	0.272	0.386	0.507	0.252	0.324	0.513
Orientation	EN-WS			WN-ES		

Fig. 6 Researchers conduct questionnaire interviews with pedestrians (self-drawn)

2.6 Data Analysis

First, this study analyzes the differences between TCV and TSV using linear regression. Second, this study will discuss the correlation between thermal equivalent temperature parameters and TSV and TCV under different age, gender, space openness, and street orientation conditions. Spearman correlation coefficient method will be utilized to test the correlation and level of significance between the relevant parameters (T_g, T_a, V_a, RH, and clothing heat resistance) and TSV, TCV, and Kruskal–Wallis test will be utilized to test the correlation between pedestrian activity level and TSV, TCV.

3 Result

3.1 *The Difference Between Thermal Perception and Thermal Comfort*

Many studies have concluded that thermal perception cannot be equated with thermal comfort [28], and this section will further confirm the difference between thermal perception and satisfaction. The TSV and TCV of the surveyed pedestrians are shown in Fig. 7a and b. There exists at least 7.3% of pedestrians who have a neutral perception of heat when faced with a hot street environment, and there exists 14.9% of pedestrians who find hot street environments comfortable, which indicates that pedestrians can feel comfortable in a slightly warmer or colder environment. The linear relationship between the TSV and the TCV ($R^2 = 0.5943$) is shown in Fig. 7c. The TCV is the same as the TCV. Overall, the TCV level will be higher than the TSV level as the environment becomes hotter, indicating that the comfort range of the interviewed pedestrians in hot outdoor environments is less than the neutral range of thermal perception, and that the requirement for thermal comfort is more stringent in hotter environments.

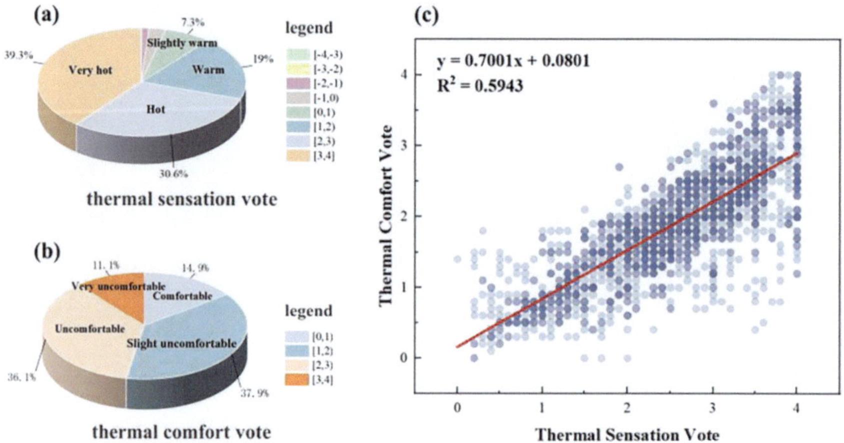

Fig. 7 **a** and **b** are TSV and TCV of pedestrians on commercial streets, **c** is linear relationship between TCV and TSV (self-drawn)

3.2 *Biological Sex Divergences in Thermal Perception and Thermal Comfort*

Table 3 and 4 addresses the differences between male and female perceptions of the street heat environment. In both men and women, TSV and TCV were positively was contingent upon T_a, T_g and V_a. However, RH demonstrated a significant negative covariance with the target variable, and as RH increased, thermal comfort elevated for both males and females, but there was no significant difference. This is consistent with a study that gender did not demonstrate a significant moderating effect on reported thermal sensations [14]. It is noteworthy that females ($p < 0.05$ for TSV, $p = 0.16$ for TCV) were slightly less sensitive to perceived changes in wind speed compared to males ($p < 0.001$ for TSV, $p = 0.37$ for TCV). Otherwise, the impact of occupant behavior patterns on subjective thermal evaluation was always significant regardless of gender. However, the effect of clothing on thermal sensation was somewhat related to gender, with males ($p = 0.003$) being able to get cooler by reducing clothing, while females ($p = 0.580$) had no significant effect on thermal sensation by reducing clothing.

As shown in Fig. 8, NPET was different for males and females, 21.33 °C ($R^2 = 0.57$) and 22.16 °C ($R^2 = 0.59$) respectively, indicating that males perceive heat first under the same thermal environmental conditions, in other words males are less heat tolerant compared to females. Comparing the slopes of the two groups, whereas when the temperature was at high temperature, the range of heat perception vote was greater for females (0.1501) than for males (0.1438), indicating that females in hot environments are more sensitive to heat.

Table 3 Differential impacts of microclimatic factors and occupant activities on TSV across gender cohorts (self-drawn)

TSV	Analyze method	Male (n = 848)		Female (n = 868)	
		C/H	p	C/H	p
T_g	Spearman	0.839	<0.001***	0.828	<0.001***
T_a		0.807	<0.001***	0.793	<0.001***
V_a		0.157	<0.001***	0.095	0.005**
RH		−0.720	<0.001***	−0.713	<0.001***
Activity	K-W test	14.060	<0.001***	31.998	<0.001***
Clothing thermal resistance	Spearman	−0.102	0.003**	0.019	0.580

Note *p < 0.10; **:p < 0.05; ***:p < 0.001). C is the correlation for the Spearman; H is the test statistic for the K-W test

Table 4 Differential impacts of microclimatic factors and occupant activities on TCV across gender cohorts (self-drawn)

TCV	Analyze method	Male(n = 848)		Females(n = 868)	
		C/H	*p*	C/H	*p*
T_g	Spearman	0.732	< 0.001***	0.719	< 0.001***
T_a		0.712	< 0.001***	0.714	< 0.001***
V_a		0.072	0.037**	-0.048	0.160
RH		-0.588	< 0.001***	-0.605	< 0.001***
Activity	K-W test	26.217	< 0.001***	49.812	< 0.001***
Clothing thermal resistance	Spearman	-0.092	0.007**	0.029	0.389

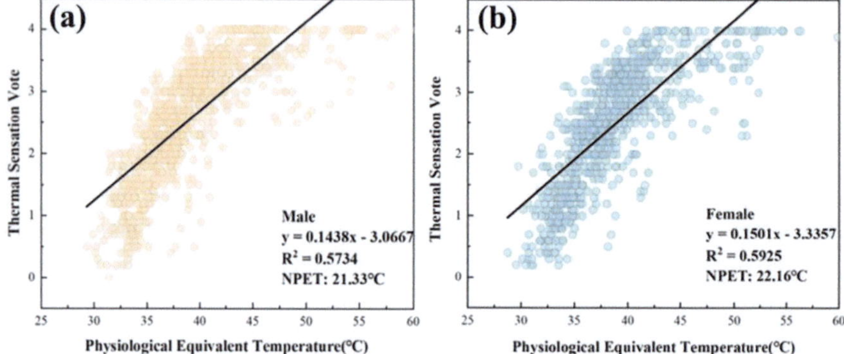

Fig. 8 Linear regression of TSV and PET. **a** is male, **b** is female (self-drawn)

3.3 *Age-Specific Differences in Thermal Perception and Comfort*

Tables 5 and 6 show the differences in thermal sensations of pedestrians in the three age groups. The directional consistency of relationships was maintained across both subjective measures: enhancing thermal parameters and activity elevated TSV/TCV, whereas increasing RH depressed these indices. In particular, the thermal sensations of adolescents and children (9–25 years old) were not significantly affected by activity ($p = 0.353$), whereas activity could significantly influence their thermal comfort. Importantly, adolescents and children could not change their behavior and clothing to achieve thermal comfort optimization outcomes, and the correlation between pedestrians' clothing and their TSV and TCV in the older group was not significant.

By regressing the TSV and PET of pedestrians, as shown in Fig. 9, the NPETs of the interviewed pedestrians were different for different age groups. The NPET of pedestrians for the 9–25-year-old cohort is obtained as 21.10 °C, the NPET of

Table 5 Age-stratified variations in TSV (self-drawn)

TSV	Analyze method	[9, 25) (n = 366)		[25, 50) (n = 1078)		[50, 77] (n = 271)	
		C/H	*p*	C/H	*p*	C/H	*p*
T_g	Spearman	0.837	<0.001***	0.892	<0.001***	0.829	<0.001***
T_a		0.787	<0.001***	0.798	<0.001***	0.816	<0.001***
V_a		0.127	0.015**	0.110	<0.001***	0.189	0.002**
RH		−0.736	<0.001***	−0.708	<0.001***	−0.716	<0.001***
Activity	K-W test	2.085	0.353	34.554	<0.001***	7.056	0.029**
Clothing thermal resistance	Spearman	0.053	0.310	−0.055	0.072*	−0.073	0.229

Table 6 Age-stratified variations in TCV (self-drawn)

TCV	Analyze method	[9, 25) (n = 366)		[25, 50) (n = 1078)		[50, 77) (n = 271)	
		C/H	*p*	C/H	*p*	C/H	*p*
T_g	Spearman	0.695	<0.001***	0.728	<0.001***	0.731	<0.001***
T_a		0.669	<0.001***	0.727	<0.001***	0.692	<0.001***
V_a		0.036	0.493	0.005	0.871	−0.003	0.959
RH		−0.591	<0.001***	−0.603	<0.001***	−0.550	<0.001***
Activity	K-W test	8.971	0.011**	52.220	<0.001***	13.925	<0.001***
Clothing thermal resistance	Spearman	0.016	0.755	−0.041	0.179	−0.008	0.897

pedestrians for the 25–50-year-old cohort is 21.48 °C, and the NPET of pedestrians in the the 50–77-year-old cohort is 23.34 °C. The younger you are the more you start to perceive heat first, in other words with age the body's sensitivity to the thermal environment decreases. This is consistent with the conclusion of kruger et al. that heat perception decreases in the elderly [14].

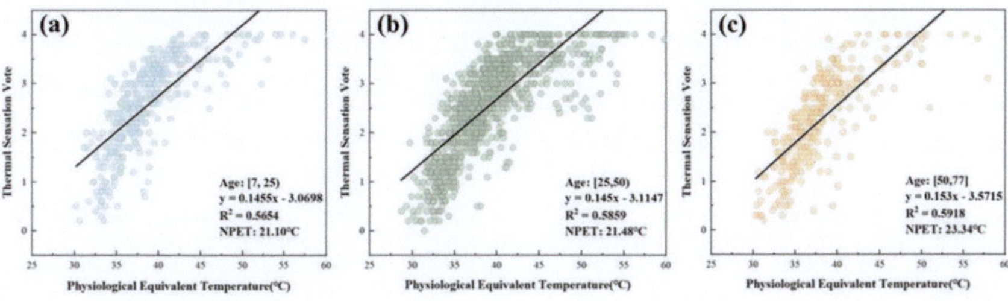

Fig. 9 Linear regression of TSV and PET. **a** is age group 7–25, **b** is age group 25–50, **c** is age group 50–77 (self-drawn)

3.4 SVF-Induced Variations in Thermal Perception and Comfort

Tables 7 and 8 quantifies the outcomes of the examination of the effect of thermal equivalent temperature elements on TSV and TCV under the three openness conditions. Consistent directional relationships were observed: T_a and T_a showed positive covariation with subjective thermal evaluations (TSV/TCV), whereas RH exhibited inverse correlations. Among them, the effect of V_a on TSV at open places was not significant ($p = 0.657$), which may be due to the fact that the thermal perception of pedestrians at open places is mainly influenced by solar radiation rather than V_a. TSV was almost unaffected by activity and clothing when pedestrians were in slightly open places. In addition, activity intensity had a notable impact on participants' thermal sensation and comfort evaluations when pedestrians were in the open ($p < 0.001$ for TSV, $p < 0.001$ for TCV) and clothing did not have a marked impact on TSV and TCV ($p = 0.621$ for TSV, $p = 0.647$ for TCV).

Table 7 Effects of thermal parameters and activities on TSV among different SVF (self-drawn)

TSV	Analyze method	[0, 0.3) (n = 593)		[0.3, 0.5) (n = 554)		[0.5, 1] (n = 569)	
		C/H	*p*	C/H	*p*	C/H	*p*
T_g	Spearman	0.805	<0.001***	0.828	<0.001***	0.869	<0.001***
T_a		0.764	<0.001***	0.783	<0.001***	0.867	<0.001***
V_a		0.217	<0.001***	0.200	<0.001***	-0.019	0.657
RH		−0.742	<0.001***	−0.745	<0.001***	−0.808	<0.001***
Activity	K-W test	19.247	<0.001***	7.585	0.023	18.703	<0.001***
Clothing thermal resistance	Spearman	−0.075	0.068*	−0.061	0.15	0.021	0.621

Table 8 Effects of thermal parameters and activities on TCV among different SVF (self-drawn)

TCV	Analyze method	[0, 0.3) (n = 593)		[0.3, 0.5) (n = 554)		[0.5, 1] (n = 569)	
		C/H	*p*	C/H	*p*	C/H	*p*
T_g	Spearman	0.677	<0.001***	0.748	<0.001***	0.760	<0.001***
T_a		0.692	<0.001***	0.725	<0.001***	0.787	<0.001***
V_a		−0.012	0.772	0.175	<0.001***	−0.053	0.209
RH		−0.611	<0.001***	−0.674	<0.001***	−0.717	<0.001***
Activity	K-W test	43.300	<0.001***	10.711	0.005**	24.976	<0.001***
Clothing thermal resistance	Spearman	−0.034	0.412	−0.066	0.123	0.019	0.647

3.5 *Orientation-Induced Variations in Thermal Perception and Comfort in Urban Street Canyons*

The study streets can be categorized into three types according to orientation: EN-WS orientation, WN-ES orientation, and E-W orientation. As shown in Tables 9 and 10, T_g, T_a, V_a, and the observed activity-TSV/TCV positive relationships and RH-TSV/TCV negative relationships remained robust when controlling for street orientation variations. However, clothing heat resistance for pedestrians was negatively correlated with TSV, TCV only in WN-ES oriented streets ($p = 0.004$ for TSV, $p = 0.005$ for TCV). Notably, the effect of V_a on TSV was strongly associated with street orientation, with the most significant effect of V_a on pedestrian thermal perception on EN-WS oriented streets ($p < 0.001$), followed by WN-ES oriented streets ($p = 0.004$), and finally E-W oriented streets ($p = 0.017$).

Table 9 Effects of thermal parameters and activities on TSV among different street orientation (self-drawn)

TSV	Analyze method	EN-WS (n = 836)		WN-ES (n = 579)		E-W (n = 301)	
		C/H	*p*	C/H	*P*	C/H	*p*
T_g	Spearman	0.817	<0.001***	0.862	<0.001***	0.872	<0.001***
T_a		0.782	<0.001***	0.849	<0.001***	0.787	<0.001***
V_a		0.13	<0.001***	0.119	0.004**	0.138	0.017*
RH		−0.701	<0.001***	−0.786	<0.001***	−0.693	<0.001***
Activity	K-W test	11.212	<0.001***	3.929	0.047**	13.856	<0.001***
Clothing thermal resistance	Spearman	0.002	0.964	−0.12	0.004**	0.018	0.751

Table 10 Effects of thermal parameters and activities on TCV among different street orientation (self-drawn)

TCV	Analyze method	EN-WS(n = 836)		WN-ES(n = 579)		E-W(n = 301)	
		C/H	*p*	C/H	*P*	C/H	*p*
T_g	Spearman	0.738	<0.001***	0.733	<0.001***	0.737	<0.001***
T_a		0.726	<0.001***	0.735	<0.001***	0.715	<0.001***
V_a		0.015	0.669	0.127	0.002**	−0.166	0.004**
RH		−0.629	<0.001***	−0.644	<0.001***	−0.476	<0.001***
Activity	K-W test	16.885	<0.001***	4.140	0.042**	9.360	0.002**
Clothing thermal resistance	Spearman	0	0.991	−0.166	0.005**	0.02	0.732

4 Discussion

A key finding was the compression of comfort boundaries relative to the broader thermal neutrality spectrum in urban walking environments. These results corroborate previous observations that Zhongyu [38] studied residential thermal comfort in Chongqing revealed a compressed comfort acceptability range compared to the broader thermal neutrality spectrum, but the difference between the two was not significant. In addition, this study found that pedestrians were more demanding of thermal comfort as the environment became hotter. Further scholarly attention ought to be directed toward the interaction between pedestrian thermal perception and thermal comfort.

Gender-based variations in thermoregulatory responses to elevated temperatures were consistently observed. At high temperatures, the TSV range changes more in females than in males. It was found that males had lower NPET than females and that males were more heat intolerant than females. This is consistent with Du et al.'s conclusion that men tend to feel more uncomfortable in hotter environments when heat is combined with still air [12, 36]. The reasons for this are that on the one hand, men have a higher metabolic rate and wetter skin [36], on the other hand, men favor lower thermal resistance in the selection of clothing materials, whereas women will choose clothing with higher thermal resistance for reasons such as aesthetics [4]. Clothing-induced thermal perception variations exhibit gender-specific patterns, and studies have shown that men prefer to achieve thermal comfort by reducing clothing. Given the gender differences in thermal comfort research, future studies should consider outdoor thermal comfort and health for men and women separately.

Pedestrians of different ages possess different thermal sensations. Physiological assessments demonstrated that thermal adaptation capacity declines with age, rendering elderly subjects more vulnerable to temperature extremes despite equivalent exposure conditions. Adolescents and children find it difficult to achieve comfort by choosing to meditate and reduce clothing, similarly, older adults find it difficult to achieve coolness by reducing clothing, however, older adults can appropriately cope with hot environments by meditating. Future research should focus on outdoor health [6] to reduce the risk of heat stroke in children and older adults.

The thermal equivalent temperature elements affect TSV and TCV differently in different SVFs. The effect of V_a on pedestrian TCV is not significant in narrower spaces (SVF = [0, 0.3)) and more open spaces (SVF = [0.5, 1)). It can be understood that a smaller SVF indicates that there is more shading in the space, in other words the less solar radiation the pedestrian is exposed to, and in this case the pedestrian's judgment of thermal comfort is mainly influenced by solar radiation rather than wind speed. In addition, López et al. [19] found that different SVF present in different neighborhoods can affect human visual comfort and solar radiation and thus overall comfort. Therefore, the study of visual perception and acoustic perception of pedestrians can be included in future research.

The effect of environmental parameters on subjective perception is related to street orientation. This is reflected as the effect of V_a on TSV and TCV, with the significance ranging from strong to weak in the order of EN-WS oriented, WN-ES oriented, and E-W oriented streets. The E-W streets are subjected to solar radiation due to prolonged periods of time, while V_a has limited effect in reducing the thermal perception of pedestrians. This coincides with the conclusion of Narimani et al. [34, 20] that pedestrians on E-W oriented streets have the worst thermal experience and need to change the orientation and plant street trees appropriately. The thermal mitigation potential of urban greenery warrants integration into parametric studies of street geometry's influence on microclimate perception.

5 Conclusion

This study employs thermal equivalent temperature (TET) as a key metric to evaluate pedestrian thermal comfort in three representative commercial streets within Chongqing's urban core. The analysis specifically addresses how thermal perception varies across (1) gender (male/female), (2) age groups (young/middle-aged/elderly), (3) street spatial openness (SVF), and (4) solar exposure patterns (N-S/NE-SW/E-W oriented). And key research outcomes include:

- The comfort range of interviewed pedestrians was less than the neutral range of heat perception. And the requirement for thermal comfort is more stringent in hotter environments.
- Under the same thermal environmental conditions, women were less sensitive to wind speed than men, and men were less heat intolerant than women.
- Aging significantly impairs heat perception capacity, resulting in attenuated thermal environment adaptability among older adults compared to younger populations.
- Wind speed fluctuations exerted negligible influence on pedestrian thermal perception, regardless of SVF magnitude.
- Street orientation in turn affects pedestrian thermal comfort by changing V_a, and the effect of V_a on pedestrian thermal perception is in the order of EN-WS toward the street, WN-ES toward the street, and E-W toward the street, from strong to weak.

Based on the research findings, summer street thermal comfort mitigation strategies are proposed: Prioritize EN-WS oriented street layouts to enhance ventilation and cooling; maintain moderate SVF to avoid excessive shading or full exposure; implement gender- and age-responsive designs including shade structures (targeting females) and rest areas (for elderly) in high-temperature zones, while improving wind speed regulation (in male-sensitive areas); optimize street tree canopy coverage through species selection, with emphasis on three-dimensional shading systems for E-W oriented streets.

However, this study still has some limitations. First, the interviewed pedestrians lacked knowledge about thermal perception and thermal comfort systems, and thus there were errors in completing the questionnaire due to individual differences.

The time of this study was summer, and the extrapolation of thermal equivalent temperature effects to other seasonal contexts remains empirically unverified.

References

1. Aboelata A (2020) Vegetation in different street orientations of aspect ratio (H/W 1:1) to mitigate UHI and reduce buildings' energy in arid climate. Build Environ 172
2. Aboelata A (2021) Reducing outdoor air temperature, improving thermal comfort, and saving buildings' cooling energy demand in arid cities—cool paving utilization. Sustain Cities Soc 68
3. Ali SB, Patnaik S (2018) Thermal comfort in urban open spaces: objective assessment and subjective perception study in tropical city of Bhopal, India. Urban Clim 24:954–967
4. Aqilah N, Rijal HB, Yoshida K (2023) Gender difference regarding thermal comfort in Japanese residential building during free running mode. Build Environ 245
5. Arias-Rosales A, LeDuc PR (2020) Comparing view factor modeling frameworks for the estimation of incident solar energy. Appl Energy 277
6. Baquero Larriva MT, Higueras E (2020) Health risk for older adults in Madrid, by outdoor thermal and acoustic comfort. Urban Clim 34
7. Cao S-H, Ming P-P, Zhao X (2021) Fuzzy comprehensive evaluation of human thermal comfort in simulating natural wind environment. Build Environ 188
8. Chatzipoulka C, Compagnon R, Kaempf J, Nikolopoulou M (2018) Sky view factor as predictor of solar availability on building façades. Solar Energy 170:1026–1038
9. Chiang Y-C, Liu H-H, Li D, Ho L-C (2023) Quantification through deep learning of sky view factor and greenery on urban streets during hot and cool seasons. Landsc Urban Plan 232
10. Chow WTL, Akbar SNABA, Heng SL, Roth M (2016) Assessment of measured and perceived microclimates within a tropical urban forest. Urban Forest Urban Greening 16:62–75
11. Chu Z, Li S, Li T, Qian H, Liu C, Yan Z (2024) Numerical simulation of layout and landscape elements on the thermal environment of urban squares. Ecol Inform 82
12. Du H, Zhao Z, Lyu J, Li J, Liu Z, Li X, Yang Y, Lan L, Lian Z (2023) Gender differences in thermal comfort under coupled environmental factors. Energy Build 295
13. Elnabawi MH, Hamza N, Dudek S (2016) Thermal perception of outdoor urban spaces in the hot arid region of Cairo, Egypt. Sustain Cities Soc 22:136–145
14. Kruger EL, Drach P (2017) Identifying potential effects from anthropometric variables on outdoor thermal comfort. Build Environ 117:230–237
15. Lai D, Guo D, Hou Y, Lin C, Chen Q (2014) Studies of outdoor thermal comfort in northern China. Build Environ 77:110–118
16. Lai D, Lian Z, Liu W, Guo C, Liu W, Liu K, Chen Q (2020) A comprehensive review of thermal comfort studies in urban open spaces. Sci Total Environ 742
17. Lai D, Liu W, Gan T, Liu K, Chen Q (2019) A review of mitigating strategies to improve the thermal environment and thermal comfort in urban outdoor spaces. Sci Total Environ 661:337–353
18. Li Z, Zhou L, Hong X, Qiu S (2024) Outdoor thermal comfort and activities in urban parks: an experiment study in humid subtropical climates. Build Environ 253
19. López CSP, Sala M, Tagliabue LC, Frontini F, Bouziri S (2016) Solar radiation and daylighting assessment using the sky-view factor (SVF) analysis as method to evaluate urban planning densification policies impacts. Energy Procedia 91:989–996
20. Narimani N, Karimi A, Brown RD (2022) Effects of street orientation and tree species thermal comfort within urban canyons in a hot, dry climate. Ecol Inform 69
21. Nguyen TT, Grote U, Neubacher F, Rahut DB, Do MH, Paudel GP, F (2023) Security risks from climate change and environmental degradation: implications for sustainable land use transformation in the Global South. Curr Opin Environ Sustain 63:101322

22. Nitidara NPA, Sarwono J, Suprijanto S, Soelami FN (2022) The multisensory interaction between auditory, visual, and thermal to the overall comfort in public open space: a study in a tropical climate. Sustain Cities Soc 78
23. Potchter O, Cohen P, Lin T-P, Matzarakis A (2018) Outdoor human thermal perception in various climates: a comprehensive review of approaches, methods and quantification. Sci Total Environ 631–632:390–406
24. Qin Z, Lu B, Jing W, Yin Y, Zhang L, Wang X, Luo W, Zhang J, Qiu J, Dong Y (2024) Creating comfortable outdoor environments: understanding the intricate relationship between sound, humidity, and thermal comfort. Urban Clim 55
25. Roshan G, Moghbel M, Attia S (2020) Evaluating the wind cooling potential on outdoor thermal comfort in selected Iranian climate types. J Therm Biol 92
26. Santamouris M (2014) On the energy impact of urban heat island and global warming on buildings. Energy Build 82
27. Sharmin T, Steemers K, Humphreys M (2019) Outdoor thermal comfort and summer PET range: a field study in tropical city Dhaka. Energy Build 198:149–159
28. Song W, Zhong F, Calautit JK, Li J (2024) Exploring the role of skin temperature in thermal sensation and thermal comfort: a comprehensive review. Energy Built Environ 6(4):762–781
29. Sun Y, Zhang C, Zhao Y, Li J, Ma Y, Zhu C (2023) A systematic review on thermal environment and thermal comfort studies in Chinese residential buildings. Energy Build 291
30. Thach T-Q, Zheng Q, Lai P-C, Wong P P-Y, Chau P Y-K, Jahn HJ, Plass D, Katzschner L, Kraemer A, Wong C-M (2015) Assessing spatial associations between thermal stress and mortality in Hong Kong: a small-area ecological study. Sci Total Environ 502
31. Wang W, Li Y, Li L, Wang R, Wang Y (2023) Study on thermal comfort of elderly in community parks: an exploration from the perspectives of different activities and ages. Build Environ 246
32. Wang Z, Yang Y, Liu C, Zhou F, Hao H (2024) Human thermal comfort model and evaluation on building thermal environment. Energy Build 323
33. Xi T, Li Q, Mochida A, Meng Q (2012) Study on the outdoor thermal environment and thermal comfort around campus clusters in subtropical urban areas. Build Environ 52:162–170
34. Xiao Q, Fan X, Guo Y, Li S, He W, Deng Y, Xiao Z, Wang P, Wu C, F (2024) Tree form characteristics as criteria for tree species selection to improve pedestrian thermal comfort in street canyons: case study of a humid subtropical city. Sustain Cities Soc 105
35. Xiong J, Cheng B, Zhang J, Liu Y, Tan X, Shi M, He X, Guo J (2023) A study of waterside microenvironmental factors and their effects on summer outdoor thermal comfort in a CFA-climate campus. J Therm Biol 117
36. Yang L, Zhao S, Gao S, Zhang H, Arens E, Zhai Y (2021) Gender differences in metabolic rates and thermal comfort in sedentary young males and females at various temperatures. Energy Build 251
37. Zhang H, Wang Y, Guo F, Zhao J, Dong J, Zhu P (2024) Factors influencing outdoor thermal comfort in a coastal park during the transition seasons in cold regions of China. Urban Clim 55
38. Zhongyu H (2020) Study on outdoor thermal comfort of existing residential areas in Chongqing in summer

BY NC ND

Spatial Correlation Study of Local Climate Zoning and Carbon Emission Performance: A Study of the Central City of Chongqing, China

Xingli Wang, Mingying Zeng, Yangli Li, and Rui Zhang

Abstract The growth of urbanization and intensive human activities has resulted in a steady increase in urban carbon emissions, posing specific repercussions and difficulties to the human environment and sustainable development. Based on the local climate zone (LCZ) system, we investigated the type and spatial distribution characteristics of the LCZ, as well as the geographic distribution traits of carbon emission performance in Chongqing's central area, and then examined the spatial correlation between the two variables. The results of the study show that: ①a total of 14 LCZ types were identified in the central area of Chongqing, including eight built landscape-type LCZ types and six natural cover type LCZ types; ②the main source of carbon emissions is the indirect carbon emissions from the construction land, and the total amount of carbon emissions in the central urban area of Chongqing is overall high; the spatial distribution of the carbon emission performance level is unbalanced, presenting a high in the central part of the spatial pattern around the low; ③overall, the LCZ and carbon emission performance in the central urban area of Chongqing have a negative spatial correlation. Among them, the built landscape-type LCZ and carbon emission performance were positively correlated spatially. Except for the two categories of LCZE rock/hardened ground and LCZF bare soil/sand, the remaining natural cover LCZ types had a negative correlation with carbon emission performance. The findings of this study can be used to guide Chongqing Municipality's green, low-carbon, and sustainable development, as well as the creation of differentiated emission reduction and efficiency strategies in other Western Chinese cities.

Keywords Local climate zoning · Carbon emission performance · Bivariate spatial autocorrelation · Chongqing

---

X. Wang · M. Zeng (✉) · Y. Li
School of Civil Engineering and Architecture, Southwest University of Science and Technology, Sichuan, China
e-mail: zmy018@hotmail.com

R. Zhang
College of Design and Engineering, National University of Singapore, Singapore, Singapore

B. He et al. (eds.), *Climate-Resilient, Low-Carbon, and Sustainable Cities*, Lecture Notes in Civil Engineering 754, https://doi.org/10.1007/978-981-95-2169-2_12

1 Introduction

In recent years, with China's rapid urbanization, the city boundary has expanded significantly, and a large number of natural areas have been replaced by high-density, impermeable urban subsurfaces, which, due to their unique geometrical configurations, have altered local climatic zones, disturbed the energy balance, and interrupted the flow of the natural water cycle, which has led to an increase in the temperature of the urban areas, and an intensification of energy consumption and carbon emissions, which has in turn given rise to a series of This has led to a series of serious urban ecological and environmental problems, posing a serious challenge to the creation of a sustainable and livable urban living environment [1]. This poses a serious challenge for the creation of a sustainable and livable urban living environment. Relevant studies have shown that [2] that carbon emissions, mainly in the form of CO_2, caused by localized land-use change, have accounted for one-third of the total emissions from human activities in the past few years. Therefore, it is of profound theoretical value and practical significance to accurately grasp the carbon emission characteristics and carbon efficiency patterns of each climate zone, in order to deepen the study of carbon emission performance and promote the scientific and effective management of carbon emissions [3, 6]. The following is a brief overview of the carbon emission characteristics and carbon efficiency patterns in each climate sub-region.

Stewart and Oke [4] 2012 proposed the theory of local climate zoning, which aims to establish a quantitative link between surface cover type and the urban heat island effect, and reinterprets the urban heat island phenomenon through a standardized terminology system [5, 7, 8] The LCZ theory is a new approach to the urban heat island phenomenon. After LCZ was proposed, numerous scholars explored the scientific validity of applying LCZ to study the urban environment. Stewart[8] et al. used temperature observations and simulations of surface atmospheric models to evaluate the conceptual division of LCZ and showed that thermal differences exist in different types of LCZ, which are mainly influenced by building height and spacing, proportion of permeable surfaces, tree density, and soil moisture. Geletič [9] et al. analyzed the relationship between LCZ and surface temperatures in the two largest Czech cities, Prague and Brno, and the results confirmed the independence of LCZ as well as the variability among LST. Wang et al. [10] analyzed the temperatures in the local climate zones (LCZ) of two arid zone cities in the United States and compared the differences between the physical quantities of the LCZ and the reference values. In recent years, there have been abundant research results on local climate zoning at home and abroad, including theoretical indicators of local climate zoning, heat island effects, model simulation, and planning applications. However, research on local climate zoning in China mainly focuses on the quantitative connection between surface cover type and the urban heat island effect, and there is a lack of research on local climate zoning in the field of carbon emission performance.

Carbon performance refers to the desired outcome that may be obtained as a result of economic activity that generates CO2 emissions. It assesses the output's efficiency for a given economic input and is an important component of environmental performance assessment [11]. It is an important component of environmental performance assessment. There has been substantial discussion in existing research on how to assess carbon emission performance, which can be classified as single-factor indicators [12]. The measurement approach divides them into single-factor indicators and full-factor indications [13]. On the other hand, researchers have steadily focused on the spatial characteristics and affecting elements of carbon emission performance [14]. On the other hand, researchers have steadily concentrated on the geographical features and affecting elements of carbon emission performance. Scholars from home and abroad have conducted theoretical and empirical studies on the optimization of carbon emission efficiency measurement, dynamic evaluation, influencing variables, and the impact of technical and green technology innovation on carbon emission efficiency. However, little research has investigated the relationship between land cover type and carbon emission efficiency from a climatic zone viewpoint, and significant issues have to be resolved and addressed.

In view of this, this study selected the downtown area of Chongqing Municipality as the study area and used the WUDAPT mapping method to map the local climate zones in the downtown area of Chongqing Municipality. It also used the spatial autocorrelation analysis method to explore the spatial correlation between the climate zones and the performance of carbon emissions, aiming at providing a scientific basis and practical reference for the formulation of climate-resilient urban development planning.

2 Study Area and Data Sources

2.1 Overview of the Study Area

Chongqing (coordinates 105°17′–110°11′E, 28°10′–32°13′N), situated in the eastern region of the Sichuan Basin, is the only municipality in central and western China and is recognized as one of the four hottest cities in the country during the summer months. Research indicates that carbon emissions from energy consumption in Chongqing Municipality have been increasing over the last twenty years, with notable variations in the spatial distribution of these emissions. The areas with the highest carbon emissions are primarily located around the central urban region of Chongqing Municipality, exhibiting a significant positive spatial clustering. Consequently, this paper focuses on the central urban area of Chongqing, which encompasses nine districts: Yuzhong, Dadukou, Jiangbei, Shapingba, Jiulongpo, Nanan, Beibei, Yubei, and Banan (Fig. 1).

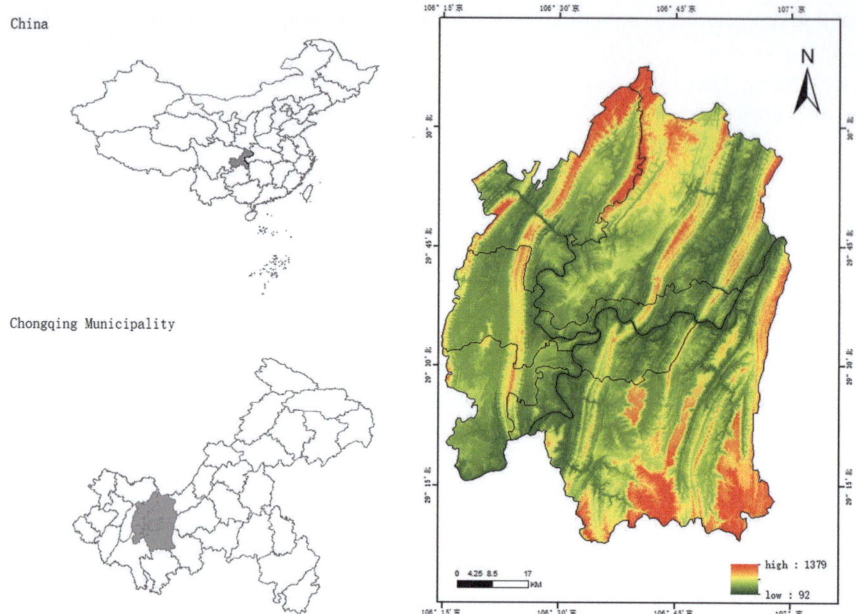

Fig. 1 Overview of the study area (Self-drawn). *Note* This map is derived from the standard map associated with review number GS (2024)0650, which was obtained from the National Geographic Information Public Service Platform-Tianmap, and it has not undergone any alterations from the original base map

2.2 Data Sources and Description

This study used data such as building data, land use data, energy data, and Landsat8 satellite remote sensing data, as shown in Table 1. The vector data about buildings was sourced from the Baidu map platform, utilizing the Baidu map interceptor BDMap-Capturer0.4Beta to acquire the raster data for building bases at level 18. This data was subsequently vectorized using the ArcScan tool within ArcGIS 10.7, allowing for the exportation of DWG files. To ascertain building heights, supplementary resources such as Google Earth, Street View, and online searches were employed. Ultimately, a foundational database of urban morphology was established in ArcGIS 10.7. Remote sensing image data were procured from the USGS website, as the local climate classification method is influenced by the background climatic conditions of the study area. Literature review indicates that these background conditions can be characterized as sunny, with minimal cloud cover and light winds. Consequently, the timing for capturing remote sensing images for this study was selected to ensure normal solar radiation levels, calm winds, and less than 10% cloud cover, alongside other specific meteorological conditions. The remote sensing images analyzed in this study, specifically those corresponding to the center of Chongqing, are designated as 39/128 and 40/128, both exhibiting less than 10% cloud cover, with the acquisition date being

Table 1 Research data sources and content. (Self-drawn)

Data type	Timing	Resolution (of a photo)	Source (of information etc.)	Source URL
Building vector data	2022	30m	Baidu's online map	https://map.baidu.com/@11657586,3,684,248,13z
Landsat 8 image data	2022	30m	USGS	https://www.usgs.gov/
Land use data	2022	30m	Wuhan University	https://zenodo.org/record/5210928#.YcZ_nWBByUk
Energy data	2022	–	Chongqing Municipal Bureau of Statistics	https://tjj.cq.gov.cn/index.html

July 5, 2022. Historical meteorological data confirm that the day was characterized by sunny weather and a light easterly wind, aligning with the identified background climatic conditions. Subsequently, the preprocessing of remote sensing images was conducted using the widely utilized software ENVI 5.6.2, following these specific steps: (1) cropping the image by the administrative boundary data in shapefile format for the study area; (2) performing atmospheric correction on the cropped image to mitigate the adverse effects of atmospheric scattering and solar radiation on the sensor, utilizing the atmospheric correction tool in ENVI 5.3 for data calibration; (3) conducting radiometric calibration using the radiometric calibration tool in ENVI 5.3 to calibrate the visible and thermal infrared bands of the remote sensing image data. Land use data for the nine districts in central Chongqing were obtained from the CLCD land use dataset released by Professor Huang Xin's team at Wuhan University, which has a resolution of 30m x 30m. This dataset encompasses nine categories of land use, including farmland, forests, shrublands, grasslands, water bodies, ice and snow, bare soil, impermeable surfaces, and wetlands. Using ArcGIS 10.7, this data was reclassified into six categories: arable land, forest land, grassland, water bodies, built-up land, and unused land, thereby generating the land use data for the year 2022. Energy data were sourced from the Chongqing Municipal Bureau of Statistics.

3 Research Methodology

3.1 Local Climate Zone

Local climate zoning theory divides local climate zoning types into two categories: built-up landscape type and natural cover type, on the basis of which they are subdivided into 17 basic types, including 10 built-up landscape type LCZ types and 7 natural cover type LCZ types (Table 2) [16]. The following table summarizes the basic types of LCZ in the city. In order to further characterize the urban local climate zoning, a large number of studies on LCZ classification methods have been carried out both at home and abroad, which are mainly divided into three methods: manual sampling methods, classification methods based on geographic information system (GIS), and classification methods based on remotely sensed images [17–19]. The first two methods are The first two methods [13] are limited in use in many Chinese cities, especially in relatively backward areas, where it is difficult to collect complete urban data, and the complexity of calculating various indicators limits the large-scale application of these two methods. In contrast, the WUDAPT method proposed by Bechtel et al. in 2015 [20] et al. proposed the WUDAPT mapping method (World Urban Database and Access Portal Tools, WUDAPT) by virtue of its easy-to-access remote sensing data (30m-resolution Landsat8 multispectral imagery), the Google Earth platform, and the open-source tool LCZ Generator with a stochastic This provides a clearer workflow for the remote sensing image classification method and reduces the user's professional background requirements in remote sensing. Compared with the other two classification methods, the WUDAPT mapping method has the advantages of easy access to data, simple operation, and high accuracy of the classification results, and the free and open data sources and software tools on the WUDAPT website greatly promote the exploration of urban climate research in data-poor areas. The free and open data sources and software tools on the WUDAPT website have greatly facilitated the exploration and development of urban climate research in data-poor regions [21, 22]. The free and open data sources and software tools on the WUDAPT website have greatly facilitated the exploration and development of urban climate research in data-poor regions, and thus the method has become a mainstream method for urban local climate zoning research and has been widely used.

Therefore, this study adopted the local climate zoning WUDAPT mapping method to produce the LCZ classification map of the central city of Chongqing. The specific steps are: (1) Satellite maps containing the study area were downloaded from the USGS website and Landsat-8 remote sensing imagery data were preprocessed. If the one-view imagery could not contain the entire study area, the satellite imagery needed to be stitched into ENVI 5.6.2. (2) Create regions of interest in Google Earth Pro 7.3.3.7699 and represent the representative regions of each local climate zoning type as training samples using polygons, and select at least 20 training samples for each LCZ for identification and classification. To avoid the influence of time, the training samples should be representative and avoid the selection of harvested farmland, construction sites, and other zones with short timeliness. Each training sample was selected by using street view images in Google Earth pro7.3.3.7699 or field research to perform preliminary statistics on the type of ground cover and morphological indicators of the area, which were then matched with the type of LCZ. (3) Landsat imagery and regions of interest in the LCZ Generator. The remotely sensed imagery was resampled to obtain the spectral signatures of the local-scale urban features. Based on the spectral similarity of training samples with the same LCZ type, local climate zoning was performed using Random Forest Classification to generate spatial distribution maps of different LCZ types. The LCZ maps generated in step 3 were exported to a KML file and opened for inspection in Google Earth 7.3.3.7699, and the areas with inaccurate classification were re-sampled and step (3) was repeated until a high level of accuracy was achieved.

Table 2 Basic types of local climate zones *source* based on Ref. s1

LCZ1 Compact High-Rise 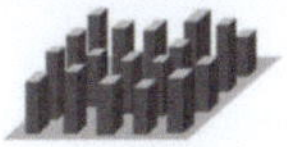	LCZ2 Compact Mid-rise 	LCZ3 Compact low-rise buildings 	LCZ4 Open high-rise building 
LCZ5 open mid-rise 	LCZ6 Open Low-rise Building 	LCZ7 Lightweight low-rise building 	LCZ8 large low-rise building 
LCZ9 Scattered low-rise buildings 	LCZ10 Heavy Industrial Plant 		
LCZA Dense Tree 	LCZB Sparse trees 	LCZC shrubs 	LCZD low vegetation
LCZE rock/hardened ground	LCZF Bare soil/sand 	LCZG waters	

3.2 Carbon Emission Accounting

Direct Carbon Accounting

The assessment of direct carbon emissions associated with land use primarily encompasses emissions originating from various land types, including cropland, forest land, grassland, watersheds, unused land, and construction land [24]. This study, utilizing existing research findings and pertinent literature, alongside the specific context of Chongqing Municipality, presents a table of carbon emission coefficients (Table 3) for distinct categories of land use [15].

Carbon Emissions Calculation Formula:

$$E_1 = \sum e_i = \sum T_i \times \delta_i \tag{1}$$

In this context, E_1 denotes direct carbon emissions; the variable i signifies the type of land, with i taking values from 1 to 6, which correspond to cultivated land, forest land, grassland, water bodies, constructed land, and unused land, respectively. e_i represents the carbon emissions produced by land type i, T_i indicates the area associated with land type i. δ_i represents the carbon emission (or absorption) coefficient

Table 3 Table of carbon emission factors (Self-drawn)

Land use classification	Carbon emission factor	unit (of measure)	Data sources
Cultivated land	0.0477	Kg C/(m2-a)	Sun Hyuk et al. [26]; Sun Hyun Bin [27]
Forest land	–0.0623	Kg C/(m2-a)	Sun Hyuk et al. [26] Zhao Rongqin et al. [28]
Grassland	–0.0021	Kg C/(m2-a)	Fang Jingyun et al. [29] Wang Gang et al. [30]
Water bodies	–0.0268	Kg C/(m2-a)	Sun He et al. [26] Hongxin Shi et al. [31]
Constructed land	–	–	–
Unused land	–0.0005	Kg C/(m2-a)	Sun He et al. [26] Hongxin Shi et al. [31]

for land type i, where positive values indicate emissions and negative values indicate absorption.

Accounting for Indirect Carbon Emissions

Indirect carbon emissions mainly refer to the estimation of carbon emissions from construction land, which are indirectly estimated through carbon emissions generated by coal, oil, natural gas, and other energy consumption. Energy-related data were collected based on the 2022 Chongqing Statistical Yearbook, which directly presents the total energy consumption at the citywide level. Additional data processing is required to obtain more detailed energy consumption data at the regional level. First, the total GDP and total energy consumption of Chongqing (in tons of standard coal) were used to calculate the energy consumption coefficient per unit of GDP (tons of standard coal per million yuan). Secondly, the total energy consumption of each region is indirectly deduced based on the GDP value of each region and this energy consumption coefficient:

$$a = \mathrm{GDP}/Q \tag{2}$$

$$E_2 = \mathrm{GDPi} \times a \tag{3}$$

where a is the energy consumption coefficient per unit of GDP; GDP is the total GDP of Chongqing; Q is the total energy consumption; E2 is the total carbon emission from construction land; and GDPi is the total GDP of the districts in the center of Chongqing.

Carbon Emission Performance Accounting

Carbon emission performance, as a key indicator for assessing the correlation between carbon emissions from the production and consumption activities of enterprises or individuals and their economic output, reflects the strength of their contribution to promoting the transition to a low-carbon economy. In this study, carbon emission performance was calculated based on the total carbon emissions, that is, by calculating the carbon emission density (the ratio of carbon emissions to the corresponding area) per unit area to quantitatively assess its performance level. The lower the carbon emission performance, the lower the spatial carbon emissions and the higher the degree of realizing carbon emission reduction targets, and vice versa. Carbon emission performance accounting aims to provide a more accurate and comprehensive perspective of carbon emission performance analysis and strong support for the formulation of targeted emission reduction measures.

$$C = E_1 + E_2 \tag{4}$$

where: C is the total carbon emissions; E_1 is the direct carbon emission; E_2 is the indirect carbon emission.

$$\mathrm{Cee} = \mathrm{C}/\mathrm{S} \tag{5}$$

where: *Cee* is the carbon emission efficiency, i.e., carbon emission performance; C is the total amount of carbon emissions; S is the area in square meters.

3.3 Spatial Autocorrelation

Spatial autocorrelation analysis serves as a method for assessing the spatial correlation and dependence characteristics of various elements, thereby elucidating the overall spatial distribution correlation among these elements. This analysis encompasses both global and local spatial autocorrelation. Global spatial autocorrelation pertains to the overall distribution of a particular phenomenon and ascertains the presence of clustering tendencies within a specified region. The global Moran's I index is the primary metric employed to evaluate this aspect, which is commonly utilized to investigate the spatial distribution patterns and the underlying causes associated with geographic elements. The specific formulations for these analyses are detailed as follow:

$$I = \frac{n\sum_{i=1}^{n}\sum_{j=1}^{n} W_{ij}(x_i - \overline{x})\left(x_j - \overline{x}\right)}{\sum_{i=1}^{n}\sum_{j=1}^{n} W_{ij}\sum_{i=1}^{n}(x_i - \overline{x})^2} \tag{6}$$

where: I represents the global Moran's I statistic, n denotes the total number of research subjects, while *xi* and *xj* refer to the observations associated with research subjects i and j, respectively. W_{ij} signifies the spatial weight matrix corresponding to research subjects i and j. The value of Moran's I range from -1 to 1; specifically, if I is greater than 0, it indicates a significant positive correlation among the observations; if I is less than 0, it suggests a significant negative correlation; and if I equals 0, it implies the absence of spatial autocorrelation among the observations, with larger absolute values of I reflecting stronger correlations in spatial distribution. To investigate local variations in the distribution of spatial objects, local spatial autocorrelation can be assessed, and local Moran's I is calculated as follows:

$$I_i = \frac{(x_i - \overline{x}) \sum_{j=1}^{n} w_{i,j}(x_i - \overline{x})}{S^2} \tag{7}$$

where: x_i is the attribute of element i; $\overline{x}$ is the mean value of the corresponding attribute; $w_{i,j}$ is the spatial weight between elements i and j; $S2 = \frac{1}{n}\sum_{i=1}^{n}(x_i - \overline{x})^2$

The analysis of spatial autocorrelation encompassed both univariate and bivariate methodologies. Univariate spatial autocorrelation serves to elucidate the overall degree of correlation and significance of the attribute values associated with a specific element within a designated region. This approach facilitates the exploration of spatial correlation patterns across various geographic locations, thereby enabling the identification of spatial agglomeration and divergence within a local context. Conversely, bivariate spatial autocorrelation is employed to examine the spatial correlation between two or more variables. In this investigation, Bivariate Moran's I was utilized to evaluate the degree of correlation between a location variable and other variables, employing the weighted average of all neighboring locations. The bivariate spatial autocorrelation method was selected for this study, utilizing GeoDa 1.22 software, to conduct a comprehensive analysis of the spatial correlation between climate zones and carbon emission performance. This method is primarily employed to characterize the spatial correlation patterns of multiple variables, with the spatial interactions and coupling characteristics being represented by the calculated Bivariate Moran's I:

$$I_{CI} = \frac{\sum_{i=1}^{n}\sum_{j=1}^{n} \mathrm{wij}(xi - \overline{x})(yj - \overline{y})}{s^2 \sum_{i=1}^{n}\sum_{j=1}^{n} \mathrm{wij}} \tag{8}$$

where I_{CI} is the bivariate Moran index, W_{ij} is the inverse distance spatial weight matrix, x_i and $\overline{x}$ are the attribute and mean values of spatial unit i, y_i and are the attribute and mean values of spatial cell i, and S^2 is the sample variance.

4 Results and Analysis

4.1 Local Climate Zone

In this paper, a total of 14 local climate zone types were identified within the study area using the WUDAPT mapping method, including eight built landscape LCZ types and six natural cover LCZ types.

To validate the LCZ classification results, this paper uses confusion matrix plots to assess the classification accuracy of each category. Four evaluation indexes of classification accuracy are used in the confusion matrix for the resulting classifications, which are overall classification accuracy, producer accuracy, user accuracy, and Kappa coefficient, and the consistent relationship between Kappa coefficient and classification accuracy is shown in Table 4. According to the confusion matrix to evaluate the accuracy of the output image classification results and the actual zoning image, the corresponding local climate zoning classification confusion matrix for 2022 is generated (Table 5). Table 6 shows that the overall classification accuracy (OA) of LCZ classification is 0.84, and the Kappa coefficient (K) score is 0.83, which shows high accuracy, indicating that the classification of the local climate zoning map is in good agreement with the actual classification and that the local climate zoning map can provide good support for the subsequent research.

The local climate zoning map of Chongqing's central urban area accurately represents the spatial distribution of urban morphology and land utilization. As indicated by the data presented in Fig. 2, the natural environment-type land cover zones (LCZ) constitute 62.14% of the central city, surpassing the built landscape-type LCZ, which comprises 37.86%. Within the natural environment LCZ categories, LCZ A, characterized by dense tree coverage, and LCZ B, characterized by sparse tree coverage, represent significant proportions, accounting for 30.71% and 26.41% of the total natural environment, respectively. These zones are predominantly located in the southeastern and northeastern regions of the central urban area. Additionally, the highest percentage of scattered low-rise buildings is observed in LCZ 9, which constitutes 21.43% of the overall built landscape type.

Table 4 Relationship between Kappa coefficient and consistency of classification accuracy (Self-drawn)

Kappa coefficients	Classification accuracy
0.0–0.20	Very low
0.21–0.40	General
0.41–0.60	Medium
0.61–0.80	Good
0.81–1	Very Good

Table 5 Confusion array of local climate zoning maps for Chongqing central city, 2022 (Self-drawn)

LCZ	1	2	3	4	5	6	7	8	9	10	A	B	C	D	E	F	G	SC	UA (%)
1	20	0	0	14	0	0	0	0	0	0	0	0	0	0	0	0	0	34	58.8
2	0	4	0	5	17	0	0	2	0	0	0	0	0	0	0	0	0	28	14.3
3	0	16	2	1	7	0	0	7	5	0	0	0	0	0	0	0	0	38	5.3
4	7	1	0	101	37	7	0	7	0	0	0	0	0	0	0	0	1	161	62.7
5	0	0	0	54	35	5	0	2	9	0	0	1	0	0	0	0	0	106	33
6	0	0	0	13	37	12	0	0	6	0	0	0	0	0	0	0	0	68	17.6
7	0	2	0	0	0	0	0	4	0	0	0	0	0	0	0	0	0	6	0
8	0	1	0	10	16	1	0	968	1	0	0	0	0	1	14	0	0	1012	95.7
9	0	0	0	3	1	4	0	6	116	0	0	13	0	15	0	0	1	159	73
10	0	0	0	0	0	0	0	22	0	0	0	0	0	1	0	0	0	23	0
A	0	0	0	0	0	0	0	1	1	0	1262	23	0	0	0	0	0	1287	98.1
B	0	0	0	0	0	0	0	0	22	0	122	140	0	0	0	0	0	284	49.3
C	0	0	0	0	0	0	0	0	0	0	0	0	0	0	0	0	0	0	
D	0	0	0	0	0	0	0	2	50	0	0	17	0	73	0	0	0	142	51.4
E	0	0	0	0	0	0	0	14	0	0	0	0	0	0	15	1	0	30	50
F	0	0	0	0	0	0	0	0	0	0	0	0	0	0	0	0	0	0	
G	0	0	0	0	0	0	0	2	0	0	0	0	0	1	0	0	642	645	99.5
SR	27	24	2	201	150	29	0	1037	210	0	1384	194	0	91	29	1	644	4023	
PA (%)	74.1	16.7	100	50.2	23.3	41.4		93.3	55.2		91.2	72.2		80.2	51.7	0	99.7		84.3
K	0.83																		
OA	0.84																		

Table 6 Scatterplot of Moran's I index of climate zoning and carbon performance for individual localities (Self-drawn)

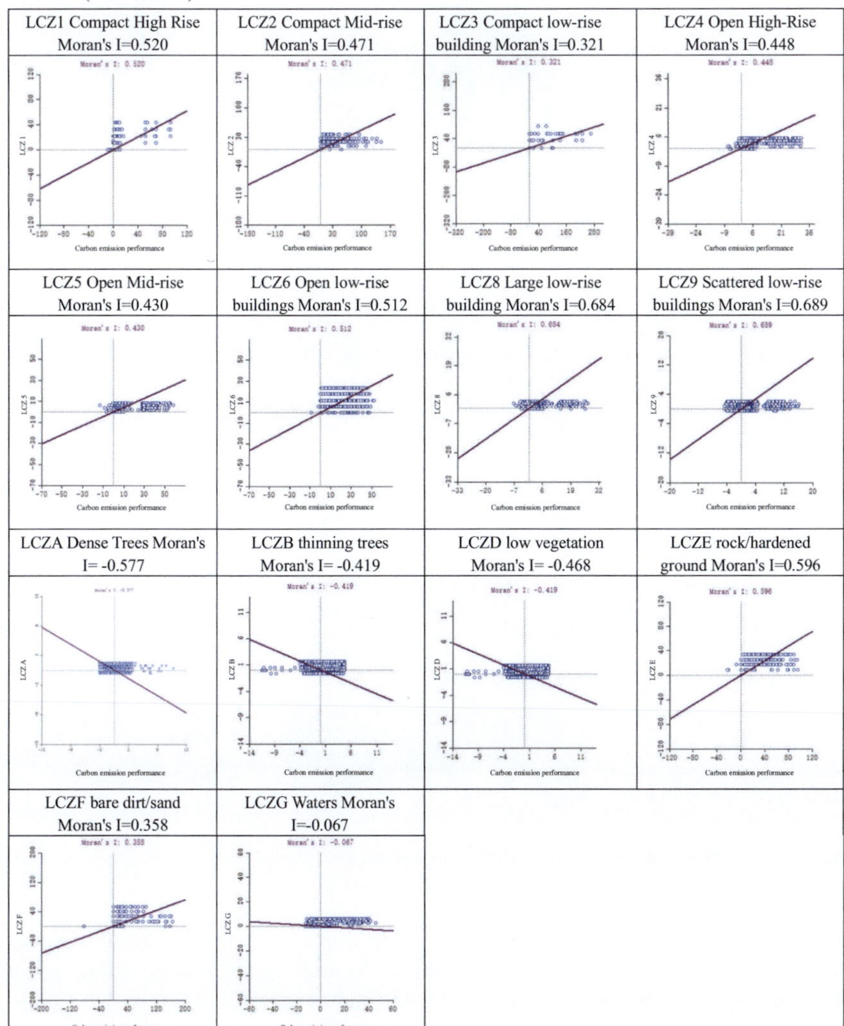

4.2 *Characterizing the Spatial Distribution of Carbon Emission Performance*

According to Eq. 1–3, the total carbon emissions for the central city of Chongqing amount to 6022.620×10^{-4} tons. Utilizing ArcGIS 10.7 software and in conjunction with the analysis provided by Eqs. 4–5, the distribution pattern of carbon emission

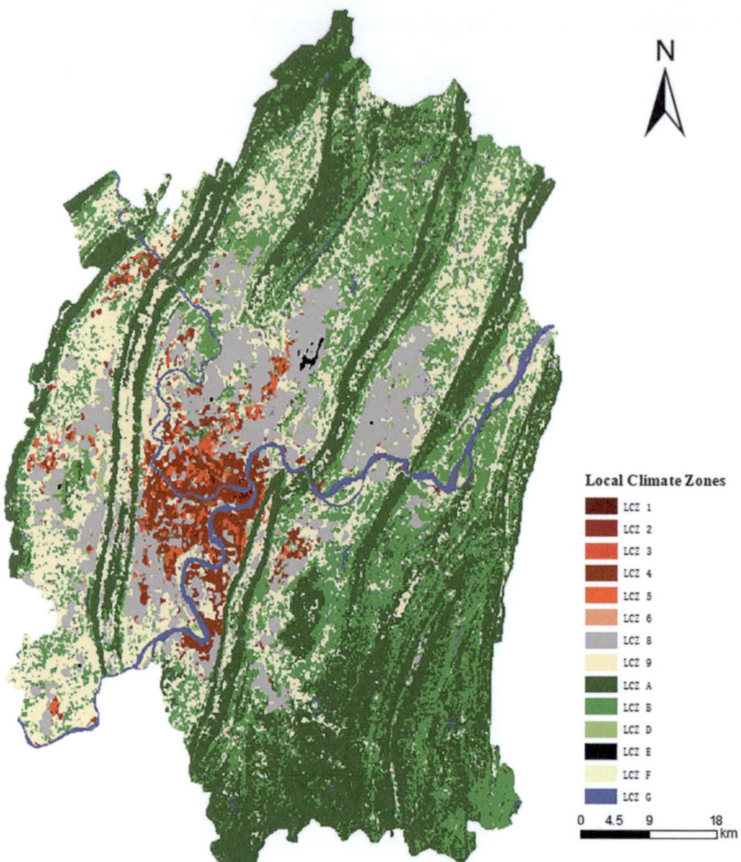

Fig. 2 Map of local climate zones in central Chongqing (Self-drawn)

performance can be elucidated (see Fig. 3). In 2022, the carbon emission performance in Chongqing's central city exhibits a spatial configuration characterized by a decline from the central core to the peripheral regions, resulting in a three-tiered distribution pattern. This spatial distribution is categorized into high-value areas, medium-value areas, and low-value areas. The high-value areas encompass Yuzhong District, Jiulongpo District, and Jiangbei District, which are situated within the core of the main city and exhibit a high degree of urbanization. These regions, marked by elevated carbon emission performance, indicate substantial spatial carbon emissions and limited success in achieving carbon reduction targets, thereby necessitating significant efforts for carbon mitigation. The medium-value areas include Yubei District, Shapingba District, and Nan'an District, located in the central urban area's core periphery. Although these areas fall within the medium-value range for carbon emission performance, their spatial carbon emissions should not be underestimated. Conversely, the low-value areas consist of Banan, Beipai Dam, and Dadukou

Districts, which are positioned on the outskirts of the central city. The spatial distribution of carbon emission performance underscores notable regional disparities in carbon emissions resulting from energy consumption within Chongqing's central urban areas. These disparities manifest in two primary dimensions: firstly, a decreasing trend from the "core of the main city" towards the "two wings" (northeast and southeast); and secondly, a reduction in carbon emissions from riverine zones to mountainous terrains, particularly highlighting a peak in carbon emissions at the core of the main city. Notably, this peak gradually diminishes towards the mountainous regions in the northeast and southeast, thereby establishing a distinct low-carbon zone. Areas exhibiting lower carbon emission performance reflect reduced spatial carbon emissions and a higher degree of success in meeting carbon reduction targets.

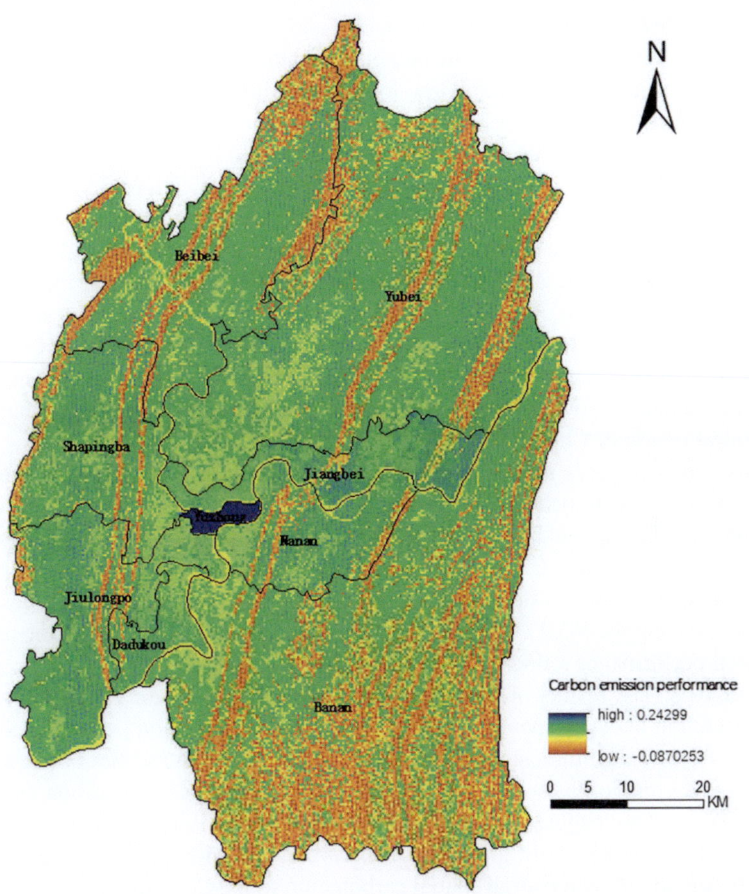

Fig. 3 Spatial distribution of carbon emission performance in central Chongqing (Self-drawn)

4.3 Spatial Correlation Analysis of Local Climate Zoning and Carbon Emission Performance

Global Spatial Autocorrelation Analysis

This study employs GeoDa 1.22 software to analyze the Local Climate Zone (LCZ) values and carbon emission performance data for the central urban area of Chongqing. The analysis utilizes LCZ values as the primary variable and carbon emission performance values, derived from Eq. 5, as the secondary variable to conduct a bivariate spatial autocorrelation analysis. The findings reveal that at a spatial scale of 100m x 100m, both the absolute value of the Z-test at a 99% confidence level and the absolute value of Moran's I are significantly elevated, with a Moran's I value of -0.175 (refer to Fig. 4) and a significance P-value of 0.001. This indicates that the spatial relationship between the local climate zones and carbon emission performance in Chongqing's central area is not attributable to random processes.The first quadrant (high-high) and the third quadrant (low-low) of the Moran's I scatterplot demonstrate a robust positive spatial correlation among the elements, suggesting that regional units with high (or low) observed values are surrounded by other regions exhibiting similar high (or low) values, thereby indicating spatial homogeneity. Conversely, the second quadrant (low–high) and the fourth quadrant (high-low) reveal a strong negative correlation, illustrating a spatial configuration where regional units with low (or high) values are encircled by areas with high (or low) values, indicative of spatial heterogeneity.As illustrated in Fig. 4, The quantity of regions located in the second and fourth quadrants consistently surpasses that of the first and third quadrants throughout all three examined periods. This pattern highlights a significant negative spatial correlation between local climate zones and carbon emission performance in the central region of Chongqing during the designated timeframe.

This study employs spatial autocorrelation analysis to investigate the intricate relationship between local climate zones (LCZ) and carbon emission performance. Initially, bivariate spatial autocorrelation analysis was conducted using GeoDa 1.22 software to evaluate the LCZ values for each type of local climate zone alongside their corresponding carbon emission performance values. The individual LCZ values were derived utilizing the WUDAPT mapping method in ArcGIS 10.7, while the carbon emission performance values were calculated in accordance with Eq. 5, also using ArcGIS 10.7. The results of this analysis, presented in Table 6, visually illustrate the correlation between specific LCZ types and carbon emission performance. The analysis of the Local Indicators of Spatial Association (LISA) scatter plot indicates a positive correlation between the built landscape LCZ types and carbon emission performance. This suggests that an increase in built landscape LCZ types within urban areas correlates with an increase in carbon emission performance, thereby contributing to elevated overall spatial carbon emissions. Consequently, this finding underscores the necessity for urban planners and policymakers to prioritize the design and arrangement of built landscapes during urbanization efforts to mitigate their adverse effects on carbon emissions.

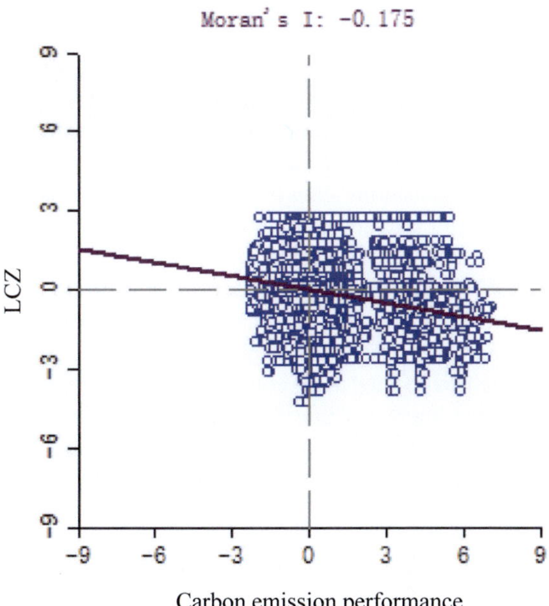

Fig. 4 Scatterplot of Moran's I index of localized climate zoning and carbon emission performance in 2022 (Self-drawn)

Among the natural-covered LCZ types, two categories—LCZE (rocky/hardened ground) and LCZF (bare soil/sand)—exhibit a positive correlation with carbon emission performance. This implies that a greater presence of impermeable urban subsurfaces, such as rock, hardened ground, and bare soil or sand, is associated with higher carbon emission performance. This observation further highlights the significance of urban subsurface types in relation to carbon emissions, suggesting that urban planning should aim to minimize the use of impermeable surfaces to curtail carbon emissions.

Conversely, other natural cover LCZ types, excluding LCZE and LCZF, demonstrate a negative relationship with carbon emission performance. This indicates that an increase in ecological elements, such as natural vegetation and watersheds, within urban environments correlates with a decrease in carbon emission performance and a reduction in spatial carbon emissions. The findings of this research offer a viable strategy for carbon emission reduction in urban settings, advocating for the enhancement of ecological components, such as natural vegetation and watersheds, to achieve carbon emission targets.

In conclusion, this paper elucidates the complex interplay between various LCZ types and carbon emission performance through bivariate spatial autocorrelation analysis. The insights gained not only furnish researchers with robust evidence for a more profound understanding of the mechanisms driving urban carbon emissions but also provide a scientific foundation for urban planners and policymakers to develop effective strategies for carbon emission reduction.

Local Spatial Autocorrelation Analysis

In order to be able to better reflect the spatial correlation and significance between neighboring regions, the 100mx100m scale with strong spatial aggregation was selected, and local bivariate spatial autocorrelation analysis was performed on the study area by GeoDa1.22 software to generate the LISA significance map (Fig. 5) and LISA clustering map (Fig. 6). From Figs. 5 and 6, it can be seen that LCZ has a significant spatial aggregation relationship with carbon emission performance, the higher the value of LCZ the lower the carbon emission performance, and the lower the LCZ located in the periphery of the central city, the lower the carbon emission performance.The areas with high one-high correlation type in 2022 are mainly scattered and clustered in the Shapingba, Nanan, Yuzhong, Yubei, and Beipei districts, all of which are located in the core area of the main city, with a high level of urbanization, and urban interact with each other and become high and high agglomeration areas; the number of low and low association type areas is higher, forming the Banan and Jiulongpo districts to the southeast of the central city, and some of these districts have low levels of industrialization and urbanization, and are mainly based on ecological protection, with forest land being the main land use type; the low and high association type areas include the Dadukou district, which has a relatively low level of carbon emissions of its own but is surrounded by cities with high carbon emissions, thus forming the low–high type.

5 Conclusion and Discussion

5.1 *Reach a Verdict*

Based on the local climate zoning system, this paper analyzes the correlation between local climate zoning and carbon emission performance in the central urban area of Chongqing, and finally draws the following conclusions: First, this paper uses the WUDAPT mapping method to obtain the local climate zoning map of the central urban area of Chongqing, and there are a total of 14 LCZ types in the map, including 8 built-up landscape types and 6 natural cover types. From the classification results of built landscape-type LCZs, the core area of the central city is mainly dominated by built landscape-type LCZs, and the edge of the core area of the central city is filled with built-landscape-type LCZs, which is more diversified than the built LCZs in the main city; from the results of the classification of natural cover type LCZs, the central city is highly fragmented in terms of the "blue-green" ecosystems. From the classification results of natural cover type LCZs, the "blue-green" ecosystems in the central city show a highly fragmented distribution, with only the Yubei District and Banan District having continuous "green" ecological patch systems, and the distribution of "blue-green" spaces is relatively independent, without forming a complete ecological network. Second, the results of carbon emission performance accounting show that the technical efficiency value of each district in the central urban area of

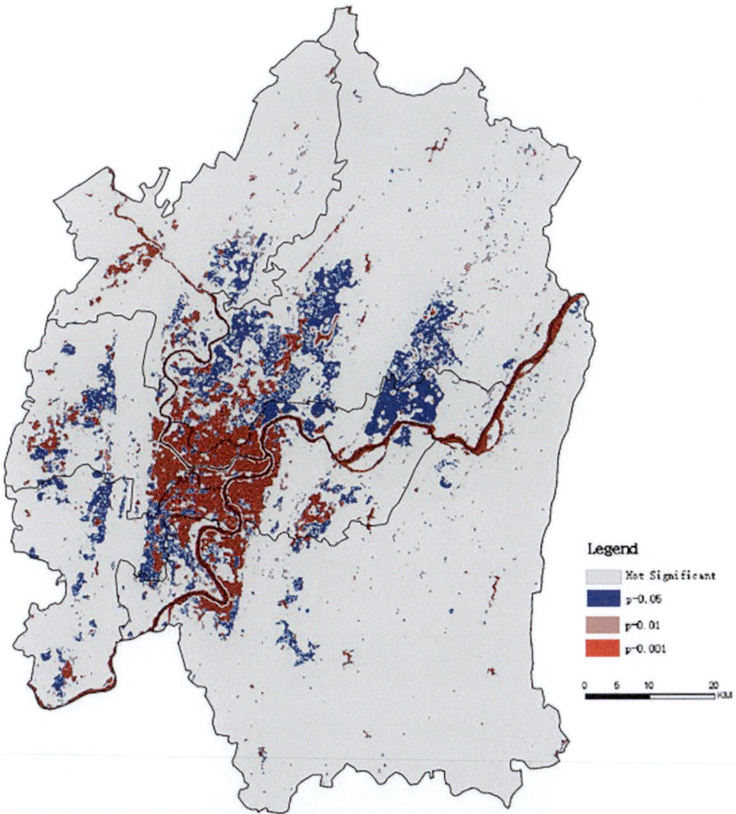

Fig. 5 LCZ versus Carbon Emissions Performance LISA Significance Plot *Source* Self-drawn)

Chongqing is generally low, among which the efficiency of the core area of the central urban area is higher, and the efficiency of the core area in the southeast and northeast directions is lower. The "technical catch-up" effect is not obvious in each region of the central area, and the economy of scale effect of land use is not significant, and the cost of high-speed economic development in some regions is the high consumption of land resources and the high emission of pollutants, and the development mode of high input and high growth needs to be further optimized. Third, the central urban area of Chongqing has a spatially negative correlation between local climate zoning and carbon emission performance and has spatial clustering characteristics. Further analysis of the correlation between each LCZ type and its corresponding carbon emission performance shows that the built landscape-type LCZ type is positively correlated with the carbon emission performance, that is, the more built landscape-type LCZ type, the higher the carbon emission performance because the continuous expansion of urban construction land will bring about an increase in engineering and

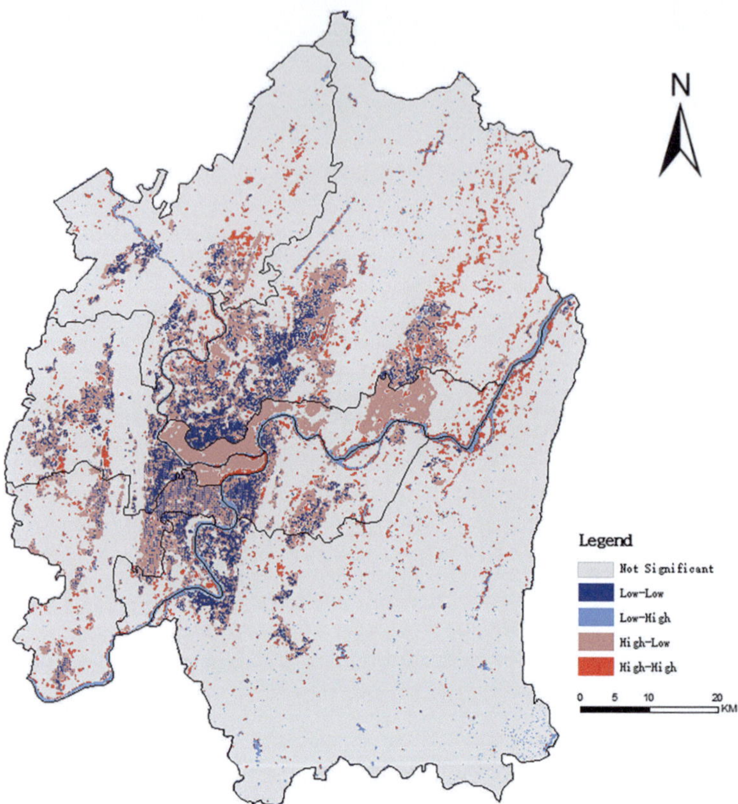

Fig. 6 Lisa clustering of LCZ and carbon performance *source* self-plotted

construction and a convergence of the population, which will increase the consumption of resources and crowd out the surrounding ecological land, arable land, and watersheds, resulting in an increase in carbon sources and a decrease in carbon sinks, thus resulting in an increase in carbon sources and a decrease in carbon sinks. In the natural cover LCZ type, LCZE rocky/hardened ground and LCZF bare soil/sand are positively correlated with carbon emission performance, that is, the more impermeable the urban sub-surface in the city, the higher the carbon emission performance. This is due to the fact that the increase in impermeability reduces the vegetation cover, which affects the carbon cycle, resulting in an increase in carbon emissions and a decrease in carbon emission performance. Except for the above two types, the other local climate zoning types were negatively correlated with carbon emission performance, that is, the more natural vegetation and watersheds, the lower the carbon emission performance. This is because the greater the blue-green space, the more carbon sinks, the lower the spatial carbon emissions, and the higher the degree of achieving the carbon emission targets.

In conclusion, the construction planning of Chongqing's central city should comprehensively consider the impact of urban expansion on carbon emission performance, control the infinite expansion of built landscape LCZ types, and expand the coverage area of blue-green spaces such as woodlands and water bodies, which will help to reduce carbon and increase sinks, and realize the green, low-carbon and sustainable development.

5.2 Talk Over

The findings from the aforementioned studies indicate that the economic development of Chongqing's urban center is achieved at the expense of significant inputs and emissions related to land resources. This suggests that the economic growth observed, which occurs under the premise of disregarding carbon emissions, may be characterized as somewhat "artificially inflated." Consequently, it is essential to expedite the transition toward a more sustainable economic development model that emphasizes low-carbon utilization of land resources in Chongqing.

Through a comprehensive analysis of empirical research, this paper posits the following arguments:

(1) Regarding land use planning, the Local Climate Zone (LCZ) theory serves as a foundational framework for evaluating the energy consumption and carbon emission performance of low-carbon eco-urban areas. This theory facilitates the identification of spatial variations in carbon emission performance and the delineation of low-emission zones when selecting such areas. Furthermore, during the planning assessment phase, it can be employed to evaluate the carbon characteristics of proposed sites. In the planning control phase, the integration of LCZ morphological characteristics can optimize urban layouts and inform adjustment strategies that align LCZ types with site attributes.

(2) In terms of index management, the urban morphology and functional data derived from LCZ theory exhibit a high degree of compatibility with low-carbon ecological policy standards. Key indicators related to energy efficiency, land conservation, material savings, ecological protection, and the creation of comfortable environments—such as ground cover, geometric form, and material composition—are effectively encompassed by LCZ theory. This data can serve as a foundational layer for evaluation and planning decision-making; during the control phase, the incorporation of LCZ morphological indicators into planning regulations will enhance the optimization of relevant indicators and their spatial arrangements.

(3) In the context of safeguarding "blue-green spaces," it is crucial to prioritize the preservation of natural landscapes and biodiversity to protect "blue-green" ecosystems. Specific strategies may include the protection of agricultural land, the establishment of ecological agriculture demonstration zones, the enhancement of forest management practices, the achievement of a balance between

harvesting and breeding, the improvement of forest quality, the prevention of soil erosion, the implementation of natural forest protection initiatives, the intensification of shrub forest cultivation, and the overall enhancement of forest quality and natural vegetation. Concurrently, adjustments to land-use patterns, comprehensive management of small watersheds, and the application of both engineering and biological measures to restore vegetation and increase forest cover should be undertaken.

This study had some limitations. First, there are temporal and spatial limitations to the research content. In the future, the LCZ system can be utilized to compare multiple cities in different periods to study the impact of the development and evolution of urban form on the urban climate. Second, this paper gives urban planning advice from the perspective of microclimate, but in the future, more complete planning advice can be given by considering other factors such as ventilation and building energy consumption.

References

1. Zhao L, Lee X, Smith RB et al (2014) Strong contributions of local background climate to urban heat Islands. Nature 511(7508):216
2. Houg H'TON KA (1999) The Anmual net flux of carbon to the atmosphere from changes in Land use 1850–19901. Telus 5l(2):298–313
3. Weng QH (2009) Thermal infrared remote sensing for urban climate and environmental studies: methods, applications, and trends. ISPRS J Photogramm Remote Sens 64(4):335
4. Stewartid, OKE TR (2012) Local climate zones for urban temperature studies. Bull Am Meteorol Soc 93(12):1879
5. Oke TR (1967) City size and the urban heat island. Atmos Environ 7(8):769
6. Patz JA, Campbell-Lendrum D, Holloway T et al (2005) Impact of regional climate change on human health. Nature 438(7066):310
7. Jiacong HU, Qijiang ZHU (2010) Progress of urban heat island research. J Beijing Norm Univ (Nat Sci Ed) 46(2):186
8. Stewart ID, Oke TR, Krayenhoff ES (2014) Evaluation of the local climate zone scheme using temperature observations and model simulations. Int J Climatol 34(4):1062–1080
9. Geletič J, Lehnert M, Dobrovolný P (2016) Land surface temperature differences within local climate zones, based on two central European cities. Remote Sens 8(10):788
10. Wang C, Middel A, Myint SW et al (2018) Assessing local climate zones in arid cities: the case of Phoenix, Arizona and Las Vegas, Nevada. ISPRS J Photogramm Remote Sens 141:59–71
11. Wang S, Gao S, Huang Y et al (2020) Spatial and temporal evolution pattern and prediction of carbon emission performance of Chinese cities based on super-efficient SBM model. J Geogr 75(6):1316–1330
12. Stretesky PB, Lynch MJ (2009) A cross-national study of the association between per capita carbon dioxide emissions and exports to the United States. Soc Sci Res 38(1):239–250
13. Wang S, Gao S, Huang Y et al (2020) Spatio-temporal evolution and trend prediction of urban carbon emission performance in China based on super-efficiency SBM model. Acta Geographica Sinica 75(6):1316–1330
14. Cheng Y, Zhang Y, Wang J (2023) Spatial-temporal evolution of provincial carbon emission performance and driving Spatial-temporal evolution of provincial carbon emission performance and driving force of technological innovation in China. Scientia Geographica Sinica 43(2):313–323

15. Suyun F, Kaixuan Z, LU L (2018) Remote sensing analysis of environmental changes in megacities along the Maritime Silk Road. J Geo-Inf Sci 20(5):602
16. Stewart ID, Oke TR, Krayenhoff ES (2014) Evaluation of the 'local climate zone' scheme using temperature observations and model simulations. Int J Climatol 34(4):1062
17. Geletic J, Lehnert M (2016) GIS-based delineation of local climate zones: the case of medium-sized Central European cities. Moravian Geographical Reports 24(3):2
18. Mitraka Z, Del Frate F, Chrysoulakis N et al (2015) Exploiting earth observation data products for mapping local climate zones: IEEE. Joint Urban Remote Sensing Event (JURSE), 20 15:1
19. Qiu CP, Mou LC, Schmitt M et al (2019) Local climate zone-based urban land cover classification fom multi.seasonalSentinel-2 images with a recurrent residual network. ISPRS Joural Photogramm Remote Sens 154:151
20. Bechtel B, Alexander PJ, Böhner J et al (2015) Mapping local climate zones for aworldwide database of the form and function of cities. Isprs Int JOf Geo-Inf 4(1):199–219
21. Bechtel B, Alexander P et al Generating WUDAPT Level 0 data -Current status of produetion and evaluation. Urban Climate 201(9)27:24
22. Ching J, Mills G, Bechtel B et al (2018) WUDAPT: an urban weather, climate, and environmental modeling infirastructure for the anthropocene. Bull Am Meteorol Soc 99(9):1907
23. Zhu Q, Mei Y, Chen Y et al Regional differentiation characteristics and optimization of the structural efficiency of land use in Hubei province based on the carbon emissions. economic Geogra
24. Han J, Zhou X, Xiang W (2016) Progress in research on land use effects on carbon emissions and low carbon management. Acta Ecologica Sinica 36(4):1152–1161
25. Sun H, Liang H, Chang X et al (2015) Land use patterns on carbon emission and spatial association in China. Econ Geogr 35(3):154–162
26. Wang H, Zeng Y (2015) Emissions in Eastern part of Qinghai Plateau. Geogr Res 34(7):1270–1284
27. Sun X (2012) Effects of carbon emission by land use patterns in Hefei's economic circle of Anhui province. J Nat Resour 27(3):394–401
28. Zhao R, Huang X, Zhong T et al (2013) Carbon effect evaluation and low-carbon optimization of regional land use. Trans Chin Soc Agric Eng 29(17):220–229
29. Fang JY, Guo Z, Piao S et al (2007) Terrestrial vegetation carbon sink in China from 1981–2000. China Series D: Earth Sci 37(6):804–812
30. Wang G, Zhang H, Xue F et al (2017) Relations between land use carbon budget and economic development at county level in Chengdu city. J Nat Resour 32(7):1170–1182
31. Shi H, Mu X, Zhang Y et al (2010) Effects of different land use patterns on carbon emission in Guangyuan city of Sichuan province. Effects of different land use patterns on carbon emission in Guangyuan city of Sichuan province. Bull Soil Water Conserv 32(3):101–106
32. Anselin L, Smirnov O (1996) Efficient algorithms for constructing proper higher order spatial lag operators. J Regional Sci 36(1):67–89

BY NC ND

Supply–Demand Matching Analysis of Urban Blue-Green Space Cooling: A Study of the Main Urban Area, Hangzhou, China

Yuhang Shi, Yijun Shi, Xinyu Zhou, Junqing Tang, and Wei Zhai

Abstract As urbanization accelerates, the thermal environment of urban areas has become a critical aspect of city planning. However, a mismatch between cooling supply and demand persists, and its underlying factors are not fully understood. This study calculates the supply–demand index for urban blue and green spaces and utilizes the LISA model to pinpoint areas with imbalanced cooling supply and demand. Key findings include: 1) Surface temperatures in Hangzhou's main urban area rose from 2016 to 2022, exacerbating thermal issues; 2) The supply–demand indices of blue and green spaces show spatial variation; 3) Areas with low supply and high demand are closely linked to socio-economic and built environments, while those with high supply and low demand are more influenced by natural factors. These results offer insights for improving urban heat and cooling strategies.

Keywords Surface temperature · Cooling correlation · Supply–demand matching · Main urban area of Hangzhou City

Y. Shi · Y. Shi (✉) · X. Zhou
College of Landscape Architecture, Zhejiang Agricultural and Forestry University, Hangzhou, China
e-mail: yijun_shi@zafu.edu.cn

Y. Shi
e-mail: 849468251@qq.com

J. Tang
School of Urban Planning and Design, Shenzhen Graduate School, Peking University, Shen-Zhen, China

W. Zhai
College of Architecture and Planning, The University of Texas at San Antonio, Texas, USA

B. He et al. (eds.), *Climate-Resilient, Low-Carbon, and Sustainable Cities*, Lecture Notes in Civil Engineering 754, https://doi.org/10.1007/978-981-95-2169-2_13

1 Research Background

Urban development has worsened global warming, leading to higher temperatures and increased heat-related issues in cities [1]. Global temperatures have risen by 1.52°C since the twenty-first century, mainly due to human activities [2]. Extreme heat events are more frequent, affecting daily life and productivity [3]. For example, the 2022 European heatwave caused 62,000 deaths [4], and high temperatures in 2019 disrupted power for 73,000 people [5]. Heat can reduce work efficiency by 10%, impacting socio-economic development [6]. Blue and green spaces, vital for cooling, are diminishing due to urban expansion, resulting in insufficient and unevenly distributed cooling [8]. Therefore.analyzing the supply–demand balance of these spaces is essential.

A comprehensive review of past research reveals that studies on the urban heat island effect and the benefits of green spaces can be divided into two main categories: research content and research methods. Regarding the content, the current consensus focuses on several key elements, including the target population, cooling mechanisms, systems, the latest cooling thresholds, and urban form These studies explore how green spaces contribute to temperature regulation and what factors determine their effectiveness.

From the perspective of target populations, such studies mainly focus on individual-level, non-systematic research, measuring thermal environments and developing vulnerability assessment frameworks for urban populations. They investigate how city surface temperatures affect both bodily and psychological well-being. By analyzing data such as age structure and socio-economic status, these studies examine the variation in surface temperatures across different times of day and night, revealing that income disparity and unequal living conditions exacerbate the negative health impacts of extreme heat on vulnerable populations. By examining factors such as age distribution and socio-economic conditions, these studies investigate the fluctuations in surface temperatures throughout the day and night. They emphasize how income inequality and poor living conditions exacerbate the negative health impacts of extreme heat on vulnerable communities [9]. Other research has shown that urban heat risks can worsen mental health issues [10]. Beyond physical and mental health, much of the research also addresses outdoor thermal comfort under high temperatures. In Hong Kong, researchers assessed the city's overall thermal comfort and identified variations in outdoor thermal comfort at different times of day [11]. In Xiamen, researchers focused on small-scale urban parks and investigated the relationship between thermal discomfort factors and outdoor space design [12].

From the perspective of impact mechanisms, research in this area primarily focuses on the cooling mechanisms of blue-green spaces. These spaces can lower temperatures through processes such as vegetation evapotranspiration and heat reflection [13]. Evidence from both research studies and real-world applications demonstrates the effectiveness of urban green spaces in mitigating heat buildup in cities

[14]. Numerous research findings suggest that the region, shape, and layout of urban green–blue areas can affect city surface temperatures [15].

From a systems perspective, this type of research focuses on analyzing cooling mechanisms under various environmental conditions. Human society is typically regarded as a large complex system, influenced by multiple factors [16, 17], but it can generally be considered in terms of three main aspects: socio-economic factors, the built urban environment, and the natural environment. For example, Liu Qing and his team employed machine learning techniques, particularly Gradient Boosting Decision Trees (GBDT), to explore the relationship between Beijing's urban landscape and the frequency of severe heatwaves. Their research identified several key factors within the city's built environment that contribute to this correlation [18]. The expansion of urban areas and the widespread presence of hard surfaces are key contributors to the urban heat island effect. In addition to the physical characteristics of the environment, socio-economic activities also play a significant role in driving fluctuations in surface temperatures. During the day, urban high temperatures strongly correlate with road density and social activities, while this relationship reverses at night [19]. Nature plays a crucial role in cooling urban environments, the amount of greenery in cities significantly impacts the cooling of surfaces. Different sizes of parks and green spaces provide varying levels of cooling, with larger areas generally offering more substantial temperature reductions to their surrounding areas [20]. Research conducted in Copenhagen found that the design and connectivity of blue-green spaces are essential to their cooling effectiveness. The study showed that green areas with complex shapes and strong interconnections are particularly effective in alleviating the urban heat island effect [21].

Furthermore, recent research on the cooling benefits of urban green spaces explores the concept of cooling thresholds and examines how urban form influences temperature regulation. This research aims to identify the critical factors that determine the effectiveness of green spaces in mitigating heat. From the perspective of cooling thresholds, In Harbin, researchers conducted a study using small urban parks and ENVI-met simulations based on observed data to investigate the real-world cooling potential of small water features and green spaces. The results revealed that different types of water bodies and green areas have varying cooling capacities. Furthermore, the combination of blue and green spaces created distinct spatial and temporal variations in cooling effects. Notably, a cooling gradient was observed, becoming more pronounced as the distance from the water bodies increased [22]. Globally, researchers have been using machine learning algorithms to analyze the factors that contribute to the cooling effects of urban parks. They have also mapped the varying distances over which these factors extend their cooling impact [23]. In Beijing's Fifth Ring, studies show spatial variations in the influence of 2D and 3D urban forms on surface temperatures throughout different daylight hours [24]. Additionally, a study in Nanjing analyzed local areas through modeling and land-use classification, finding a strong correlation between urban morphological changes and surface temperatures [25].

In conclusion, studies on urban blue-green spaces focus on their cooling potential and the thermal comfort of city residents during heatwaves However, the distribution of blue-green spaces, population density, road density, and the intensity of social activities exhibit spatial differences. As a result, some areas experience either an oversupply or a shortage of cooling from urban blue-green spaces, leading to spatial variability in the availability of urban cooling For example, areas with higher socio-economic activity, such as city centers, typically have a greater demand for cooling but often experience insufficient cooling supply, which exacerbates the urban heat island effect [26] and poses a significant risk to the urban thermal environment. Previous studies on the cooling effects of urban blue-green spaces have not considered the relationship between actual cooling demand and the supply provided by these spaces, leading to a mismatch between urban cooling and high-temperature conditions. Therefore, it is essential to incorporate a supply–demand approach into urban cooling strategies. The key to effective cooling lies in whether it can optimally and efficiently meet the urban cooling demands. These studies indicate that the supply–demand matching in urban cooling needs further exploration.

In terms of research methodology, this field predominantly relies on regression and clustering techniques. Common methods include linear and nonlinear techniques such as least squares, Pearson correlation, PCA and GWR. For clustering, popular methods include K-means, hot/cold spot clustering, coupling coordination clustering, and bivariate spatial autocorrelation clustering (Table 1).

Despite many studies delving into urban heat and the cooling effect of blue-green space, few from economic perspectives analyze supply-side cooling or uptake, and current methods have defectslimitations. Therefore, this paper fills these gaps by looking from the supply and demand angles at cooling in distinct blue-green sites

Table 1 Comparison of regression and clustering analysis methods (Self-drawn)

Model type	Model name	Advantages	Disadvantages
Regression analysis	OLS	High interpretability and concise data calculation	The exception value could not be processed
	Pearson	High interpretability and concise data calculation	Non-linear relationships cannot be detected
	GWR	Spatial heterogeneity	Computation is complex
Cluster analysis	K—means	High computational efficiency and strong adaptability	The K value setting is subjective and depends on the initial value
	Getis-Ord general G	High interpretability and concise data calculation	Spatial factors and complex big data computations
	Coupling coordination degree model	Strong adaptability	Failure to account for nonlinear relationships
	Local indicators of spatial association	Effective management of non-linear relationships and quick calculations	Relying on raw data

within modern cities. It uses surface temperature measurements and a heating and cooling supply-and-demand index. With its 500m spatial resolution analysis, this study could also have global implications.

This study lays its scientific foundation for managing urban heat through urban blue-green supply and demand. The article is organized accordingly: Chap. 2 provides the research materials and methods; Chap. 3 reports the results; Chap. 4 discusses these consequence, and finally in formal writing we discuss what has already been said up to this point as a conclusion already reached.

2 Materials and Methods

2.1 Study Area and Data

Hangzhou, the capital of Zhejiang Province, China, is becoming one of the country's "hot cities." With 66.9% forest cover, the city offers ample blue-green space. The urban area spans 3,360 square kilometers, and in the past decade, average surface temperatures have risen by 1.4 °C, with summer highs often exceeding 40 °C for 40–50 days [34]. As urbanization grows, the central area's population density has increased, raising temperatures significantly higher than in surrounding blue-green spaces. This makes Hangzhou's urban core an ideal case for studying the cooling supply and demand of urban blue-green spaces. This paper focuses on this area to explore that relationship (Fig. 1).

This study takes into account the availability and accuracy of data, and therefore plans to utilize remote sensing data for the research. Research data comes from the GEE platform, specifically the Landsat 8 C2L2 dataset, with all Remote sensing images collected from the years 2016 and 2022 (Table 2).

2.2 Grid Selection

Different grid sizes are suited to varying scenarios and data precision, so this study analyzes and compares several related studies (Table 3). When selecting an appropriate grid size, it is essential to consider not only the applicable scenario and data precision but also the analytical outcomes. For instance, analyzing the cooling capacity of blue-green areas within a 300-m grid generated landscape pattern indices that produced relatively accurate results [35]. In Hangzhou, researchers analyzed the link between water bodies, land cover, and surface temperature at 510 m, 990 m, and 1500 m scales. They found the strongest correlation at 510 m for water bodies and 990m for land cover [36]. In Beijing, researchers found that at 300 m, green spaces and water bodies had the strongest daytime impact on surface temperature, while the 600 m scale was most effective at night [37].

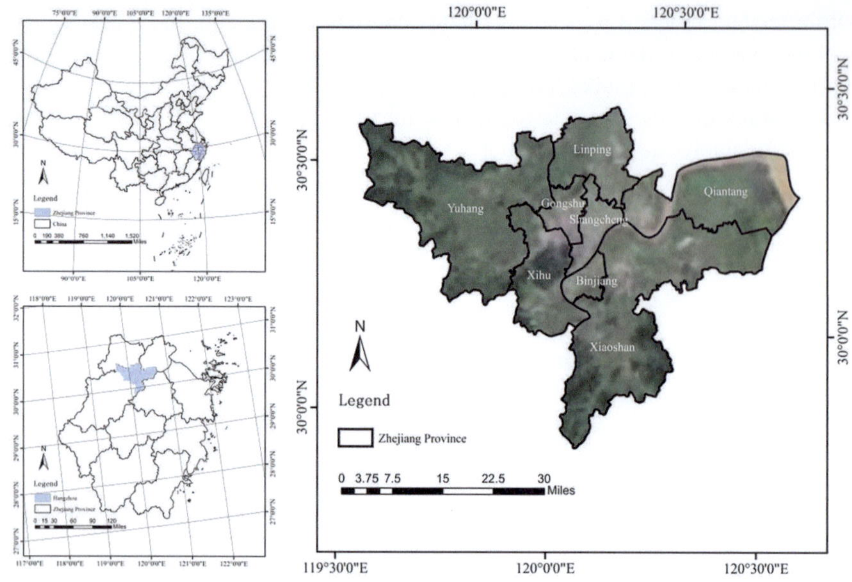

Fig. 1 Study area (Self-drawn)

Table 2 Data Sources (Self-drawn)

Data Name	Data format	Data sources	Year
LST	Tif	Google Earth Engine https://earthengine.google.com/	2016–2022
NDBI、NDWI、NDVI	Tif		
PD、COHESION、LSI	Tif		
AREA	Tif		

Table 3 Applicable scenarios for blue-green spaces of different scales (Self-drawn)

Grid size	Applicable scenarios	Data precision	Source
120	Small scale community parks and water bodies within high-density urban areas	30	Liwen [38]
300	Small scale water bodies and green spaces within cities	30	Peng [39]
500	Urban scale blue-green space	30	Xiangyi [40]
1200	Blue and green spaces in mega cities and urban agglomerations	500	Zhongjian et al. 2022

Due to the original data having a precision of 30 m, calculations using the moving window method for various landscape metrics showed that a 500 m grid size was both suitable for the analysis and the applicable scenario. Therefore, this study selects a 500 m grid scale as the research scale.

2.3 Research Methods

Examination of Influential Factors on Surface Temperature. The elements impacting urban surface temperature can be classified into three tiers: the constructed environment, natural surroundings, and socioeconomic behaviors. These various factors also interact with each other (see Fig. 2). In the socioeconomic realm, population density, nighttime light index, and functional mix can reflect the activity demands of urban populations to some extent [41]. The urban landscape is primarily shaped by factors such as building concentration, road density, the floor area ratio, and proximity to green spaces. These elements—building and road density, along with the floor area ratio—affect the reflectivity of urban areas. In contrast, the presence of nearby green spaces significantly influences the thermal environment of the surrounding urban surfaces [42]. In terms of the natural environment, key factors include the area of blue-green spaces, patch density, aggregation, and shape [43]. These elements can influence heat exchange in surrounding areas, thereby altering energy exchange dynamics.

Urban surface temperature is affected by a range of interconnected factors. This research investigates 13 influencing elements categorized into three domains: socioeconomic activities, the built environment, and the natural environment (refer to Table 4). It is important to note that there is significant overlap among the landscape pattern indicators [44].To tackle this issue, we opted for highly representative indicators by selecting one from each area, density, shape, and connectivity category. The landscape pattern indices were determined using a moving window technique, with

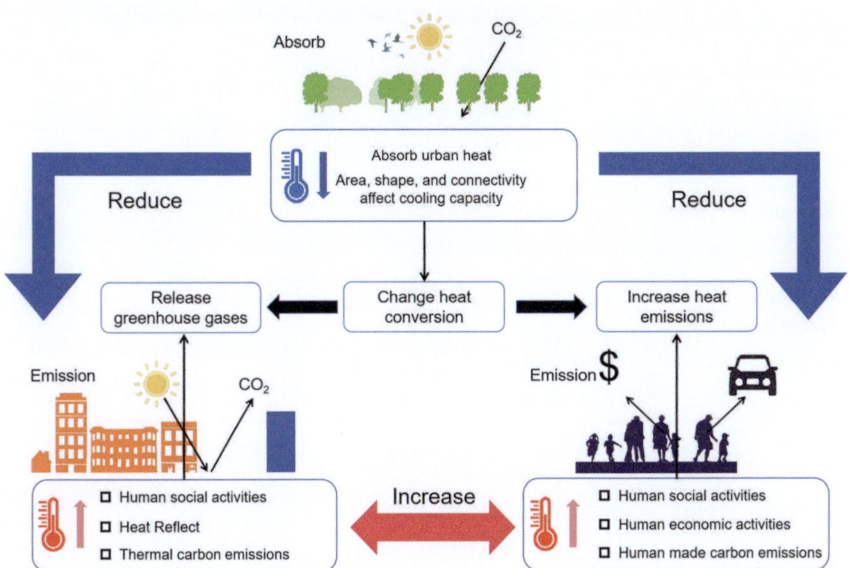

Fig. 2 Analysis of Influence Mechanism (Self-drawn)

the window size aligned with the grid. The other influencing factor indicators were analyzed through grid analysis, and the ultimate outcomes were acquired through resampling.

Methods for Measuring Thermal Environment. Current methods for calculating surface temperature include the Atmospheric correction method and single-window algorithm [45]. The radiative transfer equation method is advantageous because it can eliminate the effects of atmospheric conditions and illumination on surface reflectance [46], offering higher accuracy compared to other surface temperature algorithms. Therefore, this study utilizes the Google Earth Engine (GEE) data editor, applying median synthesis and the radiative transfer equation method. By calculating radiometric calibration, blackbody brightness radiation, and relative emissivity, we obtained the average urban surface temperatures for August 2016 and August 2022

Table 4 Calculation Indicators of CESL and CEDL (Self-drawn)

Index classification	Indicator name	Meaning of indicators	Formula
Urban natural environment	AREA	Cool—Patch area size	–
	PD	Cool—Patch density size	$PD = \frac{N_{number}}{A\_Area}$
	COHESION	Cool—Reflecting the connectivity and structural density within the landscape	$COHESION = \left[1 - \frac{\sum_{j=1}^{m} P_{ij}}{\sum_{j=1}^{m} P_{ij}\sqrt{a_{ij}}}\right]\left[1 - \frac{1}{\sqrt{AREA}}\right]$
	NDVI	Reflecting the distribution of vegetation cover	$\mathrm{NDVI} = \frac{\mathrm{NIR-RED}}{\mathrm{NIR+RED}}$
	NDWI	Reflecting the distribution of water bodies	$\mathrm{NDWI} = \frac{\mathrm{GREEN-NIR}}{\mathrm{GREEN+NIR}}$
	PLAND	Cool—Proportion of dominant plaques	$\mathrm{PLAND} = \frac{\sum_{j=1}^{n} A_{ij}}{\mathrm{AREA}} \times 100\%$
	LSI	Cool—Complexity around land use patches	$\mathrm{LSI} = \frac{0.25P}{\mathrm{AREA}}$
Social and economic activities	AREA	Heat—Patch area size	–
Urban built environment	PD	Heat—Patch density size	$PD = \frac{N_{number}}{A\_Area}$
	COHESION	Heat—Reflecting the connectivity and structural density within the lan	$COHESION = \left[1 - \frac{\sum_{j=1}^{m} P_{ij}}{\sum_{j=1}^{m} P_{ij}\sqrt{a_{ij}}}\right]\left[1 - \frac{1}{\sqrt{AREA}}\right]$
	PLAND	Heat—Proportion of dominant plaques	$\mathrm{PLAND} = \frac{\sum_{j=1}^{n} A_{ij}}{\mathrm{AREA}} \times 100\%$
	LSI	Heat—Complexity around land use patches	$\mathrm{LSI} = \frac{0.25N}{\mathrm{AREA}}$
	NDBI	Reflecting the distribution of impermeable surfaces	$\mathrm{NDBI} = \frac{\mathrm{NIR-SWIR}}{\mathrm{SWIR+NIR}}$

within Hangzhou's main urban area. The calculation method is as follows

$$L = a \times DN + b$$

$$T_b = \frac{K_2}{\ln((K/L) + 1)}$$

where, a and b represent the gain and offset, while K_1 and K_2 are predetermined constants. The brightness temperature is corrected using the surface emissivity.

$$T_s = \frac{T_b}{1 + \varphi \times T_b/\rho}$$

$$\varphi = \frac{p \times q}{c}$$

where, φ represents the emission wavelength, c is the Stefan-Boltzmann constant, p is Planck constant, and q is the photon speed.

Calculate the blue-green space cooling supply and demand level index. The blue-green space cooling supply and demand level index consists of two components: the Urban Cooling Supply level (CESL), which measures the cooling level of blue-green patches, and the Urban Cooling Demand level (CEDL), which reflects the cooling needs of urban areas. NDVI and the NDWI help identify urban green and water areas, making them useful for studying urban heat. Studies also show that the cooling capacity of blue-green spaces depends on factors like patch size, shape, isolation, and density. The NDBI is used to indicate impervious surfaces in cities. Since these indicators are spatially correlated, this study applies the moving window method, GWR, and composite index to calculate the cooling supply and demand index.

NDVI, NDWI and NDBI can be used to analyze vegetation cover, water cover and building cover, respectively. In the formulas, the near-infrared band, red band, green band, and shortwave infrared band are denoted by NIR、REDGREEN, and SWIR. PD represents patch density, while N_{number} and $A_{\_Area}$ indicate the number of patches and patch area, respectively. COHESION reflects the degree of patch aggregation, indicating the clustering and dispersion of patches across the landscape.In the formula, the area (m^2) of the j-th patch within the i-th landscape type is denoted by a_{ij}, and P_{ij} denotes the perimeter (m) of that patch. AREA represents the overall area of the landscape (hm^2).PLAND indicates the proportion of the largest patch area to the entire landscape area, where A_{ij} denotes the size (m^2) of the j-th patch in the i-th land cover type.When the percentage value of patch area approaches zero, it indicates a reduction in that patch type within the landscape. A ratio of 100 signifies that the landscape consists solely of one patch type. LSI represents the Landscape Shape Index, where N denotes the total perimeter of all landscape patches, and AREA refers to the total area covered by the patches.

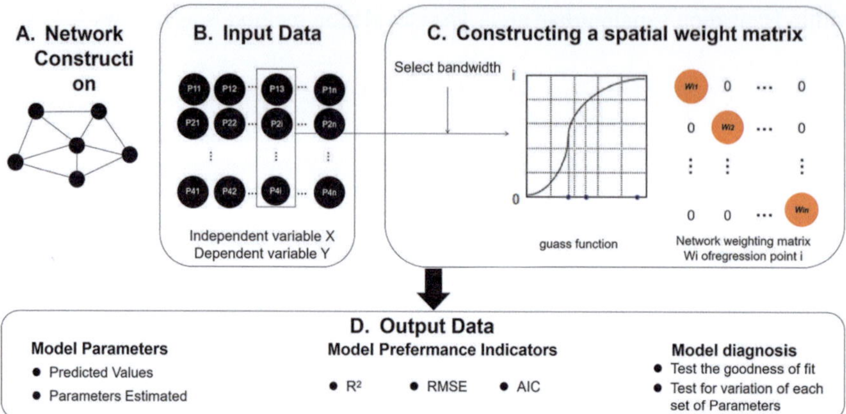

Fig. 3 Geographically weighted computation process (Self-drawn)

Landscape pattern indices exhibit nonlinear spatial relationships, and Geographically Weighted Regression (GWR) is highly advantageous for handling such spatial nonlinearity, offering greater stability. In this study, GWR is used to estimate the weights of each indicator, with GWR models being constructed using ordinary linear regression as the foundation, and in this study, Python (version 3.10) was used to conduct local regression analysis by embedding data and inputting spatial geographic locations into the model. A spatial weight matrix was constructed, and a Gaussian kernel function was applied for weight assignment (Fig. 3). The CESL and CEDL indices were then calculated using a composite index method.

Initially, a Gaussian spatial regression analysis is conducted on the random influence factor dataset. The standard equation is as follows:

$$p_i = \beta_j(x_i, y_i) + \sum_{k=1}^{q} \beta_m(x_i, y_i) n_{ik} + \varepsilon_i$$

$$i = 1, 2, 3, \ldots + \infty$$

In the method, the fitted value for the i-th pixel is represented by x_i, while y_i stands for the geographic position of the i-th sampling point. Additionally, n_{ik} refers to the k-th regression coefficient at the i-th sampling location, which is influenced by the position of that point.

Next, a spatial weight matrix is constructed, and its general formula is as follows:

$$W_{ij} = \begin{bmatrix} W_{11} & \cdots & W_{1n} \\ \vdots & \ddots & \vdots \\ W_{n1} & \cdots & W_{nn} \end{bmatrix}$$

In the method, Wij denotes the proximity between region i and region j, which can be used to express the neighboring relationship between adjacent regions.

To prevent bandwidth duplication and overfitting, the optimal bandwidth b for this study is determined using the AIC (Akaike Information Criterion). The formula is as follows:

$$AIC_c = -2\ln(\hat{\sigma}) + n\ln(2p) + n\frac{n + tr(S)}{n - 2 - tr(S)}$$

In the method, AICc is the corrected Akaike Information Criterion estimate, n is the number of sample spatial units, $\hat{\sigma}$ is the standard deviation of the error term estimate, and the remaining symbols represent constant parameters.

This study uses the Gaussian kernel function to calculate the weights, with the weight calculation formula as follows:

$$W_{ij} = \exp[-(\frac{d_{ij}}{b})^2]$$

In the equation, b represents the bandwidth, a non-negative factor that controls the relationship between the magnitude of weight and the distance. Specifically, a larger bandwidth results in a slower decline in weight as the distance between points increases, while a smaller bandwidth causes the weight to decrease more sharply with distance. Furthermore, when the bandwidth 'b' is fixed and the distance between two points is zero, the weight reaches its maximum value, which is 1.As the distance increases, the weight gradually decreases. When the distance becomes infinitely large, the weight approaches 0. Therefore, when observation points are sufficiently distant, it can be assumed that the parameter estimates of the paired regression points have negligible influence.

The composite index approach is used to calculate the CESL and CEDL indices. The respective formulas for CESL and CEDL are as follows:

$$CESL = \sum_{i=1}^{m} w_i \times x_i$$

where wi represents the weight of the i-th environmental factor, and xi is the standardized value of the i-th factor.

$$CEDL = \sum_{j=1}^{n} w_j \times y_j$$

where wj denotes the weight of the j-th environmental factor, and yj is the standardized value of the i-th degradation factor.

Matching Cooling Supply and Demand in Urban Blue-Green Spaces. Current clustering analysis methods include coupling coordination models, cold and hot spot clustering analysis using K-means, and spatial autocorrelation analysis. There is spatial consistency between CEDL and CESL; therefore, this study utilizes Moran's index to further quantify the relationship between the two.

This analysis will ultimately categorize the cooling supply and demand matching results into four types: HIGH-HIGH, HIGH-LOW, LOW–HIGH,and LOW-LOW. The equation for calculating the global Moran's Index is as follows:

$$I = \frac{m}{\sum_{q=1}^{m}\sum_{p=1}^{m} W_{qp}} \times \frac{\sum_{q-1}^{m}\sum_{p-1}^{m} W_{qp}(x_q - \overline{x})(x_p - \overline{x})}{\sum_{q=1}^{m}(x_q - \overline{x})^2}$$

In this analysis, x_q and x_p stand for the attribute values of pixels q and p, respectively. The symbol $\overline{x}$ signifies the average attribute value across all pixels within the study area. Furthermore, W_{qp} constitutes the spatial weight matrix, which delineates the spatial proximity between the target pixel and its neighboring pixels. The variable m refers to the total count of pixels encompassed within the study area.

$$I_q = \frac{(x_q - \overline{x})\sum_{p-1}^{m} W_{qp}(x_p - \overline{x})}{\sum_{q=1}^{m}(x_p - \overline{x})^2}$$

The local spatial autocorrelation statistic for pixel q, denoted as I_q, measures the relationship between the attribute values of pixel q and its neighboring pixels p. The symbol $\overline{x}$ Indicates the average value of all pixels within the research scope. W_{qp} represents to the spatial weight matrix, and m denotes the number of all pixels in the research scope. This method is similar to the global approach, but it focuses on local spatial pattern Moran's I, which measures the extent of local autocorrelation, ranges from -1 to + 1. Reflecting the spatial concentration and dispersion of elements.

3 Results Analysis

3.1 Thermal Environment Measurement Results

In this analysis, we utilized both the Atmospheric correction method and the average standard deviation technique to gauge and categorize the surface temperatures across Hangzhou's bustling downtown. As depicted in Fig. 4 and outlined in Table 5, our data reveals that the mean surface temperature in Hangzhou's urban core was a balmy 35.6 °C back in 2016. However, by 2022, that figure had climbed to a sweltering 36.9 °C, indicating a substantial rise of 1.3 °C over the six-year span. In the central urban area, high-temperature zones occupy the main urban area, covering areas of 308.4 km^2 and 336.5 km^2, respectively. The secondary high-temperature zones are located around the city center, with areas of 860.9 km^2 and 849.47 km^2. The primary reason for this temperature increase is the high intensity of urban development in these areas, which alters the underlying surface and changes the surface heat circulation,

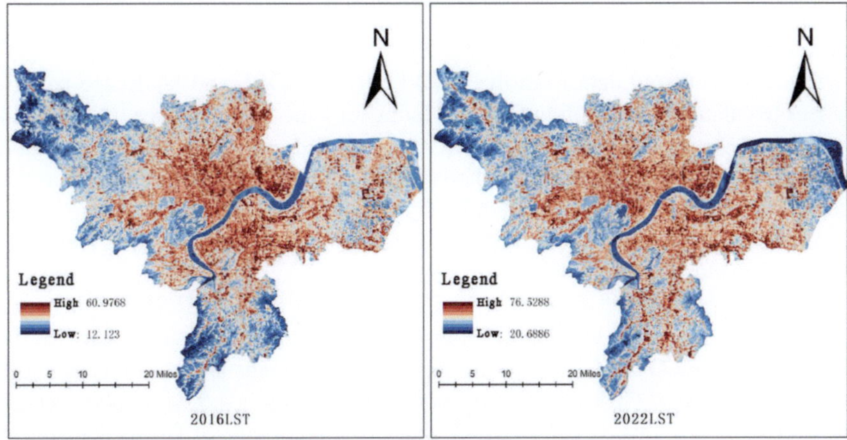

Fig.4 Surface temperature retrieval results (Self-drawn)

Table 5 Ground surface temperature classification results (Self-drawn)

Classification of surface temperature	2016		2022	
	Area (Km^2)	Rate (%)	Area (Km^2)	Rate (%)
The maximum temperature	308.48	9.18	336.52	10.01
The higher temperature	860.9	25.62	849.47	25.27
Mesotherm	942.38	28.05	907.86	27.00
The lower temperature	459.99	13.69	419.55	12.48
The lowest temperature	788.36	23.46	848.81	25.25

leading to consistently rising surface temperatures. In contrast, the lower temperature zones and cooler areas, which cover 459.9 km^2 and 419.5 km^2 respectively, the low-temperature zones, covering areas of 788.3 km^2 and 848.8 km^2, are primarily found in the northwest, southwest, and southern parts of the city. The cooler temperatures in these regions can be attributed to a higher coverage of blue-green spaces, which enhances surface cooling through temperature radiation, thus lowering the surface temperature.

3.2 Evaluation of the Cooling Supply and Demand of Urban Blue-Green Spaces

The impact indicators for the years 2016–2022 were analyzed using a Kappa consistency test. The results indicated that all indicators had Kappa values greater than 0.6, and the p-values were below 0.005.This indicates that the overall stability of the indicators during this period was relatively high, making the results highly valuable for reference (Table 6).

This study utilized data from the NDWI, NDVI, NDBI, and landscape pattern indices to calculate the Cooling Demand Index for Blue-Green Spaces (CEDL) and the Cooling Supply Index for Blue-Green Spaces (CESL) using geographic weighted regression and composite index methods (see Fig. 5). The results indicate that the factors influencing urban cooling exhibit significant spatial heterogeneity and correlation, leading to a high degree of spatial heterogeneity and clustering in the urban cooling demand and supply indices.

The findings in Fig. 5 show that the CEDL is high and the CESL is low in the urban centre. In contrast, the next locations are with a strong CESL and weak CEDL. CEDL average CEDL is 6.7 CESL average CESL is 4.5 The analysis indicates that the average value of CES initiative for each organization is 4.5, while the average value of CED initiative is 6.7. It shows that the overall level of CESL demand in the main urban area of Hangzhou is greater than that of CEDL supply. Additionally, these indicators are scattered in high-intensity activity regions and blue-green space coverage regions, to a great extent indicating the degree of correlation between the overall analysis results and its influencing factors. Additionally, the supply and demand indices exhibit significant spatial clustering. The maximum CEDL value is 96.3, primarily located in the Shangcheng District and Binjiang District. This high demand for cooling in these areas can be attributed to their diverse functions, frequent social activities, and dense population. In contrast, the maximum CESL value is 6.6, predominantly found in the western part of Yuhang District, which benefits from a high concentration of blue-green spaces, resulting in a greater supply capacity. When

Table 6 Indicator confidence test (Self-drawn)

Name	Kappa	Standard deviation	p	Confidence interval (95%)
HEAT—PD	0.7826	0.0335	0.000**	0.6954–0.8976
COOL—PD	0.7685	0.0254	0.001**	0.6857–0.8756
HEAT—PLAND	0.8623	0.0410	0.001**	0.7354–0.8645
COOL—PLAND	0.8365	0.0520	0.002**	0.7865–0.8435
HEATAREA	0.8363	0.0293	0.003**	0.7894–0.8632
COOL—AREA	0.8423	0.0320	0.004**	0.6786–0.7326
HEAT—LSI	0.7152	0.0650	0.001**	0.6836–0.8634
COOL—LSI	0.8123	0.0354	0.002**	0.6786–0.8634
HEAT—COHESION	0.7250	0.0365	0.002**	0.7936–0.8635
COOL—COHESION	0.7523	0.0354	0.006**	0.6952–0.7535
NDWI	0.8523	0.0365	0.005**	0.7230–0.7956
NDVI	0.8963	0.0245	0.004**	0.7869–0.8630
NDBI	0.8223	0.0243	0.007**	0.7936–0.8536

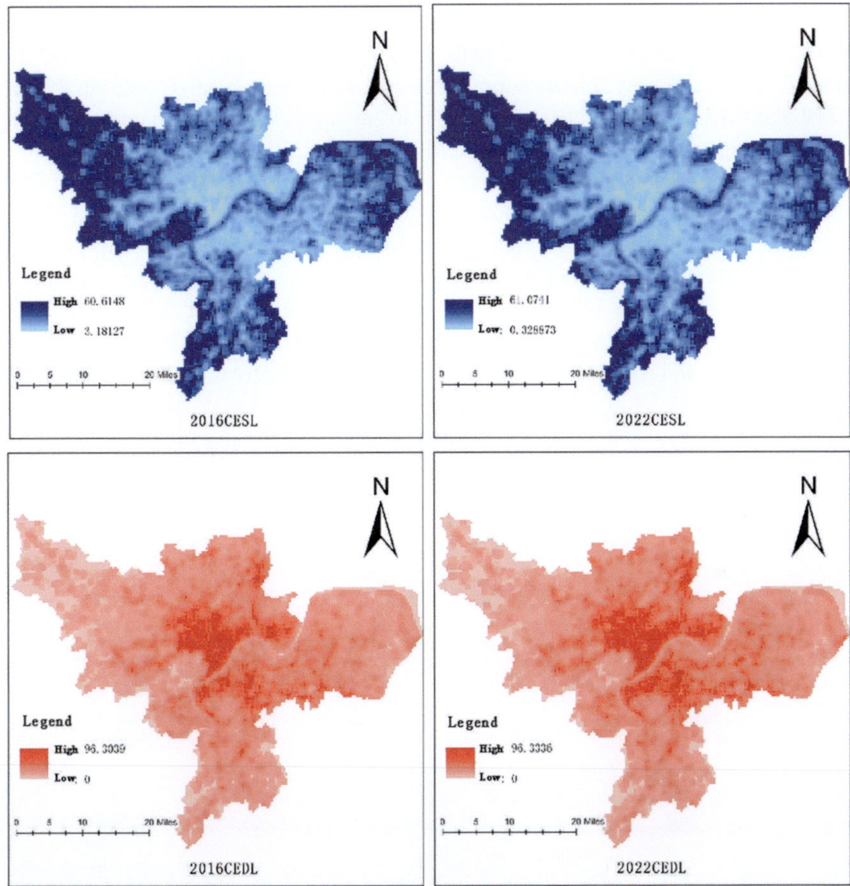

Fig. 5 Calculation results of the cooling energy supply level (CESL) and cooling energy demand level (CEDL) (Self-drawn)

local influencing factors are not considered, these analytical results provide a spatial explanation for the imbalance in the cooling supply from blue-green spaces.

3.3 Analysis of the Matching Results for Cooling Supply and Demand of Urban Blue-Green Spaces

This study employed Z-score standardization analysis and global bivariate LISA spatial autocorrelation analysis to obtain the matching results for the Cooling Supply Index for Blue-Green Spaces (CESL) and the Cooling Demand Index for Blue-Green Spaces (CEDL) for the years 2016 and 2022 (see Fig. 6). The data reveals significant variation in the cooling supply and demand match outcomes across Hangzhou's bustling city center, with a clear pattern of spatial continuity. Furthermore, the Moran's I index scores for the supply–demand alignment in 2016 and 2022 were 0.719 and 0.742, respectively, indicating a strong spatial association and clustering in the cooling supply–demand match across these years.

The matching results, with four types of clustering, HIGH–HIGH, HIGH–LOW, LOW–HIGH, and LOW–LOW, showed that the excluded areas were weakly correlated according to the spatial autocorrelation analysis. This study explores the reasons for imbalances in cooling areas for urban areas and thus requires an understanding of the regions of mismatched supply and demand for the cooling. The examination of these mismatched areas shows that, in 2016 and 2022, they represented 47.36% and 49.16% of the total study area, respectively. In 2016, the high

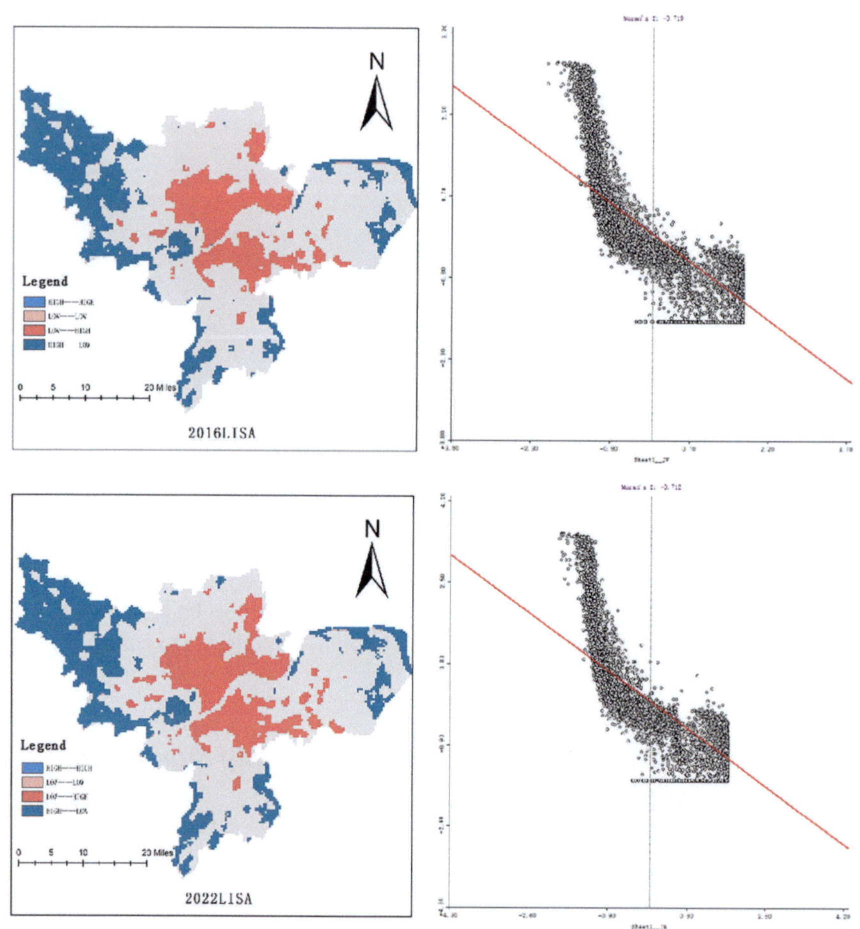

Fig. 6 Matching results of cooli6g supply and demand for blue-green spaces (Self-drawn)

supply–low demand and low supply–high demand areas measured 804.29 km^2 and 539.11 km^2, while in 2022, they expanded to 856.23 km^2 and 599.53 km^2. From 2016 to 2022, the areas with low cooling supply–high cooling demand and high cooling supply–low cooling demand grew by 60.48 km^2 and 51.07 km^2, respectively. The data indicates that the central urban sprawl in Hangzhou has exacerbated the gap between cooling availability and the need for it. Geographically, areas with limited cooling options and high demand are predominantly found in the western sectors of Qiantang, Yuhang, Linping, Shangcheng, and Binjiang Districts. On the flip side, regions with ample cooling resources and less demand are clustered in Xihu, Yuhang, the eastern part of Qiantang, and the southern Xiaoshan Districts. These trends indicate that areas with significant cooling needs are often vibrant urban centers, where growth drives increased demand. whereas areas with abundant cooling are typically located in natural settings such as along rivers and in mountainous regions, which naturally provide cooling benefits. In sum, socio-economic elements predominantly shape regions with low supply and high demand, whereas natural environmental elements have a more significant impact in areas with high supply and low demand.

4 Discussion

4.1 Further Explanation of the Urban Surface Temperature Influencing Mechanism

Multiple factors interact to give rise to the urban warming phenomenon. On the city level, a higher proportion of blue-green spaces, the density of blue-green patches, and the configuration of these patches generally contribute to a decrease in surface temperature.

It is commonly believed that the size and configuration of blue-green spaces directly increase the heat transfer surface between the patches and their adjacent surroundings [49], thus, blue-green spaces help alleviate the urban heat environment. In contrast, urban surface temperature rises with the area and density of built-up land, as impervious surfaces increase heat. Additionally, within certain ranges, a higher shape index of blue-green patches may be positively correlated with surface temperature. This may be due to the complexity of shapes, which slows down cooling rates, while increased urban activities contribute to external heat [50].This study takes a supply–demand perspective to further analyze the mechanisms and determinants of urban surface temperature at the urban scale. It explores the impact of various factors, including natural environments, economic and social activities, and the constructed urban environment, on urban surface temperature, providing a valuable reference for research in this area. By investigating the connection between urban blue-green space cooling and supply–demand, this research first offers evidence and reference for identifying high-temperature areas and assessing vulnerabilities in cities. Secondly, by examining the provision-consumption influence mechanism, the temperature-reducing potential of blue-green spaces can be improved in areas with high need and low availability. Finally, at the community level, analyzing the provision-consumption mechanism can provide reference and guidance for community-level blue-green space allocation.

4.2 The Importance of Quantifying Cooling Supply and Demand Capacities

In this research, CESL refers to the cooling supply capability of blue-green spaces, while CEDL denotes the cooling demand capacity. Quantitative analysis indicates that the primary cause of high-temperature issues in a region is the imbalance between the cooling provision and requirement of blue-green spaces. Consequently, it is crucial to assess both the cooling capacity of blue-green spaces and the cooling need capacity of the area. Previous studies have shown that urban blue-green spaces can exert a cooling effect on their surrounding environment, however, some studies also show that not all blue-green patches and urban structures meet the current cooling supply and demand capacities of the area. This discrepancy arises due to interference from their internal and neighboring environments [51]. Thus, quantitatively analyzing cooling supply and demand capacities can help us better identify blue-green patches with cooling potential. According to the identification results, some areas still exhibit inadequate cooling capacity from blue-green spaces (see Fig. 7). Primarily located in urban core regions, these areas experience frequent socio-economic activities that greatly limit the cooling capacity of blue-green spaces. Outside of mountainous forested areas, the rest of the urban core also faces a high demand for cooling. This contrasts with earlier studies, as CESL and CEDL display considerable spatial variation.

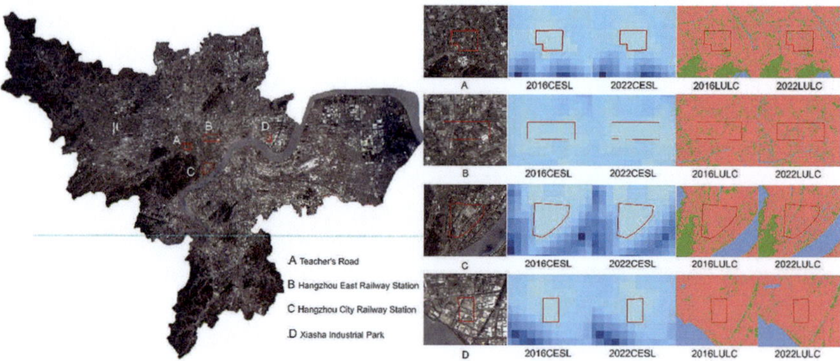

Fig. 7 Case study of blue-green space patches with insufficient cooling capacity (Self-drawn)

Based on the analysis of CEDL and CESL, we can observe that some blue-green patches in urban core areas do not have a significant impact on their surrounding environments. Similarly, built-up areas on the outskirts of the city also show little effect on their surroundings. The absence of influence could be due to factors such as the size, location, and connectivity of the patches. Overall, the spatial distribution of CESL decreases from mountainous lakes to urban center areas, while CEDL displays an opposite trend. The study area shows an uneven spatial distribution of CESL and CEDL, highlighting a significant contradiction between them. Thus, a quantitative assessment of the cooling effect availability and requirement capacities can help analyze the arrangement of blue-green spaces and serve as a basis for matching provision with need. Ensuring the adequate fulfillment of cooling needs in the urban core is crucial. A key prerequisite for this is the quantitative evaluation of CEDL and CESL, which is essential for analyzing the alignment of regional cooling provision with requirements and for balancing the cooling capacities in different areas.

4.3 Contributions of This Study Compared to Previous Research

The latest studies are focusing on three key areas: reducing temperatures in green spaces, combating the urban heat island effect, and ensuring outdoor environments are comfortable. In the quest for cooling, researchers often explore how factors such as the size, layout, and form of a space, along with its interactions with surrounding elements, can enhance its cooling capacity. The ultimate goal is to maximize enhance the cooling ability of plaques [52, 53] or to enhance their cooling efficiency [54] Second, some studies focus on how urban blue-green spaces reduce heat risks, highlighting their role in mitigating health issues, especially mental health problems in vulnerable populations [55, 56], but also disrupts normal activities of daily life. Consequently, it is suggested that controlling population density and increasing the density of outdoor blue-green spaces can effectively reduce heat risks for urban populations [57]. Finally, many researchers have focused on outdoor thermal comfort during high-temperature weather, seeking to enhance urban residents' experiences of heat. By improving building design and increasing small-scale outdoor blue-green space facilities, they aim to improve the thermal comfort of outdoor populations [58, 59] (Table 7).

Recent studies often evaluate the cooling potential of blue-green spaces and the needs of residents separately, without considering the supply–demand relationship. This can lead to areas with too much or too little cooling. Additionally, interactions between factors are frequently overlooked. To address this, it's important to analyze cooling supply and demand together using spatial models, which can better visualize these relationships and offer new insights for future studies.

Table 7 A comparison of research on the cooling effects of urban blue-green spaces. (Self-drawn)

Research perspective	Meaning	Author
Cooling perspective of blue-green space	Investigate the temperature reduction and cooling rates of various patch sizes and shapes across different greenish-blue environments, with a focus on identifying a critical point where shape, size, and cooling effectiveness converge under specific conditions	Jian [52]
	Proposing to increase the density and area of blue-green spaces can effectively improve surface temperature	Yang [53]
	It is recommended that modifying the internal form of blue-green spaces in specific research contexts could enhance their cooling effectiveness	Ouyang [54]
Perspective on reducing high temperature risk	Identify the high-temperature needs of special vulnerable populations through high-temperature brittleness assessment, and specify different blue-green space planning strategies to reduce high-temperature risks	Emma [55] Huanchu [56]
	Utilize factors such as socioeconomic status, age structure, and natural environment to conduct high-temperature risk assessments and increase blue-green space facilities	Xiaojun [57]
Thermal comfort perspective	Study the correlation between different architectural forms and the surrounding temperature of buildings, and improve the outdoor temperature of buildings	Golbarg [58]
	Exploring the impact of urban micro-parks on temperature reduction	Shivanjal [59]

4.4 Comparison with Similar Studies in Other Cities

Research on urban blue-green space cooling supply and demand is ongoing in China. Haoxin L created a framework to assess the temperature regulation services (TRS) of urban forests in Shenzhen [60]. Zhongjian Shen developed a system to evaluate cooling supply and demand in urban agglomerations in the Fujian Delta [32]. Wenning Li & Ranhao Sproposed a model that integrates vegetation cooling with high-temperature cooling demand, identifying high-risk heat areas [61]. Yasha Wang mapped the supply–demand balance in Hangzhou, prioritizing regions with low supply but high demand [45]. Yunhao F. & Liyuan Z. used a coupling coordination model in Wuhan to analyze cooling trends over a decade and factors influencing mismatches [62].These studies highlight spatial heterogeneity in blue-green space distribution. This study stands out by using geographically weighted regression and spatial autocorrelation models, focusing on spatial interactions, unlike others that use methods like principal component analysis and coupling coordination models. It provides broader insights into the connection between spatial heterogeneity and urban cooling supply and demand.

4.5 Policy Recommendations

Based on the previous analysis, there is an insufficient cooling effect in the urban core areas, which may be related to the fragmentation of blue-green spaces within these areas. In some parts of the urban core, blue-green spaces experience significant cooling deficits. Therefore, at the city level, urban planners can enhance the connectivity and area of blue-green spaces in areas with mismatched supply and demand, ensuring that these spaces can effectively contribute to cooling and meet the demand.At the community level, in areas such as Hangzhou Railway Station, Xiasha Industrial Park, commercial streets, and office parks—where land area is limited and cooling demand cannot be met—communities can optimize existing blue-green spaces and add cooling facilities. Additionally, in other areas, cities can reasonably allocate cooling resources within communities based on supply–demand matching results, maximizing the ability to meet cooling needs.

5 Conclusion and Outlook

Urban blue-green spaces have emerged as a significant research topic, attracting global attention. While many studies have explored the cooling benefits of these areas, there has been a lack of analysis from a supply-and-demand perspective, often leaving the cooling needs of hot urban areas unaddressed. This research employs Landsat 8 satellite imagery and a geographically weighted spatial autocorrelation method to thoroughly examine the spatial distribution and the balance between the supply and demand for blue-green spaces in Hangzhou from 2016 to 2022. The study identifies key factors and offers new insights into this area of research. The key findings of the study are as follows:

Surface Temperature Analysis: Through surface temperature inversion, it was found that both the overall surface temperature and the average surface temperature in Hangzhou increased from 2016 to 2022. The areas classified as urban heat zones, secondary heat zones, and cooler areas have all expanded, indicating that urban development has exacerbated the heat environment issues. Furthermore, the surrounding blue-green spaces have not effectively contributed to lowering the overall urban temperatures.

Supply–Demand Index Visualization: The analysis of the supply–demand indices for urban blue-green spaces, namely CESL (Cooling Supply Level) and CEDL (Cooling Demand Level), reveals spatial heterogeneity and clustering. This phenomenon arises from the interaction of different influencing factors. CEDL is primarily concentrated in the Upper City and Binjiang District due to the area's diverse functions, frequent social activities, and dense population, which result in high cooling demand. Conversely, CESL is mainly located in the western part of Yuhang District, attributed to the high concentration of blue-green spaces, leading to elevated supply levels. The unique spatial distribution of CEDL and CESL highlights a disparity between the supply and demand for cooling within the central urban area of Hangzhou.

Global Spatial Autocorrelation Analysis: Through spatial correlation analysis, we identified four scenarios for supply–demand matching. This study focuses on the mismatched regions, specifically the areas with low supply and high demand. These regions are characterized by high population density and frequent economic activity, yet have relatively limited blue-green spaces. In contrast, high-supply and low-demand areas are characterized by dense blue-green patches, low population density, and relatively weak economic activities. By analyzing these two mismatched regions, we can observe a certain spatial continuity between supply and demand areas.

This research explores the factors driving these dynamics, calculates the supply and demand level, and applies spatial autocorrelation techniques to assess the alignment between the supply and demand of urban green spaces for cooling. However, it is important to acknowledge that the study has its limitations. Firstly, the comprehensiveness and accuracy of indicator data need to be improved. Second, the study does not address the factors contributing to supply–demand imbalances or analyze the correlations among these factors. Lastly, the choice of a 500-m grid scale for this study may introduce statistical bias in the data for smaller blue-green patches.

In light of the research conclusions and limitations, future exploration is necessary. First, within the established study area, future research could compare and select appropriate grid sizes to enhance data precision and reduce errors. Subsequently, future research should incorporate additional factors to develop a more comprehensive index. For example, integrating high-quality GDP data with population demographics could significantly improve the accuracy of measuring demand levels. Ultimately, the matching analysis can be further refined to elaborate on the results and elucidate the roles of the influencing factors. This would greatly deepen our understanding of how urban blue-green spaces contribute to cooling, reduce heat-related risks for city residents, and improve thermal comfort in urban environments.

Acknowledgements This study is supported by the Research Project:Research Project of Zhejiang Federation of Humanities and Social Sciences (Grant Number: 2025N055) Major Humanities and Social Sciences Research Projects in Zhejiang higher education institutions, Grant Number:2024QN136

References

1. Aihaiti A et al (2021) You risk changes of compound temperature and precipitation extremes in china under 1.5 °C and 2 °C global warming. Atmospheric Research 264, Article 105838
2. World Meteorological Organization. WMO Provisional State of the Global Climate. [2023–04–21]. (2023–04–21). https://www.preventionweb.net/media/84606/download?startDownload=20241014

3. Bansal P, Quan SJ Examining temporally varying nonlinear effects of urban form on urban heat island using explainable machine learning: a case of Seoul. Building Environ 247, Article 110957. https://doi.org/10.1016/j.buildenv.2023.110957
4. Yuan Y et al (2024) Unraveling the global economic and mortality effects of rising urban heat island intensity. Sustain Cities Soc 105902. ISSN 2210–6707
5. Ballester J et al (2023) Heat-related mortality in Europe during the summer of 2022. Nat Med 29:1857–1866
6. Matteo P et al (2024) The impact of high temperatures on performance in work-related activities. Labour Econom 87:102509. ISSN 0927-5371
7. Qixin L et al (2024) Summer cooling island effects of blue-green spaces in severe cold regions: a case study of harbin, China. Build Environ 257:111539. ISSN 0360-1323
8. Fei Y et al (2023) The assessment of cooling capacity of blue-green spaces in rapidly developing cities: a case study of Tianjin's central urban area. Sustain Cities Soc 99:104918. ISSN 2210-6707
9. Prathiba A et al (2024) Assessing diurnal land surface temperature variations across landcover and local climate zones: implications for urban planning and mitigation strategies on socio-economic factors. Sustain Cities Soc 116:105880. ISSN 2210-6707
10. Huanchun H et al (2024) Urban heatwave, green spaces, and mental health: a review based on environmental health risk assessment framework. Sci Total Environ 948:174816
11. Siqi J et al (2025) A hybrid framework for assessing outdoor thermal comfort in large-scale urban environments, landscape and urban planning 256:105281. ISSN 0169-2046
12. Zhiye L et al (2024) Outdoor thermal comfort and activities in urban parks: An experiment study in humid subtropical climates. Build Environ 253:111361. ISSN 0360-1323
13. Bartesaghi Koc C, Osmond AP (2018) Peters evaluating the cooling effects of green infrastructure: a systematic review of methods, indicators and data sources. Sol Energy 166:486–508
14. Jianchun L et al (2024) Evaluation and improvement strategies for cooling performance in urban blue-green spaces. Scientia Geographica Sinica 44(08):1376–1386
15. Mangmang W et al (2024) Scenario Simulation of the urban green-blue space cooling effect under multi-dimensional collaboration: a case study based on the central city of Shanghai. South Architecture 05:95–105
16. Yijun S et al (2021) Assessment methods of urban system resilience: from the perspective of complex adaptive system theory. Cities 112:103141. ISSN 0264-2751
17. Wenxuan D et al (2024) Structural resilience disparities in regional city networks under the influence of Typhoon Fitow and COVID-19. Cities 155:105460. ISSN 0264-2751
18. Qing L et al (2024) Unveiling nonlinear effects of built environment attributes on urban heat resilience using interpretable machine learning. Urban Climate 56:102046. ISSN 2212-0955
19. Zitong S et al (2023) Modeling the intensity of surface urban heat island based on the impervious surface area. Urban Climate 49:101529. ISSN 2212-0955
20. Zhang Y et al (2024) Combined effects of urban blue–green spaces on the thermal environment: a case study of Kobe. Japan IJEPS. https://doi.org/10.1007/s42495-024-00140-4
21. Gaoyuan Y et al (2020) How can urban blue-green space be planned for climate adaption in high-latitude cities? A seasonal perspective. Sustain Cities Soc 53:101932. ISSN 2210-6707
22. Qixin L et al (2024) Summer cooling island effects of blue-green spaces in severe cold regions: a case study of harbin, China. Build Environ 257:111539. ISSN 0360-1323
23. Md Rejaul I et al (2024) Evaluating cooling effect of blue-green infrastructure on urban thermal environment in a metropolitan city: using geospatial and machine learning techniques. Sustain Cities Soc 113:105666. ISSN 2210-6707
24. Pinyang L et al (2023) How 2D and 3D built environments impact urban surface temperature under extreme heat: a study in Chengdu, China. Build Environ 231:110035. ISSN 0360-1323
25. Zihao W et al (2024) Investigating the response of urban surface thermal environment to urban morphology changes in Nanjing. J Environ Manag 370:122875. ISSN 0301-4797
26. Vujovic S et al (2021) Urban heat island: causes, consequences, and mitigation measures with emphasis on reflective and permeable pavements. CivilEng 2:459–484.https://doi.org/10.3390/civileng2020026

27. Ramírez-Aguilar EA, Lucas Souza LC Urban form and population density: Influences on urban heat island intensities in Bogota, Colombia. Urban Climate, 29, Article 100497
28. Yang L et al Dominant factors and spatial heterogeneity of land surface temperatures in urban areas: a case study in Fuzhou, China. Remote Sensing 14(5):1266
29. Gao Y et al Effects of block morphology on the surface thermal environment and the corresponding planning strategy using the geographically weighted regression model. Build Environ 216, Article 109037
30. Gamero-S et al (2024) Exploring indoor thermal comfort and its causes and consequences amid heatwaves in a Southern European city—an unsupervised learning approach. Build Environ 265:111986. ISSN 0360-1323
31. Changyeon L (2024) How do spatial scale-dependent urban and forest areas influence regionally seasonal temperatures in mid-latitude South Korean cities? J Environ Manag 360:121171. ISSN 0301-4797
32. Zhongjian S et al (2022) Examining supply and demand of cooling effect of blue and green spaces in mitigating urban heat island effects: a case study of the Fujian Delta urban agglomeration (FDUA), China. Ecological Indicators 142:109187. ISSN 1470-160X
33. Yasha W et al (2024) Matching supply and demand of cooling service provided by urban green and blue space. Urban For & Urban Green 96:128338. ISSN 1618-8667
34. Yin S et al (2024) Improvement of spatial-temporal urban heat island study based on local climate zone framework: a case study of Hangzhou, China. Build Environ 248:111102. ISSN 0360-1323
35. Qijiao X et al (2023) Regulation of water bodies to urban thermal environment:Evidence from Wuhan, China. Front Ecol Evol.11.https://doi.org/10.3389/fevo.2023.983567. ISSN=2296-701X
36. Yanfei W et al (2022) Effects of landscape patterns on the morphological evolution of surface urban heat island in Hangzhou during 2000–2020. Sustain Cities Soc 79:103717. ISSN 2210-6707
37. Yuxin Y et al (2024) Multi-scale effects of LCZ and urban green infrastructure on diurnal land surface temperature dynamics. Sustain Cities Soc 117:105945. ISSN 2210-6707
38. Liwen S et al. Study on temperature regulation function of green spaces at community scale in high-density urban areas and planning design strategies,Urban Forestry & Urban Greening,Volume 101,2024,128511,ISSN 1618–8667.
39. Peng J et al (2021) High-resolution land surface modeling of the effect of long-term urbanization on hydrothermal changes over Beijing Metropolitan area. JGR Atmospheres 126(18):e2021JD034787
40. Xiangyi D et al (2024) Characteristics of surface urban heat islands in global cities of different scales: Trends and drivers. Sustain Cities Soc 107:105483. ISSN 2210-6707
41. Yuejing G et al Exploring the spatial heterogeneity of urban heat island effect and its relationship to block morphology with the geographically weighted regression model. Sustain Cities Soc 76, Article 103431
42. Ling Y et al (2024) Is 3D building morphology really related to land surface temperature? Insights from a new homogeneous unit. Build Environ 266:112101. ISSN 0360-1323
43. Yufang J, Jing H (2022) Quantitative analysis of mitigation effect of urban blue-green spaces on urban heat island. Resour Environ Yangtze Basin 31(09):2060–2072
44. Yonghua Z et al (2022) Landscape pattern change and its correlation with influencing factors in semiarid areas, northwestern China. Chemosphere 307, Part 3, 135837. ISSN 0045-6535
45. Wang T et al (2024) Research on Surface temperature inversion and its application in Chongqing Based on split window Algorithm. Mod Inf Technol 8(15):93–98
46. Lu X et al (2023) Comparison of land surface temperature retrieval algorithms based on landsat 8 data. Arid Environ Monit 37(03):137–144
47. Bilal A et al (2024) Analysis of LST, NDVI, and UHI patterns for urban climate using Landsat-9 satellite data in Delhi. J Atmos Sol-Terr Phy 265:106359. ISSN 1364-6826
48. Yantao X et al (2024) Seasonal surface urban heat island analysis based on local climate zones. Ecological Indicators 159:111669. ISSN 1470-160X

49. Du H et al (2019) Urban blue-green space planning based on thermal environment simulation: a case study of Shanghai. China. Ecological Indicators 106:105501
50. Haoran H et al (2023) Prioritizing environmental determinants of urban heat islands: a machine learning study for major cities in China. Int J Appl Earth Obs Geoinformation 122:103411
51. Ying Z et al (2022) Is urban green space a carbon sink or source? A case study of China based on LCA method. Environ Impact Assess Rev 94:106766
52. Jian P et al (2021) How to quantify the cooling effect of urban parks? Linking maximum and accumulation perspectives. Remote Sensing Environ 252:112135
53. Yang G et al (2020) How can urban blue-green space be planned for climate adaption in high-latitude cities? A seasonal perspective. Sustain Cities Soc 53:101932
54. Ouyang W et al (2019) The cooling efficiency of urban greenery coverage in a high-density city//IOP conference series: earth and environmental science. IOP Publishing 329(1):012043
55. Emma L et al (2024) The health benefits of reducing micro-heat islands: a 22-year analysis of the impact of urban temperature reduction on heat-related illnesses in California's major cities. Sci Total Environ 949:175284
56. Huanchun H et al (2024) Urban heatwave, green spaces, and mental health: A review based on environmental health risk assessment framework. Sci Total Environ 948:174816
57. Xiaojun H et al (2024) Assessment of heat exposure risk for urban populations and spatio-temporal patterns: a perspective of urban functional zones in Xi'an, China. Urban Climate 55:101992
58. Golbarg S et al (2024) Urban microclimate analysis: residential block morphology impact on outdoor thermal comfort. In: Proceedings of the institution of civil engineers–engineering sustainability, vol 177, no 5, pp 309–319
59. Shivanjali M et al (2024) Impact of urban street geometry on outdoor pedestrian thermal comfort during heatwave in Nagpur city. Sustain Cities Soc 108:105450
60. Haoxin L et al (2022) Imbalanced supply and demand of temperature regulation service provided by urban forests: a case study in Shenzhen, China ecological indicators, vol 145, 109666. ISSN 1470-160X
61. Wenning LRS (2023) A supply-demand model of vegetation cooling for urban heatwave mitigation. Urban Climate 52:101699. ISSN 2212-0955
62. Yunhao F, Liyuan Z Exploring t& pply-dem match and drivers of blue-green spaces cooling in Wuhan Metropolis. Urban Climate 58(20):2 4,102194. ISSN 2212-0955

Exploring the Thermal Environment of Kunming, China: Based on Local Climate Zones

Zhihao Wang, Zhu He, Pengkun Zhu, and Shihan Deng

Abstract The advancement of urbanization can exacerbate the heat island effect, bringing certain impacts and challenges to human settlement environment and sustainable development. Land Surface Temperature is a crucial metric for highlighting variations in urban thermal environments. Utilizing Landsat-8 OLI/TIRS satellite remote sensing data, this research offers a detailed examination of the spatial distribution and changes in LST through the lens of Local Climate Zones. The results indicated that: (1) The area occupied by built-up types is 13.68% lower than that of natural cover types. Additionally, urbanization has led to the fragmentation of blue-green ecosystems in central urban regions. (2) The high temperature zone (24.11–27.75 °C) occupies 3.43% of the total area, mainly found in developed regions along the study area's boundary and in patches of bare soil within urban districts. In contrast, the low-temperature zone (11.64–13.60 °C) covers 13.11%, primarily including Dianchi Lake and several smaller scattered lakes. The distribution of temperature zones is closely linked to building functions and the arrangement of blue-green ecosystems. (3) Overall, the LST of developed land types is higher compared to natural cover types, while high-rise buildings with low density can achieve more effective cooling. Among natural categories, the average LSTs of dense trees and water is 5.3 °C lower than built-up areas, showing a strong cooling effect. In contrast, paved ground and bare soil have average LSTs about 2 °C higher than built-up areas. This research offers a theoretical foundation for optimizing building layout and blue-green ecosystem design, contributing to the enhancement of the thermal environment.

Keywords Local climate zones · WUDAPT · Urban thermal environment · Land surface temperature

Z. Wang · Z. He · P. Zhu · S. Deng (✉)
Faculty of Civil Engineering and Mechanics, Kunming University of Science and Technology, Kunming, China
e-mail: sdeng@kust.edu.cn

B. He et al. (eds.), *Climate-Resilient, Low-Carbon, and Sustainable Cities*, Lecture Notes in Civil Engineering 754, https://doi.org/10.1007/978-981-95-2169-2_14

1 Introduction

The rural population is gradually gathering in the cities with the continuous improvement of urbanization level and scale. Data from the National Bureau of Statistics show that the urbanization rate of China has reached 66.2% by the end of 2023 [1]. To alleviate the pressure on production activities caused by rapid population growth and concentration, the land use/land cover patterns in cities have undergone significant changes [2, 3]. The growth of urban areas has substituted natural surfaces with impervious artificial ones [4, 5], disrupting surface energy exchange and increasing anthropogenic heat emissions. Consequently, this impacts the urban thermal environment and intensifies the Urban Heat Island (UHI) phenomenon [6, 7].

The UHI effect can lead to various ecological problems in urban areas, including air quality deterioration, increased energy consumption, and a decline in biodiversity [8–10], posing certain impacts and challenges to the human living environment and sustainable development [2, 11]. Land Surface Temperature (LST) is a crucial metric for assessing and shaping the urban thermal environment. It is strongly linked to living conditions and has attracted wide attention from the academic community [12]. A wide range of studies has thoroughly investigated the influence of urban morphology on LST to alleviate urban heat, focusing primarily on the spatial patterns of single or multiple elements. Han [13] employed the Normalized Difference Vegetation Index as an assessment metric, and used the least squares regression approach to analyze the variation trends in the Yellow River Basin and its three climatic regions. Yang [14] investigated the connection between the urban morphology of central Dalian and building forms through correlation analysis and numerical simulation, utilizing urban morphology metrics like building floor area ratio, sky view factor, and absolute roughness. However, the above studies lack a unified framework to accurately characterize urban morphology, making it difficult to capture differences in the internal thermal environment of cities. The Local Climate Zones (LCZs) theory, introduced by Stewart and Oke [15], addresses the aforementioned gaps and supports urban land use planning and management by providing a scientific method for dividing urban spaces. LCZ is an innovative land cover classification system that comprehensively considers factors such as surface spatial structure, land cover types, underlying surface materials and human activities. It divides urban surface morphology into seventeen LCZ types, comprising ten building types (LCZ 1–10) and seven natural cover types (LCZ A–G). Each type has unique morphological and thermal properties that can effectively characterize the urban thermal environment. Currently, LCZ classification techniques using remote sensing imagery have become the dominant approach [16, 17]. In 2015, Bechtel proposed the World Urban Database and Access Portal Tools (WUDAPT) protocol, aimed at simplifying LCZ classification and establishing a unified framework for acquiring urban data [18]. This method constructs LCZ maps using publicly accessible remote sensing data and a random forest classifier, demonstrating high classification accuracy in applications [19, 20]. It provides foundational support for microclimate and thermal environment research.

Kunming, a mountainous city on the southwestern plateau, is distinguished by its unique geographical features and rich climatic resources. According to the Code for the *Thermal Design of Civil Building (GB50176-2016)*, Kunming is located in a region with a mild climate. In recent years, Kunming has faced frequent extreme weather events due to global climate change, highlighting growing conflicts between urban development, environmental sustainability, and energy demands [21]. This study focuses on the Kunming metropolitan area, utilizing Landsat-8 OLI/TIRS remote sensing imagery and the WUDAPT protocol to delineate the LCZ map of Kunming for the year 2023. The atmospheric correction method is used to retrieve LST and investigate the relationship and interaction between LCZ types and LSTs. The findings aim to provide scientific guidance for building a more livable and sustainable city in Kunming.

2 Materials and Methods

2.1 Study Area

Kunming is situated in central Yunnan-Guizhou Plateau, China, between 102°10′–103°40′E longitude and 24°23′–26°22′N latitude. It enjoys a mild climate, with an average annual temperature around 15 °C and annual rainfall of about 1,000 mm. The region of interest (ROI) area includes five administrative districts in Kunming metropolitan area: Xishan District (XS), Wuhua District (WH), Panlong District (PL), Guandu District (GD), and Chenggong District (CG), with a total area of 1,435.59 m^2. The study area is illustrated in Fig. 1, with its boundary coordinates as follows: Point A (N25°09′02″, E102°35′41″), Point B (N25°09′40″, E102°56′32″)), Point C (N24°47′40″, E102°56′06″), and Point D (N24°47′20″, E102°36′06″).

2.2 Data Sources

This research employs Landsat-8 OLI/TIRS remote sensing data, building data, meteorological data, and field observations as its the main data sources, as shown in Table 1. Kunming experiences clearly defined dry and rainy seasons. The high cloud cover in remote sensing images during the rainy season (May–October) reduces the accuracy of the surface temperature inversion. Considering factors such as image quality, cloud cover, and data availability, remote sensing data from January 2023 (dry season) were selected for analysis. Meteorological data were used for confirmation of weather conditions and inversion of surface temperatures. Building data, which encompasses characteristics like shape and height, were classified by height and density according to the *Uniform Standard for Design of Civil Building (GB50352-2019)* and the LCZs of theoretical framework.

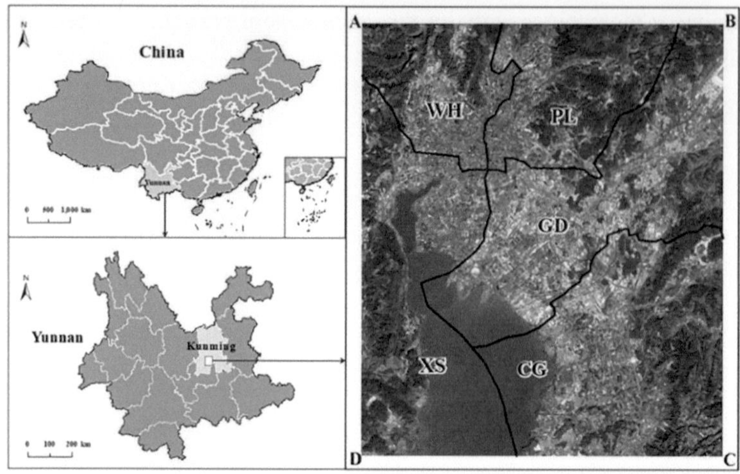

Fig. 1 Location of the study area (self-drawn)

Table 1 Data sources and description (SELF-drawn)

Data type	Source	Specification
Remote sensing images	https://glovis.usgs.gov	Landsat-8 OLI/TIRS, resolution: 30 m, cloud cover: 0.04%, date: 2023-01-26 11:35
Building data	https://map.baidu.com/	Height: high (>27 m), middle (10–27 m), low (<10 m), density: narrow (>0.6), open (0.2–0.6), sparse (<0.2)
Meteorological information	https://data.cma.cn/	Air temperature, relative humidity, air pressure and so on

3 Methods

3.1 LCZ Construction Method

This research utilizes the WUDAPT protocol to map LCZs within the study area. The process is as follows:

(1) Preprocessing of remote sensing images: The acquired remote sensing images undergo preprocessing steps such as coordinate projection transformation and clipping using SAGA GIS 2.2.0. Afterward, the images are resampled to a 100 m resolution for further analysis in SAGA GIS 3.3.0.
(2) Training area sample selection: The level 0 training template is obtained from the WUDAPT website, and 5–15 samples are chosen for each LCZ type. Given the limited samples and challenges in defining specific LCZ types for some urban areas in Kunming, the study area is eventually classified into fifteen categories

(as illustrated in Fig. 2), including eight building types (LCZ 2–6 and 8–10) and seven types of natural cover (LCZ A–G). Subsequently, the surface attributes of the samples are matched to LCZ types using Baidu Street View and field surveys.

(3) SAGA GIS classification: The preprocessed images, samples, and color schemes are imported into SAGA GIS 3.3.0. A random forest classification method is then applied to generate an initial LCZ map for the study area. For areas with significant identification discrepancies, the training samples are reselected, and the steps are repeated until a satisfactory match with actual conditions was achieved.

Fig. 2 LCZs classification basic types (self-drawn)

3.2 Surface Temperature Inversion Method

Existing studies have shown that atmospheric correction methods offer higher accuracy compared to the split-window algorithm, and the parameters required for atmospheric correction are easier to obtain. This provides an efficient and widely applicable approach for retrieving land surface temperature [22–24]. This research uses the atmospheric correction technique for surface temperature retrieval. The principle involves estimating the atmospheric influence on surface thermal radiation and then removing this atmospheric component from the total thermal radiation received by the satellite sensor, thereby converting it into the corresponding T_S. The calculation formula is as follows:

$$B(T_S) = \left[L_\lambda - L_{\mathrm{atm},i} \uparrow - \tau\left(1 - L_{\mathrm{atm},i} \downarrow\right)\right] / \tau\lambda \tag{1}$$

$$T_S = 1321.8 / \ln(774.89 / B(T_S) + 1) \tag{2}$$

In the equation, T_S represents the land surface temperature (K), and $B(T_S)$ is the blackbody radiance at temperature T_S, measured in [W/(m^2 sr μm)]. $L_{\mathrm{atm},i}$, up and $L_{\mathrm{atm},i}$, down represent the upward and downward atmospheric radiation, measured in [W/(m^2 sr μm)]. τ represents the atmospheric transmittance, τ denotes the surface emissivity, and L_λ indicates the radiant brightness received by the satellite thermal sensor, measured in [W/(m^2 sr μm)].

The parameters τ, $L_{\mathrm{atm},i}$, up and $L_{\mathrm{atm},i}$, down are calculated using the bands from the corresponding Level-2 data. ε is determined using the functional relationship between vegetation cover and NDVI proposed by Sobrino [25]. L_λ is obtained by preprocessing the imagery with radiometric calibration and atmospheric correction in ENVI 5.6.2.

3.3 Surface Temperature Classification

Classifying surface temperature categories using absolute values is challenging. The mean-standard deviation method provides a more efficient way to capture the spatial distribution and subtle variations in temperature [26]. As a result, this study applies the mean-standard deviation method to categorize temperature levels, with the classification criteria presented in Table 2. The cut-off points are derived by calculating the mean (μ) and standard deviation (std), which are subsequently used to statistical analysis of LSTs in the study area to produce a classification map.

Table 2 The classification standard of LSTs (self-drawn)

Temperature grades	Temperature classification standard
High temperature zone	$T_S > \mu + 1.5\text{std}$
Relatively high temperature zone	$\mu + 0.5\text{std} < T_S < \mu + 1.5\text{std}$
Medium temperature zone	$\mu - 0.5\text{std} < T_S < \mu + 0.5\text{std}$
Relatively low temperature zone	$\mu + 0.5\text{std} < T_S < \mu + 1.5\text{std}$
Low temperature zone	$T_S < \mu - 1.5\text{std}$

4 Results and Discussion

4.1 Spatial Distribution of LCZ

According to the comprehensively constructed LCZ map of Kunming (Fig. 3) and the statistical results of the LCZ area proportions (Fig. 4), natural cover types (LCZ A–G) occupy 56.81% of the study area, 13.68% more than built-up types (LCZ 2–6, 8–10). The built-up types are primarily characterized by open buildings (LCZ 4–6 and 9), which make up 59.45% of the area. LCZ 4 (open high-rise) and LCZ 5 (open mid-rise), mainly used for commercial and residential purposes, form block patterns in the core areas of each district, except in Guandu District, where they form strips along the east–west and north–south axes. LCZ 6 (open low-rise) and LCZ 9 (sparsely low-rise) primarily consist of villas and suburban towns, scattered across the peripheral areas of each district. Compact buildings (LCZ 2 and 3) cover 16.45% of the area. LCZ 2 (compact mid-rise), representing older settlements, is concentrated in a continuous distribution within the core of the city's second ring road, while LCZ 3 (compact low-rise) is evenly distributed in a star-like pattern throughout the city. LCZ 8 (large low-rise) and LCZ 10 (heavy industry) primarily consist of industrial sites, shopping malls, and transportation hubs. These areas are mainly concentrated in the Guandu District's Wujiaba and Airport Economic Zone, forming a continuous distribution.

Natural cover types are predominantly LCZ A (dense trees) and LCZ G (water), accounting for 38.54% and 24.75% of the area, respectively. Dense trees are primarily distributed in a patchy manner across the natural mountainous regions on the outskirts of the urban area, while water bodies are scattered throughout various districts of the main city, with the exception of Dianchi Lake, which is located in the southwestern part of Kunming. LCZ B-D cover 18.77% of the area, with LCZ B (scattered trees) distributed in patches around the urban periphery, while LCZ C (scrub) and D (low plants) are located along Dianchi Lake and around the built-up areas. LCZ E (hardened ground) and LCZ F (bare soil) account for 2.08% of the area and is mainly scattered in small patches around the outskirts of the city, corresponding to areas where urban construction activities are ongoing. The blue and green ecosystems in Kunming's central urban area show a certain level of fragmentation. It is essential to develop and protect these ecosystems wisely to achieve harmonious development

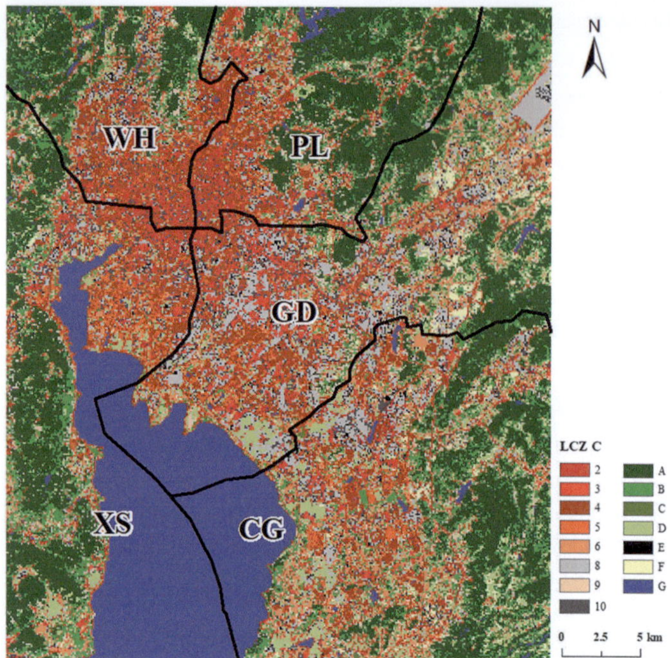

Fig. 3 LCZ map of Kunming (self-drawn)

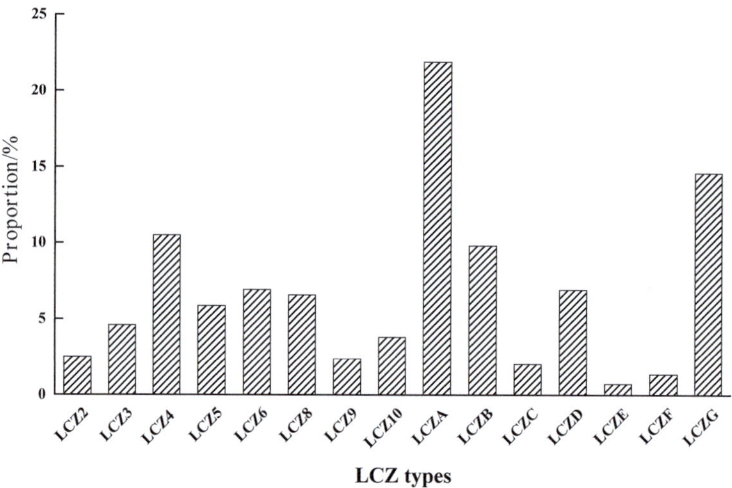

Fig. 4 Percentage of area in different LCZ zones (self-drawn)

between urban construction and the ecological environment, thereby establishing a more ecologically secure urban layout.

4.2 *Spatial Distribution of Surface Inversion Temperatures*

Surface temperature data for the study area was extracted using ENVI 5.6.2. The mean LST was found to be 18.82 °C, with a standard deviation of 3.51. The mean-standard deviation method was applied to divide the surface temperature into five categories. As depicted in Fig. 5, the medium temperature zone (17.08–20.60 °C) occupies the largest area, accounting for 41.14%. It formed a relatively even distribution around Dianchi Lake across various urban districts, primarily covering urban buildings and shrub areas. The relatively high-temperature zone (20.60–24.11 °C) is the second-largest area, covering 31.38%. It is mainly concentrated in urban areas outside the Second Ring Road, where it encompasses building-covered areas and farmland, forming a large, continuous distribution. The low temperature zone (11.16–13.60 °C) and the relatively low temperature zone (13.60–17.08 °C) accounted for 13.11% and 10.94% of the total area, respectively. The low temperature zone is primarily located in Dianchi Lake and other lakes and water bodies within the urban areas, while the relatively low temperature zone is mainly concentrated in the woodlands on the outskirts of the urban areas, along the shoreline of Dianchi Lake, and in city parks. The high temperature zone (24.11–27.75 °C) has the smallest coverage, at 3.43%, and is primarily found in built-up areas along the boundary of the study region and on bare land within each urban area, forming a scattered, point-like distribution. The distribution characteristics of the different temperature zones are closely related to building functions and the blue-green ecosystem patterns.

4.3 *Surface Temperature Variability Based on Local Climate Zones (Self-Drawn)*

The surface temperature analysis by LCZ types (Fig. 6) reveals variations in LSTs among the different LCZ categories. Generally, building types (LCZ 2–6, 8–10) exhibit higher LSTs compared to natural cover types (LCZ A–G). Among the built-up types, the average LST, in descending order, is as follows: LCZ 6 > LCZ 8 > LCZ 10 > LCZ 3 > LCZ 9 > LCZ 5 > LCZ 4 > LCZ 2. LST decreases with increasing height for both compact and open building areas. For example, the LST of LCZ 3 (compact low-rise) is higher than that of LCZ 2 (compact mid-rise). Similarly, the LST for open building areas decreases in the following order: LCZ 6 (open low-rise), LCZ 5 (open mid-rise), LCZ 4 (open high-rise). This is because taller buildings block more solar radiation, reducing the amount of direct sunlight that reaches the ground and building surfaces. The average LSTs of LCZ 10 (heavy industry) and

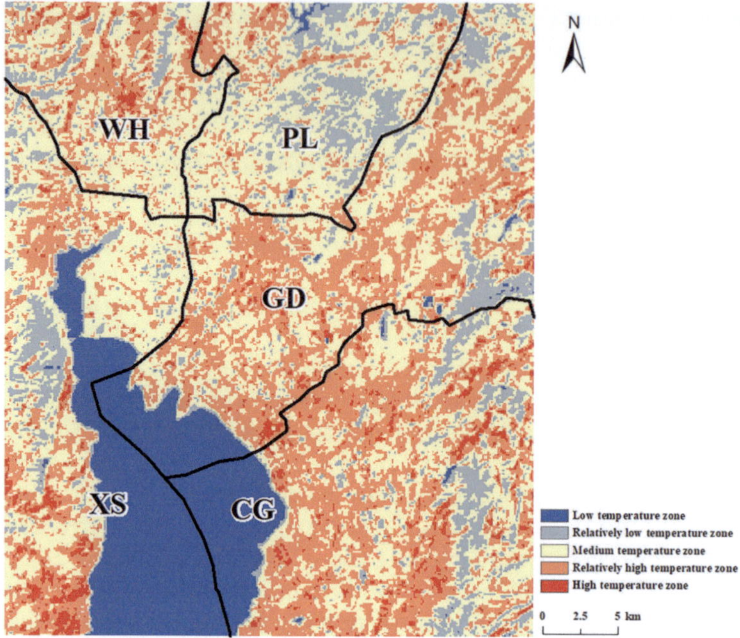

Fig. 5 Kunming surface temperature classification map

LCZ 8 (large low-rise) are significantly higher than those of other build-up types. This is because LCZ 10 involves industrial heat production, while LCZ 8 are more susceptible to solar radiation. Due to the stronger ventilation and cooling effects of low-density buildings, LCZ 9 (sparsely low-rise) have a lower LST than other buildings of the same height (LCZ 3 and 6). In summary, high-rise buildings with low density can make use of the synergistic effect of ventilation, heat dissipation and building shading, which are more conducive to achieving a desirable cooling effect.

Among the natural types, the average LST, in descending order, is as follows: LCZ F > LCZ D > LCZ E > LCZ C > LCZ B > LCZ A > LCZ G. The average LSTs of LCZ A (dense trees) and LCZ G (water) is 5.3 °C lower than that of building areas, indicating strong temperature regulation for these two types. In contrast, the average LSTs of LCZ B (scattered trees) and LCZ C (scrub) is only 0.8 °C lower than that of built-up areas, showing a less significant cooling effect. LCZ D (low plants) exhibited higher LST than the other natural cover types. This is because LCZ D in Kunming primarily consists of vegetable and flower greenhouse growing areas. The plastic material of the greenhouses reduces heat loss and increases atmospheric back radiation, providing effective thermal insulation. The average LSTs of LCZ E (hardened ground) and LCZ F (bare soil) is approximately 2 °C higher than that of built-up areas. This is due to their typically low thermal flux, which allows them to quickly absorb solar radiation, resulting in higher temperatures. Therefore, during the

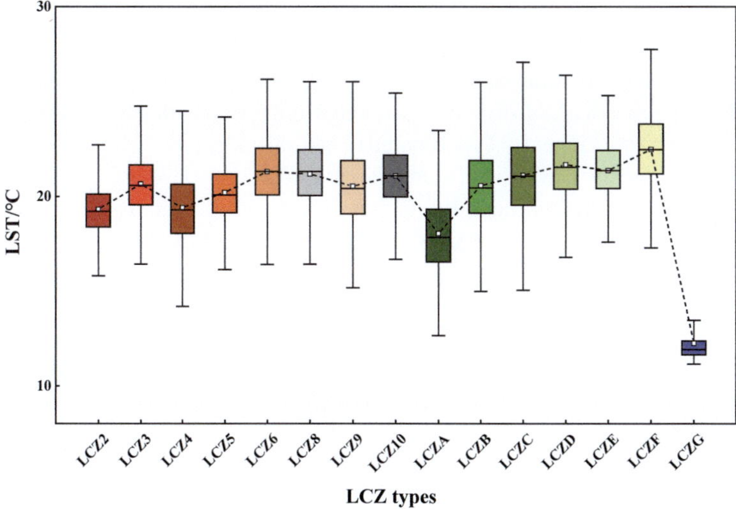

Fig. 6 Statistical map of surface temperature based on LCZs (self-drawn)

planning and construction phases of these areas, it is crucial to consider incorporating green spaces, water bodies, and permeable surfaces to achieve effective cooling.

5 Conclusion

This study utilizes 2023 Landsat-8 OLI/TRIS data to estimate surface temperature in Kunming metropolitan area, offering a refined analysis of LST spatial patterns and variations from an LCZ perspective. The key conclusions of this study can be summarized as follows:

(1) In terms of LCZ spatial distribution, the built-up types (LCZ 2–6, 8–10) cover 13.68% less area than natural cover types (LCZ A-G) in Kunming metropolitan area. The built-up types are dominated by open buildings (LCZ 4–6, 9), which are clustered throughout all urban regions. In contrast, the natural cover types are primarily composed of LCZ A and G (dense trees and water), mainly found in the surrounding natural mountainous areas of the city and the southwestern part of Kunming. Urbanization has resulted in a fragmented distribution of blue-green ecosystems in central urban areas. To achieve coordinated development between the ecological environment and urban construction, it is essential to enhance the protection system and develop land more rationally.

(2) In terms of the spatial distribution of LSTs, the high temperature zone (24.11–27.75 °C) covers 3.43% of the area and is scattered throughout the built-up areas along the study area's boundary and on bare land within each urban area. The low temperature zone (11.16–13.60 °C) makes up 13.11% of the area,

primarily found in Dianchi Lake and other lakes and water bodies in urban areas. The distribution characteristics of these different temperature zones are closely linked to building functions and the blue-green ecosystem pattern.

(3) There are distinct LST differences among LCZ types, with built-up areas generally exhibiting higher LSTs compared to natural cover types. High-rise buildings with low density can achieve more effective cooling by leveraging the combined effects of ventilation, heat dissipation, and building shading. Among natural types, the average LSTs of LCZ A (dense trees) and LCZ G (water) is 5.3 °C lower than that of built-up areas, demonstrating a significant cooling effect. In contrast, the average LSTs of LCZ E (hardened ground) and LCZ F (bare soil) is about 2 °C higher than that of built-up areas. Therefore, it is essential to plan building layouts and blue-green ecosystem configurations strategically to improve the thermal environment and create a high-quality ecological habitat for human living.

Acknowledgements This study is supported by the Yunnan Revitalization Talent Support Program (YNWR-QNBJ-2018-257) and the Scientific Research Foundation of Yunnan Provincial Department of Education (2024Y130).

References

1. Wang PP (2024) Population volume declined and high-quality population development achieved. China Information News 002
2. Rahaman ZA, Kafy AA, Saha M et al (2022) Assessing the impacts of vegetation cover loss on surface temperature, urban heat island and carbon emission in Penang city, Malaysia. Build Environ 222:109335
3. Amini S, Saber M, Rabiei-Dastjerdi H et al (2022) Urban land use and land cover change analysis using random forest classification of landsat time series. Remote Sens 14(11):2654
4. Li Y, Schubert S, Kropp JP et al (2020) On the influence of density and morphology on the urban heat island intensity. Nat Commun 11:2647
5. Shu B, Chen Y, Zhang K et al (2024) Urban engineering insights: spatiotemporal analysis of land surface temperature and land use in urban landscape. Alex Eng J 92:273–282
6. Wardeh Y, Kinab E, Escadeillas G et al (2022) Review of the optimization techniques for cool pavements solutions to mitigate urban heat islands. Build Environ 109482
7. Leal Filho W, Wolf F, Castro-Díaz R et al (2021) Addressing the urban heat islands effect: a cross-country assessment of the role of green infrastructure. Sustainability 13(2):753
8. Aslan A, Altinoz B, Ozsolak B (2021) The link between urbanization and air pollution in Turkey: evidence from dynamic autoregressive distributed lag simulations. Environ Sci Pollut Res 28(37):52370–52380
9. Karimi A, Mohammad P, García-Martínez A et al (2023) New developments and future challenges in reducing and controlling heat island effect in urban areas. Environ Dev Sustain 25(10):10485–10531
10. Harmay NSM, Choi M (2023) The urban heat island and thermal heat stress correlate with climate dynamics and energy budget variations in multiple urban environments. Sustain Cities Soc 91:104422

11. Xi C, Han L, Wang J et al (2023) How can greenery space mitigate urban heat island? An analysis of cooling effect, carbon sequestration, and nurturing cost at the street scale. J Clean Prod 419:138230
12. Tuholske C, Caylor K, Funk C et al (2021) Global urban population exposure to extreme heat. Proc Natl Acad Sci 118(41):e2024792118
13. Han Z, Wu Q, Lai R et al (2022) Spatiotemporal variations of vegetation cover and its influence on surface air temperature change over the Yellow River Basin, China. J Water Clim Chang 13(9):3239–3252
14. Yang J, Yang Y, Sun D et al (2021) Influence of urban morphological characteristics on thermal environment. Sustain Cities Soc 72:103045
15. Stewart ID, Oke TR (2012) Local climate zones for urban temperature studies. Bull Am Meteor Soc 93(12):1879–1900
16. Chung LCH, Xie J, Ren C (2021) Improved machine-learning mapping of local climate zones in metropolitan areas using composite earth observation data in Google Earth Engine. Build Environ 199:107879
17. Du R, Liu CH, Li XX et al (2023) Effect of local climate zone (LCZ) and building category (BC) classification on the simulation of urban climate and air-conditioning load in Hong Kong. Energy 271:127004
18. Bechtel B, Alexander PJ, Böhner J et al (2015) Mapping local climate zones for a worldwide database of the form and function of cities. ISPRS Int J Geo Inf 4(1):199–219
19. Demuzere M, Bechtel B, Middel A et al (2019) Mapping Europe into local climate zones. PLoS ONE 14(4):e0214474
20. Dong P, Jiang S, Zhan W et al (2022) Diurnally continuous dynamics of surface urban heat island intensities of local climate zones with spatiotemporally enhanced satellite-derived land surface temperatures. Build Environ 218:109105
21. He M, He P, Zhao L et al (2023) Analysis of thermal environment effect and its influencing factors in the main urban area of Kunming. J Environ Eng Technol 13(01):403–412
22. Wan JK, Shen ZH, Li S (2024) Difference and parameter analysis of LST inversion based on Landsat data. J Infrared Millim Waves 43(02):226–234
23. Yang L, Cao Y, Zhu X et al (2014) Land surface temperature retrieval for arid regions based on Landsat-8 TIRS data: a case study in Shihezi, Northwest China. J Arid Land 6:704–716
24. Yuan YX, Zhu XL, Hou CH et al (2022) Spatiotemporal evolution characteristics and driving forces of surface thermal environment in mining intensive areas. Sci Technol Eng 22(32):14537–14548
25. Sobrino JA, Jiménez-Muñoz JC, Paolini L (2004) Land surface temperature retrieval from LANDSAT TM 5. Remote Sens Environ 90(4):434–440
26. Chen SL, Wang TX (2009) Comparison analyses of equal interval method and mean-standard deviation method used to delimitate urban heat Island. J Geo-Inf Sci 11(02):145–150

BY NC ND

Thermal-Humid Environment Characteristics in Dry Seasons: A Study of Chenggong District, Kunming, China

Zhihao Wang, Chonghao Dai, Tong Li, Pengkun Zhu, and Shihan Deng

Abstract To investigate the thermal and relative humidity characteristics during the dry season in Kunming's Chenggong New District, this study selected three observation routes and conducted parallel mobile observations on November 27th–28th, 2022, and January 10th–11th, 2023. The routes encompassed various land use types and underlying surfaces. After processing the data, the results were imported into ArcGIS 10.6 to create band-like distribution maps of temperature and relative humidity for subsequent analysis. The results demonstrate that in the study area, daytime urban heat island intensity slightly surpasses nighttime intensity during the dry season, with an average daytime urban heat island intensity of 2.5 °C and a nighttime urban heat island intensity of 2.1 °C. A significant urban heat island effect is observed in industrial and residential areas. The rates of change for temperature and humidity are 1.3, −0.6 °C per h, and −3.9, 2.7% per h, respectively, with daytime variations being significantly greater than nighttime ones.

Keywords Thermal-humid environment · Parallel mobile observation · Heat emissions · Water bodies · Chenggong new district

1 Introduction

In recent years, China's rapid economic growth has driven substantial urbanization, resulting in a marked increase in the urban population. According to the Population Reference Bureau, as of July 2022, China's population reached 1.411 billion, with an urbanization rate of 65.22% [1], up from 36.2% in 2000, marking an annual increase of 2.71%. This rapid urbanization, however, has not come without its challenges. It has led to various environmental issues, notably the urban heat island (UHI) effect, which has become a major area of concern.

Z. Wang · C. Dai · T. Li · P. Zhu · S. Deng (✉)
Faculty of Civil Engineering and Mechanics, Kunming University of Science and Technology, Kunming, China
e-mail: sdeng@kust.edu.cn

B. He et al. (eds.), *Climate-Resilient, Low-Carbon, and Sustainable Cities*, Lecture Notes in Civil Engineering 754, https://doi.org/10.1007/978-981-95-2169-2_15

The UHI effect is a meteorological phenomenon where cities show higher temperatures than nearby rural areas, primarily due to heat from urban infrastructure, such as buildings, roads, and vehicles, as well as human activities. This heat is absorbed and retained by urban surfaces and structures, causing an increase in urban temperatures. Initial observations of UHI date back to the mid-twentieth century with Luke Howard [2] noting the warmer temperatures in urban areas. Oke [3] further elucidated the mechanisms behind UHI through comprehensive energy balance analyses, examining the exchange of energy—such as radiation, convection, and both sensible and latent heat—between urban and rural environments. Duckworth et al. [4] examined the effects of urbanization on horizontal and vertical temperature gradients. They found that urbanization significantly altered local temperature distribution, leading to distinct temperature gradients between urban areas and the surrounding rural regions. Rao [5] was a pioneer in utilizing satellite remote sensing to identify UHIs.

Currently, numerous scholars have extensively studied major Chinese cities experiencing severe UHI effects. Meng et al. [6] analyzed 12 years of remote sensing data from Beijing, proposing a method to delineate urban boundaries based on impervious surface distribution density (ISDD) and developing a novel classification for surface urban heat island intensity (SUHII). The analysis revealed that SUHII peaked during summer days and was closely associated with ISDD. However, at night, the intensity increased from the city's outskirts to its center, showing a weaker correlation with ISDD. In Shanghai, Cui and Shi [7] determined that urban expansion and population growth significantly contributed to the UHI effect, while increasing urban green spaces helped mitigate the issue. Chen et al. [8] explored the UHI effect in different local climate zones (LCZs) in Guangzhou during heatwaves, observing that UHI intensity (UHII) increased at night in densely built areas, while it decreased during the day in areas with more vegetation. Huang et al. [9] introduced the "Land Contribution Index (LCI)" in Wuhan, demonstrating that urban construction land had the greatest contribution to UHI, whereas water bodies and green spaces mitigated it. Zheng et al. [10] employed the LCZ framework to study the spatial distribution of nighttime UHI, highlighting the framework's effectiveness in understanding UHI and recommending urban planning strategies. Wang and Jun [11] identified that in Chongqing, the UHI effect was particularly strong in high-density areas during hot seasons. Luo et al. [12] utilized infrared remote sensing and the STIRPAT model to assess the impact of industrial structure on the UHI effect across the Yangtze River Delta urban agglomeration. The findings indicate that industrial upgrading significantly intensified the UHI effect, with identified threshold effects related to population size and technological advancement.

In Kunming, researchers like Li [13] have assessed the UHI effect through various determinants, including urban infrastructure, human activity, city size, population growth, and weather conditions. Although Kunming's UHI is less severe than in the aforementioned cities, the phenomenon persists. Sun et al. [14] analyzed meteorological data from 2004 to 2012 to examine Kunming's UHI effect. The analysis revealed that the UHI was more pronounced at night, with the affected area expanding after 2008 and shifting from the main urban center towards Chenggong and Shilin. Ma

[15] utilized 30 years of remote sensing data, demonstrating that land use changes in Kunming have intensified the surface UHI effect. He recommended optimizing land use and controlling heat emissions to mitigate the UHI effect. Pan et al. [16] analyzed 70 years of temperature data, finding that extreme high temperatures in summer are increasing, while extreme cold events are decreasing. The most significant UHI effects were observed in winter and spring, driven by urban population growth, urbanization, and increases in per capita GDP. Chuan et al. [17] employed high-resolution meteorological data alongside remote sensing data to examine the intricate structure of the UHI effect in Kunming. The observations reveal that nighttime UHII decreases exponentially as the distance from the city center increases. Additionally, the analysis shows that meteorological conditions significantly affect the temporal and spatial variations of the UHI, with the strongest effects occurring on autumn nights. While these studies offer valuable insights into Kunming's UHI, they are limited by low spatial and temporal resolution and do not adequately cover rapidly urbanizing areas.

Kunming has a typical subtropical monsoon climate, with distinct dry and wet seasons. The dry season, lasting from November to April, is characterized by low precipitation, relatively high temperatures, and increased evaporation. Since 1990, the UHI effect during Kunming's dry season has become more pronounced, with overall precipitation decreasing and average temperatures rising by 0.43 °C per decade. This temperature increase enhances evaporation, further intensifying the UHI effect [18]. The dry season features large diurnal temperature variations and stable weather, providing ideal conditions for long-term meteorological observations and data collection. Thus, it is an optimal period for studying the UHI effect and analyzing its thermal and humidity characteristics.

Chenggong District, an emerging area in the southeastern part of Kunming City, has experienced rapid economic development. Urbanization in Chenggong District has led to significant improvements in infrastructure and urban functions, attracting a growing population through enhanced commercial, educational, medical, and entertainment resources. However, this rapid urbanization has also introduced challenges, including the UHI effect and environmental pollution, which demand effective solutions. This study employs a parallel mobile observation method across three routes in Chenggong New District to measure temperature and relative humidity, generating band-like distribution maps. The findings provide valuable insights into the UHI effect and thermal-humid conditions, serving as a reference for future urban planning and development.

2 Testing Strategy

2.1 Methods and Routes for Testing

The mobile observation method is an effective approach for studying local climates. It involves installing meteorological instruments on vehicles to collect weather data along city routes, facilitating the investigation of UHIs. Originating in the 1930s, this method has been utilized by scholars from countries such as Germany [19], Poland [20], Austria [21], New Zealand [22], and Japan [23] to study UHI. The parallel mobile observation method, a variation of this approach, involves simultaneous observations along several parallel routes. Vehicles depart at the same time to ensure synchronized and comparable data. This method not only assesses the impact of UHII but also enables the comparison of similarities and differences across routes. By analyzing these patterns, researchers can identify the thermal and humidity characteristics of the area, contributing to the formulation of strategies aimed at mitigating UHI effects.

In the Chenggong New District of Kunming, a mobile observation was carried out along three parallel routes selected to capture diverse land use patterns and underlying surface types. Beginning at the shores of Dian Lake and extending inland to the final points, this setup was carefully designed to offer a detailed understanding of thermal and humidity variations during the dry season in Chenggong, while assessing Dian Lake's influence on the local climate. The routes are designated for clarity as Route A, Route B, and Route C, arranged from north to south as depicted in Fig. 1 (*Map Data Source* Baidu Maps).

The three observation routes start approximately 1 km away from Dian Lake and run eastward, covering the breadth of Chenggong's urban area. Each route extends around 7 km in length, with Route A positioned about 2.5 km from Route B, and Route B about 1.5 km from Route C. The distance between Route A and B is larger than between B and C due to the absence of a suitable parallel path to cover the entire district. These routes encompass diverse urban landscapes, including parks, lakes, farmland, reservoirs, densely populated high-rise and low-rise residential areas, schools, and construction sites.

Due to pandemic restrictions, the starting points for the observation routes were adjusted. Route A began at the entrance of Jiangwei Community near Dian Lake, traveling east along Shilong Road and passing through several urban areas and parks, ending near Shilongba Reservoir. Key water bodies along this route included Dian Lake and Luolong Lake. Route B started from Aochen Dianchi Xingcheng Residential Area, moving from Hailu Road to Tuo Feng Street, and ended at Bailongtan Reservoir, including stops like Wanda Plaza and several parks. Major water features on this route were Dian Lake and Muchun Lake. Route C originated from Aochen Central Square on Baozhu Street, extending east to the intersection of Liangwang Road and Lianda Street, passing through urban centers and university areas without significant water body encounters, except near the starting point at Dian Lake.

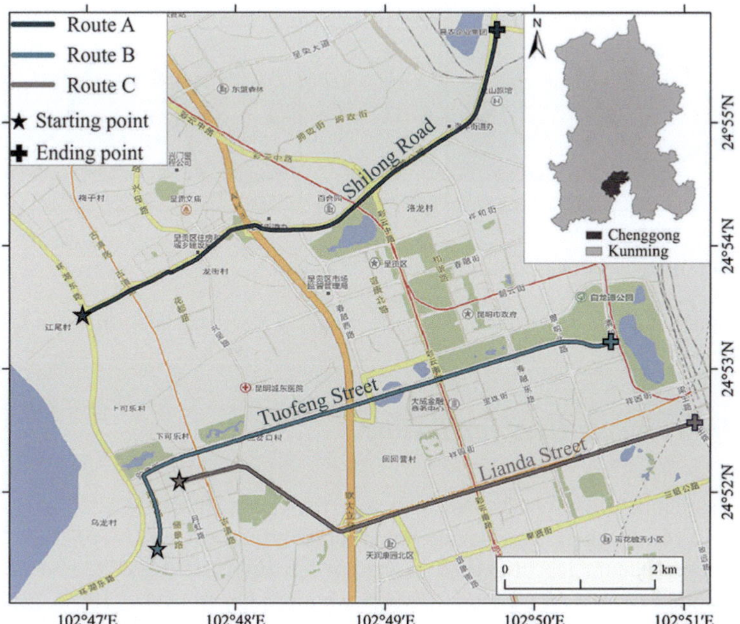

Fig. 1 Study area and test routes

2.2 *Testing Time*

To maintain the accuracy and reliability of the test data and results, and to avoid potential damage to the equipment on rainy days, it is crucial that the weather be clear with minimal or no cloud cover both 3–5 days before and on the actual days of testing. Based on these weather requirements, the testing periods have been scheduled from November 27th, 2022, until the morning of November 28th, 2022, and from January 10th, 2023, until the morning of January 11th, 2023.

As shown in Fig. 2, the full-day temperature and relative humidity data for Chenggong District, Kunming, on the day before the test dates (November 26th, 2022, and January 9th, 2023) revealed that on both days, the temperature decreased and relative humidity increased from 00:00 to 08:00, the temperature increased and relative humidity decreased from 8:00 to 17:00, and from 17:00 to 24:00, the temperature decreased and relative humidity increased again. Therefore, this study set the parallel observation periods during the temperature-decreasing, relative humidity-increasing intervals twice and the temperature-increasing, relative humidity-decreasing intervals twice, with two parallel observations during the day and two at night. This approach aimed to analyze the temperature and relative humidity variations and the thermal and relative humidity environment characteristics in Chenggong District during these intervals.

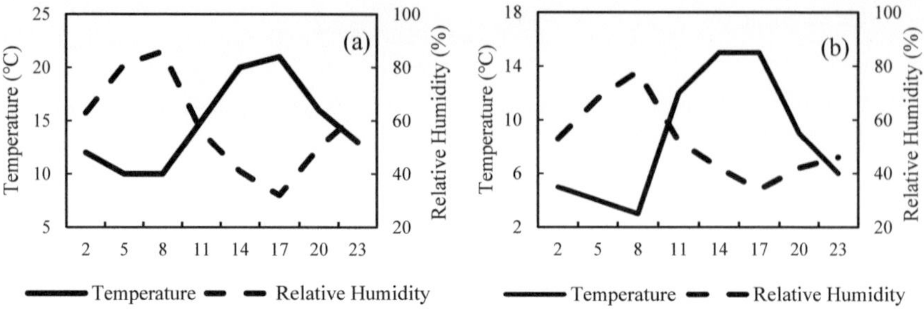

Fig. 2 Temperature and relative humidity change trend: **a** November 26th, 2022 and **b** January 9th, 2023

2.3 *Instruments Used for Testing*

The instruments utilized in this study include the Onset HOBO MX2302A Temperature/RH Data Logger. This device has a temperature range of −40–70 °C, with measurement accuracies of ±0.25 °C below freezing and ±0.2 °C above, and a fine resolution of 0.04 °C. It measures relative humidity from 0 to 100%, achieving accuracies of ±2.5% within the most common range of 10–90%, and ±5% otherwise. The Columbus P-1 Mark II high-precision GPS logger was also employed to record positional data, including latitude, longitude, speed, altitude, and other related parameters. The positioning accuracy of the GPS is 1.5 m (horizontal) at 50% CEP (Circular Error Probable) and 4.0 m at 95% CEP.

To ensure data reliability and accuracy, instruments must be calibrated before starting the test. To shield the temperature and relative humidity logger from external elements such as sunlight and wind, it is housed within a louvered cover mounted on the car's roof, about 1.7 m above the ground. Once the parallel observations commence, both the temperature and relative humidity logger and the GPS logger start recording data at one-second intervals to maintain synchronized data collection. The car is driven at a steady speed of around 40 km/h to ensure that the data are uniformly distributed spatially and to reduce the impact of wind on the measurements. During the tests, obstructions from overpasses, buildings, and trees above the car may intermittently affect the GPS signal, a factor that is monitored throughout the observation period.

3 Data Processing

After the test, latitude and longitude data from the three routes will be processed into a format compatible with AutoCAD 2019, enabling the generation of corresponding route maps. These maps, along with logger data, will be imported into ArcGIS 10.6. Interpolation analysis will match the latitude and longitude points with the corresponding temperature and relative humidity data. Kriging and extraction analyses will be applied to create temperature and relative humidity distribution maps for each route at different times, using color gradients to visualize the values. This approach facilitates the analysis of temperature and relative humidity variations within the study area. Based on the logger's specifications, the errors of temperature and relative humidity data are ±0.2 °C and ±2.5%, respectively.

In mobile observation studies, as the observation period exceeds one hour, the influence of weather changes and diurnal variations becomes more significant. Furthermore, when passing through areas with diverse urban structures or vegetation cover, which are marked by complex spatiotemporal characteristics, the data may not fully capture microclimate change patterns. To ensure accuracy, consistency, and spatiotemporal comparability, post-processing typically requires synchronization corrections and temporal alignment [24]. However, for observation periods shorter than one hour, when meteorological conditions are stable and changes in urban types or vegetation coverage have minimal effects on the microclimate, synchronization corrections are unnecessary [25]. In this study, mobile observations on different routes began simultaneously, each lasting approximately 40 min. The maximum distance between routes was only 2.5 km, and weather conditions remained stable throughout the observation, making synchronization corrections unnecessary.

4 Analysis of Thermal and Relative Humidity Conditions

The parallel observations carried out on November 27th–28th, 2022, and January 10th–11th, 2023, are categorized into daytime and nighttime sessions based on temperature and relative humidity trends—specifically, periods of temperature increase with relative humidity decrease, and temperature decrease with relative humidity increase. Each session is analyzed separately. Following this, a comprehensive analysis of all gathered data is conducted to thoroughly investigate the thermal and relative humidity dynamics in Chenggong District during the dry season. This approach is designed to provide a detailed understanding of how temperature and relative humidity interact and fluctuate, offering insights into the specific climatic conditions of the area.

4.1 Daytime Trends: Rising Temperature and Decreasing Relative Humidity

The band-like distribution maps of temperature and relative humidity, obtained during parallel mobile observations on November 27th, 2022, and January 10th, 2023, are presented in Figs. 3 and 4.

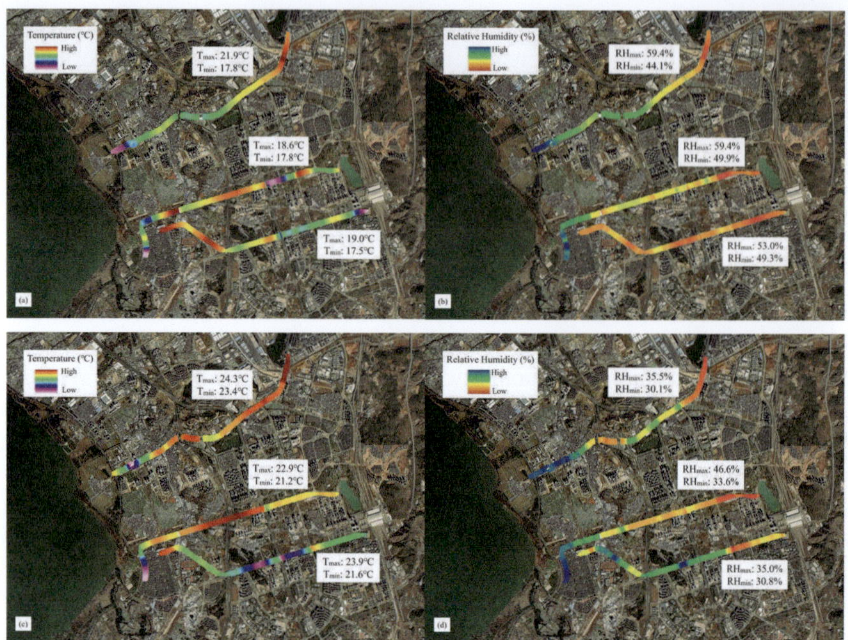

Fig. 3 Daytime temperature and relative humidity measurements at **a** and **b** 10:00 AM and **c** and **d** 2:00 PM on November 27th, 2022

At 10:00 on November 27th, 2022 (Fig. 3a and b), Route A showed the largest temperature variation of 4.1 °C and reached the highest temperature of 21.9 °C among the three routes. In contrast, Route C exhibited the smallest temperature variation of 0.8 °C and the lowest maximum temperature of 18.6 °C. Furthermore, Route A also experienced the greatest relative humidity variation of 15.3%, while both Routes A and B reached the peak relative humidity of 59.4%. In contrast, Route C had the smallest relative humidity variation of 3.7% and the lowest maximum relative humidity of 53.0%. At 14:00 (Fig. 3c and d), Route A again recorded the highest temperature of 24.3 °C, but with the smallest temperature variation of 0.9 °C. Meanwhile, Route B exhibited the lowest maximum temperature of 22.9 °C, while Route C showed the largest temperature variation of 2.3 °C. For relative humidity, Route B exhibited the greatest variation of 13.0% and the highest maximum relative humidity of 46.6%, while Routes A and C showed similar relative humidity variations and maximum relative humidity levels. These differences are attributable to the unique characteristics of each route. Route A primarily traverses residential and industrial zones characterized by high population density, heavy traffic, and low vegetation coverage. Although it passes near a large water body (Luolong Lake), the moderating effect of the water is minimal, resulting in higher temperatures and lower relative humidity levels due to insufficient heat dissipation. In contrast, Route B passes through less densely populated areas, with greater vegetation coverage, more large water bodies (Muchun Lake, Bailongtan Reservoir), wider roads, and

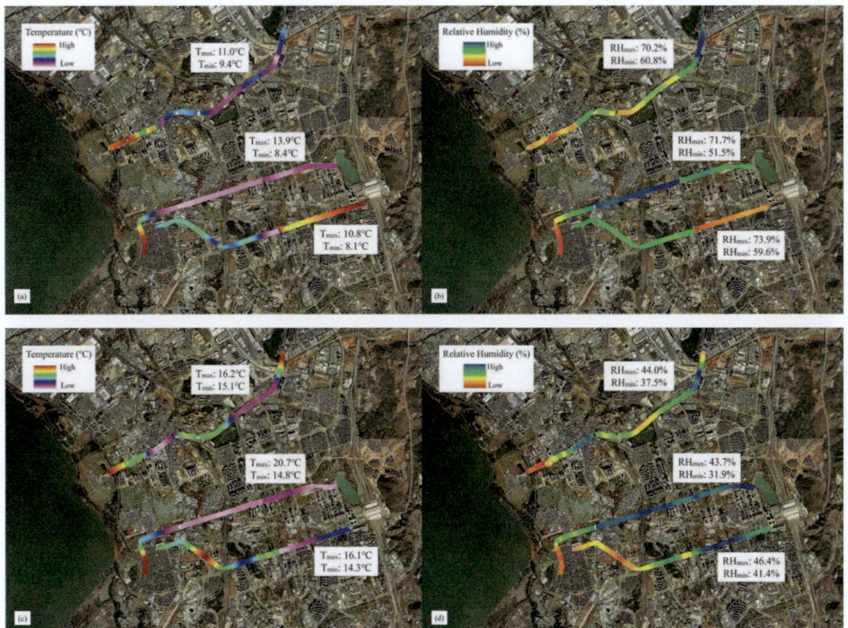

Fig. 4 Daytime temperature and relative humidity measurements at **a** and **b** 10:00 AM and **c** and **d** 2:00 PM on January 10th, 2023

less traffic. These factors facilitate better heat dissipation and moisture diffusion, leading to cooler, more humid, and more stable temperature and relative humidity levels compared to the other two routes.

At 10:00 on January 10th, 2023 (Fig. 4a and b), Route B exhibited the largest temperature variation of 5.5 °C and reached the highest temperature of 13.9 °C among the three routes. Although Route B did not reach the highest peak relative humidity of 71.7%, it displayed the greatest relative humidity variation of 20.2% and recorded the lowest relative humidity of 51.5%. At 14:00 (Fig. 4c and d), Route B once again exhibited the highest temperature variation of 5.9 °C and reached the maximum temperature of 20.7 °C. Although its peak relative humidity of 43.7% was not the highest, it experienced the greatest relative humidity variation of 11.8% and recorded the lowest relative humidity of 31.9%. Notably, Route B displayed extreme values in maximum temperature and minimum relative humidity, particularly near the starting point around the Dianchi Xingcheng Community. As the route transitions from this area into Tuofeng Street, the temperature begins to decrease, and relative humidity increases, stabilizing without significant fluctuations until the route's end. This phenomenon primarily occurs because Route B passes through a densely populated residential area near the Dianchi Xingcheng Community, where numerous restaurants and car repair shops generate considerable anthropogenic heat. Despite the proximity to Dianchi Lake, this leads to elevated temperatures and reduced relative humidity levels in the area. Once the route transitions into the wider, vegetation-rich Tuofeng Street, the temperature drops and relative humidity rises, maintaining stability thereafter.

Overall, the daytime temperature and relative humidity patterns on both days were largely consistent. Higher temperatures and lower relative humidity were observed in densely populated residential, industrial, and commercial areas, while regions near Dianchi Lake with extensive vegetation coverage and lower population density exhibited cooler temperatures and higher relative humidity levels. Relative humidity levels gradually declined as the distance from the Dianchi Lake shoreline increased. These patterns are primarily attributed to higher anthropogenic heat emissions in densely populated areas, where buildings are closely packed, vegetation and water bodies are sparse, and surfaces are predominantly composed of rough materials like cement, which impede heat dissipation and diffusion. A significant difference in temperature and relative humidity variations was observed between the two tests, particularly near Dianchi Lake. This difference is likely due to the relaxation of pandemic restrictions after December 2022, leading to increased outdoor activities. All three routes began near densely populated residential areas. During the daytime test on November 27th, 2022, the lakeside area was notably influenced by the lake-land breeze and the water's moderating effect, resulting in lower temperatures and higher relative humidity near the shore. However, in the January 10th, 2023 test, accumulated heat from human activities did not disperse quickly, leading to higher temperatures and lower relative humidity. This suggests that anthropogenic heat emissions have a greater impact on the UHI than the water body's moderating influence. The UHII for each route is determined by subtracting the lowest temperature from the highest during each time period using daytime temperature data. The average daytime UHII is found to be 2.5 °C.

4.2 Nighttime Trends: Decreasing Temperature and Increasing Relative Humidity

The band-like distribution maps of temperature and relative humidity, obtained during parallel mobile observations on the nights of November 27th–28th, 2022 and January 10th–11th, 2023, are presented in Figs. 5 and 6.

At 21:00 on November 27th, 2022 (Fig. 5a and b), Route C exhibited a larger temperature variation of 1.5 °C compared to Routes A and B, which both showed a temperature variation of only 0.4 °C. Route A recorded the highest temperature of 17.3 °C, while Route B had the lowest maximum temperature of 16.9 °C. Additionally, Route A exhibited the largest relative humidity variation of 6.2% and the highest maximum relative humidity of 59.8%, while Route C displayed the smallest relative humidity variation of 3.9% and the lowest maximum relative humidity of

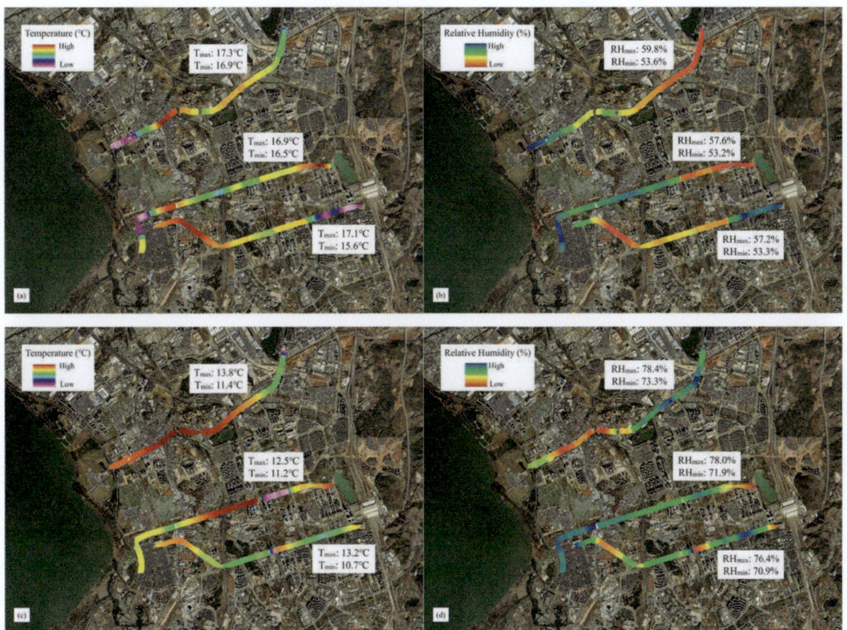

Fig. 5 Nighttime temperature and relative humidity measurements at **a** and **b** 9:00 PM and **c** and **d** 6:00 AM on November 27th–28th, 2022

57.2%. At 6:00 on November 28th, 2022 (Fig. 5c and d), Route C again exhibited the largest temperature variation of 2.5 °C, while Route A recorded the highest maximum temperature of 13.8 °C. Route B exhibited the smallest temperature variation of 1.3 °C and the lowest maximum temperature of 12.5 °C. Regarding relative humidity, Route B exhibited the largest variation of 6.1%, while Route A had the smallest variation of 5.1%, but recorded the highest maximum relative humidity of 78.4%. Route C recorded the lowest maximum relative humidity of 76.4%. The first half of Route C traverses densely populated residential and industrial areas, leading to significant anthropogenic heat emissions at 21:00, resulting in the highest temperatures and lowest relative humidity in this section of the route. The second half of Route C mainly consists of fields, schools, and unoccupied buildings, with less anthropogenic heat and more vegetation, resulting in lower temperatures and higher relative humidity compared to the first section. By 6:00 the next morning, these effects were somewhat mitigated, though the high temperature and low relative humidity conditions remained concentrated in the first half of the route. Consistent with daytime observations, Route A consistently exhibited higher temperatures than the other two routes, particularly in densely populated residential areas.

At 19:00 on January 10th, 2023 (Fig. 6a and b), Route B exhibited the largest temperature variation of 6.6 °C and the highest temperature of 19.4 °C among the three routes. The temperature variations of Routes A and C were 0.9 and 1.1 °C and their maximum temperatures of 13.9 and 14.0 °C were comparable. Additionally,

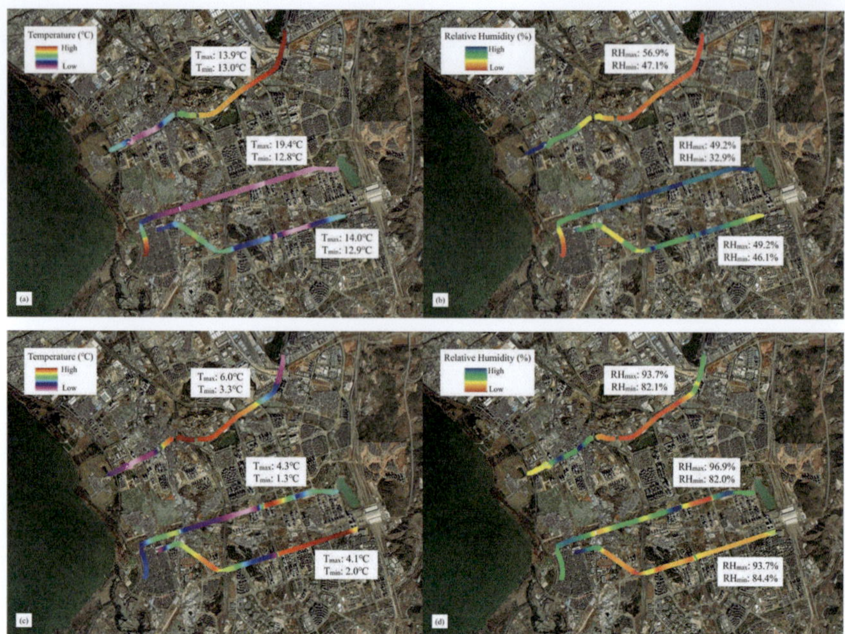

Fig. 6 Nighttime temperature and relative humidity measurements at **a** and **b** 7:00 PM and **c** and **d** 7:00 AM on January 10th-11th, 2023

Route B exhibited the largest relative humidity variation of 16.3%, whereas Route C had the smallest relative humidity variation of 3.1%. Route A recorded the highest maximum relative humidity of 56.9%, whereas both Routes B and C had the lowest maximum relative humidity of 49.2%. At 7:00 on January 11th, 2023 (Fig. 6c and d), Route B again exhibited the largest temperature variation of 3.0 °C, while Route A recorded the highest maximum temperature of 6.0 °C. Route C displayed the smallest temperature variation of 2.1 °C and recorded the lowest maximum temperature of 4.1 °C. In terms of relative humidity, Route B showed the largest variation of 14.9% and recorded the highest maximum relative humidity of 96.9%. Route C exhibited the smallest relative humidity variation of 9.3% and, along with Route A, recorded the lowest maximum relative humidity of 93.7%. At 19:00, Route B displayed a pattern similar to the daytime observations: near the starting point at Dianchi Xingchen Community, the temperature was higher and relative humidity lower, reaching 19.4 °C and 32.9%. However, after entering Tuofeng Street, the temperature dropped rapidly, and the relative humidity increased quickly, then stabilized. By 7:00, although Route B still exhibited significant temperature and relative humidity variations, this pattern was no longer evident. This change was primarily due to reduced anthropogenic heat emissions during the night in the Dianchi Xingchen Community, combined with the effects of vegetation transpiration and the lake-land breeze from Dianchi Lake, which effectively dispersed the heat accumulated during the day, resulting in lower temperatures and higher relative humidity by 7:00 in this area. By that time, densely populated residential and industrial areas had already begun daily activities, leading to elevated temperatures and reduced relative humidity in these regions.

Overall, the nighttime temperature and relative humidity patterns on both days indicated that densely populated residential, industrial, and commercial areas generally experienced higher temperatures and lower relative humidity. Conversely, areas near Dianchi Lake and those with dense vegetation exhibited lower temperatures and higher relative humidity, with relative humidity levels gradually diminishing as the distance from Dianchi Lake increased. At night, unlike during the day, Dianchi Lake releases stored heat and water vapor, increasing temperature and relative humidity in nearby shoreline areas and further intensifying the nighttime UHI. Smaller bodies of water also influence nighttime temperature and relative humidity, but their impact is less pronounced than that of Dianchi Lake. The average nighttime UHII, based on nighttime temperature data, was calculated to be 2.1 °C.

4.3 Comprehensive Evaluation

The temperature and relative humidity data from the two days of parallel tests were grouped, with data from the three routes during the same time period combined into one group, resulting in eight groups of temperature and relative humidity data, respectively. These data were visualized through box plots, as illustrated in Fig. 7a and b. The horizontal axis represents the starting time of each test, with the symbol inside the box indicating the mean value, and the whiskers extending to the maximum and minimum values.

The two-day test data presented in Fig. 7 clearly show that temperature and relative humidity exhibit opposite trends throughout the day, with higher temperatures and lower relative humidity during the day and lower temperatures and higher relative humidity at night, following the typical day-night cycle. Compared to daytime,

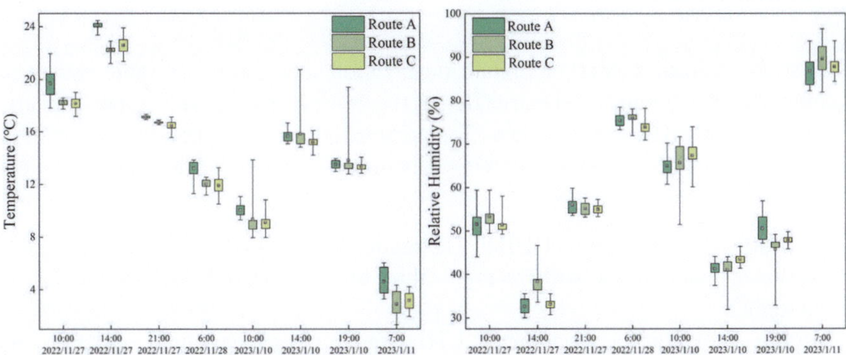

Fig. 7 The correlation between time and **a** temperature and **b** relative humidity

temperature and relative humidity on the three routes fluctuate less and are more stable at night. This stability is ascribed to the lack of solar radiation and reduced anthropogenic heat during the night, which minimize microclimate variations and result in more uniform conditions. The average temperatures for Routes A, B, and C are 14.7, 13.9, and 13.7 °C, respectively, with corresponding relative humidity values of 58.7, 59.4, and 58.5%. Route A's average temperature is significantly higher than those of Routes B and C, while the relative humidity remains similar across all routes. Table 1 presents detailed data on UHII, relative humidity variation (ΔRH), and the rate of change (ROC) for temperature and relative humidity during day and night. For Route A, the average UHII is 1.9 °C during the day and 1.6 °C at night, with a day-night difference of 0.3 °C. For Route B, the average UHII is 3.5 °C during the day and 2.8 °C at night, showing a larger difference of 0.7 °C. Route C has an average UHII of 2.1 °C during the day and 1.8 °C at night, with a difference of 0.3 °C. Smaller UHII fluctuations on Routes A and C suggest more balanced heat dynamics, while Route B's larger fluctuations indicate significant heat retention, with more heat stored during the day and less released at night. As shown in Fig. 7, Route A's overall temperature is higher than Routes B and C due to its path through industrial and residential areas, where narrow streets trap heat, slowing its release and creating broader high-temperature zones. In contrast, Routes B and C have high-temperature areas concentrated in residential and industrial zones, while most sections have dense vegetation, lower impermeability, and wider roads, contributing to a stronger UHI, especially on Route B. Relative humidity changes are small for Routes A and C but show larger fluctuations for Route B. The three routes exhibit similar patterns in the ROC for temperature and relative humidity during both day and night. The ROC are 1.3 and −0.6 °C per h for temperature, and −3.9 and 2.7% per h for relative humidity, with more pronounced daytime variations. These variations reveal spatial differences in the UHI, confirming the impact of various urban areas on the region's thermal and humidity environment.

5 Conclusions

This study utilized a parallel mobile observation method across three routes in Kunming's Chenggong New District during the dry season to record temperature and relative humidity data over two days. Temperature and relative humidity distribution maps were generated, and the following conclusions were drawn after analyzing the data:

1. In the dry season, daytime UHII in Chenggong New District, Kunming, is slightly stronger than at night, with average values of 2.5 °C during the day and 2.1 °C at night.
2. Route A exhibits a pronounced UHI, with minimal day-night variation. Its average temperature is 14.7 °C, about 1 °C higher than Routes B and C, and the day-night UHII difference is 0.3 °C. Route B has the strongest UHII, with

Table 1 The rate of change for temperature and relative humidity, UHII, and humidity variation for each route at different times

Time		Route A				Route B				Route C			
		UHII (°C)	ROC (°C/h)	ΔRH (%)	ROC (%/h)	UHII (°C)	ROC (°C/h)	ΔRH (%)	ROC (%/h)	UHII (°C)	ROC (°C/h)	ΔRH (%)	ROC (%/h)
2022/11/27	10:00	4.1	1.1	15.3	−4.7	0.8	1.0	9.5	−3.7	1.5	1.1	3.7	−4.6
2022/11/27	14:00	0.8		5.4		1.7		13.0		2.3		4.2	
2022/11/27	21:00	0.4	−0.4	6.2	2.1	0.4	−0.5	4.4	2.3	1.5	−0.5	3.9	2.1
2022/11/28	6:00	2.4		5.1		1.3		6.1		2.5		5.5	
2023/1/10	10:00	1.6	1.4	9.4	−3.4	5.5	1.6	20.2	−3.1	2.7	1.5	14.3	−4.0
2023/1/10	14:00	1.1		6.5		5.9		11.8		1.8		5.0	
2023/1/10	19:00	0.9	−0.7	9.8	3.0	6.6	−0.9	16.3	3.6	1.1	−0.8	3.1	3.3
2023/1/11	7:00	2.7		11.6		3.0		14.9		2.1		9.3	

an average temperature of 13.9 °C and a day-night UHII difference of 0.7 °C. In contrast, Route C has a relatively mild thermal-humid Environment, with an average temperature of 13.7 °C and a day-night UHII difference of 0.3 °C.

3. In the dry season of Chenggong New District, Kunming, the ROC in temperature is 1.3 °C per hour during the day and −0.6 °C per hour at night, while the ROC in relative humidity is −3.9% per hour during the day and 2.7% per hour at night. Daytime variations are more pronounced than those at night.

Acknowledgements This work was supported by the Scientific Research Foundation of Yunnan Provincial Department of Education (2024Y130).

References

1. Population Reference Bureau: 2023 World Population Data Sheet. http://www.prb.org. Accessed 20 Dec 2023
2. Howard L (1833) Climate of London deduced from meteorological observation. Harvey and Darton 1(3):1–24
3. Oke TR (1982) The energetic basis of the urban heat island. Q J R Meteorol Soc 108(455):1–24
4. Duckworth FA et al (1954) The effect of cities upon horizontal and vertical temperature gradients. Bull Am Meteor Soc 35(1):198–207

5. Rao PK (1972) Remote sensing of urban "heat islands" from an environmental satellite. Bull Am Meteor Soc 53(7):647–648
6. Meng Q, Zhang L, Sun Z et al (2018) Characterizing spatial and temporal trends of surface urban heat island effect in an urban main built-up area: a 12-year case study in Beijing, China. Remote Sens Environ 204:826–837
7. Cui L, Shi J (2012) Urbanization and its environmental effects in Shanghai, China. Urban Clim 2:1–15
8. Chen G, Chen Y, He H et al (2023) Assessing the synergies between heat waves and urban heat islands of different local climate zones in Guangzhou, China. Build Environ 240:110434
9. Huang Q, Huang J, Yang X et al (2018) Quantifying the seasonal contribution of coupling urban land use types on urban heat island using land contribution index: a case study in Wuhan, China. Sustain Cities Soc 44:666–675
10. Zheng Y, Ren C, Shi Y et al (2023) Mapping the spatial distribution of nocturnal urban heat island based on local climate zone framework. Build Environ 234:110197
11. Zhihao W, Jun L (2012) Study on heat island effect of chongqing in summer. Res J Chem Environ 16:82–85
12. Luo Y, Yang Y, He S et al (2023) Exploring the effect of industrial structure on urban heat island effect with infrared observations. Infrared Phys Technol 130:104615
13. Li Y (2015) Study on evaluation method of urban heat island effect. Chongqing University
14. Sun J, Feng J, Duan W (2015) Change in the urban heat island effect in Kunming. Clim Environ Res 20(6):645–653
15. Ma X (2023) Response mechanism of surface urban heat island to land use/land cover change in the main urban area of Kunming. Yunnan Normal University
16. Pan Y, Yang J, Li X et al (2019) Research on the impacts of the changes of temperature and urbanization in nearly 70 years in Kunming. Environ Sci Surv 38(2):29–35
17. Chuan T, Wu J, Zhao D, Yang Q, Fan W, Zhao J (2022) Fine structure analysis of urban heat island of a central city in low-latitude plateau of China. Urban Clim 44:101186
18. Huang Z (2011) Changes of dry-wet climate in the dry season in Yunnan (1961–2007). Adv Clim Chang Res 2(1):49–54
19. Kuttler WM, Barlag AB (1996) Study of the thermal structure of a town in a Narrow Valley. Atmos Environ 30(3):365–378
20. Kazimierz K, Krzysztof F (1999) Temporal and spatial characteristics of the urban heat island of Lodz. Poland. Atmospheric Environment 33:3885–3895
21. Schmidt W (1917) Zum Einfluss großer Städte auf das Klima. Naturwissenschaften 5:494–495
22. Spronken-Smith RA, Oke TR (1998) The thermal regime of urban parks in two cities with different summer climates. Remote Sens 19(11):2085–2104
23. Saitoh TS, Shimada T, Hoshi H (1996) Modeling and simulation of the Tokyo urban heat island. Atmos Environ 30(20):3431–3442
24. Zezzo LV, Coltri PP, Dubreuil V (2023) Microscale models and urban heat island studies: a systematic review. Environ Monit Assess 195:1284
25. Kousis I, Pigliautile I, Pisello AL (2021) A mobile vehicle-based methodology for dynamic microclimate analysis. Int J Environ Res 15:893–901

Assessment of Urban Thermal Environment and Analysis of Influencing Factors in Hangzhou, China

Xinyu Zhou, Yijun Shi, Yuhang Shi, Wei Zhai, and Junqing Tang

Abstract In the context of rapid urbanization and global warming, the comprehensive demands of residents for urban living environments are increasing. The urban heat island effect is a typical ecological consequence of urbanization, which not only adversely affects human health but also restricts further urban development. This paper takes the central urban area of Hangzhou as the research object and evaluates its urban thermal environment and analyzes the influencing factors based on the Landsat 8 remote sensing data of 2013, 2020, and 2024. By applying the mean-standard deviation method, the urban thermal environment of the central urban area of Hangzhou is divided into five levels, and spatial analysis methods such as the transfer area proportion matrix are used to explore the spatiotemporal evolution of the urban thermal environment in the central urban area of Hangzhou between 2013 and 2024. Pearson correlation analysis is used to quantitatively study the relationship between various influencing factors and surface temperature, and the geographic detector is employed to identify the magnitude of influence of each driver and the impact of their interactions on the urban thermal environment in the central urban area of Hangzhou. The results show that from 2013 to 2024, the urban heat island effect in the central urban area of Hangzhou is worsening day by day, and the urban high-temperature areas show a trend of spreading from the city center to the suburbs. The influence of influencing factors such as population density and land use function mixing degree on the urban thermal environment shows a gradually weakening trend. These findings can help urban planning to grasp the dynamic change characteristics of the urban thermal environment more accurately, and then allocate resources scientifically and reasonably, providing targeted reference bases for the subsequent formulation of climate adaptation plans and strategies in the city.

X. Zhou · Y. Shi (✉) · Y. Shi
College of Landscape Architecture, Zhejiang Agriculture and Forestry University, Zhejiang, China
e-mail: yijun_shi@zafu.edu.cn

W. Zhai
College of Architecture, Planning, and Public Affairs, University of Texas at Arlington, Arlington, TX 76019, USA

J. Tang
School of Urban Planning and Design, Shenzhen Graduate School, Peking University, Shenzhen, China

B. He et al. (eds.), *Climate-Resilient, Low-Carbon, and Sustainable Cities*, Lecture Notes in Civil Engineering 754, https://doi.org/10.1007/978-981-95-2169-2_16

Keywords Urban thermal environment · Radiation transfer equation method · Geographic detector · Influencing factors · Hangzhou central district

1 Introduction

On a global scale, the intertwined processes of global climate change and urbanization have conspired to exacerbate the urban heat island effect, with cities in the southern regions being especially vulnerable. In these areas, the pace of climate warming is notably accelerated, and concurrently, they are burdened with a persistently high population density. The synergy of these two elements amplifies the lethality of extreme heat events exponentially [1, 2]. The China Climate Bulletin (2023) indicates that the average temperature in China in 2023 reached the highest record since 1951. Research on urban thermal environments plays a crucial role in improving the quality of urban living environments and achieving harmonious development between urban economies and ecological environments [3]. In-depth research on urban thermal environments aids in understanding urban spatial structure and development scales, promoting the sustainable development of urban ecological environments, and enhancing the quality of residents' living environments.

The evaluation of urban thermal environments is an essential research area for analyzing and quantifying the thermal characteristics of urban areas, typically based on remote sensing data. By analyzing these data, researchers can understand the impact of urbanization on local climates. Wang Yu and colleagues, using LANDSAT 8 remote sensing images, analyzed the influence of land use patterns on urban surface thermal environments. They employed time-series Moderate Resolution Imaging Spectroradiometer (MODIS) surface temperature data and Famine Early Warning Systems Network Land Data Assimilation System FLDAS) heat flux data to study the spatiotemporal variation characteristics of urban heat islands and analyzed the mechanisms influencing urban heat island changes from the perspective of surface energy [4]. Yang and colleagues extracted surface temperature data through Landsat 8 TIRS satellite images, reclassifying them into thermal field values (TFV) to analyze the relationship between urban built-up areas, TFV, and thermal footprint environments [5]. In recent years, methods for evaluating urban thermal environments have continuously innovated. Scholars have utilized not only traditional remote sensing technology and GIS spatial analysis but also integrated research methods like numerical simulations and ecological model construction. Ke Ruipeng and colleagues, using the grey clustering evaluation method, assessed thermal environments and derived the dominant driving factors of urban thermal environments, reflecting the causes of thermal environment formation [6]. Shweta Bhati and colleagues coupled the Weather Research and Forecasting (WRF) model with the Urban Canopy Model to evaluate the impact of land use changes on model performance and estimate urban heat island intensity [7]. Researchers have revealed the spatiotemporal distribution characteristics, changing trends, and driving forces and mechanisms behind urban surface thermal environments. They focus on how urban surfaces absorb and dissipate solar radiation, maintain energy balance, and the interrelation and evolution patterns between urban surface heat island effects and atmospheric heat island effects [8, 9]. From the perspective of dynamic development, most of the existing studies have focused on the static analysis within a certain time period. There is insufficient exploration into the dynamic evolution laws of the urban thermal environment year by year, month by month or even day by day along with the urban renewal process, making it difficult to accurately predict the real-time feedback of the thermal environment after urban expansion and the construction of new districts.

Furthermore, in terms of the interaction between socio-economic factors and the thermal environment, there are seldom in-depth analyses of the complex linkage effects of urban population saturation and industrial structure adjustment on the thermal environment. It remains unclear how these emerging socio-economic phenomena quantitatively influence the thermal environment.

To explore the evolution and influencing factors of the urban heat island effect, this study selected three phases of Landsat 8 series remote sensing data from the central urban area of Hangzhou in 2013, 2020, and 2024. Using the radiation transfer equation method, surface temperatures were retrieved to comprehensively assess the urban thermal environment of Hangzhou's central urban area and identify the key driving factors affecting thermal environment changes in this region. The study will examine the relationship between land use changes, blue-green space configurations, and urban thermal environments, fill the gap in the dynamic development perspective of urban thermal environment, summarize the development trends of urban thermal environments in Hangzhou's central urban area, and explore effective thermal environment mitigation strategies, aiming to promote the sustainable development of Hangzhou's central urban area and improve the living quality and thermal comfort of urban residents.

The structure of the remaining chapters is as follows: Sect. 2 covers the study area and data sources and provides a detailed introduction to the research methods. Section 3 includes the evaluation results and analysis of Hangzhou's urban thermal environment and its spatiotemporal evolution. Section 4 explores the spatial heterogeneity of Hangzhou's LST and its relationship with influencing factors. Section 5 discusses the specific mechanisms and causes of dynamic changes in the urban thermal environment of Hangzhou, as well as the directions that urgently need to be strengthened in terms of research methods. The research conclusions were summarized in Sect. 6.

2 Study Area Overview and Research Methods

2.1 Study Area Overview

Hangzhou (118°20′ ~ 120°37′E, 29°11′ ~ 30°34′N), as the central city of Zhejiang Province, serves as an essential economic engine in the Yangtze River Delta region and a transportation hub in southeastern China. Additionally, Hangzhou is one of China's first pilot cities for carbon peaking and the host city for the 19th Asian Games. As of 2023, the urbanization rate of Hangzhou has surged to 84.2%. With the accelerated pace of urbanization, the land use patterns and the proportion of blue-green spaces in Hangzhou have undergone significant changes. These changes have profoundly impacted the urban thermal environment.

In 2023, the highest annual temperature in Hangzhou reached 42.1 °C, with 39 high-temperature days (daily maximum temperature ≥ 35 °C), ranking sixth nationwide. Hangzhou's high level of economic development and dense population make it an ideal area for this study. The study area is based on the urban scale, focusing on the central urban area of Hangzhou. The study area exhibits significant spatial heterogeneity and temporal evolution, which affects the spatial distribution of surface temperature. The selected study area primarily covers the central urban area of Hangzhou, define the study area through ArcGIS 10.7 (Fig. 1), including Shangcheng District, Gongshu District, Xihu District, Yuhang District, Binjiang District, Qiantang District, and Xiaoshan District. The central urban area has a high population mobility, a high urban built environment renewal rate, and a significant urban heat island effect. The presence of tall office buildings, extensive road networks, and numerous green parks makes it conducive to studying the spatiotemporal evolution of the urban thermal environment.

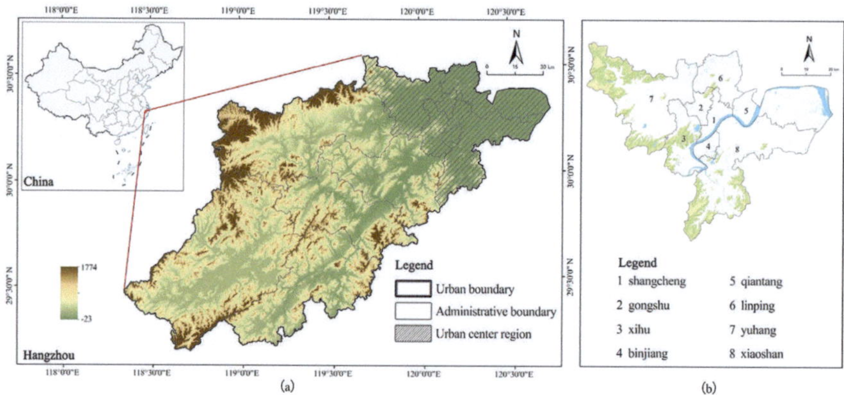

Fig. 1 Schematic diagram of the study area: **a** Topographic map of Hangzhou; **b** Administrative divisions of the central urban area of Hangzhou (Self-drawn)

2.2 Research Data and Preprocessing

Comparing surface temperature data from different years can reveal trends in urban thermal environments over time, thereby assessing the development of the urban heat island effect. Between 2013 and 2024, Hangzhou experienced rapid urbanization, significantly propelled by major events such as the G20 Summit and the successful bid for the Asian Games. These events led to significant changes in land use and increased building density, which are key factors influencing the urban thermal environment. When conducting in-depth analyses of specific periods, the year 2020 occupies a pivotal position in the development trajectory of this decade and plays a crucial role in bridging the past and the future. It wields decisive sway over the seamless unfolding of significant urban public undertakings, notably the Hangzhou Asian Games. Functioning as a linchpin stage for the unremitting propulsion of urban development, it has galvanized Hangzhou to effectuate continuous enhancements in arenas such as infrastructure and new infrastructure initiatives, thereby forging a robust bedrock for the triumphant staging of the Asian Games.

By analyzing surface temperature data from different periods over a decade, this study aims to evaluate the impact of environmental policies and urban planning on the thermal environment and reveal the thermal distribution characteristics across different urban areas, thus providing scientific evidence for urban planning. The study primarily uses thermal infrared bands from the Landsat-8 satellite, with ENVI 5.6 image processing software, based on the radiation transfer equation method for surface temperature retrieval. For analyzing the spatiotemporal evolution of urban thermal environments, the study selects remote sensing images with less than 10% cloud cover and no noise from three periods in the summer of 2013, 2020, and 2024 in Hangzhou. These images undergo key preprocessing steps, including radiometric calibration, atmospheric correction, and image cropping. Radiometric calibration and atmospheric correction are core steps in preprocessing, aimed at eliminating radiometric distortion and the impact of atmospheric reflectance on ground objects, thereby ensuring the data's high accuracy and comparability.

2.3 Mechanisms of Urban Thermal Environment

The exacerbation of urban thermal environment phenomena involves several factors related to the urbanization process, such as increased human activities, higher building density, changes in land use patterns, and climate change. These factors alter the heat exchange and distribution patterns between the urban surface and the atmosphere, leading to higher surface temperatures in urban areas [10, 11]. Over the past few decades, the urban heat island effect has increasingly impacted human health, energy consumption, and environmental quality. Its formation mechanisms are closely related to multiple factors.

Firstly, the acceleration of socioeconomic activities has led to increased human activities, significantly affecting the urban thermal environment through various channels, including population mobility, industrial production, and transportation. These activities not only foster economic growth but also generate and release large amounts of heat, causing surface temperatures to rise and forming the heat island effect [12, 13]. Specifically, during the process of urbanization, the heightened energy consumption, notably anthropogenic heat emissions, has emerged as a pivotal driving force. The expansion of urban traffic, particularly the surge in motor vehicle usage, not only emits extra heat but also intensifies air pollution and dust problems, thereby further elevating surface temperatures [14]. Secondly, the built urban environment, typified by high-rise buildings and narrow streets creating the "street canyon effect," significantly restricts air circulation, hinders heat dissipation, and increases the absorption of solar and long-wave radiation. This results in an imbalance in urban surface energy [15]. The height and density of buildings not only alter wind speed and direction but also impact heat distribution and flow. The physical composition of urban buildings, including materials like concrete and stone, typically has low albedo, absorbing more solar radiation rather than reflecting it, reducing urban surface evapotranspiration, exacerbating the heat island effect, and profoundly affecting the urban climate [16, 17]. This phenomenon not only threatens the urban ecological environment but also poses potential risks to human health.

Lastly, urbanization has led to a reduction in blue-green space coverage, while vegetation and water bodies release latent heat through evapotranspiration, which helps lower temperatures. The reduction in vegetation and water bodies means less latent heat release, forming localized high-temperature areas. Increasing green space and water body coverage and establishing urban forests and parks can effectively use urban green spaces to regulate the urban thermal environment, thereby maximizing their cooling function [18] (Fig. 2).

2.4 Methods for Evaluating Urban Thermal Environment

To effectively assess the urban thermal environment in the central urban area, this study selected the Radiative Transfer Equation method, also known as the Atmospheric Correction method. The radiative transfer equation method has remarkable advantages in the retrieval of surface temperature. Based on the physical principles such as the transmission of electromagnetic radiation in the atmosphere-surface system, it breaks free from the limitations of regional and sample statistics, accurately takes into account the atmospheric interference and the radiative characteristics of complex surfaces such as vegetation and water bodies, and achieves high-precision retrieval. Meanwhile, it efficiently utilizes multi-source remote sensing data such as thermal infrared and visible light, flexibly processes them according to different resolutions, and has both wide applicability and data adaptability. [19]. The main formula used in the surface temperature retrieval process is as follows:

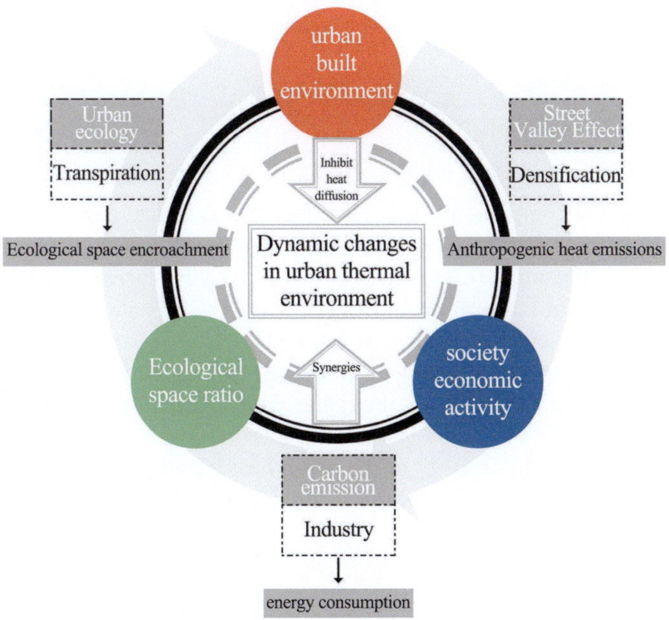

Fig. 2 Mechanisms of urban thermal environment (Self-drawn)

$$P_V = \left[(NDVI - NDVI_{Soil})/\left(NDVI_{eg} - NDVI_{Soil}\right)\right] \quad (1)$$

$$\varepsilon = 0.004P_V + 0.986 \quad (2)$$

where, P_V represents vegetation coverage; $NDVI_{Soil}$ and $NDVI_{eg}$ denote the NDVI values for bare soil and fully vegetated areas, respectively; and ε stands for surface emissivity.

Due to the discontinuation of NASA's atmospheric correction parameter calculator service for obtaining atmospheric profile parameters, this study attempts to use manual processing methods with ENVI 5.6 image processing software based on process files provided in the C2 dataset by the US Geological Survey [20]. Since the data have scaling coefficients and offsets, each band needs to be processed individually to obtain accurate results. Additionally, since the scaling coefficients for each band vary, appropriate adjustments are necessary during the processing.

$$B_\lambda(LST) = \left[L_\lambda - L_\lambda^\uparrow - \tau_\lambda \times (1 - \varepsilon_\lambda) \times L_\lambda^\downarrow\right]/(\tau_\lambda \times \varepsilon_\lambda) \quad (3)$$

where, L_λ represents the thermal infrared radiation brightness; $L_\lambda^\uparrow$ is the atmospheric upwelling radiance; $L_\lambda^\downarrow$ is the atmospheric downward radiance; τ_λ is the thermal infrared band transmittance; $B_\lambda(LST)$ is the blackbody emissivity at LST temperature; ε_λ is the surface emissivity.

The formula for calculating LST is as follows:

$$LST = K_2/\ln(\frac{K_1}{B_\lambda(LST)} + 1) - 273.15 \tag{4}$$

where, $K_1 = 774.89W/(m^2\mu_m \cdot sr)$, K_2=1321.08.

To improve the comparability of the surface retrieval data and more intuitively analyze the thermal environment changes in the study area, this study normalizes the Land Surface Temperature (LST) retrieval results, and then uses the mean-standard deviation method for uniform quantification [21]. The classification standards for surface temperature levels are shown in Table 1.

T_i is the normalized surface temperature of the i-th pixel, T_μ is the average normalized surface temperature, and T_{std} is the standard deviation of the normalized surface temperature.

To directly reflect the intensity of the urban heat island phenomenon, this study introduces the Urban-Heat-Island Ratio Index (URI) [22]. A higher URI indicates a more severe urban heat island effect.

$$URI = \frac{1}{100m}\sum_{i=1}^{n} W_i P_i \tag{5}$$

where, *URI* is the urban heat island ratio index, *m* is the regularization level index, *i* is the temperature level higher than the suburbs in the urban area, *n* is the number of temperature levels higher than the suburbs in the urban area, *W* is the weight value, take the i-th level value, and *P* is the percentage of the i-th level.

By constructing a transfer area ratio matrix, the study quantitatively analyzes the transitions between different thermal environment levels, revealing the structural characteristics of thermal environment changes [23]. In this study, spatial intersection analysis of the thermal environment distribution maps of Hangzhou's central urban area in 2013 and 2020 was conducted to calculate the area ratios of each thermal environment level transitioning to other levels at these two time points. This process includes establishing the conversion area ratio matrix, comparing and analyzing the conversion of each thermal environment level, analyzing the dynamic changes of Hangzhou's urban thermal environment in recent ten years, and further understanding the spatial structure evolution of the thermal environment in the study area. The

Table 1 Classification of thermal environment levels (Self-drawn)

Urban thermal environment level	Classification criteria
High-temperature zone	$T_i > T_\mu + T_{std}$
Sub-high-temperature zone	$T_\mu + 0.5T_{std} < T_i \leq T_\mu + T_{std}$
Medium-temperature zone	$T_\mu - 0.5T_{std} \leq T_i \leq T_\mu + 0.5T_{std}$
Sub-moderate-temperature zone	$T_\mu - T_{std} \leq T_i < T_\mu - 0.5T_{std}$
Low-temperature zone	$T_i < T_\mu - T_{std}$

specific calculation formula is as follows:

$$
S_{ij}/S_i = \begin{bmatrix} S_{11}/S_1 & \cdots & S_{1n}/S_1 \\ \vdots & \ddots & \vdots \\ S_{m1}/S_m & \cdots & S_{mn}/S_m \end{bmatrix} \tag{6}
$$

where, i represents the thermal environment level code of the reference year ($i = 1, 2, \cdots, m, m \leq 5$), j is the thermal environment level code of the target year ($j = 1, 2, \cdots, n, n \leq 5$), S_{ij} is the area transferred from the thermal environment level i of the reference year to the target year level j, and S_i is the total area of the thermal environment level i of the reference year. Therefore, S_{ij}/S_i represents the proportion of the area transferred from the thermal environment level i of the reference year to the target year level j to the total area of the thermal environment level i of the reference year.

2.5 Methods for Analyzing Factors Influencing the Urban Thermal Environment

The geographical detector is a tool based on geographic principles and spatial statistical analysis, designed to detect spatial heterogeneity and uncover its driving forces [24]. The core idea is based on the hypothesis that if an independent variable significantly influences a dependent variable, their spatial distributions should exhibit similarity [25]. The geographical detector can effectively identify and quantify the influence of various factors on spatial phenomena. Combining geographic factors and statistical models, this method can effectively analyze the spatial distribution patterns of the urban thermal environment and the spatial heterogeneity of influencing factors.

Through the geographical detector, we can identify and analyze various factors influencing the formation of the urban heat island effect, such as land use types, building density, and population density. It reveals the spatial relationships between different influencing factors, provides a deep understanding of the mechanisms behind the urban thermal environment formation, and quantitatively analyzes the contribution of different influencing factors. The specific methodology is as follows:

This study first employs the factor detector to analyze the spatial heterogeneity of land surface temperature distribution and evaluate the impact of individual factors. In the study of urban thermal environments, the explanatory power of influencing factors is represented by the q-statistic, which ranges between [0, 1]. The higher the q-statistic value, the more significant the factor's impact on land surface temperature, and vice versa. Therefore, as the q-statistic increases, geographical heterogeneity also increases. If a factor is responsible for creating stratification, its impact on land surface temperature will be more significant, and the q-value obtained by the factor detector will accordingly increase. The calculation formula is as follows:

Table 2 Types of interaction between two driving factors (Self-drawn)

Criterion	Interaction type
q(X1 ∩ X2) < Min(q(X1),q(X2))	Nonlinear attenuation
Min(q(X1),q(X2)) < q(X1 ∩ X2) < Max(q(X1),q(X2))	Single factor nonlinear attenuation
q(X1 ∩ X2) > Max(q(X1),q(X2))	Dual factor enhancement
q(X1 ∩ X2) = q(X1) + q(X2)	independent
q(X1 ∩ X2) > q(X1) + q(X2)	Nonlinear enhancement

$$q = 1 - \frac{\sum_{h=1}^{L} N_h \sigma_h^2}{N\sigma^2} \tag{7}$$

where, $h = 1, \cdots, L$ is the stratification of the influencing factors, where N_h represents the number of units in layer h, N represents the total number of spatial units in the entire study area, A and B represent the total variance and the variance of samples in sub region h. The range of the q statistic is from 0 to 1, and the larger the q value, the stronger the explanatory power of the influencing factors on the thermal environment in Hangzhou.

To further identify the interaction between different risk factors, this study employs the interaction detector. This method evaluates whether the explanatory power of two factors, X1 and X2, on the dependent variable Y, is enhanced or diminished when they act together, and whether their effect on Y is independent. The process involves calculating the q values of factors X1 and X2 on Y, denoted as q(X1) and q(X2), respectively. Additionally, the q value of the interaction between these factors, q(X1 ∩ X2), is calculated by overlaying the layers of variables X1 and X2 to form a new polygon distribution. By comparing q(X1), q(X2), and q(X1 ∩ X2), the type of interaction between the two factors can be determined [26]. The comparison results can be interpreted according to Table 2.

3 Evaluation and Analysis of Urban Thermal Environment in the Central Urban Area of Hangzhou

3.1 Results of Surface Temperature Retrieval in the Central Urban Area of Hangzhou

Based on the aforementioned research methods, this study retrieved the surface temperature of the central urban area of Hangzhou for the summers of 2013, 2020, and 2024 using remote sensing images. The vector data of the study area were used for cropping, obtaining the surface temperature data of the study area (Fig. 3).

In 2013, the surface temperature in the study area ranged from 11.60 °C to 70.57 °C, with an average of 43.68 °C. The maximum LST was 70.57 °C, with the

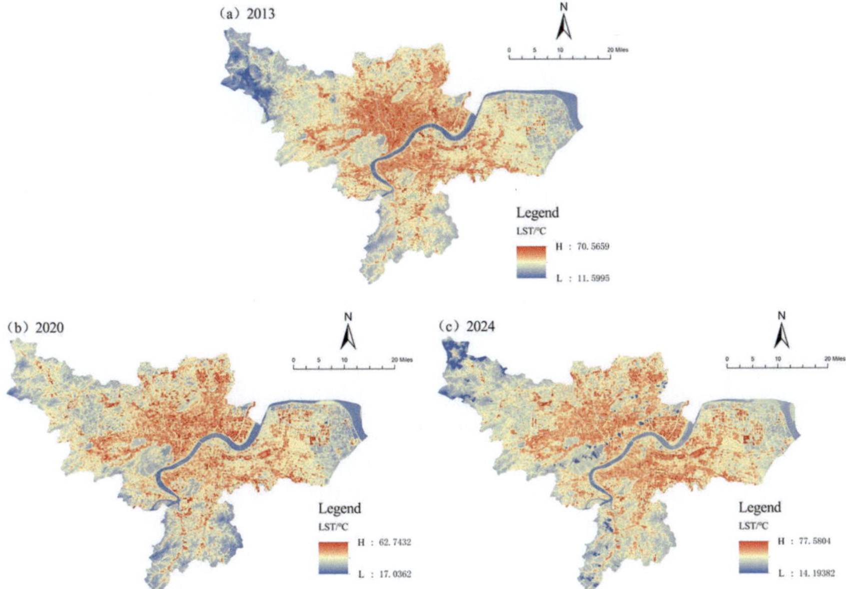

Fig. 3 Surface temperature inversion in different years: **a** 2013; **b** 2020; **c** 2024 (Self-drawn)

highest temperatures observed at urban landfill sites, due to the unique surface cover leading to extremely high urban temperatures. In 2020, the surface temperature in the study area ranged from 17.04 to 62.74 °C, with an average of 37.51 °C. Compared to 2013, the highest surface temperature in the central urban area of Hangzhou in 2020 decreased by 7.83 °C, and the average temperature decreased by 6.17 °C. This reduction could be attributed to the enhanced cooling capacity of vegetation and water bodies within the study area.

From the spatial distribution of LST, high-temperature patches within the central area were spreading, likely due to the rapid economic development and industrial expansion in Hangzhou's central urban area. In 2024, the surface temperature in the study area ranged from 14.19 to 77.58 °C, with an average of 45.96 °C. Compared to the previous periods, the highest and average surface temperatures in the central urban area of Hangzhou reached their peak in 2024.

Analyzing the retrieval results across the three periods indicates that the surface temperature in the central area of Hangzhou is significantly higher than in surrounding areas, forming a pronounced urban heat island effect. In 2013, high-temperature regions were primarily concentrated in older districts such as Shangcheng, Xihu, Gongshu, Binjiang, and the Xiasha Industrial Zone, while other peripheral areas, including some mountainous and lake regions, had relatively lower surface temperatures. By 2020 and 2024, the urban heat island effect had significantly expanded, with high-temperature areas gradually spreading to regions such as Linping, Xiaoshan, and Dajiangdong. Thus, the spatial distribution of the urban thermal environment shows an outward expansion from the old city center of Hangzhou, with the urban heat island's influence range continuously increasing, impacting the residents' living comfort to a certain extent.

To further analyze the spatial distribution changes of urban thermal environment in different periods, the retrieved surface temperatures were normalized. Using the mean-standard deviation method, the urban thermal environment of the study area was divided into five levels: low-temperature area, sub-medium temperature area, medium-temperature area, sub-high temperature area, and high-temperature area (Fig. 4). Based on the temperature level distribution map of the central urban area, the area of each heat island level and the Urban Heat Island Ratio Index (URI) for different periods, as well as the proportions of each heat island intensity level, were calculated (Table 3, Fig. 5).

It can be concluded that in terms of heat island intensity, Hangzhou's central urban area experienced a slight relief from 0.305 in 2013 to 0.295 in 2020, but then rose again to 0.312 in 2024, indicating that the urban heat island effect is worsening.

In 2013, low-temperature areas were mainly on the outskirts of Hangzhou's urban area and around the Qiantang River, covering approximately 14.88% of the study area. Sub-medium temperature zones included the West Lake Scenic Area, Lingshan Scenic Area, Xianghu Scenic Area, and the agricultural zones of Dajiangdong, characterized by high vegetation cover and proximity to water bodies, which accounted for about 14.31% of the study area. Medium-temperature areas were primarily located around the periphery of Hangzhou's central development zone, covering

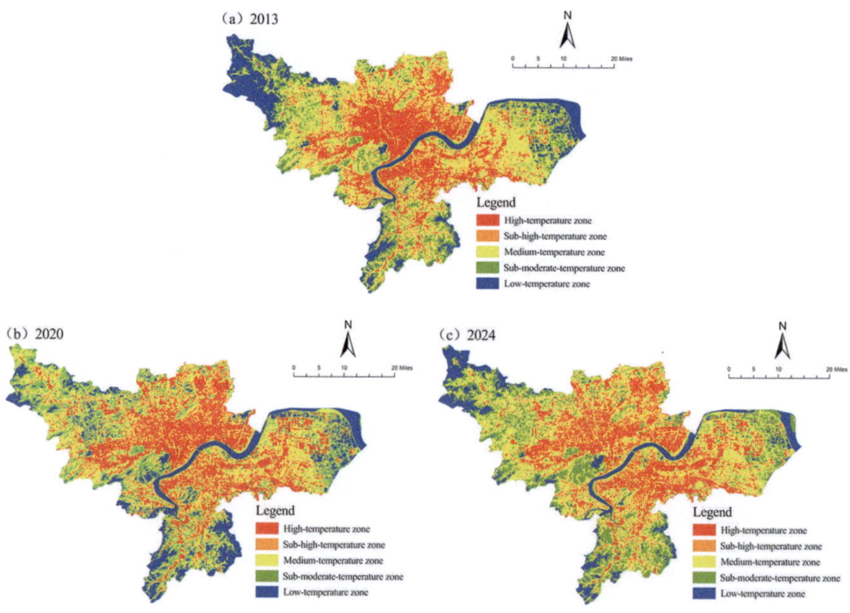

Fig. 4 Distribution of surface temperature levels by year: **a** 2013; **b** 2020; **c** 2024 (Self-drawn)

Table 3 Area of each heat island intensity level and urban heat island ratio index (Self-drawn)

Year	Urban heat island intensity level area (ha)					Urban heat island ratio index (URI)
	High-temperature zone	Sub-high-temperature zone	Medium-temperature zone	Sub-moderate-temperature zone	Low-temperature zone	
2013	62,235.36	50,244.14	125,276.39	48,045.96	49,973.83	0.305
2020	57,503.36	52,094.10	122,131.12	52,929.95	51,117.14	0.295
2024	56,998.60	59,538.34	122,566.17	62,805.09	33,867.47	0.312

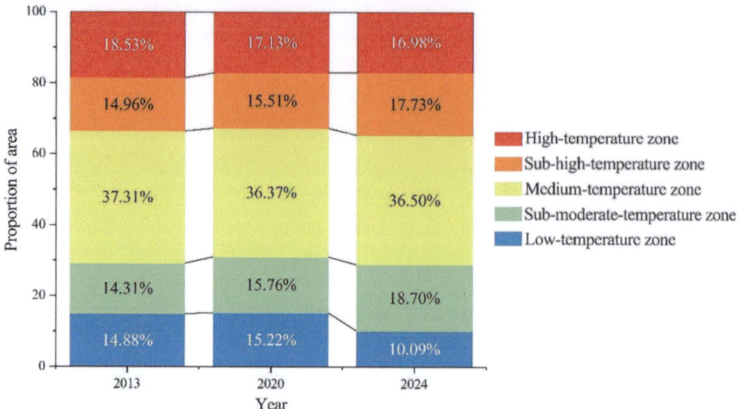

Fig. 5 Surface temperature zoning and area proportion from 2013 to 2020 (Self-drawn)

approximately 37.31% of the study area. Meanwhile, sub-high and high-temperature zones were concentrated in the core urban areas of Xihu District, Gongshu District, Shangcheng District, Binjiang District, and the Xiasha Industrial Zone, accounting for about 14.96% and 18.53%, respectively.

In 2020, the proportions of each temperature zone remained consistent with those in 2013, with slight increases in the areas of low-temperature, sub-medium temperature, and sub-high temperature zones, which accounted for approximately 15.22%, 15.76%, and 15.51%, respectively. In contrast, medium-temperature and high-temperature areas saw slight decreases, accounting for 36.37 and 17.13%.

By 2024, significant changes were primarily observed in the low-temperature, sub-medium temperature, and sub-high temperature zones. The low-temperature area dropped to 10.09%, while sub-medium and sub-high temperature zones increased to 18.70% and 17.73%, respectively. The proportions of high-temperature and medium-temperature areas remained essentially unchanged from 2020.

Overall, analysis of the spatial distribution maps of Hangzhou's central urban thermal environment from 2013 to 2024 indicates significant changes in the heat island distribution. High-temperature patches show an outward expansion from the old urban areas, with some high-temperature zones spreading from their 2013 levels. These changes were mainly observed in densely populated, commercially vibrant, and industrially concentrated areas such as major built-up areas, industrial zones, and university surroundings, which are closely linked to urban population growth and economic activity. Conversely, low-temperature areas are gradually diminishing, primarily found around water bodies and vegetated regions such as West Lake, Qiantang River, and Xianghu. Despite a decrease in the proportion of high-temperature areas within the city, many surrounding urban built-up areas are transitioning from low and medium-temperature zones to sub-medium and sub-high temperature zones. This trend indicates that the increase in sub-medium and sub-high temperature areas exacerbates the overall urban heat island effect, with urban construction and the concentration of population and commercial activities likely being major contributors to changes in the urban thermal environment.

3.2 *Spatial Transformation Analysis of Urban Thermal Environment in Hangzhou's Central Urban Area*

Based on the classification results of the urban thermal environment in Hangzhou's central urban area, a quantitative analysis of the transfer dynamics between different thermal environment levels from 2013 to 2024 was conducted. By calculating the area proportions of each thermal environment level transitioning to other levels at two different time points, the changing trends in the urban thermal environment of Hangzhou's central urban area were revealed. Figure 6 illustrates the transfer relationships between the thermal environment levels, while Table 4 presents the area proportions of the transfers between different thermal environment levels in Hangzhou's central urban area, visually reflecting the transformation of thermal environment levels.

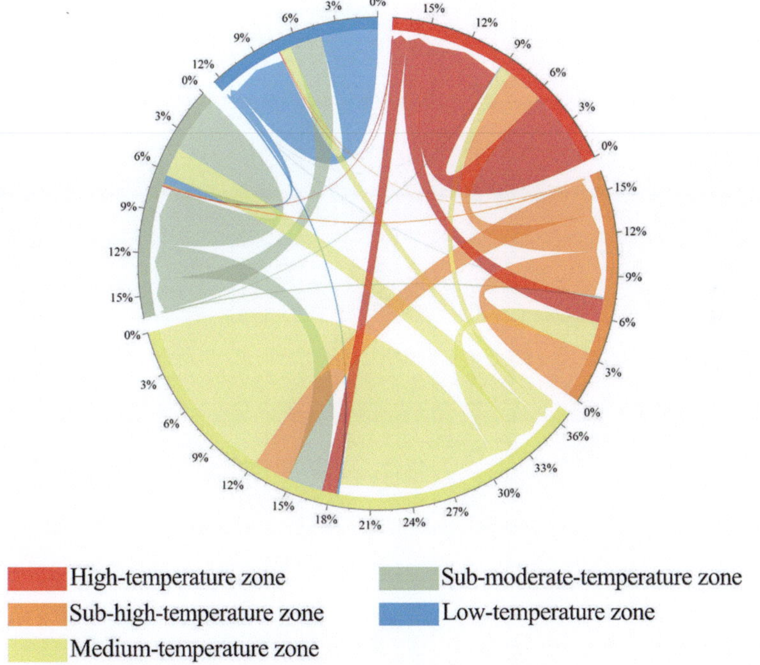

Fig. 6 Chord diagram of the transfer of urban thermal environment levels in Hangzhou's central urban area from 2013 to 2024 (Self-drawn)

Table 4 Transfer area ratio of urban thermal environment level in the central area of Hangzhou from 2013 to 2024 (%) (Self-drawn)

	Low-temperature zone (%)	Sub-moderate-temperature zone (%)	Medium-temperature zone (%)	Sub-high-temperature zone (%)	High-temperature zone (%)
Low-temperature zone	56.33	31.48	10.76	0.83	0.60
Sub-moderate-temperature zone	8.87	58.49	29.39	2.05	1.20
Medium-temperature zone	0.98	13.91	65.40	13.95	5.75
Sub-high-temperature zone	0.27	1.90	30.93	44.33	22.57
High-temperature zone	0.14	0.95	9.00	29.55	60.37

According to Fig. 6 and Table 4, frequent transitions between adjacent thermal environment levels can be observed. From 2013 to 2024, the low-temperature area decreased significantly, with most transitioning to the sub-moderate and moderate temperature zones, accounting for 31.48% and 10.76% respectively, while 56.33% of the area remained as low-temperature zones.

Within the sub-moderate temperature zone, 58.49% of the area remained unchanged, 29.39% transitioned to moderate temperature zones, 8.87% decreased to low-temperature zones, and a small portion transitioned to sub-high and high-temperature zones. In the moderate temperature zone, 65.40% of the area remained unchanged by 2024, with significant transitions mainly to adjacent sub-high and sub-moderate temperature zones, accounting for 13.95% and 13.91% respectively.

The sub-high temperature zone experienced the most significant transformations, with only 44.33% of the area maintaining its original level by 2024. Approximately 30.93% transitioned to moderate temperature zones, 22.57% increased to high-temperature zones, and the remaining small portion transitioned to lower temperature zones.

In the high-temperature zone, 60.37% of the area remained unchanged in 2024, with significant transitions primarily to sub-high temperature zones, accounting for 29.55%.

During Hangzhou's rapid urbanization, large infrastructure projects such as transportation networks, high-rise buildings, and sports facilities significantly improved the city's physical appearance. However, these constructions also intensified the urban heat island effect, resulting in noticeably higher temperatures in densely built-up areas compared to surrounding regions.

Overall, from 2013 to 2024, sub-high temperature zones exhibited the most significant changes, mainly transitioning to moderate and high-temperature zones. The area proportion of high-temperature zones remained relatively stable, with transitions primarily originating from sub-high temperature zones. Low-temperature zones, typically representing regions with vegetation and water cover, saw 43.67% transitioning to higher temperature zones, while only 10.26% of higher temperature zones transitioned to low-temperature zones. This unequal transformation suggests that the expansion of impervious surfaces and reduction of green spaces contributed to increased urban surface temperatures.

The intensification of the urban heat island effect in 2024 was primarily due to the reduction in low-temperature zones, though large-scale cross-level transitions were rare during thermal environment level changes.

4 Analysis of the Influencing Factors of Urban Thermal Environment in Hangzhou's Central Urban Area

In academic research on factors influencing the urban thermal environment, the focus is primarily on two main categories of factors: the direct impact of climatic factors on the urban thermal environment and the indirect impact of urban development changes on the thermal environment. These indirect factors include changes in land cover types, the evolution of urban spatial structures, and the increase in anthropogenic heat sources [27–29]. These factors collectively affect the urban thermal environment, causing changes in temperature distribution, which in turn affects the quality of life for urban residents and the sustainable development of the city. Based on existing research results and the specific conditions of study area, this study aims to explore the spatial heterogeneity of LST (Land Surface Temperature) in Hangzhou and its relationship with influencing factors. Using the Pearson correlation coefficient method [30], several influencing factors were identified and their correlation with LST determined. Six influencing factors, which namely, Normalized Difference Vegetation Index (NDVI), Normalized Difference Built-up Index (NDBI), Modified Normalized Difference Water Index (MNDWI), land use mix, population density, and nighttime lights, were selected and introduced into the geographical detector, a spatial statistical model.

In this model, the LST in Hangzhou's central urban area is used as the dependent variable, and various driving factors as independent variables. Under the framework of the geographical detector model, the impact of each independent variable on the spatial heterogeneity of the dependent variable is examined. Given that the geographical detector requires independent variables to be categorical, the natural break method was chosen to reduce the uncertainty caused by the selection of discretization methods. During classification, the attribute values of the six driving factors were reclassified into six categories, with the analysis results as follows.

4.1 Results of Explanatory Power Analysis of the Driving Factors of the Urban Thermal Environment

In the study of factors influencing the urban thermal environment, quantitative assessments of the impacts of six key driving factors were conducted through factor detector analysis (Table 5).The analysis results indicate that the data across the three periods all show P values less than 0.01, emphasizing the significant role these factors play in the spatial distribution of the urban thermal environment in Hangzhou's central urban area. Additionally, these factors exhibit significant differences in their impact on the spatial heterogeneity of the urban thermal environment, highlighting their importance in the urban thermal environment morphology.

In 2013, the explanatory power of the factors, ranked from highest to lowest, was as follows: NDBI, NDVI, nighttime lights, MNDWI, population density, and

Table 5 Factor detector results for 2013, 2020, and 2024 (Self-drawn)

Driving factor	2013 q-value	2013 p-value	2020 q-value	2020 p-value	2024 q-value	2024 p-value
NDVI (X1)	0.459	< 0.01	0.453	< 0.01	0.310	< 0.01
NDBI (X2)	0.463	< 0.01	0.433	< 0.01	0.288	< 0.01
MNDWI (X3)	0.320	< 0.01	0.375	< 0.01	0.252	< 0.01
Nighttime lights (X4)	0.366	< 0.01	0.163	< 0.01	0.324	< 0.01
Land use mix (X5)	0.141	< 0.01	0.196	< 0.01	0.146	< 0.01
Population density (X6)	0.237	< 0.01	0.266	< 0.01	0.159	< 0.01

land use mix. The explanatory power values for NDBI and NDVI were 0.463 and 0.459, respectively, indicating that building distribution and density, as well as vegetation, significantly impacted the spatial heterogeneity of surface temperatures in Hangzhou's central urban area. The explanatory power of nighttime lights was 0.366, showing that the density of nighttime lights in the study area was also an important factor influencing surface temperature (LST). The MNDWI q-value was 0.410, indicating that water body distribution had some impact on the spatial heterogeneity of surface temperatures. Population density and land use mix had relatively weaker explanatory power for surface temperatures in Hangzhou's central urban area, with q-values less than 0.3, specifically 0.237 and 0.141.

In 2020, the explanatory power ranking of the factors from highest to lowest was NDVI, NDBI, MNDWI, population density, land use mix, and nighttime lights. The driving factors exhibited significant differences in their impact on the spatial heterogeneity of surface temperatures in Hangzhou's central urban area, with NDBI and NDVI still being the two strongest explanatory factors for the spatial heterogeneity of surface temperatures (LST), with q-values of 0.433 and 0.453, respectively. The explanatory power of MNDWI and population density showed a slight increase, with q-values of 0.375 and 0.266. The explanatory power of the other factors was less than 0.2.

In 2024, combining the explanatory power data from 2013 and 2020, the explanatory power of the six influencing factors had decreased compared to the previous two periods. This indicates that as the urbanization of Hangzhou's central urban area continues to advance and new industries develop rapidly, the influence of factors such as building distribution, population density, and land use mix on the urban thermal environment is gradually diminishing as the built-up areas and comprehensive functions of Hangzhou's central urban area reach saturation.

4.2 Results of Interaction Analysis of Urban Thermal Environment Drivers

To analyze and reveal the comprehensive impact of different combinations of influencing factors on the urban thermal environment, the geographical detector was used to conduct interaction detection on the six influencing factors mentioned above (Fig. 7). The results showed that, for all three periods, the impact of all interaction factors on the spatial heterogeneity of surface temperature was significantly enhanced compared to individual influencing factors, with all exhibiting double-factor enhancement. This indicates that the interaction of the six influencing factors has a stronger impact on the surface temperature of Hangzhou's central urban area.

Specifically, in 2013, the interaction between nighttime lights and NDVI, with a contribution of 0.5921, and nighttime lights and NDBI, contributing 0.59196, had the greatest impact on the spatial heterogeneity of surface temperatures, both demonstrating double-factor enhancement. This was followed by the interactions between

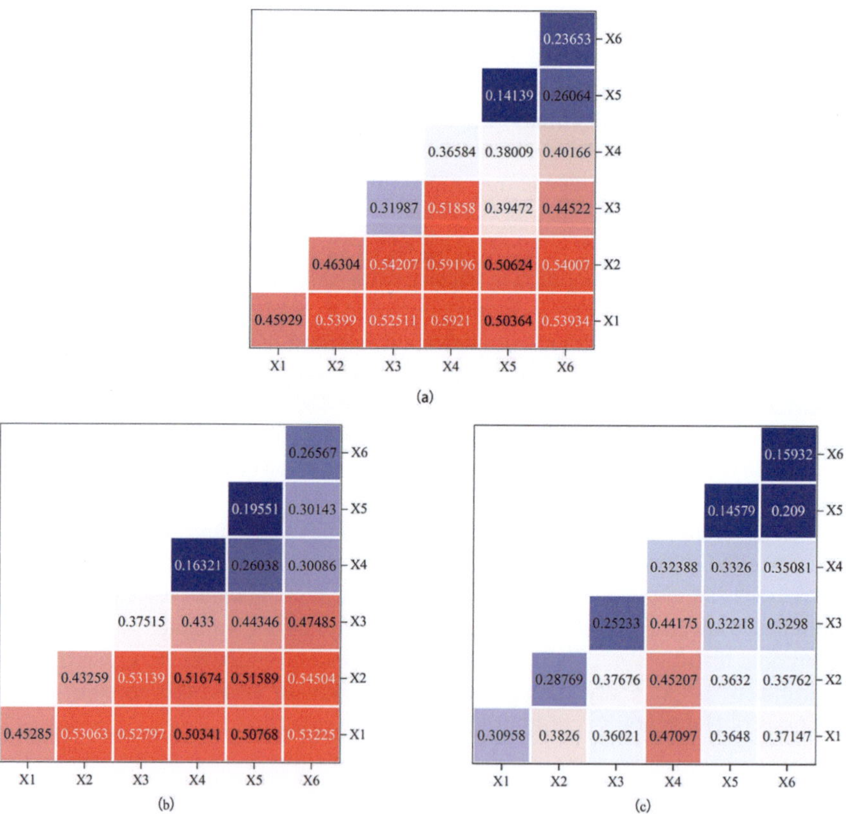

Fig. 7 Interaction detector analysis results for 2013–2024: **a** 2013; **b** 2020; **c** 2024 (Self-drawn)

MNDWI and NDBI, with a contribution of 0.54207, and population density and NDBI, with a contribution of 0.54007. The explanatory power of these combinations all exceeded 0.54, indicating their crucial role in the spatial heterogeneity of surface temperatures in Hangzhou's central urban area. In 2020, the interaction between population density and NDBI was the strongest, with a q-value of 0.54504, also showing double-factor enhancement. By 2024, all 15 pairs of double-factor combinations showed a linear synergistic enhancement effect on surface temperature, with the interaction between nighttime lights and NDVI standing out with a q-value of 0.4797, indicating that these two factors were particularly significant in their interaction within the urban thermal environment of 2024.

Further analysis of the data across the three periods revealed that the previously weaker explanatory power of factors such as population density and land use mix significantly increased when interacting with other factors like NDWI and NDVI. Additionally, although the 15 groups of interaction results in the three periods all showed a linear synergistic enhancement trend, the q-values of each factor's interaction effect decreased. This also reflects that with the accelerated urbanization process, the influence of these factors on the thermal environment in Hangzhou's central urban area is gradually diminishing due to the saturation of the built-up areas.

5 Discussion

5.1 Dynamic Changes and Cause Analysis of Urban Thermal Environment in Hangzhou

Based on the long-term observation and analysis of the thermal environment in Hangzhou, this study has revealed that the thermal environment in Hangzhou experienced a significant dynamic evolution process during the period from 2013 to 2024. Looking back to 2013, the thermal environment in some areas of Hangzhou (such as along the elevated road near the airport and the Future Science and Technology City) was prominently characterized by a relatively large proportion of medium-temperature areas. However, by 2024, its pattern had evolved into a widespread distribution of high-temperature areas. Through the meticulous interpretation and rigorous comparison of satellite imagery data during the same period, it was clearly observed that the land use types had undergone a profound transformation during this period. It gradually transitioned from the original composite land use pattern where temporary shanty buildings were intertwined with woodlands and farmlands to a construction land structure dominated by large-scale electronic industrial plants or commercial buildings. Such high-intensity urban development processes and the drastic changes in land use structures have strongly demonstrated that they have played a potent catalytic role in the deterioration of the urban thermal environment, resulting in a more pronounced temperature increase trend in local areas.

Meanwhile, the agglomeration of commercial activities has further shaped the thermal environment of Hangzhou. In the Central Business District (CBD) and the urban core areas, commercial activities presented a highly intensive situation, with frequent intersections of people and logistics, giving rise to the emergence of a large number of commercial facilities and high-rise buildings [17]. The densely packed building layout has profoundly influenced the thermal environment in two aspects. On the one hand, it has significantly enhanced the absorption efficiency of solar radiation in the region, causing more solar radiation energy to be retained by the building surfaces and the surrounding ground. On the other hand, the closely arranged buildings have seriously hindered regional ventilation, making it difficult for heat to be dissipated smoothly. As a result, the temperature in the CBD area has risen significantly, shaping a relatively independent microclimate environment with prominent high-temperature characteristics.

Overall, the dynamic evolution of urban forms and the frequent changes in land use patterns are bound to trigger adaptive changes in the urban microclimate. Especially in areas where green space resources are scarce and the building coverage rate remains high, the urban heat island effect is more pronounced, and the formation and expansion of local high-temperature areas are more significant, posing a severe challenge to the balance of the urban ecosystem and the comfort level of residents' lives.

5.2 Reflection on Limitations and Future Prospects

However, in the context of the rapid development of cities, the impacts of various urban elements on the surface temperature exhibit complex and multi-dimensional characteristics, which makes the assessment of the thermal environment in the central area of Hangzhou a highly comprehensive and challenging task. Currently, in the process of in-depth exploration of the urban thermal environment, this study still reveals certain limitations at the methodological level. In terms of the selection of driving factors, only six key elements were focused on to conduct correlation and geographical detector analyses. Although the scope involved includes conventional factors such as population density, land use type, and the proportion of blue and green spaces, many potentially crucial driving factors, such as building form, vegetation composition, and urban ventilation conditions, unfortunately, were not included in the research scope. Meanwhile, in terms of the research basis, the authoritative opinions referred to were relatively limited, mainly concentrating on some classic literature and a small number of regional case studies, lacking extensive exploration of cutting-edge academic trends and interdisciplinary theories.

Looking ahead to future research directions, in order to achieve a comprehensive improvement in the detection accuracy of the model, it is urgent to construct a more systematic and comprehensive driving factor framework. On the one hand, by fully leveraging emerging frontier technological means such as machine learning, the complex correlation patterns inherent in the city can be deeply explored [31]. For example, machine learning algorithms can be used to conduct comprehensive analysis on high-resolution remote sensing images, real-time meteorological monitoring data, and urban Internet of Things sensing data, so as to accurately capture the subtle changes in the urban thermal environment. On the other hand, widely absorbing authoritative research results and practical experiences from around the world, and starting from the interdisciplinary perspectives of architecture, ecology, and meteorology, comprehensively deconstructing the action mechanisms of driving factors, with the expectation of presenting more accurate and highly practical analyses and suggestions for the research on the urban thermal environment, and thus facilitating the sustainable development of cities.

6 Conclusion

In the context of global urbanization, the rapid expansion of cities has significantly impacted human health and ecological environments through the urban heat island effect. This study selected the rapidly developing central urban area of Hangzhou, driven by policies and major events, as a case area. Landsat 8 satellite imagery was used as the primary data source, and the Radiative Transfer Equation method was applied to retrieve surface temperatures in the area, followed by classifying the intensity of urban surface temperatures. The study examined the spatiotemporal evolution of urban heat islands, analyzed changes in Hangzhou's thermal environment spatial structure, and investigated the correlation between driving factors, including NDVI, NDBI, NDWI, nighttime lights, population density, and land use mix, with surface temperature. The geographical detector method further revealed the impact mechanisms of these factors on the urban thermal environment. The main conclusions of this study are as follows:

(1) The urban heat island effect in Hangzhou has exhibited a remarkable and deteriorating tendency. In the central urban area of this city, the variations within the low and medium temperature zones were relatively slight. However, the alterations in the sub-high temperature zones were the most prominent, with a main shift towards the medium and high-temperature zones. Regarding the patch distribution of high-temperature zones, a dispersing trend could be observed from the city center towards the outskirts. This was especially the case in core built-up areas, industrial zones, as well as the regions surrounding higher education institutions. These areas are usually characterized by dense populations, bustling commercial activities, and developed industrial sectors. It's clear that such changes are directly associated with the expansion of the urban population and the intensification of economic activities.

(2) The geographical detector analysis over three periods showed that building distribution, density, and vegetation density had the most significant impact on the spatial heterogeneity of surface temperatures in the central urban area

of Hangzhou. Interaction detector results indicated that all 15 pairs of interacting factors across the three periods showed higher influence than individual factors, with all exhibiting double-factor enhancement. This implies that the six types of influencing factors do not independently affect the surface temperature in Hangzhou. With the built-up area in the central part of Hangzhou gradually approaching a saturated state, the influence of factors that are traditionally regarded as crucial to the urban thermal environment, such as the continuous growth of population density and the increasing degree of land use function mixing, on the urban thermal environment is undergoing a gradually weakening process. From the practical perspective of urban planning and development, this implies that in the subsequent processes of urban layout optimization, functional zoning adjustment, and resource allocation decision-making, it is no longer viable to merely rely on the conventional cognitions regarding the correlations between population density, land use function mixing degree, and the thermal environment. Instead, it is necessary to re-examine the dynamic weights of various elements at different development stages.

The results of this study provide valuable insights into the spatiotemporal evolution and driving factors of the urban thermal environment in the central urban area of Hangzhou. Undoubtedly, the scope of this study focuses on the Hangzhou region, but the methodology used contains broad applicability beyond geographical limitations. The basic analysis principle of using satellite images to achieve precise tracking of the spatiotemporal trajectory of the thermal environment and identifying driving factors based on this has universal significance in the field of urban thermal environment research. Although each city exhibits different thermal environment characteristics due to its unique natural geographical conditions, socio-economic structure, and urban planning layout, the research framework constructed in this study has laid a solid foundation for further exploration of urban thermal environment in different urban contexts, and is expected to provide certain reference significance for academic research and practical applications in the field of urban thermal environment.

Acknowledgements This research was funded by the research project of Zhejiang Province Social Sciences Federation (Grant Number: 2025N055);

Major Humanities and Social Sciences Research Projects in Zhejiang higher education institutions (Grant Number: 2024QN136).

References

1. Li Y, Svenning JC, Zhou W et al (2024) Green spaces provide substantial but unequal urban cooling globally. Nat Commun 15:7108
2. He J, Shi Y, Xu L, Lu Z, Feng M, Tang Q, Guo X (2024) Exploring the scale effect of urban thermal environment through XG Boost model. Sustain Cities Soc 114:105763
3. Shi Y, Zhai G, Xu L, Zhou S, Lu Y, Liu H, Huang W (2021) Assessment methods of urban system resilience: from the perspective of complex adaptive systems. Cities 112:103141

4. Wang Y, Tang L, Zhu H et al (2021) Urban thermal environment response and attribution analysis based on multi-source remote sensing data: a case study of Shenzhen city. Acta Ecol Sin 41(22):8771–8782
5. Yang Z, Chen Y, Wu Z (2021) How urban expansion affects the thermal environment? A study of the impact of natural cities on the thermal field value and footprint of the thermal environment. Ecol Ind 126:107632
6. Ke R, Chen L, Zhang Y (2012) Spatial analysis of urban thermal environment based on grey clustering assessment method. Environ Sci Technol 35(9):185–192
7. Bhati S, Manju M (2018) WRF-urban canopy model evaluation for the assessment of heat island and thermal comfort over an urban airshed in India under varying land use/land cover conditions. Geosci Lett 5:1–19
8. Liu F, Zhang X, Murayama Y et al (2020) Impacts of land cover/use on the urban thermal environment: a comparative study of 10 megacities in China. Remote Sens 12(2):307
9. Farideh T, Abolfazl R (2021) Quantitative analysis of spatial distribution of land surface temperature (LST) in relation to ecohydrological, terrain, and socioeconomic factors based on Landsat data in a mountainous area. Adv Space Res 68(9):3622–3640
10. Georgescu M, Broadbent AM, Krayenhoff ES (2024) Quantifying the decrease in heat exposure through adaptation and mitigation in twenty-first-century US cities. Nature Cities 1:42–50
11. Li X (2018) Urban heat island effect and its impact on energy consumption and environmental quality in Hangzhou. J Clean Prod 183:301–310
12. Qian JK, Meng QY, Zhang LL, Schlink U, Hu XL, Gao JF (2023) Characteristics of anthropogenic heat with different modeling ideas and its driving effect on urban heat islands in seven typical Chinese cities. Sci Total Environ 886:163989
13. Portela CI, Massi KG, Rodrigues T, Alcântara E (2020) Impact of urban and industrial features on land surface temperature: evidences from satellite thermal indices. Sustain Cities Soc 56:102100
14. Chen XJC, Yang Z et al (2021) Spatiotemporal impact of vehicle heat on urban thermal environment: a case study in Hong Kong. Build Environ 205:108224
15. Zhang Y, Chang L (2021) Digital simulation for buildings' outdoor thermal comfort in urban neighborhoods. Buildings 11(11):541
16. Wang RF, Wang MM (2023) Multi-scale analysis of surface thermal environment in relation to urban form: a case study of the Guangdong-Hong Kong-Macao greater bay area. Sustain Cities Soc 99:104953
17. Deng XW, Cao Q, Wang LC, Wang W, Wang S, Wang SQ, Wang LZ (2023) Characterizing urban densification and quantifying its effects on urban thermal environments and human thermal comfort. Landsc Urban Plan 237:104803
18. Zhang Q, Zhou D, Xu D, Rogora A (2022) Correlation between cooling effect of green space and surrounding urban spatial form: evidence from 36 urban green spaces. Build Environ 222:109375
19. Gillespie A, Rokugawa S, Matsunaga T et al (1998) A temperature and emissivity separation algorithm for advanced spaceborne thermal emission and reflection radiometer (ASTER) images. IEEE Trans Geosci Remote Sens 36(4):1113–1126
20. Department of the Interior U.S. Geological Survey (2024) Landsat 8–9 Collection 2 (C2) Level 2 Science Product (L2SP) Guide. 6th edn. USGS, Hardeeville
21. Chen SL, Wang TX (2009) Comparative study on the delineation of urban heat islands using equidistant method and mean standard deviation method. J Earth Inf Sci 11(2):145–150
22. Xu HQ, Chen BQ (2003) Processing methods of remote sensing thermal infrared images at different times for studying urban heat island changes. Remote Sens Technol Appl 18(3):129–133+185
23. Ma YH (2021) Thermal environment study of different urban land cover types. Master's Thesis, Northwest University
24. Wang JF, Xu CD (2017) Geographical detector: principles and prospects. Acta Geogr Sin 72(1):116–134

25. Wang JF, Li XH, Christakos G et al (2010) Geographical detectors-based health risk assessment and its application in the neural tube defects study of the Heshun region, China. Int J Geogr Inf Sci 24(1):107–127
26. Han JJ, Wang JP, Chen L et al (2021) Driving factors of desertification in the Qaidam Basin, China: An 18-year analysis using the geographic detector model. Ecol Ind 124:107404
27. Yang J, Wang ZH, Kaloush KE (2016) Effect of pavement thermal properties on mitigating urban heat islands: a multi-scale modeling case study in Phoenix. Build Environ 108:110–121
28. Richards D, Fung TK, Belcher RN, Edwards PJ (2020) Differential air temperature cooling performance of urban vegetation types in the tropics. Urban For Urban Green 50:126651
29. He C, Zhou L, Yao Y, Ma W, Kinney PL (2020) Estimating spatial effects of anthropogenic heat emissions upon the urban thermal environment in an urban agglomeration area in East China. Sustain Cities Soc 57:102046
30. Hao N, Sun PX, He W et al (2022) Water resources allocation in the Tingjiang River Basin: construction of an interval-fuzzy two-stage chance-constraints model and its assessment through Pearson correlation. Water 14(18):2928
31. Zheng L, Lu WS (2024) Urban micro-scale street thermal comfort prediction using a 'graph attention network' model. Build Environ 262:111780

Matching States and Obstacle Factors of Supply and Demand of Thermal Environment Regulation Service of Blue and Green Spaces: A Case of Weifang, China

Zhongjian Shen, Chang'an Zuo, Fenghua Dong, and Tianran Wang

Abstract Blue and green spaces (BGSs) can provide thermal environment regulation service through cooling effect and consequently alleviate urban heat island (UHI) effect. This research selected Weifang urban core area (WUCA) in Shandong Province, China as the study area. An evaluation model was developed to assess both thermal environment regulation service supply capacity (TERSSC) and demand level (TERSDL), with their spatial distribution patterns being systematically analyzed. Moreover, we further analyzed the matching relationship between the TERSSC and the TERSDL and clarified obstacle factors affecting the TERSSC and the TERSDL. The results showed that (1) in the WUCA, the TERSSC, the TERSDL and their primary indicators all show a strong spatial autocorrelation effect. The TERSSC shows a decreasing trend from peripheral urban areas with dense BGSs to central built-up zones, while the TERSDL presents an opposite spatial distribution pattern. (2)The spatial correspondence between the TERSSC and the TERSDL demonstrates pronounced polarization characteristics. The area proportion of the quantitative matching states increases as the priority index value decrease. (3) Low-supply-high-demand mismatch zones predominantly coincide with densely populated urban centers. These areas require stringent regulation of human activities and urban development, including control of impervious surface ratios and building densities, coupled with strategic implementation and optimization of BGSs. Conversely, high-supply-low-demand mismatch areas primarily correspond to ecological spaces characterized by low population density and abundant vegetation-water resources. It is essential to maintain the present status, protect the original BGSs, strengthen human activities and urban construction befittingly, and further improve utilization of the thermal environment regulation service reasonably.

Z. Shen (✉) · C. Zuo · F. Dong · T. Wang
School of Civil Engineering, Shandong University, Jinan 250061, Shandong, China
e-mail: shenzj@sdu.edu.cn

Urban and Rural Human Settlement Environment Research Center, Shandong University, Jinan 250061, Shandong, China

B. He et al. (eds.), *Climate-Resilient, Low-Carbon, and Sustainable Cities*, Lecture Notes in Civil Engineering 754, https://doi.org/10.1007/978-981-95-2169-2_17

Keywords Thermal environment regulation service · Blue and green spaces · Supply · Demand · Matching · Obstacle factor · Weifang

1 Introduction

Urbanization has led to a progressively intensifying urban heat island (UHI) effect. This urban environmental problem has leaded to countless negative influences, such as reducing the quality of the ecological environment, increasing energy consumption, and causing negative effects on the well-being and health of urban inhabitants [1, 2]. A large number researches have illustrated that blue and green spaces (BGSs), such as forests, rivers, and lakes, are able to efficiently regulate the local microclimate to mitigate the negative consequences of the UHI phenomenon [3].

From an academic perspective, current researches mainly investigate the cooling effect of BGSs, assess their spatial patterns or surrounding environments, and investigate their specific cooling effects across different regions using meteorological measurements, digital simulation, and thermal infrared remote sensing technologies [4–12]. For instance, some studies have analyzed the correlation between the cooling effect and spatial characteristics, including landscape component, aggregation, connectivity, and shape [8–12]. Despite great progress in present researches, the supply–demand of cooling effect, matching relationship and specific obstacle factors are still not illuminated [1–3, 13, 14]. The concept of supply and demand is on of critical issues in ecosystem services research field [1]. The cooling effect of BGSs has been widely regarded as an important function of thermal environment regulation [1, 3]. Hence, we adapted the ecosystem services supply–demand analysis framework and clarified the definitions of the supply capacity of thermal environment regulation service (TERSSC) and the demand degree of thermal environment regulation service (TERSDL) [14]. Therefore, we define the TERSSC as the capacity of BGSs to provide thermal environment regulation service for alleviating UHI effect, and define the TERSDL as the level required by alleviating UHI effect for thermal environment regulation service.

The urban core area of Weifang city (WUCA) is one of the typical urban center areas of Shandong Province in China, and also exhibits a significant UHI effect. Accordingly, this study selected WUCA as its primary research area for investigation, using 2020 data to quantify the TERSSC and the TERSDL, and to investigate matching relationship and identify obstacle factors affecting the TERSSC and the TERSDL. This research aims to achieve three primary objectives: (1) identify spatial distribution characteristics of TERSSC and TERSDL, (2) examine the alignment between the TERSSC and TERSDL, and (3) clarify obstacle factors affecting the TERSSC and the TERSDL. The findings provide theoretical insights for urban spatial planning and UHI mitigation strategies.

2 Materials and Methods

2.1 Research Area

Weifang is a significant city in Shandong Province, located in the lower reaches of China's Yellow River. The WUCA (118°56″34′E–119°17″27′E and 36°32″41′N–36°47″65′N) was taken as the research area, covering approximately 742.54 km^2. The area proportion of impervious surfaces increased from 29.153% to 50.406% between 2000 and 2020. Because of urban area expansion and population increase, the urban overheat problem in the study area will be increasingly serious. And the study area covers not only urban built-up areas with dense population, but also substantial BGSs, such as forestland, grassland and water. Therefore, it serves as an ideal research area to investigate supply–demand of thermal environment regulation service.

2.2 Data Processing

Data source. This paper selected 2020 as the research base years. We classify the data sources of this study into four distinct types: (1) The Landsat Collection 2 Level-2 Surface Temperature data (LC2L2ST) and the Landsat-8 OLI/TIRS remote sensing images were retrieved from the geospatial cloud platform (https://www.gscloud.cn/) for the date of 20 July 2020. The LC2L2ST were used for the calculation of LST, and the Landsat images were mainly employed for the calculation of biophysical parameters, including normalized difference vegetation index (NDVI) and modified normalized difference water index (MNDWI). (2) The land cover data in 2020 were obtained from European Space Agency (https://esa-worldcover.org/en) and were employed to calculate the proportion of impervious surfaces (ISP) and the proportion of BGSs (BGSP). (3) The point of interest (POI) vector data was sourced from AMAP open data platform (https://lbs.amap.com/), which were employed for the calculation of the density of POI (POID) and the density of vulnerability facility (VPOID). (4) The population density data were obtained from gridded population data of China's seventh census [15, 16], which in 2020 were employed for the denotation of the population density (POP) and the calculation of the density of vulnerable population (VPOP). (5) The elderly ratio data were acquired from the 7th Census dataset of China, which were used of the calculation of the density of vulnerable population (VPOP) [17].

In this study, the 300 m × 300 m grid was taken as the research unit, because 300 m is a suitable unit scale for urban climate research, and is completely supported by the data used.

LST calculation and classification. The LST was calculated from the thermal infrared bands in LC2L2ST by the following conversion equation [18].

$$LST = DN \times SF + AF - 273.15 \quad (1)$$

Among them, *DN* represents the numerical value of a specific pixel. *SF* and *AF* refer to the multiplication and addition scale factors of LST products, with respective values of 0.00341802 and 149.

Based on LST values, employing the Jenks natural breakpoint classification method in ArcGIS 10.8, the LST values were categorized into nine categories: lowest, lower, low, sub-low, medium, sub-high, high, higher, highest. On basis of previous research, the areas with sub-high, high, higher, and highest were regarded as heat island, and the lowest, lower, low, and sub-low were regarded as cold island [19].

2.3 Quantitative Evaluation of the TERSSC and the TERSDL

Evaluation indicator system of the TERSSC. The TERSSC indicates the capacity of BGSs to provide thermal environment regulation services to mitigate the UHI effect, significantly influenced by cold source conditions and land cover characteristics [1, 14]. Hence, we employ two primary indicators on the basis of the previous research, including cold island supply capacity (CSC) and land cover supply capacity (LSC) (Table 1) [14].

Table 1 The evaluation indicators of the TERSSC. (Self-drawn)

Primary indicators	Secondary indicators	Indicator meaning	Data acquisition and processing
Cold island supply capacity (CSC)	LST of cold island patches (CLST)/ °C (-)	It quantifies the intensity of cold island effect	Calculation through LST data
	Cold island ratio index (CRI) (+)	It reflects the area proportion of cold island patches with low temperature	Calculation through LST data
Land cover supply capacity (LSC)	Area proportion of BGSs (BGSP) /% (+)	It reflects the coverage of BGSs in the local region	Calculation through land cover data
	NDVI (+)	It reflects the coverage of vegetation	Calculation through Landsat-8 OLI/TIRS remote sensing data
	MNDWI (+)	It reflects the coverage of water	Calculation through Landsat-8 OLI/TIRS remote sensing data

Note The " + " symbol indicates positive indicators, while " – " signifies negative ones

Evaluation indicator system of the TERSDL. The TERSDL represents the required intensity of thermal regulation services to mitigate UHI effects, primarily influenced by both physical environmental factors and socioeconomic parameters. Thus, drawing on the risk assessment framework encompassing hazard, exposure, and vulnerability proposed by IPCC, we used three primary indicators, including hazard demand level (HDL), exposure demand level (EDL), and vulnerability demand level (VDL) [3, 21, 22]. Therefore, the evaluation indicators of hazard, exposure, and vulnerability were screened, following the principles of accuracy and representativeness. Eventually, 7 indicators in three dimensions were identified (Table 2) [3, 21].

Indicator processing and weight calculation. The min–max normalization approach was employed to standardize the assessment parameters for both TERSSC and TERSDL [23]:

Table 2 The evaluation indicator system of the TERSDL. (Self-drawn)

Primary indicators	Secondary indicators	Indicator meaning	Data source and processing
Hazard demand level (HDL)	LST of heat island patches (HLST) /°C (+)	It quantifies the intensity of heat island effect	Calculation through LST data
	Heat island ratio index (HRI) (+)	It reflects the area proportion of heat island patches with high temperature	Calculation through LST data
Exposure demand level (EDL)	Population density (POP)/num · km^{-2} (+)	It reflects the human demand for thermal environment regulation service	Calculation through population density data
	POI density (POID)/ num · km^{-2} (+)	It reflects the socio-economic activities demand for thermal environment regulation service	Calculation through POI data
	Area proportion of impervious surfaces (ISP) /% (+)	It reflects the coverage of impervious surfaces	Calculation through land cover data
Vulnerability demand level (VDL)	Density of the population aged over 65 (APD) /num · km^{-2} (+)	It indicates the elderly demand for thermal environment regulation service	Calculation through population density data and the 7th Census dataset of China
	Density of the vulnerability facility (VFD)/ num · km^{-2} (+)	It reflects the people with basic diseases demand for thermal environment regulation service	Calculation through POI data of healthcare service facility

$$N_{i,+} = 0.01 + (1 - 0.01) \times \frac{V_i - V\min}{V\max - V\min}$$

$$N_{i,-} = 0.01 + (1 - 0.01) \times \frac{V\max - V_i}{V\max - V\min}$$

where $N_{i,+}$ and $N_{i,-}$ are the standardized value of positive indicators and negative indicators in the i-th research unit, respectively, V_{ij} represents the initial value of evaluation indicator, $V_{\max}$ and $V_{\min}$ are the maximum and minimum of the initial value of evaluation indicator.

In this study, subjective and objective weighting methods were utilized to derive the reasonable weight for each evaluation indicator of the TERSSC and the TERSDL. The analytic hierarchy process was employed to derive the subjective weighting values for individual evaluation indicators [14]. Simultaneously, the entropy weight method (EWM) was utilized to determine objective weighting coefficients. Subsequently, the combined weights were computed based on the minimum information entropy principle. [24]. Finally, the evaluation models of the TERSSC and the TERSDL were established:

$$TERSSCI_i = \sum_{s=1}^{n} w_s \times x_{is}$$

$$TERSDLI_i = \sum_{d=1}^{m} w_d \times y_{id}$$

where $TERSSCI_i$ and $TERSDLI_i$ represent the assessment value of the TERSSC and TERSDL in the i-th spatial unit, w_s denotes the weighting coefficient of the s-th evaluation indicator within the TERSSC framework, x_{is} indicates the standardized value of the s-th TERSSC indicator in the i-th unit, w_d represents the weight of the d-th evaluation metric in the TERSDL system, and y_{jd} signifies the normalized value of the d-th TERSDL indicator in the i-th spatial unit.

2.4 Analysis of Matching States and Obstacle Factors of the TERSSC and the TERSDL

Spatial autocorrelation analysis. Both the TERSSC and the TERSDL may be impacted by surrounding spaces, this phenomenon can be revealed by spatial autocorrelation analysis method [3]. The global Moran's I is employed to examine the spatial autocorrelation of the TERSSC and TERSDL using Geoda 1.14.0 software. Furthermore, Local Indicators of Spatial Association (LISA) is employed to analyze spatial heterogeneity patterns across neighboring areas and identify local clustering characteristics of variable distributions.

Four-quadrant division based on Z-score standardization. Using the Z-score standardization method to standardize the TERSSC and TERSDL, we assessed the spatial matching state of the TERSSC and TERSDL through a four-quadrant division in ArcGIS 10.8. The horizontal axis represents TERSSC values, while the vertical axis corresponds to TERSDL measurements. These axes divide the coordinate system into four distinct quadrants, each representing specific spatial matching patterns [3, 14]. The standardization of variables was calculated using the following formula:

$$Z_i = \frac{E_i - E_{\mathrm{m}}}{E_{\mathrm{std}}}$$

where Z_i represents the standardized values of TERSSC and TERSDL for the i-th spatial unit, E_i denotes the original values of TERSSC and TERSDL in the i-th spatial unit, E_{m} indicates the mean value of the distribution, and E_{std} signifies the standard deviation of the variable.

Priority index classification. Establishing intervention priorities is essential for optimizing the supply–demand balance of thermal environment regulation services in urban planning and management strategies [3]. In accordance with the supply–demand ratio index, the priority index was quantified to reveal the quantitative matching state of the TERSSC and the TERSDL in ArcGIS 10.8 [3, 14]. The formula of priority index is as detailed below:

$$PI_i = \frac{TERSDLI_i}{TERSSCI_i}$$

where PI_i is the priority index of the i-th spatial unit, $TERSSCI_i$ and $TERSDLI_i$ are the evaluation value of the TERSSC and the TERSDL. Employing the Jenks natural breakpoint classification method, the priority index is categorized into five categories: low priority, sub-low priority, medium priority, sub-high priority, and high priority.

Obstacle degree model. It is essential to reveal the factors that impede the TERSSC and the TERSDL so that a problem diagnosis can be made of the supply–demand of thermal environment regulation service. The obstacle degree model is generally applied for revealing the obstacle factors affecting the improvement of objects [25]. Thus, we used an obstacle degree model for revealing the chief barriers of the TERSSC and the TERSDL in the WUCA. The calculation formula is as detailed below:

$$O_{ij} = \frac{w_j\left(1 - X_{ij}\right)}{\sum_{j=1}^{m} w_j\left(1 - X_{ij}\right)} \times 100\%$$

where O_{ij} represents the obstacle degree of the j-th evaluation indicator for both TERSSC and TERSDL in the i-th spatial unit, w_j denotes the comprehensive weighting coefficient of the j-th indicator, and X_{ij} indicates the normalized value of j-th indicator within i-th spatial unit.

3 Results

3.1 Spatial Pattern of the TERSSC and the TERSDL

The comprehensive assessment value of the TERSSC, the TERSDL and their primary indicators in the WUCA are quantified respectively. The evaluation results were classified into five distinct categories using the Jenks natural breaks classification method, while spatial autocorrelation analysis method was used to investigate clustering patterns and spatial heterogeneity across neighboring units (Figs. 1, 2 and Table 3).

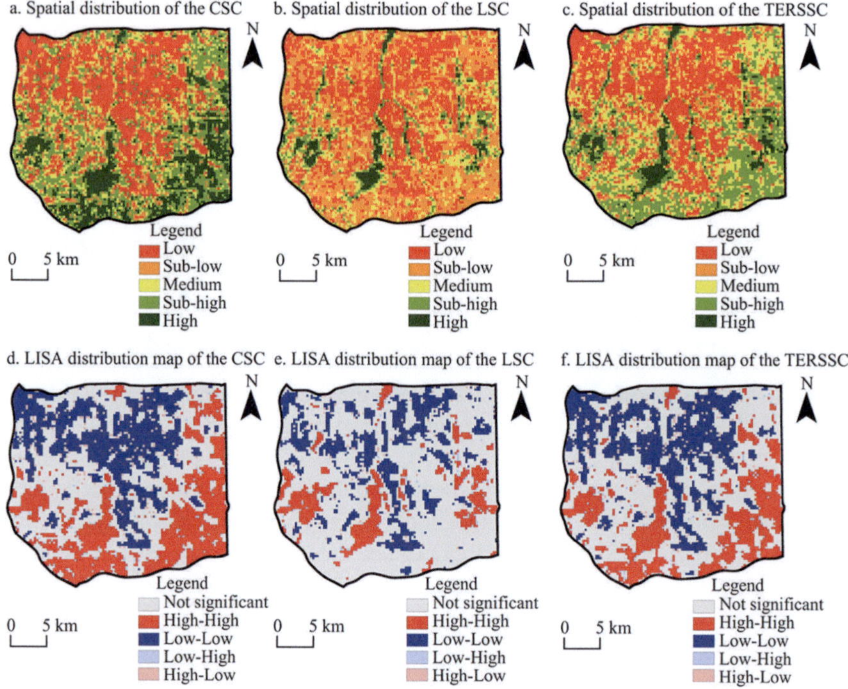

Fig. 1 Spatial distribution and LISA distribution map of the TERSSC and its primary indicators in the WUCA. **a** spatial distribution of the CSC; **b** spatial distribution of the LSC; **c** spatial distribution of the TERSSC; **d** LISA distribution map of the CSC; **e** LISA distribution map of the LSC; **f** LISA distribution map of the TERSSC. (Self-drawn)

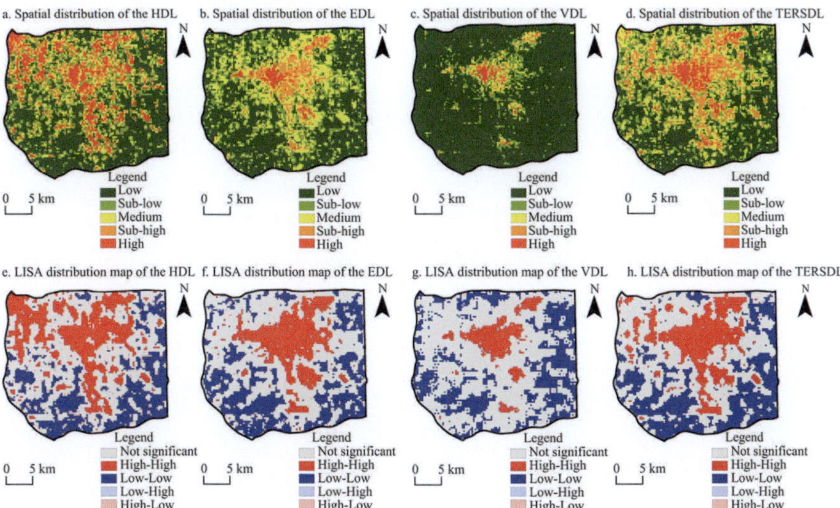

Fig. 2 Spatial distribution and LISA distribution map of the TERSDL and its primary indicators in the WUCA. **a** spatial distribution of the HDL; **b** spatial distribution of the EDL; **c** spatial distribution of the VDL; **d** spatial distribution of the TERSDL; **e** LISA distribution map of the HDL; **f** LISA distribution map of the EDL; **g** LISA distribution map of the VDL; **h** LISA distribution map of the TERSDL. (Self-drawn)

Spatial pattern of the TERSSC. As shown in Fig. 1, overall, the TERSSC is widespread but inequal in quantity for the whole WUCA and the forest, rivers and lakes demonstrate the significant thermal environment regulation capacity. Additionally, the spatial distribution of TERSSC varies extremely, the southwest and southeast show higher TERSSC values than the northwest and middle. The two primary indicators of TERSSC exhibit different spatial distribution characteristics. The spatial configuration of LSC demonstrates significantly greater heterogeneity compared to CSC. Furthermore, CSC exhibits pronounced spatial continuity and clustering patterns in both high value and low value zones, whereas LSC displays more fragmented and dispersed spatial characteristics across its value ranges. The global Moran's I of the CSC, LSC and TERSSC are 0.668, 0.596 and 0.658, respectively, indicating an obvious spatial autocorrelation. As exhibited in Fig. 2 and Table 3, the cluster results of the LISA reveal that spatial cluster of High-High mostly distributed in 26.379% of the area exhibits a High-High cluster for the CSC, and 11.485% for the LSC. When combing the two primary indicators together, for the TERSSC, 21.525% of the areas illustrates a spatial cluster of High-High. These high value areas are generally located in segments within the WUCA, generally situated around the edge zones of the southwest and southeast where human activities are infrequent. Moreover, 26.569% of the area presents a spatial cluster of Low-Low, mostly gathered in core built-up areas of middle north, and the low value areas are spatially continuous.

Table 3 Global Moran's *I* and the area proportion in LISA maps (Self-drawn)

Index		Variables						
		CSC	LSC	TERSSC	HDL	EDL	VDL	TERSDL
Global Moran's *I*		0.668^{***}	0.596^{***}	0.658^{***}	0.647^{***}	0.733^{***}	0.724^{***}	0.735^{***}
The area proportion in LSIA maps	High–high (%)	26.379	11.485	21.525	25.989	19.275	9.022	20.684
	Low–low (%)	27.279	20.270	26.569	26.735	29.931	30.038	31.234
	Low–high (%)	1.527	1.302	1.172	0.509	0.616	0.947	0.699
	High–low (%)	2.581	0.935	2.498	2.960	1.492	0.130	1.030
	Not significant (%)	42.233	66.008	48.236	43.808	48.686	59.863	46.353

Note *** denotes $P < 0.001$

Spatial pattern of the TERSDL. As illustrated in Fig. 2, spatial analysis reveals distinct TERSDL patterns across WUCA, with northern sec-tors demonstrating significantly higher values compared to southwestern and south-eastern zones. Furthermore, central urban areas show elevated TERSDL measurements relative to peripheral regions. An evident spatial heterogeneity can be observed among HDL, EDL, VDL, and TERSDL. The spatial pattern of the HDL exhibits a widespread spatial layout, while the spatial distribution of the EDL and the VDL exhibits a centripetal pattern. The integrated analysis of primary indicators demonstrates that TERSDL hotspots predominantly cluster in central built-up zones characterized by high population density and limited BGSs coverage, while coldspots are primarily located in peripheral areas featuring low population density and extensive BGSs distribution.

Consistent with TERSSC, the TERSDL exhibits significant spatial autocorrelation phenomenon. The global Moran's *I* values for the HDL, EDL, VDL, and TERSDL are 0.647, 0.733, 0.724, and 0.735, respectively. Moreover, as shown in the LISA cluster map, 25.989% of the areas shows a High-High cluster for the HDL, 19.275% for the EDL, and 9.022% for VDL. For the TERSDL, 20.684% of the areas display a high-value spatial cluster, primarily concentrated in the urban core. These areas with Low-Low cluster are mostly gathered in the southwest, southeast, and northeast ecological spaces of the WUCA.

3.2 Matching State of the TERSSC and the TERSDL

The spatial matching state of the TERSSC and the TERSDL. The spatial correspondence between TERSSC and TERSDL was analyzed through Z-score standardization and quadrant analysis, revealing their spatial coordination patterns across WUCA (Fig. 3). The analysis reveals that matched conditions between TERSSC and TERSDL, represented by Quadrant I (high-high) and Quadrant III (low-low), account for 4.44% and 9.65% of WUCA's total area respectively. Conversely, quadrant II (low supply-high demand) and quadrant IV (high supply-low demand), which represent mismatched states, encompass 40.197% and 45.714% of the WUCA. The low-supply-high-demand areas within the WUCA are predominantly located in urban built-up zones. Conversely, the high-supply, low-demand areas are primarily concentrated on the periphery of these urban built-up zones, where land cover predominantly consists of forestlands, grasslands, water bodies, and agricultural fields. This implies that because of the shortage of BGSs, most urban built-up areas are facing deficiency of thermal environment regulation service, while most urban built-up areas periphery areas have an abundance of thermal environment regulation service.

The quantitative matching state between the TERSSC and the TERSDL. The priority classification based on priority index and Jenks natural breakpoint classification method reflects the quantitative matching state between the TERSSC and the TERSDL (Fig. 4). The results show that high-priority index areas are typically

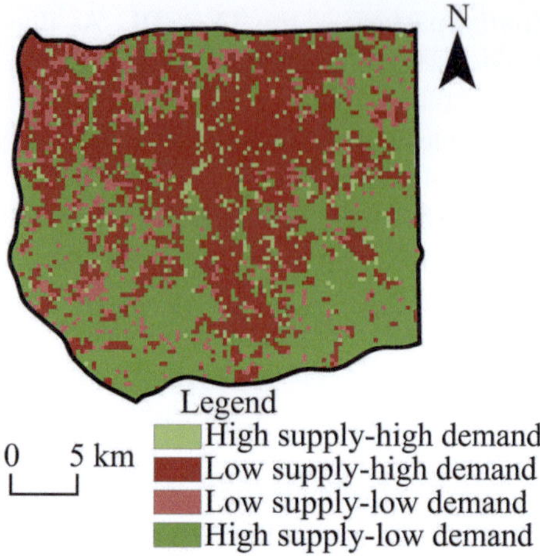

Fig. 3 A distribution map of spatial matching states of the TERSSC and TERSDL within the WUCA (Self-drawn)

concentrated in built-up regions with dense populations and limited BGSs, while the low value areas of the priority index area generally located in BGSs in the southwest and southeast. Moreover, the proportions of low priority area, sub-low priority area, medium priority area, sub-high priority area and high priority area are 41.440%, 22.614%, 19.062%, 13.403% and 3.481%, respectively. This implies that the proportion increases as the priority index value decrease.

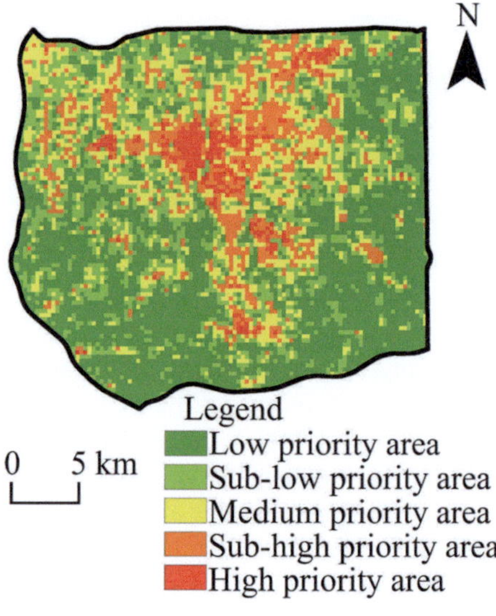

Fig. 4 Distribution map of quantitative matching states (priority index classification) of the TERSSC and the TERSDL in the WUCA (Self-drawn)

3.3 Zoning Management and Planning Suggestions of the TERSSC and the TERSDL

Zoning Management of the TERSSC and the TERSDL. According to the spatial matching states and the quantitative matching states of the TERSSC and the TERSDL, the WUCA is classified into four types and thirteen subtypes (Table 4, Fig. 5).

As shown in Fig. 5, high supply-high demand matching areas are dispersedly located in insular construction land patches close to rivers, lakes, forests, the TERSDL is high, but affected by the thermal environment regulation service in adjacent BGSs,

Table 4 Zoning management matrix of the TERSSC and the TERSDL in the WUCA (Self-drawn)

		Spatial matching states of the TERSSC and the TERSDL			
		High supply-high demand	Low supply-high demand	Low supply-low demand	High supply-Low demand
Quantitative matching states of the TERSSC and the TERSDL	Low priority area	–	–	L-L Area I (2.143%)	H–L Area I (39.297%)
	Sub-low priority area	H–H Area I (2.676%)	L–H Area I (6.086%)	L-L Area II (7.435%)	H–L Area II (6.417%)
	Medium priority area	H–H Area II (1.563%)	L–H Area II (17.428%)	L-L Area III (0.071%)	–
	Sub-high priority area	H–H Area III (0.201%)	L–H Area III (13.202%)	–	–
	High priority area	–	L–H Area IV (3.481%)	–	–

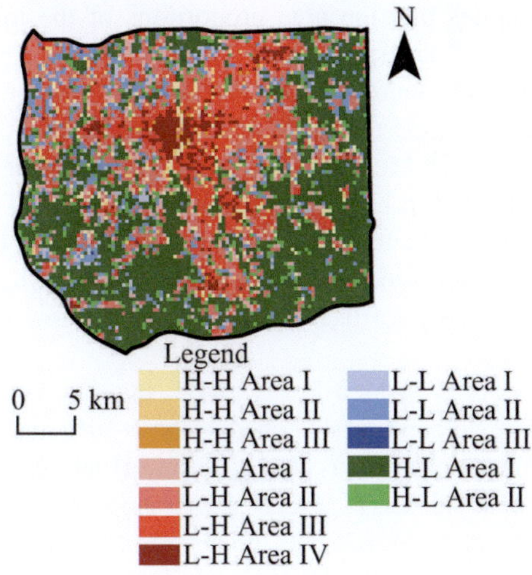

Fig. 5 Zoning management of the TERSSC and the TERSDL in the WUCA (Self-drawn)

so the TERSSC is high, too. The zoning management includes 'H–H Area I', 'H–H Area II', and 'H–H Area III'.

Low supply-high demand mismatched areas are primarily located in central urban built-up regions with dense populations and limited BGSs, where the UHI effect is severe. Hence, the TERSSC is low while the TERSDL is high. The zoning management includes 'L–H Area I', 'L–H Area II', 'L–H Area III', and 'L–H Area IV'.

Low supply-low demand matching areas are scattered in BGSs near construction lands. Due to the UHI effect of these built-up areas, these regions exhibit a low TERSSC. Additionally, with sparse impervious surfaces and low population densities, the TERSDL is also low. The zoning management includes 'L-L Area I', 'L-L Area II', and 'L-L Area III'.

High supply-low demand mismatched areas predominantly occur in ecological conservation areas characterized by sparse population distribution and rich vegetation-water resources, where TERSSC consistently surpasses TERSDL. The zoning management includes 'H–L Area I' and 'H–L Area II'.

Obstacle factors identification of mismatched zones of the TERSSC and the TERSDL. Low supply-high demand mismatched areas and high supply-low demand mismatched areas indicate a significant imbalance between the supply and demand of thermal environment regulation services, necessitating immediate attention and intervention. Hence, the obstacle degree model is applied in mismatched area to identify the barriers affecting the TERSSC and the TERSDL (Fig. 6). The results show that the distribution of obstacle degree of indicators is similarity in different low supply-high demand mismatched areas or high supply-low demand mismatched areas.

As seen the Fig. 6a–d, the key factors for low supply-high demand mismatched areas are CRI, BGSP, MNDWI, HLST, ISP, and HRI. For TERSSC, the obstacle degree of CRI is the largest generally, followed by MNDWI and BGSP. This notion implies that the area proportions of the low temperature areas and BGS are the crucial obstacle factors affecting the improvement of TERSSC. For TERSDL, the obstacle degree of POID is the largest, followed by POP, VFD and APD. By contrast, the obstacle degree of HLST is the least, followed by ISP and HRI in L–H Area I and L–H Area II, while the obstacle degree of ISP is the least, followed by HLST and HRI in L–H Area III and L–H Area IV. This implies that HLST, HRI, and ISP are dominant contribution factors forming the high demand areas. Therefore, it is recommended that the key factors, including CRI, BGSP, MNDWI, HLST, ISP, and HRI, be prioritized for adjustment in low supply-high demand mismatched areas. [26].

As illustrated in Fig. 6e, f, the distribution of obstacle degree in high supply-low demand mismatched areas is different from that of low supply-high demand mismatched areas. In H–L Area II and H–L Area I, five key indicators, including MNDWI, CLST, POID, POP, and APD, are identified. For TERSSC, the obstacle degree of MNDWI is the largest, followed by CLST. The obstacle degree of POID is the largest in TERSDL, followed by POP and APD. Therefore, the key control

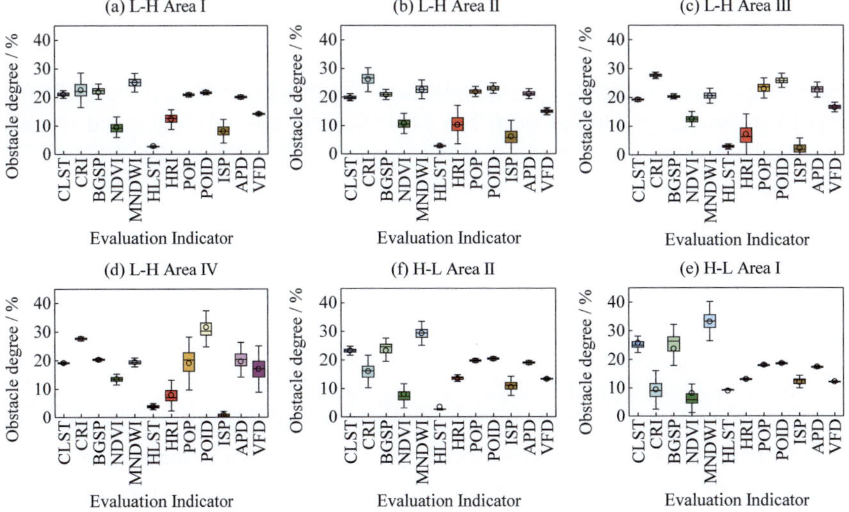

Fig. 6 Box diagram of the obstacle degree distribution of indicators of the TERSSC and the TERSDL in different management zones (Self-drawn)

indicators in high supply-low demand mismatched areas are as follows: the MNDWI, and CLST are suggested to be adjusted first, followed POID, POP, and APD.

Planning suggestions for different zones. Targeted planning intervention measures should be proposed in accordance to the obstacle degree of indicators in various management zones to improve the relationship between the TERSSC and the TERSDL.

The value of ISP in low supply-high demand mismatched areas is high, and we find that these impervious surfaces are mostly industrial areas, commercial and residential areas with high population density by using google map. Moreover, building density is also at a high level. Thus, the main issue in low supply-high demand mismatched areas is high-density buildings, leading to the serious UHI effect [14, 25]. Hence, the optimization recommendations for low supply-high demand mismatched areas focus on increasing the water coverage and reduce the impervious surface coverage and building density. Meanwhile, greening measures, including roof greening and wall greening, in high density buildings should be taken [1, 3, 25].

In high supply-low demand areas, the original forests, rivers, and lakes ought to be maintained, and gathered BGSs ought to be introduced into the unutilized land. And maximizing the thermal environment regulation effect of ecological spaces by considerably regulating exploitation and construction from surrounding forest, rivers, lawn, and lakes ought to be supported [14, 25]. Meanwhile, it is essential to appropriately strengthen human activities and urban and construction, properly mix urban built-up areas with ecological spaces and adequately utilize the thermal environment regulation service in these areas [14].

4 Conclusion

This study constructs the evaluation model to assess the TERSSC and the TERSDL, reveal the matching state between the TERSSC and the TERSDL, and identify the obstacle factors affecting the TERSSC and the TERSDL in the WUCA. The conclusions of this study are as detailed below:

(1) The TERSSC and the TERSDL show evident spatial heterogeneity. Meanwhile, the TERSSC, the TERSDL and their primary indicators all show a strong spatial autocorrelation effect. And there is a spatial imbalance between the TERSSC and the TERSDL. The TERSSC displays a clear spatial distribution, with a decreasing gradient from the urban periphery to the densely populated core areas, while the TERSDL shows an opposite trend.
(2) The spatial matching state of the TERSSC and the TERSDL demonstrates evident polarization in spatial distribution. Most urban built-up areas are facing deficiency of thermal environment regulation service, while most urban built-up areas periphery areas have an abundance of thermal environment regulation service. Moreover, the area proportion of the quantitative matching states (priority categories) increases as the priority index value decrease.
(3) The low supply-high demand mismatched areas are primarily concentrated in the core built-up areas with dense population and high impervious surface coverage, and the key control factors, including CRI, BGSP, MNDWI, HLST, ISP, and HRI are recommended to be first optimized. The high supply-low demand mismatched areas are mostly ecological spaces with sparse population and dense vegetation and water, and the key factors, including MNDWI, CLST, POID, POP, and APD are recommended to be optimized first.

This study offers a new perspective and scientific foundation for urban planning and design in mitigating UHI effect. According to the above research conclusions, the targeted planning intervention measures for improving the relationship between the TERSSC and the TERSDL are proposed in different management zones. For low supply-high demand mismatched areas, we should focus on regulating urban development intensity, controlling impervious surface ratios, moderating building densities, and implementing BGS optimization strategies. Conversely, high-supply-low-demand areas require conservation-oriented approaches, including maintaining ecological integrity, implementing controlled urban development, and improve utilization of thermal environment regulation services.

References

1. Li HX, Zhou WQ, Wang WM et al (2022) Imbalanced supply and demand of temperature regulation service provided by urban forests: a case study in Shenzhen, China. Ecol Indic 145:109666

2. Tieskens KF, Smith IA, Jimenez RB et al (2022) Mapping the gaps between cooling benefits of urban greenspace and population heat vulnerability. Sci Total Environ 845:157283
3. Wang YS, Ouyang WL, Zhang JQ (2024) Matching supply and demand of cooling service provided by urban green and blue space. Urban For Urban Green 96:128338
4. Chen M, Jia WX, Yan L et al (2022) Quantification and mapping cooling effect and its accessibility of urban parks in an extreme heat event in a megacity. J Clean Prod 334:130252
5. Du CL, Jia WX, Chen M et al (2022) How can urban parks be planned to maximize cooling effect in hot extremes? Linking maximum and accumulative perspectives. J Environ Manag 317:115346
6. Kirschner V, Macku K, Moravec D et al (2023) Measuring the relationships between various urban green spaces and local climate zones. Sci Rep 13(1):9799
7. Li WN, Sun RH (2023) A supply-demand model of vegetation cooling for urban heatwave mitigation. Urban Clim 52:101699
8. Ouyang WL, Morakinyo TE, Lee YL et al (2023) How to quantify the cooling effects of green infrastructure strategies from a spatio-temporal perspective: experience from a parametric study. Landsc Urban Plan 237:104808
9. Shi DC, Song JY, Zhong QL et al (2024) Cooling wisdom of 'water towns': how urban river networks can shape city climate? Remote Sens Environ 300:113925
10. Yao X, Ye BJ, Lan YX et al (2024) Diurnal contrast of urban park cooling effects in a "Furnace city" using multi-source geospatial data and optimal parameters-based geographical detector model. Sustain Cities Soc 114:105765
11. Wu TJ, Wang XH, Xuan L et al (2024) How to plan urban parks and the surrounding buildings to maximize the cooling effect: a case study in Xi'an. China Land 13(8):1117
12. Zhang J, Zhang HN, Qi RM (2024) A study of size threshold for cooling effect in urban parks and their cooling accessibility and equity. Sci Rep 14(1):16176
13. Fang XN, Ma Q, Wu LW et al (2023) Distributional environmental justice of residential walking space: the lens of urban ecosystem services supply and demand. J Environ Manag 329:117050
14. Shen ZJ, Zhang BH, Xin RH et al (2022) Examining supply and demand of cooling effect of blue and green spaces in mitigating urban heat island effects: a case study of the Fujian Delta urban agglomeration (FDUA), China. Ecol Indic 142:1091087
15. Chen YH, Xu CC, Ge Y et al (2024) A 100-m gridded population dataset of China's seventh census using ensemble learning and geospatial big data. Earth Syst Sci Data 16(8):3705–3718
16. Chen YH, Xu CC, Ge Y et al (2024) A 100-m gridded population dataset of China's seventh census using ensemble learning and geospatial big data. https://doi.org/10.6084/m9.figshare.24916140.v1. Last accessed 28 Sep 2024
17. National Bureau of Statistics of China (2021) The Seventh National Census. Accessed 16 Oct 2022. http://www.stats.gov.cn/tjsj/zxfb/202105/t20210510_1817176.html. Last accessed 28 Sep 2024
18. Liu YF, Zhang WJ, Liu WK et al (2024) Exploring the seasonal effects of urban morphology on land surface temperature in urban functional zones. Sustain Cities Soc 103:105268
19. Yang CB, Zhu WH, Sun JB et al (2021) Assessing the effects of 2D/3D urban morphology on the 3D urban thermal environment by using multi-source remote sensing data and UAV measurements: a case study of the snow-climate city of Changchun, China. J Clean Prod 321:128956
20. Wang JY, Meng F, Fu PJ et al (2023) Investigating the coupling of supply and demand for urban blue and green spaces' cooling effects in Shandong, China. Amosphere 14(2):404
21. Liu JP, Zhang TC, Ren YQ et al (2024) Three-dimensional evaluation framework of hazard–exposure–vulnerability for mapping heatwave risk and associated dominant dimensions in China. Int J Climatol 2024:8628
22. Huang HJ, Ma JH, Yang YF (2023) Spatial heterogeneity of driving factors for urban heat health risk in Chongqing, China: a new identification method and proposal of planning response framework. Ecol Ind 153:110449
23. Liang C, Zeng J, Zhang RC (2023) Optimizing ecological and economic benefits in areas with complex land-use evolution based on spatial subdivisions. Landsc Urban Plan 236:104782

24. Feng Y, Li GD (2020) Interaction between urbanization and eco-environment in Tibetan Plateau. Acta Geogr Sin 65(7):1386–1405
25. Sun XR, Zhou ZX, Wang Y (2023) Water resource carrying capacity and obstacle factors in the Yellow River basin based on the RBF neural network model. Environ Sci Pollut Res 30(9):22743–22759
26. Gao YJ, Zhao JY, Yu KH (2022) Effects of block morphology on the surface thermal environment and the corresponding planning strategy using the geographically weighted regression model. Build Environ 216:109037

Urban Greenway Activity Behavior Based on Thermal Comfort Evaluation: A Study in Mianyang, China

Jiayue Yin, Juanlin Fu, Li Yan, Yu Long, and Wenyan Feng

Abstract Urban greenways, as an important form of green infrastructure, play a vital role in improving the urban ecological environment and mitigating the urban heat island effect. However, research on the relationship between thermal comfort in greenways and human activity has been limited. Traditional thermal comfort studies have mainly focused on fixed open spaces, such as parks and plazas. Due to their linear layout and high mobility, urban greenways differ significantly in usage and thermal environment impacts. This study focuses on the Mianyang Riverside Greenway as the research area and analyzes the impact of thermal comfort on the behavior of greenway users based on microclimate data and questionnaire survey results. Using the Physiological Equivalent Temperature (PET) as an evaluation index, the study explores the correlation between activity intensity, duration, spatial preferences, and thermal comfort across different activity intensity groups. The results indicate that as PET increases, the average activity intensity in summer greenways decreases significantly. The demand for a suitable thermal environment varies depending on the intensity of the activities, with thermal sensitivity ranked as high-intensity > moderate-intensity > low-intensity activities. Higher-intensity activities require a more favorable thermal environment. Improving the thermal environment of greenways can enhance activity intensity and participation, thereby maximizing the health-promoting benefits of urban greenways. This study enriches the research content on thermal comfort in urban greenways from the perspective of activity type differences, providing practical evidence for achieving sustainable urban development.

Keywords Urban greenways · Microclimate · Outdoor thermal comfort · Thermal perception

---

J. Yin · J. Fu (✉) · L. Yan · Y. Long · W. Feng
School of Civil Engineering and Architecture, Southwest University of Science and Technology, Mianyang, China
e-mail: fujuanlin1227@swust.edu.cn

B. He et al. (eds.), *Climate-Resilient, Low-Carbon, and Sustainable Cities*, Lecture Notes in Civil Engineering 754, https://doi.org/10.1007/978-981-95-2169-2_18

1 Introduction

With the accelerated process of urbanization, the urban heat island effect has become increasingly prominent, and high-temperature phenomena in cities are occurring more frequently [1]. Against this backdrop, urban greenways, as important green infrastructure, have received considerable attention. Urban greenways are interconnected linear open space networks [2], not only possessing the health-promoting effects common to green spaces, such as improving the urban ecological environment and mitigating the urban heat island effect [3, 4], but also serving as ideal locations for specific physical activities like jogging and cycling due to their unique linear characteristics and accessibility. The linear layout of greenways typically makes them more suitable than traditional urban parks for meeting the diverse recreational needs of residents.

Although urban greenways offer multiple health-promoting functions, the quality of their thermal environment has a significant impact on people's thermal comfort and behavioral choices. High temperatures and high humidity within greenways can increase thermal discomfort, thereby affecting the duration of stay and the frequency of activities for users. Moreover, individuals exhibit varying levels of adaptation and sensitivity to thermal environments, which may further influence their activity levels within the greenway.

Previous studies have shown that thermal comfort significantly affects human behavior and activity. Zhao Xiaolong [5] explored the relationship between leisure physical activities in parks and thermal comfort, noting that different activity intensities correspond to varying microclimate preferences, with the duration of leisure activities linearly related to thermal comfort. Labdaoui et al. [6] employed walkability and thermal comfort indices to highlight that thermal comfort is an important factor in enhancing pedestrian activity and creating healthy environments. Rao et al. [7] analyzed the impact of thermal comfort and environmental factors such as air quality on park visitation rates, identifying temperature, humidity, and heat index (HI) as key determinants of park visitation. Their study found that park visitation is positively correlated with temperature up to a certain threshold, beyond which it becomes negatively correlated. Liu Binyi et al. [8] examined the relationship between microclimate elements, activity spaces, and human comfort, emphasizing that shading is a critical factor in improving thermal comfort during summer. In Taiwan, when the Physiological Equivalent Temperature (PET) ranges between 26 and 30 °C, the number of outdoor participants reaches its peak; as PET increases or decreases beyond this range, the number of participants declines [9]. There is a significant correlation between activity duration and thermal comfort, with a stronger correlation to air temperature compared to relative humidity, wind speed, or solar radiation [10].

Providing a suitable thermal environment for outdoor activities during the hot summer months is especially important. However, in urban planning, there is often an emphasis on the usage patterns and aesthetic functions of greenways as green spaces [11], with relatively less focus on systematically measuring the physical and microclimate environment of greenways from a human-centered perspective and analyzing their relationship with leisure activities. Therefore, research on the correlation between thermal comfort in urban greenways and human behavior holds practical significance. By gaining a deeper understanding of the relationship between thermal comfort in greenways and user behavior, it is possible to create more comfortable and pleasant urban greenway environments and further maximize the health-promoting benefits of greenways.

2 Study Area, Data, and Methods

2.1 Study Area

Mianyang is located in the northwest of the Sichuan Basin, along the middle and upper reaches of the Fujiang River. The city experiences a typical hot summer and cold winter climate. With a total area of 20,258 km^2, Mianyang had a permanent resident population of 4.911 million as of the end of 2023. The Mianyang Riverside Greenway, located in Fucheng District, was selected as the study area. The greenway is situated in the old city center of Mianyang, forming a peninsula surrounded by the Fujiang River and Anchang River. It connects various residential areas and the city's slow-moving traffic system, providing people with a green public space to engage in nature-related activities and exercise.

2.2 Study Data

2.2.1 Microclimate Data

Based on the timing of human activities and the sunrise and sunset times in Mianyang, the research period was determined to be September 4, 2024. The survey was conducted between 7:00 am and 9:00 pm. On that day, the temperature recorded at the city's meteorological station ranged from 26 to 38 °C, with the predominant wind direction shifting from northeast to no consistent wind, and wind speeds were below level 3. The perceived temperature reached 39.5 °C.

Based on the preliminary survey results, the measurement points were distributed according to the following principles: (a) The measurement points were evenly distributed as much as possible to ensure a balanced coverage of the area; (b) The points comprehensively covered user activities along the greenway, taking into account both preferred and less preferred spaces for users; (c) Different types of shading, surface materials, and other physical environmental features, as well as the microclimate environment, were considered. Instruments were set up for fixed-point microclimate measurements (see Fig. 1).

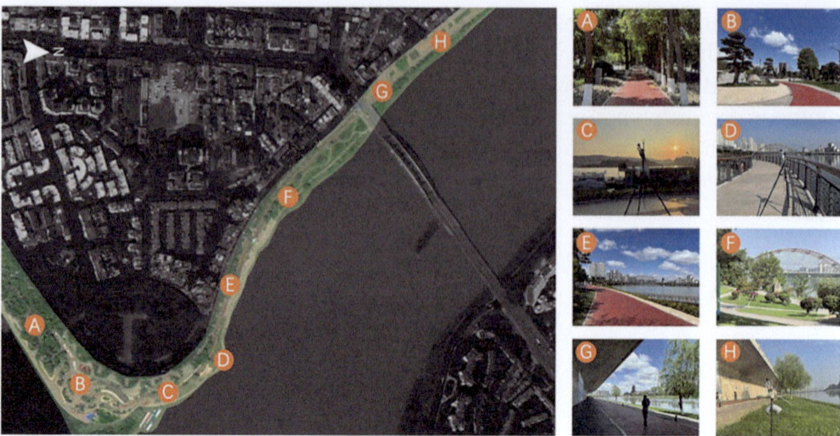

Fig. 1 Distribution of measurement points (self-drawn)

Table 1 Instrument performance parameters (self-drawn)

Measurement parameters	Instrument name	Measurement range	Accuracy
Air temperature	Testo174 temperature and humidity recorder	−20 to 70 °C	±0.5 °C
Relative humidity		0–100% RH	±3% RH
Wind speed	WWFWZY-1 wireless omnidirectional wind speed and temperature recorder	0.05–30 m/s	5%
Black globe temperature	TM-188-188D heat index tester	0–80 °C	±0.6 °C (15–40 °C)

At each measurement point, thermal environment parameters such as air temperature (T_a), globe temperature (T_g), relative humidity (R_H), and wind speed (V_a) were measured. During the on-site measurements, the instruments were fixed at a height of 1.5 m above the ground at each measurement point. All instruments were calibrated before testing to ensure accuracy. The detailed specifications of the instruments used for measurement are provided in Table 1.

2.2.2 Questionnaire Survey Data

A questionnaire survey was conducted to collect personal basic information, activity patterns, and thermal comfort conditions of participants using the greenway. During the microclimate measurement process, questionnaires were distributed to obtain direct feedback from participants on their evaluation of the thermal environment. The questionnaire consisted of three sections: The first section included basic information about the respondents, such as gender, age, height, weight, health condition, and clothing. The second section gathered information on the respondents'

Table 2 Basic information of respondents (self-drawn)

		Total number
Gender	Male	152
	Female	140
Age	<18	6
	18–40	105
	41–60	120
	>60	31

activity patterns, including duration of activity, type of activity being performed, main time periods of activity, and preferred greenway spaces for activities. The third section focused on subjective thermal comfort, including thermal sensation votes and thermal comfort votes. A total of 297 questionnaires were collected, with 292 valid responses (152 males and 140 females). The respondents covered various genders, age groups, and occupational backgrounds. Random sampling was used to ensure data representativeness (Table 2).

2.3 Study Methods

2.3.1 Thermal Comfort Evaluation Index

The thermal comfort index is a high-level evaluation metric that combines objective environmental factors with subjective perceptions, allowing for a comprehensive assessment of the thermal environment at each measurement point. The subjective thermal comfort perceptions of respondents from the questionnaire, along with the objective thermal comfort evaluation based on the Physiological Equivalent Temperature (PET), were used as the basis for the thermal comfort index. During the pilot study, various thermal comfort evaluation indices were compared, and the results showed that PET was the most closely aligned with respondents' subjective perceptions. Therefore, PET was selected as the standard for this evaluation.

(1) PET Calculation

PET is based on the energy balance of the human body, has significant thermophysiological meaning, and is highly reproducible, making it suitable for evaluating thermal conditions in different climates [12]. Using the RayMan model, various microclimate parameters collected from field measurements, as well as data on participants' age, gender, height, weight, clothing thermal resistance, and metabolic rate, were input. The statistical reference standards for clothing thermal resistance and metabolic rate were based on ANSI/ASHRAE Standard 55-2017 [13]. PET was then calculated as the objective thermal comfort evaluation index.

(2) Subjective Thermal Comfort

Table 3 Thermal sensation vote (TSV) (self-drawn)

−3	−2	−1	0	1	2	3
Very cold	Cold	Slightly cool	Neutral	Slightly warm	Warm	Very warm

Table 4 Thermal comfort vote (TCV) (self-drawn)

−2	−1	0	1	2
Very uncomfortable	Uncomfortable	Neutral	Comfortable	Very comfortable

The subjective thermal comfort of respondents was assessed using the Thermal Sensation Vote (TSV) defined by ASHRAE [14], on a seven-point scale (−3 to +3) to evaluate thermal sensation. Thermal comfort was assessed using the Thermal Comfort Vote (TCV) on a five-point scale (−2 to +2) to measure subjective thermal comfort (Tables 3 and 4).

2.3.2 Activity Intensity Classification

The data on human activities were primarily derived from the questionnaire, focusing on the sections related to physical activity. Activity type and activity duration were selected as the indicators for determining activity levels. The intensity of physical activity was classified based on Ainsworth's compendium of physical activities [15], using Metabolic Equivalent of Task (MET) as the criterion for different activity intensities [16]. The greenway users were divided into three categories: low-intensity, moderate-intensity, and high-intensity activities. The classification criteria for each intensity level are shown in Table 5.

Table 5 Classification criteria for different activity intensities

Activity type	Seated rest	Standing	Childcare	Walking	Children's play	Running	Cycling
Metabolic equivalent	1.3	1.8	2.5	3.0	5.0	≥7.0	8.0
Intensity classification	<3 METs Low-intensity			3 ≤ METs < 6 Moderate-intensity		≥6 METs High-intensity	

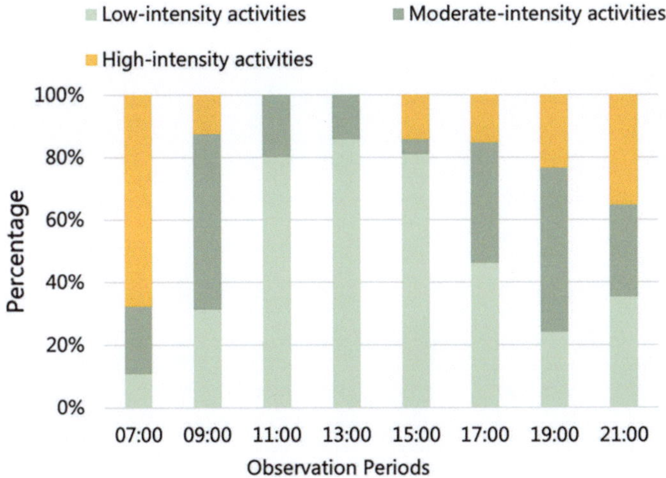

Fig. 2 Proportions of different intensity activity populations at different times (self-drawn)

2.4 *Activity Patterns and Spatial Preferences of Participants*

2.4.1 Analysis of Activity Behavior

During the experiment, the proportion of participants engaging in different activity intensities at various time intervals is shown in Fig. 2. The highest proportion of high-intensity activities occurred between 07:00 and 09:00, accounting for 68% of all activities, with an upward trend after 19:00. Moderate-intensity activities were most prominent during the 09:00–11:00, 19:00–21:00, and post-21:00 time intervals, with proportions of 56%, 52%, and 30%, respectively. The proportion of low-intensity activities increased as the temperature rose, exceeding 80% between 11:00 and 17:00. This increase was largely due to psychological factors, such as participants perceiving the greenway as having "fresh air," "beautiful scenery," and a "quiet environment," combined with the reduction in moderate- and high-intensity activities during high-temperature periods.

Overall, moderate- and high-intensity activities were more prevalent between 07:00–09:00 and after 19:00, while their proportion decreased as the temperature rose throughout the day. In contrast, low-intensity activities saw larger proportions in the late morning and afternoon, with most participants opting for shaded areas during the hottest periods.

2.4.2 Analysis of Spatial Preferences

Compared to parks and other green public spaces, greenways exhibit greater user mobility. Therefore, the spatial preference votes from the questionnaire were used

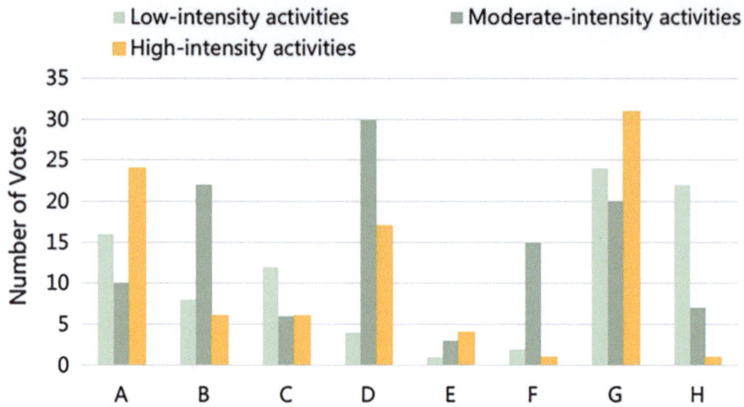

Fig. 3 Spatial preference votes of different intensity activity groups at each measurement point (self-drawn)

as the data foundation to analyze user preferences for different locations along the experimental greenway, as shown in Fig. 3. Based on the total number of votes, participants preferred measurement points A, B, D, and G more than other points, while points E and F received relatively fewer votes. Measurement point A is a shaded running track, point B is an unshaded rubber track, point D is a waterside walkway, and point G is an activity space beneath a bridge.

Low-intensity activity participants showed a strong preference for point G (the space under the bridge), with the most votes from this group. Participants engaged in moderate-intensity activities favored point D (the waterside walkway), followed by point B (the unshaded track). High-intensity activity participants voted most frequently for points D (waterside walkway) and A (shaded track).

Overall, there were significant differences in the spatial preference votes across measurement points. Both low-and high-intensity activity participants preferred shaded areas, while those engaged in moderate-intensity activities tended to favor more open and waterside spaces.

2.5 *Thermal Sensation and Thermal Comfort Voting Results*

2.5.1 Distribution Characteristics of Thermal Sensation and Thermal Comfort Votes

The results of the Thermal Sensation Vote (TSV) and Thermal Comfort Vote (TCV) for the greenway are shown in Fig. 4. Across low, moderate, and high-intensity activity groups, the highest proportion of thermal sensation votes indicated "slightly warm" (TSV = 1). For low-intensity activities, 45% of participants reported feeling

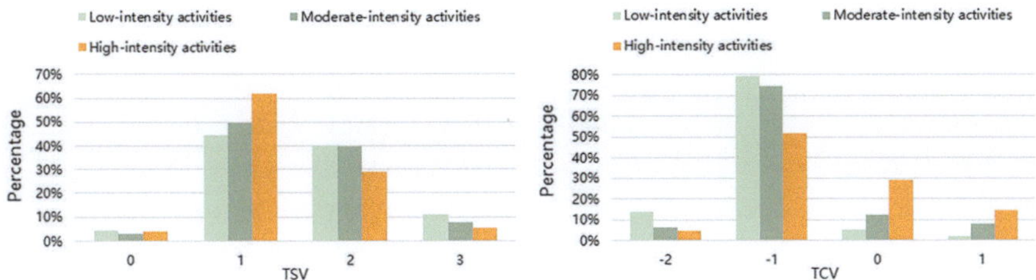

Fig. 4 Distribution characteristics of thermal sensation voting and thermal comfort voting for different intensity activities (self-drawn)

"slightly warm," while 50% of moderate-intensity participants and 62% of high-intensity participants reported the same sensation.

Regarding thermal comfort votes, participants from all activity intensity groups provided generally negative evaluations of the thermal environment in the greenway. Only a small percentage of participants gave a neutral thermal comfort rating (TCV = 0), with low- and moderate-intensity activity participants reporting neutral votes at rates of 5% and 12%, respectively. This indicates that most users found the thermal environment less than ideal, regardless of their activity intensity.

2.5.2 Correlation Analysis Between Thermal Comfort Vote and Thermal Sensation Vote

Based on the questionnaire results, a relationship between the Thermal Sensation Vote (TSV) and Thermal Comfort Vote (TCV) was established for participants engaged in activities of varying intensity levels. The analysis revealed a quadratic polynomial relationship, as presented in Table 6. When participants reached their optimal thermal comfort state (i.e., when TCV = 2), the corresponding TSV values for low, moderate, and high-intensity activity groups were −1.26, −1.21, and −0.82, respectively. This suggests that individuals engaging in activities on the greenway during the summer require a slightly cooler thermal sensation to achieve their optimal thermal comfort. However, the R^2 values for these relationships were relatively low (0.336 for low-intensity, 0.314 for moderate-intensity, and 0.483 for high-intensity). This can be attributed to the fact that thermal comfort is influenced by various factors, including not only thermal sensation but also individual physiological differences, clothing, and metabolic rate. As a result, the complexity of these factors, which may change with different levels of activity intensity, limits the model's ability to fully explain the variations in thermal comfort.

Table 6 Relationship between different intensity TCV and TSV (self-drawn)

	Expression
Low-intensity	$TCV = 0.102 - 1.182TSV + 0.256TSV^2$ ($R^2 = 0.336$) (1)
Moderate-intensity	$TCV = 0.219 - 0.972TSV + 0.126TSV^2$ ($R^2 = 0.314$) (2)
High-intensity	$TCV = 0.697 - 1.368TSV + 0.257TSV^2$ ($R^2 = 0.483$) (3)

2.6 Analysis of the Relationship Between Greenway User Activity Levels and Microclimate Thermal Comfort

2.6.1 Thermal Comfort Conditions

In summer, the microclimate of the greenway exhibits a "warming-high temperature-cooling" daily thermal comfort variation pattern. After the temperature rises, it reaches a relatively stable high-temperature state around 10:00 am, which persists until approximately 3:00 pm, during which time the microclimate thermal comfort continues to decline. During the warming phase, measurement points B, C, D, F, and H show significantly faster increases compared to points A, E, and G. In the high-temperature period, points B, C, D, and F maintain higher PET values, with point F reaching the highest value of 48.6 °C at 1:00 pm. Point H shows a significant decrease in PET from 1:00 to 2:00 pm, which is attributed to reduced solar radiation in the area due to the shading effect of nearby buildings.

According to the analysis of activity intensity proportions at different times and participants' spatial preferences, during periods of strong solar radiation, the proportion of participants choosing shaded areas for activities is significantly higher than in other locations. Measurement points A, B, and G represent different types of shading: tree shade, no shade, and building shade, respectively. As shown in Fig. 5, during the high-temperature period (10:00 am–3:00 pm), the maximum PET values for these three points are 40.0 °C, 48.2 °C, and 36.3 °C, respectively. Due to the lack of shading, point B reaches a maximum PET value of 48.2 °C, which is significantly higher than points A and G, which have shading. This clearly demonstrates that shading plays a significant role in affecting PET values. Overall, both tree and building shading can improve thermal comfort on the greenway. However, tree shading is more affected by solar radiation, meaning the thermal comfort level in tree-shaded areas is generally lower than that in building-shaded areas. Therefore, in the planning and design of greenways, careful selection of shading types is crucial to create more comfortable outdoor activity spaces.

2.6.2 Correlation Between Activity Intensity and Thermal Comfort

In this analysis, low, moderate, and high-intensity activities were assigned values of 1, 2, and 3, respectively. PET values were grouped by 1 °C intervals, and the

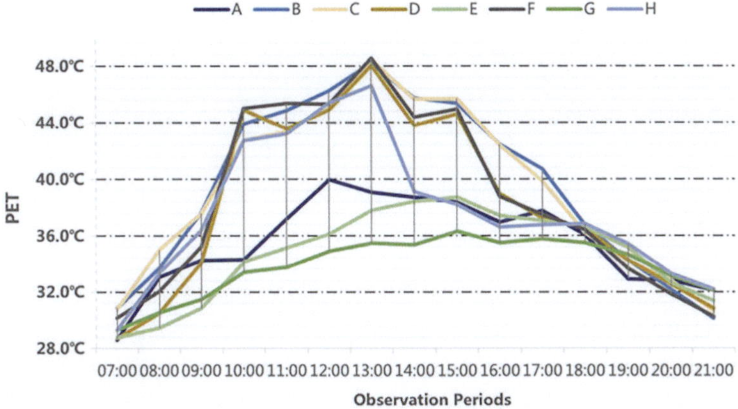

Fig. 5 Daily variation trend of microclimate thermal comfort (self-drawn)

average activity intensity for each interval was calculated. A significant linear correlation was found between activity intensity and thermal comfort, as expressed by the following equation. As the PET value increases, the average activity intensity decreases progressively.

When PET equals 30 °C, moderate to high-intensity activities are more prevalent. However, as the PET value rises to 40 °C, the average activity intensity drops to 1.19, indicating that more participants tend to opt for low-intensity activities. This negative correlation demonstrates that as PET increases, people naturally reduce high-intensity activities to mitigate health risks. This suggests that improving thermal comfort in high-temperature environments can encourage more moderate and high-intensity activities.

$$y = 5.833 - 0.116\text{PET}\ \left(\text{R}^2 = 0.917\right) \quad (4)$$

Note y is the average activity intensity, $p < 0.001$.

2.6.3 Correlation Between Activity Duration and Thermal Comfort

Correlation analysis (Table 7) shows that in summer greenways, the activity duration for different intensity levels exhibits varying relationships with temperature. For low-intensity activities, there is a positive correlation between activity duration and temperature, with a correlation coefficient of 0.535. This indicates that individuals engaged in low-intensity activities, such as sitting or resting, tend to have longer durations during high-temperature periods compared to other times.

For moderate-intensity activities, there is a negative correlation between activity duration and PET, with a correlation coefficient of −0.830. This suggests that as temperature increases, the duration of moderate-intensity activities decreases, as

Table 7 Correlation between activity duration and thermal comfort (self-drawn)

	Low-intensity activities	Moderate-intensity activities	High-intensity activities
Activity duration	0.535*	−0.830**	−0.273
Significance	0.040	0.002	0.477

Note * at 0. 05 level (two-tailed), correlation is significant; ** at 0. 01 level (two-tailed), correlation is significant

participants tend to shorten their activities in response to thermal discomfort. In contrast, high-intensity activities are usually purposeful and planned, and participants are less likely to alter their activity duration due to temperature fluctuations.

2.6.4 Comfort Thresholds for Different Activity Intensities

The relationship between thermal comfort indices and subjective votes was established to determine the thermal comfort limits and ranges for different activity types. Given the variation in sample sizes across different activity intensities and thermal environments, a statistical method of interval grouping and quantity weighting was adopted. PET values were grouped in 1 °C intervals, and groups with fewer than two samples were excluded. PET and Mean Thermal Sensation Vote (MTSV) were then used for regression analysis. The resulting regression equations all showed $P < 0.005$ and $R^2 > 0.82$, indicating that the PET index effectively reflects the thermal stress experienced by participants engaging in different activity intensities on the greenway. There were significant differences in the thermal comfort characteristics across activity intensity types. The regression coefficients for the relationship between MTSV and PET for low-, moderate-, and high-intensity activities were 0.129, 0.183, and 0.207, respectively. This means that for each 1 °C increase in PET, MTSV increases by 0.129, 0.183, and 0.207, respectively. Therefore, the thermal sensitivity ranking, from highest to lowest, is high-intensity activities > moderate-intensity activities > low-intensity activities. Activities like running and cycling, which are considered high-intensity, have higher thermal environment requirements.

According to the Thermal Sensation Vote (TSV) scale, when TSV = 0, the participant's thermal sensation is considered "neutral." Using the regression equations between PET and TSV, the PET value corresponding to TSV = 0, or the "neutral" thermal sensation, can be calculated. The neutral PET values for low-, moderate-, and high-intensity activities are 23.5 °C, 24.8 °C, and 25.0 °C, respectively. When TSV falls within the range of −1.5 to 1.5, the thermal environment is considered acceptable to the participants [17]. Therefore, the acceptable thermal environment thresholds for low-, moderate-, and high-intensity activities are 11.9–35.2 °C, 16.6–33.0 °C, and 17.8–32.3 °C, respectively. A larger range indicates that participants engaged in that intensity of activity can tolerate a wider range of thermal conditions.

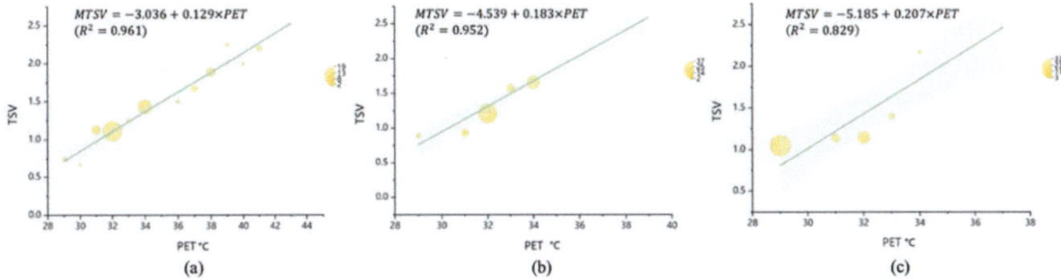

Fig. 6 a–c Correlations between TSV and PET for low, moderate and high intensity activities, respectively (self-drawn)

The results show that in summer greenways, high-intensity activity participants are more sensitive to thermal environments compared to those engaged in low- and moderate-intensity activities. This suggests that improving outdoor thermal comfort in greenways can increase the intensity of activities, thereby enhancing the health-promoting effects of greenway use (Fig. 6).

$$\text{MTSV} = -3.036 + 0.129\text{PET}\ \left(\text{R}^2 = 0.961\right) \tag{5}$$

$$\text{MTSV} = -4.539 + 0.183\text{PET}\ \left(\text{R}^2 = 0.952\right) \tag{6}$$

$$\text{MTSV} = -5.185 + 0.207\text{PET}\ \left(\text{R}^2 = 0.829\right) \tag{7}$$

3 Conclusion

This study focuses on the thermal comfort of individuals engaging in different activity intensities in urban greenways during the summer season in regions with hot summers and cold winters. The research explores the thermal comfort conditions of the greenway and the relationship between various activity levels and microclimate thermal comfort. The key conclusions are as follows:

(1) When users of the greenway achieve optimal thermal comfort during summer, the thermal sensation votes for low-, moderate-, and high-intensity activity participants are −1.26, −1.21, and −0.82, respectively. This indicates that a slightly cooler thermal sensation is necessary for optimal thermal comfort. Improving the thermal environment of the greenway can enhance the overall thermal comfort experience for users.

(2) The thermal environment requirements vary depending on the intensity of the activities. The thermal sensitivity ranking, from highest to lowest, is high-intensity activities > moderate-intensity activities > low-intensity activities. The

higher the activity intensity, the greater the demand for a suitable thermal environment. This finding is consistent with previous studies that concluded that joggers have higher thermal sensitivity compared to those resting or walking.

(3) Future greenway planning and design should consider the needs of participants engaging in different intensity activities. For example, high-intensity activity participants are more sensitive to the thermal environment and represent a larger portion of users. Therefore, during high-temperature periods, it is advisable to provide more shading facilities, such as increasing the area of tree cover and strategically planning shaded areas using buildings. This will help meet the demand for shaded spaces, improve users' thermal comfort, and make greenways more user-friendly and versatile, offering better outdoor activity spaces.

In conclusion, optimizing the thermal comfort of urban greenways not only enhances the user experience but also encourages a more diverse range of activities, ultimately improving the overall utilization of greenways and contributing to the achievement of urban health development goals.

References

1. Gulyás Á, Unger J, Matzarakis A (2006) Assessment of the microclimatic and human comfort conditions in a complex urban environment: modelling and measurements. Build Environ 41(12):1713–1722
2. Little CE (1995) Greenways for America. JHU Press
3. Huang H, Yang H, Chen Y et al (2021) Urban green space optimization based on a climate health risk appraisal—a case study of Beijing city, China. Urban For Urban Green 62:127154
4. Fang H (2018) Study on greenway planning and design in China. Beijing Forestry University
5. Xiaolong ZHAO, Qing BIAN, Yunjing HOU et al (2019) A research on the correlation between physical activity performance and thermal comfortable of urban park in cold region. Chin Landsc Arch 35(04):80–85
6. Labdaoui K, Mazouz S, Moeinaddini M et al (2021) The street walkability and thermal comfort index (SWTCI): a new assessment tool combining street design measurements and thermal comfort. Sci Total Environ 795:148663
7. Rao F, Xiao P, Zhang Y et al (2024) Unravelling key environmental factors influencing urban park visits: thermal comfort and air quality. Urban Clim 57:102096
8. Liu B-y, Mei Y, Kuang W (2016) Experimental research on correlation between microclimate element and human behavior and perception of residential landscape space in shanghai. Chin Landsc Arch 32(01):5–9
9. Lin TP, Tsai KT, Liao C et al (2013) Effects of thermal comfort and adaptation on park attendance regarding different shading levels and activity types. Build Environ 59:599–611
10. Yuan W (2015) Research of thermal comfort of microclimate on Harbin urban park based on behavior activity. Harbin Institute of Technology
11. Yi W (2022) A landscape design strategy research of typical greenway jogging space in humid and hot areas supporting thermal comfort. South China University of Technology
12. Matzarakis A, Mayer H, Iziomon MG (1999) Applications of a universal thermal index: physiological equivalent temperature. Int J Biometeorol 43:76–84
13. American Society of Heating, Refrigerating and Air-conditioning Engineers. ASHRAE Standard 55 (2017) Thermal environmental conditions for human occupancy. SHRAE Customer Service, 1791 Tullie Circle, NE, Atlanta, pp 37–39

14. 2001 ASHRAE handbook, fundamentals (SI), American Soeiety of Heating, Refrigerating and Air-conditioning, Engineers, Ine., 1791 Tullie Circle, NE, Atalanta, GA 30329
15. Ainsworth BE, Haskell WL, Whitt MC et al (2000) Compendium of physical activities: an update of activity codes and MET intensities. Med Sci Sport Exerc 32(9; SUPP/1):S498–S504
16. Zhao W, Cong L (2004) Physical activity evaluation: metabolic equivalent intensity levels and evaluation of different physical activity. J Hyg Res 02:246–249
17. Xing H, Baofeng L, Hong C (2020) Research review and evaluation framework of outdoor thermal comfort. Build Sci 36(04):53–61

Summer Thermal Comfort in Transfer Spaces of Semi-Open Rail Transit Stations in Chongqing, China

Dingran Li, Lili Dong, Xiang Cheng, Yu Luo, Mian Zhang, and Simin He

Abstract Amid rapid global urbanization, rail transit systems have expanded significantly due to economic, environmental, and efficiency benefits. However, enclosed transfer spaces within stations often suffer from poor thermal comfort due to high passenger density. This study examines Sigongli Station on Chongqing Rail Transit Line 3, analyzing thermal conditions in semi-open transfer spaces. Field data from 372 valid questionnaires and 356 environmental measurements at two test points with varying degrees of openness. The study's findings, derived from regression analysis of the collected data, yielded thermo-neutral temperatures of 23.82 °C and 22.63 °C at the two points. A comparison of the thermal satisfaction votes (TSV) regression model with the predicted mean vote (PMV) regression model revealed that the PMV model was challenging to calibrate with precision under extreme thermal conditions during summer. However, the application of the extended predictive mean vote (ePMV) modified model led to a substantial enhancement in PMV correction, with an improvement of 66.1% in prediction accuracy. This improvement was particularly notable in naturally ventilated high-temperature environments. However, under high-temperature, high-humidity, and windless conditions, the corrective effect of ePMV remains unsatisfactory. Consequently, a constant based on ePMV was introduced, and a novel PMV modified model for predicting and evaluating thermal sensation in high-temperature, high-humidity environments during summer was established: PMV' = 0.5816PMV + 0.5806. This significantly improves the prediction accuracy over the traditional PMV model. Finally, improvement strategies are proposed and reflections on this study are made in order to provide lessons and references for subsequent studies on thermal comfort in rail transportation.

Keywords Urban rail transit · Transfer space · Thermal comfort

D. Li · L. Dong (✉) · M. Zhang · S. He
School of Architecture and Urban Planning, Chongqing Jiaotong University, Chongqing, China
e-mail: dongll@cqjtu.edu.cn

X. Cheng
School of Architecture and Urban Planning, Chongqing University, Chongqing, China

Y. Luo
Faculty of Environmental Engineering, Kitakyushu University, Fukuoka, Japan

B. He et al. (eds.), *Climate-Resilient, Low-Carbon, and Sustainable Cities*, Lecture Notes in Civil Engineering 754, https://doi.org/10.1007/978-981-95-2169-2_19

1 Introduction

The acceleration of global urbanization has led to a significant expansion in the rail transit industry, which possesses distinct advantages, including economy, environmental protection, and high efficiency. In comparison with alternative forms of urban public transportation, urban rail transit possesses numerous advantages, including its substantial capacity, minimal energy consumption, safety and comfort, expeditiousness, and environmental sustainability. The enhanced accessibility, reduced commuting times, mitigation of traffic congestion, promotion of urban development, enhancement of urban continuity, and facilitation of high-density expansion through the integration of various transit stations are some of the notable benefits of efficient rail transit [1]. In recent years, environmental concerns related to high-density interchange subway stations have garnered mounting attention [2].

According to American Society of Heating, Refrigerating and Air-Conditioning Engineers (ASHRAE) 55–2017 [3], thermal comfort is defined as the subjective expression of satisfaction and evaluation of the thermal environment, which is used to define the achievement of a comfortable thermal environment. This indicates that a combination of subjective and objective perspectives is necessary in the study of the indoor thermal environment of a building. Among the factors involved in the study of the thermal environment, in addition to personal factors such as activities and clothing, are temperature, thermal radiation, humidity, wind speed, and other physical environment factors. These factors interact with each other, thereby demonstrating that the assessment of thermal comfort of an indoor environment is complex. To date, the majority of indoor thermal environment research has centered on civil buildings, particularly residential buildings and office buildings [4, 5]. However, with the recent advancements in urban rail transit, an increasing number of scholars both domestically and internationally have begun to direct their attention toward the thermal environmental comfort of rail transit stations.

Given the considerable variation in thermal environments across different areas of a given station, and the heterogeneity of factors influencing these environments, existing studies typically adopt a multifaceted approach by dividing stations into core functional units, such as carriages, platforms, and entrances and exits, thereby facilitating targeted and diversified research.

Currently, a large number of scholars have focused their research on thermal comfort in urban rail transit in conventional space units such as carriages and platforms. However, systematic studies on the thermal comfort of transfer space are relatively scarce. First, existing thermal comfort models such as PMV [6] and SET [7] are primarily developed for stable environments and fail to effectively account for the multifactorial interactions in transfer spaces, including high-density passenger flows [2], the orientation and vegetation shading of transfer spaces [8], as well as air velocity [9].

Second, the environmental control standards are insufficiently comprehensive. Existing environmental control strategies are often directly adapted from platforms design standards [10], without considering the unique minute-level environmental fluctuations characteristic of transfer space. This has led to a common contradiction in practical operations, where cooling delays coexist with redundant energy consumption [11].

The existing zoning research framework fails to address the transitional characteristics of transfer spaces, resulting in the orientation optimization proposed by Quan et al. [8] and the entrance configuration scheme validated by Omar et al. [11] not resolving the current thermal environment issues in metro transfer space, such as poor ventilation caused by narrow corridors and high temperatures and humidity during summer. This highlights the urgent need for the establishment of a research paradigm specific to transfer space.

The objective of this paper is to measure the thermal environment and conduct a questionnaire survey of representative semi-open urban rail transit stations in Chongqing. The survey results were then statistically analyzed using regression analysis to develop a linear regression model of the mean value of thermal sensory polls at the test points and the thermal neutrality temperature. This analysis provides a basis for optimizing the thermal environment in the transfer space of semi-open urban rail transit stations.

2 Research Methodology

2.1 Objects of Study

The Chongqing Rail Transit Line 3 has been identified as the metro line with the highest average daily passenger traffic in Chongqing. Furthermore, it has been determined that Line 3 is also the largest single-track line in the world. Among these stations, Sigongli Station is a noteworthy example of a semi-open interchange station on Line 3. It facilitates the interchange between Line 3 and the Circle Line, and it is situated in close proximity to the Sigongli Bus Hub, which experiences a significant passenger flow. Consequently, the location of this test was designated as the transfer space of Chongqing Railway Transportation Line 3 Sigongli Station. To ensure a comprehensive investigation, two additional locations with different degrees of openness were selected for further study, as illustrated in Table 1.

The research subject is composed of passengers traversing the transfer space, encompassing individuals of diverse genders, ages, cultural backgrounds, and other characteristics. Chongqing is situated within a region characterized by distinct thermal zones, classified according to the annual mean temperature. The month of

Table 1 Quantification of spatial patterns (Self-drawn)

Test point	Window opening frequency	Number of air changes
Test point A	0.38	2.94 times/h
Test point B	0.87	14.67 times/h

August was selected as the experimental timeframe due to its position in the local summer season, a period marked by elevated temperatures. This timing is considered sufficient to accurately reflect the prevailing climatic characteristics of the region.

2.2 *Methodology of the Survey*

Thermal environment field measurements

To expand upon the findings of the preliminary study, two additional test locations were selected for the transfer space of the selected site. These locations featured varying degrees of openings. Between August 1st and 2nd, 2024, a series of real-world tests were conducted on-site to gather more extensive data. The uninterrupted testing period was from 7:00 a.m. to 10:00 p.m. daily, with a 5-min data recording interval. The physical environment parameters measured included air temperature, relative humidity, and wind speed.

The primary experimental instruments employed were the TES 690-H temperature and humidity measuring instrument and the Kestrel 5500 handheld meteorological instrument. Prior to utilization, the meteorological instrument underwent a calibration process involving a rotary adjustment around the vertical axis. The instruments were measured by fixing them on a tripod and keeping them at a height of 1.2 m from the ground to minimize measurement errors.

Thermal Comfort Questionnaire

A subjective feeling questionnaire was used to investigate passengers' feelings about the current status of thermal comfort in the current interchange corridors. The research yielded 372 valid questionnaires, encompassing the demographic characteristics of the respondents who participated in the study. The gender distribution of the respondents is as follows: 57.03% female and 45.97% male, indicating a more balanced gender ratio. The predominant age group among the respondents was 18 to 40 years old, constituting 64.78% of the total. The questionnaire was designed to target the primary age group, yet the selection of subjects from all age categories was undertaken to ensure the representation of a comprehensive and diverse sample.

The questionnaire was utilized to evaluate human thermal sensation and comfort. The survey incorporated passengers' subjective assessments of thermal sensation, satisfaction, wind speed, and humidity levels in the indoor environment. The voting values of thermal sensation were expressed using the ASHRAE's seven-level scale, namely: very hot (+3), hot (+2), hotter (+1), moderate (0), colder (−1), cold (−2), and very cold (−3) [3].

During the survey, respondents' attire was meticulously documented, and the thermal resistance of their clothing was calculated in accordance with ASHRAE Standard 55–2017 using the clo rating (1 clo = 0.155 $m^{2-}K/W$). However, due to the impact of the summer climate, the survey participants primarily opted for summer attire, such as shorts, short sleeves, skirts, and similar garments. Consequently, the thermal resistance of bulky winter clothing, such as thick sweaters and down jackets, was not taken into account in the survey questionnaire. To ensure the reliability of the study, the clothing thermal resistance of each passenger was calculated according to the actual situation [12]. This empirical measurement method categorizes and simplifies the thermal resistance of different types of garments. According to the thermal resistance of summer garments in Table 1 and superimposed calculation, the data of the thermal resistance of garments for each sample is obtained, as shown in Table 2. The analysis reveals that the average garment thermal resistance of the respondents is 0.35clo, with a minimum of 0.14 clo and a maximum of 0.66 clo.

Concurrently, according to the criteria outlined in ASHRAE Standard 55–2017, metabolic rate was determined, with the actual activities of the respondents in the interchange corridor taken into account. The activities of the respondents and their respective metabolic rates were then categorized and enumerated according to the statistics presented in Table 3. Consequently, the metabolic rate data corresponding to each sample was obtained.

Calculation of thermal comfort indicators

Preliminary research and analysis indicate that the site experiences natural ventilation conditions throughout the year, and users of the transfer space do not remain in a stationary state for extended periods or face difficulties adjusting their attire to the outdoor temperature. Given the observed characteristics of the physical environment

Table 2 Thermal resistance of garments corresponding to different clothes (Self-drawn)

Type of clothing	Thermal resistance (clo)	Type of clothing	Thermal resistance (clo)	Type of clothing	Thermal resistance (clo)
Underwear + Socks	0.06	Woman's dress	0.12	Short-sleeved shirt	0.19
Sleeveless undershirt	0.10	Light jacket	0.25	Long-sleeved shirt	0.25
Short-sleeved shirt	0.15	Long pants/ half-length skirt	0.15	Sandals/Slippers	0.02
Long-sleeved T-shirt	0.20	Shorts/Skirts	0.06	Sneakers/ Leather Shoes	0.05

Table 3 Metabolic rates corresponding to different activity types (Self-drawn)

Type of activity	Metabolic rate	Type of activity	Metabolic rate	Type of activity	Metabolic rate
Sit quietly	1.0	Relaxed standing	1.2	Idle	1.7
Walking (3.2km/h)	2.0	Walking (4.8km/h)	2.6	Walking (6.4km/h)	3.8

and the influence of subjective factors such as psychological and movement status, the PMV model is regarded as a suitable thermal comfort evaluation index.

In this paper, the PMV indicator is the most widely used indicator for evaluating thermal environment comfort. It was developed by Prof. Fanger by combining a large amount of experimental data with physical and physiological parameters of heat production and heat dissipation of the human body. The PMV indicator considers human thermal comfort as a result of the thermal balance between the human body and the surrounding environment. The assessment of thermal comfort is predicated on the consideration of various environmental parameters, including but not limited to temperature, humidity, wind speed, and other relevant metrics. These parameters are integrated with the human metabolic rate and the thermal resistance of clothing, thereby facilitating the prediction of thermal comfort in personnel operating within a uniform and stable environment [13]. The formula is as follows:

$$\begin{aligned} PMV = & \left(0.303e^{-0.036M} + 0.0275\right) \times \{M - W \\ & - 3.05 \times [5.733 - 0.007(M - W) - P_a] \\ & - 0.42(M - W - 58.2) - 0.0173M(5.867 - P_a) \\ & - 0.0014M(34 - t_a) - 3.96 \\ & \times 10^{-8} f_{cl}\left[(t_{cl} + 273)^4 - (t_r + 273)^4\right] - f_{cl}h_c(t_{cl} - t_a)\} \end{aligned} \tag{1}$$

M is the metabolic rate of the human body, measured in watts per square meter (W/m2); W is the mechanical work done by the human body to the outside environment, generally set to zero (W); P_a is the partial pressure of water vapor in the air, measured in Pascals (Pa); Ta is the air temperature of the surrounding environment, measured in degrees Celsius (°C); and f_{cl} is the clothing area factor, defined as the ratio of the external area of the dressed human body to the surface area of the naked body; hc is the convective heat transfer coefficient, W/(m2-K); tcl is the temperature of the external surface of the clothing, °C; tr is the average radiation temperature of the environment, °C.

3 Results and Discussion

3.1 *Analysis of the Measured Results of Each Measurement Point*

Measured analysis of more confined areas

The A test point for the transfer space in the corner of the channel, in the environment of a more closed position, Fig. 1 presents the day A test point temperature, humidity, wind speed, and heat index with the time of the law of change. The test point of the measured temperature change range was from 30.0 to 33.1 °C, with an average value of 32.3 °C. The day of the temperature change was relatively small and relatively stable. During the test period, the humidity at test point A ranged from 66.2% to 77.2%, with higher humidity, coupled with poor ventilation conditions and almost zero wind speed throughout the day, resulting in the region's heat index being significantly higher than the air temperature, with a variation range of 37.3–43.8 °C. The heat index is a composite of the air temperature and the humidity, and the air temperature is also higher than the air temperature. The heat index is a composite of air temperature and air relative humidity, which is used to determine the body temperature index. That is to say, the human body experiences the heat [14] and can respond accordingly. The measured results indicate that test site A experiences elevated temperatures and humidity throughout the day, compounded by inadequate ventilation.

Measured analysis of more open areas

The B test point for the transfer space in the station hall platform by the window, in the environment of a more open position, is illustrated in Fig. 2. This figure presents the B test point temperature, humidity, wind speed, coolness, and heat index with the time of the change rule of law in a day. The measured temperature at the B test point changes within a range of 31.2–37.5 °C, with an average value of 35.1 °C, indicating significant daily temperature variations. The location in proximity to the open window has a significant impact on the natural ventilation environment during the test period. The B test point exhibited a humidity range of 40.9%-69.8%, and the wind environment for the indirect breeze exhibited a wind speed range of 0–1.8 m/

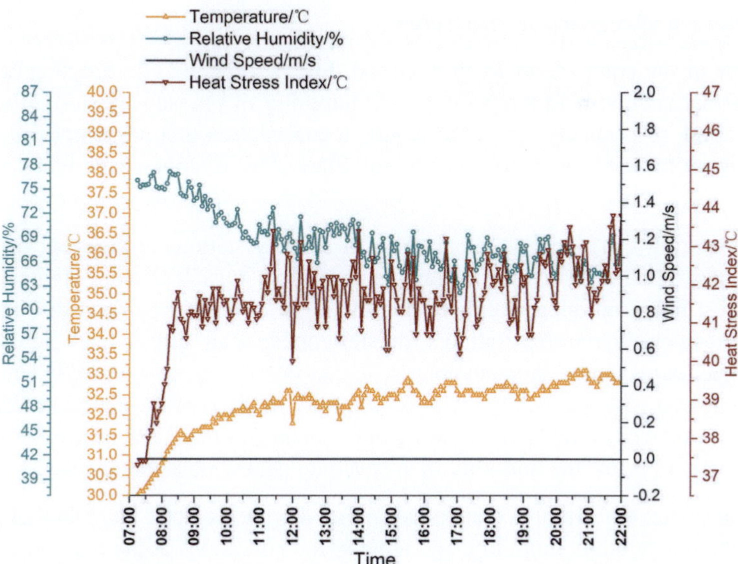

Fig. 1 Variation of thermal environment parameters throughout the day at test site A (Self-drawn)

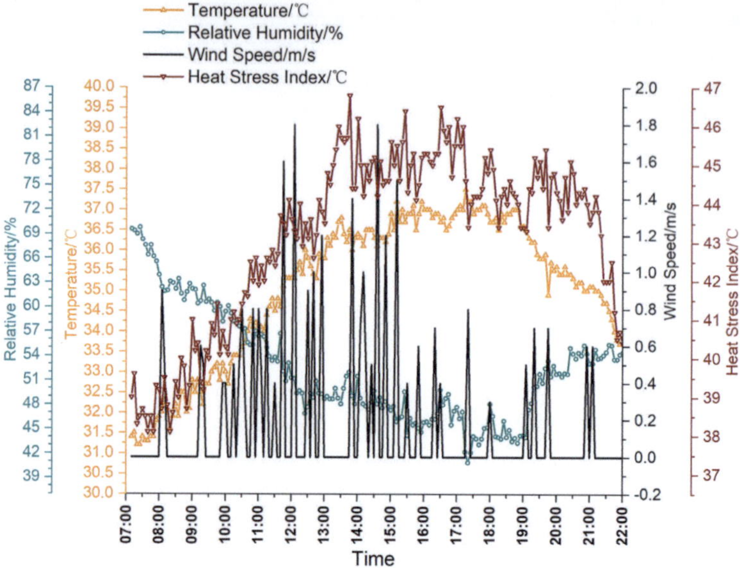

Fig. 2 Variation of thermal environment parameters throughout the day at test site B (Self-drawn)

s. The measured results indicate that the B test point exhibits a high temperature problem; however, the humidity and wind speed conditions at this location are notably more favorable than those at the A test point (Fig. 2).

Comparison of regional measurements

In terms of the comparison by time period, Fig. 3 show the box line diagrams of the thermal environment temperature and humidity of the subway station in each time period, respectively. From the charts, it can be seen that the temperature and humidity of the subway station at different times of the day show a certain pattern
:

1. Comparing the average temperatures in subway stations at different times of the day, it is found that the average temperatures in each area are obviously the lowest in the morning rush hour and the highest in the evening rush hour, and the temperatures in the station's transfer corridors are higher than the outdoor temperatures during the rush hour.
2. Compared with the average relative humidity in the subway station at different times of the day found that the air humidity in the morning peak time is obviously the largest, while the humidity in the evening peak time is the lowest.

In the context of sub-area comparisons, Fig. 4 presents a box line plot that juxtaposes the temperature, humidity, and wind speed conditions at the two test sites in the transfer space over the course of the day. The figure reveals the disparity in the physical environment between the two test points.

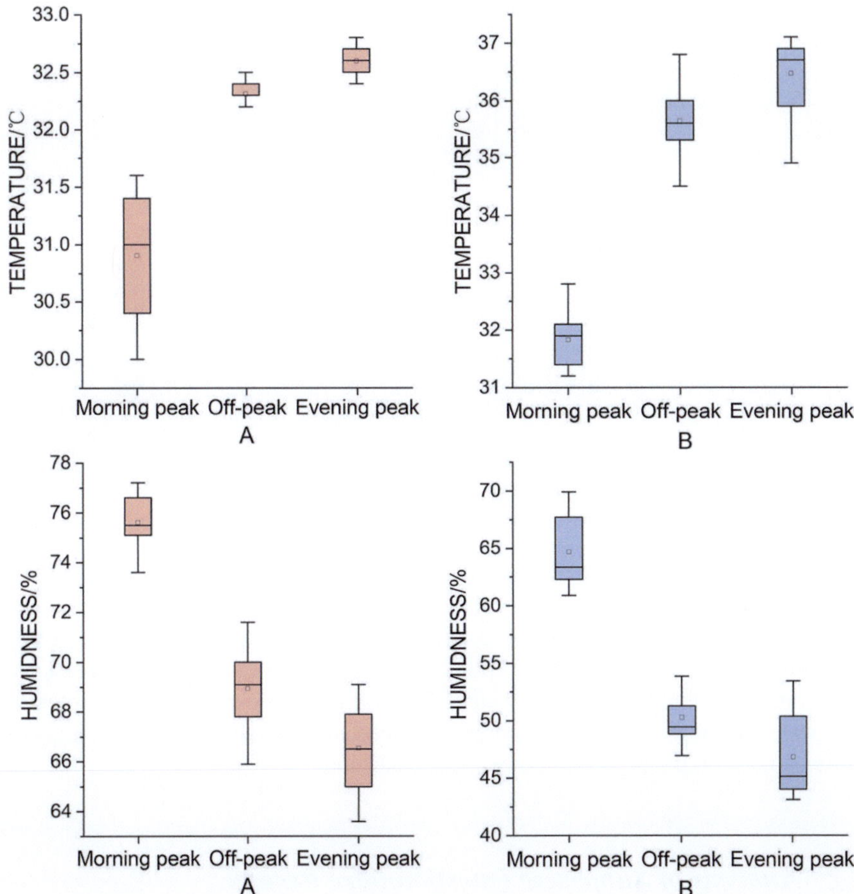

Fig. 3 Air temperature and relative humidity at different times of the day at each test (Self-drawn)

1. With respect to air temperature, test point A is 3–4 °C lower than test point B. The air temperature at test point A exhibits less variation throughout the day, while the temperature difference at test point B is more pronounced throughout the day.
2. In terms of relative humidity, the humidity at test point A is significantly higher than that at test point B, with a range of 65–70%. In contrast, the relative humidity at test point B exhibits a greater similarity to the outdoor relative humidity.
3. With respect to wind speed, test point A remained windless throughout the day, while test point B experienced intermittent, gentle winds.

A comprehensive comparative analysis of temperature, humidity, and wind speed at the two stations was conducted over different time periods and areas. The analysis yielded several key conclusions. Firstly, test point B demonstrated superior ventilation, with the area exhibiting greater openness proving to be more conducive to

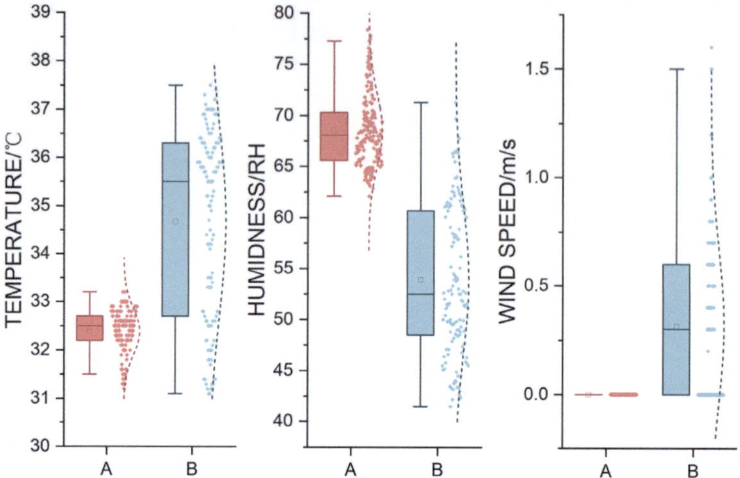

Fig. 4 Comparison of all-day physical environment parameters at each test point (Self-drawn)

ventilation compared to the other locations. Secondly, the overall thermal environment at test point A was found to be less favorable. While the temperature at this location was not as high as that of point B, the absence of wind and the presence of significantly higher humidity resulted in a stifling and pernicious waiting environment. This observation underscores the importance of considering the impact of closed spaces on the perception of waiting conditions.

3.2 Analysis of Subjective Questionnaire Results

The results of the questionnaire on the description of thermal sensation of the respondents at test point A are shown in Fig. 5 In this figure, 43.32% of the respondents thought that the thermal environment was very hot, 44.92% thought that the thermal sensation was hot, only 10.7% thought that it was slightly warm, and 1.07% thought that the thermal sensation was moderate. The findings indicate that the air temperature can be effectively reduced to enhance subjective comfort.

The results of the questionnaire administered at test point B indicated that 15.68% of the respondents found the heat to be very hot, 46.49% found it to be hot, 28.11% found it to be slightly warm, and 9.19% found it to be moderately comfortable.

Concurrently, 84.4% of the respondents indicated that traversing this interchange area would result in short-term physical discomfort, characterized by sweating, shortness of breath, and increased heart rate. Furthermore, 41.7% of respondents reported an increase in psychological pressure, characterized by feelings of unease and anxiety, with subsequent effects on mood and emotions.

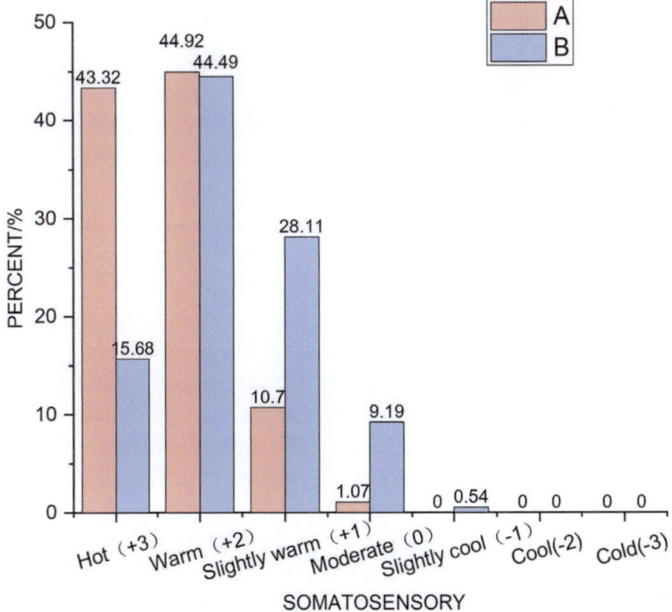

Fig. 5 Thermal sensation vote by questionnaire at each test site (Self-drawn)

According to the seven-level thermal sensation voting scale established by ASHRAE, the mean thermal sensation vote (MTS) for Test Site A was 2.30, indicating a temperature range between "hot" and "very hot." In contrast, the mean thermal sensation vote for Test Site B was 1.68, classified as "slightly warm" to "hot." The comparison of the mean thermal sensation vote values for each test point further elucidates that the thermal environment and thermal comfort are the poorest in a more confined space, such as Test Point A.

Furthermore, the majority of respondents to the questionnaire indicated that they perceived the temperature at Test Point A to be higher than that at Test Point B. However, the measured air temperature at Test Point A (30.0–33.1°C) was significantly lower than the air temperature at Test Point B (31.2–37.5°C). This discrepancy can be attributed to the elevated humidity and deficient ventilation experienced at Test Point A throughout the observation period. Consequently, the body temperature of individuals exposed to this environment exceeded the actual air temperature, thereby leading to a discrepancy between perceived sensations and the measured data. Consequently, in a hot environment, an increase in humidity will have a greater impact on people's perception of heat than an increase in temperature.

3.3 Comparison of Predicted Mean Thermal Sensation PMV and Measured Mean Thermal Sensation TSV

The Bin temperature frequency method, a technique with international application, was employed for the analysis of thermal environments. The operating temperature, designated as t0, was divided into intervals with an accuracy of 0.5 °C. The mean thermal sensation vote (TSV) within each temperature interval was designated as the dependent variable, and the center value of each interval was selected as the independent variable to construct a linear regression model [15–17]. This model yielded Fig. 6. A high R2 correlation coefficient indicates that the use of this mean thermal sensation vote regression equation can better predict human thermal sensation. Thermal neutral temperature is defined as the temperature at which the human body experiences neither cold nor hot sensations, i.e., the temperature at which thermal sensation is most moderate. When TSV = 0, the measured thermoneutral temperature at test point A is 23.82°C, and the measured neutral thermal temperature at test point B is 22.63°C, as shown in Table 4.

The physical environment factors, including air temperature, relative humidity, wind speed, metabolic rate, and garment thermal resistance, obtained from the two test points, were entered into the program and calculated to obtain the predicted mean thermal sensory value (PMV). The Bin temperature frequency method was also used

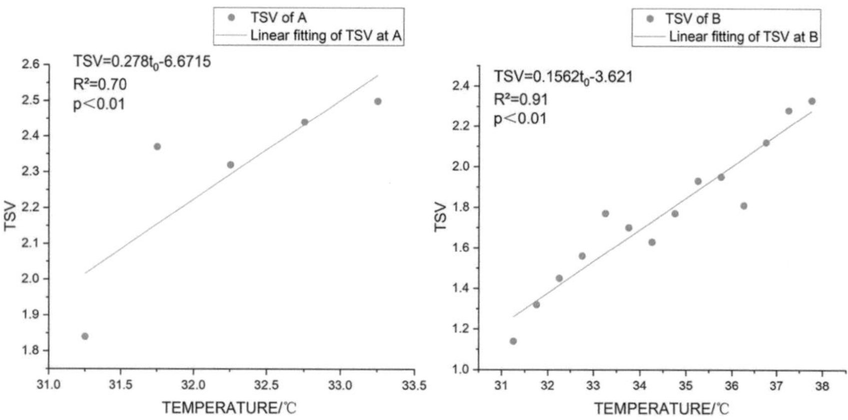

Fig. 6 Fitting curve of human subjective thermal sensation TSV versus operating temperature t_0 at each test point (Self-drawn)

Table 4 TSV regression equations and thermo-neutral temperatures for each test site (Self-drawn)

Test point	Thermal sensory regression equation	Correlation coefficient R ^{2}	Thermoneutral temperature
Test point A	TSV = $0.28t_0$-6.67	0.70	23.82°C
Test point B	TSV = $0.16t_0$-3.62	0.91	22.63°C

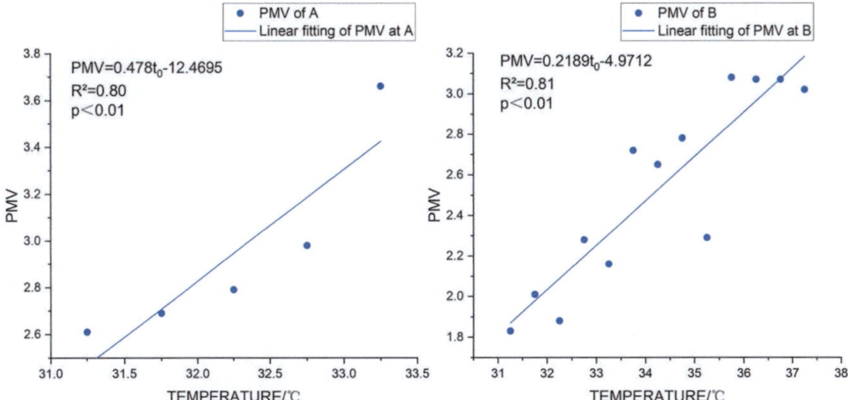

Fig. 7 Fitting curves of human objective thermal sensation PMV versus operating temperature t0 at each test point (Self-drawn)

to fit a linear regression to the operating temperature t0 to obtain Fig. 7. The predicted thermoneutral temperatures were 25.95 °C for test point A and 22.73 °C for test point B, when PMV = 0.

A comparison of the predicted PMV results and the measured TSV results of the same test point, as illustrated in Fig. 8, reveals a consistent increasing trend in both the PMV and TSV with rising operating temperature. This observation indicates that the PMV model can, to a certain extent, mirror the degree of people's thermal sensitive degree. The regression curves' slopes offer a quantitative reflection of this sensitivity to changes in operating temperature. With respect to thermoneutral temperature, a discrepancy emerges between the predicted thermoneutral temperature (25.95 °C) and the measured thermoneutral temperature (23.82 °C) at test point A. Conversely, the predicted thermoneutral temperature (22.73 °C) and the measured thermoneutral temperature (22.63 °C) at test point B demonstrate congruence.

The prediction accuracy of the PMV model for the TSV model is examined by means of the coefficient of determination. When the operating temperature is set between 30 and 40 °C, it can be concluded that the prediction accuracy of test point A is extremely low and basically unable to make a prediction, and the prediction accuracy of test point B is only 33.4%.

It has been demonstrated that in extreme thermal environments during the summer months, a discrepancy emerges between the predicted outcomes of the PMV model and the observed results. The discrepancy is particularly pronounced in environments such as Test Site A, where the thermal environment is characterized by its stability and consistently high temperatures and humidity levels. This discrepancy leads to an increase in the error of the PMV model, particularly in these specific environments.

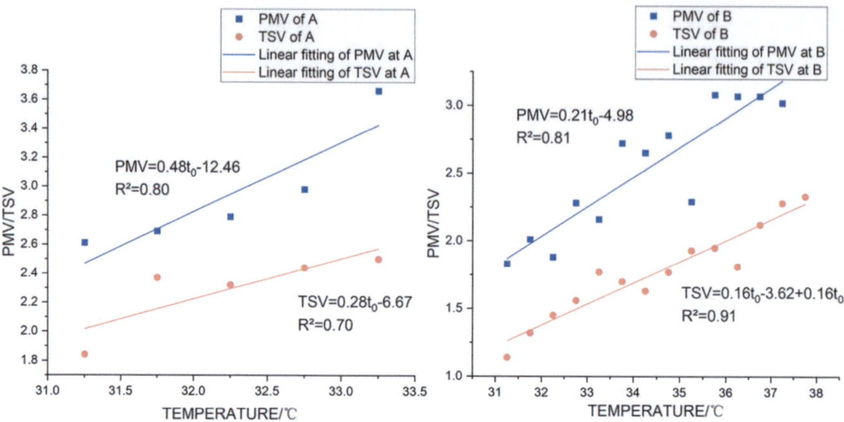

Fig. 8 Comparison of TSV and PMV at each test point (Self-drawn)

3.4 Corrections to PMV

Modified model of ePMV

As demonstrated in the preceding study, the PMV model demonstrates a substantial discrepancy in its capacity to predict human thermal sensation under natural ventilation conditions in regions characterized by high temperatures and humidity, such as Chongqing during the summer months. Professor Fanger has proposed a methodology to rectify the limitations of the PMV model. This methodology involves the incorporation of an expectation factor, e, ranging from 0.5 to 1, within non-air-conditioned buildings in warmer climates [18]. It is further posited by the same author that the value of e should be set at 0.7, a conclusion that is supported by the actual circumstances in China.

$$e\text{PMV} = \text{PMV} \times e \tag{2}$$

Given the high-temperature environment of the research site, which lacks air-conditioning, the ePMV model proposed by Professor Fanger was employed to rectify the initial PMV. The corrected ePMV regression curves were then juxtaposed with the measured thermal sensory vote (TSV) to ascertain the efficacy of the correction.

As illustrated in Fig. 9, following the application of the ePMV correction, the regression curves of ePMV and TSV at test point B demonstrate a high degree of similarity, indicating an enhanced correction effect. The coefficient of determination indicates an enhancement in prediction accuracy from 33.4% to 99.5%. Conversely, the regression curves of ePMV and TSV at test point A exhibit substantial residual errors following the correction, underscoring the necessity for alternative modified models in high-temperature and high-humidity environments.

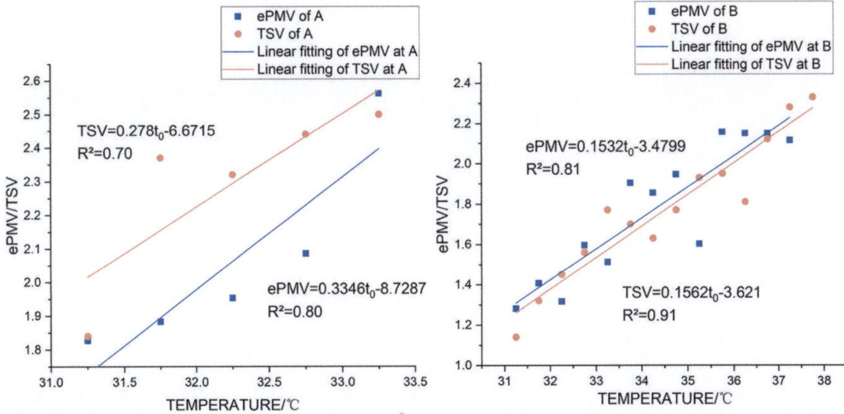

Fig. 9 Comparison of ePMV correction effect at each test point (Self-drawn)

Fig. 10 PMV' correction effect (Self-drawn)

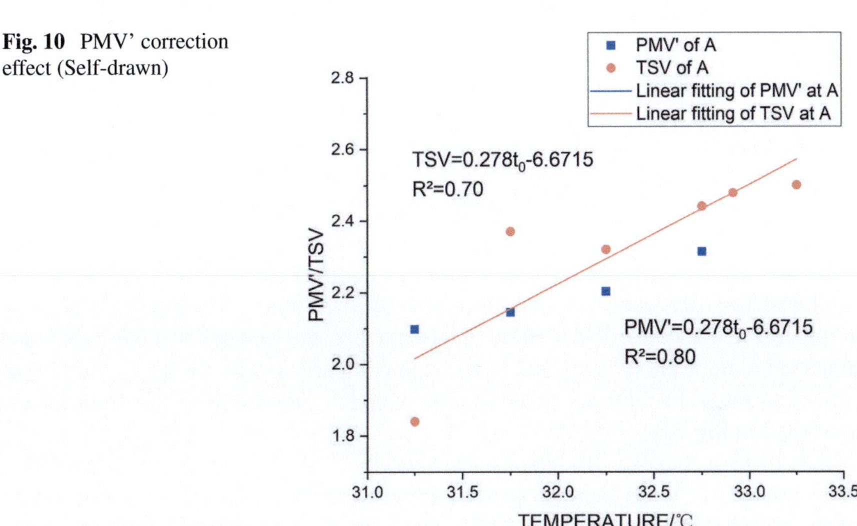

In this study, we verified that Professor Fanger's ePMV modified model has a superior correction effect on PMV in high-temperature environments with natural ventilation. However, the correction effect of ePMV in high-temperature and high-humidity environments with no wind is not optimal.

Presentation of the revised model

Exhibit a clear linear correlation with the operating temperature, thereby suggesting a corresponding linear correlation between PMV and TSV. In this paper, for the environment of high temperature and high humidity, a constant β is introduced on the basis of Prof. Fanger's ePMV modified model, and the following modified model is obtained:

$$\mathrm{PMV}' = \alpha\mathrm{PMV} + \beta = \mathrm{TSV} \tag{3}$$

The linear regression equation of PMV and operating temperature is as follows:

$$\mathrm{PMV} = at_0 + b \tag{4}$$

The linear regression equation for TSV versus operating temperature is:

$$\mathrm{TSV} = ct_0 + d \tag{5}$$

Substituting Eqs. (4) and (5) into Eq. (3) and simplifying gives:

$$\alpha = \frac{c}{a} \tag{6}$$

$$\beta = \frac{ad - bc}{a} \tag{7}$$

Based on the regression equations of TSV and PMV of test point A calculated in the previous section, the modified model of PMV in a high-temperature and high-humidity environment is obtained by Eqs. (6) and (7):

$$\mathrm{PMV}' = 0.5816\mathrm{PMV} + 0.5806 \tag{8}$$

In accordance with the aforementioned modified model, each PMV value employed for regression calculation at test point A was substituted into Eq. (8) to obtain the respective PMV value. Consequently, the updated sample data were subjected to regression fitting and analysis in comparison with the TSV, which was utilized to assess the efficacy of the modified model. The results of this analysis are illustrated in Fig. 10.

It is evident that the discrepancy between the PMV' value derived post-PMV correction and TSV has been mitigated. Furthermore, the regression curves of PMV' versus operating temperature and TSV versus operating temperature exhibit a high degree of congruence. The findings indicate that the corrected PMV' model demonstrates enhanced capability in predicting human thermal sensation under conditions of elevated temperature and humidity.

4 Conclusion and Outlook

This study employed a field experimental method that combined thermal environment testing and subjective questionnaire survey to investigate the thermal environment in the transfer space of Sigongli Station on Chongqing Railway Transit Line 3. The measured thermal sensation TSV values of the interviewed persons, the calculation of the predicted thermal sensation PMV values, and other relevant metrics were obtained through this method. The following conclusions were obtained after conducting a thorough analysis:

(1) Due to the enclosed nature of transfer spaces in rail transit, thermal comfort issues are prone to arise. Under high-temperature and high-humidity conditions, the accuracy of traditional PMV models significantly declines, while the ePMV model shows a better corrective effect in natural ventilation scenarios. However, it still exhibits limitations under windless, high-humidity conditions.
(2) PMV modified model suitable for high-temperature and high-humidity environments is proposed based on linear regression (PMV' $= 0.5816$PMV $+ 0.5806$), filling the research gap for PMV modified models in such environments and providing a reference for the establishment of PMV correction methods under other environmental conditions in the future.
(3) The proposed window optimization and dynamic ventilation strategies can effectively improve the thermal environment of the space, providing scientific evidence for the design and improvement of rail transit systems.

This study, through the analysis of the the transfer space of Chongqing Railway Transportation Line 3 Sigongli Station, exposes the key weaknesses in the current research and practice of thermal environments in transfer spaces.

Firstly, thermal comfort assessment tools are inadequate for the specific conditions of transfer spaces. Directly applying research methods based on platforms environment standards results [10] in redundant energy consumption due to mechanical ventilation in transfer space [11]. The traditional PMV model [6] introduces deviations in closed interchange corridors, failing to address the synergistic effects caused by factors such as high-density passenger flow [2] and narrow transfer spaces.

Secondly, energy consumption control strategies require improvement. The issue of spatiotemporal mismatch in environmental control strategies is prominent. Existing design standards have not sufficiently considered the transitional characteristics of transfer spaces, and the direct transplantation of platforms standards [10] to transfer spaces leads to delayed cooling effects in narrow corridors and redundant mechanical ventilation energy consumption.

Insights for urban planning and engineering design can be drawn from these findings: Firstly, the thermal environment research of transfer spaces can be made more accurate by effectively integrating SET comfort thresholds [7] with dynamic models [6]. Secondly, appropriate cooling and ventilation methods should be selected based on actual dynamic conditions. For instance, during off-peak nighttime hours, the mechanical exhaust system can be activated to reduce the thermal storage in building envelopes and mitigate the impact of thermal inertia on the following day. During peak periods with high passenger flow in the summer, pre-conditioning the air supply system can enhance thermal comfort during these peak times.

Whilst the present study provides valuable insights, it is important to note that it is not without its limitations. Firstly, the sample data for the test sites were obtained from a specific site area, and some of the conclusions may not be fully applicable to other rail transit sites. Secondly, the data collection period coincided with specific seasonal times, which may have resulted in a partial reflection of the thermal environment changes in different seasonal climates.

Insights for urban planning and engineering design can be drawn from these findings. Subsequent research will focus on addressing the following: (1) Conducting field measurements and subjective surveys for various types of stations and different seasons to validate the generalizability of these findings; (2) Simulating unstable environments under dynamic passenger flows; (3) Investigating passive energy-saving strategies for semi-open station types.

These explorations will promote the transition of thermal environment regulation in rail transit stations from reliance on energy consumption to passive energy-saving approaches, which align more closely with future trends. Additionally, they will facilitate the shift from a one-size-fits-all static platforms environment standard to dynamic ventilation control strategies more suitable for the thermal environment of transfer spaces, ultimately achieving a dual enhancement of passenger comfort and urban energy consumption optimization.

References

1. Li D, Zhang D (2018) Study on the sustainable comprehensive development along rail transit line. In: International symposium on mechanics, structures and materials science (MSMS). Tianjin, Peoples R China
2. Mao P et al (2019) Characterization of urban subway microenvironment exposure A case of Nanjing in China. Int J Environ Res Public Health 16(4)
3. Ashrae A (2017) ASHRAE Standard 55: thermal environmental conditions for human occupancy. American Society of Heating, Refrigerating and Air-Conditioning Engineers, Atlanta, GA, USA
4. Miao Y, Ding Y (2020) Indoor environmental quality in existing public buildings in China: measurement results and retrofitting priorities. Build Environ 185
5. Psomas T et al (2021) Indoor humidity of dwellings and association with building characteristics, behaviors and health in a northern climate. Build Environ 198
6. Ma Z (2023) Field study on thermal and humidity environment and thermal comfort evaluation in xian metro: a case study of metro line 2
7. Han J, Kwon SB, Chun C (2016) Indoor environment and passengers' comfort in subway stations in Seoul. Build Environ 104:221–231
8. Quan X, Yuan J (2022) Influencing factors of thermal environment at Xiamen subway station entrance. Build Energy Effic 50(11):47–52
9. Katavoutas G, Assimakopoulos MN, Asimakopoulos DN (2016) On the determination of the thermal comfort conditions of a metropolitan city underground railway. Sci Total Environ 566:877–887
10. Xu M (2020) Study on thermal environment and thermal comfort of Suzhou rail transit station—take Shantangjie station, Guangji South Road station and Mudu station for example
11. Abdelaziz OM, Cui X, Sun X (2024) Enhancing underground thermal environments in Cairo: the role of subway entrance geometry in optimizing natural ventilation. Buildings 14(9)
12. Zhu Y (2010) Built Environ
13. Fanger PO (1970) Thermal comfort. In: Analysis and applications in environmental engineering
14. Lin J (2021) Response of microclimate to plant configuration in parks along the yellow river in Ningxia

15. Cena K, de Dear RJ (1999) Field study of occupant comfort and office thermal environments in a hot, arid climate. ASHRAE Trans 105:204
16. de Dear R, Fountain M (1994) Field experiments on occupant comfort and office thermal environments in a hot-humid climate
17. Donnini G, Molina J, Martello C (1997) Field study of occupant comfort and office thermal environments in a cold climate
18. Fanger PO, Toftum J (2002) Extension of the PMV model to non-air-conditioned buildings in warm climates. Energy Build 34(6):533–536

Assessment of Outdoor Thermal Comfort of Tieniu Square in Mianyang, China

Xinyi Duan

Abstract With the development of the city and the increasing time of residents' outdoor activities, a good outdoor space environment and thermal comfort have an important impact on residents' physical and mental health. Taking Tieniu Square in Mianyang City as an example, we conducted the actual measurement and questionnaire survey of the outdoor thermal environment of the square under the hot weather in summer, and analysed the reasons for the differences in the thermal environment of each measurement point of the venue by combining the TSV (Thermal Sensation Voting) and TCV (Thermal Comfort Voting) of the respondent population, and the study showed that: (1) the correlation of the coolness index with the air temperature and the relative humidity is stronger, (2) for the wind speed, and the combination of the actual data and the questionnaire statistical result Greening and water bodies have a certain effect on thermal comfort enhancement, and (3) water bodies can improve the wind environment of the site. Through a series of current research in order to provide reference for the improvement of microclimate in the outdoor environment of Tieniu Square.

Keywords City square · Outdoor thermal comfort · Microclimate · Heat index

1 Introduction

Relevant studies have shown that the interactions between urban space and various elements of the thermal environment are multidirectional and integrated [1]. Some scholars have explored the impacts of the layout and form of greening in urban areas on the wind speed and temperature in urban areas, and have proposed adaptive strategies for cooling urban areas through greening [2].

Microclimate has a direct effect on human thermal comfort and activity. Multiple regression has been used to analyse the impact of microclimatic factors on human

X. Duan (✉)
School of Civil Engineering and Architecture, Southwest University of Science and Technology, Mianyang, China
e-mail: Sssslytherin@163.com

B. He et al. (eds.), *Climate-Resilient, Low-Carbon, and Sustainable Cities*, Lecture Notes in Civil Engineering 754, https://doi.org/10.1007/978-981-95-2169-2_20

behaviour [3]. Some scholars have conducted statistical data on microclimate elements and human behaviour to summarize the elements of microclimate influence on human behavioural activities [4].

As the main outdoor activity place for urban residents and the core node of urban external space, the city square is one of the important space types for outdoor thermal comfort research [5]. City squares in urban centers are more susceptible to weather extremes than suburbs, exacerbating microenvironmental changes. Plaza microclimates significantly impact user activities, with studies confirming temperature and solar radiation as key comfort determinants through their climatic influence. [6]Similarly, there is a causal relationship between the outdoor thermal environment and crowd activity, with improved microclimate in the outdoor environment increasing crowd attendance in summer [7].

This study analyses the current four meteorological factors: air temperature (Ta), relative humidity (RH), wind speed (W), and heat index (Heat Index) of Tieniu Square by means of literature research, field measurements, and data analysis, in which the heat index is calculated by the regression equations of air temperature and relative humidity, and the results of the analysis illustrate that Tieniu Square, Mianyang City, under the hot and high temperature weather in the summer. The study evaluates Tieniu Square's summer microclimate in Mianyang, analyzing correlations among measurement points to guide future thermal environment optimization.

2 Methods and Materials

2.1 Research Objectives

Tie Niu Square is located in Mianyang City, Binhe Road, east of Dongfanghong Bridge, the square covers an area of nearly 2 hectares, with a green space area of 1.19 hectares, is the city's residents live intensively, adjacent to the Fulingjiang River, the terrain is open, the traffic is convenient, on all sides of the road to reach the city's main traffic arteries (see Fig. 1). The square has a relatively strong sense of enclosure, featuring three sides that are enclosed and one open side. On the west side of the square, high-rise buildings provide effective shading from the afternoon sun, while the east side consists of a staggered arrangement of low-rise buildings that leave some space for ventilation in the square. Tieniu Square is located in the city center area, according to previous literature and meteorological data, the temperature in the city center area of Mianyang City is significantly higher than that in the city periphery, combined with the analysis of the square's own morphology and attractiveness to the crowd, it is concluded that Tie Niu Square is typical (see Fig. 1).

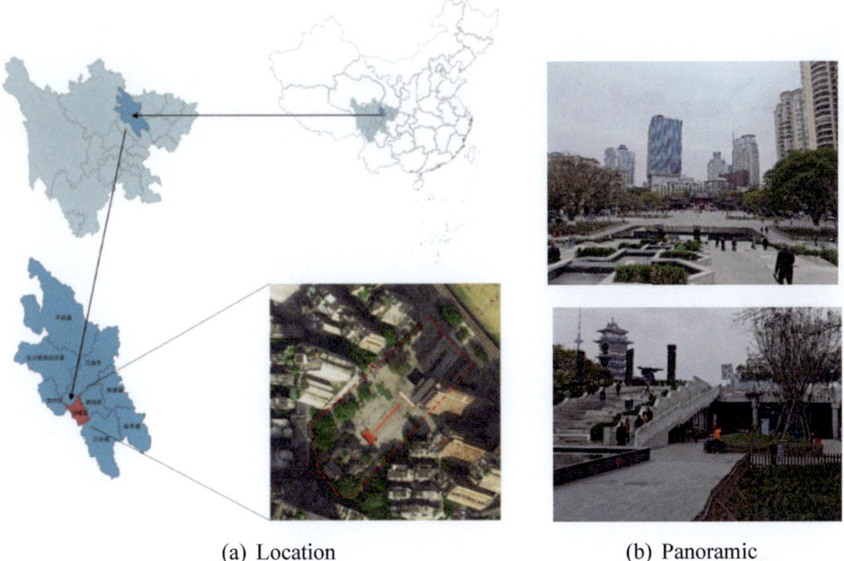

(a) Location (b) Panoramic

Fig. 1 Location and panoramic view of the square (Self-drawn)

2.2 Research Methodology

Field measurements. In this study, the time when the antecedent weather was consecutively sunny, 26 August 2024, was chosen for the experimental data measurements, and the thermal environment parameters were examined by the instruments (see Table 2) and the parameters included the air temperature (Ta), the black globe temperature (Tg), the relative humidity (RH), and the wind speed (va). And in the day from 7:00 to 21:00 every 1 min on the measurement point for measurement, measurement height of 1.5 m, the instrument automatically measured, the choice of measurement point through field research and Tieu Square landscape elements of the composition of the later divided into six measurement points, respectively, are the measurement of point 1: the square hard pavement, around the unobstructed, open space, the measurement of point 2: the crowd resting under the tree pool seating, with the shade of the trees, the measurement of point 3 Measurement point 3: the square shrubs, measurement point 4: the square fountain, near the water, surrounded by no building shade, open space, measurement point 5: the square under the tall trees, tree shade, measurement point 6 for the control group: Tieu Plaza Riverside Greenway at the open grass, surrounded by no building shade, and the square is divided into six measurement points. Location and view of measurement points (see Fig. 2) (Tables 1 and 2).

Data analysis methods. First, the air temperature, relative humidity, wind speed, and heat index at each measurement point were analyzed, of which the heat index from the U.S. National Weather Service converted the heat-sensing model into a multiple statistical regression model and renamed the heat index for issuing high

Fig. 2 Location and site photos of meteorological parameters measured at Tieniu square (Self-drawn)

Table 1 Meteorological conditions on measurement days (Self-drawn)

Test day	Maximum temperature/°C	Minimum temperature/°C	Weather conditions	Wind-chill index/°C	Sunrise/Sunset	24 h precipitation	Wind direction/velocity
August 26, 2024	38	26	Sunny day	40.5	6:33/19:33	0 mm	Northeast wind

Table 2 Instruments used for the measurement of meteorological parameters (Self-drawn)

Parameters	Instrumentation	Accuracy	Realm	Frequency	Recording method
Air temperature (T_a)	Heat-stress WBGT meter TM-188D	± 0.6	0–50 °C	1 min	Automatic
Relative humidity (RH)	Heat-stress WBGT meter TM-188D	± 3 to 5%	1–99%	1 min	Automatic
Air velocity (V_a)	WWFWZY-1 wind speed and temperature recorder	± 0.05	0.05–30 m/s	1 min	Automatic
Black bulge temperature (TG)	Heat-stress WBGT meter TM-188D	± 0.6	0–80 °C	1 min	Automatic

temperature warning information [8] The equation is:

$$H_I = c_1 + c_2T + c_3R_H + c_4TR_H + c_5T^2 + c_6R_H^2 + c_7T^2R_H + c_8TR_H^2 + c_9T^2R_H^2$$

In this equation, HI is the heat index; T is the ambient temperature; RH is the relative humidity; ci (i = 1, 2, ..., 9) is the coefficient [9].

Afterward, the average value of the measurement point in each period of time and the average value of the corresponding time period of all measurement points were to make the difference between the temperature difference, wind speed difference, relative humidity difference, and coolness index (temperature difference, wind difference, humidity difference, and heat difference), the results of the comparison, the smaller the difference in the temperature obtained indicates that the cooling effect of the measurement site is more effective, the larger the difference in the humidity obtained indicates that the humidification effect is more effective, and the larger the difference in the wind obtained indicates that the wind environment of the measurement site is more effective. The larger the heat index, the better the improvement of the thermal environment [10]. The formula for calculating each difference is as follows:

$$D_T(°\mathrm{C}) = T - T_a \tag{1}$$

$$D_H(\%) = H - H_a \tag{2}$$

$$D_W(\mathrm{m/s}) = W - W_a \tag{3}$$

$$D_I(°\mathrm{C}) = I - I_a \tag{4}$$

As shown in the equation, DW (m/s) is the difference in wind speed, W is the mean value of wind speed at a measurement point, and Wa is the mean value of wind speed at all measurement points: DH (%) is the difference in humidity, H is the mean value of relative humidity at a measurement point, and Ha is the mean value of relative humidity at all measurement points; DT (° C) is the difference in temperature, T is the mean value of air temperature at a measurement point, and DI (%) is the coolness heat index. The correlation between meteorological parameters and the thermal environment was analyzed by categorizing the environmental factors at each measurement point.

Questionnaires. The distribution of the questionnaire was conducted by randomly selecting respondents near the measurement points (see Fig. 3), A total of 100 questionnaires were distributed, consisting of sections on respondents' personal information and subjective thermal environment perceptions. The personal information included gender, age, duration of stay, primary purpose of visiting Tianniu Square,

Fig. 3 Randomly selected respondents for questionnaire survey (Self-drawn)

frequency of visits, and main activities. The subjective thermal environment perceptions were measured using the ASHRAE 7-point scale [11]. The questionnaire was divided into Thermal Sensation Voting (TSV) and Thermal Preference Voting (TPV), incorporating assessments of wind sensation, humidity sensation, and comfort voting [12].

3 Analysis of Experimental Results

3.1 Information on Respondents

A total of 100 questionnaires were distributed in this paper. Tourists from abroad may feel uncomfortable with the climate of Mianyang City, which may affect the judgment of local residents on the thermal environment needs of Tiegu Plaza, so the questionnaire design of this paper especially added the time of residence in Mianyang City as an item of investigation, of which there are 75 respondents who have been living in Mianyang City for more than one month, and there are another 25 respondents who come to Mianyang City to live for no more than One month.

The age and gender percentage of the respondents are shown (see Fig. 4). The gender ratio of men and women in this research is 50.7% and 49.3% respectively, which is approximately 1:1, and from the Mianyang Municipality Statistical Yearbook 2023, there is are 72,206 males and 68,969 females in Mianyang Municipality,

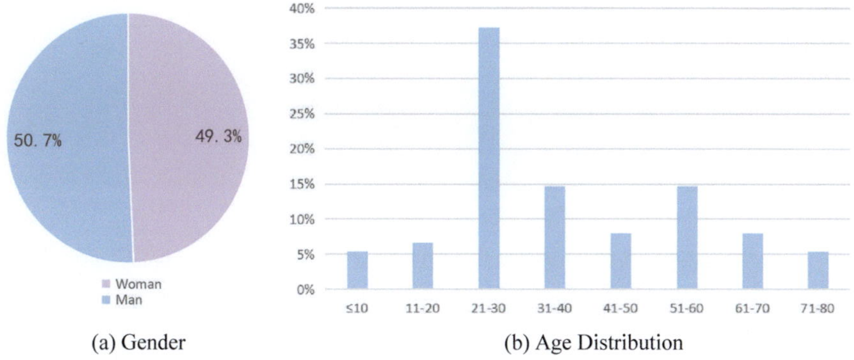

(a) Gender (b) Age Distribution

Fig. 4 Proportion of respondents by gender and age (Self-drawn)

with the ratio of 51.1% and 48.9%, respectively. This shows that the sample size of the questionnaire issued in this paper is basically in line with the gender ratio of the resident population in Mianyang City, and can objectively reflect the opinions of Mianyang City residents on the thermal environment of Tieniu Square.

The survey included a wide age distribution, ranging from 7-year-old children to 71-year-old elderly participants, with adults constituting 56% (ages 21–60) and the elderly making up 10% (over 60 years) (see Fig. 4). The majority of respondents were aged 21–30, 31–40, and 51–60. Among the respondents, 46.7% had an education level below high school, while 53.4% held a college degree or higher. Due to the presence of certain knowledge barriers in the questionnaire content, staff clarified to respondents that the focus was on their immediate feelings during the survey rather than general opinions. The frequency of visits indicated that most respondents engaged in various activities of differing intensities in the square, with 30.7% of respondents passing through only on that day, ensuring the accuracy of the collected information.

3.2 *Analysis of Thermal Environment Results*

Thermal sensory analysis. In the Thermal Sensation Voting (TSV), the most votes were for slightly warm (+1) and warm (+2), accounting for 37.3% and 41.3%, respectively, while the least votes were for slightly cool (−1), which received 0%. Overall, nearly 78.6% of the responses fell into the warmer category (TSV $\geq$ 1), indicating that the outdoor thermal environment at Tianniu Square in Mianyang was predominantly warm during the measurement period. Statistical analysis of the thermal sensation votes at five measurement points, excluding the control group, was conducted using SPSS software, and the results were illustrated in Excel (see Fig. 5). Measurement Point 5 had the most comfortable thermal sensation, with the lowest proportion of warm responses and the highest proportion of neutral responses; conversely,

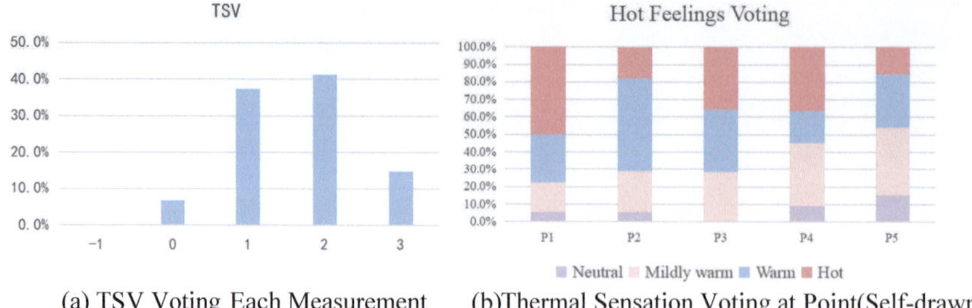

(a) TSV Voting Each Measurement (b)Thermal Sensation Voting at Point(Self-drawn)

Fig. 5 Overall thermal sensation voting in tianniu square

Measurement Point 1 was the least comfortable, exhibiting a notable proportion of neutral responses, indicating significant variations in thermal sensations throughout the day. A reasonable inference regarding the characteristics of each measurement point suggests that Measurement Point 5, located under a 15-m canopy of a parasol tree, benefits from extensive shade, which positively impacts thermal sensations both psychologically and physiologically.

Thermal acceptability analysis. In the Thermal Comfort Voting (TCV), the highest proportion of responses indicated "very uncomfortable" (−2) at 45.3%, followed by "slightly uncomfortable" (−1) at 30.7%. During the hot summer, only 1.3% of Mianyang citizens reported a thermal comfort level within the comfort zone (TCV > 0), while a significant 76% reported being in the discomfort zone (TCV < 0) (see Fig. 6). Heat preference voting situation, for humidity preference 36% of the respondents think that they prefer a wet environment, 51% of the respondents think that the current humidity is appropriate, 13% of the respondents prefer a dry environment. In terms of wind speed preference, 53% of respondents felt that the current wind environment was appropriate and did not need to change, which is consistent with the measured plaza wind speed data. In the large increase series, 78% of the respondents think that the temperature should be lower (−1), with a significant voting percentage, which means that the majority of the respondents prefer a summer with lower temperatures. For the solar radiation preference, 64% of the respondents wanted weaker solar radiation and about 36% kept the same.

3.3 Measurement Data Analysis

On August 26, 2024, average measurements of four meteorological parameters—air temperature, wind speed, humidity, and solar radiation—were taken at six measurement points in Tianniu Square, Mianyang (see Fig. 7). Throughout the measurement period, the air temperature exhibited an increasing trend, with a gradual decrease starting after 15:00; however, the overall air temperature at the end of the measurement period at 22:00 was higher than that recorded at 07:00. The variation in wind speed was notably different among the measurement points, with Measurement Point

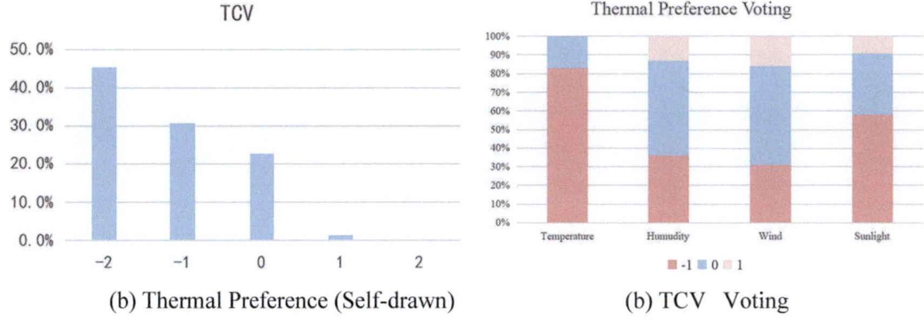

(b) Thermal Preference (Self-drawn) (b) TCV Voting

Fig. 6 Thermal sensation voting situation

4 consistently recording the highest wind speeds throughout the day. The trend in relative humidity was inversely related to air temperature, as relative humidity decreased with rising temperatures. Solar radiation, influenced by overall climatic conditions, displayed a pattern of initially increasing and then decreasing over time, with the period from 12:00 to 15:00 being the peak for solar radiation intensity.

On August 26, from 7:00 to 22:00, the air temperature varied from 29.1°C to 41.6°C, with the highest and lowest temperatures occurring at the control group's measurement point 6 (at the broad lawn); the wind speed varied from 0.09 m/s to 2.72 m/s throughout the day, with the highest wind speed occurring at the measurement point 4 (at the square fountain).(see Fig. 7)The wind environment at measurement point 4 was better than the other measurement points, with the minimum wind speed occurring at measurement point 6 (at the wide lawn); humidity varied from 33.96% to 78.3% throughout the day, with the maximum humidity occurring at measurement point 6 (at the wide lawn), and the minimum humidity occurring at measurement point 4 (at the fountain). Measurement point 2 is located under the seat of the tree pool in the plaza, and the air temperature varied less throughout the day. At the end of the measurement, i.e., at 22:00, the air temperature at this measurement point was 33.5°C, which was the highest temperature among all the measurement points, suggesting that the tree bushes would instead have a weaker heat dissipation effect due to the closure of the canopy in high temperature and heat wave weather. The range of the heat index during the measurement period was 24.2–52°C.

The wind speed at all measurement points showed a general trend of increasing, then decreasing, and then increasing, the wind speed was the highest at about 15:30 p.m. and the lowest at about 7:00 a.m. The weather on August 26 was sunny, and the wind began to blow in the plaza in the afternoon, so the wind speed increased in the afternoon hours. Measurement point 1, measurement point 4, and measurement point 6, the change rule is more similar, measurement point 1, measurement point 4, located in the center of the square space is open, with fewer trees covered, the overall wind speed is higher. Measurement point 2, measurement point 5 around the plant shade, the overall wind speed is low.

Based on the blue and green space landscape elements of Tieu Square site and crowd activities are divided into six categories of measurement points, descriptive

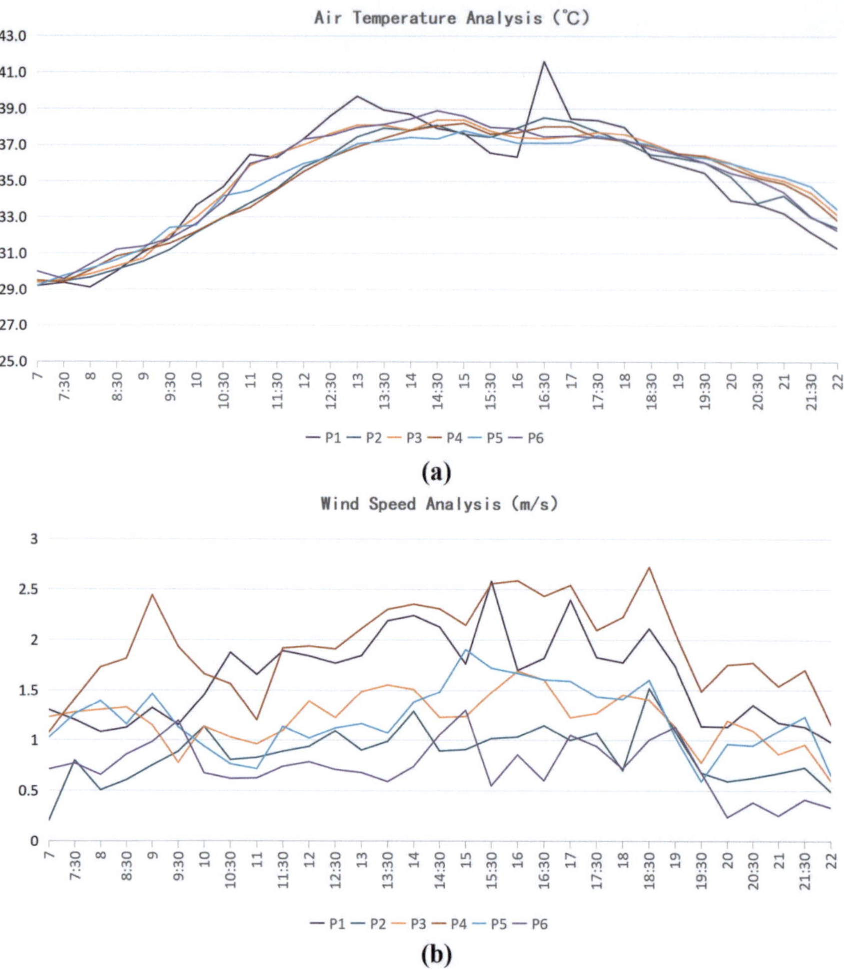

Fig. 7 Variation of meteorological factors at measurement Points from 07:00 to 22:00 at 30-min Intervals (Self-drawn)

statistics, and difference analysis for the difference of meteorological element indexes of each measurement point, first of all, by analyzing the wind environment of Tieu Square can be seen (see Fig. 8 and Table 4), Measurement point 1 and 4 are located in the scene of the wind environment difference is positive, these two types of measurement points are located in the open and unobstructed area and measurement point 1 and measurement point 4 and the plaza connected to the upper platform of the staircase in the same line, the direction is perpendicular to the Anchang River, the average value of the wind difference of measurement point 4 is the largest, 0.69 m/s, the landscape elements of the measurement point 4 contains a body of water,

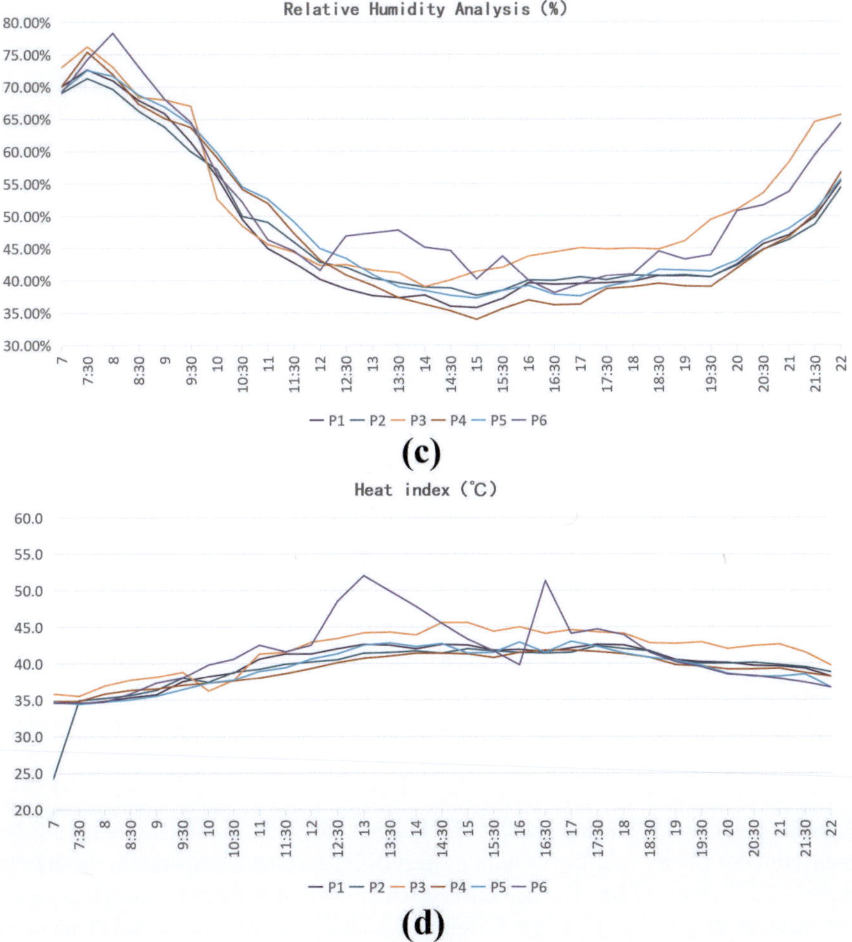

Fig. 7 (continued)

which indicates that the water elements have a positive effect on improving the wind environment。

Analyzing the outdoor thermal environment of Tieh Niu Square, the air temperature at measurement point 6 is higher than the average air temperature of the entire Square during most of the day, and the characterization of the measurement point shows that the space at measurement point 6 is more open, there is no building shade, and the radiation of sunshine is stronger, so the overall air temperature is higher during the measurement time of this measurement point, and measurement point 6 exceeds the average temperature of the Square by 3.27 °C in the afternoon time. The larger the difference in the heat index, the better the improvement of the thermal environment at the location. The maximum difference in the heat index and the maximum difference in the air temperature at point 6 in the control group were

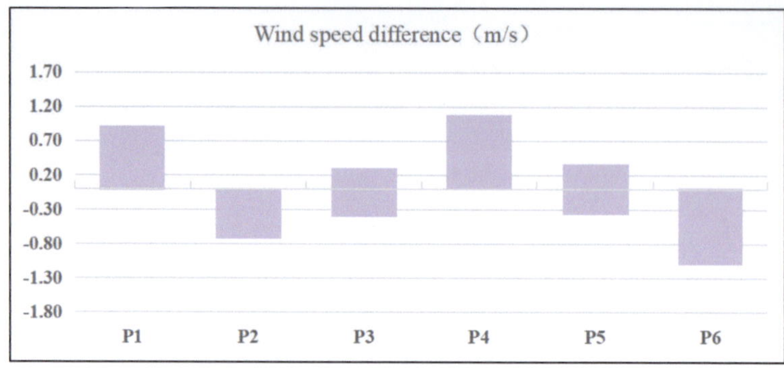

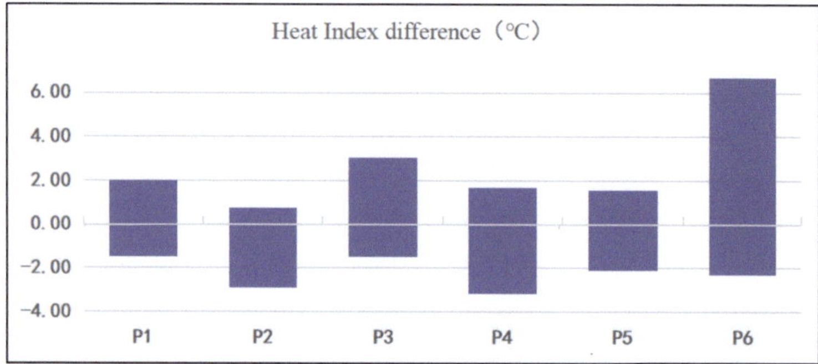

Fig. 8 Variability analysis of meteorological factors in Tieu Square by measurement points (Self-drawn)

6.68 °C and 3.27 °C, respectively, which were the highest values among all the measurement points, and the minimum heat difference and temperature difference were −2.3 °C and −1.46 °C, respectively, which were the smaller values among all the measurement points (see Fig. 8, Tables 3 and 6). The environmental element of the measurement point is a large open green space, which indicates that the presence of a large green space has a certain cooling effect, but there is a threshold for the cooling effect as the air temperature rises.

Using SPSS, Pearson correlation analysis was performed on the meteorological measurements from the six measurement points in the square. The results were visualized using Origin, where red indicates a positive correlation, blue indicates a negative correlation, and the intensity of the colors reflects the strength of the correlation (see Fig. 9). The results indicate a significant correlation ($p \leq 0.001$) between the heat index and air temperature at each measurement point. Additionally, wind speed shows a positive correlation with air temperature at all points, except for Measurement Point 6 in the control group, where the significance level is $p \leq 0.05$. Most of the relative humidity measurements exhibit a negative correlation with wind speed, and all measurement points show a significant negative correlation with air temperature ($p \leq 0.001$). During the hot summer heat period, there is a significant positive correlation between the temperature (see Figs. 8 and 9, Table 6) and the heat

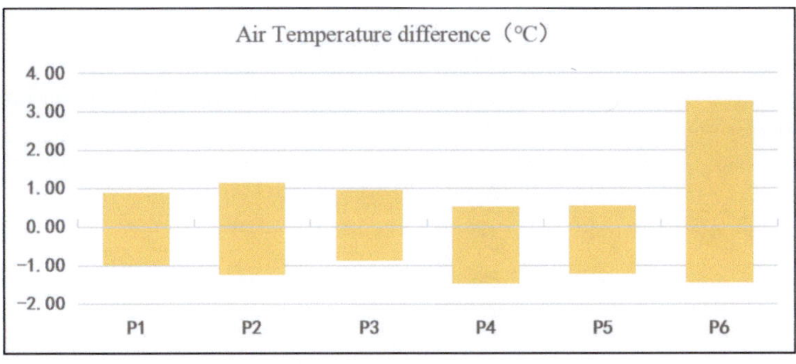

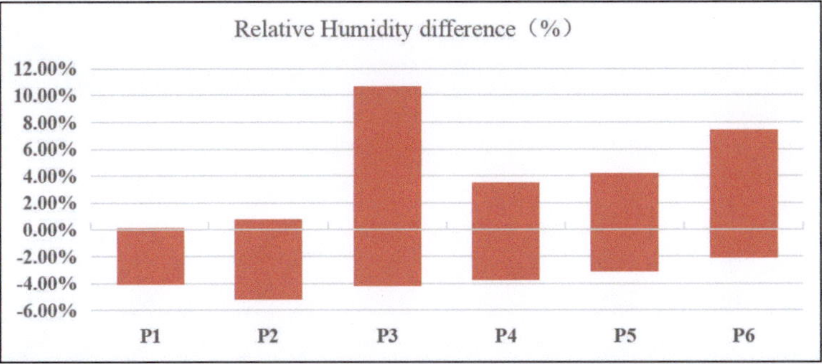

Fig. 8 (continued)

Table 3 Descriptive statistics table of temperature differences (DT) at measurement points in tianniu square (Self-drawn)

Measurement point	Point1	Point2	point3	Point4	Point5	Point6
Maximum	0.90	1.15	0.95	0.52	0.55	3.27
Minimum	−0.99	−1.25	−0.89	−1.48	−1.21	−1.46
Rang	1.89	2.40	1.84	2.00	1.76	4.72
Average	0.23	−0.06	0.20	−0.15	−0.31	0.09
Standard deviation	0.39	0.58	0.43	0.52	0.41	1.09

Table 4 Descriptive statistics table of wind speed differences (DW) at measurement points in tianniu square (Self-drawn)

Measurement point	Point1	Point2	Point3	Point4	Point5	Point6
Maximum	1.00	−0.03	0.31	1.09	0.36	0.02
Minimum	−0.03	−0.39	−0.31	0.15	−0.35	−1.44
Rang	1.03	0.70	0.71	0.94	0.71	1.45
Average	0.38	−0.39	−0.04	0.69	−0.04	−0.60
Standard deviation	0.28	0.18	0.19	0.23	0.18	0.35

index in Iron Ox Square, (see Fig. 8 and Table 5). The relative humidity at each of the Iron Ox Square measurement points has some effect on the thermal environment of the site, but the correlation is not present as a significant phenomenon (Tables 3, 4, 5 and 6).

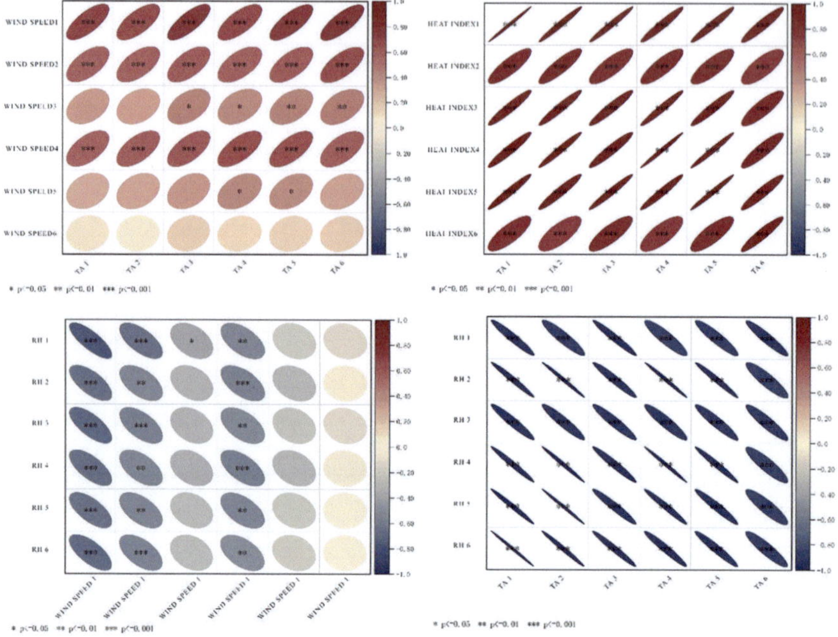

Fig. 9 Correlation heatmap analysis of meteorological factors (Self-drawn)

Table 5 Descriptive statistics table of humidity differences (DH) at measurement points in tianniu square (Self-drawn)

Measurement Point	Point1	Point2	Point3	Point4	Point5	Point6
Maximum	0.06	0.78	10.63	3.54	4.20	7.43
Minimum	−4.08	−5.24	−4.23	−3.75	−4.23	−2.10
Rang	4.15	6.03	14.87	7.29	7.37	7.37
average	−1.83	−1.20	0.22	−1.54	0.22	2.28
standard deviation	1.16	1.54	3.36	2.15	1.81	2.76

Table 6 Descriptive statistics table of heat index differences (DI) at measurement points in tianniu square (Self-drawn)

Measurement point	Point1	Point2	Point3	Point4	Point5	Point6
Maximum	2.02	0.75	3.03	1.68	1.58	6.68
Minimum	−1.48	−2.92	−1.50	−3.20	−2.12	−2.30
Rang	3.50	3.67	4.53	4.88	3.70	8.98
Average	0.22	−0.65	1.53	−0.96	−0.77	0.95
Standard deviation	0.69	0.97	1.06	1.02	0.69	2.11

4 Suggestions for Optimizing the Thermal Environment of Tieu Square Based on Thermal Comfort

4.1 *Summary of the Article*

This paper measured the microclimate factors of Tieniu Plaza in the summer at the height of pedestrians in the meteorological day, summed up the pattern of changes in microclimate factors in the square study area on the day of the measurements and correlation analysis, which resulted in the correlation of the coolness index as well as the relative humidity and the air temperature of the highest $p \leq 0.001$, followed by the wind speed $p \leq 0.05$. To ensure the credibility of the study, a questionnaire survey was conducted near each measurement point. In order to ensure the credibility of the study, a questionnaire survey was conducted near the measurement points, randomly selecting the plaza activities to distribute the questionnaire, and the results show that 78% of the residents hope that the temperature can be lowered, 51% of the residents think that the current relative humidity is appropriate, 53% of the respondents think that the current wind speed is appropriate, and the greenery in the site can play a certain role in the psychological and physiological comfort of the respondents.

4.2 *Improvement Strategies*

Creating a favorable summer thermal environment in urban outdoor spaces can encourage outdoor activities, benefiting public health while enhancing urban livability and vitality. By integrating the components of urban squares and thermal comfort evaluations, the following spatial design recommendations for urban squares are proposed:

(1) Greening system design The greening of the square thermal environment is the most obvious in this experiment because there are high-rise buildings around the square, the combination of greening and architecture, rational organization of greening arrangement, the combination of architectural shade and green shade, as far as possible for the square activities of the crowd to provide shade

throughout the day the activity space. The combination of different types of plants will have different impacts on the thermal environment, and we can try the combination of different types of trees and shrubs. Considering the arrangement of trees, a reasonable arrangement of trees will help the air circulation of the site.

(2) Water landscape design Water has good heat storage and obvious humidifying effect. At the same time in the measurement of the data obtained on the day, it can be concluded that a certain area of water can play a role in improving the wind environment of the site to the existing fountain as the main body of the linear extension to the surrounding area, by increasing the area of the water body, increase its cold island effect and improve the wind environment.

(3) Ground pavement design adopts low heat-absorbing materials to prevent excessive long-wave radiation caused by high temperatures on the ground in summer, which affects the ambient air temperature around the square.

References

1. Lu XH, Li Q, Tang PJ, Li Q, Zhao LH (2023) Effect of heat mitigation strategies on the thermal environment of hardened pavement and pedestrian thermal comfort in a hot and humid area: a case study of Guangzhou, China. J Clean Prod 427
2. Hongchi Z, Yuqiu W, Fei G (2024) Factors influencing outdoor thermal comfort in a coastal park during the transition seasons in cold regions of China. Urban Clim 55
3. Yuan Chanjuan S, Zongmao W (2024) Summer outdoor thermal comfort assessment in city squares—a case study of cold dry winter, hot summer climate zone. Sustain Cities Soc 101
4. Xiao L, Jingtang H, Ke X (2023) Identification of factors affecting public willingness to pay for heat mitigation and adaptation: Evidence from Guangzhou, China. Urban Clim 48
5. Behzad Z, Guilandoust A (2024) Enhancing outdoor thermal comfort in a historic site in a hot dry climate (Case study: Naghsh-e-Jahan Square, Isfahan). Sustain Cities Soc 102
6. Nouri AS, Costa JP (2017) Addressing thermophysiological thresholds and psychological aspects during hot and dry mediterranean summers through public space design: the case of Rossio. Build Environ 118:67–90
7. Lykoudis MN (2006) Thermal comfort in outdoor urban spaces: analysis across different European countries. Build Environ 41(11):1455–1470
8. Jianxiang H, Chaobin Z, Yanbin Z (2016) Outdoor thermal environments and activities in open space: an experiment study in humid subtropical climates. Build Environ 103:238–249
9. Bucheli J, Dalhaus T, Finger R (2022) Temperature effects on crop yields in heat index insurance. Food Policy 107:102214
10. Akkurt GK, Semsiyeci S (2024) Evaluation of nature-based solutions to improve the urban microclimate in mediterranean climate conditions: a case study of Izmir-Karsıyaka. Sustainability 16(7):2646
11. Kumar P, Sharma A (2020) Study on importance, procedure, and scope of outdoor thermal comfort—a review. Sustain Cities Soc
12. Nutsford D, Pearson AL, Kingham S, Reitsma F (2016) Residential exposure to visible blue space (but not green space) associated with lower psychological distress in a capital city. Health Place 39:70–78

Thermal Comfort of the Elderly in Different Landscape Spaces of Urban Parks in Mianyang, China

Wenyan Feng, Chunrong Zhao, Li Yan, Yu Long, and Jiayue Yin

Abstract A comfortable outdoor thermal environment is an important prerequisite for ensuring quality of life, improving urban operational efficiency, and promoting energy conservation and emission reduction. As our aging population grows, improving outdoor thermal conditions can encourage more outdoor activities among the elderly and better utilize park spaces To clarify the thermal environment characteristics of urban park spaces and the thermal comfort of the elderly, this study Takes People's Park in Mianyang as an example, analyzing and comparing the microclimate differences of various landscape space types in summer and the thermal perception of the elderly through field measurements, questionnaire surveys, and statistical methods, and establishing evaluation criteria for the thermal comfort of the elderly. The research results indicate that: (1) Black globe temperature has the greatest impact on outdoor thermal comfort for the elderly, while relative humidity has the least impact on outdoor thermal comfort for the elderly.(2) The results of the study showed a convergence between the thermal comfort vote and the thermal acceptability vote, ranked from highest to lowest as follows: corridor space-building shading > shaded walkway space-tree shading > water body space-open > square space-open.(3) The study identified the thermal neutral *PET* and comfort range for the elderly in four different types of outdoor environments. The research results can provide basic da*ta* references for the optimization design of urban spatial thermal environments.

Keywords Climate change · Elderly thermal comfort · Mianyang park · Landscape space

---

W. Feng · C. Zhao (✉) · L. Yan · Y. Long · J. Yin
School of Civil Engineering and Architecture, Southwest University of Science and Technology, Mianyang, China
e-mail: 406986617@qq.com

B. He et al. (eds.), *Climate-Resilient, Low-Carbon, and Sustainable Cities*, Lecture Notes in Civil Engineering 754, https://doi.org/10.1007/978-981-95-2169-2_21

1 Introduction

Most of the world's cities are at risk from the impacts of climate change, and these risks seriously affect sustainable urban development [1]. The 'China Climate Change Blue Book (2024)' points out that the warming trend of the global climate system is ongoing, and China is a sensitive and significantly affected area of global climate change [2]. The effective temperature increase brought by climate change has resulted in an annual increase of 0.7 days in hot and extremely hot days every decade in China, while comfortable days are gradually decreasing, making thermal comfort issues increasingly prominent. Against the backdrop of global warming, changes in urban thermal environments have become a focal point of interest for scholars [3]. The World Health Organization and various sectors are actively advocating for 'healthy aging', making the world more friendly to the elderly is an essential part of our changing demographic structure. In the context of intensified global climate change, the elderly face a higher risk of thermal health issues in their daily activities [4, 5].

The elderly are the primary user group of outdoor open spaces[6]. Existing related studies have often defined human thermal comfort standards in urban residential areas [7], park green spaces [8], and university campuses [9] through on-site measurements and surveys, analyzing the thermal comfort differences under various conditions such as different vegetation coverage [10], spatial limitations [11], visual perceptions [12], and different types of spatial landscapes [13]; however, such studies have often not Targeted the elderly population. Relevant scholars have conducted studies on the thermal comfort of the elderly in urban parks during the summer and autumn seasons [14], the thermal comfort of the elderly in cold regions during autumn [15], and the thermal comfort of the elderly in parks in humid and hot regions [16]. However, due to the influence of physiological regulation and psychological expectations, there are differences in the thermal comfort characteristics of the elderly under different climatic regions, seasonal conditions, and spatial environments, and even some conclusions are contradictory [17]. Therefore, it is necessary to explore the thermal comfort situation of the elderly in relation to local climatic conditions and different spatial types within parks, and to delineate the threshold range of thermal comfort-related indicators, making the thermal comfort characteristics of the elderly more explicit.

Urban parks are the main place for outdoor recreation for the elderly. Although much has been learned about heat perception of the elderly in urban parks, there is a lack of knowledge about summer heat perception of the elderly in urban parks in subtropical climates. Mianyang has a subtropical monsoon humid climate with hot summers and cold winters. Therefore, this study takes the People's Park in Mianyang City as an example, and analyses the thermal environment condition of the People's Park under continuous high temperature conditions, the difference of thermal perception of the elderly in outdoor space under different landscape spaces, and the main meteorological factors affecting the outdoor thermal comfort and thermal sensation of the elderly through the measurement of thermal environment in the park and the survey of the outdoor thermal comfort of the elderly, so as to investigate the appropriate temperature and thermal comfort threshold of the elderly in the Mianyang area in terms of their outdoor thermal sensation. It has certain reference significance for the scientific evaluation of summer microclimate comfort in urban parks from the perspective of the elderly and the optimisation of the thermal environment in public space. Research Area and Research Methods.

2 Research Area and Research Methods

2.1 Research Area

Mianyang City, located in southwestern China, has a typical summer hot and winter cold climate according to "Building Climate Zone Standard GB 50,178–93." This study focuses on People's Park, dividing it into four space types based on the spatial characteristics and activity paths of the elderly: G Square Space (G1 Fountain Square, G2 Liberation Monument Square), L Shaded Path Space (L1 Path, L2 Under Big Tree, L3 Under Shade), C Corridor Space (C Corridor), and S Water Body Space (Side of Artificial Lake), totaling 7 measurement points that accommodate the rich activities of the elderly. The distinctions among various landscape spaces in terms of spatial limitations are as follows: the square space is an open space; the shaded walkway is a space sheltered by trees; the corridor is a space sheltered by buildings; the water body space is an open space. The distribution of measurement points in People's Park and the surrounding environment are shown in Fig. 1 and Table 1.

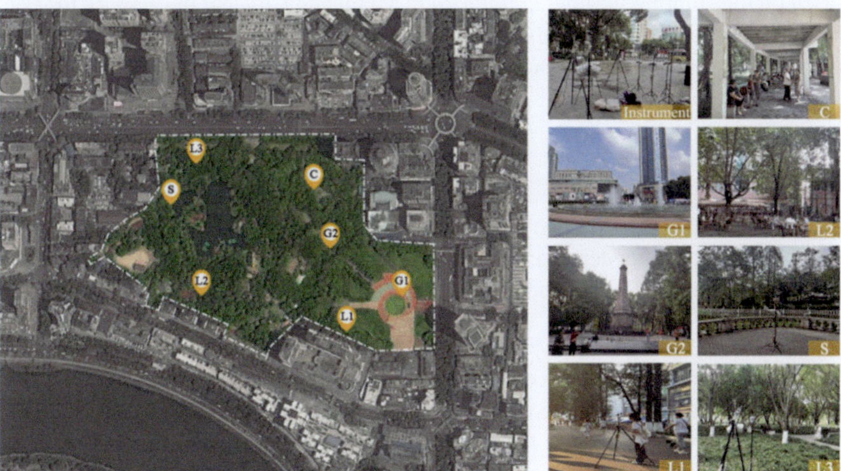

Fig. 1 Geographic location of measurement points (Self-drawn)

Table 1 Spatial characteristics of observation locations (Self-drawn)

Measurement point number	Spatial type (Classification according to different landscape spaces)	Spatial limitation (Type of space masking)	Plant community	Paving material
G1 fountain square	G square space	Open	–	Red ceramic tiles
G2 liberation memorial square	G square space	Partially shaded by trees	Tree-grass type	Light-colored floor tiles
L1 path	L tree-lined path space	Partially shaded by trees	Tree-grass type	Asphalt pavement
L2 under large trees	L tree-lined path space	Fully shaded by trees	Tree	Bluestone slab
L3 under tree shade	L tree-lined path space	Fully shaded by trees	Tree-shrub-grass type	Light-colored floor tiles
C long corridor	C long corridor space	Architectural shade	Tree-grass type	bluestone slab
S by the artificial lake	S water body space	Open	Tree-Grass Type	Asphalt Pavement

2.2 *Data Sources and Processing*

This study employed on-site questionnaire surveys of the elderly and microclimate field measurements, with the survey conducted on August 27, 2024.The questionnaire survey was conducted simultaneously with the on-site measurements of the park. Based on the summer weather conditions in Mianyang City and the activity times of the elderly population, the final determined survey period and the weather conditions on the survey day are shown in the Table 2.

2.2.1 Questionnaire Data

The study questionnaire consists of two parts: the first part collects basic background information such as gender, age, height, weight, clothing conditions, and health status of the respondents. The second part focuses on the thermal comfort evaluation of

Table 2 Basic meteorological data and research time (Self-drawn)

Date (Year. Month. Day)	Weather Conditions	Air Temperature	Relative Humidity	Average Wind Speed	Time
2024.8.27	Clear to Partly Cloudy	27~ 37 °C	40%	Northeast Wind Level 3–4 to < Level 3	7:00–21:00

Table 3 Basic information about the respondents (Self-drawn)

Designation	Classification	Percentage (%)
Genders	Males	46.29
	Females	53.71
Age	60–69 years	57.53
	70–79 years	33.79
	80 years and over	8.68
State of health	Well-being	62.56
	General	32.42
	Poor health	5.02
TSV	Neutral (0)	10.05
	A little hot (+1)	50.68
	Hot (+2)	28.77
	Very hot (+3)	10.5
TCV	Very uncomfortable (−2)	7.76
	A little uncomfortable (−1)	64.38
	Just fine (0)	19.18
	More comfortable (+1)	8.22
	Very comfortable (2)	0.46
TAV	Totally acceptable (+1)	4.57
	Just acceptable (+0.01)	42.47
	Just unacceptable (−0.01)	49.77
	Totally unacceptable (-1)	3.2

the elderly, including Thermal Sensation Voteh, Thermal Comfort Vote *(*TCV*)*, and Thermal Acceptability Vote *(TaV)*.The TSV uses a 7-level scale from very cold to very hot (−3 to + 3), the TCV uses a 5-level scale from very uncomfortable to very comfortable (−2 to + 2), and the *TaV* uses a 4-level scale from completely acceptable to completely unacceptable (+1 to −1).According to the World Health Organization's criteria for the elderly, the survey was aimed at the elderly aged 60 and above. 334 valid questionnaires were collected, and all the respondents had lived in the urban area of Mianyang City for more than one year and were able to accurately assess the outdoor thermal environment, and the basic information of the respondents is shown in Table 3.

2.2.2 Microclimate Measurement Data

Field meteorological parameter measurements were conducted by placing relevant environmental testing instruments at the measurement points for meteorological parameter testing, as shown in the Table 4. The testing content includes air temperature *(Ta)*, relative humidity (*RH*), wind speed (*Va)*, and black globe temperature

Table 4 Instrument performance parameters (Self-drawn)

Instrument name	Instrument photo	Measurement parameters	Measurement range	Accuracy	Resolution	Unit
Testo174 temperature and humidity recorder		Temperature	−20 ~ 70°C	± 0.5	0.1	°C
		Humidity	0 ~ 100%*RH*	± 3%	0.1	*RH*
WWFWZY-1 wireless omnidirectional wind speed and temperature recorder		Wind Speed	0.05 ~ 30m/s	5%	0.01	m/s
TM-188-188D heat index tester		Black globe temperature	0–80°C	± 0.6°C (15–40°C)	0.1	°C

(*Tg*).Measurements at each point were conducted continuously from 7:00 AM to 9:00 p.m. The recording interval for all instruments was 1 min, and the installation height of the instruments was 1.5 m above the ground.

2.3 *Research Methods*

2.3.1 Selection of Thermal Comfort Indicators

This paper selects Physiologically Equivalent Temperature (*PET*) as the thermal comfort evaluation indicator. Physiologically Equivalent Temperature was first proposed by Hoppe, and it is a human thermal comfort index developed based on the Munich human heat balance model [18]. It is defined as the air temperature of a typical room when the core temperature and skin temperature of a standard human in a certain environment are equal to those in a typical indoor room. It has been widely applied in outdoor thermal comfort assessments.

2.3.2 Data Analysis Methods

Based on the thermal evaluation questionnaire results and real-time meteorological data, the Rayman model was used to calculate the Physiologically Equivalent Temperature (*PET*).The elderly's height, weight, and age were set to fixed values for the calculation, with summer clothing thermal resistance assumed at 0.45 clo and a metabolic rate of 120 W as input parameters. A crucial factor in calculating *PET* is the mean radiant temperature *(TMRT)* [19]. According to ISO 7726, the formula for Tmrt incorporates the diameter of the black sphere *(Dis)*, which was set at 0.05 m in this study, with an emissivity (ε) of 0.95.

$$T_{mrt} = [(T_g + 273.15) \times 4 + \frac{1.1 \times 10^8 V_a^{0.6}}{\varepsilon D_{is}^{0.4}} \times (T_g - T_a)] \times \frac{1}{4} - 273.15 \quad (1)$$

Bivariate correlation analyses as well as multiple linear regressions of various meteorological parameters with TSV and TCV will be performed to determine the main meteorological factors affecting outdoor thermal comfort and thermal sensations of older adults. Data visualization will be performed using Origin2021 software. In SPSS, *PET* values for each landscape space will be categorized by time and mean TSV and TCV will be calculated for each time period. Linear regression analysis will be performed to establish the relationship between *PET* values and TSV and TCV to form a model for thermal perception and comfort assessment. Finally, thermal neutrality and comfort thresholds were calculated [20].

3 Results and Analysis

3.1 *Analysis of Outdoor Summer Thermal Environment*

To analyze the temporal variations among different landscape spaces and evaluate the outdoor thermal environment of these spaces, the average changes in air temperature (*Ta*), relative humidity (*RH*), wind speed (*Va*), and globe temperature (*Tg*) were calculated at each measurement point every half hour during the measurement period. Daily variation line graphs for each microclimate factor were then plotted, as illustrated in the figure.

3.1.1 Air Temperature Analysis

Statistics of air temperature values at each measurement point reveal, in conjunction with the line graphs, that the overall trend of air temperature at each measurement point first increases and then decreases. The highest *Ta* value during the testing period occurred in the G1 Fountain Square, while the lowest *Ta* value was found in the C Long Corridor. Average temperature ranking during various space testing time periods: $Ta_{\text{Square}} > Ta_{\text{Water Body}} > Ta_{\text{Tree-lined Path}} > Ta_{\text{Long Corridor}}$. The reason is speculated to be that the square is an open space, with plant characteristics having little to no shading on the upper surface, allowing for continuous sunlight exposure, resulting in the highest temperature. In S Space, the water surface creates a mirror effect, reflecting a large amount of solar radiation into the surrounding environment. Additionally, the high specific heat capacity of water allows it to store and release

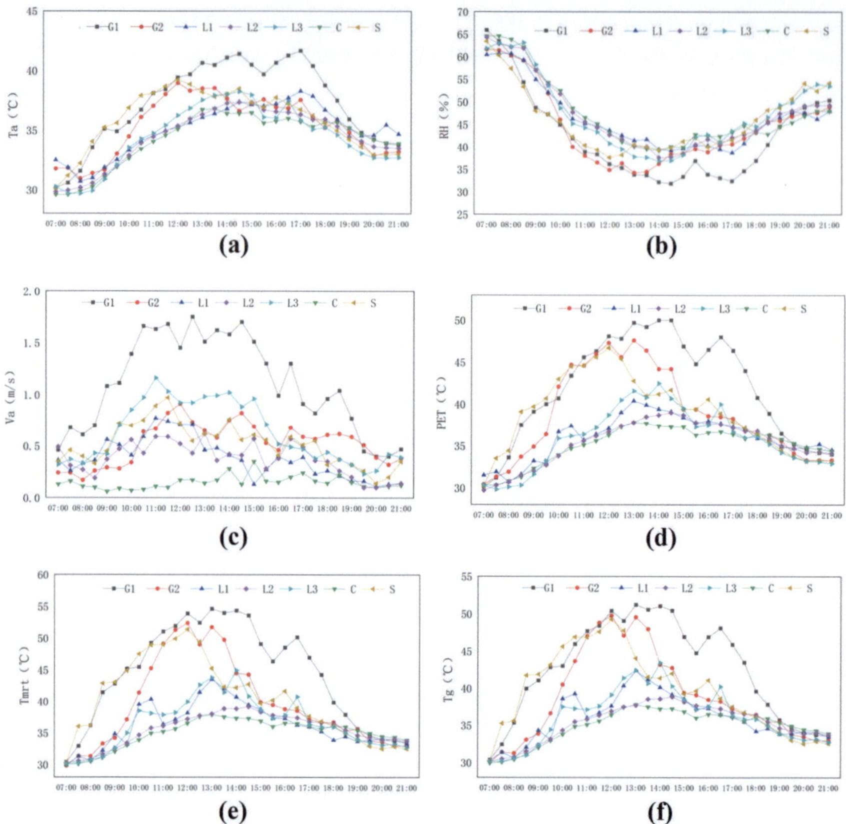

Fig. 2 Time—*Ta* (**a**); Time—*RH* (**b**); Time—*Va* (**c**); Time—*Tg* (**d**); Time—*Tmrt* (**e**); Time—*PET* (**f**) (Self-drawn)

heat slowly, maintaining a high level in the environment around the artificial lake (see Fig. 2a).

3.1.2 Relative Humidity Analysis

The daily variation trend of *RH* at each measurement point shows a pattern of first decreasing and then increasing, as illustrated in the figure. The highest and lowest values of *RH* during the testing period both occurred in G Space, followed by S Space. It is speculated that the fountain in the square is activated at 19:30 in the evening, causing the relative humidity in the space to remain relatively high. Average relative humidity ranking during various spatial testing time periods: $RH_{\text{Long Corridor}} > RH_{\text{Tree-lined Path}} > RH_{\text{Water Body}} > RH_{\text{Square}}$ (see Fig. 2b).

Table 5 Average meteorological parameters and *PET* for various landscape spaces (Self-drawn)

Space type	*Ta* (°C)	*RH* (%)	*Va* (m/s)	*Tg* (°C))	*PET* (°C)
Space G	36.33	43.50	0.81	40.71	40.31
Space L	34.65	47.06	0.46	35.80	35.76
Space C	34.23	47.50	0.15	34.87	35.10
Space S	35.87	46.10	0.52	39.65	38.87

3.1.3 Average Wind Speed Analysis

In terms of wind speed, the wind speed in the following spaces decreases in the order of: FounTain Square (G1), Liberation Monument Square (G2), by the Artificial Lake (S), under the Tree Shade (L3), Pathway (L1), Big Qiao (L2), Long Corridor (C).The average wind speed ranking from largest to smallest during various spatial testing time periods is: $Va_{\text{Square}} > Va_{\text{Water Body}} > Va_{\text{Tree-lined Path}} > Va_{\text{Long Corridor}}$ (see Fig. 2c).

3.1.4 Black Globe Temperature Analysis

The ranking of black ball temperature during various testing periods is: $Tg_{\text{Square}} > Tg_{\text{Water Body}} > Tg_{\text{Tree-lined Path}} > Tg_{\text{Long Corridor}}$. The black ball temperature is highest in G1 space, while C and L spaces have relatively lower black ball temperatures (see Fig. 2d) (Table 5).

3.2 Thermal Perception of Elderly People in Various Landscape Spaces

3.2.1 Thermal Sensation, Thermal Comfort, Thermal Acceptability Voting Distribution

There were significant differences in outdoor thermal sensations among older adults in different landscape spaces. The percentage of older adults who voted 'neutral' for thermal sensation in this survey was only 12 percent. Among various landscape spaces, the most votes for 'hot' and 'very hot' were in G Space. The most votes for 'a bit hot' were in S Space, followed by L Space. C Space had the highest votes for 'neutral' thermal sensation, accounting for 19% of its total votes. The ranking of neutral thermal sensation votes among various landscape spaces is $\text{TSV}_{\text{Long Corridor}} > \text{TSV}_{\text{Water Body}} > \text{TSV}_{\text{Tree-lined Path}} > \text{TSV}_{\text{Square}}$ (see Fig. 3a).

Due to the continuous high-temperature weather, the thermal comfort level of people throughout the park is relatively low. Among the various landscape spaces, G Space has the highest percenTage of discomfort, accounting for 69% of its total votes. L Space and S Space follow, each accounting for 61%.C Space has the

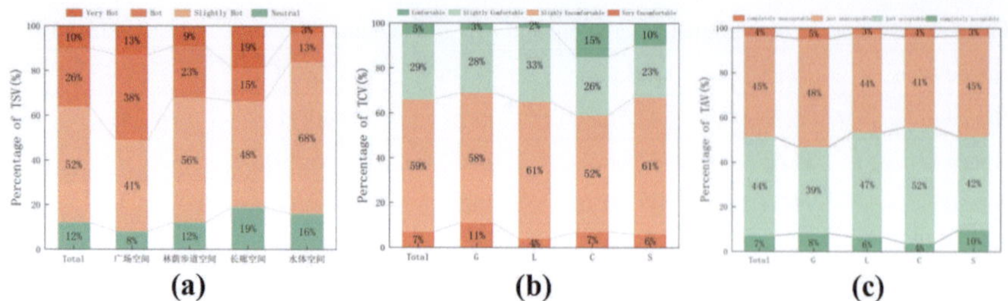

Fig. 3 Thermal sensation voting; Thermal comfort voting; Thermal acceptability voting (Self-drawn)

lowest percentage of discomfort, accounting for 59% of its total votes. The ranking of thermal comfort proportions in various landscape spaces is $TCV_{\text{Long Corridor}}$ > $TCV_{\text{Tree-lined Path}}$ > $TCV_{\text{Water Body}}$ > TCV_{Square} (see Fig. 3b).

The votes for thermal accepTability and thermal comfort tend to be consistent, with the voting ranked from high to low as $TaV_{\text{Long Corridor}}$ > $TaV_{\text{Tree-lined Path}}$ > $TaV_{\text{Water Body}}$ > TaV_{Square}. Among them, S Space has the highest proportion of 'completely acceptable'. G Space has the highest proportion of 'completely unacceptable'. Overall, older adults' acceptance of the park's outdoor thermal environment is moderate, indicating that while most individuals do not have a high sense of thermal comfort in outdoor microclimate environments, they can still accept it (see Fig. 3c).

3.2.2 Changes in Thermal Comfort Voting Over Time and Spatial Preferences of Population Activities

The testing period is divided into three phases based on temperature variations: warming phase, high temperature phase, and cooling phase. This paper considers votes of just right、fairly comfortable、 and very comfortable as comfort votes. The overall trend in thermal comfort voting across various landscape spaces first decreases and then increases over time. C Space shows the highest proportion of thermal comfort during the warming and high-temperature phases, reflecting its favorable thermal environment. In the cooling phase, G Space has the highest thermal comfort among the elderly, likely due to its open design, adequate wind speed, and the activation of a flowing fountain at night, which enhances comfort. Furthermore, G Space is preferred by the elderly during the cooling phase, while L Space is where they spend the most time overall (see Fig. 4).

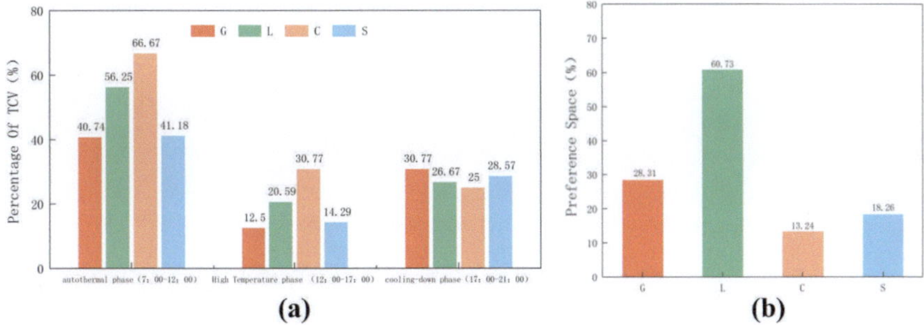

Fig. 4 Percentage of thermal "comfort" votes by landscape space over time (**a**); preferred space votes (**b**) (Self-drawn)

3.3 Discussion of Thermal Comfort Thresholds for Each Landscape Space

When TSV = 0, the thermal neutral *PET* for the elderly in G, L, C, and S spaces are 24.64°C, 28.17°C, 28.75°C, and 20.86°C, respectively. Taking TSV = (−0.5) ~ (+0.5) as the thermal sensation range, i.e., the suitable temperature range. The corresponding *PET* thermal neutrality ranges are 19.97 ~ 29.32 °C, 24.75 ~ 31.60 °C, 25.87 ~ 31.62 °C, 13.05 ~ 28.67 °C.The widths of the thermal neutrality intervals are 9.35 °C, 6.85 °C, 5.75 °C, 15.62 °C.The slope of the regression curve represents the sensitivity of the elderly to environmental changes; a higher slope indicates that the elderly are more sensitive to changes in this environment. The thermal neutrality *PET* value in water body spaces is the lowest, and the slope of the fitting formula is also the lowest. This indicates that the elderly are least sensitive to changes in *PET* in this area, with a relatively stable thermal perception. The reasons include the presence of water bodies and leisure seating, as well as the open space, which plays a significant role in maintaining the stability of the environmental microclimate (see Fig. 5a).

When TCV = 0, the thermal comfort *PET* for the elderly in the G, L, C, and S spaces are 31.44°C, 31.89°C, 32.20°C, and 32.96°C, respectively. Taking TCV = 0 ~1 as the thermal comfort range, the corresponding *PET* thermal comfort ranges for the four spaces are 20.45 ~31.44 °C, 24.59 ~31.89 °C, 25.99 ~32.20 °C, and 22.43 ~32.96 °C.The interval widths are 10.99 °C, 7.3 °C, 6.21 °C, 10.53 °C.Furthermore, the thermal comfort intervals in spaces G and S are broader, while space C has the narrowest thermal comfort interval (see Fig. 5b).

In addition, comparing the *PET* thermal neutral temperature and threshold ranges of all spaces with the standard *PET* thermal comfort range (18°C ≤ *PET* ≤ 23°C) shows that 32.96°C exceeds the Standard value by approximately 9°C.This indicates that the standards for outdoor thermal comfort perception and the acceptance threshold for high temperatures among the elderly in summer are relatively high. The reason is related to the fact that elderly individuals have a delayed thermoregulatory response to changes in environmental temperature as they age.

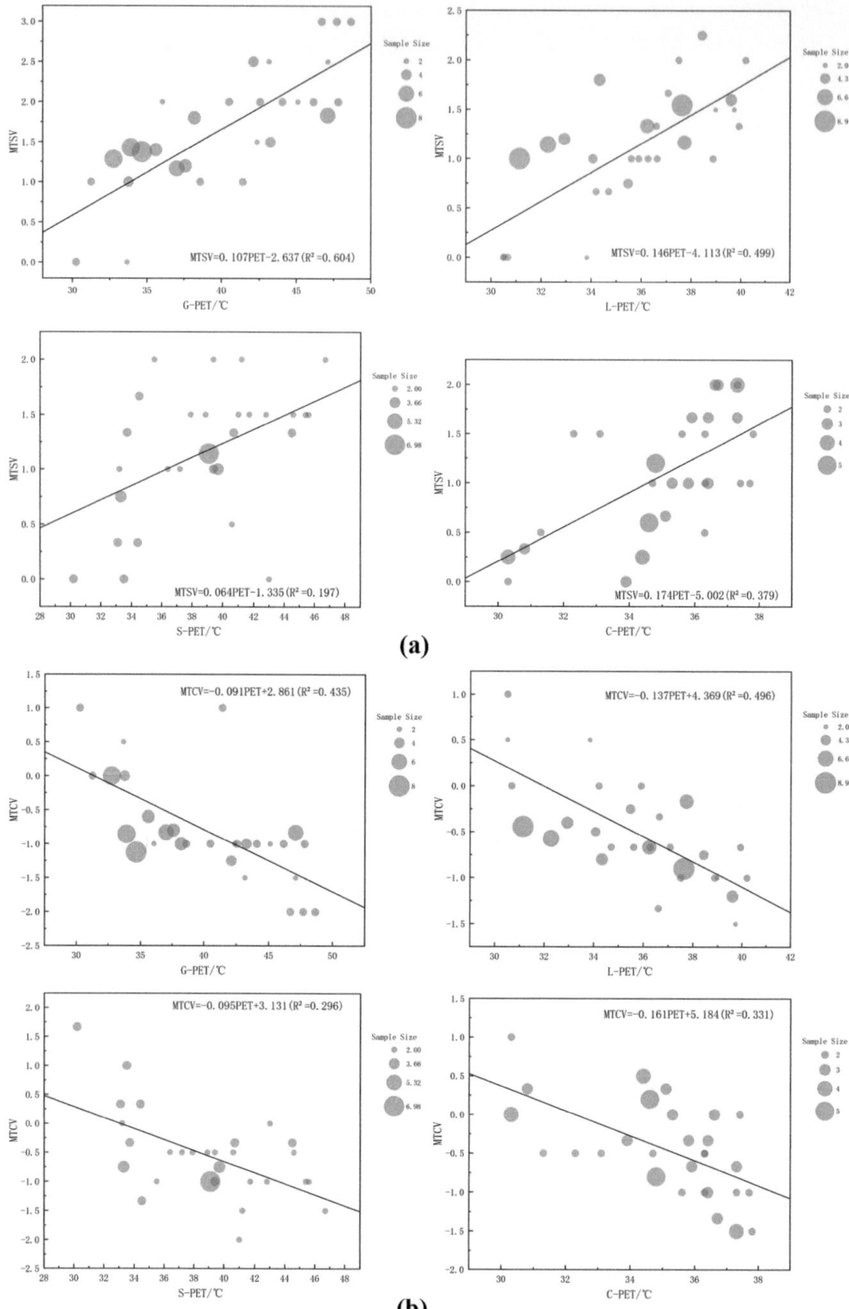

Fig. 5 TSV—*PET* across landscape spaces (**a**); TCV-*PET* across landscape spaces (**b**) (Self-drawn)

Table 6 Correlation between microclimatic factors and thermal sensory thermal comfort voting (Self-drawn)

Meteorological factors	Temperature perception TSV		Thermal comfort perception TCV	
	Correlation	Significance sig	Correlation	Significance sig.
Ta	0.639***	0.000	−0.738***	0.000
RH	−0.651***	0.000	0.759***	0.000
Tg	0.611***	0.000	−0.734***	0.000
Va	0.536***	0.003	−0.594***	0.001

Note: *** At the 0.01 level (two-Tailed), the correlation was significant.

3.4 Outdoor Thermal Comfort Evaluation Model

3.4.1 Effect of Microclimate on Thermal Sensory Thermal Comfort

Meteorological factors such as Ta, *RH*, Va, and *Tg* are the main factors affecting outdoor thermal comfort for the human body. To study the microclimate environmental factors influencing the elderly's evaluation of the current thermal environment, the research employed Pearson correlation coefficients to conduct bivariate correlation analysis between Ta, *RH*, Va, *Tg* and TSV, TCV. From the results of the charts, it can be seen that Ta, Tg, and *Va* are significantly positively correlated with the elderly's thermal sensation of the outdoor environment, while *RH* is negatively correlated with thermal sensation; Correspondingly, *Ta*, *Tg*, and *Va* are negatively correlated with the comfort level of the population, while *RH* is positively correlated with thermal comfort (Table 6).

3.4.2 Outdoor Thermal Comfort Evaluation Models

Using multiple linear regression methods, Ta, *RH*, Va, and *Tg* are Taken as independent variables, and the comprehensive comfort voting value is Taken as the dependent variable to esTablish a prediction model for outdoor thermal comfort for the elderly in Mianyang, resulting in the following relationship:

$$OCV = -0.025Ta - 0.116Tg + 1.076Va + 0.012RH + 3.497 \quad (2)$$

From the perspective of coefficients, the impact of meteorological factors on thermal comfort, in descending order, is as follows: $Tg > Va > Ta > RH$. This indicates that black globe temperature has the greatest impact on outdoor thermal comfort for elderly people, while relative humidity has the least impact on outdoor thermal comfort for elderly people.

4 Conclusion

In this study, through field measurements and thermal comfort surveys of summer microclimates in Mianyang urban parks, we analyzed the thermal comfort characteristics of the elderly in different landscape spaces and the influencing factors of microclimates, established different quantitative evaluation criteria, and came up with the following conclusions.

1. Elderly prefer shaded areas (C & L) in parks. Thermal sensation: Corridor > Water Body > Shaded Pathway > Square. Thermal comfort: Corridor > Shaded Pathway > Water Body > Square. Water improves comfort in open spaces (S > G).
2. The preferred space for elderly people in park activities is L space, which received the highest votes. Thermal comfort decreased then increased overall. C space was best in warm phases, G space in cool phases.
3. Thermal Neutral *PET* Values and Thermal Comfort Ranges for Older Adults Differ Across Landscape Spaces
4. *Ta*, *Tg* and *Va* are positively correlated with outdoor thermal sensation in the elderly and vice versa for relative humidity; Ta, *Tg* and *Va* are negatively correlated with thermal comfort in the elderly and vice versa for relative humidity and thermal comfort.

In summary, in order to improve the thermal comfort of elderly people's outdoor activities in urban parks in summer, the following suggestions are made to improve the thermal environment of the People's Park in Mianyang City: the "natural shading facilities" in the park are relatively perfect, but the "artificial shading facilities" in the park are relatively lacking. At the same time, through rational plant combinations, water features and space design, as well as regular plant pruning, different landscape spaces can be created to regulate wind speed and relative humidity to meet the requirements of the elderly for wind speed and relative humidity.

References

1. A Study on Urban Planning Response to Climate Resilient City as Exemplified in Urban Center of High-density Area at Block Scale Based on Outdoor Thermal Comfort Analysis. Shanghai Jiaotong University (2018)
2. Blue Book on Climate Change in China 2024 (2024) Energy conservation & environmental Protection (07):1
3. Rui L, Fang W, Handong M et al (2024) Research on urban thermal environment and spatial planning strategies from the perspective of climate resilience: a case study of Zhengzhou. Henan Science 1–7
4. Wang W, Li Y, Li L et al (2023) Study on thermal comfort of elderly in community parks: an exploration from the perspectives of different activities and ages. Build Environ 246:111001
5. The research on outdoor thermal comfort of the elderly in hot summer and cold winter area. Southwest Jiaotong University 2022

6. Li X, Li X, Tang N et al (2022) Summer outdoor thermal perception for the elderly in a comprehensive park of Changsha, China. Atmosphere 13(11):1853
7. Yugang GUAN, Jianwen WANG, Hong CHEN et al (2023) Study on outdoor thermal comfort and thermal environment optimization strategy of Wuhan residential area based on behavioral activities. New Arch 06:144–148
8. Pan JB, Shi C, Huang TT, Zhu DL (2023) Spatial differentiation characteristics of human thermal comfort in beijing olympic forest park. Landsc Arch 30(12):114–120
9. Huang Z, Cheng B, Gou Z, Zhang F (2019) Outdoor thermal and adaptive behaviors in a university campus in China's hot summer cold winter climate region. Build Environ 165:106414
10. Ouyang W, Morakinyo TE, Ren C et al (2020) The cooling efficiency of variable greenery coverage ratios in different urban densities: a study in a subtropical climate. Build Environ 174:106772
11. Jinhui Y (2020) Research on behavior and satisfaction of middle-aged and elderly urban parks based on thermal perception. Anhui Agricultural University
12. Lam C, Yang H, Yang X et al (2020) Cross-modal effects of thermal and visual conditions on outdoor thermal and visual comfort perception. Build Environ 186:107297
13. Min X (2019) Outdoor thermal comfort in different landscape spaces in an urban park. Northwest A&F University
14. Zhengqi P, Fu Z, Rong H (2023) Study on the difference between thermal comfort in summer and autumn of the elderly in Lanzhou Urban Park.Build Sci 39(12):285–292
15. Sihan X, Ming L, Kun W et al (2024) Analysis of autumn thermal comfort characteristics and influencing factors of elderly people in urban parks in cold regions. Sci Technol Eng 24(1):336–343
16. Haijing H, Linying W (2023) Environment factors to park thermal comfort of the elderly in hot and Humid Areas. J Chin Urban For 21(01):13–19
17. Xiong J, Ma T, Lian Z et al (2019) Perceptual and physiological responses of elderly subjects to moderate temperatures. Build Environ 156:117–122
18. Chen L, Ng E (2012) Outdoor thermal comfort and outdoor activities: a review of research in the past decade. Cities 29(2):118–125
19. Matzarakis A, Mayer H, Iziomon MG (1999) Applications of a universal thermal index: physiological equivalent temperature. Int J Biometeorol 43:76–84
20. American Society of Heating, Refrigerating and Air-conditioning Engineers. ASHRAE Standard 55 (2017) Thermal environmental conditions for human occupancy.1791 Tullie Circle, NE, Atlanta: SHRAE Customer Service, pp 37–39

Multifaceted Analysis of Green Roof Characteristics in Modulating Urban Microclimate Patterns

Qianwen Cai, Dongjin Cui, and Yutong Chen

Abstract As a proven strategy for urban microclimate enhancement, green roofs demonstrate significant potential in mitigating unpredictable meteorological events and disaster resilience. While existing research predominantly focuses on green roofs' energy-saving and local environmental benefits, their broader climate regulation mechanisms remain underexplored. This study uses the ENVI-met simulation platform to explore how green roofs regulate outdoor microclimates. The study establishes causal linkages between critical green roof parameters (LAI, substrate moisture gradients) and environmental drivers (meteorological forcing, building morphology). Comparative analysis reveals intensive systems exhibit greater efficacy in simultaneous heat mitigation, humidity buffering, and thermal comfort enhancement relative to semi-intensive/extensive systems. Increasing LAI enhances cooling and humidification, but the benefits level off after a certain point. Substrate moisture content has a significant impact during the day but less so at night. Moderate wind speeds (around 2 m/s) and clear weather conditions support green roof performance, especially during the day. The height of buildings also affects microclimate benefits, with lower buildings creating more favorable conditions. Meteorological factors have the most significant influence on green roof performance, followed by component parameters, while building height and layout have a smaller impact. Green roofs improve microclimatic conditions during the day, but their impact decreases at night, highlighting their importance for urban environmental enhancement. This study delivers the first systematic examination of green roofs' transboundary microclimate regulation mechanisms—particularly through LAI, substrate moisture, wind-buildup interactions—filling a critical knowledge gap in urban climate science.

Keywords Mitigation of urban heat island phenomenon · Vegetative roof · Urban microclimatic conditions · ENVI-met modeling

Q. Cai (✉) · D. Cui · Y. Chen
School of Architecture and Urban Planning, Shenzhen University, Shenzhen, China
e-mail: jena.djcui@szu.edu.cn

B. He et al. (eds.), *Climate-Resilient, Low-Carbon, and Sustainable Cities*, Lecture Notes in Civil Engineering 754, https://doi.org/10.1007/978-981-95-2169-2_22

1 Introduction

The intensification of urban heat islands (UHIs) has been closely linked to anthropogenic land cover modification, where large-scale conversion of natural ecosystems into built environments creates expansive impervious surfaces [1–3]. Satellite monitoring reveals a striking decline in urban vegetation coverage from 12.23% (2003) to 5.91% (2020), equivalent to an annual loss rate of 0.35% [4–6]. Concurrently, per capita accessible green space has diminished by 68% over this period—translating to a reduction of 0.51 m^2 per resident annually [7]. This phenomenon has exacerbated the UHI. It has also led to several urban problems such as increased energy consumption in buildings and reduced thermal comfort [8, 9]. Landsat satellite imagery analysis reveals Shenzhen's impervious surfaces have undergone a 15-fold expansion, escalating from 51 km^2 in 1985 to 780 km^2 in 2020, mirroring the city's meteoric urbanization trajectory over this 35-year period [10]. In high-density cities, where the supply of green space lags behind demand, traditional approaches to greening at ground level face the challenge of limited land.

Rooftop greening is an emerging form of urban greening. It improves the overall green cover of cities by making effective use of urban roof space. Similarly, green roofs can produce positive impacts that are comparable to those of ground-level green roofs [11]. For example, by reducing the albedo of building surfaces and the transpiration of plants and soil, green roofs contribute to slowing the rise in urban temperatures [12], thereby mitigating the summer heat island effect [13]. Green roofing has many benefits, including adding greenery to a city without using land resources, reducing urban carbon emissions through cooler building facades, and reducing air conditioner energy consumption [14, 15].

There are a large number of studies related to green roofs, mainly categorized into the selection of component parameters, the role of environmental improvement, and the interaction between green roofs and building energy efficiency. For example, Chen's study focused on the relationship between substrate moisture and transpiration and found that substrate moisture in the optimal range is crucial for the role of green roofs [16]. Through an integration of remote sensing and statistical modeling, Lin et al. [17] demonstrated green roofs' pivotal role in thermal regulation for urban zones deficient in expansive vegetated areas and aquatic systems. Complementing this, Bevilacqua et al. [18] quantified a 34.9% annual cooling energy reduction in buildings retrofitted with vegetated roofing systems under specific climatic regimes.

Despite valuable findings, there is a research gap in understanding how green roof components interact with meteorological factors and building characteristics to regulate microclimates. This study uses the ENVI-met simulation platform to explore these interactions, aiming to enhance our understanding of green roofs' external climate-regulating mechanisms. Field measurements and controlled experiments will evaluate the effects of various parameters, providing insights for urban planning and green infrastructure.

Focusing on the vegetation-urban climate nexus, we implement ENVI-met V5.5 to quantify green roofs' spatiotemporal thermal regulation capacities. To evaluate how well the ENVI-met model performs, empirical data collected via field observations in KLHY residential area were compared with simulation outputs. Next, we conducted three groups of controlled experiments to examine the effects of component parameters, meteorological factors, and building height on the thermal environment. By analyzing four assessment indicators, we compared the spatial distribution and diurnal variation patterns of thermal benefits across six sets of simulated data. To comprehensively assess the impact of various roof-related factors on outdoor microclimate, this study focuses on four core indicators as *Ta* (air temperature), *RH* (relative humidity), *UTCI* (Universal Thermal Climate Index), and *Ts* (surface temperature). Our goal is to thoroughly explore the effects of vegetative roofs, highlighting their potential benefits and variations across different conditions.

2 Methodology

2.1 Experimental Validation

Kaili Garden, situated in Nanshan District, Shenzhen, exemplifies a prototypical grid-form architectural configuration with a height of 24 m. The computational domain and sensor network deployment are spatially mapped in Fig. 1. The parameters measured encompassed meteorological data (*Ta*, *RH* and wind speed) in Table 1. Field measurements were conducted over an 8-h diurnal cycle (9:00–17:00 LST) on 23 July 2023, a representative summer day. All sensors were positioned at the pedestrian standard level, 1.5 m from the ground. The statistics sampled every 10 min (see Fig. 1b). A JT-IAQ-50 microclimate monitoring system (Japan Tobacco International) was deployed to capture high-resolution meteorological profiles. Instrumentation specifications are cataloged in Table 2, with spatial configuration detailed in Fig. 1a.

The input meteorological parameters in this study include hourly air temperature, relative humidity, wind speed, and wind direction from the Nanshan meteorological station. ENVI-met V5.5 typically offers spatial resolutions ranging from 0.5 to 10 m and a temporal resolution of 10 s. For this study, a finer resolution of 2 m × 2 m × 1 m was selected to enhance accuracy. As outlined in in Table 1, the simulation area covered 220 m × 200 m horizontally, divided into 110 × 100 grid cells, and extended vertically up to 72 m with 72 grid cells.

Studies have demonstrated that model validation requires the integration of 2–3 complementary metrics to ensure methodological consistency [19–21]. This study employs three validation indices:R^2 (coefficient of determination), *RMSE* (root mean square error), and *MAPE* (mean absolute percentage error), as systematically compared in Fig. 2. Generally, as R^2 approaches 1, the model is more perfect [22]. According to existing research, estimation models for air temperature (*Ta*) and relative humidity (*RH*) exhibit high accuracy. R^2 values are used by 65% of the literature as a validation metric [19]. The *RMSE* for temperature ranges from 0.28 to 4.04 °C,

Fig. 1 **a** Measuring the locations of the cases (Image map from Baidu). **b** Measuring the instruments (Self-drawn)

Table 1 Input and test parameters for KLHY validation simulation in Shenzhen (Self-drawn)

Name	Parameter	Input value
Study area	Shenzhen	22°4′N, 114°0′E
	Area size (x * y * z)	110 * 100 * 72
	Grid resolution (m)	dx = 2; dy = 2;d z = 1
Simulation time	Start time	July 23, 2023 at 03:00
	Total time	15 h
Meteorological conditions	Initial air temperature (°C)	Hourly weather station data
	Relative humidity (%)	Hourly weather station data
	Inflow direction	135°
	Wind speed at 10 m	2 m/s
Building/grounds	Ground material	Gray concrete
	Ground reflectance	0.3
	Building surface reflectance	0.3

Table 2 Accuracy specification for measuring instruments

Measuring parameters	Air temperature	Relative humidity	Black-globe temperature	Wind speed
Range	−20–120 °C	0–100%	20 °C –20 °C	0–5 m/s
Precision	±0.5 °C	±1.5%	±0.5 °C	±0.03 m/s
Interval	1 min			

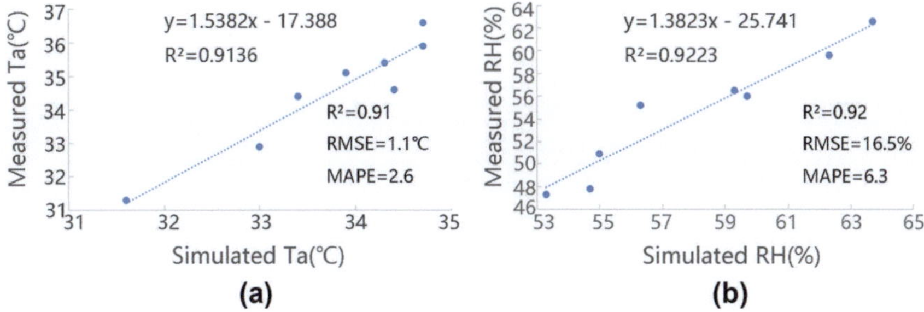

Fig. 2 Model validation: **a** Air temperature (*Ta*)—simulated versus observed; **b** Relative humidity (*RH*)—model outputs versus field measurements (Self-drawn)

while for humidity, it ranges from 2.04 to 10.20% [22, 23]. The *MAPE* is generally below 10% [24], indicating that the prediction errors of these models are within an acceptable range.

Comparative analysis revealed a high degree of consistency between field-measured and simulated *Ta/RH* values (Fig. 2), withR^2 coefficients reaching 0.91 (Ta) and 0.92 (*RH*). The error metrics further validated this alignment: *Ta* exhibited $RMSE = 1.1$ °C ($MAPE = 2.6\%$) while *RH* showed $RMSE = 16.5\%$ ($MAPE = 6.3\%$). Field-calibrated simulations achieved temperature prediction accuracy of ±1.1 °C ($R^2 = 0.91$) and humidity modeling precision of ±16.5% ($R^2 = 0.92$), meeting urban climate modeling benchmarks. These performance metrics substantiate the model's reliability in pedestrian-level environmental analysis.

2.2 *Case Parameter Configuration*

The studies used a grid size of 2 m × 2 m × 2 m to enhance simulation accuracy. In Table 3, the numerical model encompassed a horizontal area of 120 m × 120 m, with a vertical dimension of 72 m. It consisted of a grid layout with 60 × 60 cells horizontally and 24 cells vertically. The size of a single model is 40 m × 40 m × 24 m in Fig. 3, with the initialization date set as July 23, 2023 (the date of measurement). Hourly data from nearby meteorological stations are inputted selecting the "simple forcing" function. The meteorological wind speed at a height of 10 m is 2 m/s, with a wind direction of 0° (default). The simulation ran for 30 h, starting at midnight on July 23rd. This paper will provide a list of consistent generic parameters used in the simulation tests in Table 3.

Table 3 Basic parameter setting (Self-drawn)

Name	Parameter	Input value
Study area	Shenzhen	22°4′N, 114°0′E
	Area size ($x * y * z$)	60 * 60 * 48
	Grid resolution (m)	$dx = dy = dz = 2$
Simulation time	Start time	July 23, 2023, at 00:00
	Total time	30 h
Meteorological conditions	Initial air temperature (°C)	Hourly weather station data
	Relative humidity (%)	Hourly weather station data
	Inflow direction	0°
	Wind speed	2 m/s
	Cloud amount	0 (cloudless and sunny day)
Building/grounds	Ground material	Gray concrete
	Ground reflectance	0.3
	Building surface reflectance	0.3
	Building size (m)	40 * 40
	Building height (m)	24

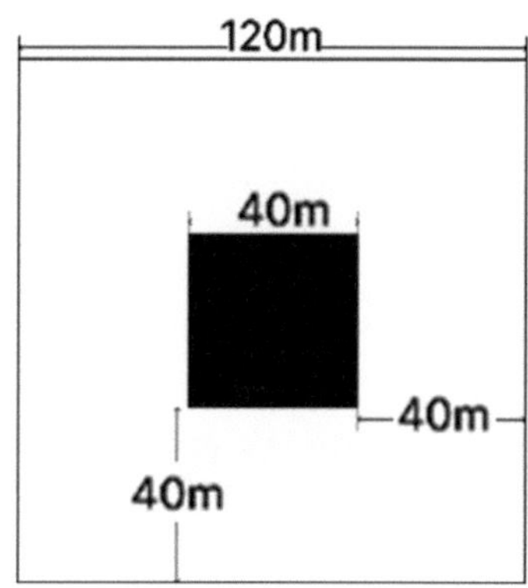

Fig. 3 Single-model planes (Self-drawn)

2.3 Green Roof Component Parameters

The structural parameters of a vegetated roof include the roof type, leaf area index (LAI), and substrate composition. Planting material selection, substrate composition, and substrate thicknesses are all critical to thermal performance [25–27]. Numerous researchers have explored the optimal selection of plant species for various types of vegetated roofs. Studies have demonstrated that LAI is pivotal in decreasing energy use in buildings across both temperate and tropical climate zones [28, 29]. Substrate composition, moisture content, and thickness are key to green roof benefits [30, 31].

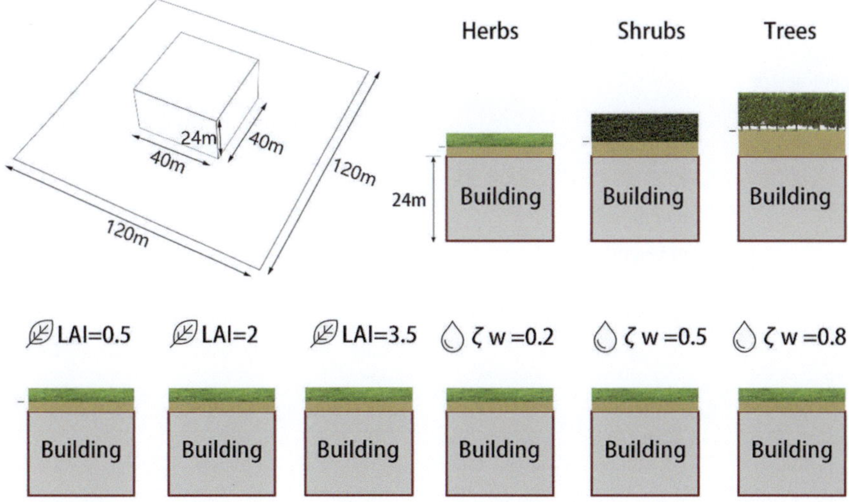

Fig. 4 Three diagrams of component parameters (Self-drawn)

In hot, arid climates, the wet soil may reduce roof surface temperatures through evaporative cooling [32].

The study explored the impact of various green roof types, leaf area indexes (LAI), and substrate moisture levels on building microclimates. The experimental setup includes three components in Fig. 4: First, the experimental groups included green roof types consisting of herbs, shrubs, and trees, with the bare roof serving as the control group. Second, the leaf area index (LAI) experimental groups are set at 0.5, 2, and 3.5, with the bare roof as the control. Lastly, the substrate moisture experimental groups are divided into dry ($\zeta w = 0.2$), moderately moist ($\zeta w = 0.5$), and wet ($\zeta w = 0.8$), with the bare roof also serving as the control group. Specific experimental parameters can be found in Tables 4, 5 and 6.

2.4 Meteorological Factors Modelling Parameters

Vegetated rooftops' efficacy is predominantly determined by regional meteorological contexts. Green roofs enhance thermal comfort and reduce cooling demand more efficiently in hot climates, while significantly decreasing heating demand in colder climates [33]. Solar radiation directly influences the temperature performance of green roof substrates and plant community composition [34, 35]. Due to varying solar radiation levels in different weather conditions, many authors have studied the performance of green roofs [13, 36, 37]. Wind speed and direction significantly affect the benefits of green roofs at both roof and pedestrian levels. Studies show that green roofs follow prevailing wind directions for cooling at pedestrian levels, with

Table 4 Green roofs with different types (Self-drawn)

Case parameters	Green roof types	Comparator	Herbs (rough)	Shrubs (semi-intensive)	Trees (intensive)
Vegetation parameters	Leaf shape	/	Grasses	Losing leaves	Losing leaves
	LAI	/	2	2	2
	Transmittance	/	0.3	0.3	0.3
	Plant height	/	0.18 m	1 m	10 m
Basal parameters	Substrate thickness	/	0.15 m	0.2 m	0.65 m
	Basic ingredients	/	Sandy loam	Sandy loam	Sandy loam
	Substrate moisture factor	/	0.5	0.5	0.5

Table 5 Green roofs with different LAI

Case parameters	Green roof types	Comparator	LAI = 0.5	LAI = 2	LAI = 3.5
Vegetation parameters	Leaf shape	/	Grasses	Grasses	Grasses
	LAI	/	0.5	2	3.5
	Transmittance	/	0.3	0.3	0.3
	Plant height	/	0.18 m	0.18 m	0.18 m
Basal parameters	Substrate thickness	/	0.15 m	0.15 m	0.15 m
	Basic ingredients	/	Sandy loam	Sandy loam	Sandy loam
	Substrate moisture factor	/	0.5	0.5	0.5

Table 6 Green roofs with different substrate humidity coefficients

Case parameters	Green roof types	Comparator	$\zeta w = 0.2$	$\zeta w = 0.5$	$\zeta w = 0.8$
Vegetation parameters	Leaf shape	/	Grasses	Grasses	Grasses
	LAI	/	2	2	2
	Transmittance	/	0.3	0.3	0.3
	Plant height	/	0.18 m	0.18 m	0.18 m
Basal parameters	Substrate thickness	/	0.15 m	0.15 m	0.15 m
	basic ingredients	/	Sandy loam	Sandy loam	Sandy loam
	Substrate moisture factor	/	0.2	0.5	0.8

Table 7 Basic parameter settings (Self-drawn)

Name	Parameter	Input value
Vegetation parameters	Leaf shape	Grasses
	LAI	2
	Transmittance	0.3
	Plant height	0.18
Basal parameters	Substrate thickness	0.15
	Basic ingredients	Sandy loam
	Substrate moisture factor	0.5

enhanced effects observed on the north and east sides of buildings [38]. Concentrated placement in upwind directions effectively reduces temperatures across the area, surpassing dispersed placement, even with the same vegetation coverage [39].

The model's grid parameters, building parameters, and simulation time are consistent with those specified in Table 3, while the vegetation and substrate parameters are shown in Table 7. This study investigates the impact of green roofs under varying weather conditions and wind speeds, revealing their effects on the climatic environment and microclimates. The experimental setup in Fig. 5 includes three meteorological conditions: sunny, cloudy, and overcast, each with a corresponding control group. For wind speed, the experiment established scenarios for $WS = 0.5$ m/s, $WS = 2$ m/s, and $WS = 3.5$ m/s. Additionally, the grid parameters, building parameters, and simulation time of the model are consistent with those specified in Table 3, while vegetation and substrate parameters are as shown in Table 7, with specific experimental parameters detailed in Tables 8 and 9.

2.5 Building Characteristics

Building characteristics exert a substantial influence on the spatial extent and distribution of green roofs. Research on their influence mainly focuses on building height and layout. Studies on building height explore differences in the adaptability and feasibility of green roofs for buildings of varying heights. Since this study focuses on a single building, the building layout is not discussed.

In high-density urban areas, green roofs may not effectively reduce street heat loads due to nearby standard building heights, limiting their cooling effects [40]. Building height governs the efficacy of rooftop greenery in altering thermal conditions within street canyons. The diminishing cooling effect of green roofs correlates with higher building structures. Research has measured the cooling effects of vegetative roofs, indicating that for buildings 10 m high, the cooling effect extends up to 10 m below the roof, while for 20-m-high buildings, it extends up to 5 m below the roof, and for 40-m-high buildings, it decreases to 0 m below the roof [39].

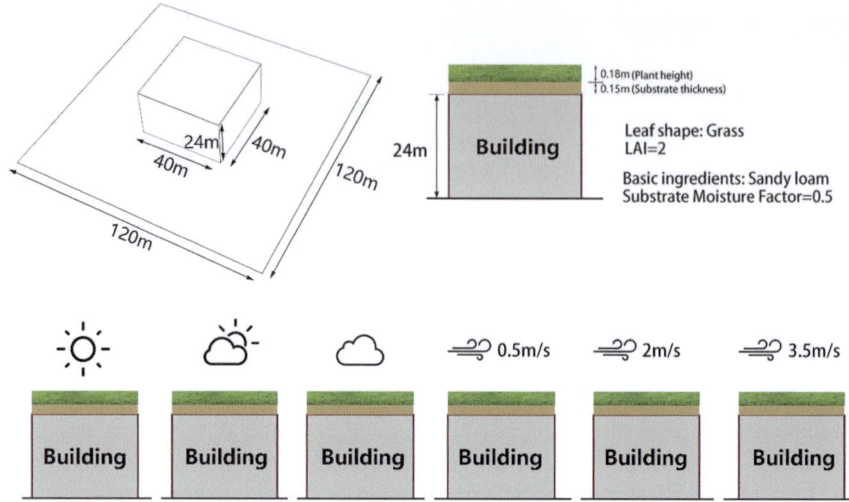

Fig. 5 Two diagrams of meteorological parameters (Self-drawn)

Table 8 Green roofs in different weather conditions. (Self-drawn)

Weather conditions	Sunny day	Cloudy day	Totally cloudy
Wind speed	2 m/s	2 m/s	2 m/s
Amount of cloud	0	4	8

Table 9 Green roofs in different wind speeds. (Self-drawn)

Case	$WS = 0.5$ m/s	$WS = 2$ m/s	$WS = 3.5$ m/s
Wind speed	0.5 m/s	2 m/s	3.5 m/s
Amount of cloud	0	0	0

Building height impacts the benefits of green roofs. The experimental group tested four elevations: 12, 24, 36, and 48 m, each with a corresponding control group in Fig. 6. Specific experimental parameters are detailed in Table 10.

3 Results and Discussion

3.1 *Effects of Different Types of Green Roofs*

The findings of the study, as depicted in Fig. 7, revealed notable disparities in the cooling and humidifying benefits among different types roofs. Tree plants performed the best with maximum cooling and humidification benefits of 1 °C and 0.15% at the roof level and 0.22 °C and 0.09% at the pedestrian level, respectively. In comparison,

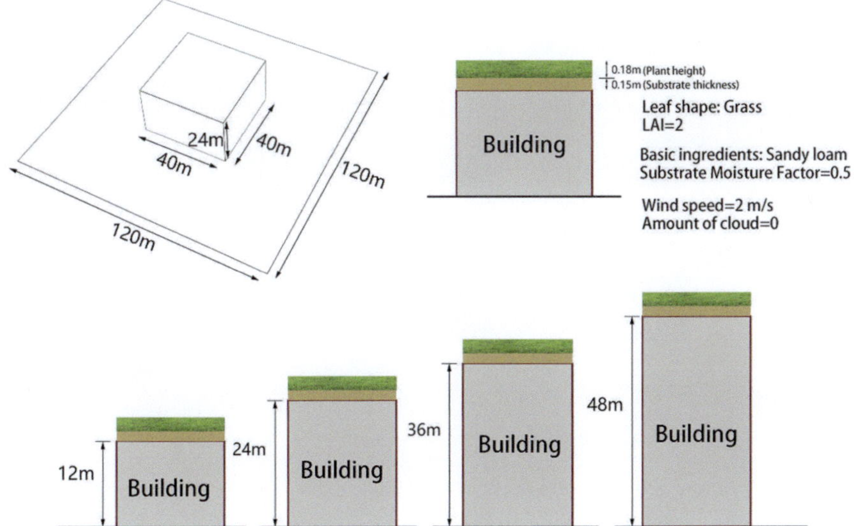

Fig. 6 Building height diagram (Self-drawn)

Table 10 Green roofs at different heights (Self-drawn)

Case	H = 12 m	H = 24 m	H = 36 m	H = 48 m
Building height	12	24	36	48
Building size	40 * 40	40 * 40	40 * 40	40 * 40

herbaceous and shrubby plants had similar cooling benefits, with maximum cooling of 0.71 °C and 0.73 °C at the roof level.

Trees also performed well in improving thermal comfort, as indicated by the *UTCI*. Specifically, the thermal comfort improvements were 0.28 °C for the pedestrian layer and 5.63 °C for the roof layer, respectively. In contrast, herbs and shrubs showed relatively smaller improvements, as illustrated in Fig. 8. In terms of surface temperature change, the maximum temperature reduction was 38.38 °C for trees and 38.47 °C and 38.42 °C for herbs and shrubs, respectively.

At night, the effects of these vegetation types on temperature and humidity remain stable, but during the day, especially from morning to noon, trees significantly outperform the others, with a midday cooling effect reaching 0.38 °C, while shrubs and herbs peak at around 0.25 °C. Trees reduce relative humidity to approximately −1.70% by 10:00 AM and reach a peak impact of 2.8 °C on *UTCI* by 2:00 PM, compared to 1.8 °C for shrubs and 0.2 °C for herbs. In the afternoon, the cooling effects of all vegetation types gradually diminish, with trees dropping to below 0.10 °C by 6:00 PM, while relative humidity effects fluctuate, with trees at about −0.30% in the evening. By night, the influences of all vegetation types converge, with trees recovering to approximately 0.4 °C on UTCI after 10:00 PM (see in Fig. 9).

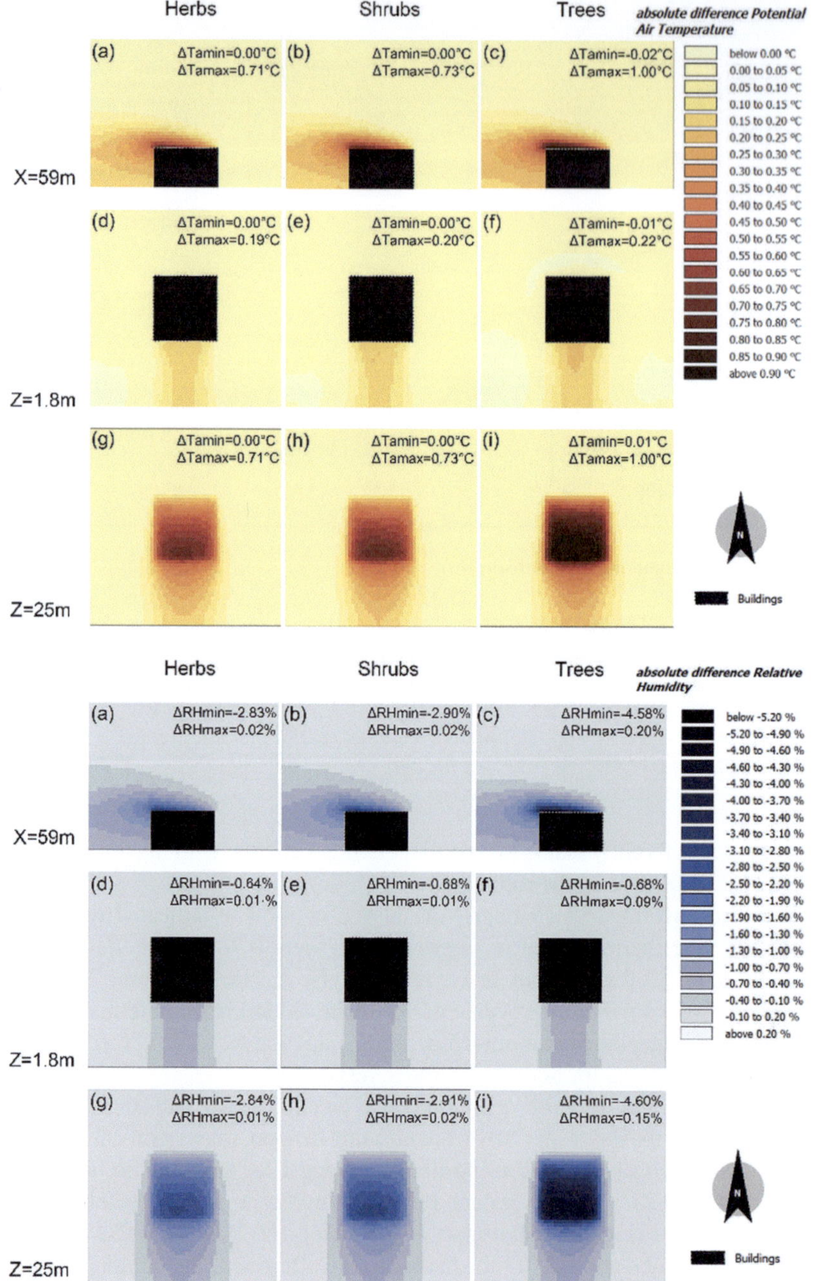

Fig. 7 Spatial distribution of different types of green roofs at 14:00 ΔTa and ΔRH (Self-drawn)

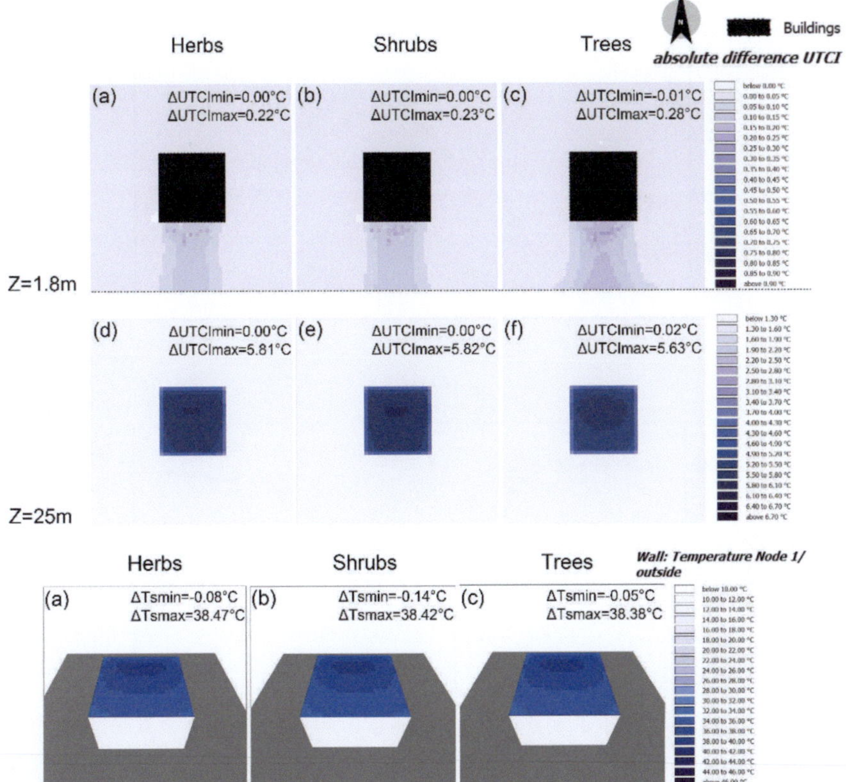

Fig. 8 Pedestrian layer distribution of different types of green roofs at 14:00 $\Delta UTCI$ and ΔTs

3.2 *Effect of Green Roofs with Different LAI*

The findings of the study, as shown in Fig. 10, indicate that green roofs with varying leaf area indices (LAIs) exhibit significant differences in their cooling and humidifying effects. Specifically, an LAI of 3.5 demonstrates the most pronounced cooling and humidifying impacts, achieving reductions of 0.76 °C and 0.13% at the roof level, and 0.20 °C and 0.04% at the pedestrian level. In contrast, the weakest effects are observed when the LAI is 0.5. The trend for cooling effects at the pedestrian level follows the order (LAI = 3.5) ≈ (LAI = 2) > (LAI = 0.5), while at the roof level, it is (LAI = 3.5) > (LAI = 2) > (LAI = 0.5).

Furthermore, green roofs with a high Leaf Area Index (LAI) of 3.5 show the greatest enhancement in thermal environment quality, as depicted in Fig. 11. These roofs result in temperature increases of 0.25 °C at the pedestrian level and 5.81 °C at the rooftop level, outperforming those with an LAI of 2. The effects of a low LAI (0.5) are notably weaker. As vegetation coverage increases, the cooling effect on

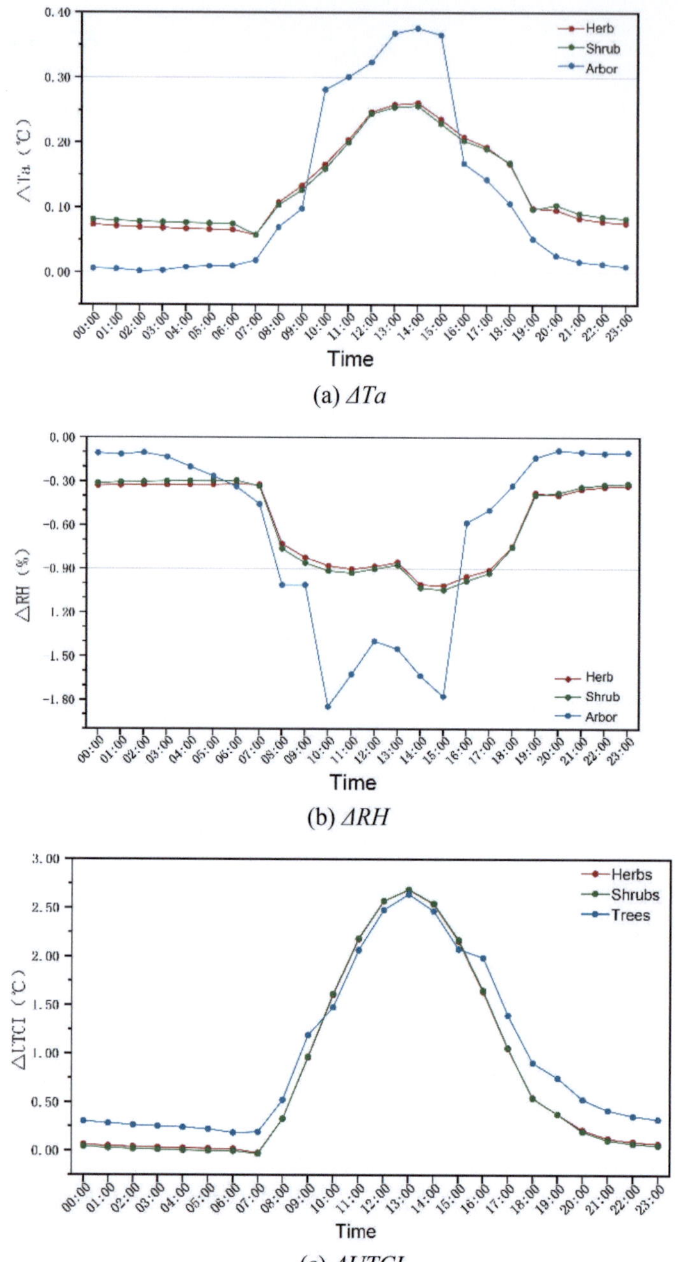

(a) ΔTa

(b) ΔRH

(c) $\Delta UTCI$

Fig. 9 Daily variation of different types of green roofs at roof level ($z = 25$ m) **a** ΔTa **b** ΔRH **c** $\Delta UTCI$ (Self-drawn)

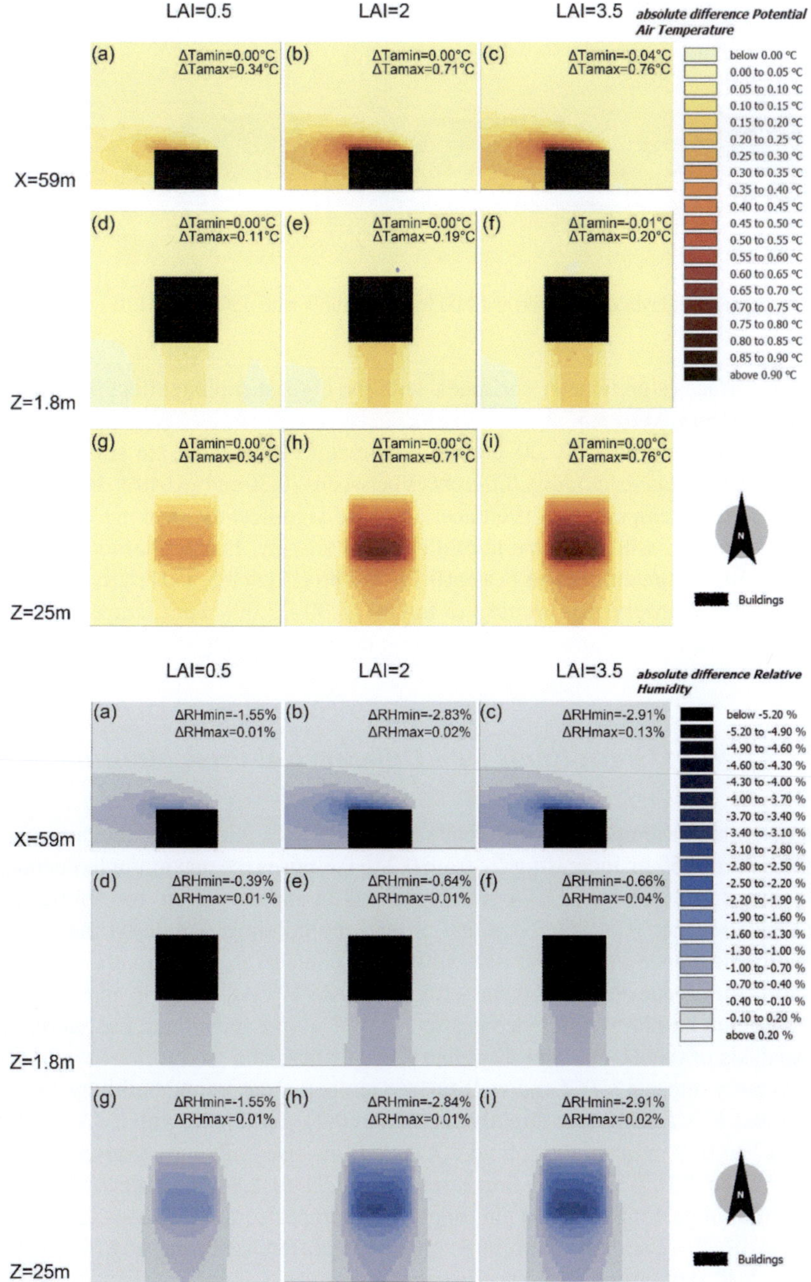

Fig. 10 Spatial distribution of green roofs at 14:00 for different LAI ΔTa and ΔRH (Self-drawn)

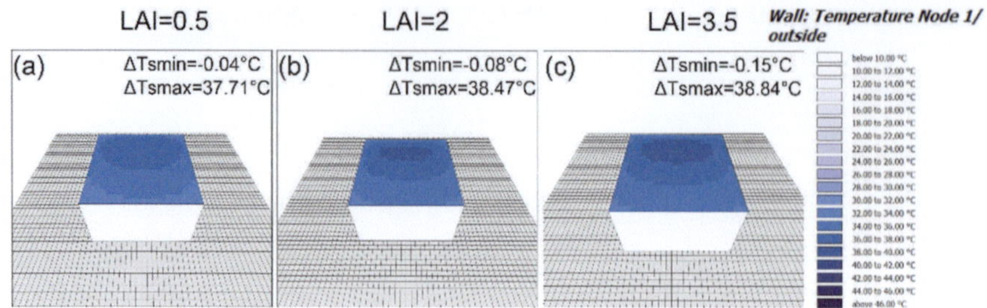

Fig. 11 Pedestrian layer distribution at 14:00 for green roofs with different LAI *ΔTs* (Self-drawn)

rooftop surfaces significantly enhances, with the highest cooling effect of 38.84 °C observed at an LAI of 3.5.

In the early morning (00:00–08:00), temperature differences are small (around 0.10 °C), with relative humidity differences between −0.20 and −0.40%. By morning (08:00–12:00), temperature differences increase significantly, peaking at 0.25 °C with LAI = 3.5, while relative humidity drops sharply. In the afternoon (13:00–17:00), temperature differences peak at 0.27 °C with LAI = 0.5, and relative humidity shows slight recovery. By evening to night (18:00–23:00), both temperature and humidity differences return to early morning levels (see in Fig. 12).

3.3 Effect of Green Roofs with Different Substrate Moisture

The potential air temperature difference (*ΔTa*) increases with water content (ζw) in Fig. 13. At $x = 59$ m, increasing ζw from 0.2 to 0.8 raises the maximum temperature difference from 0.65 °C to 0.74 °C. At $z = 1.8$ m and $z = 25$ m, the temperature difference reaches 0.74 °C above and around the building. Similarly, the relative humidity difference (*ΔRH*) also rises with water content; at $x = 59$ m, the minimum humidity difference decreases from −1.55% to −2.91%. At $z = 1.8$ m and $z = 25$ m, the minimum humidity difference is also −2.91% above and around the building.

Analysis of the *ΔUTCI* and *ΔTs* charts reveals that increased water content (ζw) significantly impacts *UTCI* and wall surface temperatures around buildings. At $z = 1.8$ m, the *UTCI* difference slightly rises from 0.21 to 0.23 °C with ζw increasing from 0.2 to 0.8, indicating minimal effects on low-rise buildings. Conversely, at $z = 25$ m, the *UTCI* difference increases from 5.71 to 5.84 °C, reflecting a more substantial impact on high-rise buildings. Wall temperature differences range from 38.33 to 38.52 °C, with the minimum temperature difference decreasing from −0.06 to −0.12 °C, demonstrating a consistent influence.

The study analyzed how temperature difference (*ΔTa*), relative humidity difference (*ΔRH*), and Universal Thermal Climate Index difference (*ΔUTCI*) change throughout the day under varying water content. From midnight to morning, *ΔTa* remains around 0.10 °C with minimal water content influence. However, from

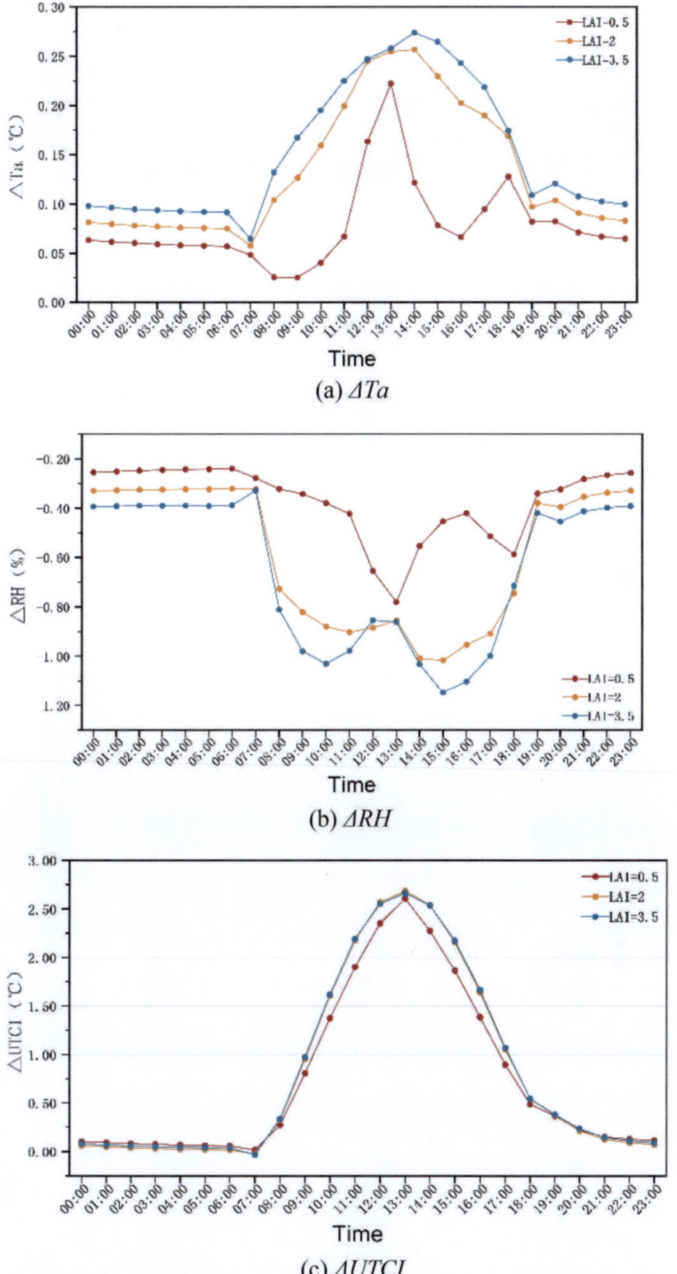

(a) *ΔTa*

(b) *ΔRH*

(c) *ΔUTCI*

Fig. 12 Daily variation of green roofs at roof level ($z = 25$ m) for different LAI **a** *ΔTa* **b** *ΔRH* **c** *ΔUTCI* (Self-drawn)

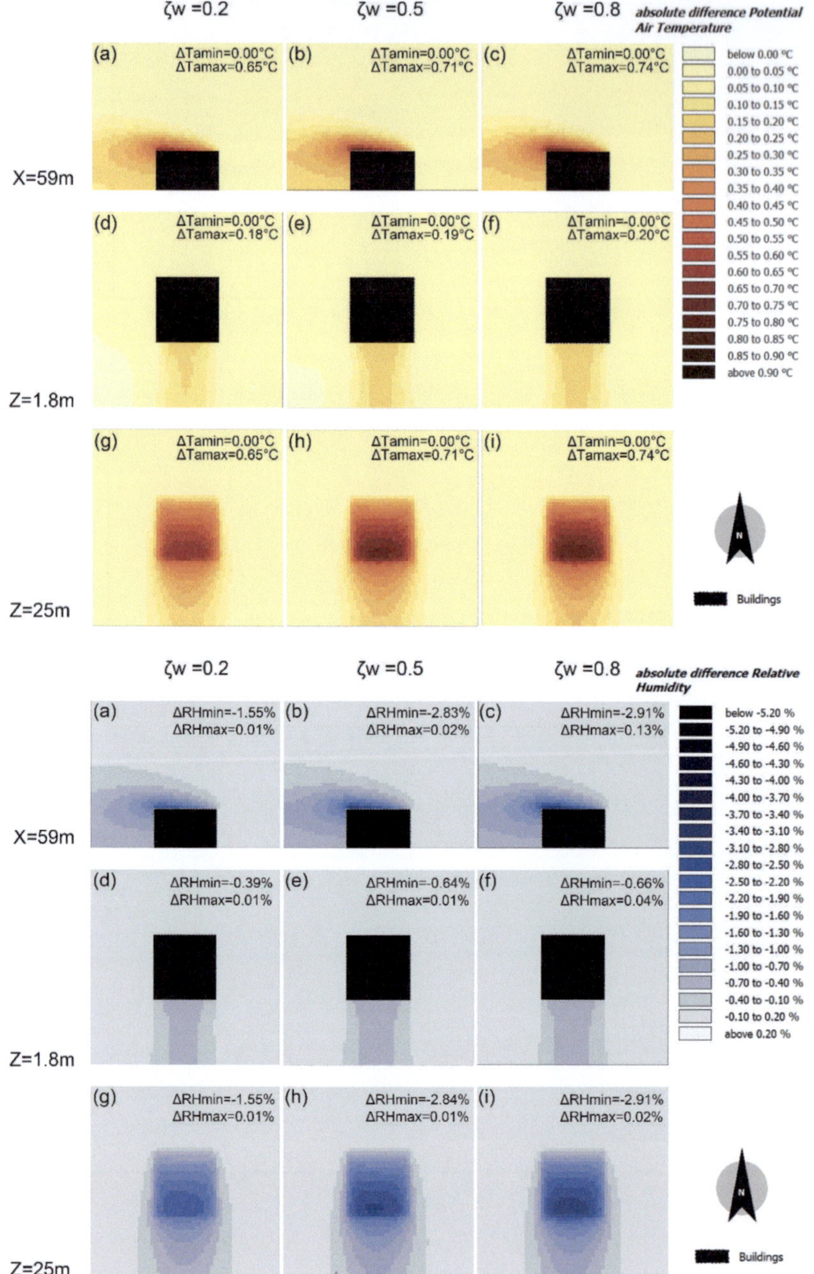

Fig. 13 Spatial distribution of green roofs at 14:00 for different substrate moisture *ΔTa* and *ΔRH* (Self-drawn)

morning to noon, ΔTa increases significantly to nearly 0.25 °C at $\zeta w = 0.8$, while ΔRH drops sharply to about −1.0% and $\Delta UTCI$ peaks at approximately 2.75 °C, indicating a strong impact of high water content.

In the afternoon, ΔTa peaks at 13:00 and then declines, with higher values under elevated water content. ΔRH slightly recovers but stays low, while $\Delta UTCI$ remains high. By evening, ΔTa returns to morning levels, and the influence of water content diminishes as both ΔRH and $\Delta UTCI$ gradually rise (see in Fig. 14).

3.4 Effect of Green Roofs at Different Wind Speeds

At heights of $x = 59$ m and $z = 25$ m, increasing wind speeds lead to greater temperature and humidity differences in Fig. 15. Specifically, at $x = 59$ m, the maximum ΔTa rises from 0.41 to 0.71 °C, and a similar trend is observed at $z = 25$ m. ΔRH also increases notably; at $x = 59$ m, the minimum ΔRH drops from −1.81 to −2.83%, and at $z = 25$ m, it decreases from −1.81 to −2.84%. In contrast, at $z = 1.8$ m, ΔTa and ΔRH show minimal changes, indicating that wind speed has a limited impact on low-rise buildings.

At $z = 1.8$ m, $\Delta UTCI$ shows minimal variation, with maximum values between 0.16 and 0.22 °C, indicating little effect on thermal comfort around low-rise buildings. In contrast, at $z = 25$ m, the maximum $\Delta UTCI$ decreases from 6.45 °C at low wind speeds to 5.17 °C at higher speeds, highlighting a substantial impact on thermal comfort around high-rise buildings. Additionally in Fig. 16, wall surface temperature difference (ΔTs) is notably affected by wind speed; under low wind speed ($WS = 0.5$ m/s), the maximum ΔTs is 44.65 °C, decreasing to 38.47 °C at $WS = 2$ m/s and 33.45 °C at $WS = 3.5$ m/s, indicating that higher wind speeds reduce wall temperature differences.

From midnight to morning (00:00–08:00), both ΔTa and ΔRH remain stable, with ΔTa fluctuating between 0.05 and 0.20 °C and ΔRH ranging from −0.80 to −0.20%. As the day progresses from morning to noon (08:00–12:00), ΔTa significantly increases, reaching around 0.25 °C at high wind speeds, while ΔRH sharply drops to −1.20%. $\Delta UTCI$ also rises rapidly, peaking at approximately 3.00 °C under similar conditions.

In the afternoon (12:00–17:00), ΔTa peaks at noon before gradually declining but remaining elevated. ΔRH reaches its lowest point at noon and then begins to recover, while $\Delta UTCI$ peaks at noon and subsequently decreases but stays high until evening. Finally, from evening to night (17:00–23:00), ΔTa returns to morning levels, ΔRH increases to about −0.20%, and $\Delta UTCI$ decreases, approaching early morning values (see in Fig. 17).

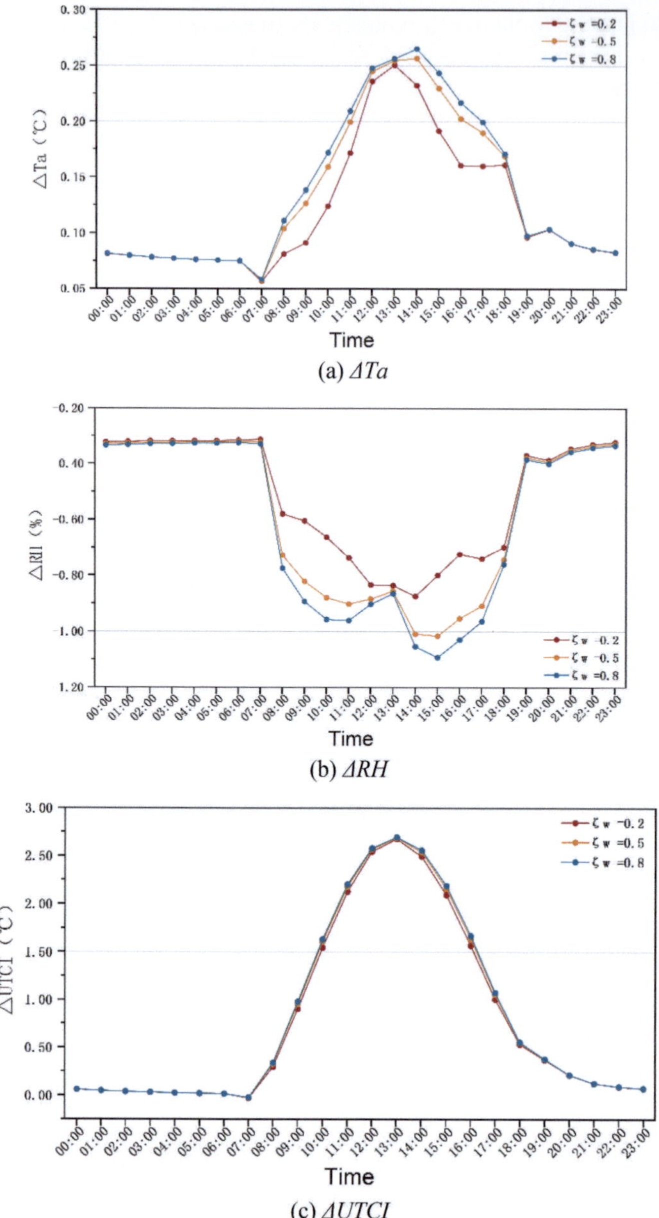

(a) *ΔTa*

(b) *ΔRH*

(c) *ΔUTCI*

Fig. 14 Daily variation of green roofs at roof level ($z = 25$ m) for different substrate moisture **a** *ΔTa* **b** *ΔRH* **c** *ΔUTCI* (Self-drawn)

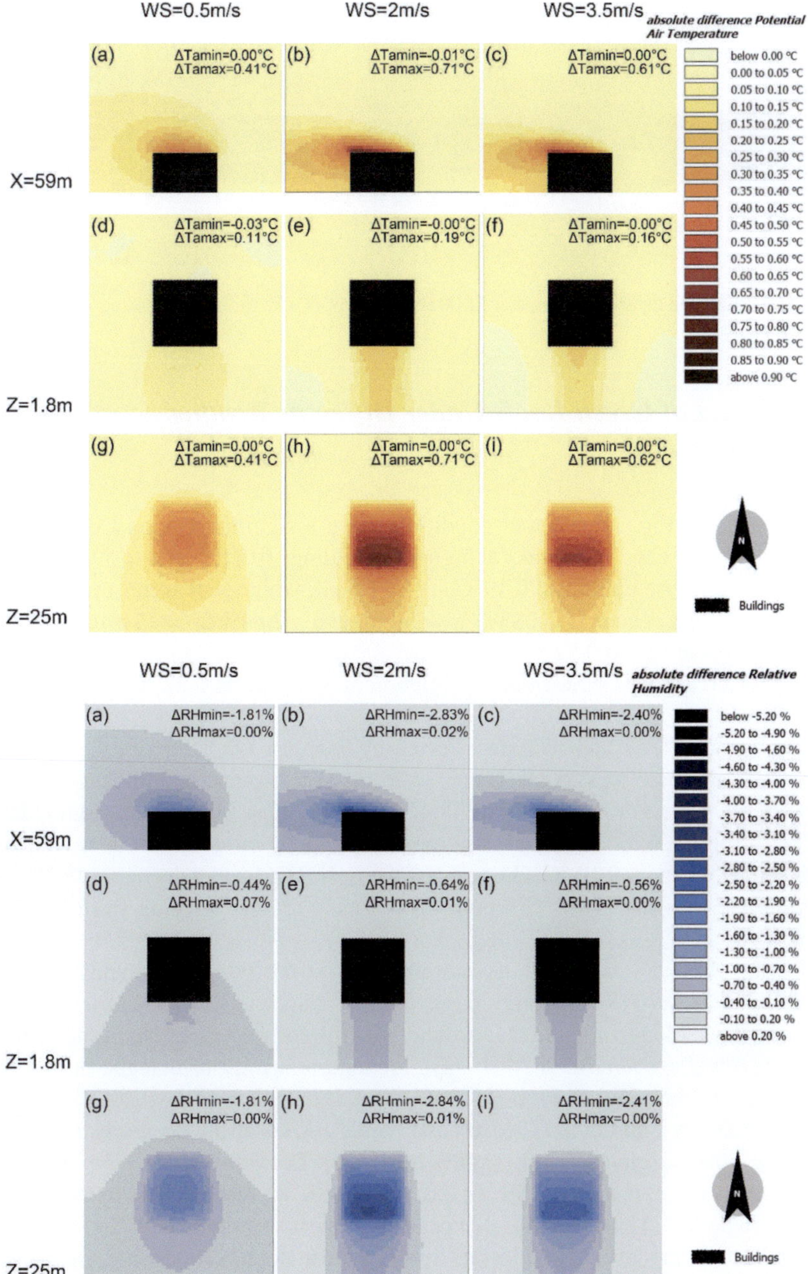

Fig. 15 Spatial distribution of green roofs at 14:00 for different wind speeds ΔTa and ΔRH (Self-drawn)

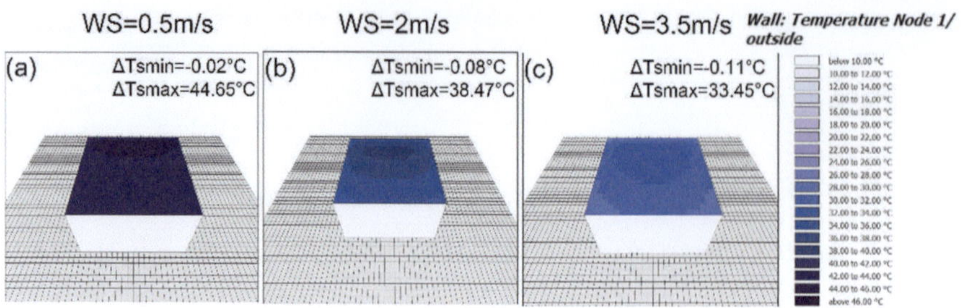

Fig. 16 Pedestrian layer distribution at 14:00 for green roofs with different wind speeds ΔTs (Self-drawn)

3.5 Effect of Green Roofs Under Different Weather Conditions

As depicted in Fig. 18, weather conditions have a significant impact on ΔTa and ΔRH. On sunny days, ΔTa peaks at high-rise buildings (0.71°C at $x = 59$ m and $z = 25$ m) and low-rise buildings (0.19°C at $z = 1.8$ m), while ΔRH is most pronounced at high-rise tops, reaching −2.83% ($x = 59$ m) and −2.84% ($z = 25$ m). On cloudy days, ΔTa decreases to 0.52 °C ($x = 59$ m) and 0.53 °C ($z = 25$ m), with ΔRH at −2.36%. Overcast days show minimal differences, with ΔTa at 0.27 °C and ΔRH at −1.56%.

Similarly, $\Delta UTCI$ and ΔTs are affected by weather. On sunny days, $\Delta UTCI$ is highest at high-rise tops (5.81 °C at $z = 25$ m) and low-rise buildings (0.22 °C at $z = 1.8$ m), with ΔTs ranging from 38.47 to −0.08 °C in Fig. 19. On cloudy days, $\Delta UTCI$ drops to 4.09 °C at high-rise tops and 0.13 °C around low-rises, with ΔTs between 29.02 and −0.10 °C. On overcast days, $\Delta UTCI$ further decreases to 2.25 °C at high-rise tops and 0.03 °C around low-rises, with ΔTs ranging from 19.11 to 0.00 °C.

During the early hours of the morning (00:00–08:00), ΔTa remains stable at approximately 0.10 °C on sunny days, 0.08 °C on cloudy days, and 0.05 °C on overcast days. Meanwhile, ΔRH is around −0.40% for *sunny and cloudy conditions and −0.20% for overcast.* $\Delta UTCI$ stays low and relatively unchanged during this period. From morning to noon (08:00–12:00), ΔTa increases to 0.20 °C (sunny), 0.15 °C (cloudy), and 0.10 °C (overcast), while ΔRH sharply decreases to −1.0% (sunny), −0.80% (cloudy), and −0.60% (overcast). The value of $\Delta UTCI$ experiences a notable increase, reaching 2.50 °C, 2.0 °C, and 1.0 °C, respectively. In the noon to afternoon period (12:00–17:00), ΔTa peaks at 0.25 °C on sunny days and gradually returns to morning levels. ΔRH hits its lowest point at noon before rising, and $\Delta UTCI$ also peaks at noon and then decreases toward midnight levels (see in Fig. 20).

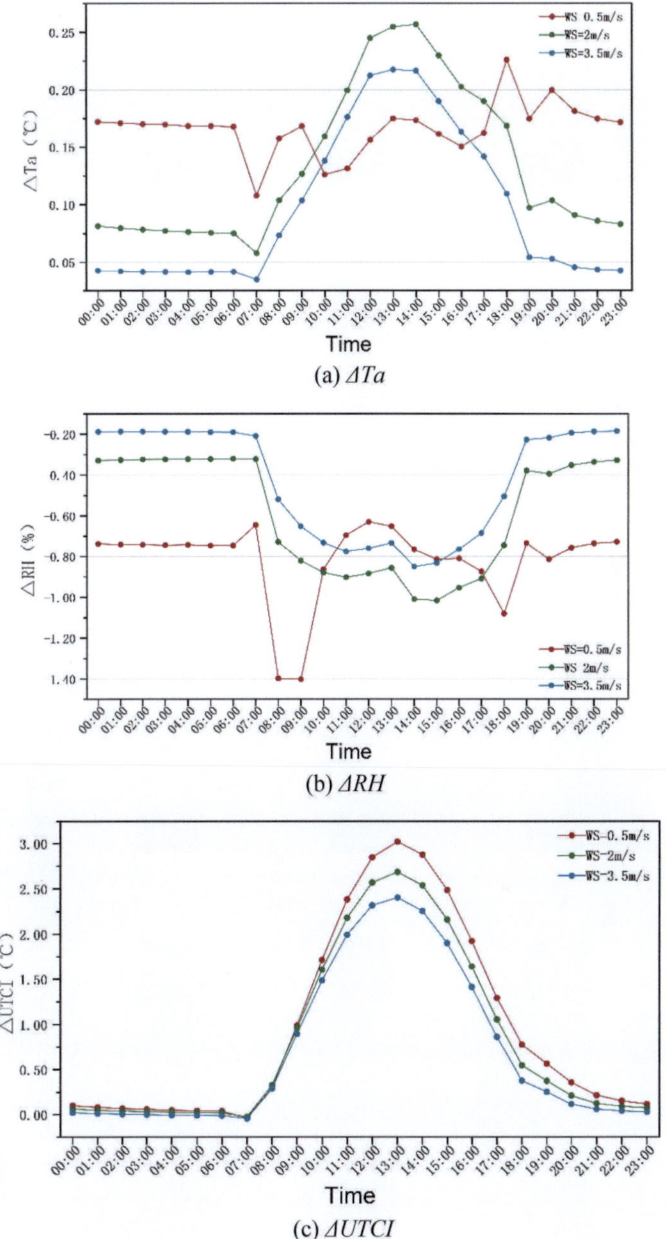

(a) *ΔTa*

(b) *ΔRH*

(c) *ΔUTCI*

Fig. 17 Daily variation of green roofs at roof level ($z = 25$ m) for different wind speeds **a** *ΔTa* **b** *ΔRH* **c** *ΔUTCI* (Self-drawn)

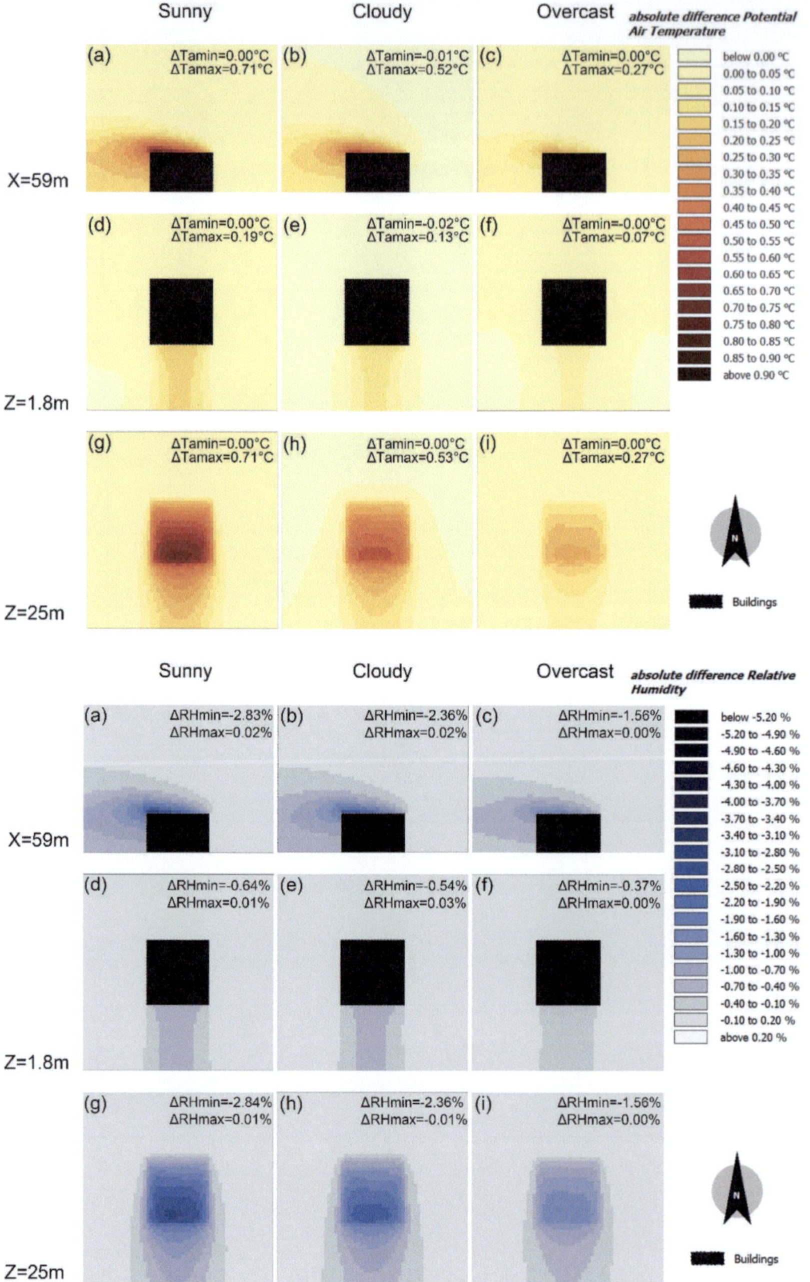

Fig. 18 Spatial distribution of green roofs at 14:00 for different weather conditions ΔTa and ΔRH (*Self-drawn*)

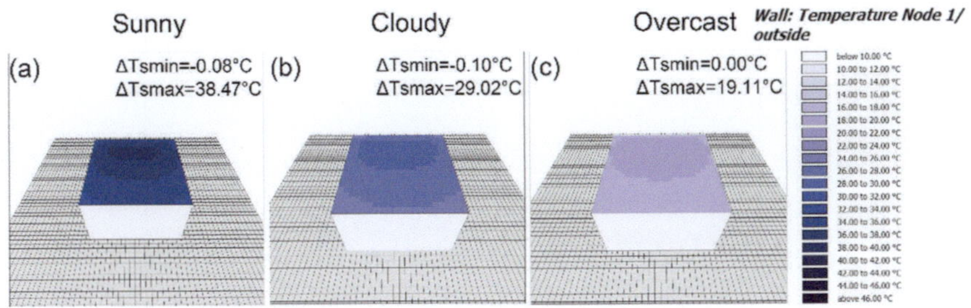

Fig. 19 Pedestrian layer distribution at 14:00 for green roofs with different weather conditions ΔTs

3.6 *Effect of Different Building Heights on Green Roofs*

As building height increases in Fig. 21, both ΔTa and ΔRH gradually decrease. At a height of 59 m, the maximum ΔTa is 0.91 °C at 12 m, decreasing to 0.56 °C at 48 m. Similarly, the minimum ΔRH is −3.60% at 12 m and drops to −2.24% at 48 m. At a reference height of 1.8 m, ΔTa decreases from 0.45 °C at 12 m to 0.04 °C at 48 m, while ΔRH declines from −1.64% at 12 m to −0.12% at 48 m. This trend is consistent across various heights ($z = 13$ m, $z = 25$ m, $z = 37$ m, $z = 49$ m), with ΔTa peaking at 0.91 °C at $z = 13$ m and $z = 25$ m, then falling to 0.56 °C at $z =$ 49 m. ΔRH also shows a decrease from −3.60% at $z = 13$ m to −2.24% at$z = 49$ m.

For low-rise buildings at a reference height of 1.8 m, the maximum $\Delta UTCI$ decreases from 0.45 °C at 12 m to 0.05 °C at 48 m. In contrast, for high-rise buildings ($z = 13$ m, $z = 25$ m, $z = 37$ m, $z = 49$ m), the maximum $\Delta UTCI$ decreases from 5.96 °C at 12 m to 5.64 °C at 48 m. Additionally, wall surface temperature differences (ΔTs) exhibit variations with building height in Fig. 22, with maximum ΔTs values recorded at 38.79 °C at 12 m, 38.47 °C at 24 m, 38.10 °C at 36 m, and 37.74 °C at 48 m. The changes in ΔTs across different heights are relatively minor.

From midnight to morning (00:00–08:00), ΔTa remains stable across heights, with values around 0.10 °C at 12 m, 0.08 °C at 24 m, 0.06 °C at 36 m, and 0.05 °C at 48 m. ΔRH is also stable, with $H = 12$ m at approximately −0.40%, while $H = 24$ m, $H = 36$ m, and $H = 48$ m range from −0.20% to −0.10%. As the day progresses from morning to noon (08:00–12:00), ΔTa increases significantly, reaching 0.30 °C at 12 m, 0.25 °C at 24 m, 0.20 °C at 36 m, and 0.15 °C at 48 m. Simultaneously, ΔRH decreases, dropping to −1.0% at 12 m and to −0.80%, −0.60%, and −0.40% at 24 m, 36 m, and 48 m, respectively (see in Fig. 23).

During the period from noon to afternoon (12:00–17:00), ΔTa reaches its peak values, with 0.35 °C at a height of 12 m, 0.30 °C at 24 m, 0.25 °C at 36 m, and 0.20 °C at 48 m. ΔRH reaches its lowest point at noon, recording −1.20% at 12 m, −1.0% at 24 m, −0.80% at 36 m, and −0.60% at 48 m. In the evening to night period (17:00–23:00), ΔTa gradually returns to morning levels, while ΔRH rises to −0.40%, −0.30%, −0.20%, and −0.10% at 12 m, 24 m, 36 m, and 48 m, respectively. Throughout the day, $\Delta UTCI$ consistently peaks around 2.5 °C at noon for all heights before gradually decreasing to near midnight levels.

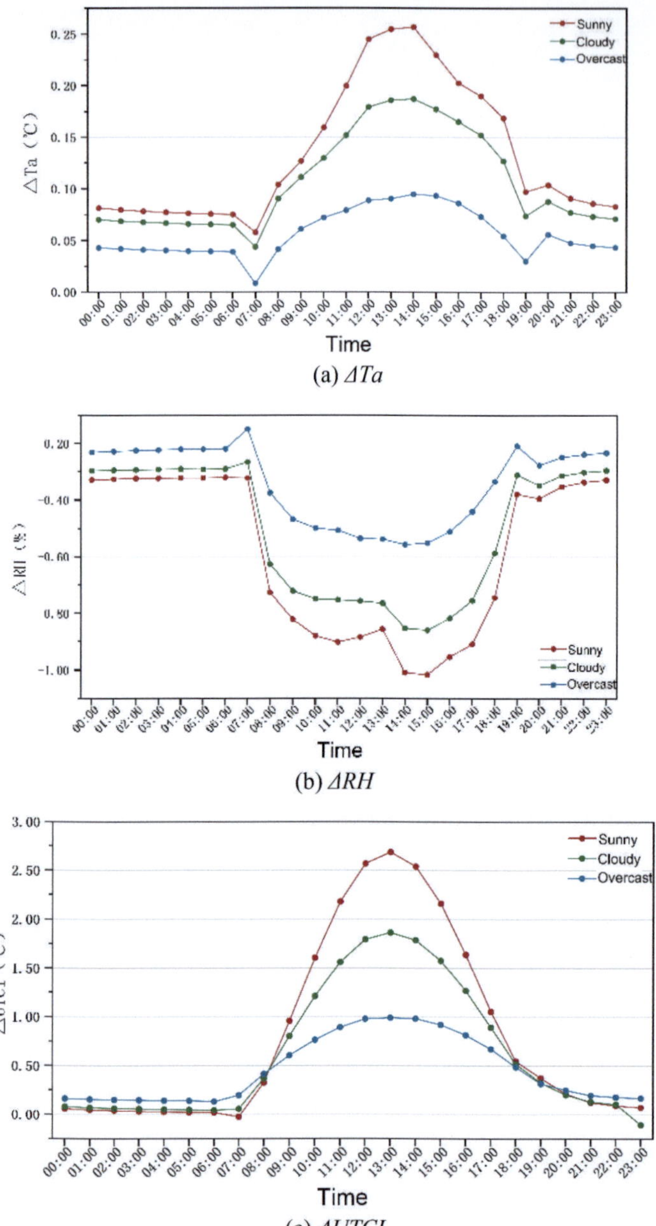

Fig. 20 Daily variation of green roofs at roof level ($z = 25$ m) for different weather conditions **a** *ΔTa* **b** *ΔRH* **c** *ΔUTCI* (Self-drawn)

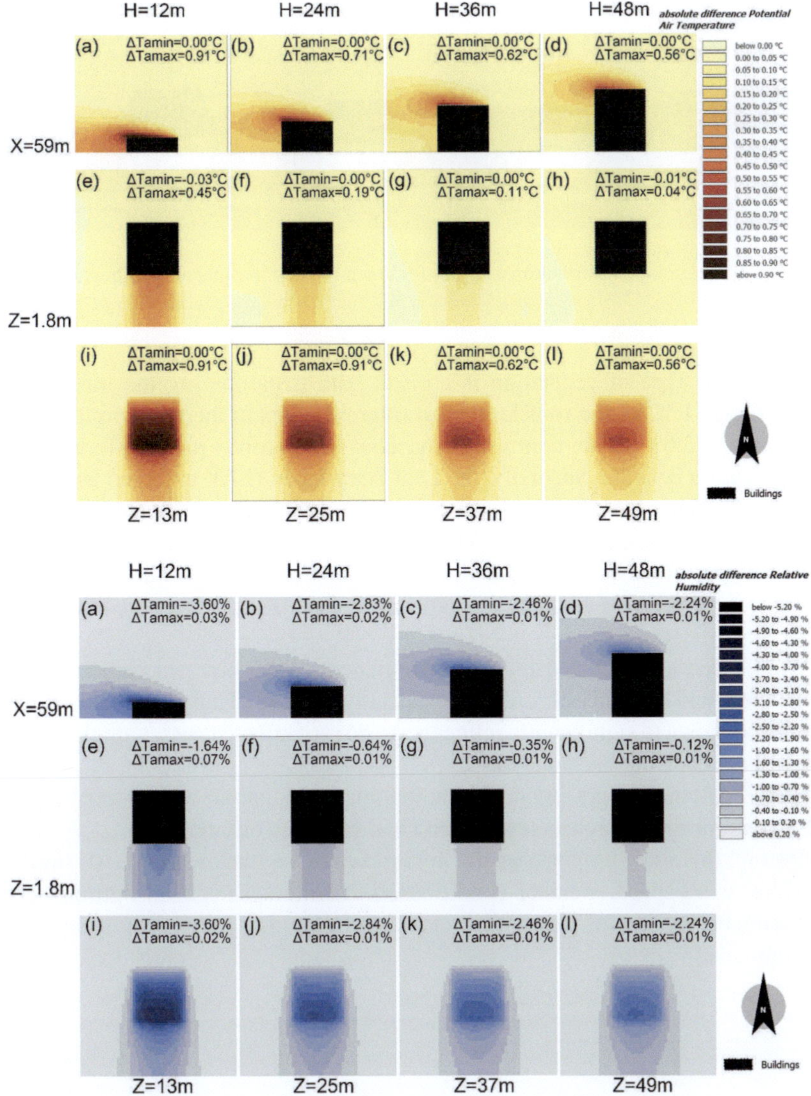

Fig. 21 Spatial distribution of green roofs at 14:00 for different building heights ΔTa and ΔRH

4 Conclusions

This research thoroughly explored the effects of various roof—related factors on the outdoor microclimate by analyzing air temperature (*Ta*), relative humidity (*RH*), surface temperature (*Ts*), and the Universal Thermal Climate Index (*UTCI*). The findings highlight the advantages of green roofs, showing notable differences based on conditions.

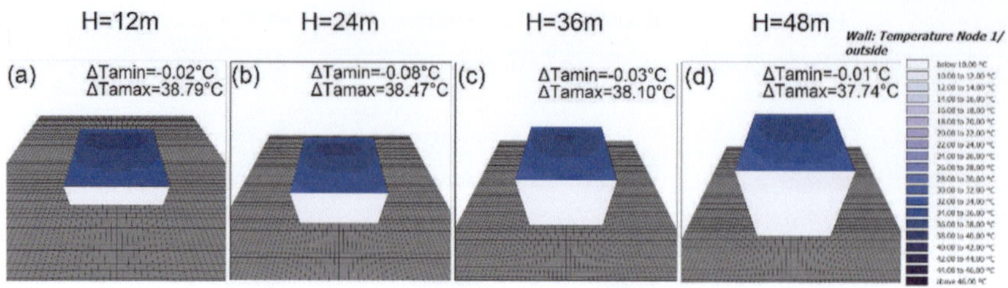

Fig. 22 Pedestrian layer distribution at 14:00 for green roofs with different building heights *ΔTs* (Self-drawn)

Firstly, the type of vegetation is critical, with trees offering the most significant cooling (1 °C at the roof level) and improvements in thermal comfort (*UTCI* reductions of 5.63 °C). In contrast, herbaceous and shrubby plants offered weaker humidification benefits. Secondly, the Leaf Area Index (LAI) emerged as a critical factor, with higher LAI (3.5) yielding maximum cooling (0.76 °C) and humidifying benefits (0.13%) at the roof level. Thirdly, substrate moisture content significantly influenced performance; increased moisture led to greater temperature and humidity differences, particularly at midday, suggesting that irrigation strategies could enhance green roof efficacy.

Wind speed also affected microclimate benefits, with higher speeds improving temperature and humidity differences, especially at elevated heights. Weather conditions were shown to significantly influence green roof effectiveness, with the greatest differences occurring on sunny days.

Lastly, building height impacted microclimate regulation, with lower buildings (12 m) exhibiting the highest temperature and humidity differences.

Overall, this study emphasizes the importance of vegetation type, LAI, substrate moisture, wind speed, weather conditions, and building height in optimizing the microclimate benefits of green roofs, guiding future design for sustainable urban environments.

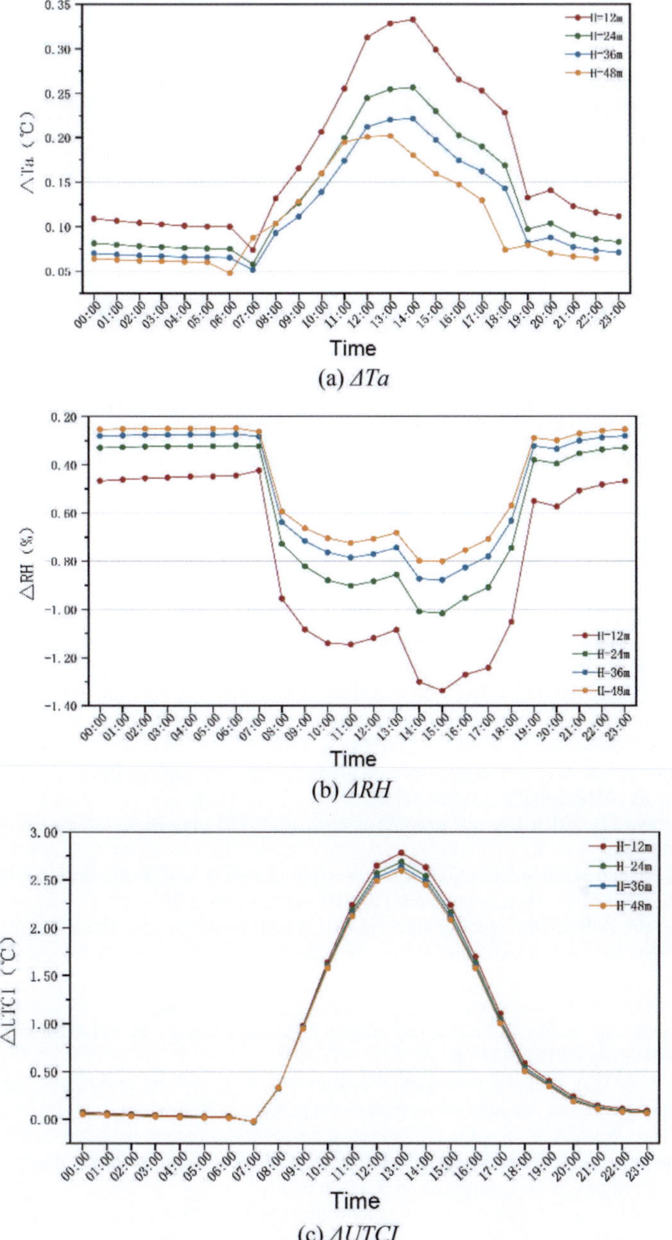

Fig. 23 Daily variation of green roofs at roof level ($z = 25$ m) for different building heights **a** *ΔTa* **b** *ΔRH* **c** *ΔUTCI* (Self-drawn)

Acknowledgements This work was financially supported by the Open Fund of the Key Laboratory of Ecology and Energy Saving Study of Dense Habitat, Ministry of Education (20220104).

References

1. Oke (1982) The energetic basis of the urban heat island
2. Rotach MW, Vogt R, Bernhofer C, Batchvarova E, Christen A, Clappier A, Feddersen B, Gryning S-E, Martucci G, Mayer H, Mitev V, Oke TR, Parlow E, Richner H, Roth M, Roulet Y-A, Ruffieux D, Salmond JA, Schatzmann M, Voogt JA (2005) BUBBLE–an urban boundary layer meteorology project. Theor Appl Climatol 81
3. Zhang N (2015) Numerical simulations on influence of urban land cover expansion and anthropogenic heat release on urban meteorological environment in Pearl River Delta
4. Tan JKN, Belcher RN, Tan HTW, Menz S, Schroepfer T (2021) The urban heat island mitigation potential of vegetation depends on local surface type and shade. Urban For Urban Green 62:127128. https://doi.org/10.1016/j.ufug.2021.127128
5. Chapma S, Watson JEM, Salazar A, Thatcher M (2017) The impact of urbanization and climate change on urban temperatures: a systematic review. Landsc Ecol 32:1921–1935. https://doi.org/10.1007/s10980-017-0561-4
6. Di Giuseppe E, D'Orazio M (2015) Assessment of the effectiveness of cool and green roofs for the mitigation of the Heat Island effect and for the improvement of thermal comfort in Nearly Zero Energy Building. Archit Sci Rev 58:134–143. https://doi.org/10.1080/00038628.2014.966050
7. Zhang P (2023) Rapid urbanization and meteorological changes are reshaping the urban vegetation pattern in urban core area: a national 315-city study in China. Sci Total Environ
8. Buchin O, Hoelscher M-T, Meier F, Nehls T, Ziegler F (2016) Evaluation of the health-risk reduction potential of countermeasures to urban heat islands. Energy Build 114:27–37. https://doi.org/10.1016/j.enbuild.2015.06.038
9. Morakinyo TE (2019) Thermal benefits of vertical greening in a high-density city_ Case study of Hong Kong
10. Cheng J, Chen M, Tang S (2023) Shenzhen–a typical benchmark of Chinese rapid urbanization miracle. Cities 140:104421. https://doi.org/10.1016/j.cities.2023.104421
11. Coutts AM, Daly E, Beringer J, Tapper NJ (2013) Assessing practical measures to reduce urban heat: green and cool roofs. Build Environ 70:266–276. https://doi.org/10.1016/j.buildenv.2013.08.021
12. Aboelata A (2021) Assessment of green roof benefits on buildings' energy-saving by cooling outdoor spaces in different urban densities in arid cities. Energy 219:119514. https://doi.org/10.1016/j.energy.2020.119514
13. Zhang K (2022) Thermal performance and energy consumption analysis of eight types of extensive green roofs in subtropical monsoon climate. Build Environ
14. Yang WD, Wang X, Fang CQ (2015) Pull-in instability of carbon nanotube-reinforced nano-switches considering scale, surface and thermal effects. Compos Part B Eng 82:143–151. https://doi.org/10.1016/j.compositesb.2015.08.044
15. Moghbel M, Erfanian Salim R (2017) Environmental benefits of green roofs on microclimate of Tehran with specific focus on air temperature, humidity and CO2 content. Urban Clim 20:46–58. https://doi.org/10.1016/j.uclim.2017.02.012
16. Chen P-Y (2022) Effects of meteorological variables and substrate moisture on evapotranspiration and thermal performance of a green roof in a subtropical climate. Ecol Eng 180:106663. https://doi.org/10.1016/j.ecoleng.2022.106663
17. Lin P, Song D, Qin H (2021) Impact of parking and greening design strategies on summertime outdoor thermal condition in old mid-rise residential estates. Urban For Urban Green 63:127200. https://doi.org/10.1016/j.ufug.2021.127200

18. Bevilacqua P, Bruno R, Arcuri N (2020) Green roofs in a Mediterranean climate: energy performances based on in-situ experimental data. Renew Energy 152:1414–1430. https://doi.org/10.1016/j.renene.2020.01.085
19. Liu Z, Cheng W, Jim CY, Morakinyo TE, Shi Y, Ng E (2021) Heat mitigation benefits of urban green and blue infrastructures: a systematic review of modeling techniques, validation and scenario simulation in ENVI-met V4. Build Environ 200:107939. https://doi.org/10.1016/j.buildenv.2021.107939
20. Alsaad H, Hartmann M, Hilbel R, Voelker C (2022) The potential of facade greening in mitigating the effects of heatwaves in Central European cities. Build Environ 216:109021. https://doi.org/10.1016/j.buildenv.2022.109021
21. Willmott CJ (1982) Some comments on the evaluation of model performance. In: Bulletin of the American meteorological society, vol 63, issue 11. https://journals.ametsoc.org/configurable/content
22. Di Giuseppe E, Ulpiani G, Cancellieri C, Di Perna C, D'Orazio M, Zinzi M (2021) Numerical modelling and experimental validation of the microclimatic impacts of water mist cooling in urban areas. Energy Build 231:110638. https://doi.org/10.1016/j.enbuild.2020.110638
23. Forouzandeh A (2018) Numerical modeling validation for the microclimate thermal condition of semi-closed courtyard spaces between buildings. Sustain Cities Soc 36:327–345. https://doi.org/10.1016/j.scs.2017.07.025
24. 2019-Influence_of_landscape_layout_on_microclimate_in_residential_area_based_on_ENVI-met_simulation.pdf
25. Morau D, Libelle T, Garde F (2012) Performance evaluation of green roof for thermal protection of buildings in reunion island. Energy Procedia 14:1008–1016. https://doi.org/10.1016/j.egypro.2011.12.1047
26. Jim CY (2014) Air-conditioning energy consumption due to green roofs with different building thermal insulation. Appl Energy 128:49–59. https://doi.org/10.1016/j.apenergy.2014.04.055
27. Chan ALS, Chow TT (2013) Energy and economic performance of green roof system under future climatic conditions in Hong Kong. Energy Build 64:182–198. https://doi.org/10.1016/j.enbuild.2013.05.015
28. Susca T (2019) Green roofs to reduce building energy use? A review on key structural factors of green roofs and their effects on urban climate. Build Environ 162:106273. https://doi.org/10.1016/j.buildenv.2019.106273
29. Zeng C, Bai X, Sun L, Zhang Y, Yuan Y (2017) Optimal parameters of green roofs in representative cities of four climate zones in China: a simulation study. Energy Build 150:118–131. https://doi.org/10.1016/j.enbuild.2017.05.079
30. Kader S, Chadalavada S, Jaufer L, Spalevic V, Dudic B (2022) Green roof substrates—a literature review. Front Built Environ 8. https://doi.org/10.3389/fbuil.2022.1019362
31. Marouli C, Savvidou P, Koutsokali M, Papadopoulou P, Misseyanni A, Tsiliki G, Georgas D (2022) Plant growth on a mediterranean green roof: a pilot study on influence of substrate depth, substrate composition, and type of green roof. Front Sustain Cities 3. https://doi.org/10.3389/frsc.2021.796441
32. Pearlmutter D, Rosenfeld S (2008) Performance analysis of a simple roof cooling system with irrigated soil and two shading alternatives. Energy Build 40:855–864. https://doi.org/10.1016/j.enbuild.2007.06.004
33. Jaffal I, Ouldboukhitine S-E, Belarbi R (2012) A comprehensive study of the impact of green roofs on building energy performance. Renew Energy 43:157–164. https://doi.org/10.1016/j.renene.2011.12.004
34. Bevilacqua P, Mazzeo D, Bruno R, Arcuri N (2017) Surface temperature analysis of an extensive green roof for the mitigation of urban heat island in southern mediterranean climate. Energy Build 150:318–327. https://doi.org/10.1016/j.enbuild.2017.05.081
35. Lee LSH, Jim CY (2020) Thermal-irradiance behaviours of subtropical intensive green roof in winter and landscape-soil design implications. Energy Build 209:109692. https://doi.org/10.1016/j.enbuild.2019.109692

36. He Y, Yu H, Dong N, Ye H (2016) Thermal and energy performance assessment of extensive green roof in summer: a case study of a lightweight building in Shanghai. Energy Build 127:762–773. https://doi.org/10.1016/j.enbuild.2016.06.016
37. Jim CY (2015) Diurnal and partitioned heat-flux patterns of coupled green-building roof systems. Renew Energy 81:262–274. https://doi.org/10.1016/j.renene.2015.03.044
38. Berardi U (2016) The outdoor microclimate benefits and energy saving resulting from green roofs retrofits. Energy Build 121:217–229. https://doi.org/10.1016/j.enbuild.2016.03.021
39. Jin C, Bai X, Luo T, Zou M (2018) Effects of green roofs' variations on the regional thermal environment using measurements and simulations in Chongqing, China. Urban For Urban Green 29:223–237. https://doi.org/10.1016/j.ufug.2017.12.002
40. Knaus M, Haase D (2020) Green roof effects on daytime heat in a prefabricated residential neighbourhood in Berlin, Germany. Urban For Urban Green 53:126738. https://doi.org/10.1016/j.ufug.2020.126738

Decarbonization of the Built Environment

Rapid Accounting Method and Project Verification of Carbon Emissions in the City Buildings Embodied Stage Based on the Emission Coefficient Method

Bing Li, Yixiang Lu, Yuanzhong Huang, and Tao Zhang

Abstract The embodied stage of city buildings has the characteristics of concentrated emission time and large emissions, which involves large volumes, many processes and complex material types, and there are few studies on carbon emission examples in the embodied stage. The team first clarified the calculation method and basic process, then investigated a large number of projects to propose a carbon emission accounting method that is different from the traditional calculation model, and finally used the two models to calculate the carbon emissions of multiple buildings in four projects. The results show that the average error between the new model and the traditional model is 3.5%, the average carbon emission intensity of the selected case is 400 kg/m^2, and the carbon emission of civil engineering accounts for more than 75% of the embodied stage, which is the most important type of project. All kinds of steel and commercial concrete account for 80–90% of carbon emissions in the embodied stage, which is the most important material type. It is of great significance to select green building materials and improve the building assembly rate to reduce carbon emissions in the embodied stage.

Keywords Building carbon emission · Embodied stage · Rapid accounting method · Carbon reduction strategies · Green building material

1 Introduction

With the acceleration of urbanization, China's construction industry is booming, but it also causes the building carbon emissions to grow year by year (as shown in Fig. 1). According to China Building Energy Consumption Research Report (2022), the total

B. Li · Y. Huang · T. Zhang
Shaanxi City Investment and Operation Group Co., Ltd, Xi'an, Shaanxi, China

Y. Lu (✉)
China Northwest Architecture Design and Research Institute Co., Ltd, Xi'an, China
e-mail: 2451007654@qq.com

B. He et al. (eds.), *Climate-Resilient, Low-Carbon, and Sustainable Cities*, Lecture Notes in Civil Engineering 754, https://doi.org/10.1007/978-981-95-2169-2_23

carbon emission of the whole process of construction in 2020 will be 5.08 billion tonnes of carbon dioxide, accounting for as much as 50.9% of the national carbon emission. The production stage of building materials accounts for the largest share of 28.2%, followed by the operation stage with 21.7% and the construction stage with 1.0% [1].

The whole life cycle of a building includes the materialization stage, the use stage and the demolition stage [2]. According to the study, the carbon emission in the materialization stage of buildings is characterized by concentrated emission time and large absolute amount, which is the key research direction of building energy conservation and emission reduction [3]. Since the reform and opening up, China's urbanization has developed rapidly, but it has also led to a significant increase in greenhouse gas emissions, which has brought about serious problems such as climate disasters, energy consumption and environmental pollution. The sources of carbon emissions are divided into three sectors: construction, industry, and transportation, with the construction sector accounting for 37% of the total carbon dioxide (CO_2) emissions [4]. In addition, unlike rural buildings, the construction process of urban buildings is large in volume, with many procedures and complicated materials, and there is not enough research on examples of carbon emissions in the urbanization stage, and there is a lack of scientific and convenient methods to measure carbon emissions. Therefore, in order to realize the ambitious goals of "2030 Carbon Peak" and "2060 Carbon Neutral", it is necessary to study the carbon emission characteristics, accounting methods and carbon reduction strategies of urban buildings.

The accounting boundaries of carbon emissions can be divided into three levels: regions (such as countries and cities), organizations (such as enterprises) and products, and different levels need to adopt different accounting methods. The temporal boundary can be categorized into pre-use, in-use and post-use, i.e., the process from "cradle to grave". The article counts and calculates the actual carbon emissions of

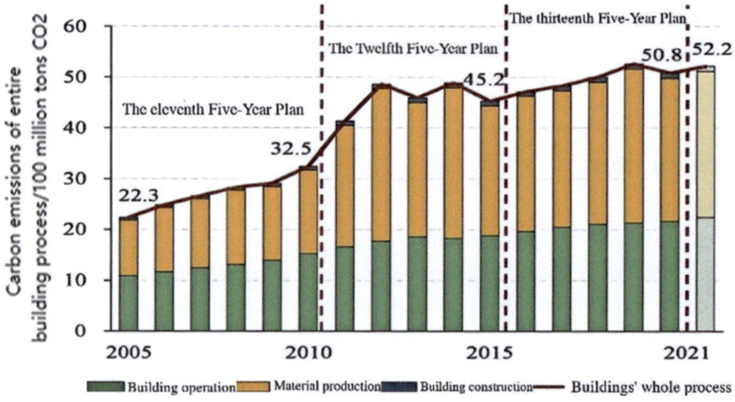

Fig. 1 Trend of carbon emissions from the whole building process in the country (2005–2021) (Self-drawn)

some cities at the stage of urbanization, which covers a wide range and a large number of samples, so the results have more analytical value.

This study focuses on the quantification of carbon emissions and the carbon reduction path in the building stage, which is divided into five parts: firstly, the applicability of the emission coefficient method is established by comparing the accounting methods, and the accounting process of the physical and chemical stage is constructed; Secondly, an empirical analysis was carried out based on more than 50 building cases in Shaanxi. Then, the composition characteristics of carbon emissions were analyzed and a rapid estimation model was proposed. Then, a phased carbon reduction strategy is proposed in combination with practice. Finally, the conclusions and prospects are summarized. By establishing a regional carbon emission database, constructing a rapid estimation method, and refining key emission reduction paths, this study provides data support for building carbon accounting, and provides methodological guidance for low-carbon design and construction management, which has important practical value for promoting the green transformation of the construction industry.

2 Calculation of the Basic Methodology and Process

2.1 Relevant Concepts

Building Carbon Footprint: This term is derived from the word "Carbon Footprint" and refers to the total amount of carbon emissions, expressed in carbon dioxide equivalents, of a product or service system over its entire life cycle [5]. A building's carbon footprint generally consists of three phases: physicalization, use and demolition of the building.

Building phase: it does not only refer to the construction phase of the building, but also includes three phases of building materials production, transportation and construction, specifically refers to the production of building materials, processing and manufacturing of components and on-site construction and installation before the building is put into use to form the engineering entity [6].

Carbon Emission Factor: A coefficient of energy and material consumption against carbon dioxide emissions to quantify the carbon emissions from activities related to different phases of the building [7].

Green building materials: building materials that can reduce the consumption of resources and mitigate the impact on the ecological environment throughout their lifespan, and that are characterized by energy saving, emission reduction, safety, health, convenience and recyclability [8].

Carbon sequestration building materials: Building materials capable of adsorbing and sequestering carbon dioxide during growth, manufacture or use [9]. For example, about 1.2 tonnes of CO_2 can be sequestered per cubic meter of bamboo.

2.2 Comparison of Methods for Calculating Carbon Emissions

By analyzing the relevant literature at home and abroad, there are three main methods for measuring carbon emissions: field measurement method, input-output method and emission factor method [10, 11] (Table 1).

Compared with the previous two methods, Although the emission factor method has the disadvantages of relying on data accuracy and dynamic update lag, the method is based on carbon emission factors, has the advantages of clear logic, simple method and wide application, which is not only suitable for the macro level of countries and cities, but also can measure the micro carbon emission of specific projects, therefore, the emission factor method is chosen for the case study.

Table 1 Comparison of common carbon emission measurement methods (Self-drawn)

	Field measurement method	The input–output method	Emission factor method
Principle	Continuous real-time detection of greenhouse gases generated in the vicinity of the product or service, using specialized equipment	Based on the principle of mass conservation, the inputs and outputs of the product or service to be measured are counted, modeled and converted to carbon emissions	Calculating carbon emissions by multiplying the activity data of the sources with the corresponding emission factors
Calculation method	Emissions are measured by means of inspection tools	Using a mathematical model of inputs and outputs to calculate the	emission factors x activity data
Characteristics	The actual measurement method is highly accurate, but it has high measurement requirements, long investment time and high cost of use	The methodology is accurate, but requires detailed in-depth analysis of specific processes, a large workload and a relatively complex process	The calculation accuracy is lower than the former two, but the calculation process is clear, the data acquisition is more convenient, and the application is relatively wide
Scope of application	Micro-computing for governmental and scientific institutions only	Applicable to macro-accounting, e.g., carbon emissions of countries, cities	Applicable at macro, micro (project, product) level

2.3 Accounting Process for the Urbanization Phase of Buildings

Each building has a huge amount of statistical data, and it takes a lot of time and effort to calculate and analyze these data. As shown in Fig. 2, in order to achieve the purpose of rapid assessment of carbon emissions, it is necessary to simplify the calculation process, which can be simplified into the following five processes.

(1) Material collection

The level of detail of the inventory greatly affects the accuracy of carbon emission calculations at a later stage, but it is difficult to collect all kinds of materials. Moreover, it was found in the carbon inventory that for buildings that have been constructed for a long time, the inventory materials are seriously missing, and there is no effective record of the energy consumption in the early stage of the building and the later stage of its use.

(2) Boundary screening

The collection of lists is so diverse in style and content that it would take a great deal of time and effort to accurately account for everything, so it is necessary to screen the accounting boundaries and discard useless lists and data.

(3) Case accounting

The carbon emission of building materials involves two key parameters: the amount of building materials and the corresponding carbon emission factor, with the amount of building materials coming from the bill of quantities and the carbon emission factor coming from relevant standards or websites. The above two stages determine the accuracy of the amount of building materials, which is then calculated for specific

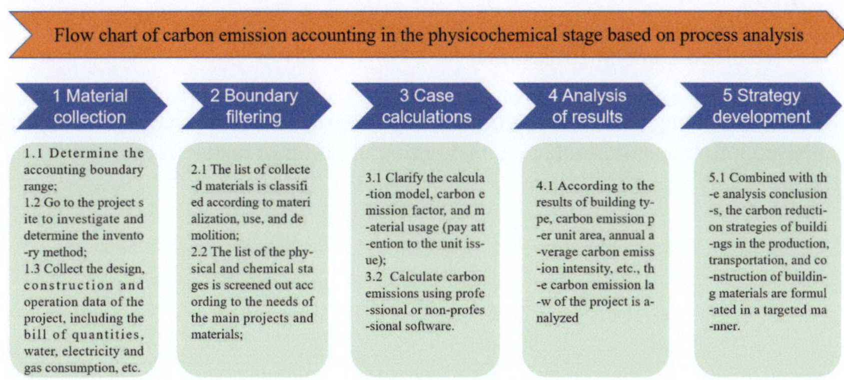

Fig. 2 Flowchart of carbon emission accounting for the buildingization phase based on process analysis (Self-drawn)

cases according to the principle of "Carbon emission = Carbon emission factor x unit quantity".

(4) Analysis of results

Analyze the carbon emission pattern and characteristics of the project based on the results of building type, carbon emission per unit area, and average annual carbon emission intensity.

(5) Strategy development

Combined with the results of the analysis, carbon reduction strategies for buildings in the production, transportation and construction of building materials are developed. The.

3 Carbon Emissions Calculation Model

3.1 Traditional Computational Models

3.1.1 Total Carbon Emissions

$$E = E_1 + E_2 + E_3 \tag{1}$$

where: E—equivalent GHG emissions from the materialization phase, $kgCO_2$ e, in the same units as below;,

E_1, E_2, E_3—Equivalent greenhouse gas emissions from the production and processing, transportation, and construction of building materials.

3.1.2 Itemized Carbon Emissions

Without considering the recycling of building materials, the carbon emissions from the production, transportation and construction of building materials are.

$$E_1 = \sum\nolimits_{i=1}^{n} Q_i \times F_i \tag{2}$$

$$E_2 = \sum\nolimits_{i=1}^{n} Q_i \times D_i \times T_i \tag{3}$$

$$E_3 = \sum\nolimits_{i=1}^{n} M_i \times N_i + E_4 \tag{4}$$

where.

Q_i—the amount of the ith building material.

F_i—carbon emission factor for the production and processing of the ith building material, kg·CO_2·e/unit.

D_i—transportation distance of the ith building material, km.

T_i—Carbon emission factor of the ith building material under a certain mode of transportation, kg·CO_2·e/(t·km).

M_i—Consumption of the ith energy source at the site of the construction machinery unit, kW·h or kg.

N_i—carbon emission factor for energy source i, kg·CO_2·e/unit.

E_4—Carbon emissions from manual activities during site construction, kg·CO_2·e.

3.1.3 Average Annual Carbon Intensity

$$I = \frac{E}{A \times T} \tag{5}$$

where: I—average annual carbon emissions per unit area, kg·CO_2·e/(m^2 ·a);

A—Gross floor area, m^2.

T—life of the building, a.

3.2 Rapid Carbon Emission Accounting Model Based on Major Building Materials

The production and construction phases have relevant standards and a large amount of data as an objective basis for calculations; while the transportation phase is highly subjective because of the variety of materials involved, the complexity and diversity of transportation conditions, and the high degree of subjectivity in transportation schemes [12]. Meanwhile, the carbon emissions of 10 types of building materials, including steel, commercial concrete, wall materials (bricks, masonry, etc.), mortar, copper-core wire and cable, building ceramics, PVC pipes, thermal insulation materials, windows and doors, and water-based paints, account for 99% of the carbon emissions of the total building materials in the building construction phase, and are the most important building materials in the physical phase of the carbon emissions [13]. The research and calculation results of this paper also prove that the carbon emissions from the production process of building materials account for 95% of the carbon emissions in the materialization stage, and only two types of building materials, steel and commercial concrete, account for about 80 to 90% of the carbon emissions in the materialization stage.

Therefore, in order to quickly assess the carbon emission situation in the early stage of the project, carbon emissions from the production of other non-major building materials as well as transportation and construction can be excluded for the time being, i.e., based on the carbon emissions from the production and processing of the major building materials (steel and concrete) and their corresponding ratios, the major works such as earthwork and civil engineering, ventilation, electricity, water supply and drainage, and heating, etc., can be selected to estimate the carbon emissions of the building in the entire physical phase, and the formula of the quick accounting model is as follows.

$$E = \frac{\sum_{i=1}^{n} Q_i \times F_i}{P} \tag{6}$$

where: E—Carbon emissions in the physical phase, kg·CO_2·e;,

Q_i—Quantity of building material type i, kg·CO_2·e/unit.

F_i—Aggregate carbon emission factor for building material i, mainly steel and commercial concrete.

P—Proportion of carbon emissions from steel and concrete production and processing in the physical phase, which in the context of this paper is taken to be 85%. This value depends on a number of factors such as the form of the building, the type of building materials, and the amount of building materials used. When the calculation includes non-major building materials such as wall materials and thermal insulation materials and their usage is large, the proportion value can be increased appropriately (maximum not more than 95%).

At present, the existing carbon emission factor is difficult to cover all types of building materials, steel, for example, the bill of quantities of steel is usually in the form of steel pipes of various calibers, reinforcing bars, combination of steel templates, supporting steel pipes and fasteners, steel plates, steel, and so on, and nails, iron, galvanized iron, connecting bushings, steel components (elbows, tees, valves, flanges, clamps, bolts, screws) carbon emissions are relatively small, the calculation of which can be appropriate to ignore the carbon emissions. Can be appropriately ignored in the calculation (Table 2).

4 Carbon Emission Distribution and Case Validation

4.1 Carbon Emissions Data Research

The most critical part of the new accounting formula is the determination of the distribution ratio P. Therefore, in order to accurately understand the carbon emission distribution characteristics of various types of buildings, the article collects and organizes a large amount of engineering data (data from books, literature, websites, etc.), and compares them with the following calculations to verify the reliability of the data. Figure 3 selects 68 residential buildings, 56 office buildings and 26 commercial buildings for comparison, and the results show that the carbon emission intensity of residential, office and commercial buildings are 570, 624 and 544 kg·CO_2·e/m^2, respectively, with the office building having the highest carbon emission intensity.

Table 2 Carbon emission factors of typical building materials [7, 14, 15] (Self-drawn)

Serial number	Typical building material names	Unit	Aggregate carbon emission factors (kg·CO_2·e/unit)	Data sources
1	Plain carbon steel (market average)	t	2050	GB/T 51,366
2	Wire mesh	m^2	1.79	DH 2.0
3	Steel pipe e.g. galvanized steel pipe DN150	m	369.36	CBCED
4	Combined steel formwork	t	3744	CBCED
5	Supporting steel pipes and fasteners	kg	2.69	CBCED
6	Steel plate e.g. steel plate 2.5 mm	t	2685.31	CBCED
7	Various steel sections	t	3351.83	CBCED
8	Commodity concrete e.g. C10	m^3	376.39	CBCED
9	Commodity concrete e.g. impermeable commodity concrete P8 C60	m^3	586.6	CBCED

The carbon emissions from the production, transportation and construction phases of the surveyed buildings accounted for 91%, 6% and 3%, respectively, with the production phase having the highest carbon emission intensity.

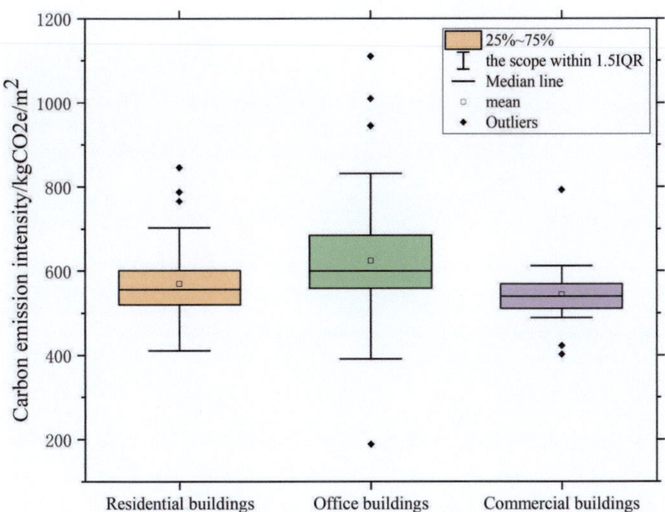

Fig. 3 Distribution of carbon emission intensity at different stages of buildingization (Self-drawn)

4.2 Case Validation Calculations

4.2.1 Calculations

In order to verify the applicability of the new model, this paper also selects four projects located in a city of multiple buildings, including commercial, residential and industrial types, project A ~ C for reinforced concrete structure, project D for steel structure, the specific information is shown in Table 3.

Figure 4 shows the real or rendered view of the project, from left to right, project A, B, C and D in order.

Table 4 is the result based on the traditional calculation model, and the carbon emissions from the production of building materials, the transportation of building

Table 3 Summary of the project (Self-drawn)

Project serial number	Building type	Gross floor area (m^2)	Useful life (a)
A	Business	240,000	50
B	Residence	450,000	100
C	Residence	600,000	50
D	Industry	36,000	50

Fig. 4 Actual or rendered view of the project (Self-drawn)

Table 4 Results of case calculations (Self-drawn)

Project serial number	Total carbon emissions ($t\cdot CO_2\cdot e$)	Carbon intensity ($kg\cdot CO_2\cdot e/m^2$)	Annual average carbon emission intensity [$kg\cdot CO_2\cdot e/(m^2\cdot a)$]	Carbon emission intensity from research ($kg\cdot CO_2\cdot e/m^2$)	Building type
A	74,759	311.49	6.23	544.24	Business
B	316,083	702.41	7.02	565.47	Residence
C	249,401	415.67	8.31	565.47	Residence
D	9284	257.88	5.16	323.64	Industry

materials and the construction process account for about 90, 6% and 4% of the carbon emissions in the materialization stage respectively, which is similar to the distribution ratio obtained from the research.

This paper also based on the proposed accounting formula (6) for the above four projects re-accounting, this paper believes that the traditional model calculated value is similar to the actual value, so Fig. 5 can be seen, the new model calculated value and the actual carbon emissions of the average error of 3.5%, the larger the carbon emissions of the project, the smaller the difference between the two percentage, it can be argued that the value of the new model is more reliable, and the new model proposed in this paper is suitable for the estimation of carbon emissions, including the larger volume of construction works, and so on. The new model proposed in this paper is suitable for the estimation of carbon emission including large construction projects.

4.2.2 Analysis of Carbon Emissions Per Unit Area

Since the types, sizes and amounts of building materials of buildings are different, it is not comparable to use only carbon emission as an evaluation index, so carbon emission per unit area is introduced to evaluate the carbon emission intensity of buildings of different sizes and types.

As shown in Table 4, except for the carbon emission intensity of Project B (702.41 $kg\cdot CO_2\cdot e/m^2$, which is obviously larger, the carbon emission intensity of the rest of the projects is lower than the average of the results of the study, which is mainly due to the fact that Project B is planned and designed to be a "century-old building", and the number of materials used is larger, and there is a significant difference in the types of materials used from traditional residential buildings, such as residential buildings. This is mainly due to the fact that Project B was planned and designed as a "century-old building", which uses a larger amount of materials and is different from traditional residential buildings. Secondly, the carbon emission intensity of projects A, C and D are 311.49, 415.67 and 257.88 $kg\cdot CO_2\cdot e/m^2$, respectively, so

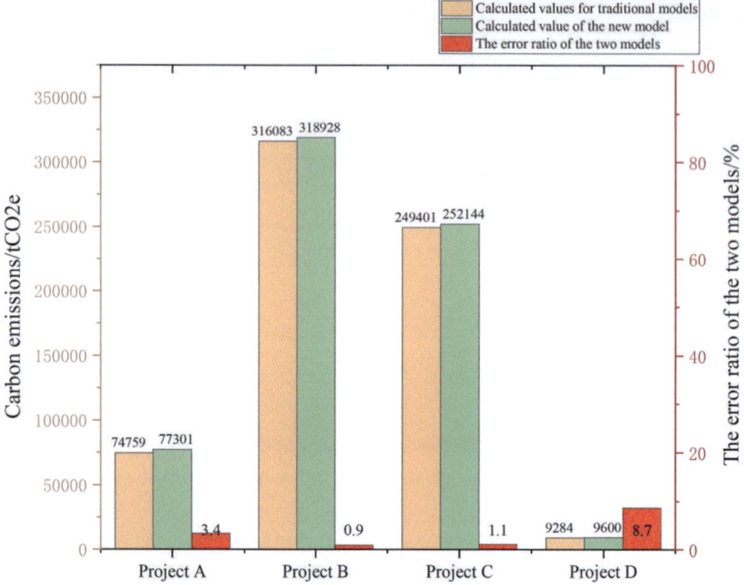

Fig. 5 Comparison of computational results between the traditional model and the new model (Self-drawn)

in the context of this paper, it can be found that residential buildings > commercial buildings > industrial buildings, which is also consistent with the carbon emission pattern obtained from the research. This is also consistent with the carbon emission pattern obtained from the research. The reason may be that the amount of building materials and the carbon emission factor of building materials in residential and commercial buildings are larger than that in industrial buildings.

4.2.3 Analysis of the Distribution of Carbon Emissions

(1) type of work

The type of construction involved in the case generally includes more than 10 projects such as heating, electrical, water supply and drainage, weak burial, ventilation and air conditioning, fire protection, civil construction, etc., of which the carbon emissions in civil construction account for the largest proportion. Take project A as an example, as shown in Fig. 6, the carbon emissions of civil engineering, ventilation engineering, electrical, water supply and drainage are 66,084, 8,454, 2,365, 1,947 t·CO_2·e, accounting for 77.12%, 9.87%, 2.76%, 2.27% respectively, accounting for a total of more than 92% of the total carbon emissions, which is the most important type of engineering at the stage of building, so in the design and construction process, attention and focus should be paid to and the construction process, which

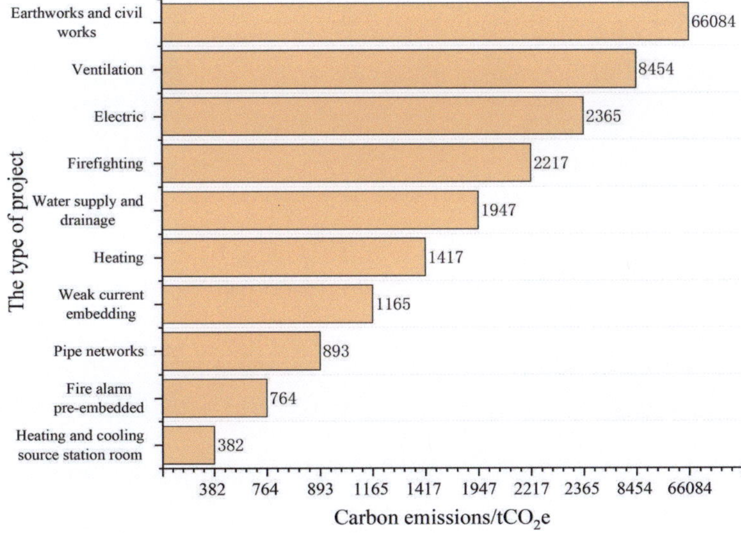

Fig. 6 Carbon emissions from different types of works (Self-drawn)

should be the most important type of engineering. Therefore, in the process of design and construction, we should pay attention to and reduce the types of materials and the amount of materials used in civil engineering, ventilation and other works, so as to reduce the carbon emissions generated in this part.

(2) Types of building materials

Building materials are divided into main structural materials, decorative materials and other materials, mainly including steel, wall materials, PVC pipes, water-based coatings and other 9 types of building materials. As shown in Fig. 7, the carbon emissions of steel and commercial concrete are 315,309 and 266,496 t·CO_2·e, accounting for 52% and 34% respectively, followed by wall materials, mortar and thermal insulation materials. In the four items from A to D, the carbon emissions of steel and concrete account for about 80% to 90% of the total carbon emissions in the materialization stage, so it can be seen that the main structural materials are the most important building materials in the building construction stage.

5 Carbon Reduction Strategies for Towns and Cities Based on the Stages of Buildings' Chemistry

Combined with the above analysis, the carbon emission of the building materialization stage accounts for about 20% of the carbon emission of the whole life cycle of the building, and in the materialization stage, the production of building materials

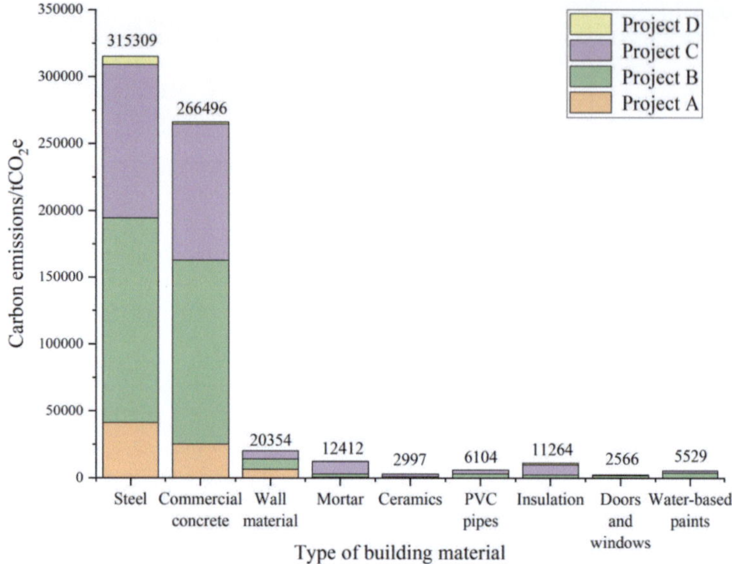

(a) Carbon emissions from different building material types

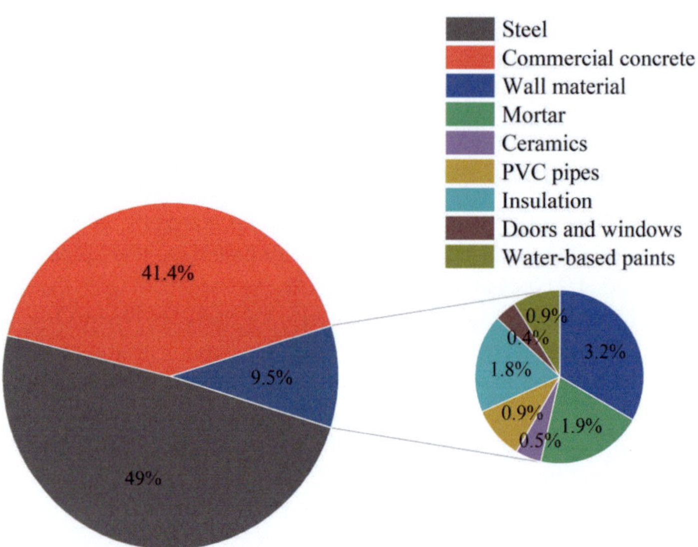

(b) the composition of different building material types

Fig. 7 Carbon emissions and composition of different building material types (Self-drawn) has the largest carbon emission, accounting for about 90% of the carbon emission of this stage. Therefore, in formulating carbon reduction strategies, the first factor to be considered is the selection and use of building materials, followed by transportation and construction, which account for about 10 per cent of the entire materialization phase.

5.1 Application of Green Building Materials

The carbon intensity of the main structural building materials varies greatly among different types of buildings, and the rational selection of building materials according to the structure and type of building can reduce carbon emissions to a large extent [14]. Specific strategies are as follows.

(1) Prioritizing the use of building materials with smaller carbon emission factors or recyclable materials, such as composite bricks instead of ordinary sintered bricks.
(2) Increase the service life of the building at the design stage, so as to reduce the average annual carbon emission intensity of the building, for example, the preparation of recycled aggregate concrete can not only reduce carbon emissions, and at the same time, due to the recycled aggregate surface mortar can be carbonized with atmospheric carbon dioxide, thus absorbing atmospheric carbon dioxide [16], and the selection of low-carbon cement or reduce the amount of cement is the most direct and effective means to reduce the carbon emissions of ready-mixed concrete [17]. The selection of low carbon cement or reduction of cement consumption is the most direct and effective means to reduce the carbon emission of ready-mixed concrete [17].
(3) Establishment of standardized components to increase the building assembly rate. Compared with the traditional cast-in-place, assembled buildings can reduce carbon emissions by 2.65 kg/m^2 [18] at the building construction stage.

5.2 Low-Carbon Transportation of Building Materials

The transportation of building materials accounts for 4% of the materialization phase, and the carbon emissions from the transportation of building materials are mainly related to the amount of building materials used, the distance between the origin of the building materials and the construction site, and the mode of transportation of the building materials [19]. Therefore, carbon reduction can be achieved by prioritizing the use of local building materials in the transportation process.

5.3 Low-Carbon Construction Management

Construction is the smallest contributor to carbon emissions throughout the physical phase, with the following main carbon reduction strategies.

(1) Reduce carbon emissions by rationalizing shift planning, i.e., reducing the number of machine shifts.
(2) Enhancement of assembly and industrialization of building construction.
(3) Adoption of the dry construction process as an alternative to the wet method.
(4) For temporary facilities such as staff quarters, existing buildings should be fully utilized for rapid assembly.

6 Conclusions and Outlook

6.1 Conclusion

The article determines the calculation method and accounting process, then obtains the carbon emission distribution pattern based on literature and field research, thus proposes the method of estimating the carbon emission of the whole materialization stage based on the main building materials, and finally carries out the validation calculations and analyses of the carbon emission of different types of buildings in a certain city using the emission factor method, which involves the residential, commercial and industrial buildings, and finally draws the following conclusions:

(1) Carbon emission accounting at the stage of urbanization is characterized by a large volume, a long cycle and many materials. As far as the accounting process is concerned, commercial concrete is mostly used in civil engineering and generally needs to be accounted for, while steel is widely used in ventilation, water supply and drainage, and fire-fighting projects in addition to civil engineering, and exists in various forms.
(2) The average carbon emission intensity of the selected cases in this paper is 400 kg/m^2, and the carbon emission of civil engineering accounts for more than 75% of the materialization stage, which is the most important type of engineering; various types of steel and commercial concrete account for about 80–90% of the carbon emission of the materialization stage, which is the most important type of materials.
(3) In the context of this paper, the production of building materials accounts for about 90% of the entire materialization phase, and the transportation and construction phase accounts for about 10%. Therefore, ignoring the carbon emissions from the production of other non-major building materials, transportation and construction, we can quickly predict the carbon emissions by counting the main structural building materials such as steel and commercial concrete.

(4) Carbon reduction strategies in the physical phase include: giving priority to the use of green building materials and carbon sequestration building materials with lower carbon emission factors, extending the service life of the building, using recycled materials to increase the recycling rate, and strengthening civil construction management, among other measures.

6.2 Outlook

It is worth noting that there are certain errors and difficulties in the application of the method, for example, there are some limitations in the carbon emission accounting for non-conventional types of urban buildings. The method is mainly used to provide data support for architectural planning, pre-project feasibility study, green and low-carbon design, as well as to provide reference for the evaluation of the project at a later stage. The universality of the ratio P in Eq. (6) for buildings of different regions, types, storeys and structural sizes has yet to be tested, and we will consider increasing the number of case samples to verify its accuracy and optimize the range of the corresponding value of P in the future.

In addition, the method of calculating carbon emissions by using the emission factor method is more mature, but the collection and screening of the inventory and the calculation by using the conventional software still require a lot of manpower and time. At present, although the commonly used carbon emission calculation software such as Donghe and CEEB support the whole life cycle calculation of carbon emission, there are still problems of missing data, non-correspondence of data forms and high modeling requirements. Therefore, in the future, there is an urgent need for a convenient and efficient calculation software that can effectively recognize all kinds of statistical inventories and cover most of the carbon emission factors.

Acknowledgements This study is supported by the Science and Technology Program (SIGC2023-KY-02) of Shaanxi Investment Group Co.

References

1. Research Report on Energy Consumption and Carbon Emission of Buildings in China (2022). Architecture (02):57–69 (2023)
2. Luo Z (2016) Research on the calculation method of building life cycle carbon dioxide emissions and emission reduction strategy. Xi'an University of Architecture and Technology
3. Luo Z, Yang L, Liu J (2014) Study on CO_2 emission in the physical phase of office buildings. Civil Const Environ Eng 36(05):37–43
4. United Nations Environment Programme (UNEP) (2021) 2021 Global status report for buildings and construction: towards a zero-emission, efficient and resilient buildings and construction sector. UNEP, Kenya
5. Carbon Footprint. What is a carbon footprint. http://www.carbonfootprint.com/carbon-footprint.html,2008-05-10.

6. Zhan X, Wang E (2015) Life-cycle assessment and control measures for carbon emissions of typical buildings in China. Build Environ 86:89–97
7. GB/T 51366–2019 (2019) Building carbon emission calculation standard. China Building Industry Press, Beijing
8. T/CECS 1555–2024 (2024) Evaluation criteria for carbon neutral buildings. China Construction Industry Press, Beijing
9. T/CSUS 60–2023 (2023) Evaluation criteria for low carbon buildings. China Building Industry Press, Beijing
10. Hu R (2023) Research on carbon emission measurement in the stage of assembled building fabrication. Guangzhou University
11. Kuan YK (2020) Research on carbon emission measurement of assembled building construction phase based on bill of quantities. Guangzhou University
12. Wang R, Gao X, Duan Y et al (2023) Research on rapid calculation method of carbon emission from building decoration. Build Sci 39(12):20–27+35
13. Kuen Y (2018) Research on carbon emission accounting methods in the stage of building chemistry. Xi'an University of Architecture and Technology
14. Li Y, Chen J (2020) Carbon footprinting in the whole life cycle of buildings. China Construction Industry Press, Beijing
15. Greenway Zero Carbon Building Network (2021). http://gdzeb.com
16. Wei M, Lin M, Xie Q et al (2024) Research on whole life cycle carbon emission analysis and calculation method (model) of recycled aggregate concrete. Concrete (2024). Online First. Accessed 30 Jul 2024
17. Pan Y, Gong J, Zhang K et al (2024) Carbon footprint calculation analysis and carbon emission reduction study of ready-mixed concrete based on full life cycle evaluation method. New Build Mater 51(04):145–149
18. Wang T, Zhang J, Zhao W et al (2022) Carbon emission modeling of assembled buildings at the stage of building fabrication and comparative evaluation of cast-in-place buildings. New Build Mater 49(10):88–91
19. Shao G, Zhao X, Gao Y et al (2012) Research on the calculation method of carbon emission from building materials in buildings. New Build Mater 39(02):75–77

Life Cycle Carbon Emissions of Lignin Asphalt Pavements: A Comparative Study on Environmental Impact

Lu He, Chongshi Xie, Tong Wu, Yue Ma, Boheng Zhang, and Saijun Cao

Abstract This study investigated the potential of lignin, a renewable biopolymer, as a partial substitute for petroleum-based asphalt in road construction. The research focused on a life cycle assessment (LCA) approach to evaluate the carbon emissions associated with lignin asphalt mixtures across various stages, including material production, asphalt mixing, transportation, and construction. Through experimental analysis and comparison with conventional asphalt, the findings demonstrated the significant environmental benefits of lignin, particularly in its ability to reduce carbon emissions. When lignin replaced 20% of the asphalt, the overall carbon footprint was notably reduced, positioning lignin as a promising alternative for sustainable pavement materials. The study highlighted that, despite the promising results in lowering emissions, further validation was required for the practical feasibility of lignin in real-world road construction projects. This research provides a comprehensive evaluation of lignin's potential in weakening the environmental impact of asphalt used in the engineering, contributing to both academic understanding and engineering application in achieving carbon neutrality goals.

Keywords Lignin · Asphalt mixture · Life Cycle Assessment (LCA) · Carbon emissions · Pavement materials

1 Introduction

As global climate change intensifies, governments and industries are increasingly implementing policies and measures aimed at reducing greenhouse gas emissions [1]. The transportation sector, particularly the carbon emissions associated with road construction, has become a crucial focus in the fight against climate change [2, 3].

L. He (✉) · C. Xie · T. Wu · Y. Ma · B. Zhang · S. Cao
Chong Qing Architectural Design Institute CO., LTD (CQADI), Chongqing, China
e-mail: He.lu2008@163.com

C. Xie
School of Architecture and Urban Planning, Chongqing University, Chongqing, China

B. He et al. (eds.), *Climate-Resilient, Low-Carbon, and Sustainable Cities*, Lecture Notes in Civil Engineering 754, https://doi.org/10.1007/978-981-95-2169-2_24

Traditional asphalt, a petroleum-based material, produces significant carbon emissions during its production, transportation, and construction phases due to its energy-intensive and pollution-heavy processes [4, 5]. Consequently, the search for alternative materials that can lower the carbon footprint of road construction has emerged as a key focus in both academic and engineering fields [6, 7].

In recent years, lignin, a natural and renewable biopolymer, has garnered significant attention [8]. As a primary component of plant cell walls, lignin is abundantly available in biomass resources such as wood and agricultural residues, making it both plentiful and sustainable. Compared to conventional petroleum-based materials, lignin generates lower carbon emissions throughout its production and use, positioning it as a potential green alternative [9]. Studies [10, 11] have demonstrated that lignin can not only serve as an additive to enhance the adhesive properties of asphalt but, under certain conditions, can also partially replace asphalt, thereby reducing overall consumption. The application of lignin as a partial substitute for asphalt in road engineering has presented a promising solution for decarbonization. As a renewable biomass material, lignin can reduce dependency on petroleum-based asphalt while contributing to carbon sequestration and lowering energy consumption and emissions during production and construction processes.

However, despite promising laboratory findings on the use of lignin in asphalt mixtures, its real-world feasibility in road construction requires further validation [12, 13]. Specifically, the full carbon footprint of lignin asphalt pavements across their entire life cycle remains underexplored. Life Cycle Assessment (LCA) offers a systematic framework to address this issue, enabling an analysis of environmental impacts across various stages, including material production, asphalt mixing, transportation, and construction [14, 15]. Through LCA, researchers can comprehensively evaluate the feasibility of lignin as an alternative to traditional asphalt and assess its true potential for reducing carbon emissions [16].

This study aimed to systematically evaluate the carbon emissions associated with the application of lignin in asphalt pavements through the LCA approach. The research encompassed the whole life cycle, from material production and asphalt mixing to transportation and construction, providing a detailed quantification of carbon emissions at each stage. By comparing these findings with conventional asphalt pavements, this paper explored the benefits and challenges of using lignin in practical engineering applications, assessing its potential as a low-carbon alternative. The results of this study offer scientific support to the road construction industry in achieving sustainable development goals and carbon neutrality policies.

2 Materials and Methods

2.1 Materials

The SBS-modified asphalt was produced by Shandong Yuanming New Material Co., LTD. As shown in Table 1, the test results of asphalt met the requirements. The coarse and fine aggregates were generated by Langfang Shikang new building materials Co., LTD. The surface of the aggregate had no cracks and no weathering phenomenon. Basic properties of aggregates were tested and listed in Tables 2 and 3. The mineral powders were purchased in Lingshou county Rong Chuang building materials Co., LTD. and basic properties met the requirements of standard in Table 4.

Table 1 Basic properties of SBS-modified asphalt (Self-drawn)

Properties	Result	Standard
Softening point/°C	68.3	ASTM D36-06
Penetration /25 °C, 0.1 mm	48.2	ASTM D5-06
Ductility /5 °C, cm	31.2	ASTM D113-07
Flash point/°C	299	ASTM D92-02
Mass loss/%	0.28	ASTM D2872-04
Wax content/%	99.1	ASTM D3344-90

Table 2 Basic properties of coarse aggregate (Self-drawn)

Properties	Result	Standard
Los Angeles abrasion test/%	22.1	T0317–2005
Crushing value/%	11.2	T0316–2005
Adhesion to asphalt	4 level	T0616–1993
Water absorption/%	0.5	T0307–2005

Table 3 Basic properties of fine aggregate (Self-drawn)

Properties	Result	Standard
Sand equivalent/%	67	T0334–2005
Soundness/%	10	T0340–2005
Mud content/%	1.7	T0333–2000

Table 4 Basic properties of mineral powder (Self-drawn)

Properties	Result	Standard
Water content (%)	0.4	T0103–2007
Plasticity index	2.9	T0354–2005
Hydrophilic coefficient	0.7	T0353–2005

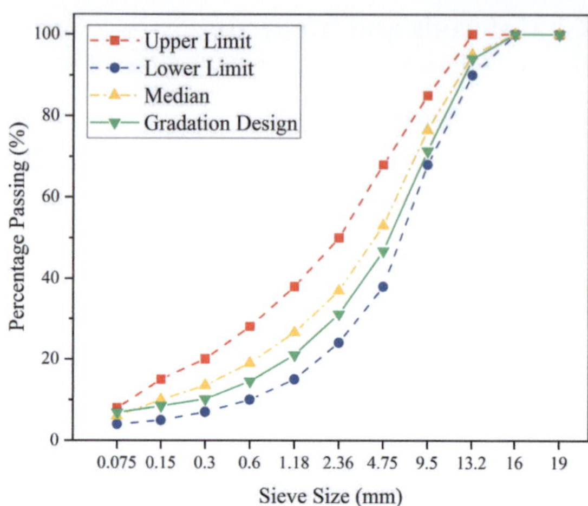

Fig. 1 Gradation design of asphalt mixture in this study (Self-drawn)

2.2 Preparation of Asphalt Binders

The preparation of asphalt employed a high-speed shearing method. First, the asphalt was heated to a specified temperature, typically between 150 and 180 °C. Then, polymeric additives or lignin were dispersed into the asphalt and sheared for 30 min at 4000 rad/min. Finally, the high-speed shearing machine was adjusted to 500 rad/min and operated for 5 min to eliminate foaming.

2.3 Gradation of Asphalt Mixtures

The gradation design of asphalt mixtures is a critical factor influencing the mechanical properties and durability of asphalt pavements. It involves selecting and proportioning aggregates of various sizes to achieve an optimal balance between strength, stability, and workability. This study utilized the Marshall design method and selected SMA-13 as the gradation for the asphalt mixture, as shown in Fig. 1. The optimum asphalt-aggregate ratio for the mixture was identified as 6.0%.

2.4 Experimental Method

This study employed three commonly used tests on asphalt mixtures to determine the optimal replacement amount of lignin for asphalt, including the rutting test, the freeze–thaw splitting test, and the low-temperature crack resistance test. These tests were conducted to assess high-temperature stability, water stability, and low-temperature crack resistance, respectively [17–19]. Additionally, the rotary viscosity

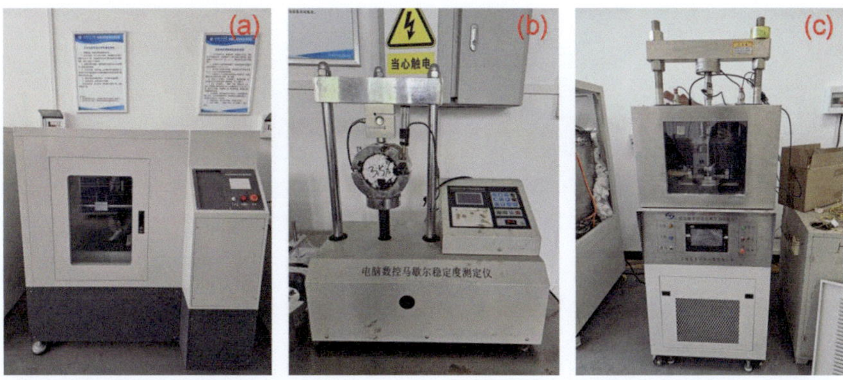

Fig. 2 Commonly used tests of asphalt mixtures: **a** rutting test; **b** freeze–thaw splitting test; **c** low-temperature crack resistance test (Self-drawn)

test was performed to measure the viscosity of the asphalt, helping to determine the appropriate mixing temperature for the modified asphalt (Fig. 2).

2.5 Methodology

In this study, system boundaries of different asphalt mixtures were divide as materials production, mixture mixing, mixture transportation and construction of three phases. LCA is a comprehensive analytical method used to assess the environmental impacts of a product, process, or service across its entire life cycle. LCA covers all stages, from raw material extraction, production, and manufacturing, through the use phase, to final disposal or recycling. By identifying and assessing environmental burdens across various phases, LCA provides scientific insights for decision-makers to optimize resource use, reduce pollution, and promote sustainable development. In this study, the system boundary was limited to materials production, asphalt mixture mixing, transportation, and construction, with the analysis primarily focused on the material characteristics and environmental impacts associated with asphalt paving. Figure 3 illustrated the system boundaries for different asphalt mixtures, including materials production, mixture mixing, transportation, and construction across three phases. One tonne of asphalt mixture was defined as the functional unit. The final output was the carbon dioxide emissions generated during the production and construction phases of the paving process.

Carbon emissions are calculated by multiplying unit emissions by their corresponding emission factors, ensuring that all relevant assumptions are explicitly stated [20]. The Life Cycle Inventory (LCI) is a crucial phase in LCA, focused on collecting and quantifying input and output data for a product, process, or service throughout its entire life cycle.

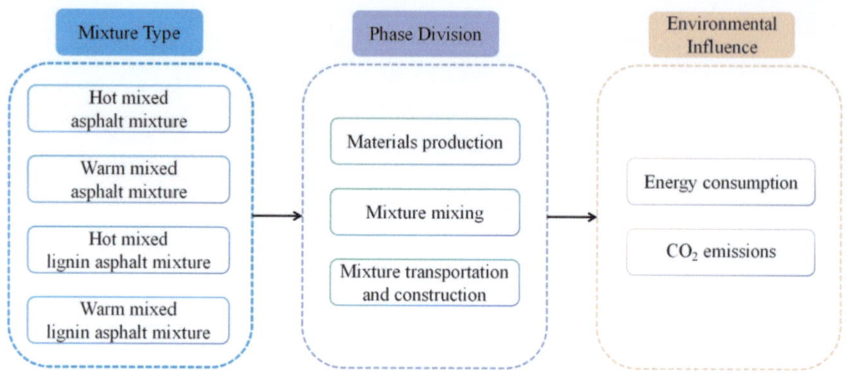

Fig. 3 System boundary of different asphalt mixtures (Self-drawn)

The primary objective of LCI is to identify the main sources of carbon emissions during each stage of pavement construction and to establish an input–output data inventory. In the raw material production stage, carbon emission resulted from material consumption, such as the production of asphalt binders, coarse aggregates, fine aggregates, mineral powder, lignin, and warm mixed additive. These materials were produced through well-established processes and techniques. The unit energy consumption and carbon emission factors for these materials were provided in Table 5.

Carbon emissions at the materials production phase were derived from the production process of various raw materials and were calculated on the basis of the following formula.

$$E_a = \sum E_i = \sum m_i \times E_{i0} \tag{1}$$

where E_a was the total carbon emissions at the materials production phase; E_i was carbon emissions from i material at the materials production phase; m_i was the amount of i material used; E_{i0} was the unit carbon emission of the production of i material.

Table 5 Carbon emission and energy consumption of raw material (Self-drawn)

	Unit carbon emission (kg/t)	Unit energy consumption (MJ/t)	References
SBS asphalt	512.50	5803.25	[21]
Coarse aggregate	2.43	31.82	[21]
Fine aggregate	8.69	58.56	[21]
Mineral powder	47.21	207.36	[21]
Warm mixing agent	112.54	621.57	[22]
Lignin	53.30	2140.21	[23]

Carbon emissions at the mixture mixing phase were derived from the heating materials, operation of the mixing plant and others. Carbon emissions from heating materials were calculated as the following formula.

$$Q_b = \sum Q_i = \sum m_i \times C_i \times (T - T_0) \quad (2)$$

where Q_b was the total heat required to heat the material; Q_i was the heat required to heat the i material; C_i was the specific heat capacity of i material; T was the mixing temperature; T_0 was the initial temperature.

$$M_o = Q_b / H_o / 0.55 \quad (3)$$

where M_o was the amount of heavy oil; H_o was the net heating value of heavy oil.

$$E_b = \sum E_j = \sum m_j \times E_{jo} \quad (4)$$

where E_b was the total carbon emissions at the mixture mixing phase; E_j was carbon emissions from j fuel; m_j was the amount of j fuel; E_{j0} was the unit carbon emission of the production of j fuel.

The mixture transportation and construction phases were carbon emissions from fuel consumed by construction and transport machinery. All asphalt mixtures had the same carbon footprint at these phases.

3 Results and Discussions

3.1 Alternative Feasibility of Lignin

To evaluate the feasibility of substituting lignin for asphalt, lignin asphalt mixtures were prepared to test their mechanical properties. According to previous research [24], lignin was mixed into asphalt at levels up to 30%, showing rheological properties similar to those of common polymer-modified asphalt. Therefore, this study selected five replacement amounts of 5, 10, 15, 20, and 25%.

To assess the effect of lignin on the high-temperature stability of the asphalt mixture, a specific rutting test was conducted on mixtures with varying lignin content. The results indicated that all lignin-containing mixtures exhibited better high-temperature stability than pure that. As shown in Fig. 4, the dynamic stability of the asphalt mixtures significantly increased as the lignin content rose, demonstrating that lignin improved the mixture's resistance to deformation at high temperatures. However, when the lignin replacement rate reached 20%, the enhancement in dynamic stability began to diminish. This improvement in high-temperature stability was attributed to the increased viscosity of the asphalt, facilitated by lignin. Additionally, a portion of the lignin, which remained largely unreacted with the asphalt

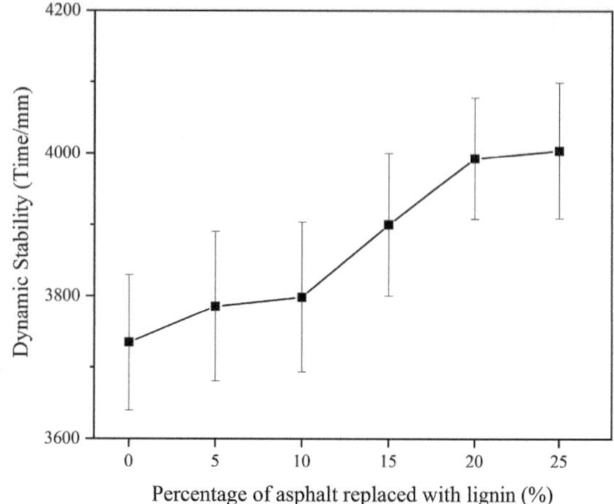

Fig. 4 Replacement effects for different content of lignin on high-temperature performance of asphalt mixtures (Self-drawn)

binders, acted as a fine filler within the mixture, filling voids and improving structural integrity. This contributed to enhanced adhesion between the aggregate and asphalt, leading to a notable improvement in the high-temperature stability of the mixtures [25].

To evaluate the influence of lignin on the water stability of asphalt mixtures, a freeze–thaw splitting test was conducted on asphalt mixtures with varying amounts of lignin. Figure 5 illustrated the variation in the tensile strength ratio (TSR) relative to the percentage of asphalt binder substituted with lignin. The TSR is an indicator commonly used to assess the moisture resistance of asphalt mixtures, indicating how well the material resists water damage. Specifically, the TSR is the ratio of the tensile strength of water-conditioned specimens to that of dry specimens. A higher TSR generally implies better resistance to moisture-induced damage, which made it a critical parameter in assessing the durability of asphalt pavements in a variety of changing circumstances. The TSR values exhibited a fluctuating trend, with the highest value recorded at 10% lignin replacement, followed by a decrease at 15%. The values slightly increased at 20% but plateaued between 20 and 25%. Notably, the inclusion of lignin positively influenced the TSR up to 10%, beyond which the effect became less pronounced, suggesting a threshold beyond which additional lignin did not significantly enhance the water stability of the asphalt mixture.

To evaluate the influence of lignin on the crack resistance of asphalt mixtures at low temperatures, a low-temperature bending test was performed on mixtures with different lignin content. As shown in Fig. 6, the failure strain and failure stiffness modulus followed similar trends as a function of the percentage of asphalt replaced with lignin. The failure strain exhibited an increasing trend as lignin replacement rose from 0 to 25%, indicating reduced ductility in the asphalt mixture. However, at 20% lignin replacement, the increase in failure strain slowed, suggesting potential alterations in the mechanical properties of the asphalt mixture at higher lignin content. In contrast, the failure stiffness modulus displayed an opposite trend. These results

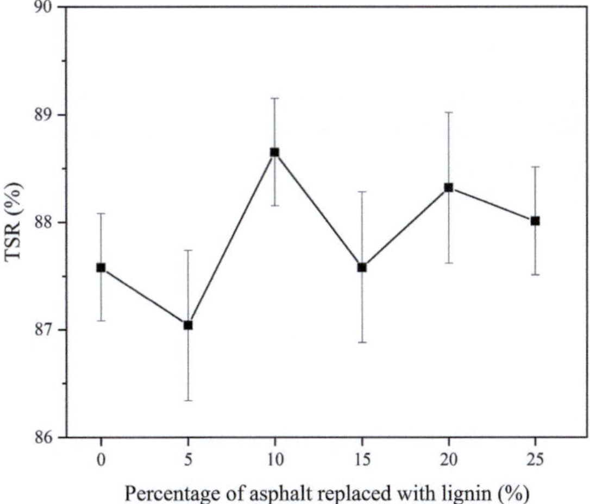

Fig. 5 Replacement effects for different content of lignin on water stability of asphalt mixtures (Self-drawn)

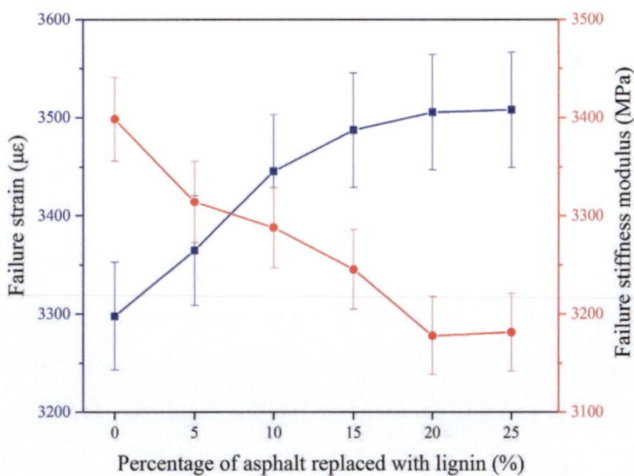

Fig. 6 Replacement effects for different content of lignin on low-temperature performance of asphalt mixtures (Self-drawn)

suggest that the introduction of lignin improved the low-temperature crack resistance of asphalt mixtures, likely due to the benefits provided by the bio-material [26, 27]. This improvement could be attributed to the development of a robust bond between lignin and bitumen, which enhanced the asphalt's ability to wrap the aggregates more effectively, ensuring durable and improved adhesion between the lignin asphalt matrix and the aggregate [28].

According to the results of the rutting test, freeze–thaw splitting test, and low-temperature bending test, a lignin content of 20% was identified as optimal for achieving desirable viscoelastic properties in the blends. Hence, this study selected 20% as the optimal replacement amount.

3.2 Functional Unit

To better assess and compare the environmental impact of asphalt mixtures partially replaced with lignin using the way of life cycle assessment, we introduced a warm mixed additive into the asphalt. It selected hot mixed asphalt mixtures (HA), warm mixed asphalt mixtures (WA), and warm mixed lignin asphalt mixtures (WLA) for comparison with hot mixed lignin asphalt mixtures (HLA). Figure 7 and Table 6 provided detailed material compositions and specific formulations of the different asphalt mixtures. The composition and ratio of the mixtures remained consistent, except that 20% of the asphalt was replaced with lignin, and the warm mixed additive was incorporated. These asphalt mixtures were chosen for the surface layer of the asphalt pavement.

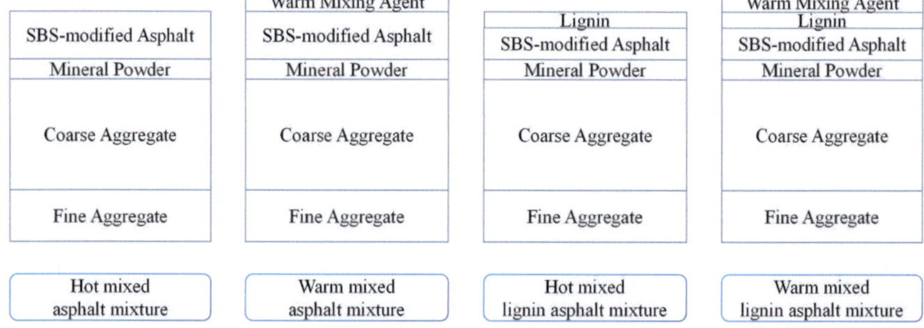

Fig. 7 Material composition of different asphalt mixtures (Self-drawn)

Table 6 Formulation of different asphalt mixtures (Self-drawn)

	SBS-modified asphalt/%	Aggregate/%	Mineral powder/%	Warm mixing agent[a]/%	Lignin/%
Hot mixed asphalt mixture (HA)	6	100	5	0	0
Warm mixed asphalt mixture (WA)	6	100	5	3	0
Hot mixed lignin asphalt mixture (HLA)	4.8	100	5	0	1.2
Warm mixed lignin asphalt mixture (WLA)	4.8	100	5	3	1.2

[a]Dosage of warm mixing agent is defined as the proportion of the total amount of asphalt binders and lignin

3.3 Determination of Mixing Temperature

In asphalt pavement construction, the mixing temperature refers to the temperature at which the asphalt binder is mixed with the aggregate. This temperature should ensure that the asphalt binders have adequate flow ability to evenly coat the aggregates. Typically, the mixing temperature is determined based on the viscosity-temperature curve of the asphalt binder. As shown in Fig. 8, the mixing temperature is usually selected as the temperature corresponding to a viscosity range of 0.15–0.19.

In this study, the temperature corresponding to 0.17 was selected as the mixing temperature. The mixing temperature and temperature difference obtained by viscosity-temperature curve were shown in the Table 7. The substitution of lignin for asphalt increased the mixing temperature, whereas the addition of warm mix additive reduced it.

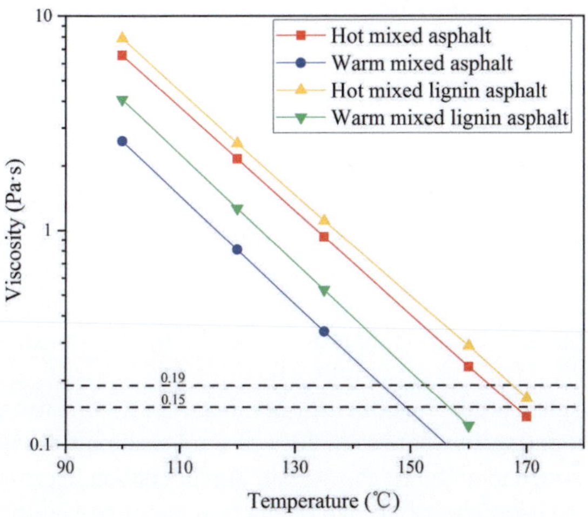

Fig. 8 Viscosity-temperature curves of different asphalt binders (Self-drawn)

Table 7 The mixing temperature of different asphalt (Self-drawn)

Type of asphalt mixture	Mixing temperature/°C	Temperature difference/°C
Hot mixed asphalt mixture	166.9	–
Warm mixed asphalt mixture	147.8	19.1
Hot mixed lignin asphalt mixture	169.2	–2.3
Warm mixed lignin asphalt mixture	154.8	12.1

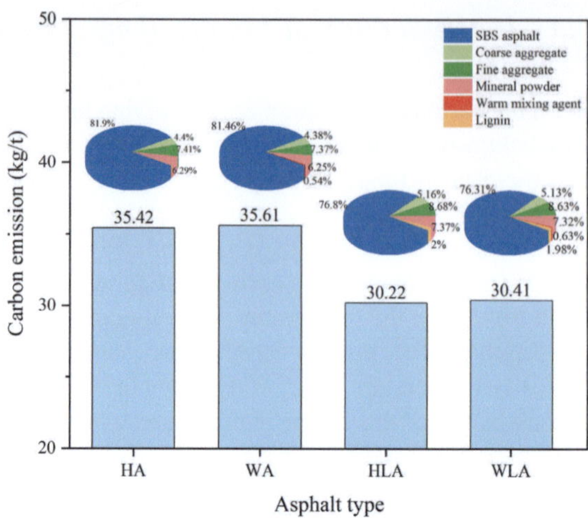

Fig. 9 Carbon emission results of different asphalt mixtures in material production stage (Self-drawn)

3.4 *Life Cycle Assessment*

In the materials production stage, the key variations among the asphalt mixtures arose from the incorporation of the warm-mix additive and lignin. As illustrated in Fig. 9, the partial replacement of the asphalt binder with lignin led to a substantial reduction in carbon emissions. Specifically, a 20% substitution of the asphalt binder with lignin resulted in a 14.68% decrease in carbon emissions. This significant reduction can be attributed to the lower carbon emission factor of lignin compared to traditional asphalt binder.

Notably, the asphalt binder remained the dominant contributor to carbon emissions in this stage, accounting for more than 70% of total emissions across all mixtures. Thus, strategies aimed at reducing the asphalt binder content in pavement mixtures, such as substituting it with materials like lignin, present a promising approach to mitigating carbon emissions in the production phase of asphalt pavements. Additionally, the limited impact of the warm-mix additive on overall emissions underscores the importance of focusing on binder substitution to achieve substantial environmental benefits.

In the mixing stage of asphalt mixtures, the primary differences among the mixtures were attributed to the mixing temperature, which was determined by the viscosity-temperature curve of the asphalt binder. As shown in Fig. 10, the carbon emission results for various asphalt mixtures during the mixing stage indicated that the addition of the warm-mix additive significantly lowered the mixing temperature, leading to a notable reduction in carbon emissions. Specifically, despite the warm-mix additive constituting only 3% of the total blend, it resulted in a 6.28 and 4.68% decrease in carbon emissions for the respective mixtures. This underscored the critical role of reducing construction temperatures in minimizing carbon emissions during the mixing phase. The reduction in temperature not only improved

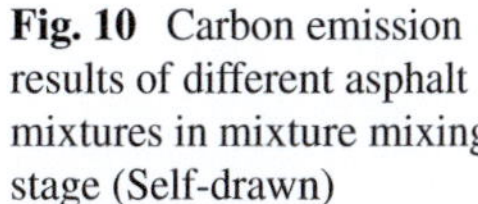

Fig. 10 Carbon emission results of different asphalt mixtures in mixture mixing stage (Self-drawn)

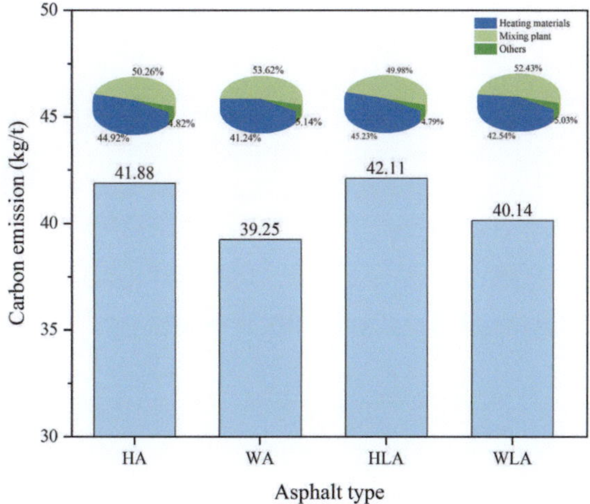

energy efficiency but also had a substantial environmental impact, highlighting the importance of temperature control in sustainable asphalt pavement practices.

Notably, the heating of materials and the operation of the mixing plant remained the dominant contributors to carbon emissions at this stage, accounting for more than 90% of total emissions across all asphalt mixtures. Despite efforts to reduce emissions through additives and temperature adjustments, the energy-intensive processes involved in material heating and mixing were the primary sources of carbon output during the mixing stage.

The transportation stage primarily involved the delivery of asphalt mixtures from the production site to the construction site. During the on-site construction phase, carbon emissions were primarily generated by construction machinery, including pavers and rollers. There were no significant differences in carbon emissions among the various asphalt mixtures during these two stages, with emissions consistently measuring 13.57 kg/t.

The carbon emissions from the production, mixing, transportation, and construction stages were summarized in Fig. 11. As shown, the substitution of asphalt with lignin and the incorporation of the warm mixed additive significantly reduced overall carbon emissions, highlighting notable environmental benefits. Specifically, the carbon emissions per ton of asphalt mixture decreased by 5.46% with lignin substitution and by 2.67% with the use of the warm-mix additive. When both lignin substitution and the warm-mix additive were applied together, total carbon emissions from the asphalt mixture were reduced by an impressive 7.42%. This reduction emphasized the potential of these sustainable practices to reduce the environmental influence during the construction of asphalt pavement. The combined effect of lignin substitution and the warm-mix additive demonstrated a synergistic relationship, achieving a greater reduction than when either technique was used individually.

Furthermore, throughout the entire asphalt production process, the material production and asphalt mixing stages contributed the largest portion of the total

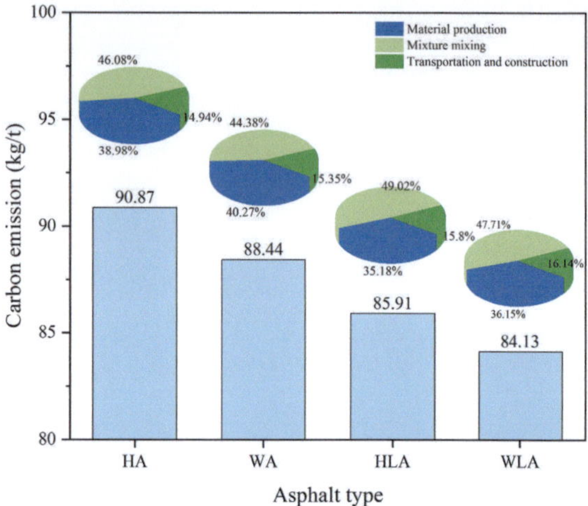

Fig. 11 Carbon emission results of different asphalt mixtures from life cycle (Self-drawn)

carbon emissions, emphasizing that these stages should be the primary targets for emission reduction efforts. The substantial environmental benefits realized through these strategies make them highly promising for improving sustainability in the road construction industry, offering an effective approach to reducing the carbon footprint of asphalt mixtures.

4 Conclusion

This study offers a thorough evaluation of the potential environmental benefits of using lignin as a partial replacement for traditional asphalt in pavement construction. By employing a LCA framework, we quantified the carbon emissions across various stages of road construction, including material production, asphalt mixture mixing, transportation, and construction. The results demonstrated that replacing 20% of asphalt with lignin significantly reduced carbon emissions, particularly during the material production phase, where a 14.68% decrease was observed. Additionally, the introduction of warm-mix additives further contributed to emission reduction by lowering the mixing temperature, resulting in a cumulative reduction of up to 7.42%. Further research is needed to explore the long-term durability and performance of lignin-based asphalt in real-world applications, as well as its economic feasibility on a larger scale. Nonetheless, this study offers valuable insights for policymakers and engineers aiming to reduce the carbon footprint of road construction and advance sustainable development goals in the transportation sector. Future work should focus on optimizing the performance and cost-efficiency of lignin asphalt while establishing industry standards and conducting long-term field trials. Additionally, interdisciplinary collaboration and lifecycle assessments are essential to accelerate its large-scale adoption and validate its environmental benefits.

Acknowledgements This study was supported by the funding of Chongqing Science and Technology Bureau (2023TIAD-GPX0066) and Chongqing Design Group CP., LTD (Research and application of key technologies regarding digitization and industrialization on municipal engineering oriented low carbon).

Declaration of Competing Interest The authors declare that they have no known competing financial interests or personal relationships that could have appeared to influence the work reported in this paper.

References

1. Becsi B, Formayer H (2024) Linking local climate scenarios to global warming levels: applicability, prospects and uncertainties. Environ Res Clim 3(4):045003–045003
2. Ding Y, Jian S, Yu L (2025) How to reduce carbon emissions in the urban transportation systems through carbon markets? Balancing the monetary and environmental benefits. Appl Energy 377(PB):124454–124454
3. Zhang X et al (2025) Establish of air pollutants and greenhouse gases emission inventory and co-benefits of their reduction of transportation sector in Central China. J Environ Sci 150:604–621
4. Qiao Y et al (2025) Stochastic analysis for comparing life cycle carbon emissions of hot and cold mix asphalt pavement systems. Resour Conserv Recycl 212:107881–107881
5. Wang J et al (2024) Carbon emission calculation and uncertainty analysis of asphalt concrete during the production stage. Constr Build Mater 447:138007–138007
6. Tiancheng L et al (2024) Contribution of climate change and traffic load on asphalt pavement carbon emissions. J Clean Prod 434:140553
7. Qi L et al (2024) Carbon emissions quantification and different models comparison throughout the life cycle of asphalt pavements. Constr Build Mater 411:134323
8. Gomes BFML, Júnior SV, Gurgel LVA (2024) Production of activated carbons from technical lignin as a promising pathway towards carbon emission neutrality for second-generation (2G) ethanol plants. J Clean Prod 450:141648
9. Yanchao Y et al (2022) A comparative life cycle assessment of asphalt mixtures modified with a novel composite of diatomite powder and lignin fiber. Constr Build Mater 323
10. Elena G et al (2022) Performance assessment of asphalt mixtures produced with a bio-binder containing 30% of lignin. Mater Struct 55(8)
11. Ahmed K et al (2021) Comparative life cycle assessment of asphalt mixtures using composite admixtures of lignin and glass fibers. Materials 14(21):6589–6589
12. Arafat S et al (2019) Sustainable lignin to enhance asphalt binder oxidative aging properties and mix properties. J Clean Prod 217:456–468
13. Tamrin, Hafizullah RA, Barita A (2023) The enhancement of mechanical, physical, and thermal properties of modified asphalt mixture with lignin from mahogany sawdust in the presence of diphenylmethane 4,4′-diisocyanate. Int J Adhesion Adhesives 125
14. Nofal H, Al-Ajarmeh R (2027) Environmental performance assessment of decentralized wastewater treatment system in Jordan using life cycle analysis. Int J Environ Sci Dev 14(6)
15. Maiti P, Meikap BC (2025) Mechanism and adsorptive removal of Pb (II) by torrefied/pyrolyzed functionalized bio-adsorbent in batch application and life cycle assessment. Sep Purif Technol 354(P7):129333–129333
16. Margalho Élida M et al (2024) Life cycle assessment of asphalt pavement with photocatalytic capability: environmental impacts from cradle to gate. Springer Nature Switzerland, pp 625–634
17. Mohammad LN, Elseifi M, Cao W, Raghavendra A, Ye M (2017) Evaluation of various Hamburg wheel-tracking devices and AASHTO T 324 specification for rutting testing of asphalt mixtures. Road Mater Pave Des 18(sup4):128–143

18. Jiang XI, Gabrielson J, Titi H, Huang B, Bai Y, Polaczyk P, Hu W, Zhang M, Xiao R (2022) Field investigation and numerical analysis of an inverted pavement system in Tennessee, USA. Transp Geotech 35:100759
19. Jiang XI, Gabrielson J, Huang B, Bai Y, Polaczyk P, Zhang M, Hu W, Xiao R (2022) Evaluation of inverted pavement by structural condition indicators from falling weight deflectometer. Constr Build Mater 319:125991
20. ISO IOfS (2006) Environmental management: life cycle assessment; principles and framework, vol 2006. ISO
21. Feng M et al (2020) Life cycle assessment of greenhouse gas emissions from asphalt pavement maintenance: a case study in China. J Clean Prod 288(prepublish):125595
22. Kenan L et al (2023) Quantitative evaluation on the energy saving and emission reduction characteristics of warm mix asphalt mixtures. Constr Build Mater 407
23. Khandelwal M (2019) Carbon footprint of Lignin modified Asphalt mix
24. Gaudenzi E, Cardone F, Lu X et al (2023) The use of lignin for sustainable asphalt pavements: a literature review. Constr Build Mater
25. Cai L, Shi X, Xue J (2018) Laboratory evaluation of composed modified asphalt binder and mixture containing nano-silica/rock asphalt/SBS. Constr Build Mater 172:204–211
26. Pérez IP, Rodríguez Pasandín AM, Pais JC, Alves Pereira PA (2019) Use of lignin biopolymer from industrial waste as bitumen extender for asphalt mixtures. J Clean Prod 220:87–98. https://doi.org/10.1016/j.jclepro.2019.02.082
27. Gaudenzi E, Cardone F, Lu X, Canestrari F (2022) Performance assessment of asphalt mixtures produced with a bio-binder containing 30% of lignin. Mater. Struct. 0123456789. https://doi.org/10.1617/s11527-022-02057-w
28. Zahedi M, Zarei A, Zarei M (2020) The effect of lignin on mechanical and dynamical properties of asphalt mixtures. SN Appl. Sci. 2(7):1–10. https://doi.org/10.1007/s42452-020-3041-4

Territorial Spatial Planning Strategy in Rural Area Under "Carbon Peaking" and "Carbon Neutrality" Goals

Xu Feng, Chunyu Li, Chuanyang Pan, and Ming Nie

Abstract Low-carbon, green, and sustainable development has become a central theme in territorial spatial planning and governance in the era of stock. Township areas, as key regions for carbon emissions in China, play a crucial role in carbon reduction and control, and are vital governance zones that are closely related to territorial spatial protection and development. However, existing research has not yet comprehensively integrated the goals of "carbon peaking" and "carbon neutrality" into territorial spatial planning, with insufficient linkage to specific spatial strategies. In particular, research at the micro-scale of rural and township areas is still nascent, hindering effective guidance for spatial optimization. This study first examines the hierarchical role of the goals within the territorial spatial planning governance system and develops a carbon assessment model for rural and township areas. Then, using Changshaoying Township, located in the ecological conservation zone of the capital, as a case study, the research delineates boundaries for protection and utilization based on ecological significance assessments. Through forest resource evaluation and visual accessibility analysis, this study proposes site selection and decision-making methods that focus on ecological restoration, planting structure optimization, and phased project implementation, all aimed at achieving carbon reduction and carbon sink enhancement. These methods are intended to promote the green transformation of urbanization in rural and township areas, advance ecological civilization, and facilitate low-carbon development. The results provide insights for micro-scale planning and development in stock regions within territorial spatial planning, contributing to the practice and development of low-carbon rural development.

Keywords Ecological conservation area · "Carbon peaking" and "Carbon Neutrality" calculation model · Township area · Chang Shaoying township

X. Feng (✉) · C. Pan · M. Nie
College of Horticulture, China Agricultural University, Haidian District, Beijing, China
e-mail: fengxu1204@cau.edu.cn

C. Li
Housing and Urban-Rural Development Bureau of Kaifeng, Kaifeng City, Henan Province, China

B. He et al. (eds.), *Climate-Resilient, Low-Carbon, and Sustainable Cities*, Lecture Notes in Civil Engineering 754, https://doi.org/10.1007/978-981-95-2169-2_25

1 Introduction

Territorial spatial planning serves as a guiding framework for national spatial development, a spatial blueprint for sustainable development, and an important pillar in the construction of ecological civilization [1]. In the context of the "dual carbon" goals, the development of low-carbon rural areas has become a key element in promoting the green transformation of territorial space. Decisions on the protection and development of territorial space aimed at reducing carbon emission and enhancing carbon sink are increasingly seen as fundamental to achieving low-carbon rural development.

Current research on urban–rural "carbon peaking" and "carbon neutrality" focuses on the following areas: First, the exploration of low-carbon urban–rural concepts [2, 3], and studies on the planning systems and transmission paths related to "carbon peaking" and "carbon neutrality". Second, the simulation of "carbon peaking" and "carbon neutrality" scenarios [4, 5], which provides a scientific basis for carbon emission management, ecological protection, policy evaluation, and urban and regional planning. Third, the development of low-carbon evaluation frameworks [6, 7] and proposed ways to optimize urban–rural spatial structures to guide the planning and development of low-carbon urban and rural environments. Fourth, research on the selection of planning schemes focused on carbon reduction and carbon sink enhancement [8, 9], which aims to promote effective carbon emission management by improving the efficiency of resource utilization and allocation. Most of the existing studies focus on larger spatial scales, such as provincial, municipal, and county levels [10], and show insufficient integration with specific spatial strategies, and fail to adequately reflect the internal carbon emission patterns and their driving factors within urban and rural areas.

In the process of achieving the "carbon peaking" and "carbon neutrality" goals, the multi-level governance framework of territorial spatial planning plays a crucial role in the hierarchical positioning within both research and practice [11]. This hierarchical positioning encompasses not only the division of policy levels but also the regional and sectoral differences encountered during implementation. By establishing clear hierarchical roles within the "carbon peaking" and "carbon neutrality" goals, governments at different levels can more effectively devise and implement relevant measures, ensuring the successful implementation of regional carbon reduction and control strategies. At the provincial level, governments primarily focus on assessing and monitoring the resource base and coordinating the balance between regional carbon reduction and carbon sink enhancement. Municipal governments, under provincial guidelines, are responsible for developing comprehensive carbon reduction strategies, including energy conservation policies and negative lists. At the county level, governments focus on optimizing the urban–rural structure and revitalizing existing spaces to formulate spatial strategies for carbon reduction and carbon sink enhancement tailored to local development needs. Township-level governments, in turn, must consider the needs of the grassroots, develop land-use control strategies in line with spatial planning, and guide decision-making for specific projects.

Guided by these hierarchical tasks, the accuracy of spatial analysis in terms of "carbon peaking" and "carbon neutrality" targets varies across levels. At the national, provincial, and municipal levels, the focus is on policy formulation, prioritizing industrial restructuring and energy conservation and emission reduction in key functional areas. Conversely, at the county, township, and village levels, the focus is on land use planning, with the goal of reducing carbon emission and enhancing carbon sink by optimizing land use patterns. Specifically, county-level planning focuses on the "three-in-one" spatial layout, village-level planning emphasizes detailed land use classification, and township-level planning is concerned with zoning and land use changes to regulate development projects related to protection and growth.

As a major contributor to carbon emission in China, township plays a critical role in carbon reduction and control, and are closely related to the protection and development of territorial spaces. Focusing research on townships not only increases policy applicability, but also provides substantial support for low-carbon rural development initiatives. Existing studies on rural areas have mainly explored the optimization of large-scale energy structures [12], changes in agricultural and rural lifestyles [13, 14], and other related topics.

Within the framework of the "carbon peaking" and "carbon neutrality" goals, several internationally accepted methods for calculating carbon emissions are used, as summarized in Table 1. These methods include the greenhouse gas inventory method, the GDP energy intensity method, direct measurement, the plot inventory method, material flow analysis, and life cycle assessment. Among these, the greenhouse gas inventory method is commonly used for urban carbon emissions accounting, which helps provide a macroscale view of greenhouse gas emissions by measuring emissions from different sectors [15, 16]. While this method is accurate, it lacks spatial correlation. The GDP energy intensity method [17] is often applied at a macro level to guide industrial policy, while direct measurement, land inventory, material flow analysis, and life cycle assessment methods are tailored to the specific needs of sectors and regions with different characteristics. The carbon measurement method [18], which is closely related to land use, provides results that are highly dependent on data accuracy.

Overall, while there has been increasing attention to the adaptive transformation of spatial planning in the context of the "carbon peaking" and "carbon neutrality" goals, the integration of these goals into territorial spatial planning remains limited, particularly at the micro scale [19]. Consequently, providing actionable guidance for rural spatial optimization and adjustment has proven challenging. Therefore, this study focuses on Changshaoying Manchu Township in Beijing's ecological reserve as a case study. Based on a thorough evaluation of the strengths and limitations of existing methods, this work develops a calculation model tailored to township-level territorial spatial planning under the goals of carbon reduction and carbon sink enhancement, and proposes technical ways for spatial optimization. The aim is to provide theoretical support for the implementation of carbon reduction and control policies.

Table 1 "Carbon peaking" and "Carbon Neutrality" calculation methods (Self-drawn)

Method	Content	Calculation formula	Characteristic
Greenhouse gas inventory (from IPCC guidelines)	Energy	$C = \sum\left(EF_{i,j,k} \times Activity_{i,j,k}\right)$ (EF: Emission factor; Activity: Consumption)	Simple, reliable data
	Industry	$E_{CO_2} = AD \times EF$ (E_{CO_2}: CO_2 emission from industrial production; AD: Output)	
	Agriculture	$E_{CH_4} = \sum EF_i \times AD_i$ (ECH4: CH4 emission; AD: Planting area or number of animals)	
	Land use and agriculture	$\triangle CAFPLU = \triangle CFL + \triangle CCL + \triangle CGL + \triangle CWZ + \triangle CSL + \triangle COL$ ($\triangle$C: Carbon; AFPLU: Land use in agriculture, forestry, etc.; FL: Forest land; CL: Crop land; GL: Grassland; WL: Wetland; SL: Settlement; OL: Other land)	
	Waste disposal	$C = E_{CH_4} + E_{CO_2}$ (E_{CH_4}: CH4 from solid waste landfill, domestic and industrial wastewater treatment; E_{CO_2}: CO_2 emission from waste incineration treatment)	
Energy consumption per unit of GDP	Measure carbon emission per unit of GDP	$C = P \times GDP/P \times I \times F$ (P: Population; GDP/P: GDP per capita; I: Energy utilization intensity; F: Carbon emission intensity)	Strong integrity
Field measurement	Measure and summarize according to emission sources	$G = K \times Q \times C$ (K: Conversion factor; Q: Air quantity; C: Gas concentration)	High sampling costs
Sample land inventory	Based on measured biomass data	$\log W = a + b \times \log(D^2H)$; $W_x = c(D^2)$ (c: Carbon sink from tree; Wx: Biomass of branch, trunk and leave; D: Diameter at breast height; H: Height; a, b: parameter)	High precision and parameter requirement
Material balance algorithm	Carbon generation and emission from production process	$G = \sum G_{input} - \sum G_{output}$ (G: Carbon emission; Ginput: Input carbon; Goutput: non-CO2 output)	Conservation of mass in production
Life cycle method	Emission of energy and material require-ments during the lifecycle	$C = \sum_i^n \sum M_j Q_j$ (C: Carbon emission of total life cycle; n: life cycle phase; Mj: J energy consumption in phase I; Qj: Energy emission factor in J)	Multiple parties involved, low accuracy

(continued)

Table 1 (continued)

Method	Content	Calculation formula	Characteristic
Carbon emission calculation method	Energy	$C_{energy} = \sum(E_j \times EF_j)$ (E: Emission from all types of commodity energy and non-commodity energy)	Accurate result, high data requirement
	Building	$C_{building} = C_n + C_t + C_c = \sum(E_m \times EF_m + G_{m,I} \times D_{m,I} \times EF_i + E_j \times EF_j)$ (Cn: Material production; Ct: Material transport; Cc: Construction site; Em: Material consumption; $G_{m,i}$: Material weight; $D_{m,i}$: Transport distance; E: Energy consumption)	
	Traffic	$C_{traffic} = \sum(V_i \times S_i \times M_i \times EFj)$ (V: Volume; S: Annual travelled distance; M: Fuel economy)	
	Agriculture	$C_{agriculture} = (N_{straw} + N_{fertilizer}) \times EF_j \times W + \sum(N_j \times \alpha_j \times 21 + N_j \times \beta_j \times 310) + \sum(E_j \times EF_j)$ (N: Biological cycle; α: Enteric fermentation from animal; β: Fecal volume; E: Fossil energy consumption in agricultural irrigation)	
	Waste	$C_{waste} = C_i + C_l + C_d + C_s = E_i \times C_C \times P \times EF_i + E_l \times \alpha_{CH4} \times (1-71.5\% + N \times BOD \times SBF \times C_{BOD} \times FTA + (E_d + E_S) \times COD \times C_{COD}$ (Ci: Emission from waste incineration; Cl: Emission from waste landfill; Cd: Discharge from domestic wastewater; Cs: Discharge from industrial wastewater)	

2 Materials and Methods

2.1 Study Area

Changshaoying Township is located in the ecological reserve of Beijing, China, in the eastern part of the Beijing-Tianjin-Hebei region. It is located between 116°30′ to 116°50′E and 39°40′ to 39°55′N. The region has a typical temperate monsoon climate with distinct seasons. Summers are hot with abundant rainfall, while winters are cold and dry. The average annual temperature ranges from 9 to 11 °C, and the annual rainfall is about 500 mm. Changshaoying Township has a population of approximately 15,000 people and covers an area of 120 km^2. The topography is relatively complex, with higher mountains in the west (the highest peaking reaches about 1,200 m) and lower hilly terrain in the east, resulting in a landscape that combines mountainous and plain areas. The township is rich in forest resources, and its proximity to the capital, along with Beijing's ecological protection policies, gives the area significant ecological value.

Located within Beijing's ecological conservation zone, Changshaoying Township is of great ecological significance, facing the dual pressure of urban–rural development and ecological preservation. The region has a large proportion of ecological carbon sink areas, demonstrating a clear advantage in carbon sequestration capacity. However, land resources are limited, and the balanced development of carbon emission reduction and carbon sink enhancement requires scientific planning and management.

2.2 Carbon Peaking and Carbon Neutrality Calculation Model in Town Level

Based on a comprehensive assessment of the advantages and limitations of existing methods, and considering the "carbon peaking" and "carbon neutrality" goals at the township level, along with the ecological conservation characteristics of Changshaoying Township, this study develops a calculation model specifically designed for land use planning at the township scale (Table 2). The model integrates the spatial interdependencies between carbon emissions and ecological restoration, while taking into account the specific resource characteristics of the region. This approach provides both theoretical underpinnings and practical pathways for facilitating the low-carbon transformation of township areas. In particular, carbon emission monitoring and carbon sink evaluation serve as essential prerequisites and benchmarks for the implementation of "carbon peaking" and "carbon neutrality" strategies [20]. These elements significantly influence the effective implementation of carbon reduction and control policies, the transition to renewable energy sources, and the advancement of low-carbon technologies, thereby contributing to the promotion of sustainable and green economic development.

The carbon calculation model for Changshaoying Township is composed of two main components: carbon emissions and carbon sinks. Carbon emissions are mainly caused by energy consumption, transportation, agriculture, and waste management. In terms of energy, the model reflects the differences in energy supply methods in different villages through energy consumption adjustment coefficients. For transportation, the model considers the volume of passenger and freight traffic on major roads. For agriculture, energy consumption for irrigation is simplified due to the limited arable land in Changshaoying Township. Waste management focuses on the incineration and landfilling of solid waste.

Table 2 Changshaoying township "Carbon peaking"and "Carbon Neutrality"calculation model (Self-drawn)

Classification		Calculation formula	Data source
Carbon emission		$C_{emission} = C_e + C_t + C_a + C_w$	Statistic yearbook, Survey data
Among	Energy	$C_e = \sum(E_j \times EF_j \times a)$	Energy consumption (electricity, gas and coal)
	Traffic	$C_t = \sum(E_i \times EF_j)$	Consumption of diesel, gasoline, etc.
	Agriculture	$C_a = \sum(E_j \times EF_j) + \sum(N_j \times \alpha_j \times 21 + N_j \times \beta_j \times 310)$	Animal species and breeding quantity, fertilizer and pesticide use, and etc.
	Waste	$C_w = \sum(MSW \times CCW \times FCF \times EF \times a)$	Waste incineration and landfill data
Carbon sink		$C_{sink} = C_f + C_c + C_g + C_w$	The third territorial survey data
Among	Forest land	$C_f = C_{forest} + C_{soil} = \sum F(W_f \times R_f \times A_f \times L_f \times CF) + S \times \varepsilon$; $\log W = a + b \times \log(D^2H)$	Forest area, dominant species (quantity, height, diameter at breast height, etc.), coverage, canopy density
	Cropland	$C_c = \sum(C_f \times D_f) = \sum(C_f \times Y_f/H_f)$	Cultivated land area, type and quantity of crop
	Grassland	$C_g = S \times \varepsilon$	Grassland area
	Wetland	$C_w = C_{vegetation} + C_{soil} = \sum(S_{i,1} \times P_i \times CF + S_{i,2} \times D_i)$	Wetland area, vegetation cover area and biomass

Carbon sink primarily stems from forests, croplands, grasslands, and wetlands. Forests are evaluated in terms of both tree biomass and soil carbon content. Carbon sequestration rates for croplands are estimated based on the yields of major crops, such as corn, wheat, and peanuts. Wetlands are categorized as rivers, reservoirs, and constructed wetlands.

3 Projects Decision in Chang Shaoying Town Under the Carbon Goal

3.1 Protection and Utilization Zoning of Territorial Space

For the ecological conservation area, ecological significance, including ecological sensitivity and ecosystem service importance level, is the core basis to separate areas between protection and utilization (Fig. 1). Firstly, the town is divided by grids of 30 m

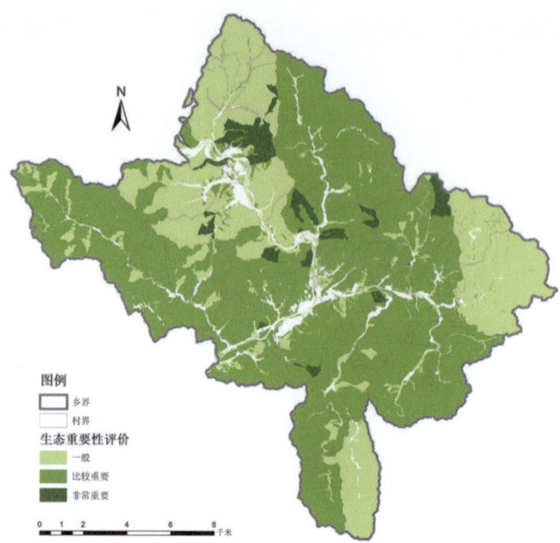

Fig. 1 Ecological importance zoning (Self-drawn)

× 30 m, and collect natural resource data, the formula required into the corresponding grid. Secondly, according to the dual evaluation of China's territorial spatial planning, formulas (1)–(6) for Chang Shaoying are extracted. Thirdly, according to the formula, the ecological significance of each grid is evaluated, and the town area is divided into three zoning types, which are ecological control zoning with strict construction activity restrictions, ecological buffer zoning with focus on vegetation conservation and ecological rehabilitation, and the ecological utilization zoning with optimized planting structure and limited construction.

$$S_{bio} = NPP_{mean} \times F_{pre} \times F_{tem} \times F_{alt} \tag{1}$$

(S_{bio}: Biodiversity maintenance function; NPP_{mean}: Net primary productivity per year; F_{pre}: Annual precipitation; F_{tem}: Temperature parameter; F_{alt}: Altitude parameter.)

$$TQ = \sum_{i}^{j} (P_i - R_i - ET_i) \times A_i \times 10^3 \tag{2}$$

(TQ: Water source conservation capacity; P_i: Precipitation; R_i: Runoff; ET_i: Evaporation; j: Ecosystem type number)

$$A = R \times K \times L \times S \times (1 - C) \tag{3}$$

(A: Water and soil conservation; R: Rainfall erosivity; K: Soil erodibility; L: Length; S: Slope; C: Vegetation factor)

$$S = \sqrt[4]{R \times K \times LS \times C} \quad (4)$$

(S: Water and soil erosion sensitivity; LS: Fluctuation)

$$D = \sqrt[4]{I \times W \times K \times C} \quad (5)$$

(D: Desertification; I: Dryness; W: Day count of sand-driving wind)

$$ES = MAX\ (\mathrm{S_{bio}},\ \mathrm{TQ},\ \mathrm{A},\ \mathrm{S},\ \mathrm{D}) \quad (6)$$

(ES: Ecological significance).

Though above analysis, Chang Shaoying's zoning is shown in Fig. 1.

3.2 *Siting and Strategy for Ecological Rehabilitation Projects within Territorial Spatial Conservation Area*

Conservation area includes ecological control zoning and ecological buffer zoning. As the forest cover exceeds 98%, the forest quality becomes an important criterion for the location of ecological restoration projects. Through expert scoring, forest health grade (0.4), canopy density (0.3), canopy cover (0.2), and age class (0.1) are selected as evaluation indicators of forest quality. After processing the total score on a hundred-point scale (every twenty points as one level), the forest quality is graded from level one to five, and it is visualized on the GIS platform (Fig. 2).

For the area with good forest quality level such as one, two, or three, as well as forest within the ecological conservation red line, only routine maintenance work is

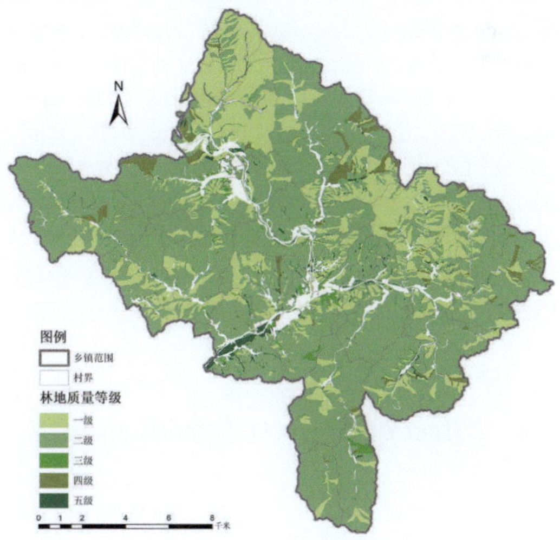

Fig. 2 Forest land quality analysis (Self-drawn)

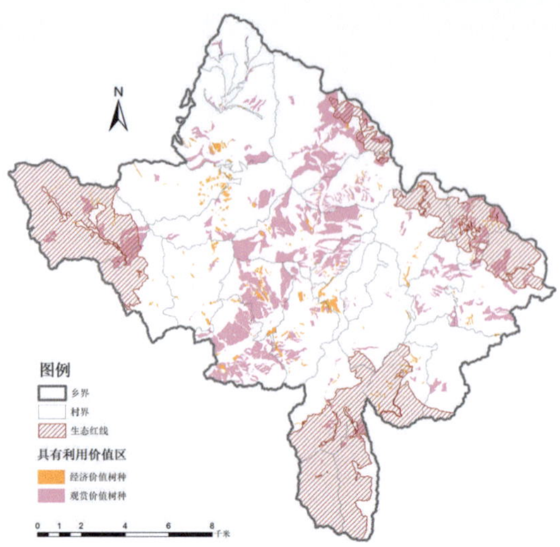

Fig. 3 Utilization projects guidance (Self-drawn)

needed. To area with poor forest quality such as four or five, density of high carbon sequestration trees will be increased to improve canopy density and coverage, or a stable structure composed of tree and shrub will be formed by increasing density of shrub-grass or replacing shrub-grass species.

3.3 Siting and Strategy for Development Projects within Territorial Spatial Utilization Area

Utilization area includes ecological utilization zoning. By analyzing the distribution characteristics of dominant tree species, visual and accessibility area, ornamental plant area and economic plant area are distinguished (Fig. 3). In the ornamental plant area, based on the surrounding ornamental tree species, the improvement of ornamental effect and carbon sink, as well as the establishment of appropriate recreational facilities, are considered as the main issue by replanting ornamental trees, partially changing high carbon sink ornamental trees partly, or matching with shrub grass. In the economic plant area, under the premise of reducing the use of fertilizer and pesticide, replanting with fruit species or developing forest-based economy will be considered to increase carbon sink while promoting rural industrial development.

3.4 Effect of Projects Introduction

Efficiency prediction of carbon reduction and sink enhancement after project implementation, as well as maintaining rural spatial integrity, are the guidance and basis for project decision. According to the calculation in the "Carbon peaking" and "Carbon Neutrality" calculation model (Table 2), Chang Shaoying will reduce carbon emissions by 835.86t and enhance carbon sinks by 6385.12t after the projects are implemented (Table 3).

3.5 Projects Arrangement

Considering costs, the above projects should be divided into short-term and long-term projects (Fig. 4). Since Chang Shaoying is not a tourist destination, in order to attract more transit tourists to stay, the short-term projects are listed within the visible and accessible area, which can achieve practical value through ecological restoration and planting structure adjustment as soon as possible.

4 Discussion and Summary

In the context of global climate change mitigation and sustainable development, low-carbon and green development has become a central issue in land use planning and governance. The goals of "carbon peaking" and "carbon neutrality," closely linked to ecological conservation, have emerged as critical drivers of contemporary development. As a major source of carbon emission, township regions bear substantial responsibility for carbon reduction and control, which is closely related to land protection and development.

This study provides an in-depth analysis of the hierarchical positioning of the "carbon peaking" and "carbon neutrality" goals within the land spatial planning and governance framework, and presents an innovative township-level "carbon peaking" and "carbon neutrality" calculation model based on ecological importance assessments. The model quantifies and identifies the potential for carbon reduction and carbon sink enhancement within township, providing accurate data-driven support for planners. This facilitates the prioritization of ecological protection in critical areas and the implementation of targeted ecological restoration measures. Using Changshaoying Township as a case study, and integrating forest resource evaluations and visual accessibility analyses, the study provides spatial decision frameworks for optimizing planting structures, conducting ecological restoration, and planning the phased implementation of land space protection and development projects. This will ensure the rational allocation of resources and promote the development of low-carbon and green infrastructure through enhanced carbon sink functionality.

Through model-based calculations and analysis, this study highlights the significant role of township regions in carbon reduction, low-carbon development, and the promotion of ecological civilization, filling a critical gap in micro-scale research

Table 3 Effect of projects introduction in Chang Shaoying town (Self-drawn)

Classification		Carbon Emissions (tCO_2)	Carbon Emission Factor (tCO_2/hm^2)	Classification		Area (hm^2)	Carbon Sink (tCO_2)	Carbon Absorption Factor (tCO_2/hm^2)
Carbon Emission	Energy	8363.55	38.35	Carbon Sink	Forest Land	22,634.70	174,739.88	7.72
					Cultivated Land	507.08	3072.91	6.06
	Waste	2824.48			Grass-land	260.95	339.24	1.30
	Transportation	3754.10	22.91		Wet-land	246.55	3286.52	13.33
	Agriculture	1639.96	49.51					
Total	–	16,582.09 (835.86 reduction)	–	Total	–	23,649.28	181,438.55 (6385.12 enhancement)	–

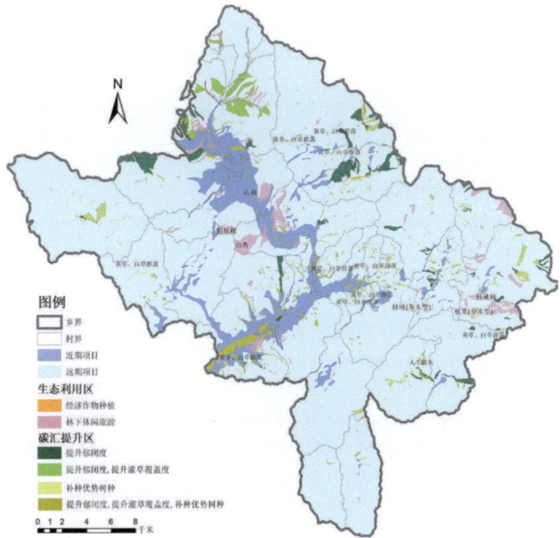

Fig. 4 Projects arrangement (Self-drawn)

on carbon emissions and ecological governance at the township level. The findings provide practical insights for low-carbon rural development in similar ecological conservation areas. Additionally, the phased implementation and site selection method proposed in this study provides local governments with a scientific approach to resource allocation, ensuring the gradual realization of low-carbon targets and promoting low-carbon, green development and ecological civilization. However, the construction of "carbon peaking" and "carbon neutrality" calculation models and the planning of protection and development projects require further optimization and refinement through extensive case studies.

Acknowledgements This study is supported by the National Natural Science Foundation of China (Grant No. 52078479).

References

1. Sun S (2020) On the transformation from urban and rural planning to territory development planning. Urban Plann Forum 04:11–17
2. Pan H, Tang Y, Wu J, Lu Y, Zhang Y (2008) Spatial planning strategy for 'Low Carbon Cities' in China. Urban Plann Forum 57–64
3. Sui X, Tang Y, Hu T (2010) Examining low-carbon rural construction in China. Soc Sci Yunnan 107–112
4. Zhu J, Tang Z, Liu X, Lu Y (2012) Applying Jiyuan City as an Example, a DPSIR model is employed to evaluate low-carbon city development. Urban Problems 42–47
5. Luo H, He Z (2015) Research on low-carbon rural-tourism evaluation based on DPSIR model—a case study of mentougou in Beijing. Sci Technol Ind 15:118–126
6. Xu X, Luo Z, Zhang X, Zhu Y, Xu W (2020) Integrating ecological information into urban master planning. Urban Dev Stud 27:9–16

7. Jiang H (2015) Transfer prices, property rights distribution, and the development of urban spatial patterns: an examination from the new institutional economics angle. Urban Plann Forum 63–69
8. Wang Z, Xiang Y, Wu Y (2016) Consensus, dilemma and strategy: a study on low-carbon rural construction in Yangtze River Delta Region. New Archit 33–39
9. Ye H, Cai Y, Dai X, Li X, YI X (2013) Guangzhou low carbon urban planning practice. Planners 29:84–88, 100
10. Xiong C, Chen S, Xu L (2020) Driving factors analysis of agricultural carbon emissions based on extended STIRPAT model of Jiangsu province, China. Growth Change 51(3):1401–1416
11. Feng X, Wang K, Mao Q, Pan C (2022) Rural spatial planning methods in the territorial planning: from an integrated perspective of planning and governance. City Plann Rev 46:21–31
12. Miao X (2017) On the current rural energy consumption and the solutions in the background of urbanization—analysis of the rural domestic energy use in Beijing. Frontiers 92–95
13. Zhao X, Song L (2018) Carbon emission from agricultural land use and its relationship with agricultural economy in hunan province. J Ecol Rural Environ 34:976–981
14. Zhang S, Wang M (2022) Factors affecting and mitigating the lifetime carbon emission of rural shandong province residents. Rural Econ Sci Technol 33:44–47
15. Li Q, Tang L, Shi L (2013) Research and compilation of urban greenhouse gas emission inventory. Acta Ecol Sin 33(02):367–373
16. Li J, Zhou P, Zhang L (2018) Analysis and prediction of changes on land-use change and forestry greenhouse gas in Guangdong province. Ecol Sci 37(06):45–51
17. Sha Z, Tong F, Zhang C (2013) Study on carbon emission model based on energy consumption per unit of GDP. Ecol Econ 10:49–52
18. Chen J, Song G Yu L (2018) An estimation and analysis method for road carbon emissions in urban agglomeration transport planning. J Highway Transp Res Dev 35(02):122–128+136
19. Xiong J (2021) Study and thoughts on territorial spatial planning under the goal of 'Carbon Emissions peaking and Carbon Neutrality'. Urban Plann Forum 74–80
20. Wang Y, Wang X, Chen Q (2025) Review of research on urban carbon emission measurement methods and its implications. City Plann Rev 1–15

Potential of Urban Greening Biomass Energy Clean Heating and the Application of Full Chain Technology Path in Xi'an, China

Ruoxi Yang, Min Zhou, Dongning Wang, Xiaoning Su, Han Song, and Jia Yang

Abstract Biomass energy, as a low-carbon energy source with high calorific value, is more suitable for the transition to clean heating in northern regions. With the ecological development of cities, the area of urban green spaces has increased, and the total amount of dead branches, fallen leaves, or maintenance and pruning materials of green plants is relatively large. However, the lack of renewable application solutions has resulted in a large amount of waste. However, the practical application of biomass energy still faces problems such as dispersed raw material resources, low technological level, and imperfect management and incentive mechanisms. This article takes Xi'an as an example and evaluates the current situation of green biomass energy in Xi'an, which shows that it has great heating potential. On this basis, a "collection addition storage operation combustion" full chain clean heating technology path with distributed collection and storage as the core was proposed, aiming to improve the efficiency of biomass energy utilization, promote clean energy heating, promote the circular utilization of urban green biomass, and reduce carbon emissions.

Keywords Urban green biomass · Heating potential · Clean heating

---

R. Yang (✉) · M. Zhou · X. Su · H. Song
China Northwest Architecture Design and Research Institute Co., Ltd., Xi'an, China
e-mail: yangruoxi_97@163.com

D. Wang
Xi'an University of Architecture and Technology, Xi'an, China

J. Yang
China Kide Engineering Corporation, Xi'an, China

B. He et al. (eds.), *Climate-Resilient, Low-Carbon, and Sustainable Cities*, Lecture Notes in Civil Engineering 754, https://doi.org/10.1007/978-981-95-2169-2_26

1 Introduction

Biomass is a kind of energy stored in biomass in the form of chemical energy from solar energy, which usually includes forest waste, agricultural waste, aquatic plants, oilseed plants, urban and industrial organic waste, animal manure and so on. As a low-carbon energy source, the utilization of biomass not only helps to reduce greenhouse gas emissions, but also realizes the sustainable development of resources through the recycling of organic waste. According to research, by 2020, biomass will account for 96% of the global renewable thermal energy production [1].

In recent years, national and local governments attach great importance to the development and utilization of biomass energy, issued a series of relevant policies and supporting measures to promote the rapid development of biomass energy industry. The National Energy Administration issued the "2024 Energy Work Guidelines" [2], pointing out that we should continue to promote clean energy substitution in key areas, promote ultra-low-emission cogeneration centralized heating and geothermal, solar energy, biomass and other renewable energy heating according to local conditions, promote the continued development of clean heating in the northern region, and encourage the gradual development of a variety of clean heating methods. Shaanxi Provincial Department of Ecology and Environment issued the "implementation of the "Action Plan for Continuous Improvement of Air Quality" implementation views (2024–2025)" [3] (draft), the development of new energy and clean low-carbon energy, and clearly stated that the new residential housing, commercial complexes, etc., must be used to clean energy heating. General Office of the Xi'an Municipal People's Government issued the "Xi'an City, pollution reduction and carbon reduction synergistic innovation pilot implementation plan" [4] emphasizes the need to vigorously promote clean energy heating, including geothermal energy, sewage energy, air energy, biomass and other energy sources complement each other clean energy security system. A series of national and local policies have been formulated in the area of biomass clean heating in cities and towns, with the aim of promoting the wide application of clean energy and optimizing the adjustment of the energy structure. The research on clean heating using biomass energy for urban greening in Xi'an city shows that from an environmental perspective, traditional heating in Xi'an relies on fossil fuels and has high carbon emissions. The carbon dioxide released from biomass energy combustion can be absorbed and balanced by plant growth, with almost zero emissions, and pollutants such as sulfur dioxide are significantly reduced, which can significantly improve air quality and help achieve the "dual carbon" goal. In terms of sustainable energy development, as a local renewable energy source, it can enrich Xi'an's energy structure, reduce dependence on external non renewable energy, enhance energy supply stability, and turn green waste into treasure, improving resource utilization efficiency. Economically, it can drive industries such as biomass energy collection, processing, and equipment manufacturing, creating employment opportunities. With the maturity of technology and the expansion of industrial scale, heating costs are expected to decrease, which has huge advantages for users and enterprises.

Although biomass energy has many advantages, it still faces some problems in the process of practical application. First, the characteristics of biomass feedstock resources limit the efficient utilization of biomass energy. Biomass raw materials are too dispersed, the collection cost is high, and there are some difficulties in transportation and storage. Secondly, the level of R&D and application of biomass energy technology in China is relatively low, with high production cost, imperfect industrial chain and weak market competitiveness. In addition, the management and incentive mechanism of biomass energy is not yet perfect, which also restricts its large-scale popularization and application.

Based on the current situation of urban greening biomass research and analysis of heating potential, this paper takes Xi'an City as an example and combines the problems in the popularization and application of urban greening biomass, and puts forward the technological path of the whole chain of clean heating "collection-addition-storage-transportation-burning".

2 Analysis of the Status of Urban Greening Biomass

2.1 Biomass for Urban Greening

Dead leaves and branches generated by the natural renewal of green plants during the growth process or plant waste materials such as tree and shrub pruning materials, lawn pruning materials, discarded flowers and weeds in gardens and flower beds generated during the greening maintenance process are called urban greening biomass [5]. Urban greening biomass mainly comes from greening plants in urban green areas, including three types of trees, shrubs and herbs.

2.2 Total Amount of Green Space and Coverage

In recent years, urban greening construction in Xi'an has achieved remarkable results, and the total amount of green space and coverage rate have been greatly increased. According to the Xi'an Statistical Yearbook 2023 (see Fig. 1), the green area of Xi'an reached 44,861 hectares in 2022, and the figure is still growing.

2.3 Characterizing the Distribution of Urban Green Spaces

(1) The distribution of green space in Xi'an is characterized by unevenness. Due to historical reasons and land use restrictions, the old urban areas have relatively less green space, while the new urban areas and suburbs have more green space,

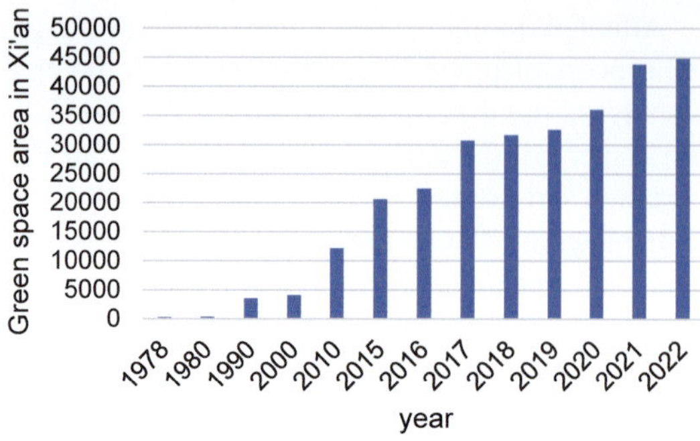

Fig. 1 Green space area in Xi'an as of 2022 statistics (unit: 10,000 square meters) (Self-drawn)

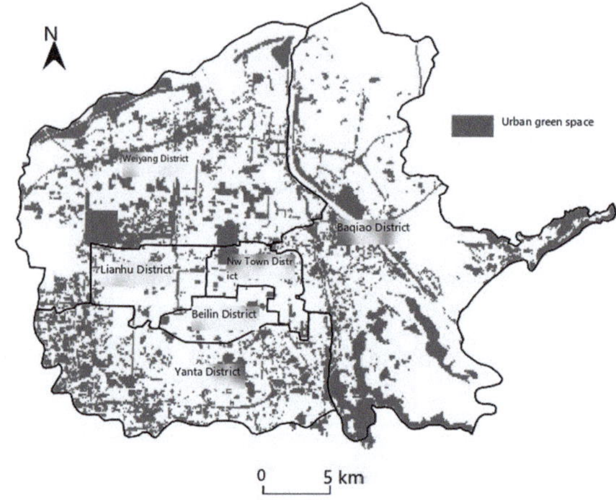

Fig. 2 Distribution of urban green space in Xi'an (Self-drawn)

according to statistics, the distribution of urban green space in Xi'an is shown in Fig. 2. Figure 2 is a reference to Dang Hui's article on the spatial pattern of urban green space ecosystem services in Xi'an based on fairness evaluation.

(2) The distribution of urban green space in Xi'an is multi-level. The plan has constructed a multi-level green space system of "city level—district level—community level".

2.4 Green Space Types and Functio

(1) Comprehensive parks

Xi'an has a certain number of comprehensive parks, such as Yannan Park, Chang'an Park, City Sports Park, etc., with an average green space rate of 75%.

(2) Community parks and street green spaces

In addition to comprehensive parks, Xi'an has also built a large number of community parks and street green spaces, which are mainly located in the vicinity of settlements, with an average green space ratio of more than 60%.

(3) Protective and productive green spaces

The city of Xi'an has set up protective and productive green spaces, such as protective green spaces on both sides of roads and rivers, as well as productive green spaces such as nurseries and flower gardens, which generally have an average green space rate of 90% or higher.

3 Assessment of Biomass Heating Potential for Urban Greening

3.1 Status of Urban Heating

Traditional heating methods (such as coal-fired heating) not only consume a large amount of non-renewable resources, but also bring serious environmental pollution problems. At the same time, due to the imperfection of the heating system, many areas have uneven heating, hot and cold phenomenon. In the process of urbanization, the area of urban green space is increasing, but at the same time, it also produces a large amount of biomass waste, and at the present stage, the treatment of green space waste is relatively single, mainly by landfill or incineration, which results in the waste of resources and secondary pollution of the environment. Through the rational use of biomass resources, we can not only realize the efficient use of energy and sustainable development of the environment, but also promote the recycling of urban waste and reduce carbon emissions.

3.2 Heating Potential

It is estimated that, under ideal conditions, Xi'an city greening biomass energy pruning amount of 1.093 million tons/year, withering amount of 176,000 tons/year. In the collection of idealized conditions, the total amount of 126.9 million tons can be used, equivalent heating potential of 17.5 to 21.5 million GJ (equivalent to 598,000 tons of standard coal), can reduce 41.9 tons of carbon emissions.

Xi'an Statistical Yearbook 2023 data pointed out that the total heat supply in 2023, including steam heating 16.82 million GJ and hot water heating 92.48 million GJ, a total of 111.1 million GJ. Calculated, the potential of urban green biomass heating is equivalent to the city of Xi'an in 2023, 0.23% of the amount of heat supply, which is equivalent to the annual centralized heating demand of 687,000 square meters of residential [6].

4 The "Collect-Add-Store-Transport-Burn" Clean Heating Technology Pathway Centered on Distributed Collection and Storage

The collection of biomass for urban greening requires a large investment in human and material resources. The establishment of a biomass heating system requires a large initial investment, including the costs of equipment purchase, installation and commissioning, operation and maintenance, and fuel storage (warehouse and land). The operating cost of biomass heating system is also relatively high, which is mainly affected by the processing of biomass raw materials, equipment operating efficiency and other factors. Although the State has introduced a series of policies and measures to encourage the development of biomass energy, the policy on the use of biomass in urban greening for clean heating is not yet complete.

4.1 Distributed Collection Methods for Urban Greening Biomass

The collection of green biomass consists of several elements, including collection methods, collection devices and collection sites. In Xi'an, green vegetation waste is diverse and scattered, and the total amount of biomass collected is greatly affected by regional and seasonal factors. According to the different sources of greening biomass, it can be categorized into street trees, parks and large-scale public green areas, residential green areas, and unit green areas [7].

There are two suggested collection methods, one mobile and the other stationary. The former refers to the use of mobile transportation vehicles to collect green waste around the point of generation, which can be driven into the living area at any time according to the need, while the latter refers to the fixed establishment of the green waste transportation point in a certain location, and the transportation staff collects the fixed point every day or every few days, which both have advantages and disadvantages. According to the distribution of green biomass in Xi'an, the corresponding collection methods are selected according to the region, as shown in Table 1.

The selection of collection sites is critical to the collection process and it is recommended that the following requirements be followed, as shown in Table 2.

In this type of scenario, the use of a collection device that meets the requirements of this type of scenario can effectively reduce the intensity of work and improve work efficiency. Since the roads in such scenes are not wide and there are often people moving around, the device should be characterized by small size, strong flexibility and low noise. For different collection scenarios, the use of appropriate collection devices can greatly reduce the work intensity and improve work efficiency.

Table 1 Collection methods in different regions (Self-drawn)

Type of region	Regional characteristics	Collection methods
Street trees	Decentralized and space-constrained	Mobile collection
Green spaces in residential areas, green spaces in flats	High demands on environmental quality (noise, vehicle emissions, etc.)	Stationary collection
parks and large public green spaces	Large and relatively concentrated (manual collection is not only intensive but also inefficient)	Mobile collection (small removal trucks) + stationary collection

4.2 *Processing of Urban Greening Biomass*

Ordinary green vegetation waste has the disadvantages of low energy density, high moisture content, irregular shape and unstable combustion, etc. The collected biomass should be processed and treated in order to effectively improve its utilization efficiency and give full play to its heating potential, and the characteristics of different processing methods are presented in Table 3.

Because of the seasonal nature of urban green vegetation waste, storage is a very important part, due to the characteristics of green vegetation waste, direct storage exists in the storage volume, low storage efficiency, quality fluctuations and other problems, so solid molding fuel is more appropriate. There are various types of solid fuels, and Table 4 shows several common biomass solid fuels and their respective characteristics.

4.3 *Distributed Storage and Transportation of Urban Greening Biomass*

The processing of biomass requires the deployment of treatment facilities, which is not practical in some regions, and therefore a "centralized + on-site" approach is used, as shown in Table 5.

As shown in Fig. 3, a storage point is set up in a certain area, and the green vegetation waste produced by nearby street trees, parks or large public green areas, green areas in residential areas and green areas of units can be centrally stored in the storage point through collection, and then centrally distributed from the storage point to each combustion point. A number of similar storage sites could be set up in Xi'an to effectively improve storage efficiency.

Urban greening biomass is usually generated within cities and can be consumed locally, avoiding the costs associated with long-distance transportation. Urban greening biomass mainly comes from the waste generated during the daily greening and maintenance process of cities, such as trimmed branches, fallen leaves, and green

Table 2 Selection of collection sites (Self-drawn)

Serial number	Specific requirements	Explanatory note
1	Feasibility of land acquisition	The replenishability of landscaping waste collection points through the land allocation process is low. Considering the scarcity of land and tight land use in megacities, the sites should be built in conjunction with parks and green spaces as much as possible. For collection points constructed in conjunction with parks and green areas, it is preferable to choose semi-underground and underground building forms [8]
2	The effectiveness of cost accounting	The location of collection points should take into account the range of waste that they can radiate for collection and transportation, and should be chosen in parks, green areas or small garbage compactor stations, mainly to reduce the cost of the site and the cost of transportation, and to facilitate the collection and outward transportation of raw materials. Where there are no conditions for new construction, facilities such as garbage rooms in parks and green areas can be modified through technological improvements [9]
3	Minimizing environmental impacts	It is preferable to choose the location of the collection point at a certain distance from residential and business office areas, preferably in an open area with a forest belt of tall trees, or a place with greenery to separate it from residential areas. Wooded areas and greenery can absorb most of the high-decibel noise generated by crushing and filter some of the dust to minimize its impact on the surrounding environment [10]
4	Collection service radius and accessibility of social services	Considering the scientific and rationality of the construction and operation of the collection points for landscaping waste, the location of the collection points should take into account the amount of green plant waste generated in a city or a region, the convenience of the transportation routes and the site conditions, etc., and combine the qualitative and quantitative aspects, while taking into account the social and environmental conditions and the limitations of the functional conditions of the facilities, so as to set up the reasonable collection points (systems) for landscaping waste. Taking into account the actual situation of the region (e.g. greening area, distance between points and construction funds, etc.), the collection points [11] are constructed in a scientific, reasonable and flexible manner

Table 3 Different processing methods for greening vegetation waste (Self-drawn)

Processing methods	Advantage	Areas of application
Solid molded fuels	Easy to store, efficient combustion	Domestic heating, industrial boilers, biomass cogeneration
Gaseous fuels	High energy density	Automotive fuels, industrial solvents
Liquid fuels	Low cost	Gas-fired power generation, gas-fired boilers, rural gas

Table 4 Biomass solid molding fuel types (Self-drawn)

Category	Dimensions	Moisture control requirements	Characteristics
Pellet fuel	0.6–1 cm in diameter	12–18%	High calorific value, low ash, good molding, high density, no coking
Lump fuel	3 to 11 cm in diameter	15–30%	High yield, moderate calorific value, easy to burn, affordable and sufficient supply

Table 5 Disposal of green vegetation waste in different regions (Self-drawn)

Region	Treatment	Remarks
Street trees	Centralized processing	Such areas have spatial limitations and are therefore treated centrally through collection
Parks and green spaces, green spaces in residential areas, green spaces in flats	In-situ treatment	In such areas, it is possible to build fully enclosed collection points with shredding and temporary storage for green vegetation waste, which can be processed on site and then stored temporarily. In areas where new construction is not possible, existing garbage rooms and other places can be remodeled to effectively improve storage efficiency

plant residues generated from landscape renovation. These biomass are usually generated on-site in cities, which is completely different from traditional energy raw materials. One significant advantage is the ability to directly consume and utilize locally. This means that there is no need to go through long and complex long-distance transportation processes like coal, oil and other energy resources. This effectively avoids a series of high costs arising from long-distance transportation, including the purchase and maintenance costs of transportation vehicles, fuel consumption costs, and possible loss costs during transportation, greatly improving the economy and efficiency of energy utilization.

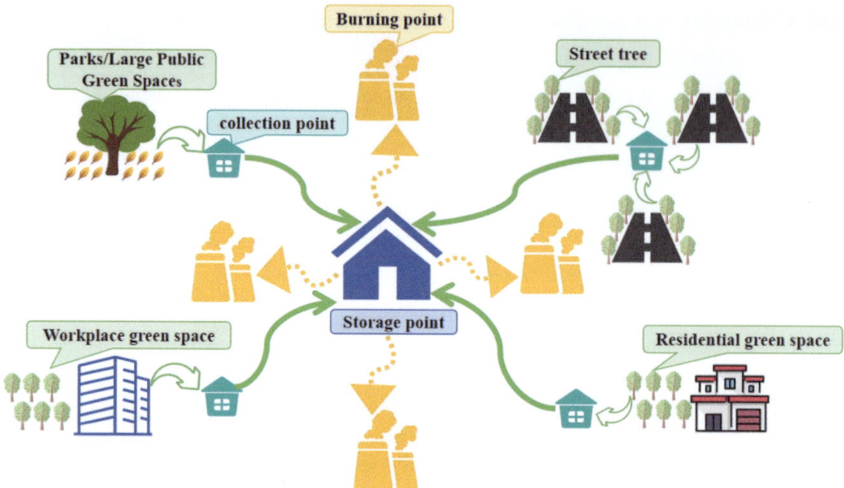

Fig. 3 Schematic diagram of the storage model (Self-drawn)

4.4 *Efficient Combustion of Biomass for Urban Greening*

(1) Optimization of biomass feedstock selection

The quality and type of biomass raw materials have an important impact on the thermal efficiency and stability of heating equipment. According to the distribution and characteristics of local biomass resources, suitable types of raw materials and processing methods should be selected, high-quality biomass fuels should be chosen, and biomass fuels should be pre-treated, e.g., crushed, screened, dried, etc., in order to improve their combustion efficiency and calorific value.

(2) Intelligent and automated control of equipment

Through real-time monitoring of the operating status of the equipment and energy consumption, rationally adjust the fuel supply, air excess coefficient and other combustion parameters, so as to make the combustion reaction more adequate and optimize the heating effect. Add combustion auxiliary equipment (such as wind pre-purifier, secondary fan, etc.) to improve the combustion process of fuel and air mixing degree, improve fuel combustion efficiency.

(3) Reduction of heat loss and waste heat recycling

Use high-quality thermal insulation materials to insulate the boiler to reduce the heat loss of the boiler itself. Regular cleaning of the boiler heating surface of the ash and ash, to maintain the heating surface of the clean and heat transfer efficiency. Reasonable control of exhaust temperature, to avoid heat loss caused by high exhaust temperature. Utilize the flue gas waste heat boiler and other equipment to recover the waste heat in the flue gas, and convert it into hot water or steam, etc.; use the waste water waste heat recovery system to recover the waste heat in the waste water of the biomass boiler, and improve the overall efficiency of heat utilization.

5 Conclusions and Outlook

(1) Urban green biomass energy is low-carbon, environmentally friendly, sustainable, and has broad prospects. Xi'an has abundant resources and significant energy utilization value. In response to the problems in its clean heating application, a full chain technology path centered on distributed collection and storage has been formed, and a systematic application plan has been proposed.

Urban greening biomass energy as a low-carbon, environmental protection, sustainable energy, has a broad development prospect. Xi'an city is rich in urban greening biomass resources, the annual pruning and withering volume can reach 1.269 million tons, equivalent to heating potential of 175,000–215,000 GJ, equivalent to 598,000 tons of standard coal, has a significant value of energy utilization. Based on the urban greening biomass in the clean heating application of biomass raw material dispersion, high collection cost, transportation and storage difficulties and other issues, the formation of a distributed collection, storage as the core of the "collection—plus—storage—transportation—combustion" full chain application technology path, through distributed collection, processing, storage and transportation and efficient combustion, the proposed urban greening biomass energy is the most effective way of energy utilization. Through distributed collection, processing, storage and transportation, and efficient combustion, a systematic application scheme of urban greening biomass energy is proposed.

(2) In the future, the promotion and application of biomass energy for urban greening needs to improve the policy system to attract capital, strengthen technology research and development to reduce costs and improve efficiency, enhance competitiveness, and promote integration with related industries to achieve resource sharing and efficiency improvement.

In the future, in the promotion and application of urban greening biomass energy, it is necessary to continue to improve the policy system of biomass energy development, increase policy support, improve the return on investment of biomass energy projects, attract more social capital to participate in the biomass energy industry. Secondly, we should strengthen the research and development and application of biomass energy technology, improve the processing efficiency and combustion efficiency of biomass raw materials, reduce production costs and improve market competitiveness. At the same time, to promote the integration of biomass energy industry and other related industries, such as the integration of landscaping, urban sanitation and other industries, to achieve resource sharing and complementary advantages, and improve overall efficiency.

Acknowledgements This study is supported by the Key Research and Development Program of Shaanxi (General Project—Social Development Field) (Program No.2023-YBSF-398); Research and application of comprehensive energy storage system regulation technology for industrial parks based on the "peak carbon dioxide emissions and carbon neutrality" goal (No. CSCEC-2023-Z-5)

References

1. World bioenergy association (2023) Global bioenergy statistics report
2. Chinese government website. https://www.gov.cn/zhengce/zhengceku/202403/content_6941170.htm. Accessed 11 Mar 2025
3. Chinese government website. https://www.gov.cn/zhengce/content/202312/content_6919000.htm. Accessed 11 Mar 2025
4. Shaanxi institute geothermal association website. http://www.sxsdrxh.com/index.php/Member/detail/id/818.html. Accessed 11 Mar 2025
5. Peng X, Xianzhong M, Yushu S et al (2017) Research on the potential of energy utilization and environmental benefits of biomass waste in Beijing-Tianjin-Hebei. Environ Sci Technol 40(12):292–299
6. Ruoxi Y, Dongning W, Xiaoning S et al (2024) Analysis of heating and carbon reduction potentials based on the current status of urban greening biomass energy resources--taking Xi'an as an example. HVAC 54(S1):44–46
7. Shaohua W (2015) Research on statistical index system and statistical method of greening residue in Beijing. Beijing Forestry University, Beijing
8. Mingjun G (2014) Site selection planning of gardening waste collection points–taking Jing'an district, Shanghai as an example. China Urban For 12(3):3
9. Huini F, Jiumin L (2010) Recycling of garden waste (II)-the establishment of collection system. Gardening (10):3
10. Ziwen L, Hailan F, Yuqing Y (2012) An analysis of green plant waste composting site construction. China Garden 28(2):4
11. Xianming S (2010) Resource utilization of gardening waste and site layout in Beijing. Beijing Forestry University
12. Hui D, Jing L, Yumeng Z et al (2021) Spatial pattern of urban green space ecosystem services in Xi'an based on fairness evaluation. J Ecol 41(17):6970–6980

Spatial Distribution of Child-Friendly Facilities and Its Influence on Low-Carbon Travel in Urban Planning

Yi Long, Mingyang Li, Panyu Peng, Yibin Ao, and Yan Wang

Abstract The keys to implementing the child-friendly concept lies in enhancing community livability, particularly by advocating for low-carbon travel. Currently, communities are still insufficiently considering children's needs, with many activity spaces and facility layouts failing to meet the requirements for low-carbon travel. Using Shuangliu District in Chengdu as a case study, this research employed both random forest and logistic regression models to examine the effects of travel distance and road environment on children's travel mode choices. The findings reveal that travel distance holds the highest relative importance in influencing children's selection of neighborhood activity travel modes, accounting for 22.48%. Additionally, a nonlinear relationship and threshold effect exist between travel distance and certain travel modes, with approximately 3 km serving as the critical intersection point for walking, cycling, and car. Furthermore, feature such as road smoothness, the availability of sidewalks, and traffic density significantly enhance the probability of children choosing to walk to child-friendly facilities. This research seeks to bridge the gap by examining how facility distribution influences travel mode choice and identifying key feature that help urban planners design child-friendly, sustainable urban environments.

Keywords Low-carbon · Child-friendly · Travel mode choice · Machine learning

Y. Long · P. Peng · Y. Ao (✉)
College of Environment and Civil Engineering, Chengdu University of Technology, Chengdu, China
e-mail: aoyibin10@mail.cdut.edu.cn

Y. Long
e-mail: longyi@stu.cdut.edu.cn

M. Li
College of Management Science, Chengdu University of Technology, Chengdu, China

Y. Wang
Department of Engineering Management, Sichuan College of Architectural Technology, Deyang, China

B. He et al. (eds.), *Climate-Resilient, Low-Carbon, and Sustainable Cities*, Lecture Notes in Civil Engineering 754, https://doi.org/10.1007/978-981-95-2169-2_27

1 Introduction

More and more cities are emphasizing child-friendly integration into urban planning and construction [1]. A child-friendly city refers to one that provides appropriate conditions, environments, and services to support the growth and development of children [2]. This requires a strong focus on the current environment for children's growth, promoting low-carbon awareness among children and teenagers, and encouraging the adoption of a lifestyle that emphasizes conservation, reasonable consumption, and green, low-carbon practices. Walking is a common mode of transportation for children, but when traveling independently, they are often at risk from traffic congestion and encroachment on pedestrian spaces [3]. Therefore, optimizing the children's travel environments is a key initiative in building child-friendly cities.

Previous research has highlighted that child-friendly facilities are not about creating spaces exclusively for children, but rather about integrating the prioritization of children's interests into urban development concepts, planning paths, and operational management. These areas should effectively address children's physical, behavioral, and emotional needs, improving the overall child-friendly atmosphere of the city. At the same time, these spatial networks can also serve other resident groups, making them sustainable, multifunctional, and intergenerational shared spaces [4]. In some Chinese cities, community life circle planning has already begun to place necessary children's facilities within 5, 10, and 15-min walking distances, aligning with the daily activity patterns and behavioral characteristics of children. These measures not only embody the concept of child-friendliness and raise residents' awareness of children's spatial needs, but also play a significant role in advancing the development of child-friendly urban environments. International studies has underscored the necessity of including child-friendly amenities into urban development to foster sustainable and inclusive communities [5, 6]. As a result, many scholars have increasingly focused on public facility planning and the factors influencing children's travel choices to these spaces.

First, children, through their cognitive exploration and behavioral needs, often reshape public spaces to suit their activities, exerting a degree of initiative. Conversely, public spaces can positively impact children by providing leisure facilities and social interaction venues, fostering opportunities for exercise, building moral values, and increasing their awareness of their environment. Such interactions promote physical and mental development. Secondly, the child-friendliness of activity spaces is also reflected in the availability of diverse venues for various activities, satisfactory tree coverage, clean and hygienic surroundings, a high proportion of soft pavement, and safe, scaled spaces designed to accommodate children's activities. These features help prevent accidents and allow children to reach these spaces in the shortest time, shortest distance, and through the simplest routes.

Existing research indicates that internal space facilities are crucial as material support elements in community public spaces, which serve as the primary platforms for children's independent activities [7]. Additionally, studies have shown that children's activity facilities located within residential areas are typically the most accessible, enabling children to reach them on foot [8]. Such layouts allow children to move freely and safely within the community, improving their overall safety when using these facilities independently. Meanwhile, a number of factors, including the population density and accessibility to transit, impact the placement of public amenities outside residential areas, which may make them less child-friendly. It is important to consider travel distance, time consumption, and equity [9]. In a study of public cultural facilities in Zhejiang Province, found that museums are increasingly prioritizing child and parent–child services. However, more attention is needed on factors like children's comfort and ergonomics, as well as creating interactive parent–child experiences [10].

Concerns over children's safe travel began to gain attention in the public health sector as early as the 1920s. Research in this field indicates that encouraging children to choose active, sustainable travel modes such as cycling, walking, or taking the bus can mitigate the negative impacts of urban travel environments. Studies have shown that the travel environment affects children's travel behavior, for example, reducing traffic density at intersections increases the probability of children choosing to walk or cycle [11]. Among the factors affecting school-age children's commuting patterns, distance emerges as the foremost determinant in their selection to walk to school [12]. In a similar study, McMillan revealed that built environment characteristics, including pedestrian infrastructure, dedicated lanes, and speed bumps in front of schools, enhance walking and cycling behaviors, with neighborhood safety being a significant influence on these choices [13]. Children's cognitive development and parents' employment status also influence their travel choices. Parental accompaniment is particularly important during childhood, with parents often choosing nearby destinations that cater to their children's needs and allowing children to influence travel decisions.

In summary, although prior research both domestically and internationally recognizes the significance of developing supportive, child-friendly environments, there remains a distinct gap in understanding how the spatial arrangement of child-friendly facilities directly impacts children's travel mode choices and their ensuing low-carbon behaviors. Therefore, the aim of this study is to quantify this relationship, identify key feature that influence mode choice, and propose actionable urban planning and policy recommendations. This study set out to: (1) examine the relative importance of features in influencing children's choice of mode of travel; (2) examine the nonlinear relationship of mode of travel and travel distance; (3) Investigate the influence of road environment on whether children walk. The main tasks include conducting case studies in the two-stream zone, collecting and processing survey data, applying machine learning models (random forests), and interpreting survey results to inform policy.

2 Data and Methodology

2.1 Data

The research process is divided into four specific parts: establishing the research objectives, formulating the research plan, conducting the research, and analyzing the results and feedback. The survey obtained ethical approval from the College of Environment and Civil Engineering, Chengdu University of Technology. The study focused on 118 communities located in Shuangliu District, Chengdu. Shuangliu District of Chengdu is one of the first pilot projects of child-friendly community construction in Chengdu, which has the background of child-friendly city construction. This district provides a suitable context for examining how spatial distribution of facilities influences children's travel mode choices and can offer insights applicable to other developing urban areas.

The research work included the following: in the pre-survey phase, a preliminary analysis of the distribution of child-friendly resources and facilities, as well as the daily travel patterns of children, was conducted. During the formal research phase, we carried out surveys with children and communities, conducted interviews with government officials, and performed fieldwork in the Shuangliu District of Chengdu, Sichuan Province, to gain a comprehensive understanding of child-friendly, low-carbon travel patterns. The main process of the investigation is shown in Fig. 1.

The survey was conducted over a nearly four-month period (2022.08–2022.11). Throughout this timeframe, 3,332 questionnaires were distributed and collected. Following the exclusion of responses with polygraph inconsistencies and other unreliable data, 1,127 questionnaires were deemed valid. This process yielded a total of

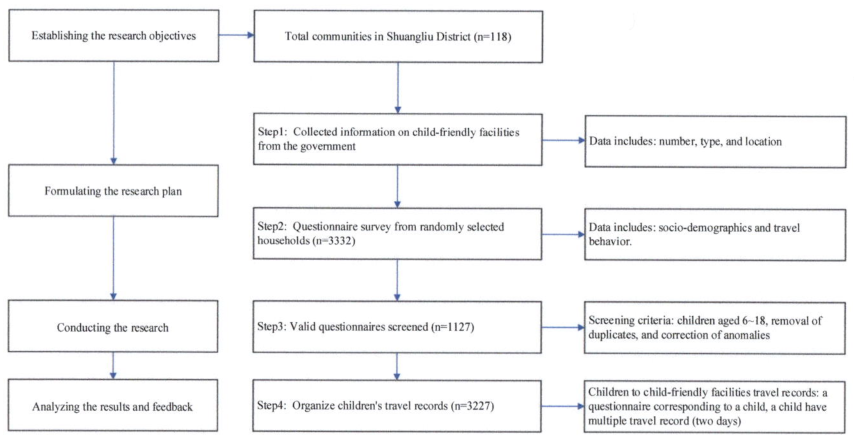

Fig. 1 The main process of the investigation (Self-drawn)

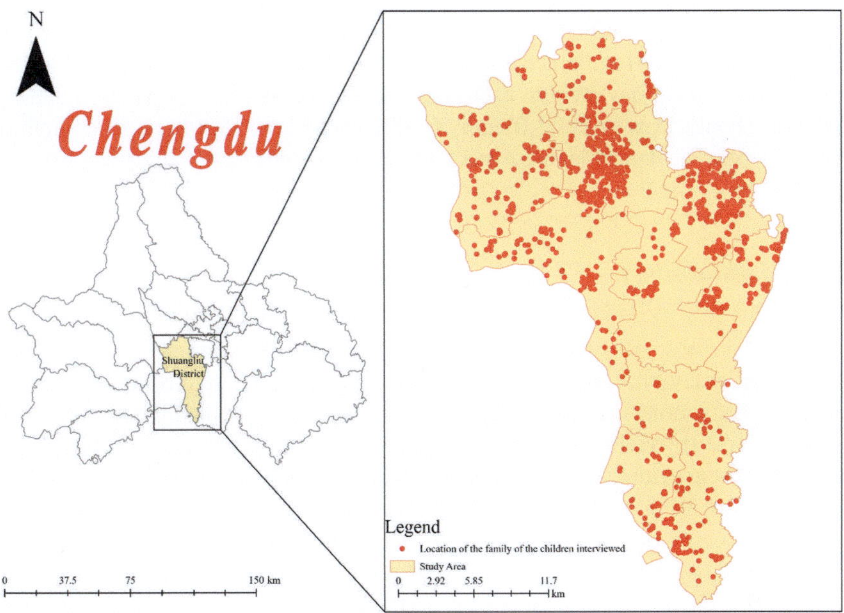

Fig. 2 Family location and spatial distribution of interviewed children. (Self-drawn)

3,227 actionable data points concerning children's travel behaviors. Figure 2 illustrates the distribution of the family locations of the surveyed children in the study area. The Fig was drawn using ArcGISPro (3.0).

To obtain children's travel distance, this study utilized the Amap Open Platform's "Route Planning API." By inputting the latitude and longitude coordinates of the origin and destination, along with the selected travel mode, into a pre-written program, we were able to output information for various routes. To accurately estimate the distance from origin to destination, the length of the first-choice route was employed, considering real-world travel paths and prevailing road conditions.

The research team identified 24 categories of child-friendly facilities by referencing the Child-Friendly Community Construction Standards (CFCCS) [14] and incorporating the requirements observed during field surveys. These facilities were categorized into five main groups based on their function, including children's homes, open outdoor play/exercise/leisure areas, and children's psychological counseling rooms [15].

The data on population density was collected through WorldPop (https://www.worldpop.org) and then revised through the 7th National Census. Road density information was obtained from OpenStreetMap (https://www.openstreetmap.org). In order to adapt built environment variables to the needs of children, the widely recognized "5Ds" model [16] was used and made consistent with the specific research objectives of this study. These variables include population density, land use mix, accessibility to child-friendly destinations, distance to transit stops, and road density. In this study, the built environment around children's settlements was measured using buffer zones with a radius of 1000 m, in accordance with the 15-min living circle planning and construction guidelines. Within each buffer zone, we assessed population density and road density. Data on child-friendly facilities were collected through community governments. The data of land use mix and bus stops are from Amap Open Platform.

2.2 Methodology

The random forest (RF) model was used in this study. The principle of RF model is based on ensemble learning. The data obtained through random sampling is fed to many weak learners (decision trees), and the category with the highest number of votes is selected by voting [17]. This method offers several advantages over traditional linear models, providing better predictive performance when addressing mode choice problems [18]. RF models have been widely adopted to solve issues in travel mode choice [19], commute time prediction [20], and nonlinear correlation analysis [21].

In this paper, we also introduce the Partial Dependence Plots (PDP) method to visualize the nonlinear relationship and assess the importance of explanatory variables in children's travel mode choices. The PDP demonstrates the incremental effect of the independent variables on the dependent variable. PDP do not rely on prior assumptions and provide a more accurate representation of the relationships between variables compared to traditional statistical models. The partial dependence of characteristic variables is calculated as follows:

$$\widehat{\phi}(Z_{i1}) = \frac{1}{T}\sum\nolimits_{i=1}^{T} f(Z_{i1}, Z_{i2}, \ldots, Z_{ij})$$

where $f(z)$ stands for the dependent variable, and $\widehat{\phi}(Z_{i1})$ is the partial correlation function for the feature variable Z_{i1}, T stands for the total number of instances in the collection, and $(Z_{i1}, Z_{i2}, \ldots, Z_{ij})$ shows the actual observed value of the j th feature of sample i. This article employs PDP to elucidate the impact of independent variables, such as travel distance, on children's mode of travel choosing.

When optimizing the RF model, a grid search method was used to determine the most effective combination of these parameters [22], The grid search process requires defining a specific range for each parameter—the depth of the tree from 1 to 20, the number of features per tree from 2 to 10, and the number of trees from 10 to 1000 (in increments of 10). A total of 18,000 ($20 \times 9 \times 100$) different parameter configurations were evaluated, using out-of-pocket error rates to evaluate the performance of the model [23]. After testing, the optimal model of parameters: max_depth = 2, n_estimators = 700, max_features = 2.

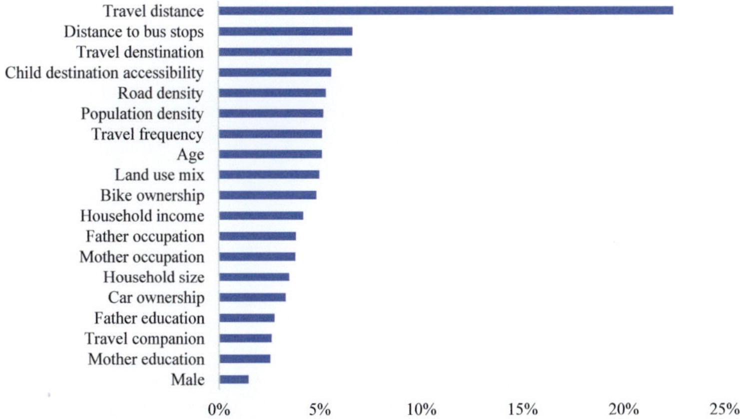

Fig. 3 Relative importance of independent variables (Self-drawn)

3 Results and Discussion

3.1 Feature Importance Analysis

After fitting the data, the random forest model provides a measure of the importance of the data attributes, which in sklearn is represented by the feature importances parameter. This parameter returns a numpy array that reflects the importance of each feature in the training process, where a higher value indicates that the corresponding feature is more important for the prediction accuracy. The visualization of the variable importance is shown in Fig. 3.

As illustrated in Fig. 3, travel distance emerges as a significantly more influential feature than other variables in determining children's choices of travel modes (relative importance: 22.48%). This highlights the critical role that distance plays in shaping how children decide to travel. Nevertheless, this observation does not elucidate the specific ways in which distance impacts each distinct mode of transportation. Consequently, it is imperative to conduct a comprehensive analysis for each travel mode, specifically examining walking, cycling (including bicycles and electric bikes), bus, and car travel.

3.2 Nonlinear Effects of Travel Mode Influences

In Fig. 4, the X-axis represents the travel distance, in km; The Y-axis represents the probability values of four travel modes. The dashed line illustrates the observed data

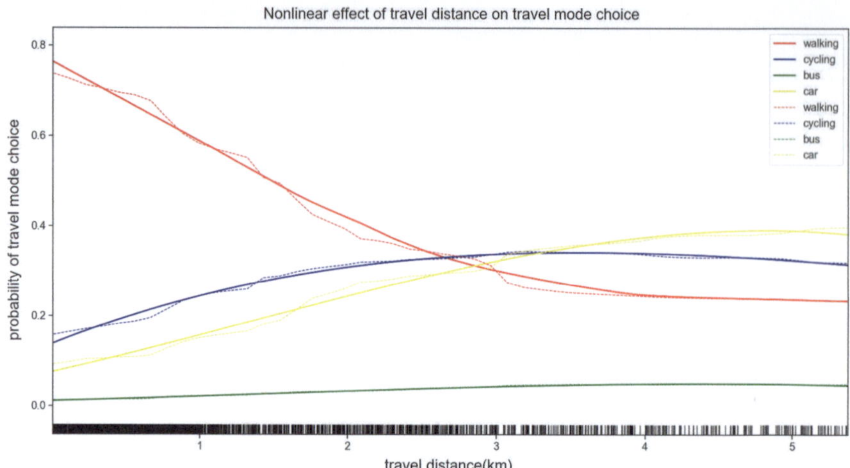

Fig. 4 The nonlinear relationship between children's travel mode choice and travel distance (Self-drawn)

points, while the solid line depicts a smoothed trend curve based on the methodology of Tao [24]. This smoothing technique helps to mitigate data noise and more accurately delineate the relationship between the variables.

Figure 4 illustrates that the relationship between travel distance and travel mode choice is nonlinear, exhibiting a distinct threshold value. For walking, travel distance is negatively associated with the probability of choosing to walk when it is within 3.07 km. Beyond this distance, the influence of travel distance on walking remains relatively stable, and the likelihood of walking remains around 0.2. In contrast, the probabilities of cycling and car usage increase as travel distance grows, indicating that travel distance has a positive effect on both these modes. Within 3.28 km, the probability of cycling surpasses that of car usage, but beyond 3.28 km, car usage probability exceeds that of cycling. Notably, the probability of choosing the bus is consistently the lowest across all distances, despite a gradual upward trend. This finding suggests that public transportation may not serve as the primary mode of travel for children in the community, warranting further investigation to clarify the underlying reasons.

Another key observation is the critical distance around 3 km, where there is a shift in the probability of choosing between walking, cycling, and cars. Within 2.78 km, walking is more likely than cycling or car usage. Beyond 2.78 km, cycling becomes the most likely mode, until private cars surpass both walking and cycling in probability at 2.97 km. After 3.17 km, car usage becomes more likely than cycling as well. Overall, in community activities, children's primary travel modes are walking, cycling, and private cars, with a very low probability of choosing public transportation.

These outcomes are consistent with the '1-3-5' low-carbon travel framework, which encourages walking for trips within 1 km, cycling for distances up to 3 km, and the use of public transportation for travel extending to 5 km. Understanding how travel distance influences mode choice behavior can assist urban planners in designing more child-friendly communities and transportation systems. Consequently, it is advisable for planners to prioritize the enhancement of children's transportation infrastructure within a 2 to 4.5 km radius surrounding residential neighborhoods. Targeted approaches encompass the establishment of community bus routes, the expansion of public transportation services, and the improvement of cycling infrastructure.

3.3 Analysis of Children's Low Carbon Travel Road Satisfaction

In this study, independent variables such as pavement type, road smoothness, road cleanliness, sidewalk provision, non-motorized lane setup, traffic lights, rest facilities, crosswalks, lighting conditions, speed reduction/limit signs, speed bumps, traffic density, street layout, street trees, and landscape features were selected for binary logistic regression analysis, with the dependent variable being whether children chose to walk to child-friendly facilities.

The aforementioned variables were incorporated into the equation, and binary logistic regression was performed using SPSS. The Logistic regression results are presented in Table 1.

several key factors significantly influence children's decision to walk to child-friendly facilities. Road smoothness has a significant positive influence on whether to choose walking ($p = 0.002$), with each unit increase in smoothness increasing the probability of children walking by approximately 35.2% ($OR = 1.352$). Sidewalk availability also plays a crucial role ($p < 0.001$), where the presence of sidewalks increases the probability of walking by 72.2% ($OR = 1.722$), underscoring the importance of pedestrian infrastructure.

Conversely, road cleanliness shows a significant negative association with walking ($p < 0.001$). Specifically, improved road cleanliness decreases the probability of children walking by 35.5% ($OR = 0.645$). This counterintuitive finding may suggest that cleaner roads are associated with higher traffic speeds or volumes, which could deter children from walking due to safety concerns. Additionally, the presence of non-motorized lanes significantly reduces the probability of walking ($p < 0.001$, $OR = 0.697$), possibly because children may opt for cycling when such facilities are available.

Traffic management features also impact walking behavior. The presence of traffic signals ($p < 0.001$, $OR = 0.745$) and speed bumps ($p = 0.002$, $OR = 0.747$) negatively affect the probability of children walking, perhaps due to increased complexity or perceived safety risks at intersections and roadways. Interestingly, traffic volume exhibits a significant positive relationship with walking ($p = 0.001$, $OR = 1.239$), suggesting that in areas with higher traffic density—often urban environments with closer amenities—children are more inclined to walk despite the traffic.

Table 1 Logistic regression results (Self-drawn)

Variable	Coefficient	Standard error	Wald	P	OR (Odds ratio)
Road width	0.044	0.077	0.336	0.562	1.045
Pavement type	0.071	0.102	0.481	0.488	1.073
Road smoothness	0.302	0.096	9.811	0.002	1.352
Road cleanliness	−0.438	0.079	30.801	0.000	0.645
Sidewalk	0.544	0.088	38.199	0.000	1.722
Non-motorized	−0.361	0.076	22.495	0.000	0.697
Traffic signals	−0.294	0.077	14.557	0.000	0.745
Recreational facilities	0.107	0.067	2.535	0.111	1.113
Crosswalks	0.105	0.094	1.235	0.266	1.111
Street lights	0.089	0.074	1.438	0.23	1.093
Speed limit signs	0.013	0.101	0.016	0.899	1.013
Speed bumps	−0.292	0.094	9.715	0.002	0.747
Traffic volume	0.214	0.062	11.902	0.001	1.239
Shops	0.031	0.082	0.14	0.708	1.031
Street trees	−0.081	0.083	0.963	0.327	0.922
Landscape pieces	0.019	0.074	0.065	0.798	1.019

4 Conclusion

This research utilizes survey data from Shuangliu District in Chengdu and applies RF and Logistic model to examine the nonlinear relationship between the community travel environment and children's low-carbon travel behaviors. The primary objective is to enhance the low-carbon travel setting by prioritizing children's perspectives.

First, the relative importance analysis indicates that travel distance is the most significant factor. Furthermore, the PDP reveal the presence of distinct thresholds and a nonlinear relationship between travel distance and certain travel modes. In particular, travel distance exerts a negative influence on walking, but positively affects cycling and car. An intersection points for walking, cycling, and cars emerges around 3 km, while the probability of choosing the bus remains consistently low. Based on these findings, we recommend walking as the preferred mode within 1 km and cycling within 1.5 km. To encourage greater reliance on buses, additional stops and routes should be established.

Understanding how travel distance influences the travel mode choices of children is crucial for urban planners working to create more child-friendly neighborhoods and transit systems. As such, it is recommended that planners focus on enhancing transportation options for children within a 2 to 4.5 km radius of residential areas. To achieve this, key strategies include introducing community bus routes, expanding public transportation services, and improving cycling infrastructure. The goal is to encourage children to use sustainable travel modes like walking, cycling, or public transit.

Walking is a well-recognized active mode of transport, known for being both the most affordable and environmentally sustainable low-carbon option. Our survey findings reveal that certain factors significantly increase the likelihood of children walking to child-friendly locations, such as the smoothness of the road, the availability of sidewalks, and lower traffic density. In particular, children are more likely to walk when the roads are smooth, sidewalks are well-maintained and clearly marked, and traffic is less congested. On the other hand, factors like road cleanliness, the presence of non-motorized lanes, traffic lights, and speed bumps tend to negatively impact the chances of children walking.

From an urban planning and design perspective, these findings underscore the importance of following clear principles when adding child-friendly facilities. Planners should adhere to accessibility standards, ensuring that essential facilities are placed within walking or cycling distance. Designing mixed-use neighborhoods with integrated child-friendly amenities, safe pedestrian and cycling paths, and well-maintained public spaces can significantly enhance children's active travel modes. Moreover, these principles should be integrated into local urban design guidelines, zoning regulations, and comprehensive community planning to create neighborhoods that inherently support low-carbon, child-friendly mobility.

This study primarily focuses on children as a homogenous group, but the needs of different age groups vary significantly. Younger children may have different mobility requirements and safety concerns than older children. Future research should incorporate age-segmentation analyses to understand how facility distribution and travel distance thresholds vary across age ranges. Additionally, while this study concentrates on a single district in Chengdu, comparative analyses with other international case studies would provide broader insights and help validate the generalizability of these findings. Expanding the analysis to include nighttime travel patterns, seasonal variations, and digital navigation aids could also further enhance the robustness of future research.

Acknowledgements This study is supported by the National Natural Science Foundation of China (72171028), the China Postdoctoral Science Foundation (2022T150077, 2022M710496), the Project of Chengdu Park City Demonstration Area Construction Research Center (GYCS2023-YB009), the Project of Chengdu Green and Low Carbon Development Research Base (LD2024Z06), the Project of Chengdu Population and Development Research Base (2023PD01), the Project of Sichuan Children Protection and Development Research Center (ETBH2023-YB002), the Project of Chengdu Philosophy and Social Sciences' Comprehensive Education Reform Experimental Base (CDZSQY-2024(3)), the Project of Sichuan Mental Health Research Center (XLJKJY2401A), the Project of National Research Center of Early Childhood Physical Education Development, Chengdu University (YETY2023A02).

References

1. Nan F (2020) Policy innovation on building child friendly cities in China: evidence from four Chinese cities. Child Youth Serv Rev 118:105491
2. Riggio E (2002) Child friendly cities: good governance in the best interests of the child. Environ Urban 14(2):45–58
3. Southworth M (2005) Designing the walkable city. J Urban Plan Dev 131(4):246–257
4. Peng W (2020) Child's infrastructure in cities—urban planning and design to guarantee child rights. Landsc Architect Front 8(02):100–109
5. Cordero-Vinueza VA, Niekerk FF, Van Dijk TT (2023) Making child-friendly cities: a socio-spatial literature review. Cities 137:104248
6. Jansson M, Herbert E, Zalar A et al (2022) Child-friendly environments—what, how and by whom? Sustainability 14(8):4852
7. Zheng Y (2021) Evaluation and optimization strategies of child-friendliness of public spaces in urban residential areas: a case study of district 6 of Lianhua New Village. Suzhou University of Science and Technology, Suzhou City
8. Xu Z (2022) Research on landscape design of child-friendly residential public space from the perspective of ERG theory. Wuhan University of Science and Technology
9. Liu C (2021) Study on the strategy of optimizing the traveling environment of the child-friendly public cultural facilities in Dalian. Dalian University of Technology
10. Wang Y (2018) Research on design of public faclities innovation service system based on children-friendly city. Ind Des 03:71–72
11. Lin J-J, Yu T-P (2011) Built environment effects on leisure travel for children: trip generation and travel mode. Transp Policy 18(1):246–258
12. Panter JR, Jones AP, Van Sluijs EM et al (2010) Neighborhood, route, and school environments and children's active commuting. Am J Prev Med 38(3):268–278
13. Mcmillan TE (2007) The relative influence of urban form on a child's travel mode to school. Transp Res Part A Policy Pract 41(1):69–79
14. Xu M, Tim G, Mao P et al (2021) International certification mechanism of "child friendly cities (communities)" and related practices and theoretical developments of European and American cities. Urban Plan Int 36(1):1–7
15. Long Y, Ao Y, Li H et al (2024) Nonlinear effects of children's daily travel distance on their travel mode choice considering different destinations. J Transp Geogr 118:103921
16. Ewing R, Cervero R, Trb et al (2001) Travel and the built environment-a synthesis. Land development and public involvement in transportation: planning and administration 87–114
17. Rhodes JS, Cutler A, Moon KR (2023) Geometry- and accuracy-preserving random forest proximities. IEEE Trans Pattern Anal Mach Intell
18. Zhao XL, Yan X, Yu A et al (2020) Prediction and behavioral analysis of travel mode choice: a comparison of machine learning and logit models. Travel Behav Soc 20:22–35
19. Wang B, Gao LJ, Juan ZC (2018) Travel mode detection using GPS data and socioeconomic attributes based on a random forest classifier. IEEE Trans Intell Transp Syst 19(5):1547–1558
20. Liu J, Xiao L (2023) Nonlinear relationships between built environment and commuting duration of migrants and locals. J Transp Geogr 106
21. He M, Pu L, Liu Y et al (2022) Research on nonlinear associations and interactions for short-distance travel mode choice of car users. J Adv Transp
22. Stuke A, Rinke P, Todorovic M (2021) Efficient hyperparameter tuning for kernel ridge regression with Bayesian optimization. Mach Learn-Sci Technol 2(3):14
23. Cheng L, Chen X, De Vos J et al (2019) Applying a random forest method approach to model travel mode choice behavior. Travel Behav Soc 14:1–10
24. Tao T, Wang J, Cao X (2020) Exploring the nonlinear associations between spatial attributes and walking distance to transit. J Transp Geogr 82

BY NC ND

Retrospect and Prospect of Research on Flat Micro-Heat Pipe Technology

Bingbing Leng, Zeyi Wang, Jiatong Zhang, Tianyang Zhang, Taiguan Song, and Yutao Sun

Abstract Heat pipes are widely used in microelectronic cooling, heat recovery in heat exchangers, and a heat transfer material for solar collectors. We look back on the exploration and research of micro-plate heat pipe by domestic and foreign scholars in recent years, on the basis of analyzing the structure and working principle of the micro-plate micro thermal tube, this paper summarizes the research progress of the micro-channel structure shape size, filling quality and its filling rate and working state affecting the thermal transmission performance of the micro-plate micro thermal tube. This paper summarizes the development of the structural design and theoretical prediction of the microtubule heat pipe model, and looks forward to the development trend of the microtubule heat pipe microchannel.

Keywords Heat transmission performance · Micro-channel structure · Work quality and filling rate · Working status · Compatible

1 Introduction

Flat plate micro heat pipe is through the capillary force in a small space to drive the evaporative cooling of the work material to complete the phase change heat transfer device. In 1965, Cotter was the first to put forward the complete heat pipe theory, and in 1984 he was in the Fifth International Heat Pipe Conference and put forward for the first time conception of "miniature heat pipe", laying the theoretical foundation for the subsequent performance analysis and design research of the heat

B. Leng (✉) · Z. Wang · Y. Sun
Changchun Power Supply Company, State Grid (Jilin Province) Electric Power Co., Ltd., Changchun, China
e-mail: 1904521705@qq.com

J. Zhang · T. Song
School of Municipal and Environmental Engineering, Jilin Jianzhu University, Changchun, Jilin, China

T. Zhang
Graduate School of Environmental Engineering, The University of Kitakyushu, Fukuoka, Japan

B. He et al. (eds.), *Climate-Resilient, Low-Carbon, and Sustainable Cities*, Lecture Notes in Civil Engineering 754, https://doi.org/10.1007/978-981-95-2169-2_28

pipe. Design research has laid a theoretical foundation [1]. There are mainly two kinds of definitions for micro heat pipe. One is that Cotter and G.P. Peterson define it as the heat pipe whose hydraulic radius is greater than or equal to the radius of curvature of the liquid curved meniscus in the heat pipe [2], Will be defined as when the number of Bond less than or equal to the heat pipe 2. In 1994, Khrustalev et al. [3] proposed a detailed mathematical model to study the heat and mass transfer process in micro heat pipes (MHP), and studied the heat transfer process in a channel core heat pipe with high heat flow density. A detailed mathematical model was proposed to study the heat and mass transfer process of the micro heat pipe (MHP) and investigate a channel core heat pipe with high heat flow density, It was determined that when the heat pipe's operational temperature attained 120 °C, it could sustain a radial heat flow density of 200 W/cm^2 and an axial heat flow density of 150 W/cm^2. A mathematical model to predict the minimum radius of the curved moon face and the maximum heat transfer in a triangular channel was proposed by Peterson et al. [4]. The findings indicated that the heat transfer capability of the micro heat pipe was significantly influenced by the liquid-artery apex angle, the liquid-flow contact angle, the length of the heat pipe, the characteristics and rate of vapor flow, as well as the tilt angle. Meanwhile, the structural parameters of the micro heat pipe for obtaining the maximum heat transfer capacity, the hydraulic diameter of the triangular notch is equal to 0.484 mm and the notch reaches 6.79 W/cm^2.

Owing to its superior heat transfer capabilities and consistent temperature distribution, the micro heat pipe is extensively utilized in areas such as electronic device cooling, solar air collectors, and high-efficiency heat exchangers. Based on the specific application, it is generally necessary to assess the heat pipe's heat transfer efficiency, startup characteristics, and isothermal behavior. The critical heat flux serves as the primary metric for evaluating the heat transfer performance of micro heat pipes. The maximum heat transfer capability of the micro heat pipe is defined by the lowest value among the heat transfer limits [5]. The heat transfer limit mainly has the following five kinds: capillary limit, continuous flow limit, condensation limit, boiling limit, entrainment limit. There are many factors affecting the heat transfer limit, directly related to its own structure, including the size of the heat pipe, shape, evaporation section and condensation section length, tube work, working temperature and the inclination angle of the heat pipe, etc., and secondly, such as capillary force, boiling and condensation phenomenon, pressure drop, and so on. With the increase of heating power, the thermal resistance of the heat pipe decreases, and the main thermal resistance of the heat pipe consists of two parts: the evaporating section and the condensing section [6]. Based on this, this paper reviews and summarizes the performance research results of micro heat pipe technology in recent years, and looks forward to the hot research issues that may be concerned in the future development of this field.

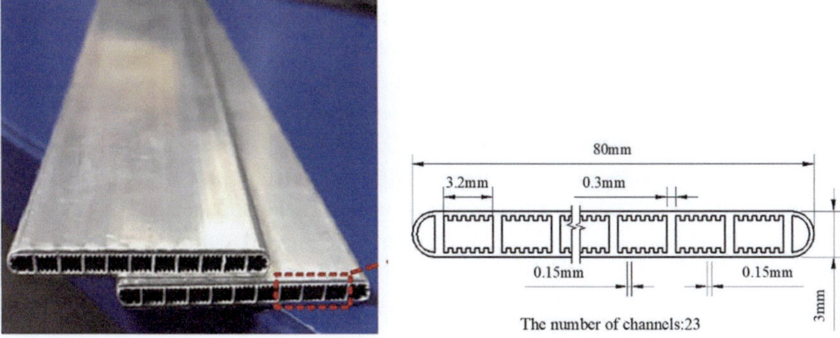

Fig. 1 The physical picture of the flat micro heat pipe (Self-drawn)

2 The Architecture and Operational Mechanism of Flat Plate Micro Heat Pipe

Flat plate micro heat pipe array (MHPA) technology combines microscale heat and mass transfer, phase change heat and mass transfer, microscale complex flow under high-pressure conditions, and other basic theories with materials and processing, packaging technology, and other technological applications. Flat plate micro heat pipe array as a super-conducting performance of the thermal conductivity of the element, compared with conventional heat pipe has obvious advantages: low thermal resistance and high thermal conductivity, surface contact heat transfer area, high reliability, high pressure capacity and other characteristics. A flat plate micro heat pipe system consists of an aluminum shell and a small amount of internally charged working fluid, the array is arranged inside a number of independent operation of the micro heat pipe, each micro heat pipe has the micro-fin structure to strengthen heat transfer, the micro-channel parallel composition of the micro heat pipe array. Such a structure increases the heat pipe direct heat and heat absorption area, due to the micro heat pipe hydraulic diameter of only 1.0 mm or so, the wall of the pressure capacity is very high, so it is not easy to leak. Furthermore, since each micro heat pipe functions independently, even if one section of the heat pipe is damaged, the remaining micro heat pipes can continue to ensure the normal operation of the heat exchanger, thus maintaining high reliability [7] (Fig. 1).

3 Examination of Thermal Transfer Efficiency in Flat Plate Micro Heat Pipes

In contrast to other heat pipes, flat plate micro heat pipes exhibit distinct characteristics, such as their compact dimensions, the absence of a liquid absorption core within the tube, and ease of manufacturing, which facilitates their application. The capillary limit and boiling limit are two crucial constraints for heat pipes, with the primary limitation of heat transfer in flat plate micro heat pipes being attributable to the capillary limit. The capillary force and the amount of working fluid have a significant impact on its thermal performance, while the latter two are determined by the flat plate micro heat pipe structure. The following section summarizes the study of the influencing factors on the heat transfer performance of flat plate micro heat pipe from the aspects of the channel structure shape and size of the heat pipe, the charging medium and its charging rate, and the working condition, and gives an outlook on the performance prediction study.

3.1 Influence of Micro-Slot Structure

The capillary pressure, which is a driving force for fluid reflux in a miniature heat pipe, can be used as an indicator to assess the heat transfer performance of a flat plate micro heat pipe, and the greater the capillary pressure inside the tube, the better the micro heat pipe heat transfer performance [8]. The capillary force required for reflux is provided by the pressure difference in the axial channel, which is generated by the sharp corners of the inner wall of the tube or the meniscus surface of the liquid seam around the steam channel. It is recalled that the design of previous microtank structures was generally based on factors such as external dimensions, length subject to gravity, and total heat transfer. As the heat transfer cycle process of micro heat pipes is mainly driven by the capillary pressure of the inner channel, its size is related to the effective capillary radius and solid–liquid contact angle. It is recalled that the design of previous microtank structures was generally based on factors such as external dimensions, length subject to gravity, and total heat transfer. The subsequent section primarily concentrates on these elements and evaluates the associated experiments (Figs. 2 and 3).

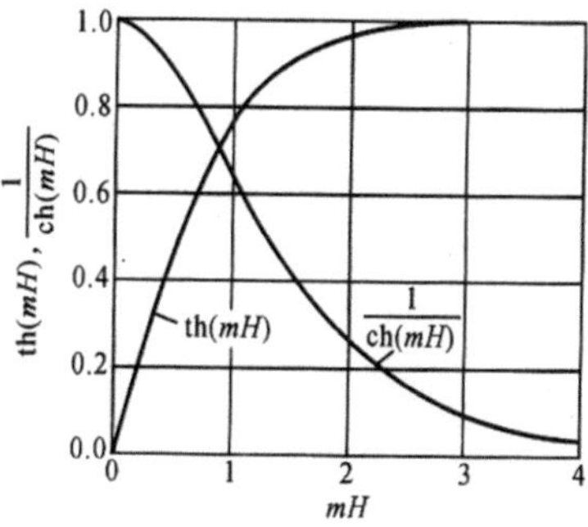

Fig. 2 Diagram of fin efficiency

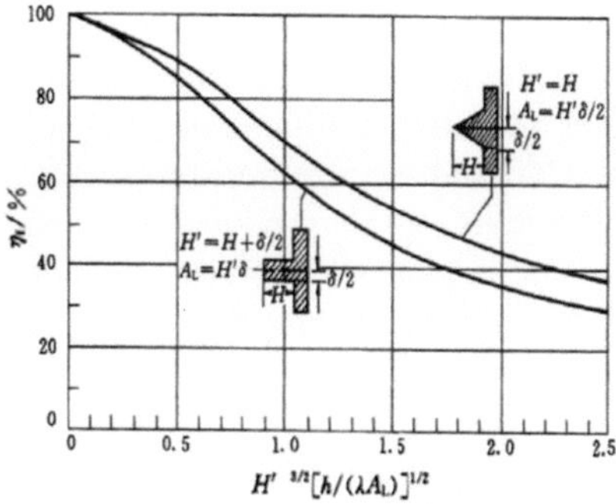

Fig.3 Fin analysis diagram (Self-drawn)

3.2 Study of Micro-Slot Structure

The capillary force, differential pressure gradient at both ends, surface roughness of the wall material, solid–liquid contact angle and wettability, etc. are all related to the flow and heat transfer characteristics of the mass inside the tube. Capillary force, differential pressure gradient at both ends of the tube as a gas–liquid flow of the "driving force", directly affecting the heat pipe heat transfer, tube length, tube wall, axial heat of the three factors play a role in the critical heat transfer, the increase in the effective heat transfer area of the tube heat transfer performance of microtubes has been improved. According to the theory of enhanced convective heat transfer proposed by Academician Guo et al. [9], we know that the thermal boundary layer of the flow of the mass of the velocity vector and the temperature gradient between the angle of the synergy can be realized enhanced heat transfer, the degree of synergy can be based on the synergy of the number of discrimination, so we need to clarify the connection between the above parameters in the future.

3.3 Influence and Study of Filling Masses and Their Filling Rates

The varying physical characteristics of distinct working fluids differ, and the choice of an appropriate working fluid can significantly influence the heat transfer efficiency of the flat plate micro heat pipe. The selection of the working fluid will depend on factors such as the operational environment, working temperature, and the inherent properties of the fluid itself, among others.

Zhao et al. [10] investigated the effects of four working fluids (methanol, ethanol, acetone, and R141b) on the heat flux of a flat plate heat pipe with microstructure. The results showed that the heat dissipation effect of the flat-plate micro heat pipe array was good, and the maximum heat flux and the total heat transfer capacity of the heat sink using methanol as the work material could reach 102 W/cm^2 and more than 102 W, respectively. They tested the heat transfer performance of the composite hollow heat pipe under the conditions of different working materials (distilled water, anhydrous ethanol, potassium dichromate solution) and filling rates. The results show that the optimum filling rate of the composite hollow heat pipe with distilled water is 33%, and the heat transfer performance of the composite hollow heat pipe with anhydrous ethanol is optimal when the filling rate is 33%.

Through the above study, we know that methanol, acetone, ethanol are excellent thermal conductivity of the working fluid, generally preferred methanol, followed by acetone or ethanol can be selected in accordance with the actual working condition of the environment, which is inextricably linked with the quality factor of the work material [11]. The rate at which a tank is filled with fluid stands as a crucial element influencing the heat transfer efficiency of microtubes. The performance of flat plate microtubes in heat transfer varies with alterations in the filling rate, with a typical optimal range falling between 20 and 30%. The optimal liquid filling ratio of the working material is also subject to the heat pipe heat transfer limit. When the liquid filling rate is too low, it will lead to insufficient liquid mass reflux, and the heat in the evaporation section is not transmitted in a timely manner, so that the temperature of the evaporation section rises; When the liquid filling rate is too high, the liquid in the condensing section is clogged, so that the temperature of the section is significantly reduced [12]. In addition, the liquid filling rate is too high to reduce the steam space in the tube, resulting in increased resistance to the flow of steam, leading to deterioration of heat transfer performance. Both of these conditions lead to an increase in the thermal resistance of the heat pipe and a decrease in the equivalent thermal conductivity of the micro heat pipe. The presence of a vapor cavity reduces the resistance to flow in the gas–liquid circulation of the work material. In order to facilitate the engineering application, we need to firstly clarify the compatibility of the working fluid and the shell material in the working environment, and secondly, the role of the circulating state of the working fluid with different filling rates in the micro heat pipe and the residual volume in the pipe on the heat transfer performance of the micro heat pipe. Therefore, the representative microheat tube array of R141b, acetone and ethanol were selected for testing. The specific parameters are shown in Table 1.

As shown in Fig. 4., the impact of varying external ambient heat temperatures on the rate of heat transfer per unit time is illustrated. As can be seen from the figure, when the heating sheet temperature of R141b working medium micro-heat pipe array is 50 °C, when the external temperature is low, the heating power changes greatly with the external temperature. When the heating sheet temperature is 25, 30 and 40 °C, the heating power changes less with the external temperature, and

Table 1 Parameters of the microheat pipe array for experiments. (Self-drawn)

Serial number	Size (mm)	Working fluid	Filling rate (%)	Ribs
1	800 × 80 × 3	R141b	20	0
2	800 × 80 × 3	Acetone	20	0
3	800 × 80 × 3	Ethanol	20	0

the heating power decreases with the increase of the external temperature. R141b The performance of the working medium microheat tube array is similar to that of acetone when the heating sheet temperature is 25, 30 and 40 °C. When the external temperature is higher than −5 °C, the heating power changes gently with the external temperature, and the heat transfer performance is poor.

The temperature diagram shows that after the release from the evaporation section, it is easier to form the condensation section, and the middle adiabatic section is short. Although the thermal resistance is large, the absorbed heat will be better transmitted to the released heat due to the increase of the condensation section, and the heat release capacity efficiency is higher (Fig. 5).

The influence of heat exchange at varying environmental temperatures is illustrated in Fig. 6. When the external temperature drops below −11 °C, the working medium's temperature surpasses the threshold limit, leading to a minimal degree of phase transformation within the tube.

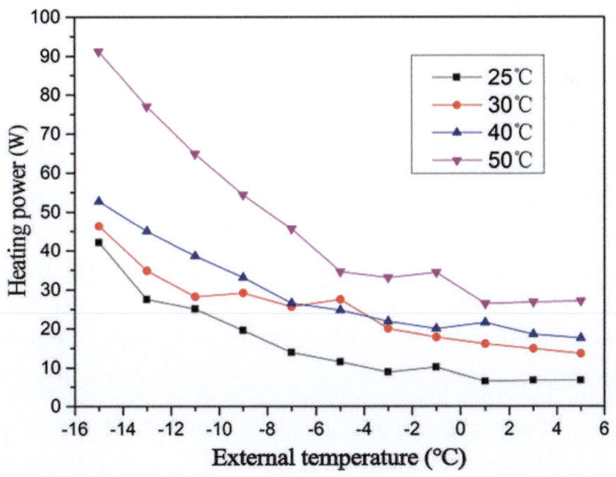

Fig.4 Effect of flat microheat pipe with R141b fluid on heat transfer per unit time at different ambient temperatures (Self-drawn)

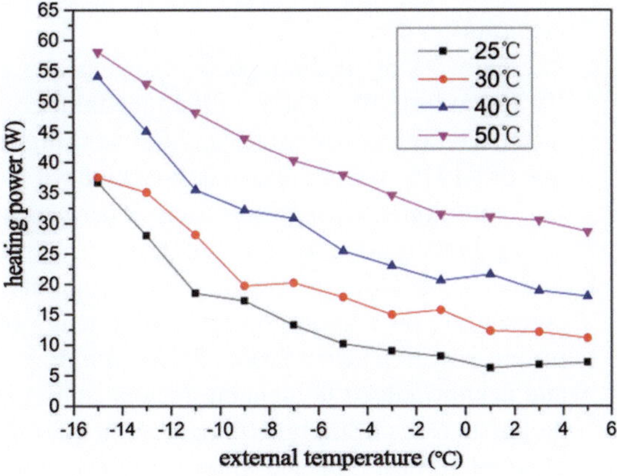

Fig.5 Effect of flat microheat pipe with acetone fluid on heat transfer per unit time at different ambient temperatures (Self-drawn)

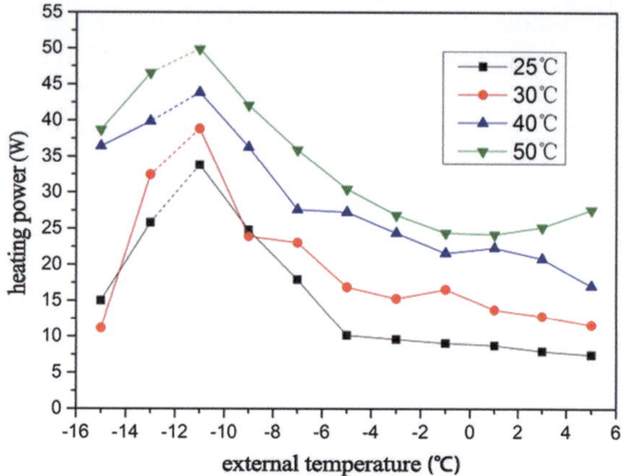

Fig.6 Effect of flat microheat pipe with ethanol fluid on heat transfer per unit time at different ambient temperatures (Self-drawn)

4 Conclusions and Outlook

In this paper, by reviewing the research results of domestic and foreign scholars on the flat plate micro heat pipe in recent years, the research progress on the heat transfer performance of the flat plate micro heat pipe is summarized from the aspects of the structure and shape dimensions of the microchannel, the charging medium and the liquid charging rate, and the working condition, and the selection of the micro heat pipe material and the compatibility relationship are summarized on the basis of the above, the following summary is made as well as the research outlook on the future development of the flat plate micro heat pipe is put forward:.

(1) Through the research results of a large number of scholars, we know that the shape and size of the groove structure of the flat plate micro heat pipe, the depth of the groove and the groove of the wettability of the tube greatly affects the role of capillary force, a comprehensive measure of the heat transfer performance of the heat pipe should also take into account the relative size of the effective heat transfer area.

(2) Internal working medium phase change heat transfer plays a leading role in the flat micro heat tube, choose filling medium according to the working environment, working temperature, and the working medium properties, 20% filling rate of R141b, acetone, ethanol three kinds of working medium at low temperature conditions have high heat transfer performance, including acetone can form a long condensation section, the strongest heat transfer capacity.

(3) In the future, additional studies on flat plate micro heat pipe technology should concentrate on refining the internal structure of this heat pipe type and identifying superior heat transfer fluids. Building upon current research findings and acknowledged influencing factors, subsequent research endeavors should be pursued accordingly. To improve the thermal transfer efficiency of the flat plate micro heat pipe and to investigate the practical applications of the micro heat pipe in real-world scenarios.

Acknowledgements This study is supported by the project of Changchun Power Supply Company of State Grid Jilin Electric Power Co., Ltd., entitled "Development and Application of Moisture Dissipation and Dehumidification Technology of Composite Capillary Core Structure Micro Heat Pipe Array Elements for Metal-enclosed Air Insulation Equipment [13]", which was used for the development and application of metal-enclosed air-insulated dehumidification technology (SGJLCC00KJJS2402034).

References

1. Cotter TP (1985) Principles and prospects for micro heat pipes. In: Proceedings of the 5th international heat pipe conference, pp 328–335
2. Peterson GP (1992) Investigation of micro heat pipes fabricated as an integral part of silicon wafers. In: Proceeding of the 8th international heat pipe conference. Beijing, pp 385-395
3. Khrustalev D, Faghri A (1995) Heat transfer in the inverted meniscus type evaporator at high heat fluxes. Int J Heat Mass Transf 38(16):3091–3101
4. Peterson GP, Ma HB (1996) Theoretical analysis of the maximum heat transport in triangular grooves: a study of idealized micro heat pipes. Heat Transfer 118:731–739
5. Peteron GP Heat pipe modeling and simulation. AIAA 23rd aerospace sciences meeting
6. Gong C (2019) Research on laser etching process and heat transfer performance of micro heat pipe with composite grooves capillary structure. Zhejiang University
7. Guangtao H, Yifeng Y (2012) Research and practice on recover boiler waste heat by reduce the exhaust gas temperature. Energy Conserv Technol (4):295–298
8. Xu K, Xin W, Yang Y, Xiaozhu X, Wei H (2018) Research on the influencing factors on the heat transfer performance of grooved micro heat pipes. Cryog Supercond 46(03):58–63
9. Zengyuan G, Suyi H (2004) Principle of field synergy and new technology of enhanced heat transfer. China Electric Power Press, Beijing
10. Yaohua Z, Hongyan W, Yanhua D, Xinyue W, Yuechao D (2011) Heat transfer characteristics of flat micro-heat pipe array. CIESC J 62(02):336–343
11. Wang G, Quan Z, Zhao Y et al (2019) Performance of a flat-plate micro heat pipe at different filling ratios and working fluids. Appl Therm Eng 146:459–468
12. Wu Y, Jia J, Tian D et al (2020) Heat transfer performance of microgroove back plate heat pipes with working fluid and heating power. J Therm Sci 29(4):982–991
13. Yang J, Feng Z, Liu Z, Song D (2024) Study on the performance of oil fume heat recovery heat exchanger based on micro heat pipe array. Energy Rep 11:240–248

Steam Curing Process Parameters of Prefabricated T-Beam Based on Intelligent Construction

Hongyu Hu, Hao Cao, Dahai Yang, and Zhitian Wang

Abstract In order to systematically explore the steam curing process of T-beam construction, the influences of water-cement ratio, sand ratio and admixture content on the compressive strength of concrete were investigated by orthogonal test and range analysis method. After selecting the optimal mix ratio of prefabricated T-beams, the effects of static holding time, heating rate, constant temperature, constant temperature time and cooling rate on the steam curing system were tested to provide data support for the optimization of the steam curing system. The results show that in the steam curing process of prefabricated t beam, the significant degree of influence of each factor on the compressive strength of concrete is in the order of content b > water-cement ratio > sand ratio. The optimal steam curing time of prefabricated T-beams is recommended to be 3 h, the constant temperature is 50 °C, and the constant temperature time is 6 h. The recommended heating rate is 4 °C /h, and the recommended cooling rate is 4 °C /h in practical applications.

Keywords Assembled T-beam · Steam curing process · Fitting ratio · Polar analysis · Experimental study

1 Introduction

Concrete traditional curing methods are inefficient, unrefined, unscientific, and the process is not standardized and other drawbacks, mainly reflected in the traditional curing methods are prone to cause low concrete strength and temperature (shrinkage) cracks, further affecting the service life of the concrete, the existing curing is mostly natural curing, the temperature and humidity can not be adjusted and controlled, which affects the speed of the concrete strength enhancement, the molds can not be

H. Hu · H. Cao
Anhui Province Transportation Holdings Group Co., Ltd., Hefei, Anhui, China

D. Yang · Z. Wang (✉)
Anhui Province Transportation Planning and Design Institute Co., Ltd., Hefei, Anhui, China
e-mail: 3301361089@qq.com

B. He et al. (eds.), *Climate-Resilient, Low-Carbon, and Sustainable Cities*, Lecture Notes in Civil Engineering 754, https://doi.org/10.1007/978-981-95-2169-2_29

released in time, the turnover efficiency is low. Artificial sprinkling is not standardized; inaccurate grasp of temperature and humidity; insufficient or excessive water consumption for curing; low curing efficiency. All these problems will cause huge economic losses to the construction.

Steam curing can improve the early strength of concrete, shorten the curing time of girders, enter the storage area after early tensioning, and reduce the occupation time of the pedestal [1–7]. Nowadays, some girder factories abandon the traditional covering spraying curing process, draw on the mature experience of curing similar projects in the industry, and adopt modern intelligent variable temperature steam curing system [8–13]. The steam curing flow operation area consists of steam curing shed and variable temperature steam curing equipment: the steam curing shed adopts the standard construction of experimental constant-temperature curing room, with heat-insulating and thermal-insulating layers on the outside; the variable-temperature steam curing machine can real-time adjust the temperature and humidity of the steam curing room along with the pre-embedded temperature sensing data of the girder body, which will efficiently improve the strength and elasticity modulus of the concrete and meet the tensioning conditions as early as possible. However, it was found in the research that the following problems still occurred in the process of promoting intelligent steam curing in the girder factory: (1) The energy consumption of intelligent steam curing of components is large, the steam curing quality and efficiency need to be further improved, and the steam curing process needs to be optimized. (2) The maximum temperature of concrete steam curing and holding temperature and time on the quality of the girder in the end how much to be verified by the exact data, the steam curing system of the girder factory to be established [14–17].

In this paper, Anhui Jiaocheng Industrial Construction Co., Ltd. is taken as an example. Firstly, the influence of the ratio of specimens and the steam curing environment on the beam quality of prefabricated components in the steam curing process is studied, and the parameters such as the amount of mineral admixture and sand rate are optimized to find the mix ratio of prefabricated T-beam that is more suitable for steam curing is found. Finally, the effects of static time, heating rate, constant temperature, constant temperature time and cooling rate on the steam curing system were studied to provide data support for the optimal design of the steam curing system.

2 Project Overview

The Hefei to Zongyang section of the Dezhou to Shangrao Expressway project connects to the north of the Dezhou to Shangrao Expressway from Huainan to Hefei, and to the south it connects to the Chizhou Yangtze River Highway Bridge section. The total length of the project route is 134.158 km, featuring full interchanges, fully enclosed, and high-level expressway standards. The design speed is 120 km per hour, and it is designed as a four-lane bidirectional highway. The bridge and culvert structures are designed to meet the Highway Class I vehicle load. Due to the large number

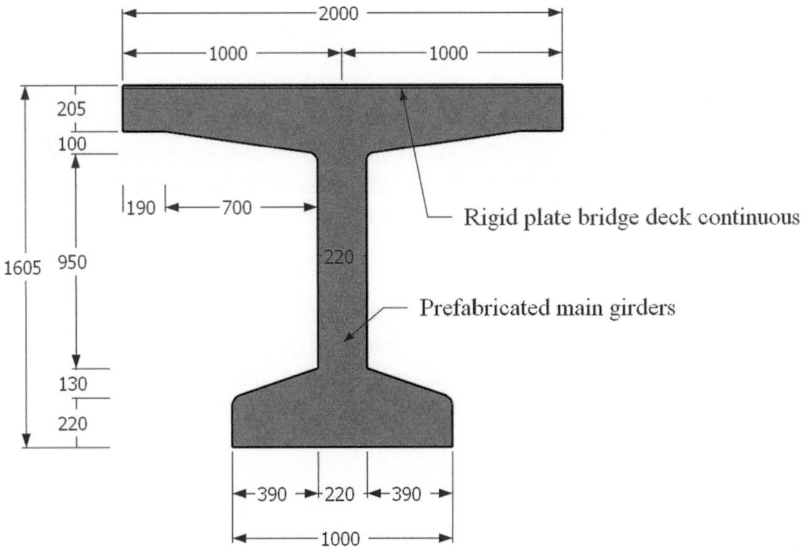

Fig. 1 Schematic diagram of cross-sectional dimensions of lightweight T-beam (Unit: mm) (Self-drawn)

of bridges along the entire route, prefabricated components such as lightweight T-beams are used extensively. All prefabricated T-beams along the entire route have a span of 25 m and are cast with C50 concrete. The specific dimensions of the T-beams are shown in Fig. 1.

3 Experimental Study on the Design Parameters of Steam-Cured Concrete

3.1 Experimental Design

The steam curing method is a measure to accelerate the hydration reaction of concrete, enabling the concrete to reach the required compressive strength in a short period, which can significantly reduce the construction period. However, due to the various influencing factors of the steam curing process, such as the dosage of mineral admixtures, sand ratio, water-cement ratio, admixtures, etc., different parameters have different effects on the quality of steam-cured concrete for precast T-beams. Therefore, it is necessary to scientifically design the concrete mix ratio and ensure a mix suitable for steam curing.

Based on field investigations at the beam factory and literature review, several important indicators of the concrete mix ratio, including water-cement ratio (A), sand ratio (B), and admixtures (C), are selected as the variables for this experiment.

Using the original mix ratio at the beam factory as the basis, a three-level three-factor orthogonal experiment is designed for the mix ratio. The experiment adopts L9(34) and arranges 9 sets of experimental combinations, with the levels and factors detailed in Table 1. To prevent experimental errors caused by human factors, each set of experiments includes six 150 mm × 150 mm × 150 mm concrete test blocks. After the concrete mixing is completed, the cohesion, water retention, and slump of the test blocks are tested to describe their workability. Subsequently, the test blocks are cured in a steam curing chamber for three days, and after the curing, their compressive strength is tested.

Meanwhile, the range analysis method is used to analyze the experimental results, aiming to obtain the optimal parameter combination for the quality analysis of precast T-beams of different sizes and materials produced by the intelligent beam factory, and to determine the extent of the influence of each parameter on the strength of steam-cured precast T-beam concrete.

3.1.1 Original Mixing Ratio and Raw Materials

All the raw materials for precast T-beams are provided by Anhui Jiaokong Group. The cement used is Conch brand ordinary Portland cement (P.II 52.5R), and its physical and mechanical properties as well as chemical composition are shown in Table 2. The coarse aggregate used mainly consists of two sizes of gravel, with the mass ratio of gravel with particle sizes of 10–20 mm at 60% and gravel with particle sizes of 5–10 mm at 40%. The fine aggregate is medium sand, and the fly ash is produced by Huaneng Beijing Thermal Power Plant, with a grade of S95. The mineral powder is of S95 grade. The water-reducing agent used is the ZY-HPWR series high-performance standard water-reducing agent produced by Hunan Zhongyan Building Materials, with a water reduction rate of 15%.

3.1.2 Specimen Preparation and Molding

The dosage of mineral admixtures, sand ratio, water-cement ratio, and admixtures have different effects on the quality of steam-cured precast T-beams. The original mix ratio from the beam factory should be used as the control group. The raw materials are mixed according to the designed mix ratio, and after mixing in the mixer for 1 min, the water-reducing agent and water are added and thoroughly mixed. This mixture is then used to make 150 mm × 150 mm × 150 mm cubic specimens for compressive strength testing. According to the requirements of different test items, the specimens are placed in different molds (150 mm × 150 mm × 150 mm) and kept for steam curing. The concrete is steam cured using a concrete steam rapid curing box according to the "Standard Test Method for Mechanical Properties of Concrete" (GB/T 50,081–2019). The specific process of the specimens is shown in Fig. 2.

Table 1 Experimental design for optimization of mix ratios (Self-drawn)

Serial number	W/C	Sand rate (%)	Water reducing agent dosage (%)	Amount of material used (kg/m^3)							
				Clinker	Coal ash	Mineral powder	Sand	Gravel (5 mm-10 mm)	Gravel (10–20 mm)	Water	Water-reducing agent
1	0.36	0.41	1.0	375	48	59	716	287	762	172	5
2	0.36	0.37	0.8	386	50	62	644	300	796	180	4
3	0.36	0.34	1.2	364	46	56	606	322	855	168	5.5
4	0.38	0.41	0.8	367	47	56	751	296	786	180	3.8
5	0.38	0.37	1.2	346	44	53	691	322	854	168	5.1
6	0.38	0.34	1.0	356	45	54	636	338	897	172	4.4
7	0.41	0.41	1.2	310	42	50	788	310	825	168	4.8
8	0.41	0.37	1.0	338	43	51	704	328	871	172	4.3
9	0.41	0.34	0.8	349	38	52	648	344	915	180	3.5

Table 2 Cement physical properties (Self-drawn)

Fineness/ %	Condensation/ min	Congeal/min	Stability	28dflexural strength /MPa	28dcompressive strength /MPa
3.9	171	259	Eligible (voter etc.)	9.2	54.5

Fig. 2 Flowchart depicting specimen preparation process (Self-drawn)

3.1.3 Test Methods

To analyze the significance of the effects of pre-curing time, admixtures, water-cement ratio, constant temperature, dosage of mineral admixtures, and sand ratio on the concrete strength, a range analysis was conducted. Based on the results of the orthogonal experiment, the significance of the influencing factors was analyzed to identify the optimal solution for each factor. Ultimately, the optimal combination of parameters for this experiment was determined, as shown in Fig. 3.

3.2 Analysis of Test Results

3.2.1 Compressive Strength Test Results

Following the experimental design in Table 1, tests were conducted, and the compressive strength of the specimens was analyzed to obtain the average compressive strength, as shown in Table 3.

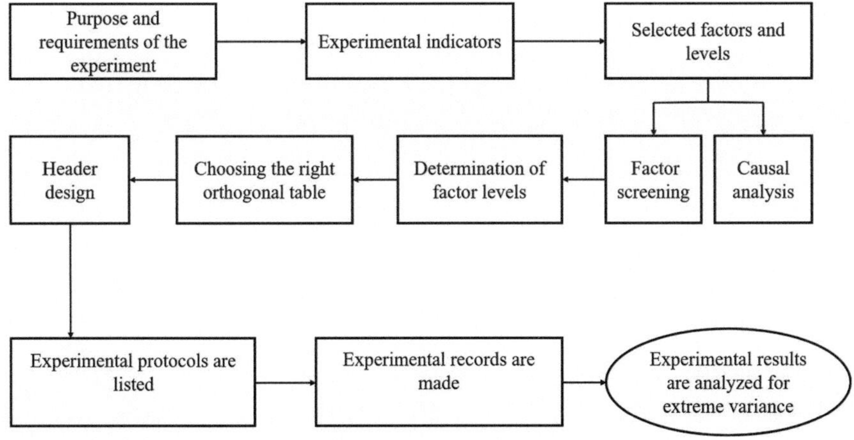

Fig. 3 Experimental analysis process (Self-drawn)

Table 3 Record of results of orthogonal test program for concrete (Self-drawn)

Serial number	Compressive strength (MPa)						Average strength
	Test block1	Test block2	Test block1	Test block1	Test block1	Test block1	
1	45.6	46.7	38.3	47.7	42.5	41.9	50.45
2	41.4	43.2	45.2	44.1	44.9	46.6	49.23
3	46.5	39.0	45.9	44.7	46.5	40.8	48.90
4	41.4	42.3	40.3	39.8	38.3	41.1	48.87
5	47.8	47.5	40.8	42.6	41.3	36.2	41.03
6	46.9	44.2	45.2	37.3	39.8	42.3	39.28
7	43.3	45.9	43.4	41.5	42.2	40.5	39.47
8	42.4	43.2	44.6	46.7	39.0	42.9	48.13
9	37.3	44.4	39.7	42.1	38.6	41.7	49.13

3.2.2 Polarization Analysis

By visually analyzing the compressive strength data of the 9 groups of specimens, a table of extreme variance analysis was obtained respectively, as shown in Table 4. K_1, K_2, K_3 is the sum of the test results under all levels of each factor, $\overline{K}_1$, $\overline{K}_2$, $\overline{K}_3$ indicates the average of the experimental results under all levels of each factor, and indicates the extreme difference. The larger the extreme difference, the more significant the effect of the factor on the indicator.

Based on the above results, we compared the R_f sizes, and since $1.9 > 1.77 > 1.53$, we can infer the significant impact of each factor on compressive strength: dosage of water-reducing agent > water-cement ratio > sand ratio. At the same time, it is

Table 4 Compressive strength extreme difference analysis table (Self-drawn)

Considerations	Water-to-cement ratio (A)	Sand rate (B) (%)	Water reducing agent dosage (C) (%)
K_1	131.8	127.1	125.3
K_2	127.4	131.6	129.4
K_3	126.5	127.0	131.0
$\overline{K}_1$	43.9	42.4	41.8
$\overline{K}_2$	42.5	43.9	43.1
$\overline{K}_3$	42.2	42.3	43.7
R_f	1.77	1.53	1.9
Horizontal number	3	3	3
Number of repetitions at each level	3	3	3

also derived from Fig. 4: A1 is the optimal solution of the factor, B2 is the optimal solution of the factor B, C3 is the optimal solution of the factor C, and the resulting optimal combination is A1B2C3.

To examine the significance of the impact of each factor and analyze the influence of experimental errors, a variance analysis was conducted on the compressive strength test results to obtain the F-test value, as shown in Table 5.

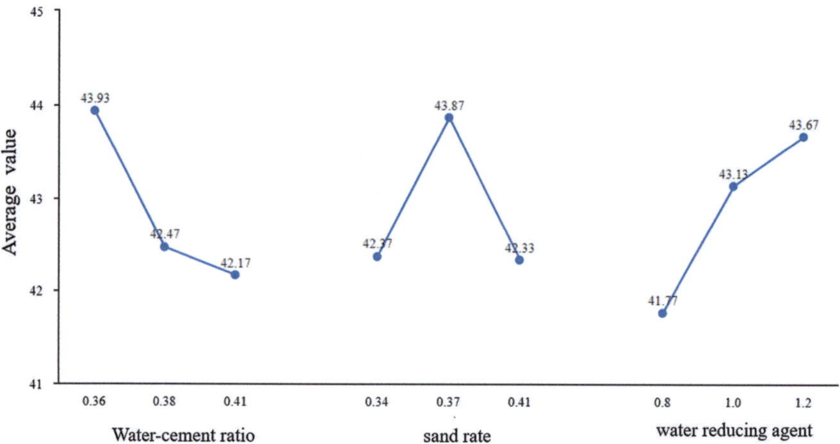

Fig. 4 Relationship between compressive strength and different levels of each factor (Self-drawn)

Table 5 ANOVA for compressive strength

Factor	SS	df	MS	F	p
Intercept	16,529.39	1	16,529.39	44,943.96	0.000
Water-cement ratio	5.362	2	2.681	7.29	0.121
Sand ratio	4.602	2	2.301	6.257	0.138
Water reducing agent dosage	5.762	2	2.881	7.834	0.113
Test error	0.736	2	0.368		

4 Research on the Establishment and Optimization of T-beam Steaming System

4.1 Parameter Impact Analysis and Optimized Design of the Autoclave System

The entire experiment was conducted in accordance with the "Technical Specification for Prefabricated Concrete Structures" (JGJ-2014). The optimal mix ratio obtained in the experiment was used as the baseline mix ratio, and the pre-curing system of the beam factory was optimized through one-on-one comparative experiments based on the reference pre-curing system.

The study investigated the effects of static dwell time, heating rate, constant temperature, dwell time at constant temperature, and cooling rate on the pre-curing system, providing data support for the optimization design of the pre-curing system. The parameters of the specific experimental program are shown in Table 6.

Group 1: The pre-curing times for the experimental specimens were set at 1, 2, 2.5, 3, and 4 h. The heating rate was adjusted to 5 °C/h, with a constant temperature of 50 °C for 8 h and a cooling rate of 5 °C/h. After steam curing, the specimens were immediately demolded, and the demolding strength was measured. Subsequently, the specimens were further cured for 3 days, 7 days, and 28 days in a standard curing environment, and their strength was measured at a certain age.

Group 2: Under the same mix ratio and static dwell, constant temperature, and cooling regimes, the experiment compared the demolding, 1-day, and 3-day compressive strengths of concrete specimens under different heating rates, set at 4, 4.5, 5, 5.5, and 6 °C/h.

Group 3: Under the same mix ratio and other steam curing process parameters, five different constant temperature groups were established, with temperatures set at 40, 45, 50, 55, and 60 °C, to determine the suitable constant temperature for concrete compressive strength, assuming the optimal heating rate determined from the previous experiment to be 4 °C/h.

Group 4: Based on the optimal temperature determined from the previous experiment, the experiment compared the strength variation of concrete under different

Table 6 Experiments on the effect of static stopping specimens on the strength of concrete (Self-drawn)

Experiment No	Static stopping time/h	Rising temperature rate/°C/h	Constant temperature/ °C	Constant temperature time/h	Cooling rate/°C/h	Stage
B1	1	5	50	8	5	Group 1
B2	2	5	50	8	5	
B3	2.5	5	50	8	5	
B4	3	5	50	8	5	
B5	4	5	50	8	5	
B1	3	4	50	8	5	Group 2
B2	3	4.5	50	8	5	
B3	3	5	50	8	5	
B4	3	5.5	50	8	5	
B5	3	6	50	8	5	
B1	3	4	40	8	5	Group 3
B2	3	4	45	8	5	
B3	3	4	50	8	5	
B4	3	4	55	8	5	
B5	3	4	60	8	5	
B1	3	4	50	6	5	Group 4
B2	3	4	50	7	5	
B3	3	4	50	8	5	
B4	3	4	50	9	5	
B5	3	4	50	10	5	
B1	3	4	50	9	6	Group 5
B2	3	4	50	9	5.5	
B3	3	4	50	9	5	
B4	3	4	50	9	4.5	
B5	3	4	50	9	4	

constant temperature times, assuming the optimal constant temperature determined from experiment 3 to be 50 °C.

Group 5: Due to challenges in controlling the cooling rate during actual experimental operations, the experiment compared the compressive strength of concrete under different cooling rates, while maintaining the same mix ratio and the consistency of the first four steam curing parameters.

These are the different experimental groups and their respective parameters, conducted in adherence to the scientific writing style.

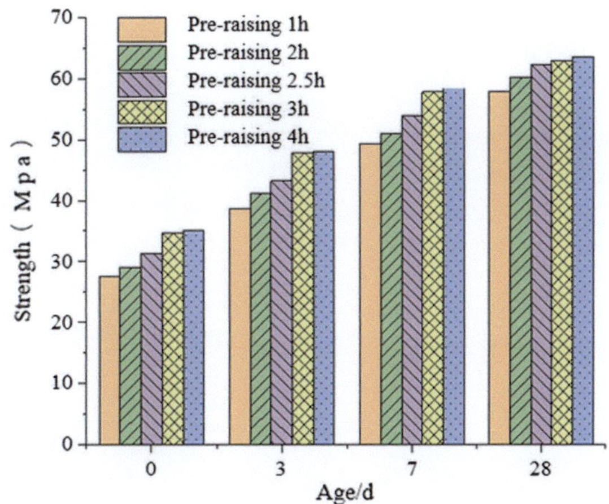

Fig. 5 Effect of preconditioning time on concrete strength (Self-drawn)

4.2 *Analysis of Test Results*

4.2.1 Effect of Static Stopping Phase on Concrete Strength

From Fig. 5, it can be observed that the influence of the static dwell stage on the strength of concrete shows an increasing trend with the extension of the pre-curing time at various ages. The demolding strength, as well as the strengths at 3 days, 7 days, and Fig. 5: Effect of preconditioning time on concrete strength28 days, increased by 26.2%, 23.8%, 19.1%, and 8.8% respectively when pre-cured for 3 h compared to pre-curing for 1 h. These strengths were only slightly lower by 1.4%, 0.4%, 1.7%, and 1.0% respectively when compared to pre-curing for 4 h. Although the demolding strength after pre-curing for 3 h only reached 69.4% of the design strength, the strength at 3 days had already reached 95.8% of the design strength, meeting the requirements. This is because water evaporates faster in short-term curing, and the surface layer is prone to drying cracks, while a large number of unreacted cement particles and residual voids cannot be filled, resulting in a decrease in the overall compressive strength. And although long-term maintenance can make up for these defects, the prolonged maintenance leads to a decrease in the efficiency of the beam factory and lacks advantages in terms of economy. Therefore, from the consideration of both product quality and production efficiency, the optimum pre-conditioning time should be about 3 h with reference to the experimental comparison.

4.2.2 Effect of Rate of Warming on Concrete Strength

From the above Fig. 6, it is evident that the faster the heating rate, the lower the strength of the concrete at various ages. While heating can accelerate the hydration reaction of cement, the process also accelerates the conversion of water vapor,

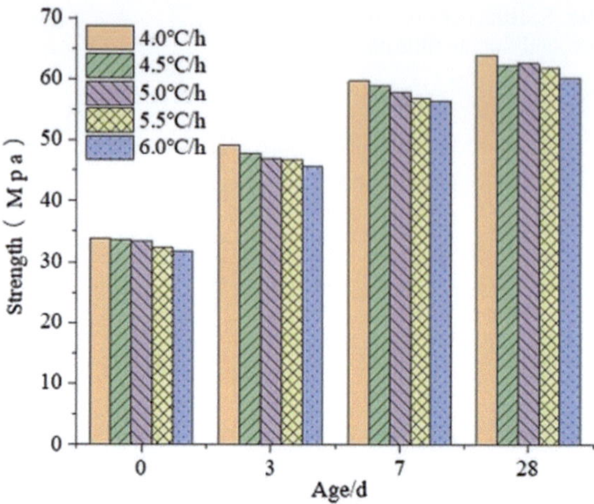

Fig. 6 Effect of rising temperature on the strength of concrete (Self-drawn)

exacerbating the damage to the pores from expansion pressure. A faster heating rate leads to a greater number of cracks, which adversely affects the later-stage strength development and durability of concrete. Therefore, in the production process, while ensuring production efficiency, it is advisable to minimize the heating rate and heat up slowly. Under these conditions, a heating rate of 4 °C/h is recommended.

4.2.3 Effect of Constant Temperature on Concrete Strength

The experimental results in Fig. 7 indicate that under the same steam curing parameters, the demolding strength, as well as the strengths at 3 days, 7 days, and 28 days of concrete, initially increase with the rise in the constant temperature until reaching 50 °C, after which they exhibit a decreasing trend. This is because as the constant temperature increases, although the degree of hydration increases, it also exacerbates the expansive effect of water vapor conversion on the pore structure, leading to a reduction in later-stage strength. The number and width of microcracks generated are much greater than in the case of lower constant temperatures. Therefore, it is recommended that the optimal constant temperature be 50 °C.

4.2.4 Effect of Constant Temperature Time on Concrete Strength

From the above Fig. 8, it can be observed that during the constant temperature stage, as the constant temperature time extends, the strength at various ages increases. The demolding strength at 6 h of constant temperature increased by 19.1% compared to 4 h. However, when the constant temperature time exceeds 6 h, the demolding strength, as well as the strengths at 3 days, 7 days, and 28 days, gradually decrease. The analysis suggests that on one hand, the longer the constant temperature time, the greater

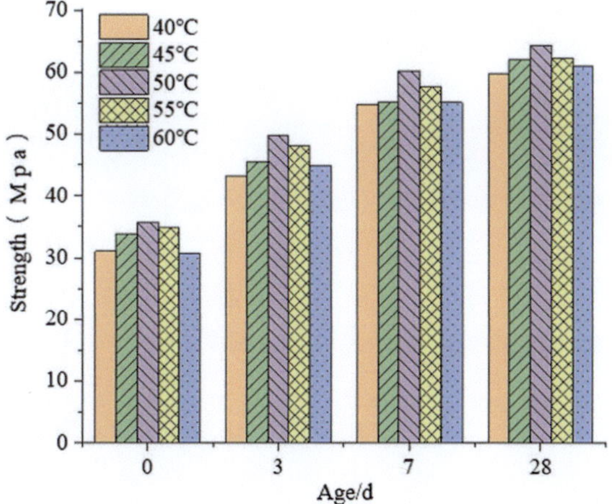

Fig. 7 Effect of constant temperature on concrete strength (Self-drawn)

the damage to the concrete structure, which may have an adverse effect on the later-stage strength of the concrete. On the other hand, the longer duration in the constant temperature stage leads to a higher degree of aggregation of hydration products, resulting in uneven distribution and hindering further hydration, thereby affecting the development of later-stage strength. Moreover, prolonging the constant temperature time can impact the turnover speed of molds, reducing production efficiency. Therefore, considering these factors, it is advisable to use a constant temperature time of 6 h.

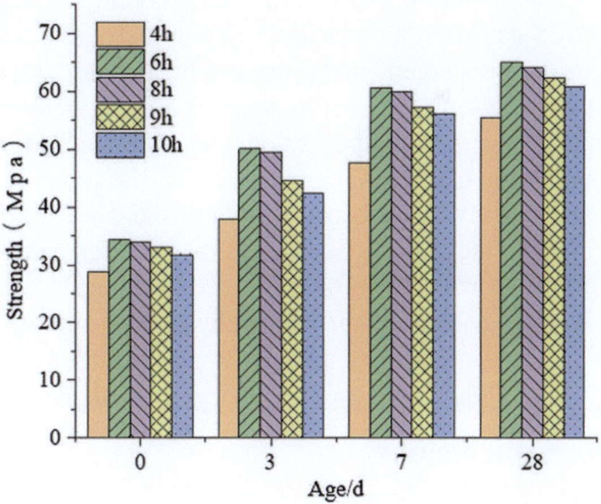

Fig. 8 Effect of constant temperature time on the strength of concrete (Self-drawn)

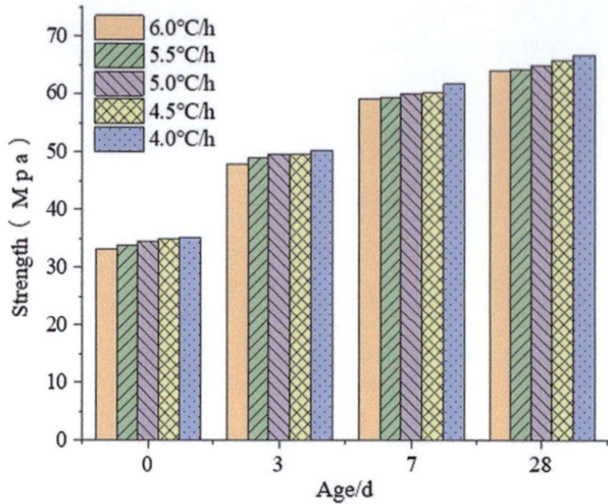

Fig. 9 Effect of constant temperature time on concrete strength (Self-drawn)

4.2.5 Effect of Cooling Rate on Concrete Strength

Based on the conclusions drawn from the previous four sections regarding the optimal steam curing parameters, the experiment aimed to compare the compressive strength of concrete under different cooling rates, while keeping the mix proportion and other steam curing parameters constant, in order to determine the optimal cooling rate. The experimental results are presented in Fig. 9.

From Fig. 9, it can be observed that during the cooling stage, the slower the cooling rate, the higher the strength of the concrete at various ages. When the cooling rate decreased from 4.5 to 4 °C/h, the demolding strength only increased by 0.3%, and the 28-day strength increased by only 1.2%. It is evident that reducing the cooling rate does not significantly promote the increase in concrete strength, as the concrete strength during the cooling stage is already able to meet the required strength. In actual engineering applications, an excessively low cooling rate can also impact production efficiency. However, a very fast cooling rate is also not advisable, as it can have adverse effects on strength and can lead to cracking on the surface of the concrete due to the greater temperature difference between the interior and exterior of the concrete. Therefore, it is recommended to select a suitable upper limit for the cooling rate, and in practical applications, a cooling rate of 4 °C/h is recommended.

5 Conclusions

This article conducted research on the control of steam curing parameters for prefabricated T-beams from the perspectives of optimizing concrete mix design and optimizing steam curing processes. The following conclusions were drawn, providing valuable references for the future production of prefabricated bridges using steam curing:

(1) The significance of various factors in the impact on the compressive strength of concrete in prefabricated T-beams through steam curing is as follows: admixture dosage > water-cement ratio > sand ratio. Simultaneously, A1 represents the optimal solution for factor A, B2 represents the optimal solution for factor B, and C3 represents the optimal solution for factor C. The optimal combination obtained is A1B2C3.
(2) Taking into account both product quality and production efficiency, the recommended pre-curing time is 3 h. It is suggested that the optimal constant temperature be 50 °C, and the optimal constant temperature time be 6 h.
(3) During the production process, while ensuring production efficiency, efforts should be made to minimize the heating rate and adopt a slow heating approach. Under the conditions of this experiment, a heating rate of 4 °C/h is recommended, and in practical applications, a cooling rate of 4 °C/h is advisable.
(4) The optimization conclusions of steam curing parameters obtained in this study can not only provide guidance for the production of precast T-girders, but also provide a reference for the optimization of steam curing process for other types of precast concrete components (e.g., box girders, hollow-core slab girders, etc.), especially when production efficiency and component quality are taken into account.
(5) Future research can further explore the mechanical properties of different types of concrete materials (e.g., high-performance concrete or fiber-reinforced concrete) under similar steam curing processes. At the same time, it is recommended that in-depth studies on the effects of different climatic environments and member cross-sectional dimensions be conducted in conjunction with actual production conditions in the field, in order to realize the wider promotion and application of the steam curing process.

References

1. Zhao L (2020) Exploration and research on T-girder prefabrication plant of steam curing flow operation_Lei Zhao. Value Eng 39(28):240–242
2. Chunming X, Zhe Z (2023) Fully automatic temperature-controlled steam curing precast T-beam technology and its application_Xiao Chunming. Sichuan Cement (04):186–187, 190
3. Zhigang H (2022) Research on key technology of factory prefabrication of prestressed concrete T-girders for highways_Huang ZG. Hebei University of Technology
4. Han G (2022) Fully automatic hydraulic formwork closed steam curing T-beam construction technology_Ge Han (1). Build Mater Technol Appl (04):70–72
5. Abdulla AI, Khatab HR (2014) Behavior of multilayer composite ferrocement slabs with intermediate rubberized cement mortar layer. J Eng Appl Sci 39:5929–5941
6. Hao C, Mengmeng G, Jun S (2024) Research on key technology of automatic production of concrete panel. Highway 69(12):254–258
7. Fayu S (2024) Smart factory prefabricated T beam production optimization measures research. Transp World (33):163–165 + 172. https://doi.org/10.16248/j.carolcarrollnki. 11–3723/u2024.33.034

8. Tays¸I N, Abid S (2015) Temperature distributions and variations in concrete box-girder bridges: experimental and finite element parametric studies. Adv Struct Eng 18:469–486
9. Abdullah QN, Abdulla AI (2022) Flexural behavior of hollow self compacted mortar ferrocement beam reinforced by GFRP bars. Case Stud Constr Mater 17:e01556
10. Xing Z (2022) Research on complete set of intelligent steam curing technology for prefabricated girder yard with dual temperature control_Zhang Xing. Railw Constr Technol 12:125–129
11. Jingsen BAO (2022) Application of intelligent hydraulic formwork system and automatic vaporization system in girder yard_Bao Jingsen. China High-Tech 03:45–46
12. Bin YANG (2020) Research and application of intelligent vaporization control system for precast concrete tube sheet of metro shield_Yang Bin. Build Mater Decor 18:266–267
13. Hao C, Eryu Z, Zhengwei B (2022) Research on optimization of concrete proportion for CRTS III track slab based on intelligent manufacturing_Chen Hao. Constr Technol (in Chinese and English) 51(03):103–106, 133
14. Yang LIU, Jidong YUAN, Mingyang GAO et al (2021) Establishment and optimization of steam curing system for PC components_Liu Yang. Sichuan Constr 41(04):217–219
15. Effect of steam curing system on strength of precast concrete box girders_Zeng Mingle
16. Jing ZHAO, Yiming YU, Wuju XUN et al (2020) Study on the effect of autoclaving system on compressive strength and impermeability of concrete_Zhao Jing. Low Temp Constr Technol 42(07):36–39
17. Hao CAO, Zhitian WANG, Xiaowu ZHOU (2023) Experimental study on optimization of steam-curing system of C50 concrete for precast components_Cao Hao. Sichuan Build Mater 49(05):17–20

Unite! Poster Competition Review Research on Architectural Energy for Low-Carbon to Sustainable Development

Bin Li, Yuqing Zhang, and Qipeng Mu

Abstract Unite! Poster Competition is a worldwide student competition to review the sparks on architectural energy for low-carbon to sustainable development. First, the competition was launched based on global environmental problems such as climate change and related policy guidance such as a series of guidance from the United Nations. Then, 12 posters from 10 universities worldwide were collected. The authors initially sorted out the related low-carbon strategies, and the related SWOT analysis were also summarized. The third section analyzed the evaluation contents of 15 judges. The jury characteristics, evaluation procedure and related comments are studied. Finally, the research got the ranking list of the posters. The fundamental analysis of the posters by authors compared with the scores and comments from professional judges to verify the rationality of the results. Then, more profound low-carbon revelations are sorted out. In this research, there are also some limitations, such as the wording problem of the "poster" use and overly broad requirements for the call for posters. However, the research is conducive to researching and promoting low-carbon development and also realizes closed-loop research on mitigating background problems.

B. Li (✉) · Y. Zhang
School of Architecture, South China University of Technology, Guangzhou, China
e-mail: bin.li@polito.it

Department of Architecture and Design, Politecnico di Torino, Torino, Italy

Q. Mu
College of Architecture and Urban Planning, Qingdao University of Technology, Qingdao, China

B. Li
Energy Saving Technology Research Institute, South China University of Technology, Guangzhou, China

Architectural Design and Research Institute Co., Ltd., South China University of Technology, Guangzhou, China

B. Li · Y. Zhang
State Key Laboratory of Subtropical Building Science, South China University of Technology, Guangzhou, China

B. He et al. (eds.), *Climate-Resilient, Low-Carbon, and Sustainable Cities*, Lecture Notes in Civil Engineering 754, https://doi.org/10.1007/978-981-95-2169-2_30

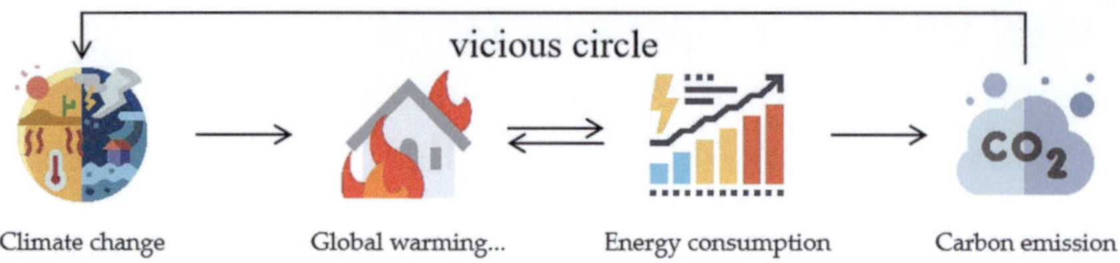

Fig. 1 Vicious circle of environmental problems. (Self-drawn)

Keywords Low-carbon · Sustainable development · Architectural energy · Unite! poster competition

1 Competition Background

1.1 Environmental Problems

Climate change has been a persistent global problem. The impacts of climate change come with increasing environmental problems, such as global warming, air pollution, bushfire and flooding [1]. Correspondingly, more energy consumption, especially for fossil energy use to deal with these problems, has contributed to more carbon emissions and the energy crisis. Therefore, a vicious circle was formed (Fig. 1).

1.2 Policy Guidance

To address these challenges, the United Nations has introduced various policy frameworks to foster sustainable development. The journey began in 1972 with the United Nations Conference on the Human Environment (UNCHE) in Stockholm, the first global gathering to highlight international concerns regarding global warming, air pollution, and so on [2]. In 1987, the World Commission on Environment and Development (WCED), a sub-organization of the United Nations, published "Our Common Future," a seminal work that sought to establish a consensus-based definition of sustainability for global development. The publication emphasized the need to reduce carbon dioxide emissions, a major driver of global warming [3]. In 1992, the United Nations Conference on Environment and Development (UNCED), also known as the Earth Summit, occurred in Rio de Janeiro. This landmark event led to the adoption of the Rio Declaration on Environment and Development and Agenda 21 and introduced the Framework Convention on Climate Change (UNFCCC). These initiatives were designed to promote sustainable development through specific actions, including efforts to reduce carbon emissions [4]. In 1997, the Kyoto Protocol was adopted at the 3rd United Nations Climate Change Conference. As the first quantified legal document under the UNFCCC, it aimed to advance global energy conservation and emissions reduction efforts in pursuing sustainable development [5]. In 2002, the World Summit on Sustainable Development was convened in South Africa. This summit sought to build on the momentum generated by the 1972 United Nations Stockholm Conference on the Environment, the 1992 Rio Earth Summit, and the 2000 Millennium Declaration to accelerate the development of sustainability initiatives [6]. In 2009, the 15th United Nations Climate Change Conference was held in Copenhagen, Denmark. The primary objective of this conference was to negotiate a global emissions reduction agreement from 2012 to 2020, following the expiration of the first commitment period under the Kyoto Protocol [7]. In 2015, the United Nations Summit on Sustainable Development was held in New York, where an ambitious new agenda for sustainable development was formally adopted. This agenda, titled "Transforming Our World: The 2030 Agenda for Sustainable Development," includes 17 Sustainable Development Goals (SDGs), such as Goal 7, "Affordable and Clean Energy," and Goal 11, "Sustainable Cities and Communities" [8]. Also, in 2015, the 21st United Nations Climate Change Conference occurred in Paris, France. The negotiations culminated in adopting the Paris Agreement, which has been the guiding framework for climate change mitigation efforts since 2020 [9]. In 2023, the Intergovernmental Panel on Climate Change (IPCC), the United Nations body responsible for assessing climate change science, released the Synthesis Report of the IPCC Sixth Assessment Report (AR6). This report revealed that the global surface temperature had risen by 1.1 °C above the 1850–1900 average during the period from 2011 to 2020. It also unequivocally attributed global warming to human activities [10]. Furthermore, in December 2023, the 28th United Nations Climate Change Conference was held in Dubai, UAE. One of the key objectives of this conference was to conduct the first evaluation of the Paris Agreement [11]. Alarmingly, the world is currently far from achieving its climate goals and is on a trajectory towards a temperature increase of 3 degrees by the end of this century [12]. UN Secretary-General António Guterres has called for a complete phase-out of fossil fuels to limit the temperature rise to 1.5 °C. He emphasized the need for countries to commit to tripling renewable energy capacity, doubling energy efficiency, and ensuring access to clean energy for all by 2030 [13]. These policy directives serve as the foundation for our research on sustainable development (Fig. 2). Clearly, to disrupt the aforementioned vicious cycle, research into low-carbon development is imperative.

1.3 Competition Origin

Therefore, the Unite! Poster Competition focuses on the Architectural Energy for Sustainable Development was launched. After almost half a year application from Unite! Seed Fund, the Poster Competition proposal was finally accepted on June 17, 2024. This project is related to SDGs 7 & 11, which aim to collect advanced ideas from bachelor, master and PhD students to contribute to sustainable development for architectural energy, especially in low-carbon fields. These practical and advanced

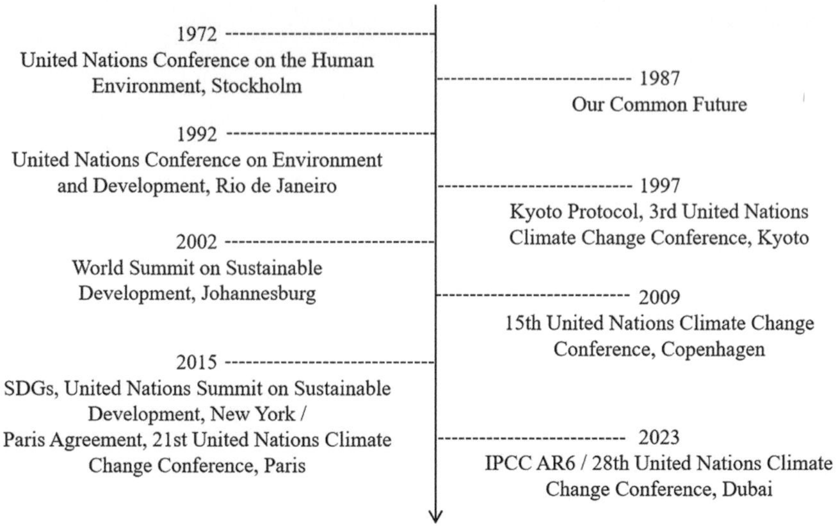

Fig. 2 UN policy guidance. (Self-drawn)

ideas will promote academic research and practical application, letting more people know that for a sustainable future. The competition principle is not to compete but to achieve an open, inclusive, and free exchange of cutting-edge knowledge under the same goals (SDGs), to deeply integrate our views on how building energy can achieve sustainable development in different ways. The main topics include architectural design with energy focus, green buildings and health, low-carbon demonstration projects, sustainable building design, advanced methods for architectural energy, but not limited to these [14]. The winning posters will be awarded with one 1st, three 2nd and five 3rd Prizes chosen by the jury members, plus a Special Prize selected by the public via an online voting poll. The award-winning posters will be part of the first Unite! online exhibit. Moreover, the poster must be anonymous and shown in a vertical one-page A1 (841 mm × 594 mm) size JPG, no less than 300dpi, no more than 50 Mb (CMYK Mode), and no layout requirements. The language used should be English.

2 Posters Analysis

2.1 Collected Posters

After one month of collecting, 12 qualified works have been received from 18 students in 10 universities worldwide. These posters are listed in Fig. 3, which clearly shows two styles. In one style, the content is compact and contains a large amount

Fig. 3 Collected posters. (Self-drawn)

of information, while the others are indeed full of design sense to reflect poster connotation.

2.2 Low-Carbon Strategies

From these posters, the main low-carbon strategies have been sorted out in Table 1. Whether for content-intensive posters or posters that lean towards design styles, low-carbon design strategies that respond to the theme are reflected from different points of view. All low-carbon design strategies can be summarized as 1. solar panel application, 2. natural ventilation system, 3. ecological material use, 4. high insulation installation, 5. rainwater use, 6. grey water system, 7. energy-efficient system, 8. waste reduction, 9. air–water heat pump, 10. bamboo/wood structure, 11. triple-glazed window, 12. energy storage system, 13. smart grid, 14. AI adjustment, 15. heat recovery ventilation, 16. smart energy management, 17. modular construction, 18. greenery, 19. inter-modal transport. It can be seen that there are at least 19 low-carbon strategies related to architecture, even transportation, management, and so on. These low-carbon strategies are helpful to the public to retrofit their buildings in the future or use some of the strategies immediately indoors, such as greenery, grey water usage, and so on. From the architecture discipline, these low-carbon strategies for sustainable development can be divided into passive design, active design, and renewable production in three main ways [15]. Passive design aims to reduce the building energy demand from the source, active design aims to save energy use with high energy efficiency during the building operation phase, and renewable production aims to produce green energy to cover the inevitable building energy use and even produce more for other use.

Table 1 Unite! Poster competition low-carbon strategies. (Self-drawn)

Number	Theme	Low-carbon strategies
1	No theme with AI generation	Solar panels on green roof, natural ventilation system, ecological materials with high insulation, rainwater harvesting system, grey water usage, use of renewable energy sources
2	No theme	Energy-efficient system, renewable energy sources, sustainable material, water conservation system, recycling system of water, waste reduction
3	Passive lakehouse	Photovoltaic system, air–water heat pump, cellulose fiber insulation, airtightness, renewable energy use, wood frame construction, foam glass gravel insulated floor, triple-glazed window, energy storage system
4	No theme with AI generation	Sustainable materials, natural light, natural ventilation, renewable energy, smart grid, AI technologies, advanced energy storage system, heat recovery ventilation
5	Johannesburg gold mining repair	Arundo donax material used for building panels and generating green energy, bamboo plywood, bamboo wooden poles, solar panels, recycled shed tiles
6	Vertical urban farm	Smart energy management, water recycling system, modular construction, roof water use, vertical planting
7	Shape-optimized photovoltaic modules on facades	Thin-film photovoltaic system, building integrated photovoltaics (BIPV)
8	Sustainable building with transportation	Rainwater system, renewable production, low impact storage, local utilization of waste/recovery, greenery, inter-modal transport
9	From grey to green	Water cycle regulation, biodiversity, ground-bound green facade, wall-bound green facade
10	No theme	Local materials
11	2030	Not applicable
12	Refurbishment with energiesprong	Prefabrication, digitalization (BIM), PV system on roof, basement insulation

2.3 SWOT Analysis

Based on these low-carbon strategies, SWOT analysis is used primarily for applying for that. For the Strengths, some posters focus on the specific issue, even only for the research of facade shapes for PV installation, such as poster 7. However, this poster wonderfully demonstrates the depth of research for one aspect of the competition topics. Some posters focus on multiple points simultaneously, which is more systematic. This provides people with an in-depth or comprehensive understanding of how to apply strategies for different situations. Regarding weaknesses, few posters present in-depth insights into specific issues or engage in extensive systematic summaries. There are also some cons in terms of aesthetic design for posters. This will give people a bad impression and even affect their understanding or use of relevant strategies. The opportunities show that only 19 low-carbon strategies are found. However, there should be more useful low-carbon strategies for future development. The threats show that if everyone uses the same low-carbon strategies, such as for building retrofit. All the buildings may look the same. This may destroy the diversity of the living environment. More comments can be seen from judges.

3 Evaluation Analysis

3.1 Jury Characteristics

For this competition, 11 professional judges in the architecture discipline and 4 professional judges in the energy discipline are invited from 11 universities worldwide. Their research focuses on architectural energy, especially for low-carbon strategies. Even though this competition started in Europe, 4 Chinese professors were invited to join as judges to enhance the internationalization of this competition (Fig. 4). This may lead to attracting 4 Chinese students who submitted their posters in a one-month short-term submission period out of 18 submitters to some extent.

Moreover, an online voting award from the global public gathered more than 2000 voters from October 22 to November 1, 2024. The voting website is https://take.supersurvey.com/poll5306566xB8744F5a-160.

3.2 Evaluation Procedure

For jury review, the evaluation procedure included four steps, as shown in Fig. 5. To ensure a fair evaluation of this competition, all the judges were required to detect plagiarism on all the posters as the first step. In fact, the organizers detected ahead when receiving that in case of any academic misconduct. More importantly, all

Fig. 4 Judges by country. (Self-drawn)

contributors are responsible for ensuring that their works are original and not plagiarized [16]. The second step is to indicate the relationships between judges and posters in advance to avoid any unfair evaluation. In this competition, one professional judge encouraged his students to submit four posters. They are numbers 3, 7, 9 and 12. However, the data shows that even if all the scores of this judge are removed, the final results are the same, proving the fairness and rationality of his professional scoring. Step 3 is to indicate that the judges only need to score the best 9 posters because this competition only has 9 jury rating awards. Scoring all of them may lead to the same rating results, affecting the setting of the awards. Moreover, referring to similar competitions, if a large number of works are collected, it is hard to score all of them. However, some judges neglected this notice and scored all the posters. After the unified requests, the judges scored again only for the best 9 posters. However, some judges gave more than three posters a score of 0, which means that in their professional opinion, some works had no chance of winning the award. This does not affect the fairness and rationality of the results. As for the specific scores from each judge, no one valued the poster with 1, 2, 3, 4, 5, 6, 7, 8, 9. This ensures consistency of results to a certain extent. For step 4, all judge were informed that their decisions would be announced to the public. It is intended to suggest and remind that professionalism should be exercised fairly and impartially. Overall, all the steps aimed to obtain a fair and credible result on the sustainable development of building energy, especially for the low-carbon aspect.

Blind Review of Posters

Dear judge,

Thank you a lot for your support for the Unite! Poster Contest.
Please review the 12 posters we collected and value them below.
You can modify this document according to your comments.
We sincerely thank you for your **fair evaluation!**

Step 1. Do you detect plagiarism?
If so, which one(s)? The one(s) do not need to be reviewed.
No.______

Step 2. Do you know the author of any submitted posters? Or are you the tutor of any posters?
If so, which one(s)?
No.______

Step 3. Please value the top 9 posters with an integer value from 1-10.
1 means pass. 10 means the best. The values can be 1,2,3,4,5,6,7,8,9 or 3,5,5,5,7,8,8,7,8 as yo wish!
Then, make comments on all the valued posters.

Step 4. We will collect all judges' reviews and calculate the value for ranking, then announce the results to the public.

I declare that my review is a fair evaluation based on my professional knowledge.

Signature:
______
Date:
______

Fig. 5 Evaluation procedure. (Self-drawn)

3.3 *Evaluation Comments*

After more than a month of jury evaluation, the blind review scores from the judges were received (Fig. 6). It can be seen that almost all the scores are significant differences on the same poster. Each poster has a minimum score of 0. At the same time, there is at least one maximum value of no less than 7 points. On the one hand, this is due to the judges' different backgrounds and scoring thresholds. On the other hand, the two styles of the submitted posters analyzed in 2.1 also led to different scores among the judges. Most judges made text comments on posters to explain how they evaluated them. Such as for reviewers 1 & 4, they gave posters 5, 6, 7 lower scores. The main comment is that there are too many graphics, which makes it too hard to read. This can be seen as a perspective-oriented poster design evaluation. For most other posters, the research content on low-carbon sustainable design mainly affected the evaluation results. Although the evaluation results also showed two tendencies, they also reflected the inclusiveness of the call for posters and the comprehensiveness of the evaluation.

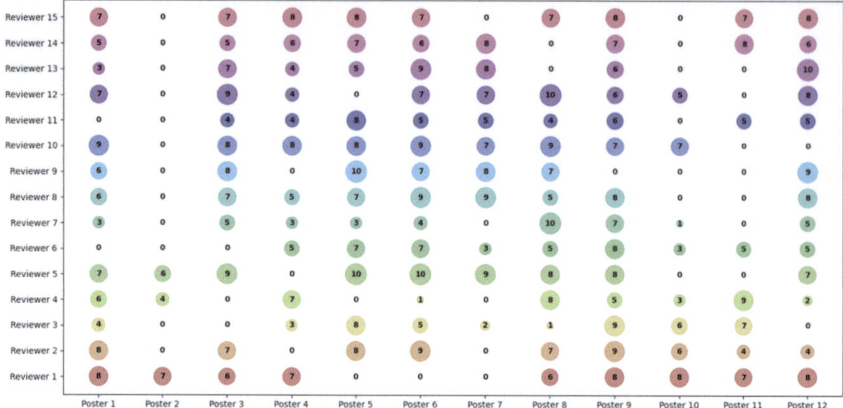

Fig. 6 Jury scores (Generated by Python 3.11). (Self-drawn)

4 Research Results

4.1 Ranking List

Finally, the ranking list by jury was calculated with each poster's average score after subtracting one highest and lowest scores separately from the total score (Fig. 7). To verify the accuracy of the results, the average of the total scores of each work was also used in the ranking calculation. The ranking results shown by the two calculation methods are the same, even though the average scores are different. At the same time, for judge 12 mentioned above, who encouraged his students to submit four posters, removed all the ratings of this judge and then calculated the average score, the ranking result is consistent with the above two results. This proves the rationality and effectiveness of the judge rating to some extent. Therefore, poster 9 got the first prize, posters 6, 5, and 8 got the second prize, and posters 12, 3, 1, 7, and 4 got the third prize.

For the public vote, voting closed at 23:59 on October 31, 2024. Finally, poster 8 won the Special Prize selected by the public (Fig. 8). It votes accounted for 30% of the total, while poster 7 was very close to first place with 29%.

4.2 Comparison Between Sects. 2 and 3

Based on the ranking list by jury, the analysis of the second and third sections can be compared and analyzed to reveal the rationality of the result. Poster 9, which got the first prize, shows the harmony between the aesthetics of poster design and the in-depth research questions. It balanced the two styles of posters and got the favor of

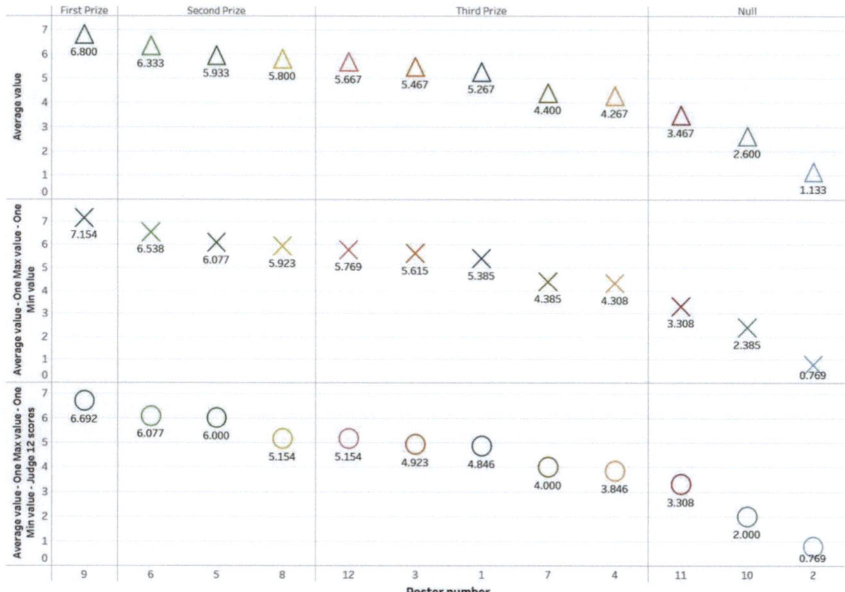

Fig. 7 Ranking list (Generated by Tableau Public 2022.4). (Self-drawn)

Fig. 8 Online voting results. (Self-drawn)

most two fields of judges. Judges 1, 2, 3, 6, and 15, one-third of reviewers gave poster 9 the highest score. They made comments showing: The contrast between gray and green colors highlights sustainable design's advantages in energy conservation and pollution reduction, offering depth and creativity; One of the best posters presented, with a clear message style and idea attractive; The conception is quite interesting; A good balance; and so on. Poster 8 seems to use a similar way to present the work. This may be one of the reasons why it won the first prize in the online voting. However, half of the panel focuses on green transportation, which weakens the response to the theme. The other second prize, posters 6 & 5, are all compact research-oriented posters. For the third prize posters, posters 12, 3, and 7 are also research-oriented. Posters 1 & 4 are more perspective-oriented with AI generation. From the comparative analysis of Sects. 2 and 3, the ranking results reflect to a certain extent that the winning works are well-deserved.

4.3 Low-Carbon Revelations

Through a series of research and analysis of the Unite! Poster Competition, the low-carbon revelations can be sorted out. First, renewable energy production is essential nowadays for low-carbon development. It is the green energy that opens the source for building energy use. Second, green measures, including greening and natural material use, are also vital for low-carbon development. Vertical planting helps improve the building's performance over the seasons. Using green materials for building structures or enclosure materials can effectively achieve carbon sequestration. Third, the energy management system in buildings is also useful. It generally combines with AI technologies to achieve real-time energy regulation and improve building energy efficiency. More low-carbon strategies can be found in Table 1. In short, these revelations correspond to the three main aspects of renewable production, passive and active design. But it is difficult to determine which technology is better. There is no good or bad technology, only reasonable use [8].

5 Discussion

So far, the review research of Unite! Poster Competition on architectural energy for low-carbon to sustainable development is almost done. However, there exist some limitations. First, the wording used in the poster may not be quite appropriate. The word "poster" gives people the illusion of preconceived results, which leads some judges and students to focus on pure layout design from an aesthetic perspective. Although aesthetic literacy may be more important and harder, it weakens the focus of this research on low-carbon development. This problem will be solved in the future with more accurate language use. Second, the requirements of the call for posters were too broad, leading to receiving too many different content posters, including case studies, system summaries, and graphic designs. This brought challenges to the evaluation for judges. Correspondingly, it affected the priority judgment of this research on low-carbon design strategies. These limitations will be avoided in future competition with clearer competition targets, goals, principles, and more. The next competition will collect more low-carbon strategies, expand the number of submissions, and further improve the quality of selected works to promote the sustainable development of the competition effectively. The improvement of applying low-carbon strategies should be shown in the works, especially from practical engineering. Therefore, the next competition may mainly call for practical engineering applications to show the improvement of low-carbon strategies to some extent. The designers' capacity should be considered compared to the first competition; only the better ones should be considered for awarding, attempting to further achieve sustainable development on the existing foundation.

6 Conclusions

Based on the background analysis of environmental problems and policy guidance, this research reviewed the Unite! Poster Competition as a case study introduced the competition framework. 12 collected posters worldwide are used to sort out the low-carbon strategies. The related SWOT of these posters are analyzed. Meanwhile, the jury evaluation on these posters is also analyzed based on the jury characteristics, evaluation procedure, and related comments in three aspects. Then, the ranking results are obtained according to the jury's scores. The basic posters analyzed in Sect. 2 were compared with the evaluation from judges in Sect. 3 to verify the reasonableness of ranking results. Moreover, the low-carbon revelations based on these valuable results are sorted out. Finally, this research came to conclusions after the necessary discussion. The research process is shown in Fig. 9. The conclusions are:

1. Based on the Unite! Poster Competition operation, 12 posters from 10 universities worldwide were collected. These posters demonstrate different low-carbon development thinking and current major low-carbon strategies.
2. At least 19 low-carbon development strategies have been summarized from these 12 posters.
3. The first-prize poster, poster 9, achieved a balance between low-carbon technologies and poster aesthetic design. The low-carbon technologies mainly include greening on buildings, which is also reflected in the second prize, poster 6, as the main low-carbon strategy.
4. For low-carbon development in architectural energy fields, passive, active design and renewable production are the three main ways to achieve that. Passive design focuses on the building itself, reducing energy consumption through

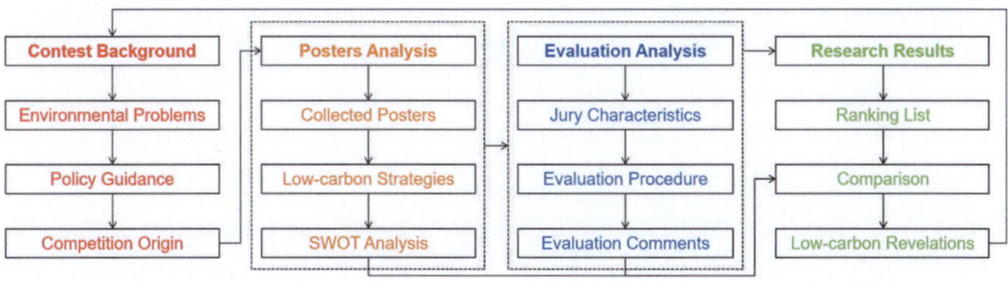

Fig. 9 Research process. (Self-drawn)

greening, recyclable materials, high-performance construction, etc. The active design combines AI management, efficient equipment, etc., to achieve energy conservation. Renewable production opens the source to generate green energy through solar panels and so on to achieve green energy use.

5. Tools are neither good nor bad, only for fair use. There are 19 low-carbon strategies in these posters. However, it is difficult to determine which strategy is more important than others. In particular, sustainable development in buildings needs to be adapted to local conditions and time. Appropriate use of the above strategies is encouraged.
6. After systematic research, based on the Unite! Poster Competition, this study achieved a closed-loop research from background problems to low-carbon development, aiming to contribute to sustainable development, especially in the architectural energy aspect.

Acknowledgements This study is supported by the China National Arts Fund (Grant No. 2024-A-05-098-610), Unite! Seed Fund initiative, and Ministero degli Affari Esteri e della Cooperazione Internazionale (MAECI) scholarship (Prot. 6160). We also thank all the judges for evaluating the posters and the students from around the world for joining this competition. Moreover, the organizers from the Unite! universities are also sincere thanks!

Declaration of Competing Interest The authors have declared no conflict of interest.

Data Availability The organizers have the right to exercise the right of multi-terminal information network dissemination of the entries, especially for academic research purposes only, not for commercial use, according to https://www.unite-university.eu/unitenews/draw-your-vision-of-our-sustainable-future. The scores data collected from 15 judges were only used for anonymous analysis.

References

1. He BJ, Wang JS, Liu HM, Ulpiani G (2021) Localized synergies between heat waves and urban heat islands: implications on human thermal comfort and urban heat management. Environ Res 193:110584. https://doi.org/10.1016/j.envres.2020.110584
2. United Nations Conference on the Human Environment, 5–16 June 1972, Stockholm. Homepage. https://www.un.org/en/conferences/environment/stockholm1972. Accessed 17 Dec 2024
3. World Commission on Environment and Development (1987) Our common future. Oxford University Press, Oxford, UK
4. Haines A, Alleyne G, Kickbusch I, Dora C (2012) From the Earth summit to Rio+20: integration of health and sustainable development. Lancet 379(9832):2189–2197. https://doi.org/10.1016/S0140-6736(12)60779-X
5. Li B, Guo WH, Liu X, Zhang YQ, Russell PJ, Schnabel MA (2021) Sustainable passive design for building performance of healthy built environment in the Lingnan area. Sustainability 13(16):9115. https://doi.org/10.3390/su13169115
6. Schnoor J (2002) Examining the World Summit on sustainable development. Environ Sci Technol 36(21):429A-430A. https://doi.org/10.1021/es0224611
7. Lau LC, Lee KT, Mohamed AR (2012) Global warming mitigation and renewable energy policy development from the Kyoto Protocol to the Copenhagen Accord-A comment. Renew Sustain Energy Rev 16(7):5280–5284. https://doi.org/10.1016/j.rser.2012.04.006

8. Li B, Zhang Y, Caneparo L, Guo W, Meng Q (2024) Energy use in residential buildings for sustainable development: the fifth Solar Decathlon Europe revelations. Heliyon 10(9):e30701. https://doi.org/10.1016/j.heliyon.2024.e30701
9. Gunfaus MT, Waisman H (2021) Assessing the adequacy of the global response to the Paris agreement: toward a full appraisal of climate ambition and action. Earth Syst Governance 8:100102
10. Synthesis Report of the IPCC Sixth Assessment Report (AR6). Homepage, https://report.ipcc.ch/ar6syr/pdf/IPCC_AR6_SYR_SPM.pdf. Accessed 17 Dec 2024
11. United Nations Climate Change conference. Homepage. https://en.wikipedia.org/wiki/United_Nations_Climate_Change_conference. Accessed 17 Dec 2024
12. COP28 talks open in Dubai with breakthrough deal on loss and damage fund. Homepage. https://news.un.org/en/story/2023/11/1144162. Accessed 17 Dec 2024
13. COP28 in Dubai enters final week, negotiations ramp up on emissions cuts, fossil fuels. Homepage. https://news.un.org/en/story/2023/12/1144542. Accessed 17 Dec 2024
14. Draw your vision of our sustainable future! Homepage. https://www.unite-university.eu/unitenews/draw-your-vision-of-our-sustainable-future. Accessed 14 Dec 2024
15. Li B, Guo W, Liu X, Zhang Y, Caneparo L (2022) The third solar decathlon China buildings for achieving carbon neutrality. Buildings 12(8):1094. https://doi.org/10.3390/buildings12081094
16. Call for Posters! Homepage. https://mp.weixin.qq.com/s/GNbLVLj1jpKhFDDQ96Djmg?token=855743472&lang=en_US. Accessed 15 Dec 2024

Relationship Between Urban Form Indicators and Building Energy Consumption from the Perspective of Urban Design

Mengyue Mei and Tongyu Sun

Abstract The increasing challenges of climate change and resource shortages have led to a global focus on sustainable development and energy consumption. Cities and buildings, as major energy consumers, play a key role in addressing these issues. Energy-efficient urban and architectural design has become a central focus for researchers and designers. While much attention has been given to building energy performance, buildings are also affected by their surrounding urban environment, meaning that district-scale urban form also impacts energy consumption. Previous studies have focused on elements such as density, floor-area ratio (FAR), and building height, but these alone do not fully describe the urban space created by buildings. Block-scale indicators should also be considered. These indicators affect public space quality and pedestrian experience, making it necessary to explore their relationship with building energy performance. This study proposes key public space indicators, such as alignment ratio, effective street length, and street space density, based on urban design guidelines and practice. By designing six urban form types with the same FAR but different public space indicators, the study simulates energy consumption for each type. The research results indicate that courtyard-dominated form types outperform tower and slab types in terms of both low energy consumption and high public space contribution. The difference between the best and worst energy performance is 9%, while the difference in urban design form indicators can be as high as 11.5 times. Additionally, for morphological types with similar energy consumption, the differences in urban design form indicators are approximately 1.7 times and 4.6 times, respectively. These results suggest the existence of urban form types that can simultaneously optimize energy performance and public space contribution. Even morphological types that perform slightly worse in one aspect have the potential for balance and optimization, thereby offering new possibilities and inspiration for early-stage considerations of morphological types in practical urban design processes. The study compares the results to examine the relationship between urban form elements, particularly design indicators, and building energy performance, aiming to offer a design perspective that balances low-energy districts with high-quality public spaces.

M. Mei (✉) · T. Sun
College of Architecture and Urban Planning, Tongji University, Shanghai, China
e-mail: mengyue_mei@163.com

B. He et al. (eds.), *Climate-Resilient, Low-Carbon, and Sustainable Cities*, Lecture Notes in Civil Engineering 754, https://doi.org/10.1007/978-981-95-2169-2_31

Keywords Urban form · Building energy consumption · Urban design

1 Introduction

About 68% of people are estimated to live in urban settlements by 2050 and the urbanization process has already become a global ongoing phenomenon and will continue over the next three decades. The transition means an increase of 2.2 billion urban residents, mostly in Asia and Africa [1]. The rapid urbanization and migration of people require more buildings to be built and used, which leads to higher energy consumption and emissions. According to International Energy Agency (IEA), the operations of buildings account for 30% of global final energy consumption and 26% of global energy-related emissions [2]. Two-thirds of the emissions is attributed by electricity generation for use in buildings and although the sector's energy intensity per square meter improved, its emissions increased more than 25% since 2000. This reflects a 65% increase in floor area since then, growing demand for energy services and rising electricity consumption. Electricity use in buildings grew five times faster than improvements in the carbon intensity of power generation since 2000 [3]. From a global view, the Paris Agreement also requires that the emissions need to be reduced by 45% by 2030 and reach net zero by 2050 [4]. The urgent and pressing goals make the study of energy consumption in urban environment particularly important.

Building energy consumption is typically categorized into operational energy consumption and life-cycle energy consumption. In a narrower sense, it refers to operational energy use such as heating, cooling, and lighting. According to data from the International Energy Agency (IEA), space cooling is one of the primary contributors to the increasing demand for electricity, with approximately 2 billion air conditioning units currently in operation globally, of which residential units account for nearly 70% of the total. Furthermore, heating for both space and water accounts for nearly half of global energy use in buildings. Effective strategies to reduce operational energy consumption go beyond improving the energy efficiency of heating and cooling systems. They also rely on passive, nature-based, and alternative solutions, including the enhancement of building and district design.

Building energy consumption is influenced by a variety of factors. Previous studies have primarily focused on internal building factors, such as surface area-to-volume ratio, envelope materials, and building orientation. However, in an urban context, buildings exist within a complex external environment, rather than in isolation. Therefore, it is crucial to examine the impact of external environmental factors, particularly the surrounding urban block, on building energy consumption.

A series of studies have demonstrated that urban form elements have a significant impact on building energy consumption. Ratti et al. identified four principal factors influencing energy performance: user behavior, energy system efficiency, architectural design, and urban form [5]. Subsequent studies have further elucidated the relationship between urban form and building energy consumption. For instance, Delmastro et al. demonstrated that variations in urban form could account for an 11%

difference in heating energy use, which increases to 16% when considering solar gain [6]. Similarly, Silva et al., through a case study of Porto, Portugal, established that urban form could explain up to 78% of the variance in total energy consumption, including heating, cooling, and mobility [7]. Therefore, when discussing building energy consumption, it is essential to conduct an in-depth exploration of the urban form elements at the block scale.

On the other hand, block form elements are also an integral part of urban design. In the early stages of block design, the form of the block not only influences building energy consumption but also affects the quality of public spaces and the pedestrian experience. Therefore, identifying key block form elements and exploring their multi-dimensional impacts is of paramount importance. While existing research often focuses on building energy consumption, the results typically lack effective guidance on public space and pedestrian experience in the block context.

Therefore, urban morphological elements can be considered to have a dual impact: they influence both building energy performance and the quality of public spaces and pedestrian experiences in cities. Existing research often focuses on only one aspect, yet in practical urban design practice, both aspects need to be considered, and the design process should integrate multiple factors. Thus, unlike studies that concentrate on a single aspect, this research attempts to simultaneously address both energy consumption and public space contribution, particularly emphasizing the latter from an urban design perspective rather than a purely technical one.

On the other hand, previous studies exploring the relationship between urban morphology and energy consumption have predominantly focused on building-scale morphological elements. However, urban morphological elements should also include neighborhood-scale indicators, as these indicators similarly influence building layout, solar access and shading, microclimate conditions, and other factors that ultimately affect building energy performance. Additionally, this study sets the same FAR as a prerequisite for comparative experiments, reflecting real-world urban design practices where designers often need to explore and develop morphological solutions under given FAR constraints. This approach is also one of the innovative aspects of this research.

This study highlights a series of indicators from an urban design perspective, including alignment ratio, effective street length, and street space density. By designing a set of urban forms with identical FAR conditions, energy consumption simulations are conducted to explore whether there exists an urban form that strikes a better balance between energy performance and high-quality public spaces. This approach provides valuable insights for designers in the early stages of urban planning. On the other hand, it encourages future research to not only focus on urban forms with high energy efficiency but also to place greater emphasis on human-centered urban public space quality.

2 Literature Review

Unlike the building scale, the mechanisms by which urban form elements at the block scale influence building energy consumption are more complex. This is because block form not only directly affects the building's passive solar gain and heat dissipation but also indirectly influences building energy performance through its impact on the microclimate. Therefore, it is essential to consider the impact of urban form on the microclimate.

On the other hand, the description and selection of urban form elements that influence building energy consumption can vary. However, existing research has primarily focused on several categories of form elements, which can generally be classified into: density measures, geometric measures, land use and land cover measures, and typologies [8]. These classifications are derived from the identification and categorization of morphological factors identified in prior research. The three types of measures—density measures, geometric measures, and land use and land cover measures—are more easily analyzed and described through data. In contrast, typology-based measures are more effective in capturing the complexity of urban form. As a result, many studies on urban morphology integrate these types or expand upon morphological typologies by incorporating specific indicators relevant to the research. This integrated approach to description allows for both the quantitative analysis of relationships and provides a practical means for urban and architectural designers to understand real-world scenarios.

However, when attempting to synthesize findings regarding energy consumption variations for each individual factor, it becomes apparent that most studies do not facilitate direct comparisons of results within the context of a single morphological factor.

Nevertheless, through a review of existing studies, common indicators can be identified. The Table 1 below presents some of the commonly used indicators regarding the influence of urban form on building energy consumption.

From the perspective of the objects being described, the morphological factors identified in existing studies can also be classified into two categories. The first category includes factors related to the building form and its relationship with the environment, such as Floor Area Ratio (FAR), building density (BD), and building coverage ratio (BCR). The second category consists of factors related to non-physical spatial attributes, such as open space ratio (OCR), obstruction angle, and sky view factor (SVF). Both categories of factors influence building energy consumption, but it is evident that when considering the quality of public space and pedestrian experience, the second category has a more direct impact.

It is worth noting that although many studies focus on the impact of urban form at the block scale on energy consumption, the indicators used to describe urban form tend to prioritize the characteristics of building form, often lacking relevant indicators from an urban design perspective. Taking density and floor area ratio (FAR) as examples, using only these two types of indicators to describe urban form can lead to vastly different outcomes due to variations in building form and layout. These

Table 1 Morphological indicator choices of some previous research

Year	Researchers	Location	Indicators
2016	Quan et al. [9]	Shanghai, China	Building density (BD), building height (BH), floor area ratio (FAR)
2018	Javanroodi et al. [10]	Tehran, Iran	Building density (BD), height to width ratio (H/W)
2019	Lima et al. [11]	Maceio, Brazil	Building shape factor (BSF), building height (BH), building density (BD), building orientation (BO)
2021	Mangan et al. [12]	Istanbul, Turkey	Building typology (BT), height to width ratio (H/W), number of floors (NS), building orientation (BO)
2021	Ahmadian et al. [13]	London, UK	Plan depth (PD), number of floors (NS), building density (BD)
2021	Zhang and Gao [14]	Nanjing, China	Floor area ratio (FAR), surface area ratio (SAR), sky view factor (SVF)
2023	Liu et al. [15]	Jianhu, China	Building density (BD), open space ratio (OSR), shape coefficient (SC), average perimeter-area (PAR)

differences not only manifest in the buildings themselves but also affect the public space configurations influenced by building forms. Relying solely on indicators like open space ratio may fail to comprehensively represent the actual urban form and neglect the pedestrian experience from an urban design perspective.

Existing indicators for describing urban form often overlook the importance of public space form, which plays a crucial role in urban design. Given this context, obtaining answers to how public space form influences building energy consumption becomes especially important. This is particularly true when the design objective is to create urban forms, which not only require energy efficiency but also aim to foster more human-centered, functional public spaces.

Urban design is one of the key approaches to creating and optimizing urban public spaces. During the urban design process, elements that are often less addressed or entirely overlooked in overall planning and architectural design are taken into consideration, such as the continuity of building facades, support for public spaces, the formation and shaping of streets, and street density, among others. Therefore, research from an urban design perspective can effectively fill the gap in understanding how these elements influence energy consumption.

The selection of public space morphological factors can be guided by existing street design and urban design guidelines. For idealized high-quality urban spaces, urban design manuals typically provide clear directives and design recommendations, often including specifications for street space morphology. For instance, the Central Melbourne Design Guide [16] emphasizes that improving both the quantity and quality of public spaces is central to urban design. It specifically highlights that urban development and renewal should prioritize supporting walking as the primary mode of transport.

At the urban morphology level, this is primarily achieved through the development of laneways and pedestrian connections. For example, the guidelines stipulate that pedestrian connections should be no more than 70 m apart and have a minimum width of 6 m. Building volumes should align with streets to form an effective street interface, except where setbacks are used to create plazas. The guide encourages attractive colonnades with a height that is twice their depth. Additionally, street walls should be lower along laneways and streets less than 10 m wide. Buildings with a street frontage longer than 25 m should be divided into smaller sections to provide modulation in the street facade.

Regarding ground floor levels, the guidelines specify that at least 80% of the ground-level frontage of buildings facing streets and laneways should consist of entrances or windows, thereby ensuring active frontage. Finally, the Central Melbourne Design Guide defines the average urban block length as the mean distance between opposite sides of an urban block, measured from laneways and streets.

This comprehensive approach underscores the importance of integrating public space quality considerations into urban design, ensuring that energy-efficient, walkable, and people-centered spaces are prioritized in city planning.

Similarly, the Shanghai Street Design Guidelines [17] emphasize that street spaces should be orderly, comfortable, and pleasant. This is achieved through the regulation of building control lines and alignment ratios, which ensure a cohesive and continuous spatial interface. By maintaining overall consistency, these guidelines aim to establish a foundation for functional spaces that support public life and create a walkable urban scale.

3 Method

To further investigate the relationship between urban morphological elements and building energy performance from an urban design perspective, this study introduces a set of indicators derived from urban design guidelines that capture the characteristics of external building spaces. A comparative analysis is then conducted through a series of simulation experiments. The overall research methodology and workflow are illustrated in the diagram below (Fig. 1).

The study is divided into three main stages: Block Parametric Modeling, Energy Performance Simulation, and Relationship Analysis. The first stage focuses on modeling urban forms, which serve as the foundation for energy performance simulations in the second stage. The outcomes of these simulations provide the basis for the relationship analysis conducted in the third stage.

The primary goal of the first stage is to construct abstract urban forms and develop their parametric models. A key requirement in this stage is ensuring that all comparative forms share the same Floor Area Ratio (FAR). This constraint is implemented for three reasons:

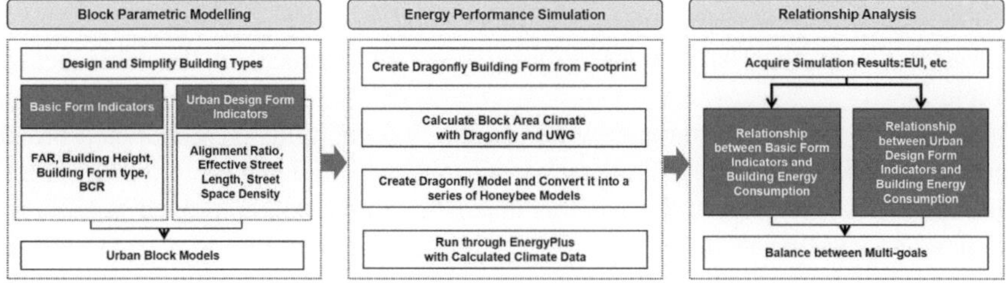

Fig. 1 Workflow of the research process with three stages

- In practical design processes, urban designers and architects often receive explicit and fixed FAR requirements at the early stages of planning.
- Controlling variables is essential for isolating effects; FAR is commonly used as a standard indicator in studies examining the relationship between urban form and energy performance.
- Despite the uniformity of FAR, the inherent complexity of urban form allows for significant variations in external spatial configurations, making these differences critical components of design decision-making.

To maintain a consistent FAR, this study employs a typical residential unit as a base model. The typical residential unit has a length of 15 m, with of 10. 5 m and height of 3 m. By altering building heights, typologies, and spatial arrangements, six distinct block models with varying external environmental characteristics are designed, as illustrated in the accompanying figure (Fig. 2).

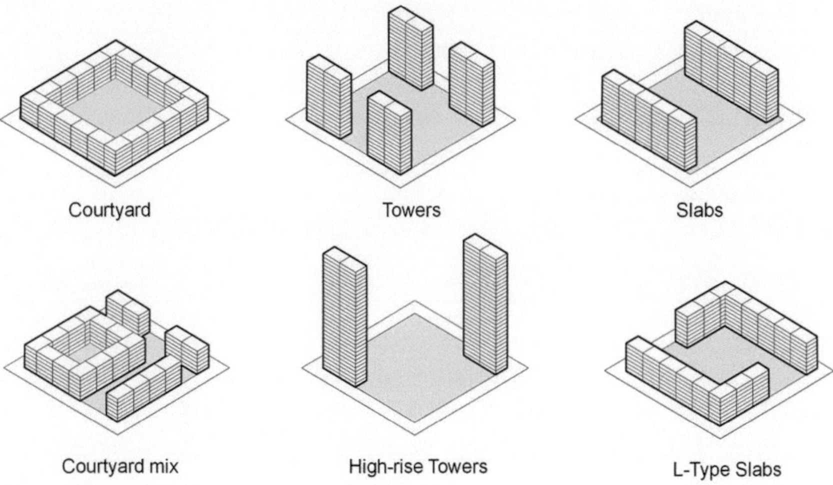

Fig. 2 Residential unit and six block form types

On the same designated site, these block models share identical FAR values. However, their other basic form indicators are summarized in the table below (Table 2).

Parametric modeling and energy performance simulations are conducted based on the defined block forms. While numerous tools are available for building energy simulation, it is crucial to assess whether the methods and tools employed in previous studies are appropriate for block-scale analyses. This study adopts a workflow centered on Ladybug Tools (Fig. 3). This workflow operates primarily within the Rhinoceros and Grasshopper platforms, utilizing EnergyPlus as the core simulation engine for energy performance analysis.

Developed by the U.S. Department of Energy, EnergyPlus provides comprehensive simulations of a building's thermal performance, humidity levels, airflow, and lighting. It accounts for key factors such as thermal conduction, air infiltration, solar radiation, lighting, and HVAC system performance. EnergyPlus fundamentally operates through a thermal balance physical model for simulations. Compared to engines like DOE-2 and DeST, EnergyPlus not only ensures thermal balance within rooms but also goes beyond linear time-invariant systems to handle surface convective heat transfer coefficients. Crawley et al. validated the high precision and accuracy of EnergyPlus in building energy consumption simulations through multiple test cases, confirming the reliability of its core algorithms [18].

Table 2 Basic form indicators of six types

No	Form type	Site area (m^2)	Total building area (m^2)	Building footprint area (m^2)	Number of floors
C-1	Courtyard	11,130.25	18,900	3150	6
C-2	Courtyard mix	11,130.25	18,900	3150	6
T-1	Towers	11,130.25	18,900	1260	15
T-2	High-rise towers	11,130.25	18,900	630	30
S-1	Slabs	11,130.25	18,900	1575	12
S-2	L-type slabs	11,130.25	18,900	2520	7, 8

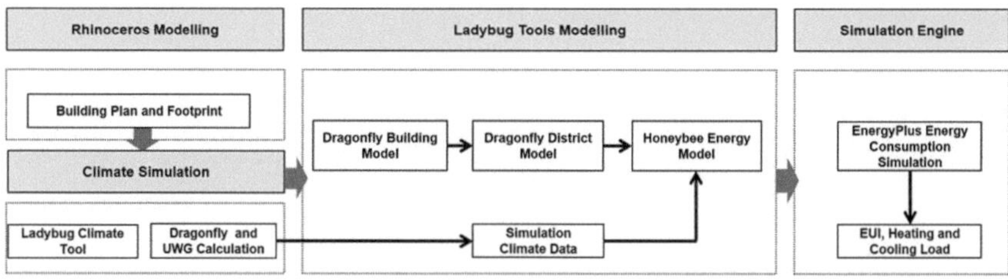

Fig. 3 Energy performance simulation workflow

To improve usability and practical application, platforms such as OpenStudio, DesignBuilder, and Ladybug Tools have incorporated the EnergyPlus engine into their user-friendly interfaces.

The Grasshopper platform facilitates visual programming, integrating a suite of user-friendly plugin tools, with Ladybug Tools being a central component.

Ladybug Tools comprises four plugins—Ladybug, Honeybee, Butterfly, and Dragonfly—each catering to specific analytical domains. Ladybug focuses on climate and meteorological analysis with advanced visualization capabilities. Honeybee supports energy performance and daylighting simulations, while Butterfly specializes in wind environment analysis. Dragonfly is designed for studying urban microclimates, macroclimates, and energy performance at the urban scale. In this study, Honeybee and Dragonfly are primarily utilized to conduct performance simulations.

Compared to tools like OpenStudio and DesignBuilder, which also utilize the EnergyPlus engine, Ladybug Tools offers several distinct advantages. First, as a parametric platform, Ladybug Tools provides greater comprehensiveness and adjustable flexibility. By combining tools such as Ladybug, Honeybee, and Dragonfly, it enables a simulation workflow that considers multiple factors simultaneously. As a result, Ladybug Tools is widely used in research related to solar potential, microclimate, daylight, energy consumption, and their connections to building and urban morphology. For example, Rostami et al. employed a Ladybug Tools workflow integrated with the UMI tool to study the solar potential and energy consumption of urban canyons and blocks under the influence of urban morphological elements in semi-arid regions [19]. Second, as a Grasshopper platform plugin, Ladybug Tools offers superior usability: compared to other environmental performance simulation software, Rhino & Grasshopper is already integrated into common design workflows in the urban design field, further reducing the learning and usage costs and facilitating performance-optimized urban and architectural design operations. This tool has also been widely adopted by researchers studying the relationship between urban morphology and building energy consumption [15, 20, 21] Therefore, this study also employs EnergyPlus as the core engine and utilizes Ladybug Tools, based on the Rhino and Grasshopper platforms, as the building energy simulation tool.

To account for local urban climate conditions, the study employs the Urban Weather Generator (UWG), a tool integrated with Dragonfly. The UWG tool has been validated across various cities for its effectiveness in adjusting urban weather data [22]. Using the modified climate data, the Honeybee plugin invokes EnergyPlus simulations to model and evaluate the energy performance of the block configurations.

The simulation generates data on Energy Use Intensity (EUI) and annual heating and cooling loads. The selection of simulation software emphasizes result reliability, workflow consistency, and the ability to perform parametric form manipulation. The simulation process utilizes the Grasshopper parametric design platform within Rhino to define pre-set building site plans, establish building volumes through floor height and floor count, and configure daylighting parameters based on a window-to-wall ratio (WWR). A Dragonfly model is then developed and converted into a series of Honeybee models for energy simulations, with a maximum WWR of 0.35 for residential buildings.

To ensure consistency, all block models are placed in the same urban context, minimizing external interference. Shanghai is selected as the experimental city due to its representative hot summer and cold winter climate, classified as 3A under ASHRAE standards (warm and humid). EnergyPlus simulations, incorporating Shanghai's climate parameters, produce annual EUI data and monthly heating and cooling load outputs.

Before comparing the energy performance data, it is essential to define four urban design form indicators specific to the identified typologies (Table 3). These indicators primarily focus on the external spaces shaped by buildings, aiming to evaluate whether these forms create sufficient and high-quality public spaces.

The first indicator is the Alignment Ratio, defined as the percentage of total building street wall length relative to the total length of the building setback line. This metric is frequently emphasized in urban design guidelines across various cities, as it reflects the degree to which buildings align with the designated setback lines and contribute to the definition of the street interface. The Alignment Ratio is a critical urban design indicator, as it captures the extent to which building placement shapes and reinforces the spatial character of urban blocks and streetscapes.

One of the key indicators is Effective Street Length, which refers to the total length of street walls within the block that contribute to the definition of functional and high-quality street spaces. Streets lacking supporting building interfaces are less likely to generate engaging and effective public spaces, making this metric critical for evaluating urban form.

Another important metric is Effective Street Space, representing the total street area enclosed by effective building interfaces. This indicator highlights the spatial extent of well-defined streetscape areas that support public activities.

Table 3 Urban design form indicators

Indicator	Formula	Connection to high-quality public spaces
Alignment ratio	$AR = \left(\frac{\sum \text{Street wall Length}}{\sum \text{Building Setback Line Length}}\right) * 100\%$	High AR helps to create continuous and enclosure street interfaces, thereby promoting pedestrian activity [23]
Effective street length	$ESL = \sum \text{Street wall Length}$	Longer street wall length can enhance the vitality of the street attract more pedestrians to stay [24]
Effective street space	$ESS = \sum ESL * \text{Pedestrian Space Width}$	Higher effective street space provides more spaces for activities [25]
Street space density	$SSD = \frac{\sum \text{Non-Urban Space Occupied By Building}}{\text{Site Area}} * 100\%$	Higher street space density means more accessible pedestrian spaces for vitality and more pedestrian-friendly environment [26]

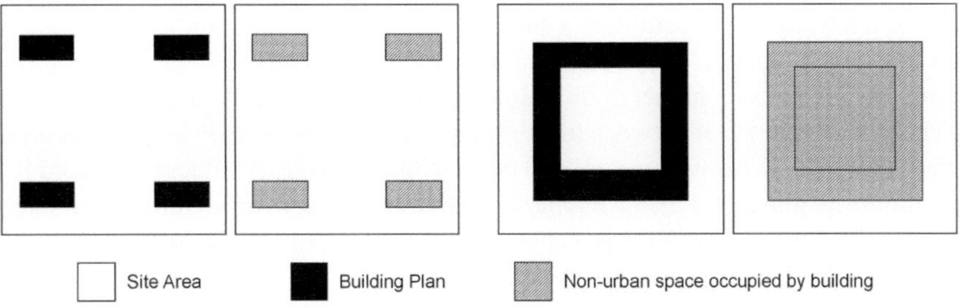

Fig. 4 Definition of street space density

Table 4 Urban design form indicators of six types

No	Form type	Effective street length (m)	Effective street space (m^2)	Street space density (%)	Alignment ratio (%)
C-1	Courtyard	342	3820	65.7	100
C-2	Courtyard mix	231.5	3025	65.7	67.7
T-1	Towers	120	1200	11.3	35.1
T-2	High-rise towers	60	600	5.7	23.7
S-1	Slabs	150	1500	14.2	43.9
S-2	L-type slabs	261	2810	65.7	76.3

Lastly, Street Space Density, a novel indicator introduced in this study, requires further elaboration. Street Space Density is defined as the proportion of non-building urban space in relation to the total site area (Fig. 4). This metric captures the spatial configuration of urban blocks, with higher street space density indicating more compact, pedestrian-friendly urban environments. It serves as a crucial measure to assess the urban morphology's capacity to support walkability and public space usability.

The following table (Table 4) shows the urban design form indicators of each block form type.

4 Results and Discussions

The energy performance simulation results for the six urban form types are presented as follows (Table 5). Overall, the C-1 type exhibits the lowest Energy Use Intensity (EUI) value, while the T-1 type has the highest EUI. Meanwhile, the T-type form exhibits a higher EUI compared to the C-type and S-type forms. The cooling and heating load differences vary among the types, with the smallest fluctuation observed in C-2 and the largest in S-1. From these results, it can be inferred that the cooling and heating load performance of C-2 is relatively balanced. This notable difference may be attributed to the complexity and mixed nature of its morphology. Differences in cooling and heating loads may have been influenced by building orientation; for example, S-1 shows a higher cooling load and a lower heating load compared to C-1.The difference in EUI between the highest (T-1) and lowest (C-1) is 9%, indicating relatively close performance across all types. Notably, the EUI values of S-1 and S-2 are almost identical. T-1 and T-2 types are also very similar. This similarity could be attributed to the control of several shared variables, such as assumed environmental conditions and the fixed Floor Area Ratio (FAR).

A ranking analysis, from best (6) to worst (1), was conducted based on energy performance and urban design indicators (Fig. 5). The results are illustrated in the figure below, which provides a clear visual representation of the performance across five key dimensions.

When urban design form indicators are considered, the C-1 type clearly outperforms the others, demonstrating the best overall performance in both public space quality and energy efficiency. In contrast, The T-2 type performs the worst in all aspects except for the EUI value, and closely following it is the T-1 type, which has

Table 5 Simulation results of six block types

No	Form type	EUI (KW/m^2)	Cooling load (KW/m^2)	Heating load (KW/m^2)
C-1	Courtyard	167.591	52.496	43.411
C-2	Courtyard mix	174.148	53.915	48.267
T-1	Towers	182.791	59.248	50.700
T-2	High-rise towers	181.936	60.094	49.745
S-1	Slabs	170.102	54.854	42.971
S-2	L-type slabs	170.641	54.001	44.555

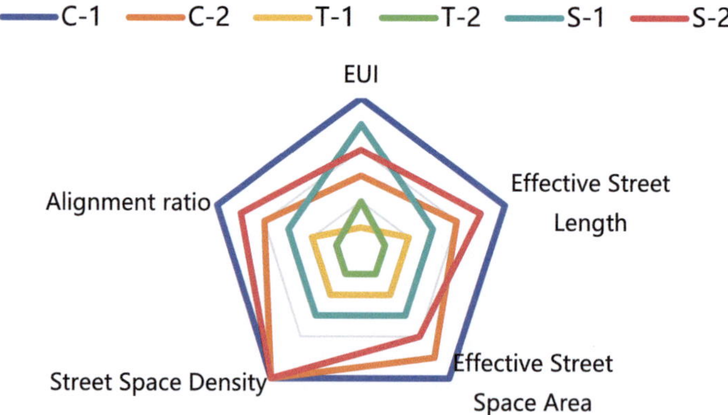

Fig. 5 Comparation rank with urban design indicators

the highest energy consumption and relatively poor contributions to public space. Although S-1 and S-2 types exhibit similar energy performance, S-2 demonstrates superior urban design performance, as it generates longer effective street and larger public spaces compared to S-1.Comparing the three major morphological types—Courtyard, Tower, and Slab—it becomes evident that the Courtyard type performs the best overall, achieving relatively lower energy consumption and better public space support to some extent. In contrast, the Tower type shows the opposite trend, while the Slab type demonstrates potential for improvement.

The comparative results of relative rankings provide insights into how different morphological types balance energy performance and contributions to public space.

First, under the same FAR (Floor Area Ratio), different morphological types have a significant impact on building energy consumption. This also implies that when energy consumption is roughly the same, there can be different morphological type options, such as S-1 and S-2, or T-1 and T-2.

Second, when morphological types with similar energy performance emerge, their contributions to public space can vary significantly. For example, S-2 provides better support for street interfaces and creates more pedestrian-friendly public spaces compared to S-1. This suggests that when selecting morphological types for design, attention must be paid to urban design form indicators.

Third, the overall experimental results indicate that there are morphological types that can offer both optimal energy performance and the best support for public spaces. However, this is not a unique optimal solution, as real-world design scenarios involve more complex choices. Nevertheless, these results demonstrate the possibility of achieving a balance between energy performance, morphological type, and public space support. In other words, when energy performance requirements are a priority, the chosen morphological type should not only address energy efficiency but also consider its contribution to public space, as this balance is achievable.

This study has certain limitations, primarily in the following areas.

First, the urban form types used in the simulations were simplified compared to real-world urban environments. This simplification was necessary to clarify the research focus and reduce the computational time required for energy performance simulations.

Second, the study was conducted within a single, controlled environmental setting and focused on a relatively small scale, lacking a broader exploration of street-level morphological types.

Third, this study primarily focuses on the comparative analysis of relative results among different morphological types. Due to the limitations in the variety of types and data, the absolute relationships between the four urban design form indicators and energy consumption remain to be explored in subsequent research.

Fourth, the urban building function selected in this study is residential. Therefore, the results are not applicable to other functions. Future work should explore the impact of urban design form indicators on blocks with different functions.

Despite these limitations, the findings provide valuable insights for urban designers. Specifically, they suggest that during the early stages of urban form planning, designers should prioritize configurations that achieve a balance between improved energy performance and high-quality public spaces. Importantly, such configurations are less influenced by variations in FAR, offering a flexible approach to sustainable urban design.

5 Conclusion

In most studies on factors influencing building energy performance, building characteristics are often the primary focus. However, in the context of real-world urban and architectural design processes, the early stages of urban design involve not only building forms but also the external public spaces shaped by buildings and other elements. These external spaces are a core component of the urban experience for users. To create orderly, comfortable, and sustainable urban environments, it is crucial to revisit the relationship between urban morphology and building energy consumption from the perspective of urban design.

This research employed controlled-variable energy simulation experiments for a set of comparative morphological types. The results reveal that urban form elements and typologies significantly impact building energy performance, while also exhibiting notable differences in their capacity to create public spaces. The C-1 type demonstrates both the best energy performance and the highest public space support indicators, while the T-type forms perform poorly in both aspects. Among the S-types, energy consumption is generally consistent, but there are notable differences in urban design form indicators. These findings suggest the possibility of identifying morphological types capable of achieving multi-objective optimization, which could be preliminary addressed in the early stages of urban design. This study aims to provide a fresh perspective and a set of novel indicators to inspire future research and design practices.

The findings offer meaningful insights for urban designers. It is suggested that during the early stages of urban design, it is feasible to prioritize morphological types that achieve both improved energy performance and high-quality public spaces, with minimal dependency on FAR. The design strategies for reference include the following:

First, under the design condition of a given FAR (Floor Area Ratio), priority should be given to morphological types that not only exhibit lower building energy consumption but also provide high-quality public spaces, such as the Courtyard type.

Second, when morphological types are constrained, efforts should be made to explore types that can balance energy performance and public space contributions within the given limitations.

Third, it is crucial to avoid the worst-case morphological types—those that result in higher building energy consumption while offering minimal support to streets and public spaces.

Future research directions could expand the scale of study blocks, include a broader range of program types, and investigate correlations to identify urban design form indicators most strongly associated with energy performance. Addressing these challenges would contribute to the development of more precise and sustainability-oriented urban design guidelines in the future.

References

1. United Nations Habitat (2022) World cities report 2022. https://unhabitat.org/sites/default/files/2022/06/wcr_2022.pdf
2. International Energy Agency. https://www.iea.org/energy-system/buildings
3. International Energy Agency: The critical role of buildings: perspectives for the clean energy transition. https://iea.blob.core.windows.net/assets/026bff1b-821d-48bc-8a0e-7c10280c62bc/Perspectives_for_the_Clean_Energy_Transition_2019.pdf
4. United Nations: The Paris agreement. https://www.un.org/en/climatechange/paris-agreement
5. Ratti C, Baker N, Steemers K (2005) Energy consumption and urban texture. Energy Build 37(7):762–776
6. Delmastro C, Mutani G, Pastorelli M, Vicentini G (2015) Urban morphology and energy consumption in Italian residential buildings. In: 2015 International conference on renewable energy research and applications (ICRERA). IEEE, Location, pp 1603–1608
7. Silva M, Oliveira V, Leal V (2017) Urban form and energy demand: a review of energy-relevant urban attributes. J Plan Lit 32(4):346–365
8. Quan SJ, Li C (2021) Urban form and building energy use: a systematic review of measures, mechanisms, and methodologies. Renew Sustain Energy Rev 139:110662
9. Quan SJ, Wu J, Wang Y, Shi Z, Yang T, Yang PPJ (2016) Urban form and building energy performance in Shanghai neighborhoods. Energy Procedia 88:126–132
10. Javanroodi K, Mahdavinejad M, Nik VM (2018) Impacts of urban morphology on reducing cooling load and increasing ventilation potential in hot-arid climate. Appl Energy 231:714–746
11. Lima I, Scalco V, Lamberts R (2019) Estimating the impact of urban densification on high-rise office building cooling loads in a hot and humid climate. Energy Build 182:30–44
12. Mangan SD, Oral GK, Kocagil IE, Sozen I (2021) The impact of urban form on building energy and cost efficiency in temperate-humid zones. J Build Eng 33:101626
13. Ahmadian E, Sodagar B, Bingham C, Elnokaly A, Mills G (2021) Effect of urban built form and density on building energy performance in temperate climates. Energy Build 236:110762
14. Zhang M, Gao Z (2021) Effect of urban form on microclimate and energy loads: case study of generic residential district prototypes in Nanjing, China. Sustain Cities Soc 70:102930
15. Liu K, Xu X, Zhang R, Kong L, Wang W, Deng W (2023) Impact of urban form on building energy consumption and solar energy potential: a case study of residential blocks in Jianhu, China. Energy Build 280:112727
16. Central Melbourne Design Guide. https://participate.melbourne.vic.gov.au/amendmentc308/melbourne-design-guide
17. Shanghai Street Design Guidelines. https://hd.ghzyj.sh.gov.cn/zcfg/zhl/201610/t20161019_696909.html
18. Crawley DB, Lawrie LK, Winkelmann FC, Buhl WF, Huang YJ, Pedersen CO, Glazer J (2001) EnergyPlus: creating a new-generation building energy simulation program. Energy Build 33(4):319–331
19. Rostami E, Nasrollahi N, Khodakarami J (2024) A comprehensive study of how urban morphological parameters impact the solar potential, energy consumption and daylight autonomy in canyons and buildings. Energy Build 305:113904
20. Feng W, Chen J, Yang Y, Gao W, Zhao Q, Feng H, Yu S (2024) The impact of building morphology on energy use intensity of high-rise residential clusters: a case study of Hangzhou, China. Buildings 14(7):2245
21. Ibrahim Y, Kershaw T, Shepherd P, Coley D (2021) On the optimisation of urban form design, energy consumption and outdoor thermal comfort using a parametric workflow in a hot arid zone. Energies 14(13):4026

22. Bueno B, Norford L, Hidalgo J et al (2013) The urban weather generator. J Build Perform Simul 6(4):269–281
23. City of New York, Department of City Planning: Active Design Guidelines: Promoting Physical Activity and Health in Design (2011). https://www.nyc.gov/site/planning/plans/active-design-guidelines/active-design-guidelines.page
24. Mehta V (2014) Evaluating public space. J Urban Des 19(1):53–88
25. City of Vancouver: Downtown Eastside Urban Design Guidelines (2013). https://guidelines.vancouver.ca/guidelines-downtown-eastside.pdf
26. Gehl J (2003) Life between buildings: using public space. Van Nostrand Reinhold

Effect of Wood Fiber on Strength and Microstructure of Alkali-Activated-Slag-Based Low Carbon Concrete Containing Recycled Clay Brick Powder

Saijun Cao, Lu He, Dawei Lv, Yue Ma, and Tong Wu

Abstract This study investigated the performance of an innovative low-carbon concrete (LCC) produced by partially replacing Portland cement with recycled wood fiber (WF), RCBPand AA-GGBFS. The primary aim was to evaluate the effects of various material ratioson on the microstructure and mechanical properties of LCC by experiments. By setting different control groups, they were subjected to curing at ambient temperature for durations of 7 and 28 days, respectively, to enable thorough testing and analysis, encompassing compressive strength and hydration products,, microstructure, thermogravimetric analysis. The findings indicated that the optimal substitution rate for RCBP was determined to be 20%, with a corresponding AA-GGBFS substitution rate of 60%, and a mass ratio of WF to RCBP was 1: 4. At these levels of substitution, the produced LCC paste exhibited commendable compressive strength. From the perspective of environmental and economic analysis, The carbon dioxide emissions per unit of concrete produced across all experimental groups were markedly lower than those associated with conventional Portland cement concrete, and the highest effective carbon reduction efficiency was as high as 70%. Overall, this study has further deepened the utilizationof RCBP to promote the sustainability of raw materials in tunnel engineering.

Keywords RCBP · GGBFS · Wood fiber · Alkali-activated concrete

Nomenclature

AA	GGBFS Alkali-activated Ground Granulated Blast Furnace Slag
CA	Coarse basalt Aggregate
FA	Fly Ash
LCC	Low-carbon Concrete

S. Cao · L. He (✉) · D. Lv · Y. Ma · T. Wu
Chong Qing Architectural Design Institute Co., Ltd. (CQADI), Chongqing, China
e-mail: He.lu2008@163.com

B. He et al. (eds.), *Climate-Resilient, Low-Carbon, and Sustainable Cities*, Lecture Notes in Civil Engineering 754, https://doi.org/10.1007/978-981-95-2169-2_32

OPC	Ordinary Portland Cement
PS	Polycarboxylate Superplasticize
RCBP	Recycled Clay Brick Powder
RS	River Sand
SF	Silica Fume
SP	Slag Powder
WF	Recycled Wood Fiber
WGP	Waste Glass Powder

1 Introduction

High-quality development is the core imperative for comprehensively building a modern socialist country. This is a strategic choice based on the transformation of the principal contradiction in our society and an important path to achieving Chinese modernization. Therefore, in the process of promoting modernization, we should adhere to high-quality development as the core while actively practicing the concept of green development, promoting efficient utilization of resources, reducing environmental pollution, and laying a solid and green foundation for the comprehensive construction of a modernized country.

Tunnels serve as a crucial component of the transportation infrastructure, significantly contributing to traffic safety and addressing challenges posed by unique topographical conditions. Nevertheless, the construction of tunnels is associated with substantial carbon emissions; specifically, the average carbon output per kilometer for highway tunnels is approximately 2247.01 tons [1], while for railway tunnels, it is around 2556 tons [2]. The carbon emission of open-cut cast-in-place subway stations is about 55,075.28t [1]. Based on the average carbon trading price of 68.15 yuan/t in China in 2023 [3], the carbon cost of tunnels and underground projects alone can reach millions of yuan. At the same time, relevant carbon emission assessment studies show that about 80% of carbon emissions come from the material production stage [4, 5], among which, concrete and steel bars account for the largest proportion of carbon emissions [6], mainly because they are used most in tunnels and underground engineering. It takes about 10,000 tons of concrete and 45 tons of steel to build 1 km of tunnel [7]. Therefore, the use of high-strength steel bars and low-carbon concrete in tunnels and underground projects is an effective measure to reduce carbon emissions [8].

Blast furnace slag serves as a ubiquitous supplementary cementitious material in the manufacture of low-carbon concrete. In certain existing studies, Pan et al. [9] have established that the performance of low-carbon concrete is optimized when specific parameters are met: water consumption is adjusted to 35%, cement content is maintained at 10%, and the ratio between steel slag and GGBS, as well as between fly ash and GGBS, is equalized at 1:1. Furthermore, this methodology promotes the efficient use of steel slag characterized by low reactivity and elevated doping concentrations. Sara et al. [10] employ fine particles derived from construction and demolition (C&D) waste as an alternative to natural sand and utilize GGBFS as a cementitious substitute. There is also a body of scholarly research [11] dedicated to the development of low-carbon concrete through the utilization of supplementary cementitious materials such as SF, FA and WGP, amongst others. Consequently, the adoption of alkali-activated slag as a substitute for cement in the production of concrete is both suitable and warranted. In addition, there are also scholars [12] who use recycled materials to work together with industrial waste to replace cement, thereby reducing CO_2 emissions. Chen et al. [13] undertook an investigation into an innovative formulation of low-carbon concrete that incorporates RCBP and GGBFS, with a potential cement replacement ratio of up to 20%. However, due to the volcanic ash reaction, the strength of concrete with higher substitution rate will be reduced. Previous research has suggested that incorporating recycled clay brick powder at concentrations ranging from 5 to 15% is appropriate, with a relatively low cement replacement rate [14]. Furthermore, studies on the integration of wood fiber into slag concrete are limited, and the underlying mechanisms of its effects remain unclear.

The novelty of this paper lies in the adoption of a specialized mixing technique to incorporate wood fiber into an existing ternary blend of slag-brick powder-cement concrete, as well as in investigating whether recycled wood fiber can enhance the application of RCBP. The principal objective of this research is to ascertain the feasibility of incorporating wood fiber into concrete compositions comprised of alkali-activated slag and RCBP, with the ultimate objective is to optimize the utilization of RCBP and reduce carbon emissions.

2 Materials and Methods

2.1 Raw Materials

The cement utilized in this research was classified as P · O 42.5 grade, while the slag consisted of S95 mineral powder produced by Henan Bangneng Metal Materials Co. Ltd. The alkali activator utilized in this study was sodium silicate, commonly referred to as water glass., characterized by a modulus of 1.65, sourced from a company in Henan Province, with the chemical formula $Na_2O \cdot SiO_2 \cdot 9H_2O$. Sodium hydroxide (NaOH) was employed as a modulating agent to modify the modulus of the water glass. The waste clay brick powder (RCBP) was obtained from commonly discarded red bricks generated from construction activities. These bricks were initially cleaned using a high-pressure water jet and then dried in an oven. Subsequently, they were crushed and sieved to achieve the desired particle size. Subsequently, the aggregate is subjected to grinding in a ball mill to yield a functional brick powder. Figure 1 and Table 1 illustrate the particle size distribution, chemical composition, and physical properties of the gelling materials, while Fig. 2 depicts the particle size distribution of RCBP.

Fig. 1 Morphology of raw materials (OPC, GGBFS, RS, RCBP, CA and WF) (self-drawn)

Table 1 Chemical composition of main cementitious materials (self-drawn)

Cementitious materials	SiO_2	AL_2O_3	Fe_2O_3	CaO	MgO	Na_2O	K_2O	TiO_2	SO_3	LOSS	Others
Cement	24.99	8.26	4.03	51.42	3.71	0.27	0.71	–	2.51	3.31	0.79
RCBP	59.88	22.34	6.84	1.26	2.12	1.57	2.66	–	0.63	–	2.7
GGBFS	36.23	9.76	1.99	39.4	10.5	–	–	0.7	1.42	–	–

Fig. 2 Particle size distribution of RCBP (self-drawn)

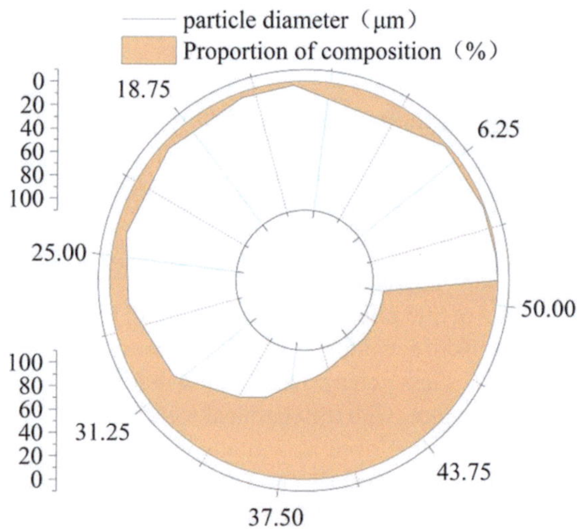

The compositional analysis of RCBP, conducted using XRD testing, is illustrated in Fig. 3. The primary crystalline component of Recycled Concrete Block Powder is quartz, with minor occurrences of feldspar and hematite. Semi-quantitative Rietveld refinement, employed as part of the XRD analysis methodology, indicated that the

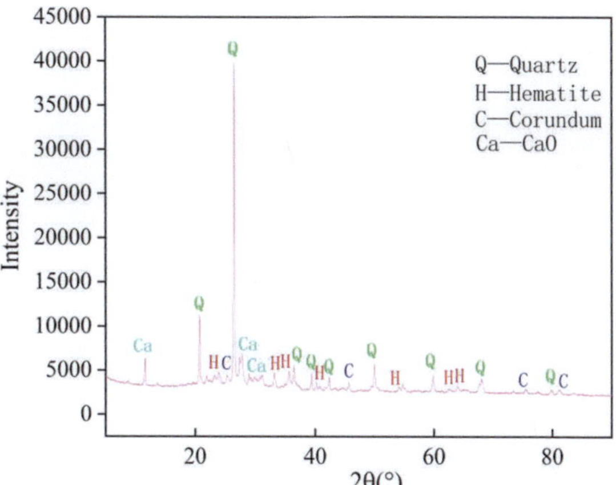

Fig. 3 Mineral composition of RCBP (self-drawn)

amorphous (glassy) phase comprises approximately 60% of the RCBP. Furthermore, in alignment with the national standard GB/T 1596, the reactivity index of the RCBP was ascertained to be 73.6%.

The coarse basalt aggregate exhibited a particle size distribution ranging from 5 to 10 mm. For concrete mixing, natural river sand was employed, which exhibited an apparent density of 2.65 g/cm^{3}and a fineness modulus of 2.4. The wooden fiber, sourced from recycled paper scraps, was obtained from Zhejiang Baofeng Mineral Processing Factory and had a fineness of 2.02 dtex. The water-reducing agent utilized in this study was a liquid polycarboxylate superplasticizer.

2.2 *Preparation of LCC*

GGBFS and RCBP were the main cementitious materials. According to the basis of the existing research [13] and the purpose of this study, a control group was set up, and no other admixtures were only prepared by pure Portland cement. The mass ratio of water to solid was 0.12 [15], and the alkali content of the activator was quantified as 4% based on the mass ratio of Na_2O to the binder [16]. In addition, five groups of experimental groups were set up, a total of five concrete test blocks, and R20-G60 was used to represent the proportion of RCBP and GGBFS in gel materials (self-drawn) (Table 2 and Fig. 4).

All raw materials were combined at ambient temperature, specifically 24 ± 2 °C. Figure 5 depicts the procedure for specimens preparation: dried OPC and RCBP were blended, then mixed with the prepared alkali activator and GGBFS, soaked for 5 min. Subsequently, the aforementioned composite mixture was placed into the mixing apparatus and subjected to agitation for a duration of two minutes. Subsequently, the superplasticizer (PS) along with the remaining aqueous component was added, and the mixture was subjected to blending for an additional duration of ten minutes

Table 2 Proportions of components in concrete mixtures(kg/m^3)

Samples	RCBP	GGBFS	WF	Cement	Water	River sand	Coarse aggregate	Polycarboxylate superplasticizer
OPC	0	0		400	264	680	1120	0.168
R0-G80	0	320	0	80	264			
R20-G60-1	80	240	0	80	264			
R20-G60-2	80	240	10	80	264			
R20-G60-3	80	240	20	80	264			
R30-G60	120	240	30	40	264			

Note R30-G60 indicates that RCBP accounts for 30% of cementitious materials and GGBFS accounts for 60%

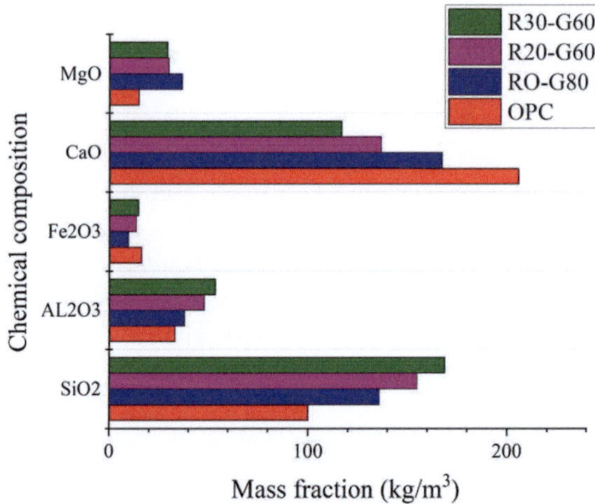

Fig. 4 Mix proportions of LCC (weight ratio percentage) (self-drawn)

until a fluid consistency was attained. Thereafter, the pre-measured coarse and fine aggregates were incorporated into the mixture and subjected to mixing five minutes. Finally, WF was incorporated into the mixture, and the blending process persisted until an optimal dispersion of the fiber was achieved. The resulting slurry was then poured into molds measuring 50 × 50 × 50 mm. After completing a three-day curing phase, the molds were disassembled, and experimental evaluations were conducted at curing durations of 7 and 28 days, respectively (Fig. 6).

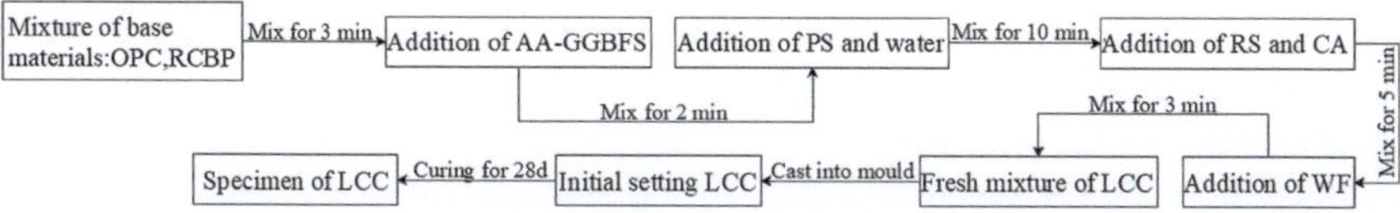

Fig. 5 Mixing procedure for LCC specimens (self-drawn)

Fig. 6 Morphology of LCC specimens (self-drawn)

2.3 *Test Methods*

Compressive Strength

Concrete samples in the form of 400 mm cubes were utilized for this test, employing a 300 kN electro-hydraulic servo-controlled compression apparatus operating at a velocity of 1 kN per second. The curing duration for the test samples was established at 7 and 28 days, respectively. The strength of the samples discussed in this study is represented as the mean value derived from three individual specimens.

Scanning Electron Microscopy (SEM)

The microstructural characteristics of each series of fragments were analyzed utilizing a German ZEISS GeminiSEM 300 scanning electron microscope. Before conducting the analysis, the specimens were thoroughly dried. A small quantity of the sample, whether in the form of a block or a thin film, was affixed to a conductive adhesive. Subsequently, the sample underwent a gold-plating process for a duration of 45 s. Subsequent to this preparatory phase, the ZEISS GeminiSEM 300 was utilized to obtain images of the sample's morphology and to conduct energy spectrum mapping. During the imaging of the morphology, an acceleration voltage of 3 kV was utilized, while for the energy spectrum mapping, an acceleration voltage of 15 kV was applied.

X-ray Diffraction (XRD)

At the specified curing age, hydration arrest of the samples was achieved by substituting the solvent with anhydrous ethanol. Subsequently, the specimens were carefully pulverized into a fine powder utilizing an agate mortar, while anhydrous ethanol was utilized to prevent carbonation. The powdered specimens were then thoroughly dried for a minimum duration of five days before being sieved through a 75 μm sieve. The scanning range for the subsequent analysis was established between 5° and 90°, with a step increment of 0.02°. The analytical instrument utilized was a Rigaku Ultima IV, sourced from Japan. This instrument utilized Cu-Kα radiation as the incident beam, functioning at a tube voltage of 40 kV and a tube current of 40 mA. Ultimately, the quantitative determination of the product phases was conducted using the Rietveld refinement technique within the Topas software framework [15].

TG Analysis

The hydration products were analyzed by Swiss Mettler TGA/DSC1 thermogravimetric analyzer. The procedure for preparing the TG powder sample is identical to that utilized for the XRD sample. In a nitrogen-rich atmosphere, the specimen was heated from ambient temperature 30–500 °C.

3 Results and Discussion

3.1 Compressive Strength

The compressive strength test was utilized to evaluate the variations in the strength among different concrete mixtures. The findings are illustrated in Figs. 7 and 8, which present the compressive strengths of each specimen and the compressive strength ratios of each specimens relative to ordinary Portland cement concrete, respectively. As indicated in Fig. 7, the compressive strengths of all tested samples exhibited a gradual increase with advancing curing age. Notably, the concrete mixture containing 80% Ground Granulated Blast-Furnace Slag (GGBFS) alone did not achieve strength levels comparable to those of the concrete mixture that included 20% Recycled Concrete Brick Powder (RCBP) and 60% GGBFS. The addition of Wood Fiber (WF) was found to enhance the later-age strength of concrete with a consistent RCBP content, although it appeared to diminish the early-age strength. In comparison to the R20-G60-1 mixture, an increase in the mass ratio of WF to RCBP to 1/4 resulted in improvements in both early-age and later-age strengths, with increases ranging from 5 to 15%. By preserving the mass ratio, an elevation in the replacement level of Recycled Concrete Block Powder (RCBP) was associated with a reduction in the strength of the specimens, indicating an inverse relationship. In the early phases, RCBP predominantly served as a filler material; consequently, as the percentage of cement substituted with RCBP increased, the overall strength of the concrete decreased. At ambient temperature, the reactive silica (SiO_2) and alumina (Al_2O_3) present in the early-added RCBP reacted with calcium hydroxide (CH) and water, resulting in the synthesis of silicate hydrates. As the curing period progressed, the pozzolanic activity of RCBP became increasingly effective, and the incorporation of WF improved the adhesion between hydration products and aggregates, which subsequently led to an enhancement in the strength of the concrete at later ages.

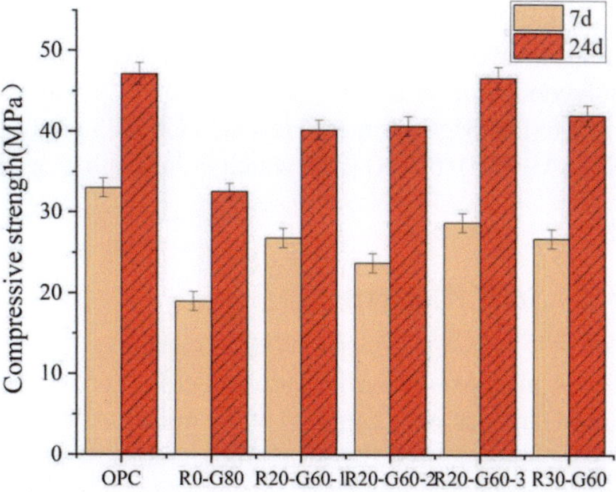

Fig. 7 Compressive strength of concretes at 7 and 28 d (self-drawn)

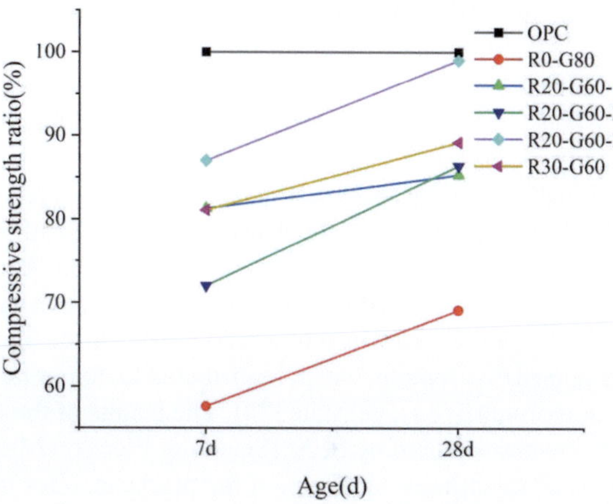

Fig. 8 Compressive strength ratios of concretes at 7 and 90 d (self-drawn)

From the perspective of the chemical composition depicted in Fig. 3, RCBP characterized as a low-calcium aluminosilicate material, predominantly comprises SiO_2 and Al_2O_3. When NaOH and Na_2SiO_3 serve as alkaline activator solutions, NaOH holds a pivotal role in the dissolution of Si^{4+} and Al^{3+} ions, as well as in the alkaline activation process [17] (with the specific process outlined in Fig. 13). Conversely, Na_2SiO_3 functions to release silicate molecules into the system, thereby contributing to the formation of sodium aluminum silicate hydrate (NASH) gel [18]. In relation to the hydration of gelling materials in an alkaline milieu, research [19] has demonstrated that, when the CaO fraction is set at 0.1 or 0.2, an augmentation in the Si/Al fraction demonstrates a pattern of initially elevating and subsequently diminishing the compressive strength of LCC. Regarding the impact of the SiO_2/Al_2O_3 ratio on the mechanical robustness of the resultant product, the optimal proportion falls within the range of 3–4 [20]. The alkaline hydration process of GGBFS, which is commonly employed as an industrial waste material for the reduction of carbon emissions, involves several stages, including the initial disintegration of GGBFS and the polycondensation of the resultant reaction products. Importantly, the high magnesium oxide (MgO) content present in GGBFS promotes the synthesis of hydrotalcite ($Mg_6Al_2(OH)16CO_3$) [21], which in turn enhances the early-age strength of LCC.

3.2 SEM Analysis

To visually observe the hydration process of concrete, Scanning Electron Microscope (SEM) images were captured at the micro-scale.

Figure 9 showed the SEM micrographs of ordinary silicate concrete and other experimental groups after adding different WF. Comparing Fig. 9a and b, the surface of OPC mainly presents a stacked morphology, which is related to the accumulation of calcium carbonate. Since more Calcite plays a positive role in the structure after OPC hydration [22], there are flocculent structures and shale states on the surface of Fig. 9b, which is similar to some research results [23]. In Fig. 9c and d, there are dense network structures, but small pores can be observed. Especially, as shown in Fig. 9c, larger pores are also observed, but with the increase of WF, the pores gradually decrease and the structure tends to be dense.

To examine the elemental distribution of the synthesized products, Energy Dispersive Spectroscopy (EDS) analysis was performed. As shown in Fig. 10, Ca and Si are the primary phases, with calcium and silicon phases widely distributed and exhibiting an accumulative state. In contrast, Al and Mg in the figure are present in a punctate and dispersed distribution, which is attributed to their relatively low content and manifestation primarily in a gel state [24]. The results of the EDS analysis corroborate with the images captured by SEM (Scanning Electron Microscope). For specific qualitative and quantitative analyses of the products, refer to the XRD (X-ray Diffraction) analysis.

3.3 XRD Analysis

To investigate the composition of hydration products among various groups in the past, XRD experiments were conducted on each group.

Figure 11 previously illustrated the XRD patterns corresponding to the experimental groups. The principal hydration products identified within the concrete of each group were Albite, Hatrurite, and Gismondine. In addition, minor amounts of unhydrated C3S and C2S were detected at approximately 32°, aligning with findings reported in some previous studies [25]. It was noteworthy that Quartz in the quartz phase was observed in the XRD patterns irrespective of whether RCBP was incorporated, with the highest diffraction peak intensity occurring near 28°. As the replacement rate of Recycled Ceramic Brick Powder (RCBP) increased, there was

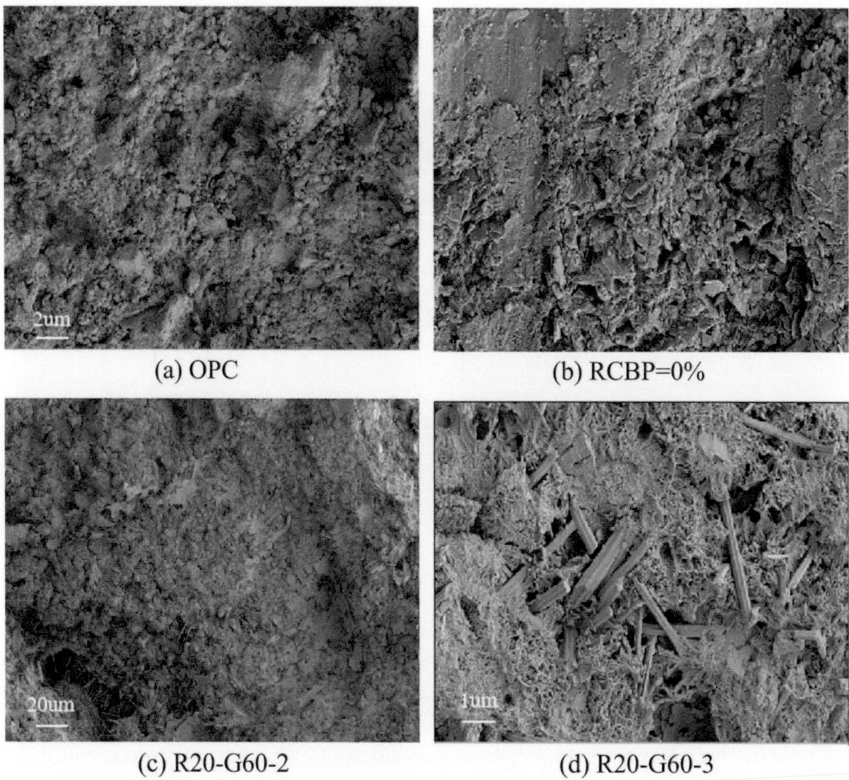

(a) OPC (b) RCBP=0%

(c) R20-G60-2 (d) R20-G60-3

Fig. 9 SEM morphology of the experimental group (self-drawn)

a corresponding rise in the intensity of the diffraction peaks related to the quartz phase. The incorporation of RCBP resulted in an elevated concentration of Albite, as previously illustrated in Fig. 3, which can be ascribed to the higher levels of Corundum present in RCBP. Conversely, with the introduction of Wood Fiber (WF), a decline in the diffraction peak intensity of Albite was observed, while the intensity of the calcium phase exhibited a gradual increase. This suggests that WF may have a facilitating effect on the hydration process. To quantitatively assess the differences in hydration products of slag-based pastes containing RCBP after the incorporation of WF, the rietveld refinement method [26] was employed in previous studies to calculate the hydration product content for each experimental group, as illustrated in Fig. 12.

On the seventh day, compared to other experimental groups, the OPC exhibited the highest content of Calcite, which accounted for its higher early-age strength [27]. As the cement was progressively replaced by GGBFS and RCBP [28], the hydration products gradually transformed into N-A-S-H gel [29]. This product exhibited a high porosity coupled with a low elastic modulus [30], which contributed to a decrease in the early-age strength of LCC. The incorporation of wood fibers (WF), owing to

Fig. 10 EDS morphology of the experimental group (self-drawn)

their capacity for water absorption, extended the setting time of the concrete and facilitated a more thorough diffusion of alkaline-reactive materials. As the curing period progressed, there was a corresponding increase in the C-A-S-H gel, resulting in a progressive elevation of the calcium-to-silicon (Ca/Si) ratio within the system. The calcium-rich gel was characterized by low porosity [31] and, in combination with N-A-S–H, provided excellent mechanical properties to the structure [32], making the structural network more stable.

Figure 13 showed the movement of ions during the dissolution process of gel materials under alkaline conditions. From left to right, the initial step involved the exchange of H^+ with Ca^{2+} and Na^+ present in the system. Following this, the hydrolysis of Al–O–Si bonds disrupted the depolymerized network, ultimately resulting in the release of Si and Al, which then formed hydrated anions that combined with cations to produce a gel. As the curing age progressed, the proportion of calcium-rich gel gradually increased, leading to the formation of a dense structural network [33].

3.4 TG Analysis

Figure 14 showed the TG results of three groups of R20-G60-1, R20-G60-3 and R30-G60. Upon examination of the figure, it was observed that the initial structural loss in the R20-G60-3 group was minimal due to the more dense structure of its hydration products, which correlated with the previously analyzed role of wood fiber (WF). However, at temperatures ranging from 290 to 300 °C, the lightweight cementitious

Fig. 11 XRD patterns of pastes (self-drawn)

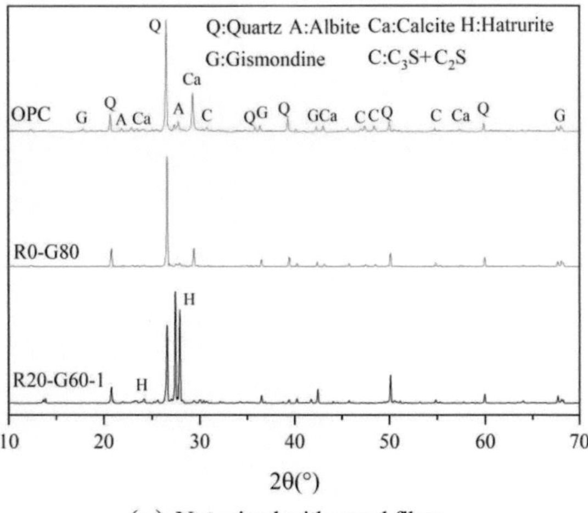

(a) Not mixed with wood fiber

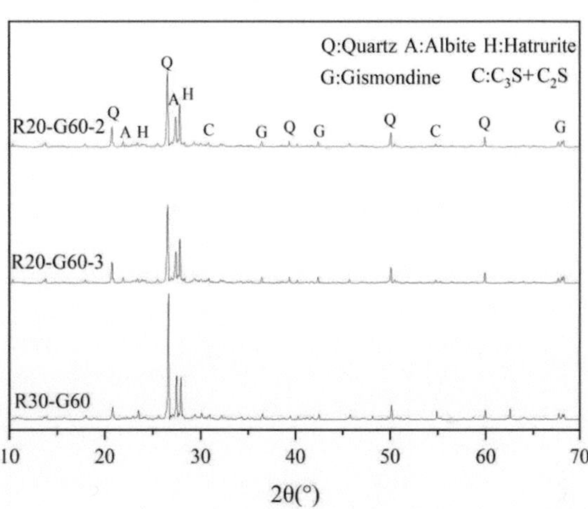

(b) Mixed with wood fiber

composites (LCC) with added WF exhibited more pronounced fluctuations, potentially related to the loss of calcium hydroxide (CH) partially bound to the WF [34]. Additionally, the figure also revealed that incorporating up to 30% of recycled crushed brick powder (RCBP) remained a challenge to the thermal stability of the structure. Based on the thermogravimetric analysis findings, the optimal cement replacement ratio, without compromising the structural stability, was determined to be 80%, with RCBP comprising 20% and the remaining portion consisting of GGBFS, forming a ratio of cement, GGBFS, and RCBP at 1:3:1.

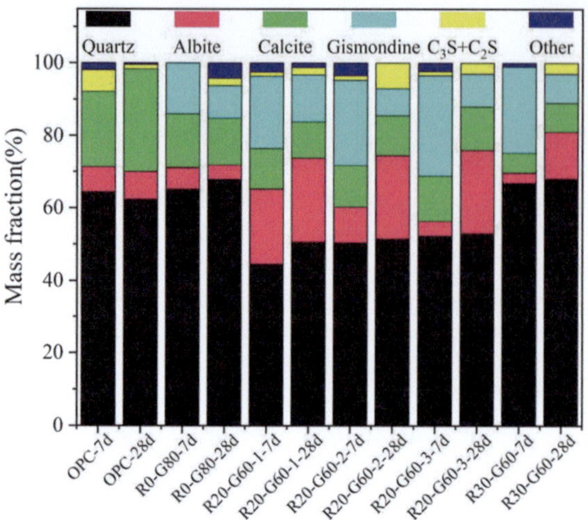

Fig. 12 Quantitative examination of the progression of hydration products (self-drawn)

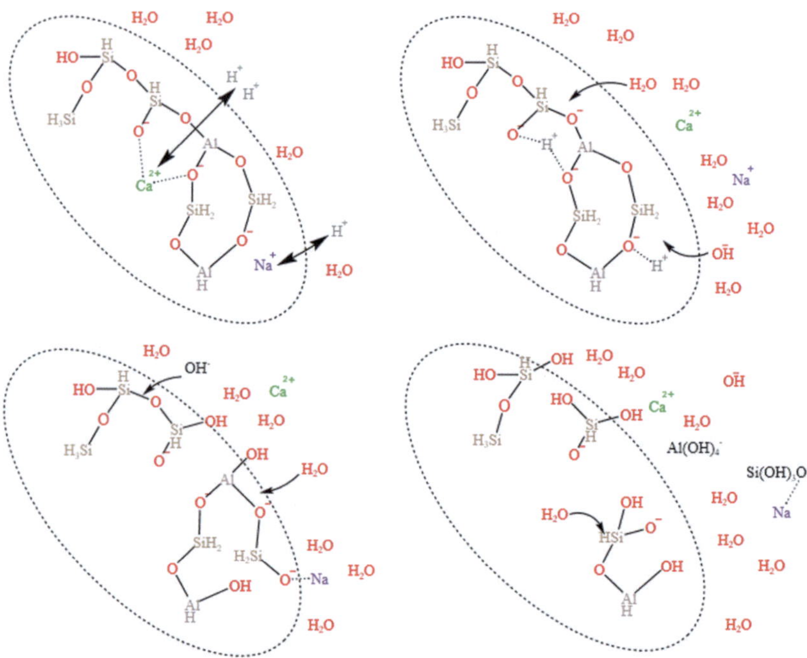

Fig. 13 Hydration ion movement of gel materials under alkali activation state (self-drawn)

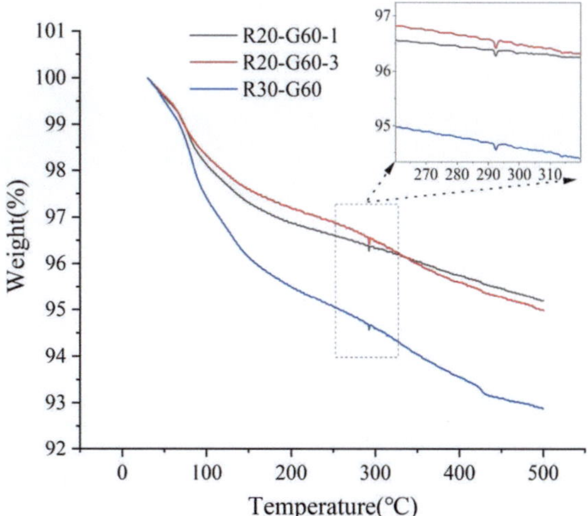

Fig. 14 TGA analysis of pastes at 28d (self-drawn)

3.5 Cost and Carbon Emissions Analysis

Through cost and carbon emission analyses, a comprehensive assessment of the viability of employing RCBP and GGBFS in the formulation of mortar/concrete can be performed, while concurrently determining the optimal cement replacement ratio. In this study, concrete with varying admixtures of RCBP and GGBFS was selected as the research subject for the purpose of conducting cost and carbon emission analyses. All expenses/carbon emissions were assessed in relation to the production of one cubic meter of concrete.

3.5.1 Cost Analysis

The cost analysis primarily took into consideration the prices of raw materials and transportation costs. According to the data published by the Price Bureau of Chongqing in 2024, the cost of cement was recorded at 460 yuan per ton, while the prices of standard sand and gravel, blast furnace slag, water, and waste cardboard in the same year were 240, 100, 3.5 yuan/t. As previously mentioned, RCBP being obtained from waste, was provisionally priced at 0 yuan. The transportation costs using a 10-ton gasoline truck and a diesel truck were both set at 0.6 yuan per ton per km [35]. The transportation distance was assumed to be 30 km [36].

The raw material prices can be calculated as follows [37]:

$$C_{\mathrm{R}} = \sum_{i=1}^{i=8} (M_i \times R_i) \tag{1}$$

The transportation costs can be calculated as follows [37]:

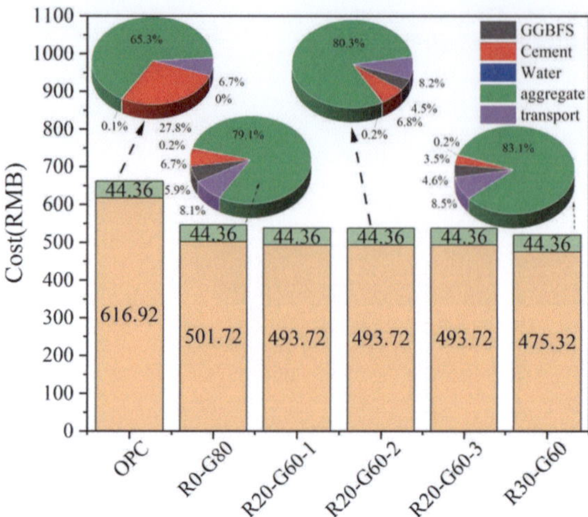

Fig. 15 Cost of experimental composition (self-drawn)

$$C_Y = \sum_{i=1}^{i=8} (M_i \times D_i) \tag{2}$$

where C_R is the total cost of raw materials, M_i is the consumption quantity of the *i*th material, R_i signifies the unit price of the *i*th material, C_Y is the transportation cost of raw materials, and D_i is the unit price per unit weight per kilometer of transportation distance.

Figure 15 showed the cost expenditures of the various experimental groups established in this study. In terms of total cost, OPC incurred the highest expense, whereas R30-G60 incurred the least. It was observed that the cost of aggregates was the highest, accounting for 27.8% of the cost of OPC cement, whereas the cost of gelling materials in the other groups was lower than the transportation cost. When the cement replacement rate was established at 80%, the cost of producing concrete using different low-carbon materials was 18–20% lower than that of OPC. When the replacement rate attained 90%, the cost could be decreased by 23%. However, the overall benefits of utilizing materials such as RCBP as partial substitutes for cement in concrete preparation should not solely have been focused on cost savings but should also have considered the changes in concrete strength and other properties for assessment [38].

3.5.2 Carbon Emissions Analysis

The emissions of carbon dioxide were calculated according to the current national standards GB/T 24,040, "Principles and Framework for Life Cycle Assessment of Environmental Management," and GB/T 24,044, "Requirements and Guidelines for Life Cycle Assessment of Environmental Management."

Table 3 Carbon emission coefficient of raw materials (self-drawn)

Materials	Carbon emission factor ($kgCO_2/kg$)	References
Slag	0.0520	[39]
Cement	0.7350	[37]
Water	1.68×10^{-4}	[37]
RCBP	0.075	[40]
Wood fiber	1.43×10^{-3}	[41]
Natural river sand (f = 2.4)	2.51×10^{-3}	[37]
basalt coarse aggregate	5.08×10^{-3}	[37]
Polycarboxylate superplasticizer	0.72	[42]

The emissions from various types of concrete during production can be calculated as follows [39]:

$$C_F = \sum_{i=1}^{i=8} (M_i \times CO_{2i}) \tag{3}$$

where C_F denotes the carbon dioxide (CO_2) emissions associated with concrete, CO_{2i} represents the CO_2 emissions per unit mass originating from each constituent of the concrete (referred to as the Carbon Emission Factor, as detailed in Table 3). Sustainable concrete consists of water, cement, slag, recycled crushed brick particles (RCBP), fine aggregate, wood fiber, coarse aggregate, and a water-reducing agent, specifically Polycarboxylate superplasticizer.

Figure 16 showed the quantities of carbon emissions associated with the production of one cubic meter of concrete across various experimental groups. The graphical representation clearly indicated that the predominant contributor to carbon dioxide emissions was the application of Portland cement [43]. The incorporation of low-carbon materials, including slag, RCBP, and WF, demonstrated potential for substantially mitigating these emissions [44]. In the context of this investigation, the utilization of pure Portland cement for the production of one cubic meter of concrete yielded CO_2 emissions as substantial as 303.56 kg. Two distinct cement replacement ratios, namely 80 and 90%, were established within the experimental groups, yielding efficiencies of 72 and 80% respectively in terms of reducing carbon dioxide emissions.

3.5.3 Comprehensive Analysis

To evaluate the specific feasibility of the experimental groups, a comprehensive consideration of three factors—compressive strength, cost-effectiveness, and carbon emissions—was undertaken [36]. Using OPC as a reference, the changes in these influencing factors across different experimental groups were calculated to obtain

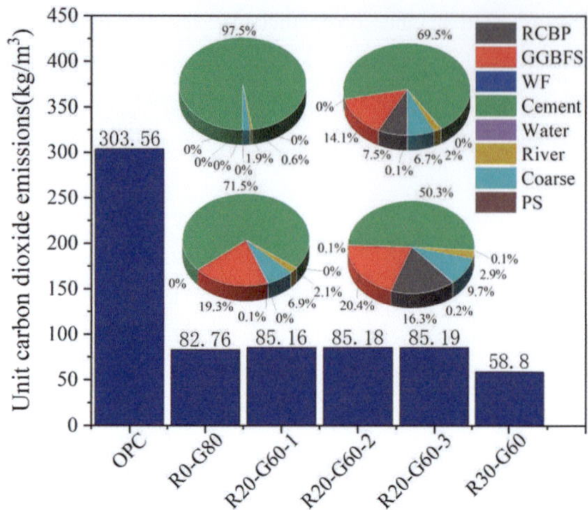

Fig. 16 Unit carbon emissions of each experimental group (self-drawn)

a low-carbon concrete mix that met both strength requirements and environmental benefits.

As depicted in Fig. 17, negative values were assigned to the changes in compressive strength due to their detrimental effects on the system, while positive values were assigned to the positive impacts of cost savings and reductions in carbon emissions achieved through the adoption of low-carbon materials. Subsequently, an overall evaluation of the system was conducted. The figure indicates that R20-G60-3 exhibited the most favorable overall indicators, followed by R30-G60. The group with only GGBFS added showed the least favorable results. The remaining groups deviated by approximately 30% from the optimal solution. These results align with earlier analyses of mechanical property and microstructural analysis results, providing mutual corroboration.

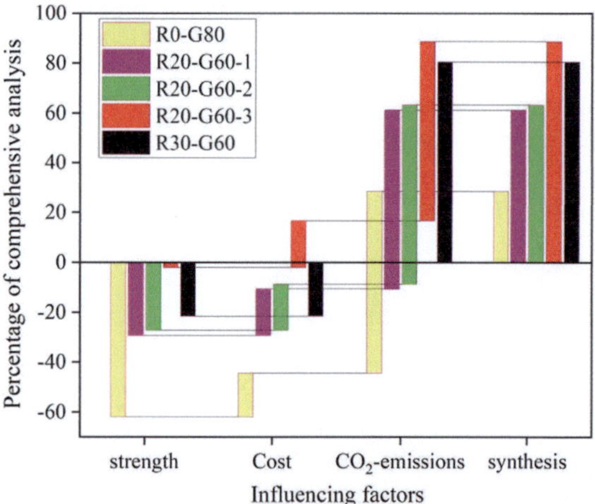

Fig. 17 Comprehensive analysis results of experimental group (self-drawn)

4 Conclusion

This study provided a comprehensive evaluation of the alkali-activated-slag-based concrete mixed with wood fiber and RCBP, focusing on its microstructure, mechanical properties and environmental impact. The incorporation of waste fibers (WF) has the potential to prolong the solidification duration of low-calcium cement (LCC), which is beneficial for the transport of alkali-activated materials. An increase in the content of recycled concrete block powder (RCBP) correlates with an enhancement in the early compressive strength of LCC; however, its influence on the later strength is minimal. This study identifies an RCBP content of 20% as optimal. Achieving a higher substitution rate is constrained by the physical properties of RCBP, which is rich in quartz, leading to an increased presence of quartz phase substances in the LCC that do not participate in the reaction. The primary hydration products within the alkali-activated slag system are identified as albite and gismondine. The complementary properties of these minerals enhance the structural strength. The appropriate incorporation of WF can further improve the structural compactness and facilitate the development of calcium-rich gel. Furthermore, the materials utilized in this study, namely WF, RCBP, and GGBFS, are characterized as low-carbon alternatives, the experimental group with the highest feasibility can decrease CO_2 emissions by as much as 70%, which is extremely beneficial for environmental protection.

Acknowledgements This study is supported by the funding of Chongqing Science and Technology Bureau (2023TIAD-GPX0066) and Chongqing Design Group CP., Ltd. (Research and application of key technologies regarding digitization and industrialization on municipal engineering oriented low carbon).

Declaration of Competing Interest The authors affirm that they possess no identifiable financial conflicts of interest or personal relationships that may have had the potential to affect the research presented in this manuscript.

References

1. Zeng H, Sun W, He W et al (2023) Study on the carbon emission prediction model for railway tunnel construction based on machine learning. Mod Tunn Technol 60(6):29–39
2. Liu M, Jia S, He X (2018) A quota-based GHG emissions quantification model for the construction of subway stations in China. J Clean Prod 198:847–858
3. Zhang X (2023) National Carbon Market to Transact $14.444 bn annually in 2023, with Gradual Expansion of Trading Scale (31 December 2023) [11 April 2024]
4. Yao J (2024) Research on carbon emission accounting & estimation indicators during the construction period of shield tunnels on urban roads. China Municipal Eng (1):94–98+103
5. Wang D, Bi Y, Wang C et al (2024) Accounting and analysis of carbon emission in metro TBM tunnel: a case study of Qingdao Metro. Build Energy Effi 52(2):136–142
6. Chen K, Chen Xiangsheng, Wang L et al (2023) Greenhouse gas emissions of open-cut metro station in construction phase: a case study of a metro station on Shenzhen urban rail transit line 12. Tunn Constr 43(S1):189–197
7. China Concrete & Cement-Based Products Association (2023) China's low carbon concrete in action: concrete "Shield" for China's Metro "Sinews" (8 October 2023) [1 March 2024]

8. Chen X, He Q, Wu Z et al (2024) Low-carbon sustainable development status and thinking of tunnel and underground engineering. Mod Tunn Technol 61(02):5–17
9. Pan H et al (2024) Preparation of low-carbon and environmentally friendly non-sintered ceramsite (NSC) employing steel slag, GGBS and fly ash: experiment and performance regulation. Constr Build Mater 411:134438
10. Sara B et al (2023) Elaboration of a self-compacting mortar based on concrete demolition waste incorporating blast furnace slag. Constr Build Mater 366:130165
11. Yan P et al (2024) Study on mechanical properties and microstructure of green ultra-high performance concrete prepared by recycling waste glass powder. J Build Eng 82:108206
12. Santorsola J, Butler LJ (2023) Material behaviour and flexural performance of low carbon concrete beams containing very high quantities of recycled and secondary materials. Constr Build Mater 407:133350
13. Chen G, Li S, Liu C et al (2024) Rheological behavior and early-age hydration performance of blended cement pastes incorporating recycled brick powder and slag. Constr Build Mater 436:136925
14. He Z et al (2021) Research progress on recycled clay brick waste as an alternative to cement for sustainable construction materials. Constr Build Mater 274:122113
15. Ye H et al (2024) Mechanical performance, impact behavior and environmental assessment of coal furnace slag based low-carbon ultra-high performance engineered cementitious composites (UHP-ECC). J Clean Prod 450:141921
16. Zheng X et al (2024) Performance evaluation and multi-objective optimization for alkali-activated slag concrete with recycled aggregate. Constr Build Mater 411:134318
17. Hwang C-L et al (2019) Performance evaluation of alkali activated mortar containing high volume of waste brick powder blended with ground granulated blast furnace slag cured at ambient temperature. Constr Build Mater 223:657–667
18. Thunuguntla CS, Gunneswara Rao TD (2018) Effect of mix design parameters on mechanical and durability properties of alkali activated slag concrete. Constr Build Mater 193:173–188
19. Xu S et al (2021) Development and preliminary mix design of ultra-high-performance concrete based on geopolymer. Constr Build Mater 308:125110
20. Garcia-Lodeiro I et al (2014) Alkaline activation of synthetic aluminosilicate glass. Ceram Int 40(4):5547–5558
21. Fu Q et al (2023) Hydration characteristics and microstructure of alkali-activated slag concrete: a review. Engineering 20:162–179
22. Sun C, Zhang J, Yan C et al (2022) Hydration characteristics of low carbon cementitious materials with multiple solid wastes. Constr Build Mater 322:126366
23. Zhang X, Li H, Wang H et al (2024) Properties of RCA stabilized with alkali-activated steel slag based materials in pavement base: laboratory tests, field application and carbon emissions. Constr Build Mater 411:134547
24. Esteves LP (2011) On the hydration of water-entrained cement–silica systems: combined SEM, XRD and thermal analysis in cement pastes. Thermochim Acta 518(1–2):27–35
25. Yao X et al (2009) Geopolymerization process of alkali–metakaolinite characterized by isothermal calorimetry. Thermochimica Acta 493(1–2):49–54
26. Chen G et al (2023) Hydration and microstructure evolution of a novel low-carbon concrete containing recycled clay brick powder and ground granulated blast furnace slag. Constr Build Mater 386:131596
27. Sirotinkin V, Bush A, Kozlov V (2023) Structural study of ceramic samples of the $PbTiO_3$–$BaTiO_3$–$BaZrO_3$ system with a high $PbTiO_3$ content studied by the Rietveld method. Zeitschrift für Kristallographie-Crystalline Materials 238(1–2):39–46
28. Li L, Cao M, Yin H (2019) Comparative roles between aragonite and calcite calcium carbonate whiskers in the hydration and strength of cement paste. Cement Concr Compos 104:103350
29. Vyšvařil M, Rovnaníková P, Keppert M (2018) Rheological properties of alkali-activated brick powder based pastes: effect of alkali activator and silicate modulus. Solid State Phenom 276:185–191

30. Fang G, Zhang M (2020) Multiscale micromechanical analysis of alkali-activated fly ash-slag paste. Cem Concr Res 135:106141
31. Zhu Y, Longhi MA, Wang A et al (2022) Alkali leaching features of 3-year-old alkali activated fly ash-slag-silica fume: for a better understanding of stability. Compos B Eng 230:109469
32. Ismail I et al (2014) Modification of phase evolution in alkali-activated blast furnace slag by the incorporation of fly ash. Cem Concrete Compos 45:125–135
33. Rafeet A et al (2019) Effects of slag substitution on physical and mechanical properties of fly ash-based alkali activated binders (AABs). Cem Concrete Res 122:118–135
34. Deboucha W, Leklou N, Khelidj A et al (2017) Hydration development of mineral additives blended cement using thermogravimetric analysis (TGA): methodology of calculating the degree of hydration. Constr Build Mater 146:687–701
35. Peng Y, Zhao ZX, Xiao YP (2023) Collaborative distribution route optimization of truck and subway based on data-driven (In Chinese). Electron Des Eng 31:43–148+154
36. Zhang H et al (2023) Recycling fine powder collected from construction and demolition wastes as partial alternatives to cement: a comprehensive analysis on effects, mechanism, cost and CO_2 emission. J Build Eng 71:106507
37. GB/T 51366-2019 (2019) Building carbon emission calculation standard. China Building Industry Press, Beijing (in Chinese)
38. Wu Y, Liu C, Liu H et al (2022) Pore structure and durability of green concrete containing recycled powder and recycled coarse aggregate. J Build Eng 53:104584
39. Li B, Yu ST, Gao BH et al (2023) Effect of recycled aggregate and steel fiber contents on the mechanical properties and sustainability aspects of Alkali-activated slag-based concrete. J Build Eng 66:105939
40. Dai T et al (2024) Effect of waste clay brick powder on microstructure and properties in blended oil well cement pastes at HTHP conditions. Geoenergy Sci Eng 237:212823
41. Ayodele TR, Alao MA, Ogunjuyigbe ASO (2018) Recyclable resources from municipal solid waste: assessment of its energy, economic and environmental benefits in Nigeria. Resour Conserv Recycl 134:165–173
42. Shi Y, Long GC, Ma C et al (2019) Design and preparation of ultra-high performance concrete with low environmental impact. J Clean Prod 214:633–643
43. Miller SA, John VM, Pacca SA et al (2018) Carbon dioxide reduction potential in the global cement industry by 2050. Cem Concr Res 114:115–124
44. Şenol AF, Karakurt C (2024) High-strength self-compacting concrete produced with recycled clay brick powders: rheological, mechanical and microstructural properties. J Build Eng 88:109175

Sun-Tracking Solar Intelligent Shading Systems with Embedded Photovoltaics for Enhancing Building Energy Efficiency: A Case Study of UK Educational Buildings

Elliot Machin, Deo Prasad, Baojie He, Xiaohong Chen, and Siliang Yang

Abstract Given the escalating impact of climate change coupled with rising carbon emissions, there is an urgent need for emerging architectural technologies that redefine the design of sustainable building solutions. Solar energy represents a ubiquitous source of sustainable and renewable energy, while shading facades represents a passive cooling strategy by reducing solar heat gain. By investigating the intricate interplay between photovoltaic electricity generation, intelligent shading and solar tracking, this paper investigates how these intelligent systems affect solar gain, human comfort and productivity, and environmental impact. This study comparatively analyses the effect on building energy and photovoltaic loads by analysing key differences of building performance without shading, with fixed overhang shading and intelligent shading embedded with photo-voltaic cells with sun-tracking capabilities. There were significant performance differences that identified the benefits of using sun -tracking photovoltaic shading devices (PVSDs). The study focused on the running simulations within the canteen area of an educational building and the results

E. Machin
School of Built Environment, Engineering and Computing, Leeds Beckett University, Leeds, UK

D. Prasad
School of Built Environment, University of New South Wales, Sydney, Australia

B. He
Centre for Climate–Resilient and Low–Carbon Cities, School of Architecture and Urban Planning, Ministry of Education, Chongqing University, Chongqing, China

Key Laboratory of New Technology for Construction of Cities in Mountain Area, Ministry of Education, Chongqing University, Chongqing, China

School of Architecture, Design and Planning, University of Queensland, Brisbane, Australia

X. Chen
Build Planet Zero, Northumbria University, Newcastle Upon Tyne, UK

S. Yang (✉)
Mott MacDonald, Leeds, UK
e-mail: siliang.yang@mottmac.com

B. He et al. (eds.), *Climate-Resilient, Low-Carbon, and Sustainable Cities*, Lecture Notes in Civil Engineering 754, https://doi.org/10.1007/978-981-95-2169-2_33

proved the sun-tracking PVSD compared with no shading, reduced internal temperatures up to 24.15% through mitigating solar gain. Furthermore, a reduction of overall cooling load requirement from 6.36 to 4.05 kW to maintain a setpoint temperature of 23 °C in summer. Additionally, embedding photovoltaics onto the shading devices enhanced daylight control whilst ensuring minimum daylight factors of 2% could be achieved. Overall, the shading devices on the case study building contributed to a total energy generation of 42.54 MWh annually.

Keywords Building physics · Shading · Solar technologies · Building energy performance · IES VE

1 Introduction

Building facades are crucial for reducing building energy consumption [1]. Intelligent facades are generally associated with dynamic shading systems, solar technologies, sensors, controls, and sustainability in conjunction with energy efficiency [2, 3]. The major focus of these systems is to reduce energy consumption, optimise building performance, whilst most of all promoting human comfort [4]. Natural daylight through building envelopes boasts positive impacts on the productivity of occupants within buildings especially for educational environments [5]. One investigation study by Heschong Mahone Group found that students with the most daylighting within classrooms progressed 26% faster on reading tests and 20% faster on mathematic tests with those with the least; generally, implementing natural light into spaces increased test scores by up to 20% [6]. Considering this, it is evident to install sufficient glazing while limiting solar gain without the obstruction of reducing natural light to prevent overheating [7]. Limiting solar gain can be achieved through higher G-values of glazing but this reduces natural lighting levels and alters the appearance of the building [8].

In recent years, there are novel shading types that buildings can use, such as fixed-shading and sun-tracking shading [9–11]. Fixed-shading devices (FDs) are strategically designed to reduce solar gain while allowing natural light to penetrate into the interior of the building through glazing [12]. However, the main disadvantage of FDs is that natural lighting can be limited due to over-excessive solar protection [13]. Sun-tracking capability is possible through MPPT (maximum power point tracking) by using an algorithm to find the maximum power curve through the use of the embedded photovoltaic panel [14, 15]. This innovative emerging technology (that is, sun-tracking solar intelligent shading) stands as a cornerstone in the industry of building and construction that offers a multifaceted range of benefits with limited re-search [16]. A key benefit of sun-tracking solar intelligent shading (STSIS) is its contribution to enhancing building energy efficiency [17]. This can be achieved through the combination of shading functions with solar energy generation [18]. The integrated system can effectively manage solar heat gain whilst harvesting renewable electricity [19–21], which is known as building-integrated photovoltaic [22].

Although building-integrated PV may affect the urban temperature [23], the STSIS still has notably positive environmental impact, for example it improves indoor comfort through offering better control on internal temperatures, reducing internally absorbed solar gain and creating high levels of indoor comfort [24, 25]. Moreover, integrating shading devices with photovoltaic panels represents an innovative blend of technology and architectural design, which has not been well studied [26].

The core emphasis of this paper investigates the STSIS, therefore the creation of an integrated photovoltaic shading device (PVSD). With continuous revisions to the UK Building Regulations Part L (Conservation of Fuel and Power) and the shift toward sustainable buildings, new intelligent emerging technologies are required to stay in line with new demands [27]. There are six sections in this paper. Section 1 (Introduction) specifies the context of the study and the research aims. Section 2 (Materials and Methods) spells out the research methods used in the study. Section 3 (Results) presents the results obtained which are discussed in Sect. 4 (Discussion). Section 5 (Conclusions) summarises the paper, while limitations and future research of the study are highlighted in Sect. 6.

2 Materials and Methods

This study, through the use of computational models, is intended to provide analytical information within the rarely explored sector of the PVSD. The computational models simulate the effects of STSIS on a new educational build (the case study building) in the rural location of London in the UK. The system parameters underwent adjustments aimed at optimising the operational efficiency of the PV shading devices whilst harvesting solar energy through embedded photovoltaic panels [28].

2.1 Building Descriptions

To explore the operational parameters of intelligent solar shading devices for improving building energy efficiency, the architectural model consisted of a significant amount of glazing. This deliberate inclusion of glazing aimed to accentuate the effect of solar gain, thereby facilitating an in-depth assessment of the anticipated impacts and efficacy of the STSIS system. To avoid the interference of other factors within this study, auxiliary ventilation, natural ventilation, infiltration and heat gain from internal adjacent spaces were not included in thermal load calculations. As illustrated in Fig. 1, the case study building was a single-story educational building with teaching spaces, meeting rooms and toilet facilitates, with the largest space being the canteen which is southern facing. The details of the building geometry and thermal properties are given in Tables 1 and 2, respectively. It should be noted that the proposed case study only looked at the canteen area as a pilot study.

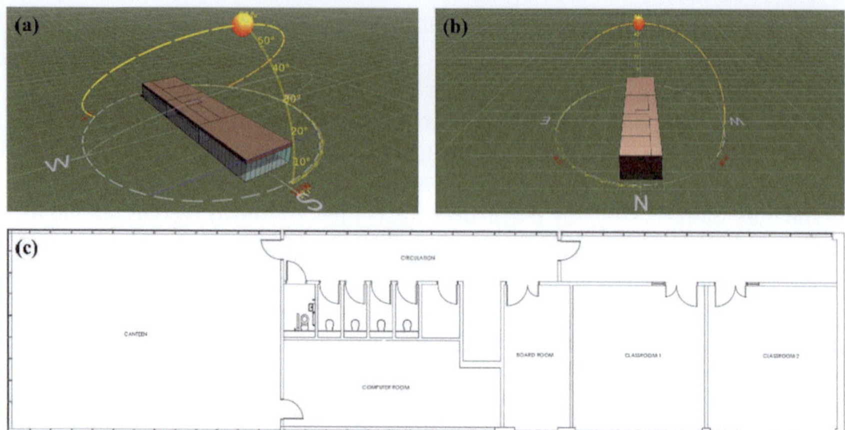

Fig. 1 The case study building model and floor layout: **a** South-west façade, **b** North façade, **c** Floor layout. (Self-drawn)

Table 1 Geometric data of the case study building. (Self-drawn)

Description	Characteristics
Gross floor area (including toilets)	388.28 m^2
Internal ceiling height	3.1 m
Ceiling void height	0.7 m
Area of canteen	127.22 m^2
Area of circulation 1	39.89 m^2
Area of computer room	44.61 m^2
Area of board room	23.81 m^2
Area of classroom 1	46.57 m^2
Area of classroom 2	48.42 m^2
Area of circulation 2	34 m^2

Table 2 Fabric thermal properties. (Self-drawn)

Description	U-value ($W/m^2 \cdot K$)
Roof (flat)	0.18
External wall	0.26
External floor	0.18
Glazing/Windows	1.4

2.2 *Applying PVSD Concepts*

Photovoltaic systems have been widely applied on buildings and serving as a part of the building component. Building-applied photovoltaic (BAPV) is fixed onto a building component or structure. Building-integrated photovoltaic (BIPV), rather than BAPV, represent integrated PV systems into the buildings structure [29]. Both BAPV and BIPV systems inherit the established merit of an on-site renewable energy source characterised by noise-free operation and emission-free functionality whilst benefiting from minimal maintenance requirements across the anticipated lifetime of 20–25 years [15]. For the purpose of the PVSDs, the main focus of this study falls under the BAPV category that allows dynamic movement through the use of sun-tracking capabilities to maximise its benefits of solar shading and energy generation.

2.3 Applying PVSD Concepts

Sun/Solar tracking technology, a pivotal technology in various solar energy applications involves orienting devices toward the sun's movement, also known as the sun path [30]. Typically, solar tracking systems can be achieved through either relying on weather data derived from factors such as latitude, altitude, sunset and sunrise times [31]. However, this method often falls short as this does not factor weather changes, changes of temperature and nearby obstructions such as shading from buildings [32]. As such, the proposed study involved optical sensors that track the brightest point in the sky, which considers the atmospheric conditions such as lighting fluctuations caused by weather variations [33].

Because the intelligent shading devices are installed with embedded PV, a more accurate method without involving additional sensors would be controlling the shading devices through the MPPT (maximum power point tracking). The MPPT is an algorithm that finds the maximum power curve of PV panels implemented with a simple BEMS (Building Energy Management System) and motor system, while the performance of the shading device can be altered for optimal performance during increased solar gain [12]. This systematic approach ensures the intelligent sun tracking shading façade maximised shading whilst concurrently optimising solar energy production. Specifically, the proposed study used the MPPT for tracking based upon the sun path with an automated microcontroller-driven solar tracker which employs a hybrid algorithm. This merges both active and chronological algorithms, using mathematical models and sensors to precisely locate the sun's position to adjust the multiple slatted louvres. As shown in Fig. 2, this simplified process consists of tracking the sun path whilst intricately moving the louvre arm.

2.4 Computational Modelling

IES VE v.2023.5.2.0 (Integrated Environmental Solutions Virtual Environment) simulation software was used in this study as it is a practical and accessible tool pivotal to this study. The initial 2D building geometry was designed in AutoCAD

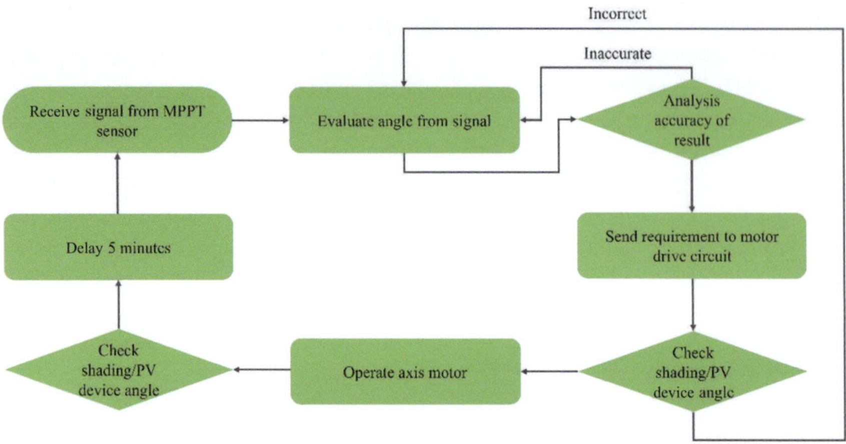

Fig. 2 The microcontroller working procedure for sun path tracking (Self-drawn)

with no BAPV solar shadings and subsequently the case study building's 2D geometry was imported in DXF format. This facilitated the creation of the building's geometry in ModelIT (the modelling interface in IES VE), allowing for the configuration of input data such as the building dimensions and properties. Once the building geometry had been built and the site rotation added, the weather file could be selected based on ASHRAE design weather database v7.0. To ensure accurate solar gains are measured a SunCast simulation within IES VE was initially run with no shading as a baseline case. Figure 3 visually represents the sun's path across the sky during different times within a year, which provides an understanding on maximum solar exposure of the specific location of the case study building.

2.5 *Modelling PVSD*

Designing the optimum size and overhang of the solar shading devices with embedded PV is important for a well-designed system, which reduces solar gain within summer when cooling is required and admits solar gain in winter to benefit from 'free' passive heating from the sunlight. Solar gain alongside the requirements for natural daylight was provided through the slats of the louvre shading system. This shading device was used to help modulate internal temperature by allowing more or less natural light to be received by the windows of the building, in turn this increases or decreases solar heat gain to prevent fluctuations of internal temperatures. This paper involves simulating three shading models (Fig. 4), that is, one model without PVSDs, another model with fixed overhang PVSDs, and the primary focus of the study—multiple movable PVSDs. Figures 5 and 6 illustrate the PVSDs details and 3D geometry in the IES VE modelling, respectively.

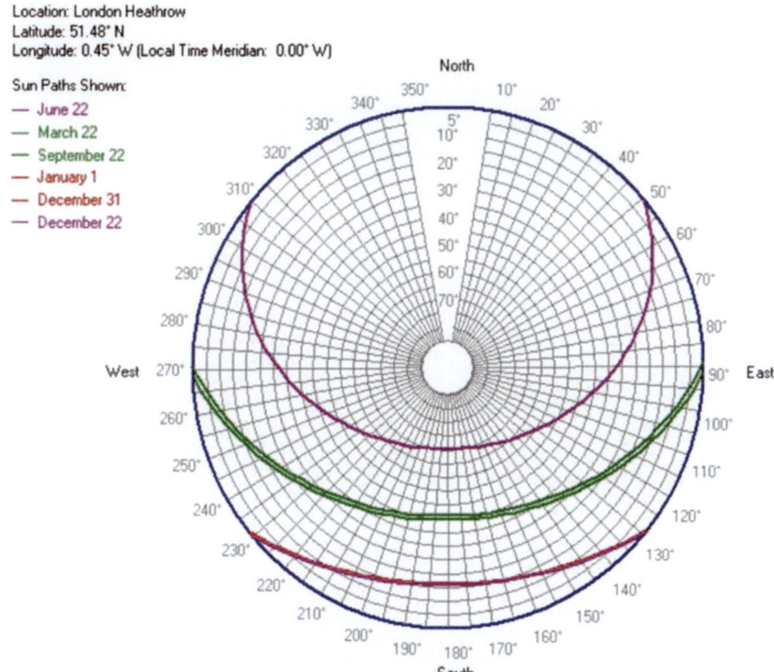

Fig. 3 IES VE produced sun path diagram (Self-drawn)

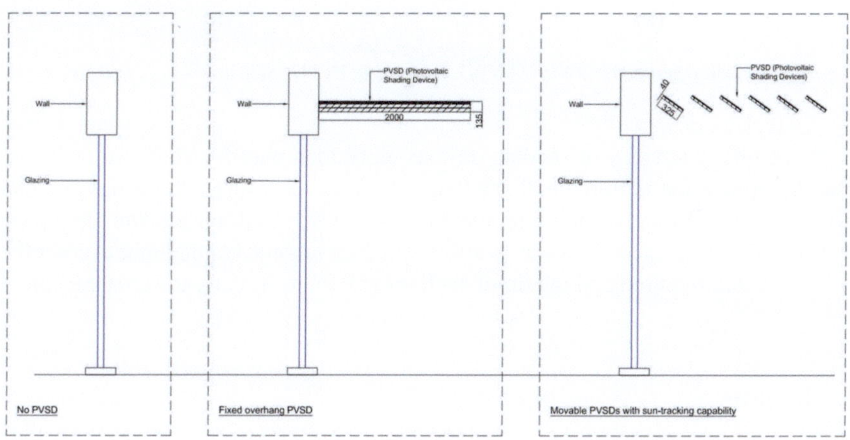

Fig. 4 PVSD types for the proposed study (Self-drawn)

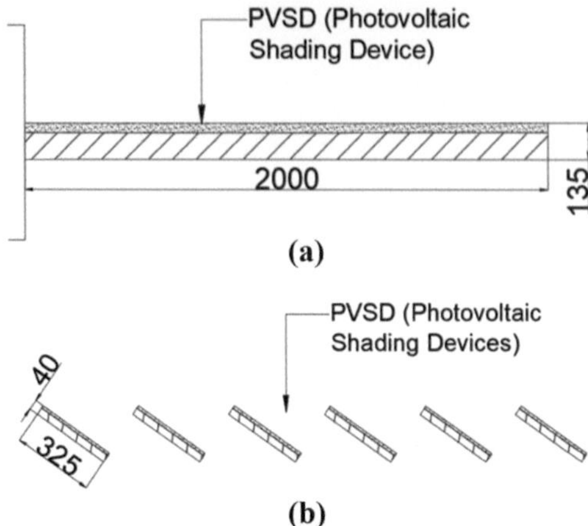

Fig. 5 The details of the proposed PVSDs: **a** Fixed PVSD, **b** Movable PVSD (Self-drawn)

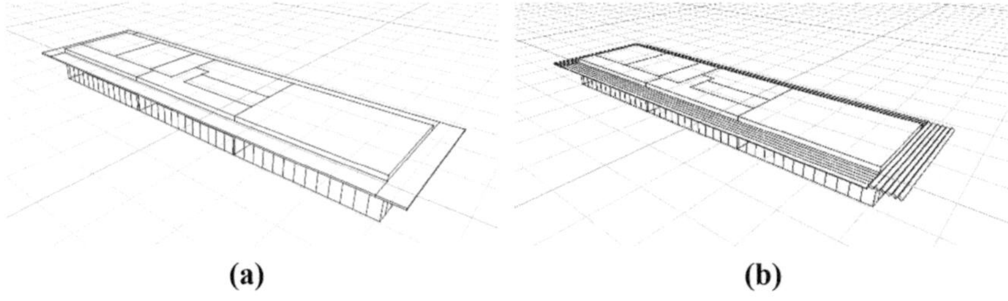

Fig. 6. 3D shading geometry: **a** Fixed PVSD, **b** Movable PVSD (Self-drawn)

Based on the strategy of shading devices embedded with PV, the most optimum shading orientation is horizontal shading. Although to ensure natural lighting can still be harvested into the building, the movable PVSD used five smaller slats at an inclination of 35 degree. This was to allow optimum amounts of daylight to enter the building when required and the tip of each louvre points directly towards the sun.

3 Results

3.1 Internal Air Temperature

To obtain internal room temperatures both heating and cooling systems were switched off. By doing this the simulation was run based on using no forms of heating and cooling so it does not interfere with internal room temperatures. The three models (no PVSD, fixed PVSD, and movable PVSD) were simulated using the IES VE

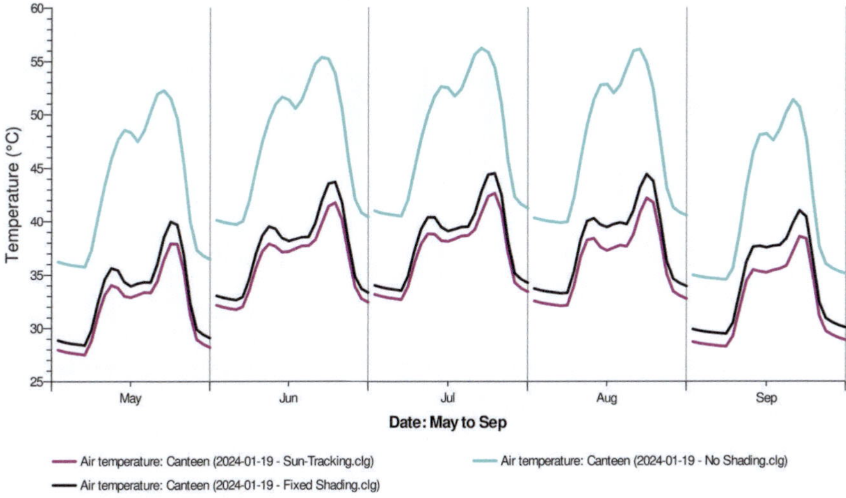

Fig. 7 Canteen maximum indoor air temperatures (Self-drawn)

simulation engine—ApacheSim. Figure 7 extracted from the Apache simulation results illustrates the maximum internal room temperatures during the warm periods within the canteen.

As can be seen, the room without PVSD has had a significantly higher room temperature than that of the two PVSD scenarios from May to September. Specifically, the canteen with no PVSD reaches a maximum temperature of 56.2 °C approximately. In comparison, the fixed PVSD device performs well by reducing maximum temperatures to 44.48 °C, whilst the best forming PVSD device was the dynamic moveable PVSD. Using the movable PVSD device, internal temperatures rose to 42.63 °C which is a 24.15% reduction of internal temperature. Through minimising the internal room temperature using dynamic movable PVSD, a reduction of cooling demand can be foreseen and therefore leading to a more energy efficient building and consequently the fuel usage and carbon emissions will be reduced.

3.2 *Solar Gain*

As mentioned previously, internal air temperatures are significantly lower whilst using dynamic movable PVSD louvres. This occurs because the PVSDs prevent primary transmittance of solar gains into the building through the window. As per Fig. 8, the dynamic movable PVSD louvres perform the best to reduce solar gain for the canteen area in any month and therefore decrease the required cooling loads accordingly. Specifically, the highest solar gain to the Canteen is 8.61 kW for no PVSD, 6.06 kW for fixed PVSD and 5.52 kW for dynamic movable PVSD.

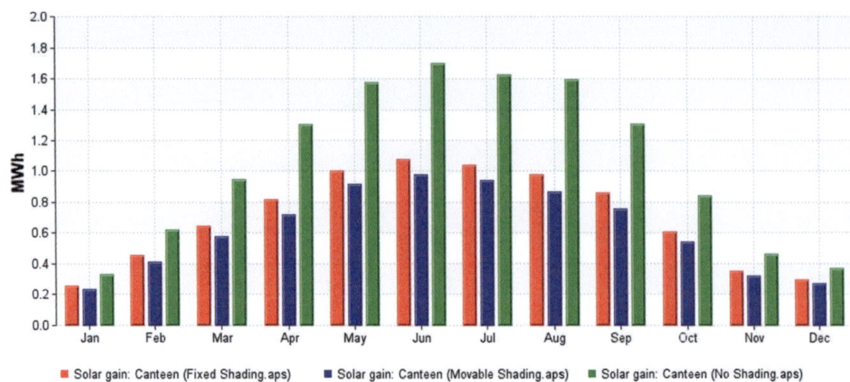

Fig. 8 Solar gain comparison between the different PVSD types (Self-drawn)

Table 3 Sensible cooling load requirement for different shading scenarios. (Self-drawn)

Scenario	Minimum value (kW)	Minimum time	Maximum value (kW)	Maximum time	Mean value (kW)
Fixed PVSD	–4.5	18:30, 17 Jul	0	0.18	–0.72
Movable PVSD	–4.05	18:30, 17 Jul	0.013	0.26	–0.62
No shading	–6.36	18:30, 17 Jul	0	1.4	–1.36

3.3 Sensible Loads

As can be seen in Table 3, the no shading scenario requires a sensible cooling load of 6.3639 kW to achieve the maximum setpoint temperature of 23 °C. Introducing fixed shading acts to reduce heat gain to 4.4965 kW and maintain a more comfortable internal temperature. This reduction supports the principle of energy-efficient design which emphasizes importance of energy efficient design solutions through the use of passive shading strategies. Furthermore, employing movable PVSDs results in a further decrease of cooling load requirements within the canteen to 4.0471 kW. This supports that the use of dynamic and movable shading devices allow for an even higher effectiveness over solar heat gain which in turn optimises the energy usage for the building.

3.4 Daylight Analysis

In addition to thermal loads, daylight analysis calculations were also produced to understand the daylight factor within the building for the sake of overall indoor comfort. This is considered the ratio of indoor illuminance to the external horizontal

Table 4 Daylight calculations—room attributes input data (Self-drawn)

Room attributes	Data
Working plane height	0.7 m
Surface maintenance factor	0.9

Table 5 Daylight calculations—internal reflectance (Self-drawn)

Internal reflectance	Data
Floor	0.25
Wall	0.8
Ceiling	0.8

illuminance. According to the Code for Lighting by the Society of Light and Lighting [34], the average daylight factor should be at least 2% for the canteen area, while a daylight factors of at least 5% would indicate internal electrical lighting would not normally be required during the daytime, subject to uniformity requirements.

In the proposed study, the case study building does not have any adjacent buildings or structures that would interfere with natural light, so the calculations were based on the CIE (Commission Internationale de l'Éclairage) overcast sky as the average condition. In this case, the daylight conditions with the sky completely covered with clouds indicate an overcast sky. The UK usually experience overcast sky conditions, and thus the CIE overcast sky has been used as part of the daylight calculations as this best suit the conditions of the case study building. The design input for the daylight calculations, as per the Code for Lighting, can be concluded in Tables 4, 5 and 6.

Daylight Analysis—No Shading. In the IES VE simulation, spatial daylight autonomy (sDA) is used to describe annual sufficiency of daylight levels in an interior environment. Through simulating a sDA using the given data input (Tables 4, 5 and 6) with a working plane grid of 1 m. In Fig. 9, it can be found that all points upon the working plane achieved the minimum threshold of 300 lx (as per the BREEAM Visual Comfort for Health and Wellbeing [35]) at 50% of the occupied hours per year.

The illumination levels within the canteen area typically reach more than 300 lx, suggesting this would be a well-lit space through the use of natural daylight. Furthermore, as shown in Fig. 10 and Table 7, the area-weighted average daylight factor was 12.6% for the room with no PVSDs. This significantly achieves the 2% requirement of daylight factor.

Table 6 Window properties (Self-drawn)

	Normal transmittance	Frame width (m)	Maintenance factor	Visible light transmittance (%)
Windows	0.71	0.03	0.92	68

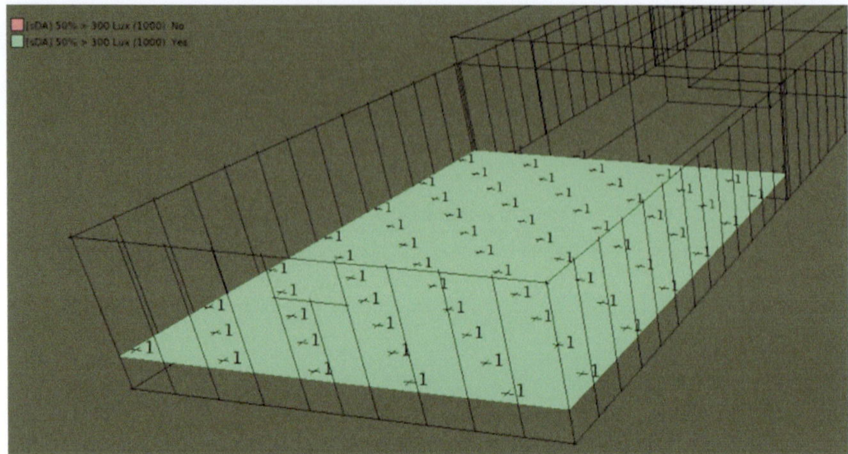

Fig. 9 SDA results across the working plane of the canteen room (Self-drawn)

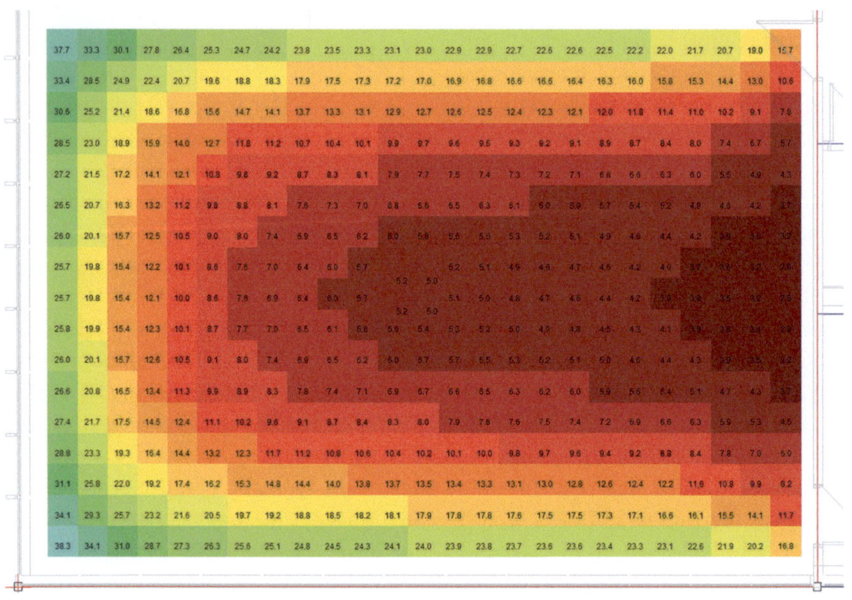

Fig. 10 Canteen daylight factors—no shading (Self-drawn)

Table 7 Daylight thresholds—no shading (Self-drawn)

Total floor area (m^2)	Total floor area above threshold (m^2)	Percentage floor area above threshold (%)	Area-weighted average daylight factor (%)	Area-weighted average illumination (lux)
105.36	105.36	100	12.6	1541.21

It should be noted that, given the primary emphasis of this research paper on sun-tracking photovoltaic shading devices, the fixed shading system was not simulated as part of the daylight calculations.

Daylight Analysis—Sun-Tracking Movable PVSD. As per Fig. 11 and Table 8, the area-weighted average illumination levels achieved 606.02 Lux which is a reduction of 935.18 Lux with the use of movable PVSDs. Additionally, the area-weighted average daylight factor has seen a reduction from the previous results (no shading scenario) with a daylight factor of now 5%.

It is also important to note that the movable shading devices are in a fixed position, due to the limitations of IES VE and the time and resources for this research paper the shading devices do not automatically move in orientation with sun-tracking capability so it would be likely the moveable shading devices would achieve greater results if there was the possibility to activate this within simulations.

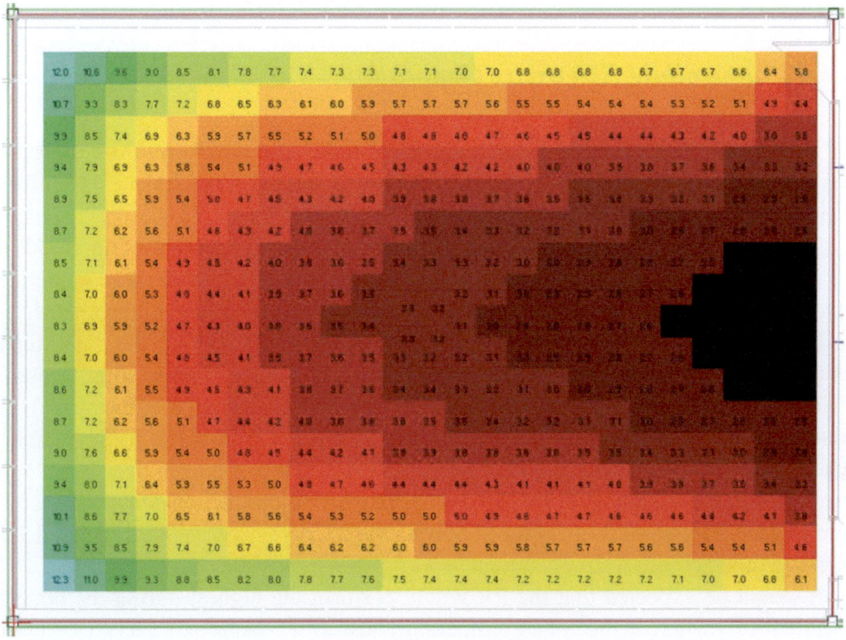

Fig. 11 Canteen daylight factors—sun-tracking movable PVSD

Table 8 Daylight thresholds—sun-tracking movable PVSD (Self-drawn)

Total floor area (m^2)	Total floor area above threshold (m^2)	Percentage floor area above threshold (%)	Area-weighted average daylight factor (%)	Area-weighted average illumination (lux)
105.36	105.36	100	5	606.02

Fig. 12 Photovoltaic shading devices modelled in IES VE (Self-drawn)

3.5 Photovoltaic Electricity Generation

As part of the study the aim was to embed photovoltaic panels into the movable shading devices. IES VE does have limitations to what can be achieved therefore sun-tracking capability cannot be achieved through the results. However, the single axis sun-tracking PV panels (Fig. 12) can see an energy production of about 30%. To achieve the most accurate results in this scenario, the PV panels were modelled at a horizontal inclination and a correction factor of 1.3 was added to the results.

Basically, the simulation uses fixed photovoltaic panels within the case study location of London, and the type of panels used are monocrystalline silicon based. Figure 13 shows the electricity generated by the PVSDs throughout the year. It is found that the PV electricity production was significant from April to September, when the monthly electricity generated exceeded 3 MWh and even over 4.5 MWh from May to July (detailed in Table 9). Although the PV production declined notably during the colder months (all below 2 MWh), the STSIS systems with embedded PV panels would achieve a summed total of 42.54 MWh over the year.

It would be expected that the PV would perform the highest during the hottest month of July where in the hottest temperature was 31.5 °C approximately. However, in this case, the highest PV electricity production is found in June, this is because PVs are less efficient in hotter conditions. In theory, PV performance generally decreases by 0.34% for each degree that rises above 25 °C [36]. However, using sun-tracking capabilities through MPPT could mitigate the issue by altering the inclination during hot weather to achieve the highest performance of PV generated electricity through not facing directly at the sun and allow heat loss from the back of the panel to reduce its surface temperature.

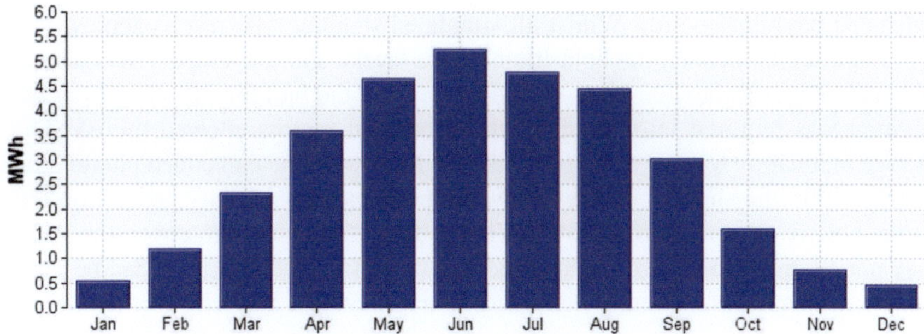

Fig. 13 PV generated electricity from the PVSDs (Self-drawn)

Table 9 Tabulated PV generated electricity (Self-drawn)

Months	PV generated electricity (MWh)	U-value (W/m^2·K)
Jan	0.55	0.72
Feb	1.2	1.56
Mar	2.32	3.02
Apr	3.6	4.68
May	4.65	6.05
Jun	5.25	6.83
Jul	4.79	6.23
Aug	4.45	5.78
Sep	3.03	3.94
Oct	1.61	2.09
Nov	0.79	1.02
Dec	0.48	0.62
Summed total	32.72	42.54

4 Discussion

This paper explores the performance and enhancement of the building energy efficiency using sun-tracking PV shading device (PVSD). Through computational analysis, the study supports the usage of the innovative PVSD strategy in public buildings in the UK. The results proved a successful relationship between the usage of sun-tracking PVSDs compared with the scenario not using PVSDs. Basically, there were significant improvements in thermal condition within the pilot study area (canteen). It was found that the movable PVSDs effectively performed the best in reducing internal room temperatures up to 24.15% due to mitigating of excessive solar gain. This, however, reduced the cooling load requirements and lowered the overall energy consumption of cooling within the canteen from 6.36 to 4.05 kW which was a reduction of approximately 36.32%. The PVSD system was also found benefitting from enhanced daylight control. Whilst all simulated shading type results achieved the minimum recommended daylight illuminance levels, sun-tracking PVSDs particularly offered more flexible control on daylight into the building. In conditions that daylight was required into the space the controlled motors allowed the PVSD to change inclination to control daylight factors. Moreover, the embedded photovoltaic panels within the shading devices allowed photovoltaic electricity generation with a total energy generation of 42.54 MWh throughout the year. Moreover, the PVSD may not be preferable from the architectural design perspective, as it can be rather conspicuous and consequently detracting from the overall aesthetic of a building.

5 Conclusions

This study demonstrated that the innovative technology of sun-tracking solar intelligent shading systems with embedded photovoltaics effectively improve the performance of buildings through reducing internal room temperatures especially in summer, while during winter months the use of the movable PV shading devices (PVSDs) can be altered to allow solar gain for the benefit of passive heating through the solar energy. However, the main issue with new buildings especially for those within London is that of overheating, which causes discomfort and thus leads to increased energy costs because of the increased cooling demand. Through implementing sun-tracking PVSD it was found that cooling loads can be passively reduced. Specifically, movable PVSDs effectively outperformed no shading and fixed PVSD options by achieving a substantial 24.15% reduction of internal temperatures through mitigating unwanted solar gain. The reduction of solar gain consequently reduced the cooling load requirements within the canteen from 6.36 kW to 4.05 kW. This represents a noteworthy decrease of approximately 36.32% of cooling energy demand. On the other hand, it can usually be found that implementing shading devices on buildings causes an unwanted reduction of daylight factors, however it has been proved within the case study building that minimum daylight factors can still be achieved through using the movable PVSD, as it offers a more flexible approach to controlling daylight entering the building. When daylight is desired, the controllable motors allow the PVSD to adjust its inclination which influences daylight factors. Further, it has been proved that a total annual energy generation of 42.54 MWh from the embedded PV panels on the movable shading devices without the need of taking up often wanted flat roof space for building services equipment requirements.

6 Limitations and Future Research

Overall, the limitations of this study stem from the simulation software, preventing definite conclusions to the performance of sun-tracking capabilities. To enhance the robustness of the future research it would be recommended to replicate these simulations either with another software or conduct multiple simulations with certain degree alterations of inclination of the PVSD.

References

1. Hepple R et al (2024) Investigating the effects of PCM-integrated walls on thermal performance for UK residential buildings of different typologies. Buildings 14(11)
2. Attia S et al (2022) Comparison of thermal energy saving potential and overheating risk of four adaptive façade technologies in office buildings. Sustainability 14(10)
3. Sureshkumar Jayakumari SD et al (2024) Energy and daylighting performance of kinetic building-integrated photovoltaics (BIPV) Façade. Sustainability 16(22)
4. Yang S et al (2023) Optimising design parameters of a building-integrated photovoltaic double-skin facade in different climate zones in Australia. Buildings 13(4)
5. Roberts F et al (2023) The application of human-centric lighting in response to working from home post-COVID-19. Buildings 13(10)
6. Heschong Mahone Group (1999) Daylighting in Schools. In: An investigation into the relationship between daylighting and human performance
7. Scibetta J et al (2023) Thermal and optical performance of semi-transparent BIPV windows in high-rise office buildings: a case study in the UK. In: International conference on urban climate, sustainability and urban design. Springer
8. Turner L, Yang S, White M (2022) Study of energy saving potential of solar shading devices in various climates. In: International conference on building energy and environment. Springer
9. Alwetaishi M et al (2021) An investigation of shading devices in a hot region: a case study in a school building. Ain Shams Eng J 12(3):3229–3239
10. Heidari A, Taghipour M, Yarmahmoodi Z (2021) The effect of fixed external shading devices on daylighting and thermal comfort in residential building. J Daylight 8(2):165–180
11. Yang S et al (2018) Study of building integrated photovoltaic/thermal double-skin facade for commercial buildings in Sydney, Australia. In: Final conference of COST TU1403 "Adaptive Facades Network". Lucerne, Switzerland
12. Mangkuto RA et al (2021) Design optimisation of fixed and adaptive shading devices on four façade orientations of a high-rise office building in the tropics. Buildings 12(1):25
13. Kim H-J, Yang C-S, Moon HJ (2019) A study on multi-objective parametric design tool for surround-type movable shading device. Sustainability 11(24):7096
14. Baouche FZ et al (2022) Design and simulation of a solar tracking system for PV. Appl Sci 12(19):9682
15. Jhumka H et al (2023) Assessing heat transfer characteristics of building envelope deployed BIPV and resultant building energy consumption in a tropical climate. Energy Build 298
16. Javed K, Ashfaq H, Singh R (2020) A new simple MPPT algorithm to track MPP under partial shading for solar photovoltaic systems. Int J Green Energy 17(1):48–61
17. Ragb O, Bakr H (2023) A new technique for estimation of photovoltaic system and tracking power peaks of PV array under partial shading. Energy 268:126680
18. Yang S et al (2017) Studies on optimal application of building-integrated photovoltaic/thermal facade for commercial buildings in Australia. In: Proceedings of SWC2017/SHC2017, pp 1–10
19. Obukhov S et al (2020) Optimal performance of dynamic particle swarm optimization based maximum power trackers for stand-alone PV system under partial shading conditions. IEEE Access 8:20770–20785
20. Roberts F et al (2023) Effect of semi-transparent a-Si PV glazing within double-skin façades on visual and energy performances under the UK climate condition. Renew Energy 207:601–610
21. Yang S et al (2020) Performance assessment of BIPV/T double-skin façade for various climate zones in Australia: effects on energy consumption. Sol Energy 199:377–399

22. Yang S (2020) Studies on the performances of building integrated photovoltaic/thermal double-skin facade for commercial buildings in Australia. In: School of built environment. University of New South Wales, Sydney, p 299
23. Elhabodi TS et al (2023) A review on BIPV-induced temperature effects on urban heat islands. Urban Clim 50
24. Hwang R-L, Fang P-L, Chen W-A (2023) Impact of solar radiation on indoor thermal comfort near highly glazed façades in a hot-humid subtropical climate: an experimental evaluation. Build Environ 243:110725
25. Yang S et al (2018) Numerical simulation study of BIPV/T double-skin facade for various climate zones in Australia: effects on indoor thermal comfort. Build Simul 12(1):51–67
26. Minelli F et al (2023) PhloVer: a modular and integrated tracking photovoltaic shading device for sustainable large urban spaces—preliminary study and prototyping. Energies 16(15):5786
27. Hepple R et al (2023) Sustainability and carbon neutrality in UK's district heating: a review and analysis. e-Prime—Adv Electr Eng Electron Energy 4
28. Asfour OS (2018) Solar and shading potential of different configurations of building integrated photovoltaics used as shading devices considering hot climatic conditions. Sustainability 10(12):4373
29. Yang S et al (2021) Numerical simulation modelling of building-integrated photovoltaic double-skin facades. In: Bulnes F, Hessling JP (eds) Recent advances in numerical simulations. IntechOpen, London, United Kingdom, pp 61–75
30. Awasthi A et al (2020) Review on sun tracking technology in solar PV system. Energy Rep 6:392–405
31. Amelia A et al (2020) Technologies of solar tracking systems: a review. In: IOP conference series: materials science and engineering. IOP Publishing
32. Yang S et al (2021) A sensitivity analysis of design parameters of BIPV/T-DSF in relation to building energy and thermal comfort performances. J Build Eng 41
33. Jung S-Y et al (2022) Development of a solar tracking-based movable louver system to save lighting energy and create a comfortable light environment. Buildings 12(11):2017
34. The Society of Light and Lighting (2012) The SLL Code for Lighting
35. Building Research Establishment (2018) BREEAM UK New Construction in Non-Domestic Buildings Technical Manual
36. Idoko L, Anaya-Lara O, McDonald A (2018) Enhancing PV modules efficiency and power output using multi-concept cooling technique. Energy Rep 4:357–369

Climate-Responsive and Sustainable Planning

Spatial Patterns of Flood Risks and Their Economic Linkages: A Preliminary Study in the Fujian Delta Urban Agglomeration, China

Qianwen Wang, Guangyu Zhao, Xin Tu, and Huijun Yang

Abstract The rapid economic expansion has catalyzed urbanization within urban agglomerations, thereby increasing flood risks due to changes in surface characteristics and atmospheric boundary layer properties. This research takes the Fujian Delta urban agglomeration as an example to explore the connections between flood risk and gross domestic product (GDP) growth across 29 districts during the period from 2010 to 2020. Our analysis reveals that flood risk distribution is spatially clustered at the scale of urban agglomerations, with clustering patterns strongly associated with levels of urbanization and economic development. Specifically, in highly urbanized districts, high-risk zones are dispersed throughout the urban areas. In contrast, in moderately to highly urbanized districts, although high risks are randomly distributed within urban areas, there are significant 'high-high' risk clusters at the urban edges. In less urbanized districts, these high-risk clusters predominantly occur within the urban centers. Drawing on these spatial correlations, the study proposes adaptive flood risk management strategies that consider the diverse urbanization and economic statuses within the agglomeration. This research not only sheds light on the intricate dynamics between urbanization, economic growth, and flood risks but also lays the foundation for future studies aimed at incorporating more extensive socio-economic factors to bolster urban resilience against climate-induced hazards.

Keywords Urbanization · Urban economy · Flood risk · Urban agglomerations · Spatial distribution

---

Q. Wang (✉) · X. Tu · H. Yang
Department of Geography, Tianjin Normal University, Tianjin, China
e-mail: qwwang@tjnu.edu.cn

G. Zhao
Tianjin University Research Institute of Architectural Design and Urban Planning Co., Ltd., Tianjin, China

B. He et al. (eds.), *Climate-Resilient, Low-Carbon, and Sustainable Cities*, Lecture Notes in Civil Engineering 754, https://doi.org/10.1007/978-981-95-2169-2_34

1 Introduction

The process of urbanization plays a essential role in the socio-economic development observed worldwide [1]. The United Nations estimates that by 2050, 68% of the world's population is expected to reside in urban areas [2]. This shift in urban population is anticipated to have a profound impact on multiple aspects of society and economic activities. It simultaneously presents both opportunities and challenges for sustainable urban development [3].

Urbanization has a considerable influence on regional hydrometeorological conditions [4, 5]. Extensive research has investigated the increasing flood risks confronting metropolitan areas, highlighting this issue as a persistent and pressing challenge in the context of continuous global urbanization. Recent studies further demonstrate that urbanization markedly alters the physical characteristics of precipitation, affecting cumulative rainfall, intensity, frequency, and the progression of storm systems [6]. This outcome is attributed to the interaction between urbanization and critical urban meteorological systems, including heat island effects, urban canopy structures, aerosol concentrations, and changes in the spatial distribution of rainfall [7]. The modifications previously discussed indirectly affect flood disasters associated with urbanization. In contrast, Urbanization's influence on flood risk is directly associated with transformations in land use and land cover, which subsequently affect hydrological response mechanisms [8, 9]. While urbanization generates positive feedback loops that promote and sustain development, it also produces a greater number of negative feedbacks that impose constraints and pose significant threats to the ecosystem [10, 11]. While accelerated socio-economic development linked to urbanization can supply technical and financial resources to address flood disasters [12], the intensification of economic concentration and unchecked urban sprawl can further exacerbate the disruption of urban ecosystems [9]. For example, The expansion of impervious surfaces in urban settings diminishes the infiltration capability of surface runoff [13]. Additionally, rapid urbanization has resulted in the excessive exploitation and encroachment on urban water bodies and additional aquatic ecosystems, leading to a considerable decrease in water area and thus weakening their regulatory capacity against floods [14]. These factors significantly escalate the likelihood of flood disasters during periods of heavy precipitation. Consequently, the urban agglomeration scale, characterized by a rapidly developing economy and high levels of urbanization, is at a much higher risk of flooding [15, 16].

Furthermore, increasing evidence demonstrates that there is spatial heterogeneity in the distribution, extent, and impact of flood disasters based on varying levels of urbanization [17, 18]. It is essential to take into account not only urban meteorological feedback but also the spatial patterns of urbanization, rainfall, and flood risk [19]. **However, key questions remain unresolved by scholars: Does flood risk in urban agglomerations exhibit a random spatial distribution? Additionally, are there disparities in the spatial patterns of flood risk among towns, which vary by levels of urbanization and economic development status?**

In this study, the Fujian Delta urban agglomeration is used a case in point for analysis to investigate the spatial patterns of flood risk and their relationship with economic development. **This study presents a novel approach by integrating multi-dimensional scales that include both urban agglomerations and individual towns, addressing the limitations of previous research that often considers urban agglomerations as uniform regions. It adopts an exploratory method to examine the spatial distribution patterns of flood risk, meticulously categorizing spatial heterogeneity by evaluating levels of urbanization and urban economic development across various towns, thereby enabling a precise, zone-specific, and tiered adaptation of disaster prevention strategies within urban agglomerations.** The research aims to comprehensively grasp the urbanization-flood risk relationship. From a regional management perspective within urban agglomerations, it seeks to develop tailored flood risk management strategies for towns varying in their levels of urbanization.

2 Method

2.1 Study Area

The Fujian Delta (116°53′–119°01′E, 23°33′–25°56′N), located on China's southeastern coast, comprises 29 administrative districts and counties under Xiamen, Zhangzhou, and Quanzhou. Despite covering only 16% of Fujian Province's land area, it accounted for 45.9% of the province's GDP in 2023, underscoring its critical role as an economic powerhouse.

The region has a subtropical humid monsoon climate, with an annual precipitation of 1,000–1,200 mm and recurrent flood events that challenge urban infrastructure and resilience. Given its rapid urbanization and pronounced flood risks, the region serves as an exemplary case for examining the interplay between economic growth and flood risk dynamics within a coastal urban agglomeration.

2.2 Study Design

A systematic analysis process for this study was organized as follows. Firstly, we collected meteorological, hydrological, and geographical data relevant to the flood risk system, and assessed the level of flood risk using the backpropagation (BP) algorithm. Secondly, we obtained land use and socio-economic data for the period 2010–2020. We identified the concentrated construction zones in each of the 29 districts within the Fujian Delta in 2020 and calculated the incremental gross domestic product (GDP) value per square kilometer over the same period. Finally, we used the spatial autocorrelation model to determine the spatial distribution pattern of flood

Table 1 Data sources in the study

Data	Description	Source
Land use	The 30 m annual land cover dataset (CLCD) in China [21]	https://essd.copernicus.org/articles/13/3907/2021/
Topography	Geospatial Data Cloud GDEMDEM 30M resolution digital elevation dataset	http://www.gscloud.cn/
Meteorology	World weather database and China surface climate information daily value dataset (V3.0)	http://data.cma.cn
Hydrology	Hydrological data published by the local government water agency	http://sl.xm.gov.cn/zfxxgk/; http://slj.quanzhou.gov.cn/zwgk/ghjh/; http://slj.zhangzhou.gov.cn/cms/html/zzsslj/tjxx/index.html
Soil	Harmonized world soil database (HWSD)	http://westdc.westgis.ac.cn
Ecosystem Service Value (ESV)	Spatial distribution dataset of terrestrial ESV in China [22, 23]	http://www.resdc.cn/
Road	OpenStreetMap	https://www.openstreetmap.org/
Population and economy	The 2021 Statistical Yearbooks of Xiamen, Zhangzhou, and Quanzhou	http://www.xm.gov.cn/; http://tjj.quanzhou.gov.cn/; http://tjj.zhangzhou.gov.cn/

risk across the urban agglomeration and investigated its correlation with levels of urbanization and economic development.

The primary source of research data was gathered from various publicly accessible sources, including the Fujian Delta government websites, the National Earth System Science Data Center, the China Meteorological Science Data Center, and the Geospatial Data Cloud, etc. (Table 1).

2.3 *Calculation of Flood Disaster Risk Using the BP Algorithm*

The overall flood risk level in the Fujian Delta was evaluated by integrating natural and artificial flood risks through a superimposition technique. The risk levels were assigned values ranging from 1 to 10, with higher values indicating more dangerous conditions and lower values indicating safer conditions. In this setting, we developed a 3-layer structure of the backpropagation algorithm for low error identification of flood risk, i.e., input layer, hidden layer, and output layer.

The input layer of the BP algorithm is composed of several indices used to evaluate flood risk. Among them, the natural flood risk indicators include surface factors (surface roughness and normalized difference vegetation index), which influence the landscape's ability to absorb or slow down runoff. Topographic factors (slope and elevation) determine water flow direction and accumulation points, while hydrological factors (surface runoff) directly represent water movement patterns. Meteorological factors (rainfall conditions) are critical in assessing flood risk due to their role in triggering flood events, and water source factors (river buffer range) measure proximity to flood-prone areas. The artificial flood risk indicators were selected for their ability to represent human vulnerability and exposure to flood risks. Socio-economic factors such as population density and economic activity reflect potential human and economic losses. Ecological service values, including agricultural productivity, climate regulation, and gas exchange functions, help capture the loss of ecosystem services during floods.

In alignment with the spatial scale of the Fujian Delta urban agglomeration, we partitioned the data for each indicator into units and spatially linked the attributes based on a raster size of 1 km × 1 km. This approach ensured uniform analytical precision across the dataset and facilitated the construction of a robust dataset for predicting flood disaster risk levels. We allocated 60% of the total samples to the training set and 40% to the testing set. The neural network toolbox in MATLAB was employed for network development, training, and testing. To optimize model performance, training parameters were fine-tuned, resulting in a training error of 0.0000004, a learning rate of 0.1, and a maximum of 1000 iterations. The credibility of the flood risk predictions generated by this network was assessed by achieving at least 90% accuracy on the validation set of flood risk levels (Fig. 1). These validation procedures collectively confirm the reliability and accuracy of the BP algorithm in predicting flood risks within the Fujian Delta.

2.4 Calculation of Flood Risk Distribution Patterns Using Spatial Autocorrelation Models

The global Moran's I index was utilized to analyze the overall spatial autocorrelation of flood risk (Eq. (1)). Furthermore, the spatial distribution patterns of flood risk were evaluated by measuring the correlation between neighboring and local units using the Local Indicators of Spatial Association (LISA) (Eq. (2)).

$$Moran^{'}sI = \frac{N\sum_{i=1}^{n}\sum_{j=1}^{n} w_{ij}(x_i - \overline{x})\left(x_j - \overline{x}\right)}{\sum_{i=1}^{N}\sum_{j=1}^{N} W_{ij}\sum_{i=1}^{N}(x_i - \overline{x})^2} \tag{1}$$

$$LocalMoran^{'}sI_i = Z_i\sum_{j}^{n} w_{ij}Z_j \tag{2}$$

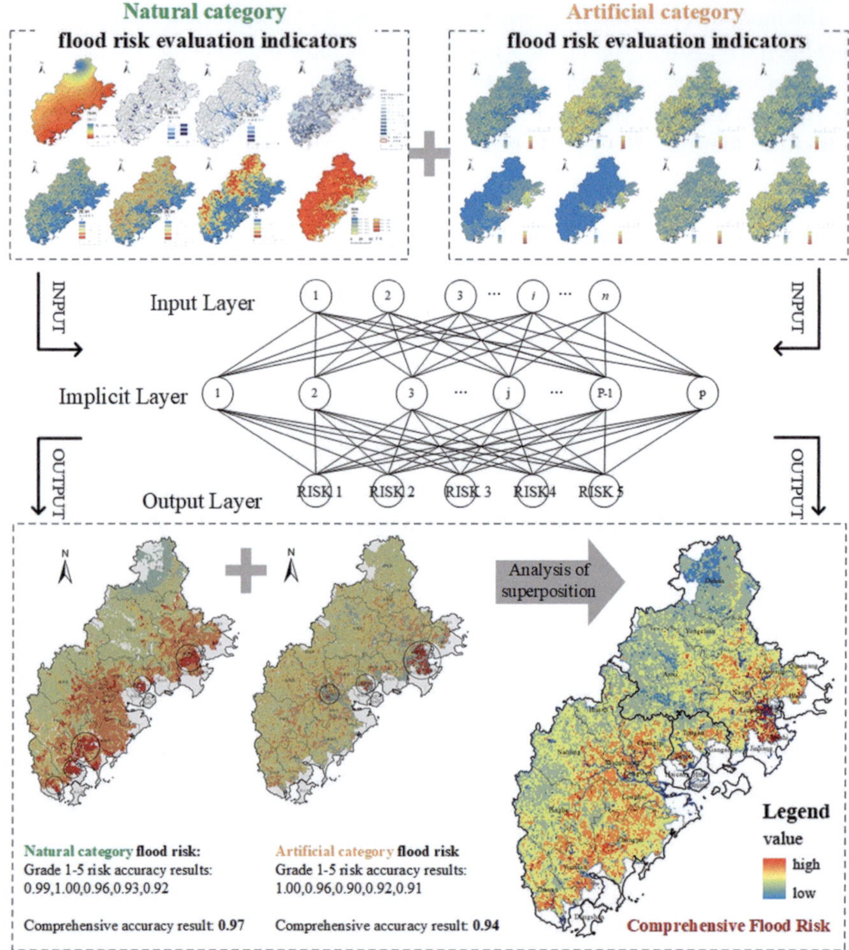

Fig. 1 The process for calculating flood risk (Self-drawn)

where x_i and x_j are the risk values at positions i and j of the spatial grid, $\overline{x}$ is the average risk value, w_{ij} is the spatial weight matrix, $Z_i = \frac{(x_i-\overline{x})}{\sqrt{\frac{1}{n}\sum_{i=1}^{n}(x_i-\overline{x})^2}}$, $Z_j = \frac{(x_j-\overline{x})}{\sqrt{\frac{1}{n}\sum_{j=1}^{n}(x_j-\overline{x})^2}}$ are the normalized form of the flood risk value for each spatial grid.

The *Moran's I* index ranges from -1 to 1, where values close to 1 indicate strong positive spatial autocorrelation, values near –1 signify strong negative spatial autocorrelation, and values around 0 suggest a random spatial distribution.

A 'high-high' cluster indicates that the grid's risk value and that of its adjacent grids are both high. A 'high-low' outlier indicates that the grid's risk value is high, but the adjacent grids have low-risk values. A 'low–high' outlier indicates that the grid's risk value is low, but the adjacent grids have high-risk values. A 'low-low' cluster indicates that the grid's risk value and that of its adjacent grids are both low.

3 Result and Discussion

3.1 Spatial Distribution of Flood Risk

The results reveal that flood risk in the Fujian Delta is primarily concentrated in the western regions, where high-risk zones dominate, while the eastern regions display predominantly low-risk areas. This distribution pattern demonstrates point-like aggregations (Fig. 2). Notably, the zones of highest risk are primarily situated within the central urban areas of Xiamen, Quanzhou, and Zhangzhou, as well as the central regions of several southern coastal counties including Zhangpu, Yunxiao, and Zhaoan. The regions identified share common characteristics, including gently rolling terrain adjacent to river systems, shaped by the unique geographical and environmental context of the Fujian Delta. These areas also exhibit either a high level of urban development or rapid expansion of urbanized areas. These observations suggest a potential correlation between heightened urbanization levels and increased flood disaster risk.

3.2 Spatial Distribution of Economic Levels

A positive correlation between urban land size and urban economy has been confirmed [20], which also holds for the Fujian Delta. The central construction zones in Xiamen, Zhangzhou, and Quanzhou constitute over 80% of the region's consolidated construction activity. Notably, Xiamen stands out as the first city in the Fujian Delta to reach the advanced stage of urbanization, with an urbanization rate of 89.2% by 2020. Over the past decade, however, its economic growth has shown signs of deceleration. In contrast, Zhangzhou and Quanzhou, the central cities, exhibit an average urbanization level of 85.2%. These cities have maintained a strong financial foundation and recorded the highest economic growth rates from 2010 to 2020. Meanwhile, other regions in the area continue to exhibit low to medium urbanization levels, characterized by smaller concentrated construction zones. Their economic growth rates fall between those of Xiamen and the combined rates of Zhangzhou and Quanzhou, as illustrated in Fig. 2.

3.3 Flood Risk Patterns in Relation to the Economic Levels

The Moran's I index calculation confirms that the distribution of flood risk within the Fujian Delta is not random, but rather demonstrates a clustered pattern. An analysis of the flood risk distribution in conjunction with urban socio-economic development revealed three distinct spatial patterns of high flood risk (Fig. 3).

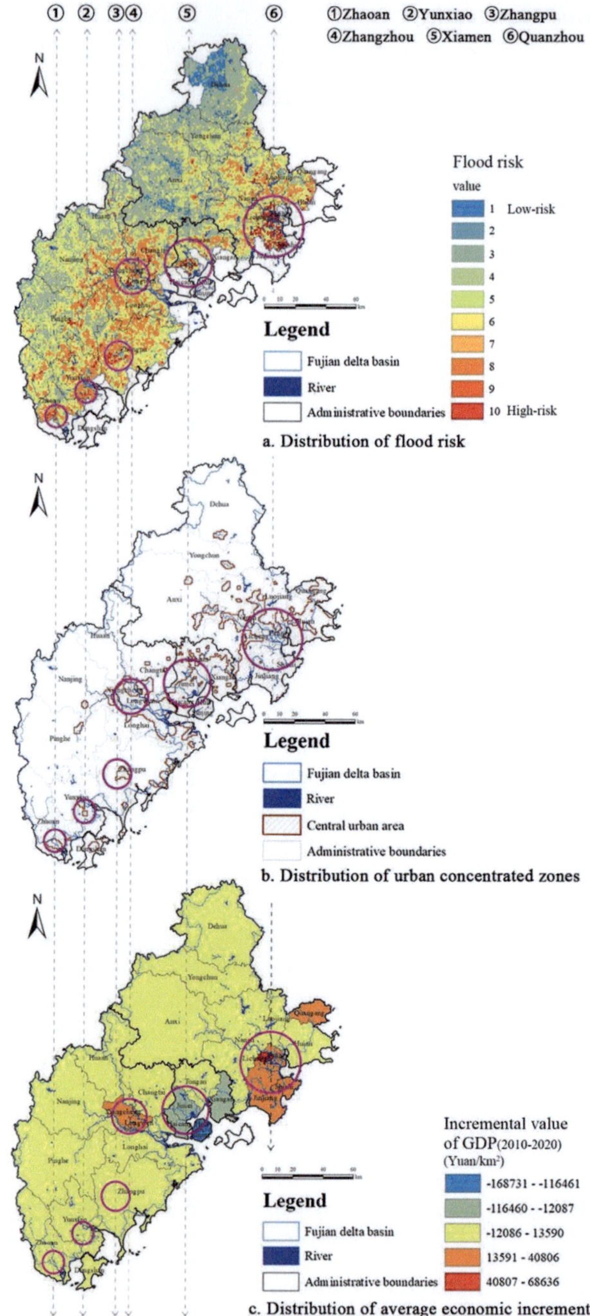

Fig. 2 Distribution of flood-risk, urban concentrated areas and average economic increment (Self-drawn)

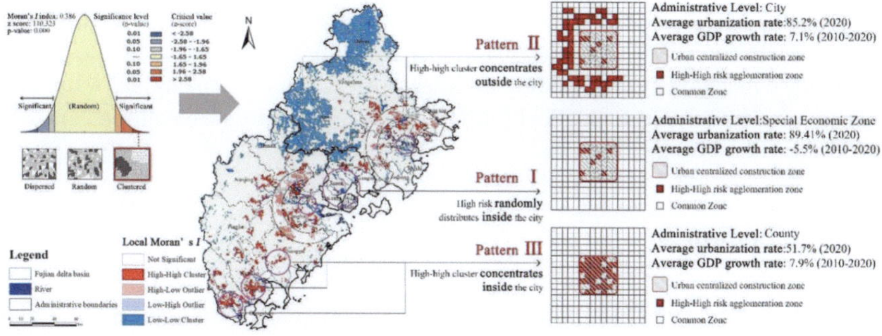

Fig. 3 Different flood risk distribution patterns (Self-drawn)

Pattern I is observed in the densely urbanized areas, where high-risk is randomly distributed within the city, exemplified by the central construction zones of Xiamen. In this region, floods are predominantly locally generated, and the distribution of high-risk areas appears random. Xiamen, serving as the principal city in the Fujian Delta, boasts a robust urban economy but experiences slow economic growth. It is plausible that during advanced stages of urbanization, areas of high risk are associated with the deterioration of local infrastructure. However, urbanization also coincides with the development of advanced facilities, refined spatial planning, and enhanced societal responses, all of which contribute to mitigating flood risks. Consequently, these areas should not only focus on micro-scale risk diagnostics, early warnings, and optimizations but also on enhancing urban resilience and reducing reliance on engineered solutions.

Pattern II is identified in moderately to highly urbanized areas where high-risk is randomly distributed within the city, yet clusters of high-high risk are concentrated on the urban fringe. This pattern is exemplified by the central construction areas of Quanzhou and Zhangzhou, which are emerging growth poles of the Fujian Delta. Interestingly, the distribution of high flood risk in this pattern is similar to that of Pattern I in the urban center, but it is clustered at the urban fringe. This suggests that while rapid urbanization may begin to mitigate flood risks similar to Pattern I, the condition of the flood hazard system at the urban fringe significantly influences risk levels within the urban center. Consequently, this pattern is particularly sensitive to the dynamics between urban development and flooding. Effective strategies should focus on enhancing the stability of the hydrometeorological system and managing water resources at the urban fringe, while carefully weighing the trade-offs between socio-economic progress and flood risk.

Pattern III characterizes the low to moderately urbanized areas, where the clusters of high-high risk are concentrated within the city. This pattern is exemplified by the central regions of Zhangpu, Yunxiao, and Zhaoan, which are typical counties of the Fujian Delta. Despite having a relatively poor economic base, these areas show growth potential. However, the high risk in the concentrated built-up area and its adjacent areas suggests that the rough urbanization pattern is closely linked to poor facility support, incoherent flood prevention engineering, and a flawed disaster response system. Therefore, it is imperative for districts exhibiting this pattern to prioritize defense mechanisms by seamlessly integrating risk management and watershed management into the frameworks of urbanization and economic development.

4 Conclusion

This study conducted a preliminary investigation into the spatial interrelationships between flood risk, urbanization, and economic development within the urban agglomerations of China's Fujian Delta. The findings reveal that the distribution of flood risks across these urban agglomerations is not random but is systematically linked to adjacent hazard conditions. Additionally, certain regions exhibit distinct 'high-high' or 'low-low' risk clustering. Our analysis compared the spatial characteristics of areas with 'high-high' flood risk clustering against their respective levels of urbanization and economic development. This comparison identified varied patterns of flood risk distribution, which are influenced by differing degrees of urbanization and economic growth within the urban clusters. These patterns are categorized into three main groups: (1) In highly urbanized areas, where high risks are randomly distributed across cities, disaster prevention strategies should focus on enhancing micro-scale risk diagnostics and reducing dependence on engineering interventions. (2) In medium-to-highly urbanized areas, where high risks are also randomly distributed within cities but with 'high-high' risk clusters at the urban peripheries, disaster mitigation efforts should prioritize improved flood management at the urban–rural interface. (3) In low-to-medium urbanized areas, where high risks are concentrated in inner cities, efforts should aim to elevate flood prevention thresholds and improve the quality of urban infrastructure.

The findings of this study hold considerable implications for urban agglomerations prone to frequent flooding, particularly those with varying degrees of urbanization. It is crucial for policymakers to comprehend the spatial distribution of flood risks within these urban clusters and tailor disaster prevention strategies accordingly. Such an understanding will not only enhance the efficiency of flood management but also facilitate the optimal allocation of disaster prevention budgets and the strategic implementation of preventative measures. By prioritizing areas most at risk and aligning resource distribution with the specific needs and vulnerabilities of different urban zones, policymakers can significantly improve resilience and reduce the overall impact of flood events on urban populations.

However, this study still has some limitations. Although it primarily focuses on mapping the spatial patterns of flood disaster risk and economic development within the urban agglomeration, it does not thoroughly explore the specific mechanisms that drive these spatial variations. Additionally, the impact of policy interventions, governance structures, and urbanization patterns on the distribution of flood risk has not been thoroughly analyzed. Future research should integrate multiple factors, such as governance capacity, land-use planning strategies, and infrastructural vulnerabilities, to reveal the processes contributing to the spatial differentiation of flood risks.

Acknowledgements This study is supported by the Youth Fund for Humanities and Social Sciences Research of the Ministry of Education of China [24YJCZH308].

References

1. Hou MY, Deng YJ, Yao SB (2022) Coordinated relationship between urbanization and grain production in China: degree measurement, spatial differentiation and its factors detection. J Clean Prod 331:129957
2. United Nations (2019) Department of Economic and Social Affairs, Population Division: World urbanization prospects: The 2018 revision. United Nations, New York
3. Zhang X, Han L, Wei H, Tan X, Zhou W, Li W et al (2022) Linking urbanization and air quality together: a review and a perspective on the future sustainable urban development. J Clean Prod 346:130988
4. Sheng JF, Wilson J (2009) Watershed urbanization and changing flood behavior across the Los Angeles metropolitan region. Nat Hazards 48:41–57
5. Zhong S, Qian Y, Zhao C, Leung R, Yang XQ (2015) A case study of urbanization impact on summer precipitation in the Greater Beijing Metropolitan Area: Urban heat island versus aerosol effects. J Gerontol Ser A Biol Med Sci 120:10903–10914
6. Luo ZR, Liu JH, Zhang SH, Shao WW, Zhou JN, Zhang L et al (2022) Spatiotemporal evolution of urban rain islands in China under the conditions of urbanization and climate change. Remote Sens 14:4159
7. Kusaka H, Nawata K, Suzuki-Parker A, Takane Y, Furuhashi N (2014) Mechanism of precipitation increase with urbanization in Tokyo as revealed by ensemble climate simulations. J Appl Meteorol Climatol 53:824–839
8. Feng BY, Zhang Y, Bourke R (2021) Urbanization impacts on flood risks based on urban growth data and coupled flood models. Nat Hazards 106:613–627
9. Yin J, Ye MW, Yin Z, Xu SY (2015) A review of advances in urban flood risk analysis over China. Stoch Env Res Risk Assess 29:1063–1070
10. Liu Y, Huang XJ, Yang H (2022) An integrated approach to investigate the coupling coordination between urbanization and flood disasters in China. J Clean Prod 375:134191
11. Yang C, Zeng W, Yang X (2020) Coupling coordination evaluation and sustainable development pattern of geo-ecological environment and urbanization in Chongqing municipality, China. Sustain Cities Soc 61: 102271
12. Huang GW (2109) A revisit to impact of urbanization on flooding. In: Urban planning and water-related disaster management, pp 43–56
13. Onate-Valdivieso F, Onate-Paladines A, Collaguazo M (2022) Spatiotemporal dynamics of soil impermeability and its impact on the hydrology of an urban basin. Land 11:250
14. Kundzewicz ZW, Su BD, Wang YJ, Xia J, Huang JL, Jiang T (2019) Flood risk and its reduction in China. Adv Water Resour 130:37–45
15. Sharma SE (2023) Urban flood resilience: governing conflicting urbanism and climate action in Amsterdam. Rev Int Political Econ 30(4):1413–1435
16. Zhou F, Xu YP, Chen Y, Xu CY, Gao YQ, Du JK (2013) Hydrological response to urbanization at different spatio-temporal scales simulated by coupling of CLUE-S and the SWAT model in the Yangtze River Delta region. J Hydrol 485:113–125
17. Li CJ, Chai YQ, Yang LS, Li HR (2016) Spatio-temporal distribution of flood disasters and analysis of influencing factors in Africa. Nat Hazards 82:721–731

18. Lu M, Yu ZH, Hua J, Kang CX, Lin ZX (2023) Spatial dependence of floods shaped by extreme rainfall under the influence of urbanization. Sci Total Environ 857:159134
19. Yang L, Tian FQ, Niyogi D (2015) A need to revisit hydrologic responses to urbanization by incorporating the feedback on spatial rainfall patterns. Urban Clim 12:128–140
20. Bai XM, Chen J, Shi PJ (2012) Landscape urbanization and economic growth in China: positive feedbacks and sustainability dilemmas. Environ Sci Technol 46:132–139
21. Yang J, Huang X (2021) The 30m annual land cover dataset and its dynamics in China from 1990 to 2019. Earth Syst Sci Data 13:3907–3925
22. Costanza R, d'Arge R, de Groot R, Farber S, Grasso M, Hannon B et al (1997) The value of the world's ecosystem services and natural capital. Nature 387:253–260
23. Xie GD, Zhang CX, Zhen L, Zhang LM (2017) Dynamic changes in the value of China's ecosystem services. Ecosyst Serv 26:146–154

Evaluation of Youth Belonging Based on Place Attachment Theory: Implications for the Optimization of Public Space in Urban Villages

Lu Wei, Siyin Zheng, Yanling He, Minghong Yu, Jing Xie, and Zhuixin Guan

Abstract Young people are the future of urban development, but the lack of a sense of belonging among young people has led to frequent movements of talent, which is not conducive to urban development and construction. Therefore, the state has introduced many policies to provide hard security for young people to settle down and live in urban villages, but the effect is limited in solving the problem of young people's psychological sense of belonging. Based on the theory of place attachment, this research explores the influencing factors of place attachment and their interactions in the public space of urban villages. Analytical results demonstrate that place attachment exhibits significant linkages with young adults' behavioral patterns and the perception of spatial environment, in which commercial activity, green landscape quality and richness of facilities all positively and significantly affect the place dependence, and personal background show negligible influence on the place dependence; the mean value of the place attachment in Yulin Area is higher than that in Workers' Village, which is mainly affected by the function and culture of the public space, and the different types of public space will also form different degrees of place attachment. This article concludes by offering recommendations for urban village spatial optimization and a policy foundation for addressing youth spiritual needs and fostering youth groups' social cohesion.

Keywords Youth belonging · Public space · Place attachment

1 Introduction

Aggregation of resources with increased urbanization in recent years [1]. The influx of a large number of young people into large cities, however, due to the lack of economic strength of the majority of young people and the pressure of life, it is

L. Wei · S. Zheng · Y. He (✉) · M. Yu · J. Xie · Z. Guan
School of Civil Engineering and Architecture, Southwest University of Science and Technology, Mianyang, Sichuan, China
e-mail: yanling.he@swust.edu.cn

B. He et al. (eds.), *Climate-Resilient, Low-Carbon, and Sustainable Cities*, Lecture Notes in Civil Engineering 754, https://doi.org/10.1007/978-981-95-2169-2_35

difficult to afford the high prices of housing in first- and second-tier cities, resulting in the majority of young people having difficulties in settling down and renting a home [2]. The phenomenon of youth apartments. Therefore, convenient location, lower rent and cost of living have become popular choices for the majority of youth groups, such as youth apartments, urban villages, old and dilapidated neighborhoods and so on.

The State has formulated policies such as the Medium- and Long-Term Youth Development Plan (2016–2025) (2017), Opinions on Accelerating the Development of Guaranteed Rental Housing (2021), and Opinions on Launching a Pilot Program for Building Youth-Developmental Cities (2022) in order to address the housing difficulties of new citizens and young people. These policies have made significant progress in enhancing the hard needs of youth for urban settlement, but have not been as effective as expected in addressing the psychological problems of youth [3]. Public space, as a vital component of urban life, is crucial in addressing the spiritual needs of youth and creating a sense of belonging. The theory of "place attachment" is frequently linked to the emotional belonging of public space in order to shape a public space that embodies identity, belonging, and comfort [4].

Place attachment is a classic theory in the study of place emotion in environmental psychology and behavioral geography. Through constant contact with places, people develop individual perceptions, preferences, or judgments about places, or an emotional belonging and connection, i.e., place attachment [5]. The Place Attachment Scale was created by Williams et al. to gauge how attached a person is to a location. They introduced the idea of place attachment and noted that it has two components: place identity and place dependence [6]. Its research interests mainly include rural place attachment [7], as well as the connection between customer place attachment and service scenarios [8]. International studies have focused on emotional ties to rural environments, such as land conservation [9], urban sprawl [10], rural revitalization [11], and other areas, while domestic studies have focused on sense of place, place attachment, landscape gardening [12]. The research scope mainly focuses on scenic spots and urban parks. The scope of research is mainly on scenic spots and urban parks, but there is no research on the application of place theory to the spatial optimization of urban villages [13].

This study, which is based on the "place attachment" theory, examines how public space features affect young people's sense of belonging in Chengdu's urban villages and examines the causes of the varying degrees of place attachment that are produced by various public space kinds. Finally, the study suggests ways to optimise public spaces in urban villages in terms of function and culture in order to enhance the sense of belonging of youth groups.

2 Research Methodology

2.1 Overview of the Study Area

This research selects two areas in Chengdu, Workers' Village and Yulin Area. Yulin Area is located in the middle of Wuhou District, starting from Renmin South Road in the east, connecting to Nijiaqiao Road in the south, Xinguang Road in the west and Yulin East Road in the north, with a total area of 2.7 square kilometers, as shown in Fig. 1. Workers' Village is located on the west side of Love Lane Block at the intersection of Caojia Lane and Ma'anxi Road in Jinniu District, with a total area of 0.6 square kilometers, as shown in Fig. 2. Yulin area is one of the successful cases of urban village transformation in Chengdu, which has formed a relatively perfect and pleasant public space. The Workers' Village area is a typical representative of the old workers' housing complex, which is still under construction.

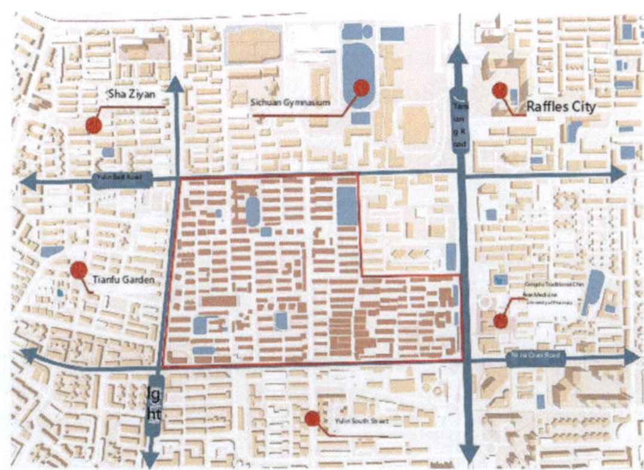

Fig. 1 The scope map of Yulin area in Chengdu (Self-drawn)

Fig. 2 The scope map of workers village in Chengdu (Self-drawn)

2.2 The Survey Sample and Data Collection

The survey was divided into two parts: pre-survey and formal survey. The pre-survey was conducted in April 2024 in the Workers' Village and Yulin districts, with 94 questionnaires distributed and 87 valid questionnaires recovered, with a validity rate of 92.55%. The formal survey was conducted in May 2024, 214 questionnaires were distributed, and 192 valid questionnaires were recovered. The study was analyzed by SPSS 27.0 data statistical software.

The participants of this survey were young people aged 18–35 years old. The questionnaire consisted of five parts: demographic characteristics, recreational behavior, perception of spatial environment, measurement of place attachment, and spatial reason scale, and the demographic data of the respondents are shown in Table 1. All of the study's variables were developed and modified using established, mature scales to guarantee the survey's high reliability and validity. Using the categories of place identity and place dependence, the measure evaluates respondents' emotional and physical ties to urban village public spaces [14]. The questionnaire was measured using a five-star Likert scale (1 for totally disagree, 2 for disagree, 3 for unsure, 4 for agree, 5 for totally agree), aiming to measure the level of youth's attachment to public space, as shown in Table 2.

Table 1 Demographic characteristics of participants (Self-drawn)

Name	Content	Count	Percentage (%)
Gender	Female	124	64.6
	Male	68	35.4
Age	18–25 years old	50	26.1
	26–30 years old	97	50.5
	31–35 years old	45	23.4
Career	Private business owners	12	6.3
	Freelancer	32	16.7
	Public servant	36	18.8
	Institutional person	20	10.4
	Company employ	67	34.9
	Student	25	13.0
Degree of education	Less than 1 year	72	37.5
	1–3 years	48	25.0
	4–7 years	39	20.3
	More than 7 years	33	17.2

Table 2 Place attachment questionnaire scale (Self-drawn)

Serial number	Place identity	Place dependence
1	I think this is a representative public space	I'm used to moving around here and not changing places
2	I feel like it's become an indispensable part of my life	I feel like the conditions here are the best for doing what I want to do
3	I have a deep affection for this place	I'll come back here whenever I have time
4	When I stay here, I feel like I'm part of the city	I'll always think about what's going on here
5	When I'm here, I'm very relaxed and can do whatever I want	I miss it when I can't come here for quite some time

2.3 Data Analysis Methods

Based on the questionnaire, the data were analyzed using SPSS 27 software, and the data results were plotted and processed using Origin 2024 and Photoshop 2021.

3 Data Analysis and Results

3.1 Reliability and Validity Analysis

The findings demonstrated that the Place Attachment Scale and the Spatial Reasons Scale had Cronbach's alpha coefficients of 0.807 and 0.859, respectively, both of which were higher than 0.8, suggesting that the scales were reliable. Furthermore, the spatial reasons scale's KMO value was 0.812, the place attachment scale's KMO value was 0.733, and the Bartlett's spherical test's significance p-value was less than 0.001. The two communal factors were retrieved after doing exploratory factor analysis on each scale independently, and the contribution rate of the variance of the place attachment scale was 56.2%, and that of the spatial reasons scale was 72.234%, which was higher than 50%, indicating that the degree of interpretation was good.

3.2 Descriptive Analysis

(1) Overview of the characteristics of population activities and information on places in public space

In terms of activity time, youths in the Workers' Village mainly concentrated on irregular hours (50%) and evening to night time (46.67%). In the Yulin community, evening and nighttime were the main activity periods, accounting for 42.42%. As for the frequency of short-term activities, the frequency of activities every 1–2 h was the highest in both places (39.39%), while the frequency of long-term activities was the highest in 3–5 times per month (50%), followed by 3–5 times per week (20%), and the lowest in every day (3.33%).

In terms of the perception of public space environment, the mean value of overall satisfaction in Workers' Village is 3.91, indicating that the youth's overall evaluation of public space is relatively positive. However, in terms of property maintenance and management, the mean satisfaction was lower, at 3.06. In contrast, the youth were more satisfied with the accessibility of transportation (M = 3.94), living atmosphere (M = 3.87) and neighborhood interaction (M = 3.83). The mean value of overall satisfaction in the Yulin community was 3.82, and the youth had the highest satisfaction with the living atmosphere (M = 4.0), as well as superior commercial activity (M = 3.93) and neighborhood interaction (M = 3.8). However, the satisfaction level of leisure facilities in public space was lower, with a mean value of only 2.7.

(2) Overall place attachment status

Based on the scores of the Place Attachment Scale, the mean values of the indicators in the Workers' Village and Yulin Area were calculated as follows: Place Identity, 2.74 in the Workers' Village and 3.51 in the Yulin Area; Place Dependence, 2.72 in the Workers' Village and 3.45 in the Yulin Area, indicating that the sense of belonging to the Yulin Area is significantly higher than that of the Workers' Village. In Yulin Area, the mean value of place identity is higher than that of place dependence, showing a state of "high identity and high dependence", while the mean value of place attachment in Workers' Village is lower, showing a state of "low identity and low dependence".

From the Place Identity Measurement Scale, the Workers' Village performs better on Indicator 5, but scores lower on Indicator 3, which shows that the type of youth activities is recognized, but the weaker cultural atmosphere of the community and the lack of interactive experiences limit their sense of integration. The Yulin community scored lower on indicators 2 and 3, which shows that the weaker cultural atmosphere of the community and the lack of interactive experiences limit the sense of integration of the youth.

In the Place Dependence Measurement Scale, Workers' Village scored the lowest on Indicator 3, indicating that its public space lacks character and is easily substituted. The Yulin community also scored low on indicators 2 and 3, indicating that the lack of functional facilities makes the Yulin community also face a high substitutability problem (Tables 3 and 4).

3.3 Correlation Analysis

Based on the results of the questionnaire, Pearson's correlation coefficient was chosen to test the correlation between place attachment and personal background, activity

Table 3 Workers' village place attachment measurement scale (Self-drawn)

Category	Evaluating indicator	Strongly disagree/ %	Disagree/ %	Neutral/ %	Agree/ %	Strongly agree/%	Mean
Place identify M = 2.74	1	0.00	22.20	77.80	0.00	0.00	2.78
	2	0.00	55.60	27.80	16.7	0.00	2.61
	3	0.00	44.40	55.60	0.00	0.00	2.56
	4	16.70	16.70	44.40	11.10	11.10	2.83
	5	0.00	27.80	50.00	22.20	0.00	2.94
Place dependence M = 2.72	1	5.60	16.70	38.90	38.90	0.00	3.11
	2	0.00	55.60	38.90	5.60	0.00	2.89
	3	0.00	27.80	61.10	0.00	11.10	2.50
	4	0.00	27.80	55.60	16.70	0.00	2.56
	5	0.00	61.10	22.20	16.70	0.00	2.55

Table 4 Yulin community's place attachment measurement scale (Self-drawn)

Category	Evaluating indicator	Strongly disagree/ %	Disagree/ %	Neutral/ %	Agree/ %	Strongly agree/%	Mean
Place identify M = 3.51	1	0.00	0.00	33.3	56.70	10.0	3.77
	2	10.00	13.30	40.0	33.30	3.30	3.07
	3	0.00	13.30	53.30	30.00	3.30	3.23
	4	0.00	3.30	40.00	43.30	13.30	3.67
	5	0.00	3.30	20.00	63.30	13.30	3.83
Place dependence M = 3.45	1	0.00	20.00	10.00	66.70	3.30	3.53
	2	3.30	10.00	36.70	50.00	0.00	3.33
	3	0.00	10.00	53.30	36.70	0.00	3.27
	4	0.00	6.70	43.30	43.30	6.70	3.50
	5	0.00	10.00	30.00	50.00	10.00	3.60

characteristics and perception of spatial environment, as shown in Fig. 3. According to the statistics, location attachment, transit convenience, and personal history did not significantly correlate. A substantial positive correlation was found between the activity features of visit frequency, stay duration, and place attachment. This suggests that the higher the frequency of visits and the longer the stay, the higher the degree of place attachment. A better spatial environment in the public space can help to foster youth's dependence on the place, as evidenced by the significant positive correlation found between place dependence and commercial activity, green landscape quality, facility richness, and other aspects of the perception of the spatial environment. Place identity and perception of the spatial environment are substantially positively correlated, suggesting that the more favorable the spatial environment, the more a

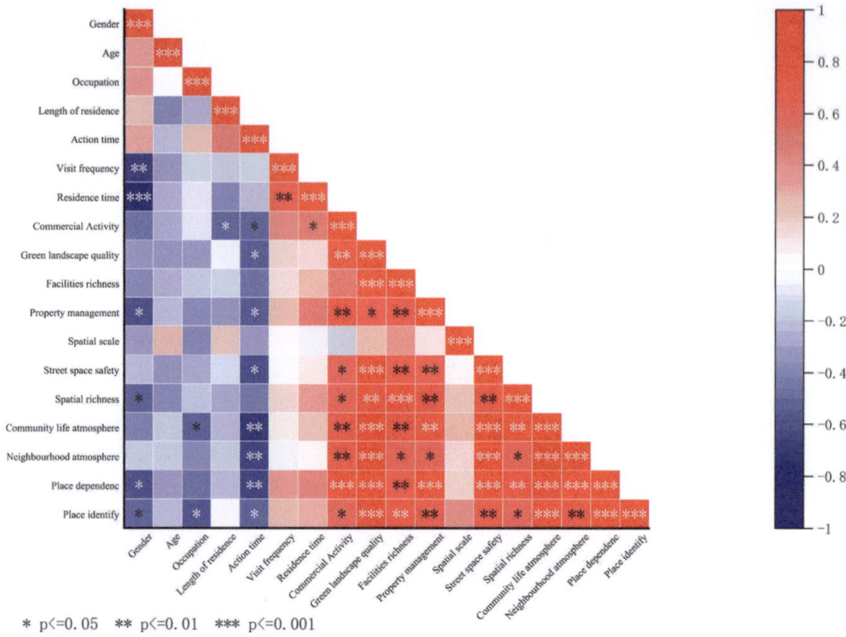

Fig. 3 Correlation analysis between place attachment and various influencing factors (Self-drawn)

young person will identify with the location. Place reliance and place identification have a strong positive correlation in Yulin Community and Workers' Village, meaning that users would have more a feeling of identity with the place when their dependence on it is high.

3.4 Exploration of Differential Place Attachment

This study assessed the attractiveness, satisfaction, and irreplaceability of space for users from the functional level of space using the Spatial Problems Scale, and examined the awareness, feelings, and self-integration of users from the cultural level of space, and explored the importance of spatial function and culture in different public spaces through the Pearson's correlation test.

The analysis results show that the mean value of spatial functions in Yulin is 3.68, which is higher than that of Workers' Village (3.4), and both of them only satisfy the basic needs of the users without significant functional activities or environmental characteristics, which makes the users' dependence on these spaces lower. In contrast, Yulin is significantly better than Workers' Village in terms of spatial culture, indicating that Yulin has better emotional communication with users in terms of cultural activities, which enhances emotional identity. On the other hand, the Workers' Village

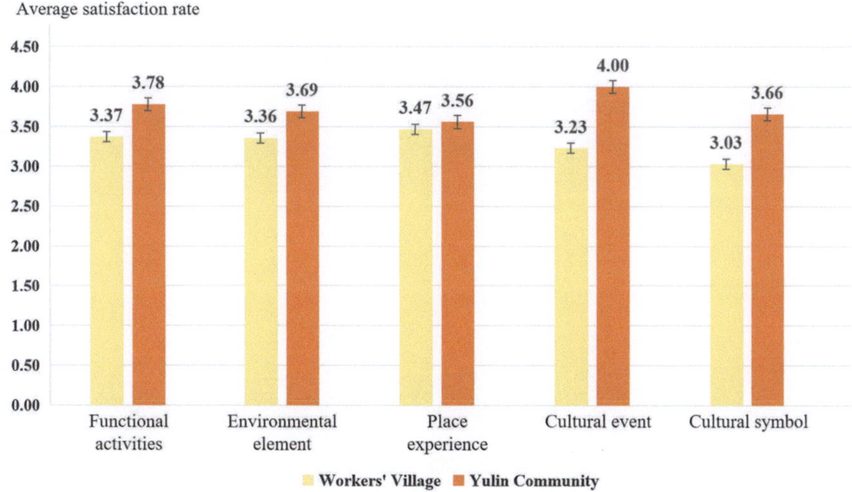

Fig. 4 Spatial functional cultural satisfaction mean (Self-drawn)

lacks the display of local characteristics and culture, and the interaction between space and emotion is insufficient, making it difficult to establish a better emotional relationship with the users, and the inability to realize spatial self-integration leads to a lower sense of belonging (Fig. 4).

The correlation analysis between the feedback of the youth group on the functional and cultural aspects of the space and the attachment of the youth group to the space was conducted, and it was found that the feedback of the youth group on the functional and cultural aspects of the space and the attachment of the youth group to the space were basically in a significant positive correlation among the six different types of public spaces, and there are significant differences in the hierarchy of demand for different categories of public space at the functional and cultural levels (e.g. Table 5). At the spatial-functional level, the order of importance of the influence of inter-house, street, entrance, and community public space is environmental elements, functional activities, and place experience; the place experience of plaza space has the greatest influence on place attachment, followed by environmental elements and functional activities; and park space focuses on functional activities. In the spatial culture level, the importance of the influence of inter-house and plaza public space is ranked as cultural symbols, place interpretation and cultural events; cultural events in street space have the greatest influence on place attachment, followed by place interpretation and cultural symbols; cultural symbols in entrance and exit space are significantly more important than cultural events and place interpretation; and community public space is dominated by place interpretation, followed by cultural symbols and cultural events. Community public space is dominated by place interpretation, followed by cultural symbols and cultural events. The above results show that the enhancement of place attachment in public space should respond to environmental suitability and

Table 5 Correlation of Place Attachment Generated by (Self-drawn) Youth Groups with Feedback on Functional Cultural Use of Space

		Inter-house public space						Community public space					
		Functional activities	Environmental element	Place experience	Cultural event	Cultural symbol	Place Interpretation	Functional activities	Environmental element	Place experience	Cultural event	Cultural symbol	Place Interpretation
Place attachment	Pearson Correlation	0.718*	0.915**	0.490	0.551	0.886**	0.776**	0.512*	0.582*	0.476	0.454	0.577*	0.745**
	Significance (two-tailed)	0.019	0.000	0.150	0.099	0.001	0.008	0.036	0.014	0.053	0.067	0.015	0.001
Place attachment	Pearson Correlation	0.551**	0.687**	0.183	0.585**	0.292	0.457*	0.460	0.874**	0.923**	0.612	0.831*	0.798*
	Significance (two-tailed)	0.006	0.000	0.404	0.003	0.176	0.028	0.251	0.005	0.001	0.107	0.011	0.018
Place attachment	Pearson Correlation	0.491*	0.513**	0.349	0.574**	0.654**	0.157	0.563*	0.338	0.549*	0.574*	0.667**	0.730**
	Significance (two-tailed)	0.011	0.007	0.081	0.002	0.000	0.444	0.023	0.201	0.028	0.020	0.005	0.001

Note *At the 0.05 level (two-tailed), the correlation is significant, **Significant at the 0.01 level (two-tailed)

activity carrying capacity, and the cultural design should focus on symbolic representation and event participation, and the differentiated ranking characteristics also provide priority design references for spatial optimization.

4 Policy Recommendations

This study of urban villages in Chengdu (Workers' Village and Yulin Area) is based on the theory of place attachment. It explains the relationship between place attachment, activity characteristics, and the perception of the spatial environment in public spaces in urban villages. It concludes that the function and culture of the space will determine the degree of place attachment that is formed in different types of public spaces. The results of the study show that the activity characteristics and environmental perception of public space are important factors affecting youth's place attachment, in which personal background does not affect the activity characteristics, but different behavioral characteristics will lead to differences in the perception of space, that is, behavioral activities play a role in the physical environment, and the environment in turn affects the activities, forming place identity and dependence, thus generating place attachment. The type of activities, frequency of visits, and length of stay are all significantly influenced by the spatial environment during this process. In particular, place identify and dependence influence one another and together shape attachment, while activity features are strongly influenced by the spatial environment, which in turn influences spatial perception and place attachment. The stability and fit offered by the spatial function, as well as the intimacy and irreplaceability of the spatial culture, account for the variations in place attachment among various public space kinds. The study's findings have led to the following policy recommendations being put out.

(1) Optimizing spatial functions, improving spatial environmental conditions, and forming places where behavioral dependence meets the needs of youth

The study shows that spatial functions (e.g., accessibility, convenience richness, etc.) are significantly and positively correlated with place dependence, while the mean value of convenience satisfaction in Yulin Area is significantly lower than that of Workers' Village. Improve the function of public space in urban villages, such as optimizing the transportation network of public space in workers' villages to improve the accessibility and relevance of the places; secondly, add more interesting and interactive places to enhance the opportunities for exchanges among young people, so as to make their interaction more diversified. In the Yulin community, first of all, it should pay attention to the improvement of basic activity facilities and leisure facilities to improve the comfort of the place; in addition, it should also optimize the public space between houses, and enhance the relationship between neighbors and form the place dependence by beautifying the green landscape and installing landscape sketches and so on.

(2) Shaping competitive functional activities and differentiated spatial environments to enhance place identity

Workers' Village has high spatial substitutability due to the lack of activity and interaction, while Yulin Area improves irreplaceability by integrating cultural landscape elements. Competitive local activities, such as Chinese New Year bazaars, online red landmarks, cultural performances, traditional handicraft activities and non-heritage experiences, should be integrated into the micro-remodeling and renewal of public spaces, and some enclosed or fragmented spaces can be managed through demolition and remodeling. In response to the problem of community infrastructure, differentiated greening landscapes should be added to the available areas such as street corners and building facades, and combined with community building. Increasingly, landscape nodes should be established to allow residents to participate in landscape revitalization.

(3) Shift from patterned construction to the construction of distinctive cultural features to enhance the city's spiritual attachment

Yulin Area has a higher mean place identity than Workers' Village due to the abundance of cultural activities, and the influence of cultural symbols on place attachment is greater in all types of public places. Under the landscape design environment of avoiding homogenization of public space, starting from the perspective of urban area, let the pattern construction turn to the construction of characteristic cultural landscape, discovering the historical elements of streets and lanes, integrating the emerging technology, evolving into the material carrier of public space through the topographical elements, reproducing the landscape memory, and guiding the community residents to look back at the memory of time and space. In addition to offering the required area for relaxation and leisure, residents should be encouraged to take part in its development and upkeep. This will allow cultural awareness to be incorporated into residents' everyday lives, improve their sense of experience, and foster a sense of cultural identity, pride, and belonging.

References

1. Yanfen H, Li D (2013) An analysis of social problems in the process of urbanization in China. Hebei Acad J 33(01):116–121
2. Jie L (2023) Face to face to solve the problem of college graduates renting. China Youth Daily (007). https://doi.org/10.38302/n.cnki.nzgqn.2023.000598
3. Yan Z (2024) Strategy research on youth demand orientation and high-quality youth development city construction. Grade Classic (06):76–78 + 132
4. Meichen W (2023) Reconstruction of community belonging based on place attachment—a case study of Wenchong Village, Guangzhou. Intell Buil Smart City (02):73–75. https://doi.org/10.13655/j.cnki.ibci.2023.02.020
5. Scannell L, Gifford R (2010) Defining place attachment: a tripartite organizing framework. J Environ Psychol 30(1):1–10

6. Williams DR, Roggenbuck JW (1989) Measuring place attachment: some preliminary results. In: San Antonio: Proceeding of NRPA symposium on leisure research
7. Junjie H (2022) The influencing factors and promotion strategies of the attachment to township public space:a case study of Yaopu Town, Chuzhou City, Anhui Province. J Anhui Agric Univ (Social Science Edition) 31(03):26–32. https://doi.org/10.19747/j.cnki.1009-2463.2022.03.004
8. Meizhen Z (2019) Research on the effect of servicescape on consumers' scene attachment and online behavioral intention. Wuhan University. https://doi.org/10.27379/d.cnki.gwhdu.2019.000444
9. Cundill G, Bezerra CJ, Vos DA et al (2017) Beyond benefit sharing: place attachment and the importance of access to protected areas for surrounding communities. Ecosyst Serv 28140–28148
10. Aśkiewicz MB et al (2018) Place attachment and collective action tendency. Soc Psychol Bull 13(4):e25612
11. Kastenholz E, Marques PC, Carneiro JM (2020) Place attachment through sensory-rich, emotion-generating place experiences in rural tourism. J Destination Mark Manag 17
12. Liu QY, You D, Zhu LY et al (2017) The relationship between place attachment and restorative perception of tourists visiting Fuzhou urban parks. Resour Sci 39(7):1303–1313
13. Guangsi L, Ange W, Keyi C (2019) Research on place attachment: concept, progress and trend. Chinese Garden 35(10):63–66. https://doi.org/10.19775/j.cla.2019.10.0063
14. Williams DR, Patterson ME, Roggenbuck JW (1992) Beyond the commodity metaphor: Examining emotional and symbolic attachment to place. Leisure Sci 14(1). https://doi.org/10.1080/01490409209513155

Analysis of Dynamic Changes and Driving Factors of Small Wetlands Around Chaohu Lake, China: Based on Remote Sensing Images

Jing Jin, Jiaqi Zhou, Mengmin Lu, and Qing Deng

Abstract Small wetlands, are vital for water resource management, environmental regulation, biodiversity conservation, and as habitats for migratory species. However, they are increasingly threatened by climate change and urbanization. In this study, small wetlands and other land use types around Chaohu Lake were extracted and validated from 2015 to 2021. The extraction and validation process involved the use of high-resolution remote sensing imagery and three classification methods: random forest (RF), support vector machine (SVM), and maximum likelihood estimation (MLE). ArcGIS was employed to analyze changes in wetland areas and key driving factors over six years. Results indicated that RF achieved the highest classification accuracy. Throughout the course of the study, the area of small wetlands increased from 9,114.42 hectares to 10,706.84 hectares, while their numbers declined from 22,279 to 21,338. The findings of the study indicated a substantial interaction between built-up land, average annual precipitation and elevation, which exerted a significant influence on wetland dynamics. The investigation further revealed that the combined effect of these factors exceeded that of a single factor when considered in isolation. The findings underscore the effectiveness of high-resolution remote sensing images and proper classifier selection in improving the accuracy of wetland research. To preserve the ecological functions of small wetlands, reducing anthropogenic disturbances is essential. These insights provide the foundation for scientific decisions that are instrumental in achieving the long-term, continued development of small wetlands.

Keywords Small wetlands · Land use transformation · Random forest · Ecological protection

J. Jin · J. Zhou · M. Lu · Q. Deng (✉)
College of Forestry and Landscape Architecture, Anhui Agricultural University, Hefei Anhui, China
e-mail: dwq@ahau.edu.cn

B. He et al. (eds.), *Climate-Resilient, Low-Carbon, and Sustainable Cities*, Lecture Notes in Civil Engineering 754, https://doi.org/10.1007/978-981-95-2169-2_36

1 Introduction

Wetlands are essential for maintaining ecological balance, serving as biodiversity reservoirs and critical components of natural ecosystems [1]. It is evident that small wetlands exhibit higher species diversity and perform unique ecological functions. These functions include water purification, flood mitigation, irrigation and water supply, production of fishery resources, habitat connectivity and climate change mitigation [2]. However, these wetlands are frequently excluded from inventories, leading to neglect and inadequate protection [3].

Recent studies have shifted focus from large wetlands to smaller ones, recognizing their vital role in ecosystem stability [4]. Notably, the Chinese government launched the "Small Wetland Conservation and Management" initiative at the COP13 preparatory meeting of the International Convention on Wetlands, and this initiative attracted considerable attention from the international community [5]. Concurrently, advancements in remote sensing technologies, particularly China's Gaofen (GF) satellite series, have enabled fine-scale identification and classification of small wetlands through enriched multispectral data. Besides, the employment of machine learning algorithms, including RF, SVM and MLE, has further enhanced mapping precision and efficiency, rendering them effective tools for monitoring small wetlands.

Despite these advancements, significant research gaps persist. Small wetlands remain underrepresented in inventories, and limited studies have explored their dynamic distribution, area changes, and interactions with other land-use types [6]. Traditional remote sensing approaches, such as the Landsat series with 30 m resolution, lack the precision required for detailed mapping of small wetlands. Moreover, the relative impacts of natural drivers and anthropogenic issues on wetland degradation and transformation are not well understood [7].

This study focuses on addressing critical research gaps in the Chaohu Lake region, a key wetland area in China. The objectives include utilizing high-resolution GF-1 and GF-6 satellite imagery with machine learning classifiers (RF, SVM, and MLE) to improve classification accuracy [8], monitoring changes in wetland area and quantity with field survey validation, and applying the Geoprobe model to identify natural and anthropogenic drivers of wetland changes [9]. Ultimately, the study aims to establish a theoretical framework for sustainable management, restoration, and planning of small wetlands, contributing to broader conservation efforts regionally and globally.

2 Material and Methods

2.1 Research Region

The present research concentrates on the geographical area encompassing Chaohu Lake (Fig. 1). The research area corresponds to the projects "Master Plan for the Wetland Park Group around Chaohu Lake," encompassing a lake, two municipal

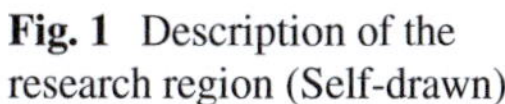

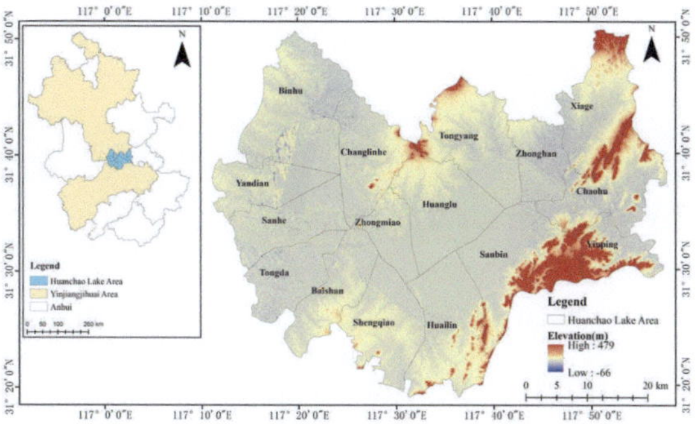

Fig. 1 Description of the research region (Self-drawn)

regions, and fourteen towns. Rapid urbanization has caused significant loss of cropland, forest land, and small wetlands, leading to degraded ecological functions. Consequently, it is paramount to map the distribution of small wetlands and to undertaking a quantitative analysis of their dynamics.

2.2 *Data Collection Source*

The spatial resolution, temporal coverage, and sources of the remote sensing data utilised in this study are detailed in China Centre for Resources Satellite Data and Application. This study utilized data collected by the GF-1 and GF-6 satellites, equipped with PMS1 and PMS2 sensors, during April 2014, April 2015, October 2018, May 2019, and March to April 2021. These sensors deliver high-resolution imagery with a spatial resolution of 2 m for panchromatic data and 8 m for multi-spectral data, ensuring the level of precision necessary for detailed and accurate analysis.

The elevation data, which were obtained from the Geospatial Data Cloud, were used alongside population density data sourced from WorldPop. Meanwhile, the temperature, GDP, and administrative boundary data, provided by the Resource and Environment Science and Data Center of the Chinese Academy of Sciences, were incorporated into the analysis. Precipitation data, collected from the National Earth System Science Data Center, and policy documents, retrieved from the official websites of the respective township governments, were also included in the study.

Following the Technical Specifications for Investigation and Evaluation of National Ecological Status—Field Observation of Wetland Ecosystems (HJ1169-2021), field investigations were conducted in the Chaohu Lake area in July and December 2022, and February 2023. A total of 191 small wetland and water body sample sites were surveyed, including 31 rivers, 71 reservoirs, 7 pit ponds, 26 ditches, 9 paddy fields, 24 aquaculture ponds, and 23 urban landscape water bodies. The geographic coordinates of each site were recorded using handheld GPS, and the wetland types were documented to provide training samples for subsequent data interpretation.

2.3 Research Approach

The methodological framework of this research consisted of three main stages: (1) Data preparation: High-resolution satellite imagery and on-site survey datasets collected in the years two thousand and fifteen, two thousand and eighteen, and two thousand and twenty-one underwent processed. Acquisition of land use data by means of RF, SVM and MLE algorithms with the ENVI5.6 platform, with accuracy verified using a confusion matrix and Kappa coefficient. Small wetland patches around Chaohu Lake were extracted using ArcGIS 10.8. (2) Analysis of wetland changes: ArcGIS 10.8 was employed to evaluate the extent, patch count, and change rate of small wetlands, as well as unidirectional transitions in specific wetland types across townships. (3) Driving factor analysis: Seven environmental and human-induced driving variables were identified based on feasibility, representativeness, and temporal consistency to investigate the changes in small wetlands within the research region.

Image processing for remote sensing. *GF Remote Sensing Image Preprocessing*. The processing of the GF series remote sensing images was undertaken with the utilisation of ENVI15.6 software. This process involved a number of steps, including radiometric calibration, atmospheric and geometric corrections, followed by image mosaicking, to produce a comprehensive remote sensing map of the study area [10] (Fig. 2).

*Confusion matrix***.** A confusion matrix evaluates classification performance by comparing true and forecasted categories. In this research, five metrics were selected for the purpose of evaluating accuracy: overall accuracy, producer accuracy, user accuracy, misclassification error and omission error.

Fig. 2 Pre-processed GF-1 and GF-6 remote sensing images in the years two thousand and fifteen, two thousand and eighteen, and two thousand and twenty-one (Self-drawn)

Overall Accuracy; The ratio of samples that have been correctly categorised out of the total number of samples. It is calculated by the formula:

$$P_{OA} = \frac{\sum_{i=1}^{n} P_{ii}}{N} \tag{1}$$

Producer's Accuracy; The proportion of reference samples in a given category that are correctly classified during the classification process. The formula is:

$$P_{PA} = \frac{P_{ii}}{P_{+i}} \tag{2}$$

User's Accuracy; The ratio of samples that have been classified into a particular category and which, on further investigation, are found to belong to that category, is determined by reference data. The formula is:

$$P_{UA} = \frac{P_{ii}}{P_{i+}} \tag{3}$$

Random Forest (RF). The proposed methodology is an ensemble-based machine learning approach that employs categorical regression trees to evaluate the effect of an independent variable on a dependent variable in a complex dataset with numerous unknown features. It aggregates multiple independent decision trees without interdependence. RF has been widely applied to complex geographical issues, including wetland remote sensing classification, urban ecological spatiotemporal patterns, and driving factor analysis.

Dynamic analysis of the speed and trend of change in small wetlands. To quantify variations in small wetlands and spatial disparities in their change rates, Eqs. 4 and 5 were employed to compute the proportion of small wetland areas within each administrative district and the dynamics of specific wetland types across townships.

The formula is as follows:

$$W_i = \frac{S_i}{P_i} \times 100\% \tag{4}$$

where W_i (%) indicates the percentage of small wetlands within each administrative region S_i (hm^2) denotes the total surface area of small wetlands in each region; i refers to the specific administrative regions; and P_i represents the overall area of each corresponding administrative region.

The formula is as follows:

$$K = \frac{U_b - U_a}{U_a} \times \frac{1}{T} \times 100\% \tag{5}$$

where K indicates the direction of change during the study timeframe, U_a represents the baseline area at the beginning of the analysis, U_b denotes the final area at the end of the observation period, and T refers to the length of the study duration.

Geoprobe Model. The Geodetector method is a spatial statistical tool based on the principles of spatial variance analysis. Of particular note is the factor and interaction method within Geodetector, which has been identified as a highly effective technique for quantitatively assessing ecological impact factors.

The formula is:

$$q = 1 - \frac{1}{N\sigma^2}\sum_{h=1}^{L} N_h\sigma_h^2 \tag{6}$$

where: h =1,..., L denotes the number of layers of factor X; N and N_h denote the total number of units in the overall domain and the h-story, respectively; σ^2 and σ_h^2 correspond to the variability of variable Y across the whole area and within the h-th layer; the parameter q lies within the range [0, 1], where a larger q value indicates that Y is more effectively explained and influenced by X.

The interaction analysis identifies the connection linking two distinct factors (Xi and Xj) as well as their joint impact on the dependent variable (Y). The strength of these interactions is assessed by comparing q(Xi ∩ Xj) with q(Xi) and q(Xj). Interactions are categorized into five types: nonlinear weakening, single-variable attenuation, independence, dual-variable enhancement, and nonlinear amplification.

3 Results

3.1 Acquisition and Analysis of Information on Small Wetlands

Production of land use maps. Small wetlands are mainly concentrated in rural and remote areas, where water bodies constitute their fundamental components. This study concentrated on aquatic ecosystems and rice paddies with sizes ranging from 0.06 to 8 hm^2. Due to constraints in acquiring high-resolution images and the study's objectives, single-month satellite data from the steady water phase between 2015 and 2021 were employed to investigate small wetlands in the vicinity of Chaohu Lake. Three classification algorithms—RF, SVM, and MLE—were utilized to analyze 2 m resolution remote sensing imagery from 2015, 2018, and 2021. The outcomes are illustrated in Fig. 3.

Accuracy evaluation and comparative verification. The classification results derived from the three methods were evaluated in terms of accuracy using the confusion matrix and Kappa coefficient. In 2015, RF achieved the highest OA (87.66%) and

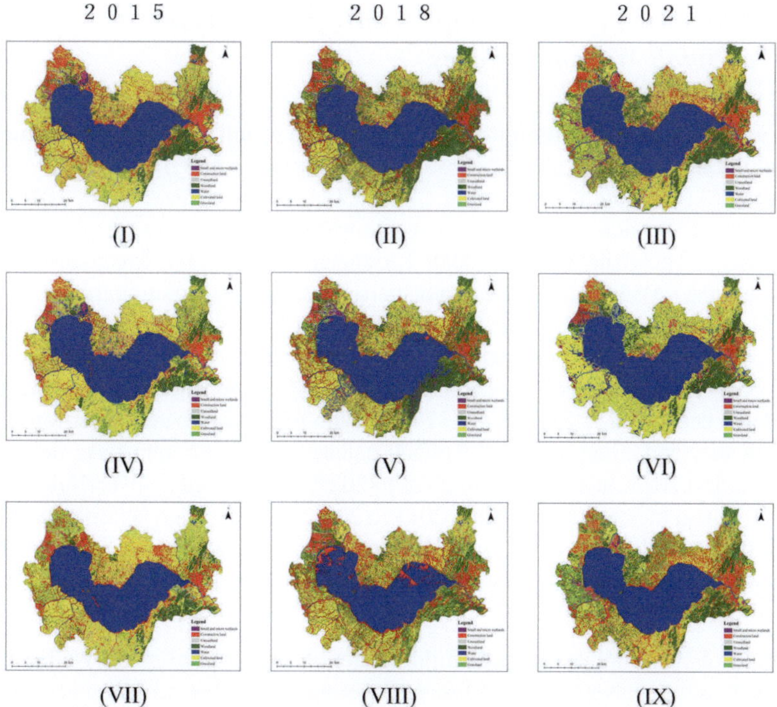

Fig. 3 Visualization of classification results generated by the Random Forest model (I, II, III), Support Vector Machine (IV, V, VI), and Maximum Likelihood Method (VII, VIII, IX), in Two thousand fifteen, Two thousand eighteen, and Two thousand twenty-one. (Self-drawn)

Kappa (0.84), followed by SVM (OA: 85.52%, Kappa: 0.81), while MLE showed weaker performance (OA: 75.97%, Kappa: 0.69). In 2018, RF remained superior (OA: 78.97%, Kappa: 0.73), outperforming SVM (OA: 77.79%, Kappa: 0.71) and MLE (OA: 70.28%, Kappa: 0.62). By 2021, RF continued to lead (OA: 82.32%, Kappa: 0.78), whereas SVM dropped to an OA of 70.89% and Kappa of 0.62, and MLE declined further to an OA of 66.38% and Kappa of 0.57. Among the three methods evaluated—RF, SVM, and MLE—the RF method achieved the highest levels of overall precision and the Kappa coefficient, and the RF classifier exhibited relatively lower misclassification and omission errors for small wetlands, thereby significantly enhancing detection precision. And its classification results for the years 2015, 2018, and 2021 satisfied the required accuracy thresholds, underscoring their reliability for subsequent analysis. Hance, among the three classification methods, RF was identified as the most effective approach, fulfilling the methodological and research requirements of this study.

3.2 *Evaluation of Alterations in the Area and Quantity of Small Wetland Patches*

The overview of spatiotemporal distribution of small wetlands surrounding Chaohu Lake. The RF algorithm was implemented in ArcGIS to identify small wetlands (0.06–8 hm^2) surrounding Chaohu Lake for the years 2015, 2018, and 2021. The area, patch count, and percentage were calculated, and the spatial distribution map is shown in Fig. 4. The spatial arrangement of small wetlands surrounding Chaohu Lake (2015–2021) exhibits a dispersed, star-shaped configuration. In 2015, 22,279 wetlands covered 9,114.42 hm^2 (2.58% of the research region). By 2018, this decreased to 17,595 wetlands and 7,375.75 hm^2 (2.09%). In 2021, it increased to 21,338 wetlands covering 10,706.84 hm^2 (3.03%). Wetland areas in townships fluctuated between 287.28 and 777.32 hm^2, with 685–1,919 patches per township. The quantity and extent of small wetland patches around Chaohu Lake decreased from 2015 to 2018, increased from 2018 to 2021, and showed an overall decline in number but an increase in area from 2015 to 2021. This variation is attributed to the development of fisheries in 2021, particularly in Tongyang, Tongda, and Zhonghan Townships, where arable land was converted into breeding ponds, increasing wetland area (Fig. 5).

Analysis of the change rate and unidirectional variation of small wetlands. A comprehensive analysis of 32 datasets, covering a six-year period, revealed that the proportion of small wetlands within townships remained comparatively low, at less than three percent. Further analysis of the dynamics and direction of change reveals substantial differences in the pace of transformation of small wetlands among various townships surrounding Chaohu Lake. This finding highlights the spatial heterogeneity of small wetland area changes at the township scale.

Between 2015 and 2018, construction land expansion caused the largest reduction in small wetlands in Binhu New Area (–13.77%) and Xiage Township (–12.99%). From 2018 to 2021, Three Rivers Township saw rapid wetland growth from 279.91 to 726.14 hm^2 due to new polders. The overall dynamic change rate of small wetlands from 2015 to 2021 was two point nine one percent, with Zhonghan Township exhibiting the most significant variation (24.82%), increasing from 295.03 hectares to 42,716.76 hectares. The directional trend of small wetlands shifted from –6.23% to 2.91%, indicating overall growth and development. To clearly illustrate temporal

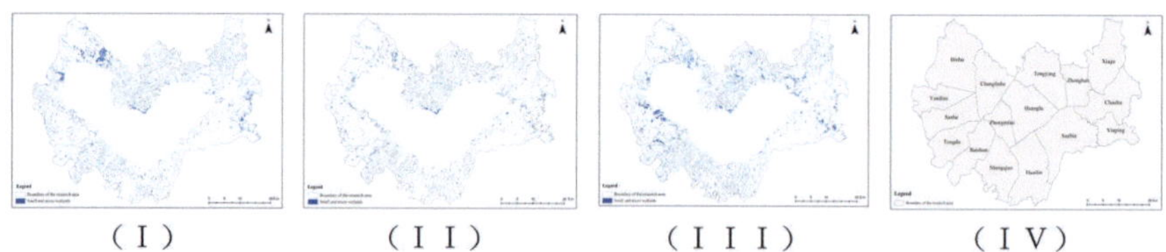

Fig. 4 Spatial distribution maps and administrative boundary maps of small wetlands for 2015 (I), 2018(II), 2021(III)

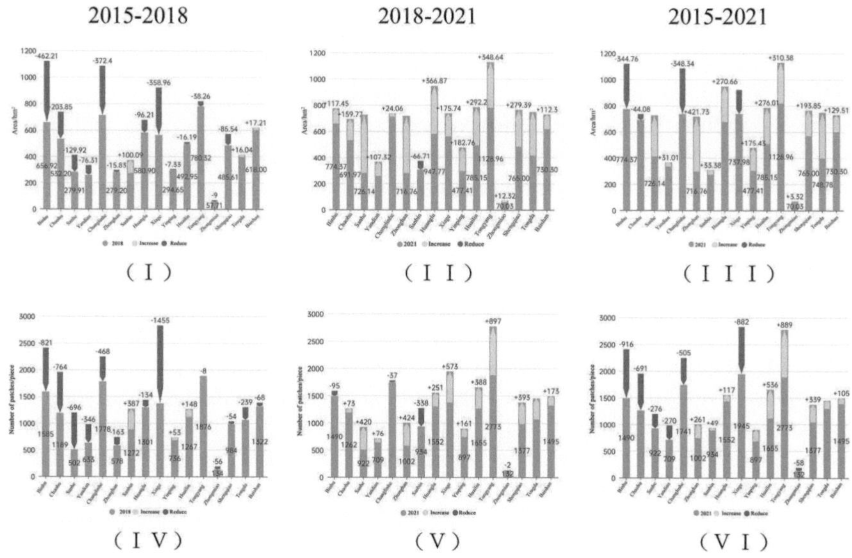

Fig. 5 Trend graphs of small wetland areas (I, II, III) and patch counts (IV, V, VI) from 2015 to 2021 (Self-drawn)

differences in small wetland distribution across townships, their area and proportions for 2015, 2018, and 2021 were depicted, compared, and examined. In 2015, the top three areas for small wetland area were Binhu New Area, Changlinhe, and Xiage, with the highest proportions in Changlinhe, Binhu New Area, and Xiage. In 2018, Tongyang, Changlinhe, and Binhu had the largest areas, while Baishan, Changlinhe, and Tongyang had the highest proportions. In 2021, Tongyang, Huanglu, and Huailin led in area, with the highest proportions in Zhongcong, Tongda, and Tongyang.

3.3 *Correlation Analysis of Drivers*

Qualitative Analysis of Driving Factors. In this research, spatial distribution maps, which include mean annual precipitation, temperature, GDP, population density, and topographic elements, were acquired from the National Earth System Science Data Centre for the years 2015, 2018, and 2020. The research results indicate that precipitation exhibited a fluctuating trend: an increase followed by a decrease during 2015–2018, and a decrease-increase–decrease pattern during the period from 2018 to 2021. The precipitation center of gravity shifted eastward to westward, then southward, corresponding to the spatial variations in small wetlands and the migration of their center of gravity around Chaohu Lake. The mean temperature in Anhui Province gradually increased from 2015 to 2021, stabilizing at 16.6 °C during 2017–2020 before peaking at 17.2 °C in 2021, the highest in six years. In socio-economic terms, the population across 13 townships experienced steady growth from 2015 to 2021, except for Yinping Township, which exhibited a decline followed by an increase. GDP in the region rose significantly, increasing from 222,469 million Yuan in 2015 to 456,123 million Yuan by 2021, driven by accelerated regional development. Topographically, the northeast and southeast are characterized by mountainous terrain, including Bushan and Jimaoyan, while the west, south, and north are relatively flat. Due to unavailable 2021 data, 2020 data were used as a substitute.

Quantitative Assessment of Driving Factors. Using the GeoDetector factor detection module, q-values of seven driving factors affecting small wetlands in 2015 and 2018 were calculated (Table 1). All q-values were >0, indicating the combined effects of natural and human-induced factors on wetland distribution. As presented in Table 2, seventy-one point four percent of the results were statistically significant, with notable differences in factor contributions between the two years. In 2015, annual precipitation had the highest contribution (q = 0.123), while in 2018, elevation was the dominant factor (q = 0.121), with wetlands more aggregated at lower elevations. Construction land and slope also showed high q-values, emphasizing their role in spatial differentiation. In general, topography, construction land, and precipitation were identified as key factors driving spatial differentiation in small wetlands. Among human-influenced factors, GDP and population density exhibited relatively low q-values but still made notable contributions to spatial variability.

Table 1 Factor detector Q-values for 2015 and 2018 (Self-drawn)

Factor indicators	MAP	MAT	GDP	PD	Building site	Altitude	Elevation
2015	0.123***	0.069**	0.053*	0.031*	0.120***	0.118***	0.095***
2018	0.034*	0.065***	0.030***	0.082*	0.098***	0.121***	0.095***

Note ***$p < 0.01$; **$p < 0.05$; *$p < 0.1$; MAP: Mean annual precipitation; GDP: Gross Domestic Product; MAT: Mean annual temperature; PD: Population density

Table 2 Results of interactive detection for spatial differentiation drivers of small wetlands in 2015 (Self-drawn)

Year	Impact factor	MAP	MAT	GDP	POP	Building site	Altitu-de	Elevation
2015	MAP	0.123*						
	MAT	0.235*	0.069*					
	GDP	0.277*	0.259*	0.053*				
	PD	0.231*	0.293*	0.290*	0.031*			
	Building site	0.375*	0.257*	0.242*	0.300*	0.120*		
	Altitude	0.307*	0.337*	0.234*	0.224*	0.281*	0.118*	
	Elevation	0.236*	0.254*	0.250*	0.182*	0.248*	0.224*	0.095*

Note *Represents a nonlinear enhancement between the two factors

Table 3 Results of interaction detection for spatial differentiation drivers of small wetlands in 2018 (Self-drawn)

Year	Impact factor	MAP	MAT	GDP	POP	Build-ing site	Altitu-de	Elevat-ion
2018	MAP	0.034*						
	MAT	0.286*	0.065*					
	GDP	0.239*	0.234*	0.134*	0.082*			
	PD	0.277*	0.254*	0.242*	0.336*	0.098*		
	Building site	0.286*	0.385*	0.284*	0.323*	0.351*	0.121*	
	Altitude	0.273*	0.314*	0.244*	0.222*	0.285*	0.225*	0.095*

In this research, an interaction detector approach, which was utilised to investigate the interactions among seven driving factors and their effects on the space distribution of small wetlands, was employed. Permutations of the seven factors yielded 28 driver pairs (Table 2). Comparing Tables 1 and 2, no mutual independence or weakening effects were observed. Instead, all interactions showed nonlinear enhancement, with q-values of factor pairs exceeding those of individual factors, indicating strong synergistic effects. As shown in Table 2, in 2015, the interaction between annual rainfall and building site exerted the most significant influence on small wetland distribution around Chaohu Lake ($q = 0.375$). In 2018, as shown in Table 3, altitude and annual temperature had the greatest impact ($q = 0.385$). Analysis of single-factor interactions revealed that precipitation, altitude, and construction land were the primary drivers of small wetland changes. These factors require comprehensive consideration in future planning and ecological restoration efforts.

4 Conclusion

This research analyzed the characteristics, trend and driving mechanisms of small wetlands in the vicinity of Chaohu Lake between 2015 and 2021. Key aspects included wetland area, patch count, spatial distribution, one-way changes, and influencing factors. High-resolution remote sensing images and classification algorithms (RF, SVM, and MLE) were utilized, with RF consistently performing the best. The results revealed a general trend of an initial decrease in wetland extent, followed by subsequent growth, resulting in a net gain of 1,592.42 hectares. Meanwhile, the total number of patches was found to have decreased by 901, indicating a transition in the wetland landscape from fragmentation to greater connectivity and structural integration. The spatial distribution followed a "small aggregation, multi-core dispersion" pattern, reflecting both global wetland fragmentation trends and the positive impacts of conservation efforts, particularly during 2018–2021.

Driving factor analysis revealed that annual precipitation, elevation, and construction land are the primary drivers of wetland changes, with significant nonlinear interactions demonstrating strong synergistic effects. These findings underscore the intricate interplay between natural and anthropogenic factors, where conservation policies, as key human interventions, have played a pivotal role. Policies implemented by the Chinese government significantly promoted wetland recovery, emphasizing the importance of integrated conservation strategies. However, the study faces several limitations, including inconsistent temporal remote sensing data, neglect of seasonal dynamics, and underestimation of policy impacts, which constrained its scope and accuracy. Additionally, strict criteria (0.06–8 hectares) and limited consideration of interannual wetland variability posed challenges in identifying small wetlands.

The findings align with global observations, such as wetland fragmentation in India, declines in Vietnam's Mekong Delta, and reclamation-driven losses in Egypt's Lake El-Brurus. These cases underscore the universal pressures on wetland ecosystems and the necessity for coordinated conservation and restoration initiatives. Despite its limitations, this research establishes a solid foundation for analyzing the temporal and spatial dynamics of small wetlands, offering valuable insights for the ecological management of wetlands around Chaohu Lake.

Building on these findings, this study proposes strategies to optimize the ecological environment of small wetlands, structured across macro, meso, and micro levels. At the macro level, a "one network, two cores, three zones" spatial conservation framework is developed to optimize the regional ecological network. At the meso level, township-level protection and restoration strategies are formulated based on the intensity of wetland dynamic changes to enhance ecological functions. At the micro level, targeted measures are proposed, including bank revetment design, vegetation configuration, cultural function integration, and wetland cluster construction, addressing key drivers and landscape design elements.

In addition, this study puts forward a conservation and recovery strategy for small wetland ecosystems, encompassing three levels: an overarching conservation and restoration pattern, a zoning conservation and restoration strategy, and a conservation and restoration design practice in the Chaohu Lake. The overarching protection and restoration is based on a macro pattern, which is refined into a zonal protection and restoration strategy, dividing the study area into five functional zones: comprehensive management region, ecological restoration region, ecological enhancement, ecological improvement region, and ecological conservation region. Furthermore, targeted protection and restoration countermeasures are developed in combination with the influencing factors identified in the study. These strategies aim to guide the sustainable management and restoration of small wetlands in Chaohu Lake and similar regions, bridging theoretical research and practical applications.

Future research should address current gaps by incorporating seasonal and long-term dynamics, refining classification methods, and expanding datasets to capture additional natural and anthropogenic drivers. Integrating policy analysis and improving temporal matching of data will further enhance the understanding of wetland changes. Ultimately, this study establishes a solid foundation for promoting ecological security and developing targeted strategies for the conservation and rehabilitation of small wetlands in Chaohu Lake and similar regions.

Acknowledgements This work was supported by General project of Humanities and Social Sciences Rsearch, Ministry of Education in China (No.22YJC630047), Anhui Province of China social science innovation development research subject (No.2021CX194) and Quality Engineering Project of the Anhui Provincial Department of Education (2023jyxm0215).

References

1. Gxokwe S, Dube T, Mazvimavi D (2022) Leveraging google earth engine platform to characterize and map small seasonal wetlands in the semi-arid environments of South Africa. Sci Total Environ 803:150139
2. Blackwell MSA, Pilgrim ES (2011) Ecosystem services delivered by small-scale wet-lands. Hydrol Sci J 56:1467–1484
3. Kim B, Lee J, Park J (2022) Role of small wetlands on the regime shift of ecological network in a wetlandscape. Environ Res Commun 4
4. Liao K, Song Y, Xie S, Luo Y, Liu Q, Lin H (2022) Quantitative analysis of the fac-tors influencing the spatial distribution of Benggang landforms based on a geograph-ical detector. ISPRS Int J Geo Inf 11:337
5. Richardson SJ, Clayton R, Rance BD, Broadbent H, McGlone MS, Wilms-hurst JM (2015) Small wetlands are critical for safeguarding rare and threatened plant spe-cies. Appl Veg Sci 18:230–241
6. Riley JW, Calhoun DL, Barichivich WJ, Walls SC (2017) Identifying small depres-sional wetlands and using a topographic position index to infer hydroperiod regimes for pond-breeding amphibians. Wetlands 37:325–338
7. Ulrich U, Lorenz S, Hörmann G, Stähler M, Neubauer L, Fohrer N (2022) Multiple pesticides in lentic small water bodies: exposure, ecotoxicological risk, and contami-nation origin. Sci Total Environ 816:151504
8. Zheng Y, Liu H, Zhuo Y, Li Z, Liang C, Wang L (2019) Dynamic changes and driving factors of wetlands in Inner Mongolia Plateau, China. PLoS One 14:e0221177
9. Feng Y, Wang S (2022) Study on dynamic changes of land use in Ganzhou District and Linze County from 2005 to 2020. Henan Sci Technol 41(07):116–120
10. Long T, Jiao W, He G, Wang G, Zhang Z (2023) Domestic optical satellite orthorecti-fied image products and automatic generation algorithms. J Remote Sens 27:635–650

Raster-Based Techniques for Efficient and Pertinent Evaluation of Morphological Urban Ventilation Potential

Yuli Fan, Guangzhong Chen, Huizi Zhang, Wei Xue, and Qipeng Liao

Abstract Morphological properties of buildings and open spaces are important aspect for evaluating the ventilation potential in urban space. Current techniques for calculating frontal area density, road orientation consistency and space openness are mostly based on vector objects, which inherently limits its accuracy and location specifity due to non-negligible size of each object relative to the phenomenon in question. Also, reliance on geometric calculations can be time-consuming when facing large data volumes. In light of this, we developed a series of workflows that exploit features and techniques such as orthogonal indexing, sliding-window filtering and mathematical morphology. Using these techniques: (1) time and memory consumption for FAD calculations can be reduced by over 60%; (2) road orientation windward consistency can be calculated locally rather than for each road segment, ruling out inaccuracy on curving roads without significant increase in cost. (3) semi-enclosed open space can be accurately detected and described when designated with an opening width threshold, either in isotropic or anisotropic way. These workflows either improves the calculation of existing indicators or provide new indicators that are closely related to the mechanisms of near-surface wind flow, which could be utilized in the construction of surrogate ventilation models.

Keywords Ventilation potential · Mathematical morphology · Frontal area density · Orientation consistency · Semi-enclosed space

---

Y. Fan · G. Chen · Q. Liao
School of Public Administration, China University of Geosciences Wuhan, Wuhan, China

H. Zhang
PowerChina Zhongnan Engineering Co., Ltd., Changsha, China

W. Xue (✉)
School of Digital Construction and Blasting Engineering, Jianghan University, Wuhan, China
e-mail: 2019102090003@whu.edu.cn

Hubei Key Laboratory of Industrial Fume and Dust Pollution Control, Wuhan, China

B. He et al. (eds.), *Climate-Resilient, Low-Carbon, and Sustainable Cities*, Lecture Notes in Civil Engineering 754, https://doi.org/10.1007/978-981-95-2169-2_37

1 Introduction

Ventilation potential assessment provides background knowledge for the construction of urban ventilation corridors, thus facilitating the mitigation of urban heat islands and the improvement of outdoor thermal comfort [1, 2]. In practice ventilation potential is mostly assessed from two aspect: the morphological aspect, and the thermodynamical aspect [3]. Morphological potential is closely connected with terrain and buildings, which are major morphological obstacles to natural ventilation in urban environment, and furthermore with major roads and greenspace, serving as open space that are spatially continuous [4, 5]. Therefore morphological indicators are basically designed around three key features: (1) spatial openness, indicating whether there is adequate spaciousness for the flow of natural wind; (2) directional consistency, indicating whether road or other open spaces are directionally consistent with natural wind, thus facilitating its smooth movement; and (3) obstacle density, indicating the density of windward obstacles on both vertical and horizontal dimensions that may hinder the introduction of natural wind [5].

Technically, these indicators are commonly calculated in a vector or object-based fashion: buildings and roads are organized as vector objects in geodatabases or shapefiles, and initial parameters such as orientation, height, etc. are calculated and recorded for each building or each road segment in advance. In calculating frontal area density, or FAD, some would organize the building data into grid datasets, yet still calculating the respective normal vector for each grid cell, making it essentially still an object-based technique [6]. This raises several issues: (1) the volume of calculation grows rapidly with the number of buildings or roads in question, or the fineness of the original vector data. For example, vector-based calculation of FAD requires the interception of projection lines and building outline, which is a consumptive topological operation. To simplify the original vector data would introduce more complexity and workload. (2) the objectification process does not necessarily serve the purpose of ventilation potential assessment. For example, road segments may be continuous curves in many instances, in which case the orientation of the whole segment does not fully reflect its ventilation potential. (3) the dependence on GIS platforms. The low compatibility of vector shapefiles restricts most tools for calculating morphological ventilation potentials to GIS platforms, which are not originally designed for large-scale calculations. This inflexibility limits the potential of pertinent improvement on algorithms and workflows.

With the above in mind, we introduce a series of grid-based methods for faster, less consumptive and more pertinent calculation of morphological ventilation potentials. These techniques conceal the concept of individual buildings, road segments, etc. and regard them as aggregations of grid cells that are identical in horizontal shape, thus utilizing convenient techniques such as morphological transformation, sliding window filtering and mathematical morphology [7]. It also allows us to develop software applications on matrix-focused platforms such as MATLAB, further increasing calculation efficiency and enabling ordinary planners and analytics to perform larger-scale analysis on less demanding hardware. In this paper, we will introduce the principles and processes of these techniques in detail, and provide comparison with existing methods in terms of performance and accuracy.

2 Frontal Area Density

The existing calculation methods for FAD primarily fall into two categories: vector calculation models and grid calculation models. (1) In vector-based model, three-dimensional vector building data is used as the data source. The map area would then be separated into square blocks, most commonly 50 by 50 m to 500 by 500 m. A projection line perpendicular to a specific wind direction is determined in that direction, and the position information of the two endpoints of this projection line is obtained. Then, for each building block, the length of the windward line for each building is determined by searching for the relationship between the nodes of the building outline and the projection line vector perpendicular to the wind direction. This is combined with the building height to obtain the windward area. Finally, the windward area density is calculated as the ratio of the total windward area on the plot to the plot area. (2) In grid-based model, a 2.5 dimensional grid matrix is created, with each grid representing the building height at the corresponding location. Standard image processing algorithms are used to obtain the unit normal vector of each pixel on the DEM surface, where the determination of the normal vector is centered on the pixel, and a triangular surface is constructed by the cross-hatch method. The normal vector of the current pixel center point is obtained by weighted averaging the normal vectors of multiple triangular surfaces around the pixel. Then the dot product of the normal vector of the pixel and the horizontal unit vector in the given wind direction is calculated to obtain the projection of the pixel in the given direction. Therefore, the negative results of the dot product of all pixels in the given direction are accumulated to obtain the windward area density in that direction.

Either way, the existing methods involves respective processing of each and every object, whether the object is the outline of a building or a matrix grid. By exploiting the features of raster grid data, we established a rotation-segmentation workflow rather than the original segmentation-projection workflow, which largely reduces the time consumption of geometrical calculations. Here we introduce this new workflow based on global morphological transformation, with example code on MATLAB platform:

a. Similar to the original grid-based model, transform the building data into a 2.5 dimension raster data ***Bh***, in which the column and row of each grid indicates its location and the value on that grid indicates the building height at the location. Import ***Bh*** into MATLAB as geo-raster matrix. Let k be the raster resolution in meters. Here we have $k = 2$ considering the balance between data volume and the integrity of building outline.

```
[Bh, GeoRef] = geotiffread('Bh.tif');
GeoInfo = geotiffinfo(Bh);
```

b. Split the grid space into $m \times m$ square blocks by creating a label layer ***Lb*** with identical number of columns and rows with ***Bh***, where subsequent integer block identifiers are assigned into each raster grid. Here we have $m = 100$ by referring to most known exercises (Fig. 1).

```
[rows, cols] = size(Bh);
Lb = zeros(rows, cols);
n = 1;
for i = 1:ceil(rows/m)
    for j = 1:ceil(cols/m)
        Lb((i-1)*m+1:min(rows, i*m), (j-1)*m+1:min(cols, j*m)) = n;
        n = n+1;
    end
end
```

c. Let α be the incoming wind direction represented in degrees. By using morphological function imrotate, rotate ***Bh*** into $\boldsymbol{Bh}_\alpha$ and ***Lb*** into $\boldsymbol{Lb}_\alpha$ so that the windward direction faces downside. With fine raster resolution there would be negligible accuracy loss during this transformation (Fig. 2).

```
Bh_a = imrotate(Bh, a);
```

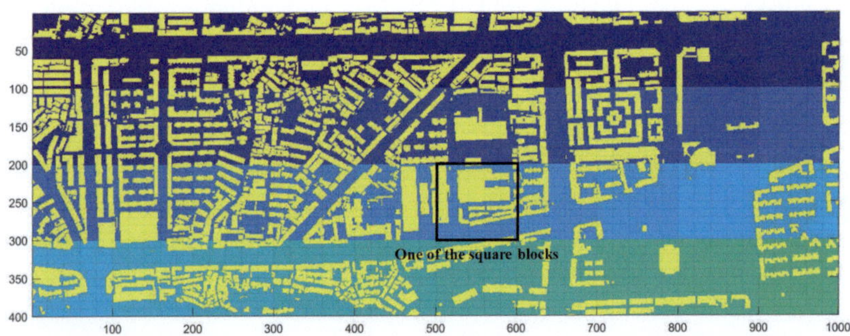

Fig. 1 Demonstration of 100×100 grid (200×200 m) square blocks. Building outlines overlayed in yellow for clearer presentation of spatial scale (Self-drawn)

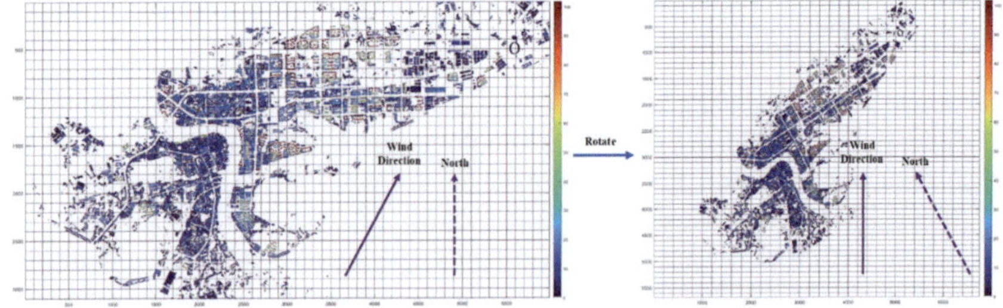

Fig. 2 Original and rotated building height raster (Self-drawn)

Lb_a = imrotate(Lb, a);

d. Create a k row integer list *R_list* to store the FAD results in each block, where k is the total number of blocks.

R_list = zeros(n, 1);

e. Subsequently extract the rotated building height data in each block by dot multiplication; in the extracted raster data ***Bl***, calculate the maximum values in each column *Bl_max*, which indicate the maximum building height of a segment projected to the windward direction. Sum up these maximum values into *Bl_sum*; *Bl_sum* $\times k$ would thus be the projected area of buildings in this block in the windward direction. Store the value of *Bl_sum* $\times k$ in the corresponding row of *R_list*. Parallel calculation can be utilized in this process to further improve performance, but to do so *R_list* must be transformed into a cell array rather than a raster matrix (Fig. 3).

```
for i = 1:n
   Bl = Bh_a.*(Lb_a==i) ;
   Bl = Bl(find(sum(Bl)>0,2),find(sum(Bl)>0)) ;
   Bl_max = max(Bl) ;
   Bl_sum = sum(Bl_max) ;
   R_list(i) = Bl_sum ;
end
```

f. Create a new empty matrix ***Result*** with the same size as ***Bh***. Assign FAD values in *R_list* to corresponding raster blocks in ***Lb***, thus restoring the rotated map into the original orientation and creating an FAD map under wind direction α (Fig. 4).

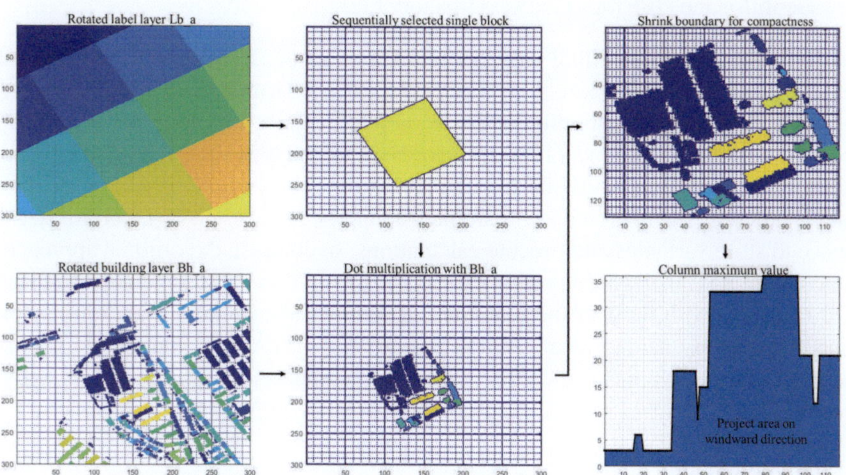

Fig. 3 Process of calculating the FAD of each block by using the rotated data (Self-drawn)

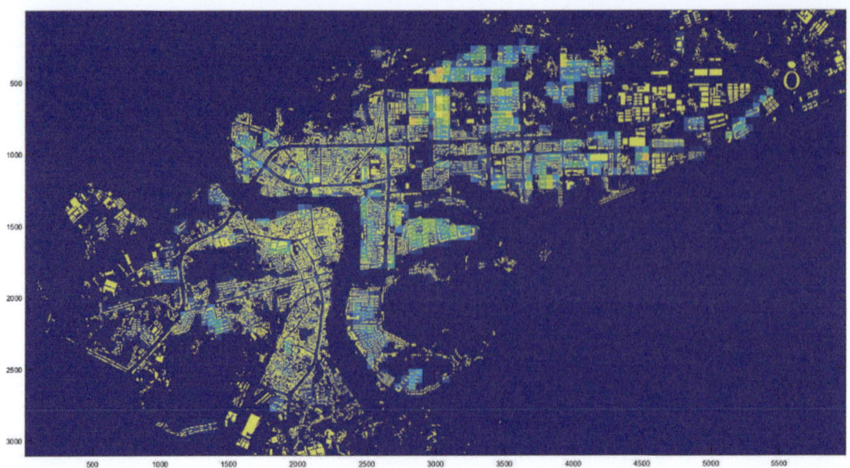

Fig. 4 Example of FAD calculation results. Building outlines overlayed in yellow for clearer presentation of spatial scale(**Self-drawn**)

```
Result = zeros(rows, cols)
for i = 1:n
    Result (Lb==i) = R_list(i) ;
end
```

g. Export ***Result*** to a geo-tiff according to the original spatial reference.

```
geotiffwrite('Result.tif', Result, GeoInfo, 'GeoKeyDirectoryTag',
GeoInfo.GeoTIFFTags.GeoKeyDirectoryTag);
```

Here we take the building data in Chibi City, Hubei as example. With a spatial extent of 6.2 km by 12.0 km, the data consists of 18,511 individual building object. Under the raster grid resolution of 2m and the block size of 100 m by 100 m and using a single Intel Core i7 11700 k CPU, the preprocessing, calculating, and value assigning processes cost 201.07 s, with block-by-block calculations taking 156.72 s and result assignment taking 37.12 s. The whole process takes up to 4.8 GB of RAM space to store variables and execute calculations. In contrast, the original approach, calculating FAD using the same data at the same resolution, cost 543 s and uses 16.5 GB of RAM, with most of the time used on geometric calculations and projection line creation. By using global rotation and converting FAD calculations to simple sums of matrix columns, we conserve 63% of computation time and 71% of RAM consumption.

3 Windward Orientation Consistency

Currently, the windward orientation consistency of roads is generally calculated by the orientation of each road segment in the polyline vector shapefile of roads in the area of interest. In most datasets, one road segment is defined by consistent road name. However, it is not unusual that a road can be bended along its path, resulting in two consequences: (1) The true orientation of the straight part of the road could be neglected. A road that consists of long straight lines along the windward direction can be easily underscored if there are sharp turns at the beginning or the end of the aforementioned straight line, and a potential corridor would be neglected. (2) A winding road could be mistakenly marked as effective ventilation corridor. If the orientation connecting the starting point and ending point of a however curving road points at the windward direction, it would be wrongfully overscored, whereas the curving nature would indeed deem it most unsuitable (Fig. 5).

Behind this problem is the object-oriented nature of the vector-based workflow. With each road segment regarded as a polyline object, most GIS platforms provide only the coordinates of the starting and ending point of the polyline in their toolboxes or attribute tables, such as Calculate Geometry in ArcGIS. ArcGIS Engine provides the program interface for acquiring the coordinate of each node on the polyline, the IPointCollection function; but on curving road segment, node-by-node calculations can bring the spatial resolution beyond the actual need in ventilation analysis scenario and thus demand extra storage and computing power.

With this in mind, we develop a new approach utilizing the local-based nature of sliding-window algorithms, and realize it by a raster-based workflow on the

Fig. 5 Demonstration of inconsistency between actual and calculated orientation using a segment of actual road data (Self-drawn)

MATLAB platform. Sliding window filters provide easy and efficient comparison with local values and known patterns; in this case, the pattern would be three raster grids representing roads aligned on the windward direction or a direction with certain degree of deviation, indicating the local road orientation around the center grid in question. The specific workflow is presented as follows.

a. Rotate the road vector data in ArcGIS platform by Data Management Tools/ Rotate tool. 13 new road vectors, $\overline{Rt_1}$ to $\overline{Rt_{13}}$, are created, respectively rotated by $(WindwardDirection \pm 15 \times n | n \in \{0,1,2,3,4,5,6\})$ degrees. Rotation is executed on original vector data rather than converted raster matrixes to maintain original directivity and continuity.
b. Convert the rotated vectors to raster matrixes and export them to.tif format. Grid size should be set small enough to distinguish neighboring roads but not too small to result in large file size. Considering the common interval of urban branch roads, we set the grid size to 10 m.

Import all rotated raster matrixes into MATLAB as elements in a cell array ***Rt***.

```
Rt = cell(13,1);
for i = 1:13
    filename = strcat('Rt_', num2str(i), '.tif');
    Rt{i} = imread(filename);
end
```

c. Create a 3 by 1 filter *c*, where all elements are assigned the value 1. During the sliding window filtering process, the first row of this filter will cover the downwind neighboring grid of the center grid, and the last row of the filter will cover the upwind neighboring grid of the center grid.

```
c = [1;1;1];
```

d. Use imfilter function to acquire filtered raster matrixes, and store them in a new cell array ***Filtered***. Dot multiply these filtered raster matrixes with the original raster grids in ***Rt*** to rule out none-zero values outside of roads. Grids with the value 3 indicate that the local orientation at the position of the grid is consistent with the rotation of the corresponding raster matrix; those with the value 2 indicate that either the upwind neighbor grid or the downwind neighbor grid is not a road grid. Then, recover the original orientation of each filtered raster matrix. Like in Step e of FAD calculations, these raster matrixes are cropped to recover its original extent (Fig. 6).

```
Filtered = cell(13,1);
for i = 1:13
    f = imfilter(Rt(i), c).*Rt(i);
    f = imrotate(-(wd-90+15*i));
    %wd is wind direction in degrees
    f = f(find(sum(f)>0,2),find(sum(f)>0)) ;
    Filtered{i} = f;
end
```

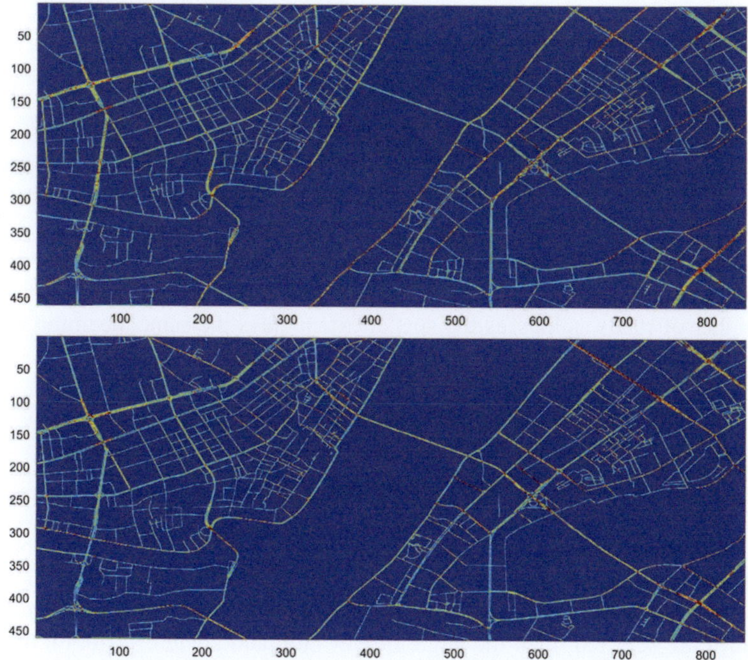

Fig. 6 Example of filtered data under different rotation angles. Data resampled for clearer visual presentation (Self-drawn)

e. Create a result matrix ***Result*** to store windward orientation consistency score of each raster grid. For a grid at the position (row, col), there is $n \in [1, 13] \cap N$ which ensures $Rt_n(row, col)$, then the consistency score is determined by the deviation indicated by n.

```
[rows, cols] = size(Rt{7});
% Acquire the original extent of unrotated raster matrix.
Filtered_3d = zeros(rows, cols);
for i = 1:13
   Filtered_3d(:,:,i)=Filtered{i};
end
```

% Aggregate recovered filtered raster matrixes in a 3-dimensional matrix, so that finding local maximum value would be convenient.

% cell2mat function is not usable in 3-dimensional conversions.

Max_1 = find(Filtered_3d = = max(Filtered_3d, [], 3), 'first');

Max_2 = find(Filtered_3d = = max(Filtered_3d, [], 3), 'last');

Max = (Max_1 + Max_2)/2;

% Acquire the position of first and last maximum value on each 3-d column, to cover the scenario where the orientation around a raster grid is near the mean value of two rotation angles.

Result = zeros(rows, cols);

Result = 100-abs(Max-7)*10;

% The closer the maximum filtered value is to the no-rotation raster matrix, the higher the consistency score.

Using the road network data of Wuhan, our approach cost 23 s in conversions, exports and calculations; whereas the traditional method cost 17 s with the same hardware. Costing 35% more time, we have taken the calculations of road windward orientation consistency to more data-adaptive and location-specific criteria. Also, the time and memory consumption of the raster-based approach will not change with the number of roads as long as the spatial extent stays the same, deeming it more adaptive for more complex networks.

4 Spatial Openness

Commonly, urban spatial openness is characterized by Sky View Factor, or SVF. This is realized by geometric calculations of building vector data: for every location in question, which usually are equidistantly sampled on roads and squares, buildings close to it are selected; then the angle of elevation and angle of azimuth between the location in question and points representing the location and height of buildings within range are calculated. Then, a hypothetical skyline is formed by connecting the largest-angle points at each direction, and SVF is calculated by the angular proportion of the space beyond the skyline to the hemisphere upon horizon.

SVF provides very useful general knowledge on the openness of a specific location; however, it does not provide many scalar details that are vital to the actual ventilation effect, especially for semi-enclosed structures surrounded by buildings which are very common in urban space: (1) is a given point in a semi-enclosed structure that may bring structural change to near-surface winds? (2) If so, how is its aspect ratio and what is its orientation? (3) How wide is its horizontal opening, is it wide enough for introducing mild natural wind? It is not easy in SVF calculations to sort out the spatial relationship among neighboring cells, and therefore so is it to obtain such indicators.

This section introduces a mathematical morphology-based approach for new indicators in characterizing spatial openness for urban ventilation evaluations. Opening width is obtained by gradual increase of dilation operator sizes until formation of independent regions; and orientation, size, and other features of any semi-enclosed structure is based on the morphological and geometrical properties of such region. The specific workflow is presented as follows.

a. Converse the original building data into raster grids as in Sect. 2a; let the created raster data be ***Bh***. Assign value 1 to raster grids that are not buildings or are buildings below a specific height that can be deemed effective in obstructing air flow.

   Bh = Bh < 1;

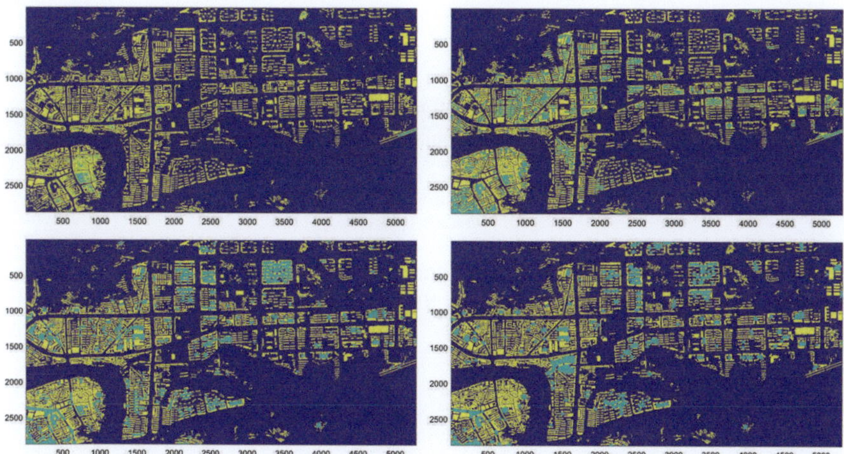

Fig. 7 Semi-enclosed regions extracted under different presumed opening width (Self-drawn)

b. Create a set of circular operators stored in cell array ***C***, which serve as detectors of morphological features at different scales. The size of the operators are set in range that near-surface air flow can be sensitive to. Here, the shape of the operator can be altered to ellipses etc. to indicate the anisotropy under known wind direction (Fig. 7).

```
C=cell(12);
for i = 1:12
C{i} = fspecial('disk', (i-1)*2+1)>0;
end
```

c. Conduct morphological opening operations on ***Bh*** using each of the operators; store the results in a new cell array Filtered. Use 'bwareaopen' function reversely to rule out the overly large regions, which are background open spaces formed by roads, greenspace etc.

```
Filtered = cell(12);
for i = 1:12
    filtered = imdilate(imerode(Bh,C{i}),C{i});
    background = bwareaopen(filtered, 10^6);
    filtered = filtered-background;
    Filtered{i} = filtered;
end
```

d. Use 'regionprops' function to label the area, orientation, etc. of the extracted semi-enclosed open spaces. Store the results in data tables in a created cell array ***Props***. These properties can be stored or displayed on the map by taking steps like in Sect. 2f.

```
Props = cell(12);
for i = 1:12
   table = regionprops(logical(Filtered{i}),'Centroid', 'MajorAxisLength', 'MinorAxisLength', 'ConvexArea', 'Extent', 'Orientation', 'Solidity');
   % Conversion to logical data for calculating binary region property
   Props{i} = table;
end
```

%% Example: Display orientation and solidity of semi-enclosed regions for a specific opening width on map (Fig. 8).

```
filtered = Filtered{7};
props = Props{7};
% Obtain scale-specific data
map1 = 0.*filtered;
map2 = map1;
% Create empty maps
label = bwlabel(filtered);
% Create label map for correspondence between map regions and property items
for i = 1:max(label(:))
   map1(label==i) = props(i).Orientation;
   map2(label==i) = props(i).Solidity;
end
```

In this way, we have obtained specific morphological information including size, orientation, solidity, etc. on semi-enclosed open-space structures in urban areas. Anisotropy can be further achieved by design of operator kernel or respective rotation of the original raster matrix. With significant connection with near-surface air flow, surrogate models could be designed for evaluation of ventilation potentials without numeric simulations.

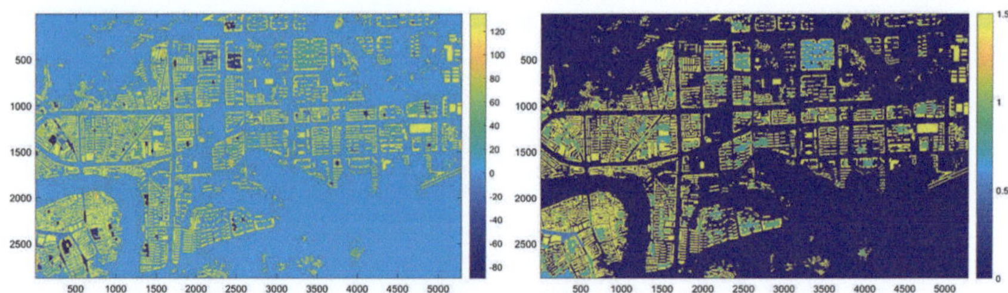

Fig. 8 Semi-enclosed regions orientation(left) and solidity(right) when opening width is 30 m (Self-drawn)

5 Discussion and Conclusion

In calculating three different morphological ventilation potential indicators, we have achieved different progresses by raster-based techniques: (1) Significantly less time and memory consumption in calculating FAD; (2) Overcoming inherent inaccuracy and realizing location-specificity in calculating windward orientation consistency of roads; (3) Introducing entrance spaciousness to the evaluation of spatial openness, especially for commonly seen semi-enclosed open spaces. Thus, the workflow or systemic accuracy of some existing indicators are improved, and some new indicators pertinent for complex built environment are created for potential use in evaluations or surrogate models [8, 9].

In a broader sense, the clever utilization and organic combination between object-based and location- or time point-based techniques is of great importance in establishing a proper workflow in geo-analysis [10]. In establishing the approaches in this paper, some key features of raster grid data are exploited; for various other issues, these features can also be put to use: (1) Spatially indexed by orthogonal directions, calculating intersections, maximus, sums etc. at a universal angle can be very efficient after global rotation transformation, as they are conversed from solving geometric equations to arithmetic operations on certain matrix columns. This is usually a solid presumption in ventilation evaluations, as background natural wind are basically homogeneous in a small area. (2) Building height, elevation etc. are stripped from affiliated objects (buildings, roads etc.) and are represented as the property of rasterized location. Thus, the results will not be interfered by the irrelevant overall characteristics of these objects. (3) Properties such as orientations and spatial openness can be regarded as image patterns, allowing morphological operations for their local-based calculations.

Acknowledgements The authors disclose receipt of following financial support for the research of this article: The Open Project Fund of Guangdong Provincial Key Laboratory of Intelligent Urban Security Monitoring Urban Security Monitoring and Smart City Planning (GPKLIUSMSCP-2023-KF-04), the Open Project Fund of Hubei Key Laboratory of Industrial Fume and Dust Pollution Control (HBIK2023-01), Youth Project of Hubei Provincial Natural Science Foundation (2024AFB110), Natural Science Foundation of Wuhan (2024040801020335), and Guiding Project of Scientific Research Program of the Hubei Provincial Department of Education (B2023254)

References

1. Ng E, Yuan C, Chen L, Ren C, Fung JC (2011) Improving the wind environment in high-density cities by understanding urban morphology and surface roughness: a study in Hong Kong. Landsc Urban Plan 101(1):59–74
2. Jasinski MF, Crago RD (1999) Estimation of vegetation aerodynamic roughness of natural regions using frontal area density determined from satellite imagery. Agric Forest Meteorol 94(1):65–77. https://doi.org/10.1016/S0168-1923(98)00129-4

3. Dang B, Liu Y, Lyu H et al (2022) Assessment of urban climate environment and configuration of ventilation corridor: a refined study in Xi'an. J Meteorol Res 36(6):914–930. https://doi.org/10.1007/s13351-022-2035-0
4. Qingming Z, Wanlu O, Zhicheng J et al (2015) RS and GIS based ventilation potential study and planning. Planners
5. Yin J, Yin J, Yin J et al (2021) The ventilation efficiency of urban built intensity and ventilation path identification: a case study of Wuhan. Int J Environ Res Public Health 18(21):11684. https://doi.org/10.3390/ijerph182111684
6. Juanjun S, Xinfa Q, Yongjian HE et al (2017) Study on comparison of vector/raster calculation model of frontal area density of urban buildings. J Geo-Inf Sci
7. Adolphe L (2001) A simplified model of urban morphology: application to an analysis of the environmental performance of cities. Environ Plann B Plann Des 28(2):183–200
8. Wu Y, Zhan Q, Quan SJ et al (2021) A surrogate-assisted optimization framework for microclimate-sensitive urban design practice. Build Environ 195(1–3):107661. https://doi.org/10.1016/j.buildenv.2021.107661
9. Zhang M, Guo X, Li J et al (2024) Urban-canopy airflow dynamics: A numerical investigation of drag forces and distribution for generic neighborhoods, and their relationships with breathability. Sci Total Environ 926. https://doi.org/10.1016/j.scitotenv.2024.171836
10. Fan Y, Zhan Q, Tang L et al (2021) Temporal characterization of minute-level PM2.5 variation within a local monitoring network using DWT-DTW. Build Environ 205:108221. https://doi.org/10.1016/j.buildenv.2021.108221

Land Use Regulation Strategies from the Perspective of Reducing Sources and Increasing Sinks

Danfeng Liu, Yang Du, and Junhui Hu

Abstract The dual carbon goal is a solemn commitment made by China to deal with global climate change contained in the territorial space system. As an important means of ecological civilization construction, territorial space regulation is of great significance for the implementation of the two-carbon goal. On the basis of systematic review of relevant studies, the connotation of source reduction and sink increase and the internal logic of territorial space response to source reduction and sink increase are expounded, and the concept of land space use control is clarified. The regulation strategies are proposed from the aspects of refining the control zoning of land space use, optimizing the transmission function of carbon reduction and carbon fixation in five levels and three categories of land space, establishing a control platform, exploring the measurement of carbon sinks and carbon emissions as the basis for the delineation of three regions and three lines, establishing a flexible transmission system and a control system for reducing sources and increasing sinks, etc., in order to provide references for the governance of land space.

Keyword Dual carbon goals · Territorial spatial planning · Land use control

1 Introduction

To address climate change on a global scale, the Chinese government has put forward the "double carbon target" of "2030 Carbon peak-2060 Carbon neutrality", and land space as an important carrier of "reducing sources and increasing sinks", carrying high-intensity land space development activities. Especially since the opening up and reform, various parts of China are faced with diversified problems such as low efficiency in the use and development of natural resources, climate change situation and extremely severe ecological environment problems, which, objectively, calls for a shift in land use control from safeguarding formerly farmed land to managing the use of national land space while taking all relevant circumstances into account. It was

D. Liu · Y. Du (✉) · J. Hu
Henan University of Science and Technology, Luoyang, China
e-mail: 1679265323@qq.com

B. He et al. (eds.), *Climate-Resilient, Low-Carbon, and Sustainable Cities*, Lecture Notes in Civil Engineering 754, https://doi.org/10.1007/978-981-95-2169-2_38

made explicit that "the use control of all territorial spatial zoning and classification shall be implemented on the basis of territorial spatial planning" when the state started to modernize the planning system in 2019. Use control is a crucial tool for carrying out territorial space planning, and it may enhance the effectiveness and capacity for governance of territorial space growth, provide a favorable institutional guarantee for optimizing the layout of territorial space, and become the focus of attention of all sectors of society. However, based on the huge potential of carbon sinks and carbon emission reduction in territorial space, there has not been any systematic results in the research on territorial space use regulation. Therefore, this study starts with the mechanism of land space response to source reduction and sink increase, and establishes the implementation path of land space use control by subdivision to provide support for deepening the theoretical discussion and institutional practice of land space use control.

2 Review and Analysis of Relevant Research

2.1 Research on the Control of Land Space Use

Since the promulgation of the "Several Opinions", the control of land space use has received extensive attention, and relevant researches have focused on the analysis of the connotation of control [1, 2], the reference of foreign experience [3, 4], the construction of control system and system [5], the process of control research and policy design [6], and the control of ecological space, with relatively rich results. It provides theoretical support for the relevant research of territorial space use regulation. However, due to the profound influence of the long-term multi-factor and multi-department classification control, the traditional control rules have been difficult to adapt to the internal requirements of the new era's territorial spatial planning framework, resulting in many problems such as the conflict of map spots of controlled land types, the difficulty of coordinating the use control line, the strong rigidity, and the weak control of ecological space. However, the control system of global all-factor coordination which is targeted to solve this problem is not perfect, and has not formed a unified theoretical framework and method system. In practice, based on the technical means of "double evaluation" and "three zones and three lines" demarcation in the preparation of territorial space planning, more practical exploration has been carried out in the establishment of a land use classification system and the acknowledgment of territorial space use control zoning. While improving the scientific implementation of use control, the theory lags behind practice.

2.2 Relevant Research on "Carbon Reduction and Carbon Sequestration" in National Space

Territorial spatial planning affects the process of space production and the realization of space value [7], and it should strengthen the implementation of the "double carbon" goal [8, 9]. To this end, scholars have investigated and studied spatial planning. based on carbon reduction and sequestration from the aspects of preparation ideas [10, 11], spatial planning methods [12, 13], preparation framework [9], preparation strategy [14, 15], assessment of territorial carbon emissions [16], renewal of industrial heritage and park design [17, 18]. Additionally, Scholars have discussed the integration methods and paths and the establishment of a transmission system [19], so as to provide a theoretical foundation for the implementation of the "dual carbon" layout of the upper-level planning, but have achieved few results.

In the field of general planning, detailed planning and village planning of national spatial planning, Copenhagen, Denmark, has always emphasized the concept of "carbon reduction and carbon fixation" into the planning scheme in the calendar edition of the urban master planning. Wei [20] and Xun [21] proposed the planning path and technical field of integrating general regulations, detailed regulations and special planning into the "double carbon target" based on the multi-level governance idea. He [12] proposed a low-carbon planning method in the overall planning stage based on differentiated thinking [12]; Jiaqiang [22] proposed low-carbon ideas for rural planning from the perspectives of industry, land use, residential buildings and energy facilities [22]. Community planning and building design carbon control projects have also attracted much attention, such as the zero carbon community design in Paris, France, the "Green Print" pilot program in Singapore, the building emission reduction renovation plan in New York, the United States, the building carbon emission monitoring guidelines in London, and the "zero carbon unit" design in Hangzhou, China. According to the above analysis, scholars have discussed the implementation of the "double carbon target" in various planning types, but the overall results are still few.

It can be seen that there is an inherent logical relationship between territorial spatial planning and "double carbon target" (see Fig. 1), and it is helpful to "reduce source and increase sink" by choosing reasonable planning modes, paths and methods. The regulation of land space use is a crucial tool for optimizing the land space layout, which can establish a coordinated coupling relationship with the combined control and control of the flexibility and rigidity of the whole area and all elements in the stages of planning, project implementation permit and supervision. Therefore, "reducing sources and increasing sinks" and territorial space use control also have an internal transmission mechanism, which also has great research potential.

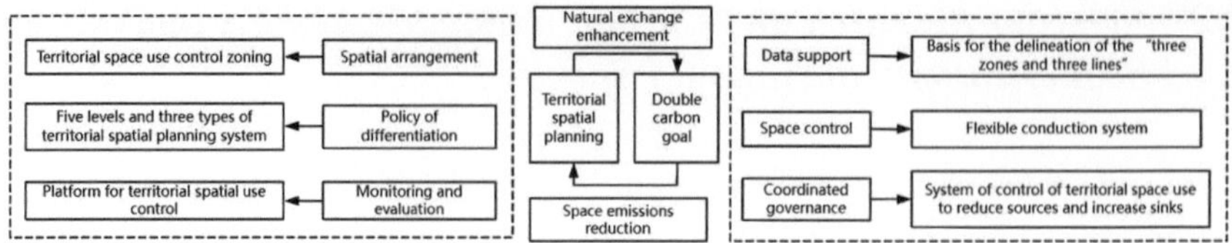

Fig. 1 Logic diagram of territorial spatial planning and "dual-carbon" objectives (Self-drawn)

3 The Internal Logic of the Territorial Space Operation of "Reducing Sources and Increasing Sinks"

3.1 The Concept of "Reducing Emissions and Increasing Carbon Sinks"

The Chinese government has put forth a "two-carbon goal" in response to climate change. As the concentration of carbon sink and carbon emission, national space is the key region to implement carbon reduction and carbon sequestration. "Reducing sources and increasing sinks" is to reduce carbon emissions and increase carbon sinks from the perspective of territorial spatial planning and design, so as to explore the spatial planning response mechanism for realizing the "double carbon goal" in the whole domain, the whole space and the whole elements, so as to make the execution of the "double carbon goal" in the field of territorial spatial planning more specific and targeted. Therefore, "reducing sources and increasing sinks" is conducted as a research perspective.

3.2 The Logic of Decreasing Sources and Increasing Sinks in Response to Territorial Space

"Reducing source and increasing sink" is an important way to mitigate climate change, and all human activities related to carbon emission and carbon absorption occur in a specific territorial space [23]. From the perspective of element composition, territorial space contains two important characteristics: territorial elements and spatial scale [13]. Land elements are the land elements that carry all kinds of development activities in the space, which can be separated into two categories of land elements in the built area and outside the built area. Spatial scale refers to the regional scope of the development intensity of human activities on land elements. In other words, the macro area outside the built area has weak construction intensity, low population concentration, and a large number of blue resources such as water systems, forests, grasslands and farmlands, which are more concentrated carbon sink resources, but there are also carbon emissions, such as agricultural production and breeding activities. The medium and micro areas in the built area have large construction intensity, high population concentration, high concentration of high carbon emissions from construction, transportation, industry, etc., and are more concentrated places for carbon emissions, but there are also carbon sink resources such as parks and green Spaces.

From the perspective of carbon reduction and carbon sequestration, the preservation and expansion of carbon sink resources outside the built-up area should be prioritized, and strict spatial management and control systems such as resource development and protection strategies and ecological restoration should be formulated. The built area places a high value on low-carbon guidance, and carries out fine management of carbon emission reduction in regulation coordination from the aspects of multi-center compact structure shaping, slow traffic system building, green building renewal and transformation, low carbon renewal design of historic districts, building layout design, green public space creation, green infrastructure network remodeling, and low carbon transformation of industrial parks (see Fig. 2).

Therefore, strengthening territorial spatial planning and control, focusing on differentiated control and control in resource development, buildings, transportation roads, energy and other fields, and realizing long-term carbon storage, carbon absorption and carbon emission reduction in territorial space play an important role in realizing "reducing source and increasing sink".

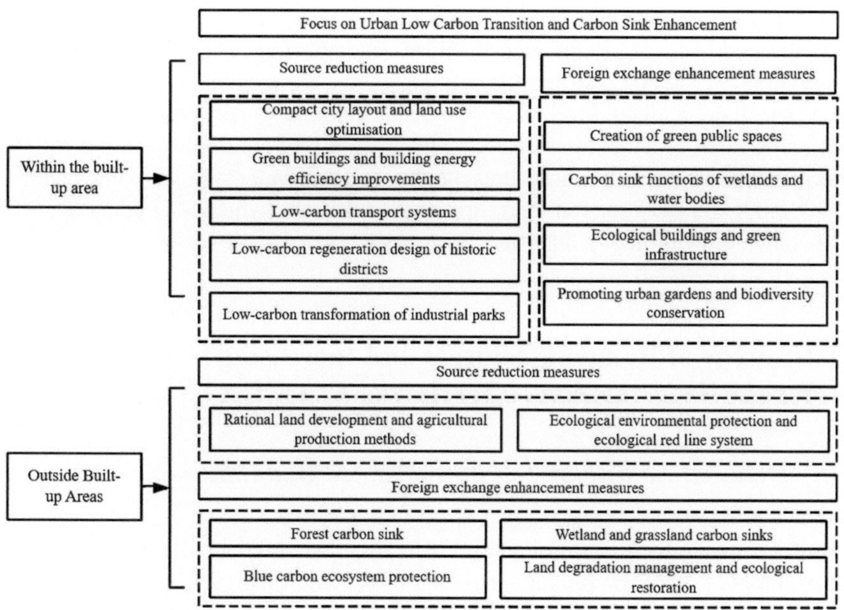

Fig. 2 The internal logic diagram of the response of land space to reduce sources and increase sinks (Self-drawn)

4 The Concept of Territorial Space Use Control

Territorial spatial use control centers on promoting the rational allocation of land resources, with the aim of promoting the coordinated and sustainable development of the regional environment, economy and society. Internationally, many countries have developed relatively mature control systems, such as the land zoning systems of the United States, Japan and Canada, the land planning permit system of the United Kingdom and the construction and development permit systems of France and the Republic of Korea, which provide effective reference for spatial resource control.

The extension of land use management with an emphasis on protecting farmed land is known as land space use control. The whole-cycle control of all resources in the entire area, as well as the control of all land use and space for development and construction activities covering urban and rural ecology, production, and life, are the main focuses of this approach, which is based on the idea of the life community of mountains, rivers, forests, fields, lakes, and grasses and on territorial space planning. The essence lies in the sustainable utilization and optimal arrangement of land resources, including elements such as space scope, regulation content and regulation behavior. Scope elements include urban, ecological, agricultural and other geographical Spaces and boundaries; The content of control includes the determination of space use and scale, the form and pattern of elements, the conversion and change of use, the determination and change of development intensity, permission and approval, etc. Regulation includes development, construction, protection, restoration, etc. China clarified the primary body responsible for controlling land space use at the national level in 2018 and stated unequivocally that the Ministry of Natural Resources should exercise the responsibility of controlling the use of land space, including cultivated land, forest land, grassland, water system, mountain field, sandy land, and construction land, in a unified manner, which is conducive to the sustainable utilization of land space resources.

5 Strategy of Reducing Sources and Increasing Sinks for the Transmission of Territorial Space Use Control

5.1 Detailed Zoning of Territorial Space Use Control

In order to make the texture of source reduction and sink increase more clear, the three spaces of "ecology, production, and life" are carefully refined using land space as the carrier of source reduction and sink increase in conjunction with the "five levels and three categories" system for planning land spaces, which is based on the source perspective. The "Decision on Several Major Issues on Comprehensively Deepening Reform" makes it apparent that the following must be done: establish the boundaries of development control, implement use control, and delineate production, living, and ecological space. Ecological space is a national area that serves ecological purposes and offers ecological goods and services [24]. It is divided into nature reserves (forest land, grassland), water source reserves, ecological conservation areas, etc., and the core area, buffer zone and transition area are defined, which is more reflected as the carbon sink function; Living space to fulfill the daily needs of people who live there, including rural settlements, urban construction concentration areas, flexible development and special use areas in addition to large-scale industrial park space, centralized implementation of carbon emission reduction measures, while urban green infrastructure, parks and green space to implement carbon sequestration measures; The production space is the space for carrying out various production activities, including agricultural planting area, economic forest area, orchard area, breeding area, industrial park, mining area, rural historical and cultural relics area, etc. Carbon sequestration measures should be implemented in agriculture, forestry, orchard, rural historical and cultural relics area, etc. Carbon emission measures will be implemented in farming, industry and mining. Based on this, we should consider the protection and development as a whole, coordinate the control needs of various departments, and strengthen the refinement of the "production, living and ecological" space for key uses in combination with the analysis results of regional "double evaluation", the development needs of cities and industries, and regional resource endowments. At the same time, ensure that the control measures are accurately corresponding to the implementation of space control.

Taking the streets of a county in a city as an example (see Fig. 3) in the delineation of the 'three living spaces', the ecological space is refined into a multi-level and multi-functional integrated forestry system, which improves the microclimate of the region through the carbon-consolidation and water purification functions of the waterfront greening and landscape belt and the wetland vegetation; and creates a 'blue vein and green network' type of ecological space relying on the forest belt of the agricultural field and the forest belt of the mechanic road to enhance ecosystem connectivity and improve vegetation cover and carbon sink efficiency. Relying on the farmland forest belt and the forest belt of farm roads, it creates a 'blue vein and green network' type of ecological space, enhances the connectivity of the ecosystem, and improves the vegetation coverage rate and carbon sink efficiency. The living space is designed to improve the living environment, reduce the waste of resources and the duplication of infrastructure construction caused by decentralised living, and thus reduce the consumption of construction materials and energy. The production space is developed according to regional characteristics, with primary planting as the main industry and farming as a supplement, to build a rural idyllic tourism industry system and enhance the efficiency of land use; for a small amount of industrial land, carbon emission control measures are strictly implemented. Through the scientific refinement of the 'three living spaces', the functional coordination and spatial optimisation of production, living and ecology will be achieved.

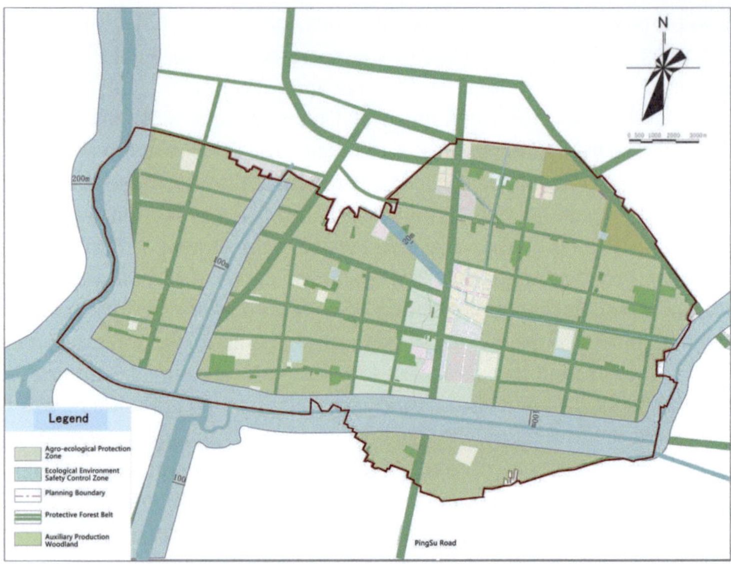

(a) Ecological spatial planning map of a street

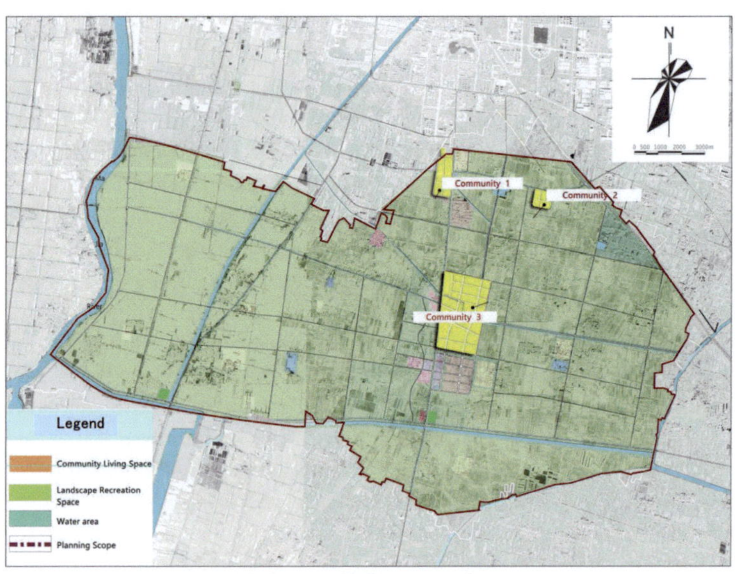

(b) A street living space plan

Fig. 3 Schematic diagram of "ecological-living-ecological space" of a street (Self-drawn)

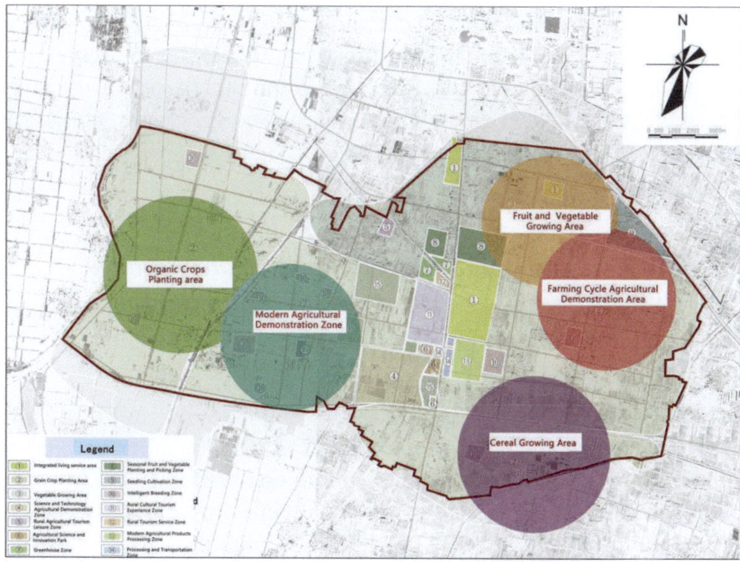

(c) A street production function layout plan

Fig. 3 (continued)

5.2 *Strengthen the Transmission of Carbon Reduction and Carbon Sequestration Regulation by Five-Level and Three-Level Territorial Spatial Planning*

National space planning should strengthen its primary leadership position in the field of carbon sink and carbon emission. The national territorial spatial planning comprehensively determines the macro-strategic positioning of the control of the use of "ecology, production and life", focuses on the creation of pertinent policies, and reflects the strategy; Based on the characteristics of the provincial resources, the territorial spatial planning of the province determines the binding indicators of carbon sink and carbon emission of various Spaces, focusing on the coordination role of the country and cities and counties. City, county and even township level land spatial planning to reduce carbon sink as the starting point to determine various types of land development intensity, land structure and layout, land mixed use degree to determine carbon sink, carbon emission reduction indicators, focusing on implemensibility. The overall plan of national space should optimize the pattern of national space development in terms of the overall distribution of carbon sink resources, the adjustment of the layout of important industries, the clarification of ecological protection objectives and bottom line management, and the identification of ecological restoration projects, and implement the dual-carbon goal. The detailed planning controls the boundary and scale of land use, construction behavior and development intensity, unit carbon emission and carbon sink index of the area, and achieves the balance between carbon emission and carbon sink from the micro level. The special plan focuses on carbon reduction in industrial structure adjustment, energy structure optimization, transportation network remodeling and other aspects, and gives play to carbon sinks in ecological environment, forests and grasslands, agriculture and other fields.

5.3 Establish a Territorial Space Use Control Platform to Achieve Dynamic Supervision of Reducing Sources and Increasing Foreign Exchange

We will expedite the creation of a "five levels and three categories"of land space use control map in compliance with the Party Central Committee's "unified base map, unified standards, unified planning, and unified platform." In this process, a blueprint for street space development in a county in a province (see Fig. 4) is used as a typical case to provide feasible practical basis and ideas. First of all, actively promote the construction of the use control data system, especially the standardization of data standards, establish differentiated spatial management and control standards, build a step-by-step, class-by-class global all-factor information supervision platform, and open the smart territory management and control. This street, for example, in the process of delineating the "three living spaces" through in-depth analysis of resource endowment and demand orientation, the whole factor system covering ecological, living and production Spaces is constructed, which provides an empirical basis for the formulation of differentiated space management and control standards. Secondly, an integrated assessment model of source reduction and sink increase is established to timely feedback the status quo of carbon sink and carbon emission reduction of various elements of national space, so as to facilitate dynamic early warning. Third, artificial intelligence and massive data technologies are used to build a territorial spatial governance information model, mine spatial governance information and data in real time, improve the informatization level of spatial governance, and provide a basis for highlighting the people-oriented spatial governance model. For example, a dynamic monitoring and feedback mechanism can be introduced on the basis of the formulation of a blueprint for spatial development to track in real time the status of carbon sinks and emissions in various types of space, which provides effective technical support for optimising the control of uses and achieving the goal of carbon neutrality. Fourth, establish a data sharing platform and sharing mechanism, explore the distribution mechanism of data acquisition income, improve the application coverage and utilization level of data, and truly play the socialization and normalization of territorial space control.

Fig. 4 Blueprint for street space development in a county (Self-drawn)

5.4 Explore the Use of Carbon Sinks and Carbon Emissions as the Basis for the Delineation of "Three Regions and Three Lines"

At present, the state clearly requires that "three zones and three lines" be delineated in the territorial space planning as a crucial foundation for the control of the use of territorial space. However, in the implementation process, the principle of leading functions should be highlighted in principle, and adjacent functional zones can penetrate each other to reflect the flexibility of spatial zoning. However, in practice, due to the game of different subjects' interests. There are more conflicts in the third zone and the third line in different levels, and there are more difference map spots, which leads to low efficiency of space control. From the perspective of carbon emissions and carbon sinks, a scientific and reasonable carbon measurement system should be gradually established from the logic of land space responding to source reduction and sink increase, and the delineation of "three regions and three lines" of land space should be carried out according to this, which is in line with the princsiple of fine and scientific control, and the efficiency of land space governance should be improved.

5.5 Establish a Flexible Transmission System

The rigidity of traditional space regulation is strong, which can not be implemented or bring more prominent environmental problems. Based on the dynamic nature of carbon sources and emissions, the territorial space regulation from the perspective of reducing sources and increasing sinks takes into account the demands of various spatial stakeholders, and considers the uncertainty of urban and rural spatial development. Based on the perspective of elasticity, an established constraint system with related rules and multiple authorities determined from the control methods, principles, behavioral regulations, zoning access, and control processes, as well as a differentiated and effective use control system, ensures the diversity and flexibility of spatial control.

5.6 *Formulate a Highly Enforceable Control System for the Reduction of Sources and Increase of Foreign Exchange for the Use of National Space*

System is the basis to ensure that all kinds of construction activities proceed smoothly according to the requirements of control. From the perspective of reducing sources and increasing sinks, the first step is to establish and improve the policy and legal system for the control of territorial space to achieve the dual carbon goals. Under the premise of satisfying the dual carbon goals, it is necessary to actively explore the mixed use of non-construction space functions within the built-up area, and revitalize existing land use. Second, explore incentive policies for carbon emission reduction and carbon sink increase in land supply, finance, taxation, and supporting facilities, and encourage social capital to participate in ecological restoration and governance. Third, explore the establishment of a dual carbon evaluation system for territorial space to provide a quantitative basis for reducing sources and increasing sinks of territorial space. Fourth, the Guide to Urban Design of Territorial Space Planning points out that urban design is an important method of general planning, detailed planning and special planning. From the perspective of the dual carbon target contained in the land space planning system, urban design should be actively explored as a policy guarantee mechanism for the management of land space use to reduce sources and increase sinks.

6 Conclusion

An essential first step and management tool for the development of the ecological civilization is the control of territorial space utilization. Territorial space is a significant carrier of carbon sources and sinks under the national strategy of carbon peak and carbon neutrality, and space use regulation should actively play the spatial governance function of reducing source and boosting sink. This paper's research is merely a preliminary effort in the fields of spatial planning for territory and dual-carbon strategy. The theoretical framework of land space use control under the "dual-carbon target" needs to be further explored and improved as a crucial supporting system for optimizing urban and rural human settlements in light of the future changes in the major national strategic background and the modernization of spatial governance. Furthermore, the "dual carbon" goal of regulating land space use necessitates extensive research in a number of areas, including the use of new technologies, innovative policy, multi-scale collaborative governance, social participation, theoretical framework innovation, and reference to international experience. In addition to contributing Chinese knowledge and solutions to the achievement of the global "double carbon" objective, we will advance the development of land space use regulation in a more intelligent and scientific direction through multidisciplinary and multi-field integration.

Acknowledgements This study is supported by the Humanities and Social Sciences Research Project of Henan Provincial Department of Education, No. 2024-ZZJH-249; Key project of Luoyang Social Science Federation (2024A077); Luoyang Social Science Federation Project (2024B452).

Danfeng Liu, 34, Master, research direction: Low-carbon urban and rural planning; Yang Du, 23, Master candidate, 1679265323@qq.com, Research interests: Quantitative methods of urban design; Junhui Hu, 38, Ph.D., 530307448@qq.com, Research interests: Urban sustainable design.

References

1. Zhuangzhuang Y, Yuan Y, Yahua W et al (2020) Territorial space use control in the context of ecological civilization: connotation cognition and system construction. China Land Sci 34(11):1–9
2. Yanping LI, Dahai LIU (2019) Thoughts on the construction of territorial space use control system. China Land 3:27–29
3. Zihui D, Guofang Z (2020) Japan's experience and enlightenment in the regulation of territorial space use. China Land Sci 34(5):33–42
4. Zhiyuan S (2017) Effective Implementation of Natural resource use regulation and its institutional Guarantee: American experience and Chinese Strategy. China Soft Sci 9:1–9
5. Zhifeng J, Xiaolei Z (2020) The basic framework of provincial territorial space use control. China Land 3:31–34
6. Shuangqing T, Lei C, Hai J (2020) From land use control to territorial space use control: evolution, track characteristics and policy implications. Econ Syst Reform (4):12–18
7. Meiqiu C, Ge Y (2021) Thinking on constructing low-carbon land spatial pattern. China Land 11:9–11
8. Jinming Y, Jianjun W (2021) Consideration on the preparation of Guangzhou's territorial Spatial Planning under the dual-carbon Goal. In: 2020/2021 China Urban planning annual conference and 2021 China urban planning academic season. China Society of Urban Planning, Chengdu, pp 525–532
9. Xiong J, Lu K, Jiang ZY et al (2021) Research and thinking on national spatial planning under the goal of "carbon peak and carbon neutrality." Urban Plann Forum 4:74–80
10. Xiaodong S, Dan Z, Qiwen C (2020) Response path of national space planning under the "dual carbon" target. Sci Technol Rev 40(6):20–29
11. Xianjin H, Anlu Z, Rongqin Z et al (2020) Carbon peaking, carbon neutrality and the implementation mechanism of territorial spatial planning. Modern Urban Res (1):1–5
12. He Z, Rui W, Ding Y (2021) First-class. Research on low-carbon planning method of county land space based on differential carbon control. Urban Plann Forum (5):58–65

13. Jialin S, Xinhai Y, Fei L (2002) Existing problems and optimization paths of township territorial space under the "dual carbon" goal. J Plann 38(1):41–48
14. Shuangcheng L (2021) How national spatial planning contributes to carbon neutrality. Contemp Guizhou (26):78
15. Chunhui L, Lin H, Xiaoning W et al (2022) Urban and rural planning Strategy based on "double carbon" goal. Planners 38(1):41–48
16. Xiaochun Z, Zijun W, Qun S (2023) Evaluation of China's carbon emission policy under the dual-carbon target. Stat Decision (2):167–172
17. Han H (2021) The path of urban industrial transformation and heritage renewal under the goal of "dual carbon" : from the perspective of industrial heritage conservation and renewal. J Shanghai Normal Univ (Philosophy and Social Sciences Edition) 6:88–94
18. Chen LJ (2021) The, "double carbon" goal guides the green development path of China's industrial parks. China Environ Manag 6:5–6
19. Chan L, Baixia X (2022) Research on the transmission path of "carbon peaking and carbon neutrality" in Guangdong Province in territorial spatial planning. J Planners 7(38):66–71
20. Wei W, Wei Z, Guobiao Z et al (2022) Spatial planning path and governance mechanism of urban agglomeration under the "dual carbon" goal. Environ Prot 50(Z1):64–69
21. Xun L (2022) Thinking on the path of urban development under the "dual-carbon" strategy. Urban Dev Res 29(08):1–11
22. Jiaqiang W, Xianghua W, Li W (2022) Research on the application of low-carbon technology in rural planning. Urban Constr Theory Res (36):4–6
23. Zhang ZL (2022) Legal response to the "dual carbon" target in territorial spatial planning. China Popul Resour Environ 32(12):11–20
24. Maoji C, Xiuhua T, Yan Z (2020) Research on systematic construction and differentiated management of ecological space. Planners 36(02):48–53

Carbon Sink Balance and Planning Indicators for Rural Collective Operational Construction Land: A Microscopic Approach to Achieving Sustainable Development

Fan Li

Abstract Coordinating economic development with environmental protection has been a significant global challenge. Currently, China's rural revitalization is progressing rapidly. Adjusting and improving relevant indicators in the planning of rural collective commercial construction land based on the concept of carbon sink balance will contribute to climate and environmental improvement. This paper adopts a micro perspective, constructs carbon sink measurement indicators for the planning and design of rural collective commercial construction land before and after implementation, and utilizes the emission factor method for hypothesis testing. The study explores how controlling indicators such as green space can achieve carbon sink balance in planning and design. The conclusions of the study are as follows: under the existing urban green space regulations, the carbon sink of grassland in the original rural collective commercial construction land will increase by 656 t/hm^2·a; for cultivated land, carbon sink balance requires the total green space and rooftop greening area to reach 50%; for forest land, the average green space rate of the entire collective commercial construction land must reach 40% to achieve balance. The study emphasizes the importance of carbon sink standard measurement of rural green space, which is helpful to low-carbon construction in rural areas.

Keyword Microscopic approach · Carbon sink assessment · Rural collective operational construction land

1 Introduction

Global warming is one of the most pressing environmental issues facing humanity today. Many scholars believe that climate change poses irreversible risks, with the expansion of economic scale being a primary driver of the rapid growth in carbon

F. Li (✉)
School of Architecture, Southwest Jiaotong University, Chengdu, China
e-mail: fanli@my.swjtu.edu.cn

B. He et al. (eds.), *Climate-Resilient, Low-Carbon, and Sustainable Cities*, Lecture Notes in Civil Engineering 754, https://doi.org/10.1007/978-981-95-2169-2_39

emissions. Therefore, achieving a balance between carbon footprints and economic development has become a central issue in the process of improving economic standards, and it has been a focal point of research both domestically and internationally in recent years.

In China, the country is currently facing a severe international and domestic situation regarding energy and climate. With the introduction of the Central Document No. 1 on rural revitalization, development is extending from urban to rural areas. This includes transitions from "broadening the applicability of collective operational construction land to the market in 2017" to "ensuring and regulating land for the integrated growth of various industries in rural locales by 2022." Such policy direction indicates that industrial integration has consistently served as a developmental driver in rural regions, necessitating support and assurance regarding land resources.

Unlike urban construction land, rural collective operational construction areas host considerable natural carbon sinks. If urban development standards are applied to the allocation of green space in these rural areas, it could result in the depletion of these carbon sinks. Thus, the planning and design of rural collective construction land remain in a phase of exploration and innovation, requiring modeling and direction based on indicators from multiple viewpoints.

Currently, studies on low carbon are at the national (provincial) and urban scales, with a primary focus on examining the spatial variability of carbon emissions and the impacts of population, economy and other factors on carbon emissions.For instance, Wang et al. [1] employed the top-down life cycle approach to calculate the carbon footprints of various Chinese provinces, revealing significant differences in both overall and per capita footprints across these provinces. In a separate study; Wu et al. [2] utilized the INVEST model to investigate the temporal changes and spatial disparities in carbon stocks within the Guangdong, Hong Kong, and so on. Additionally, researchers [3, 4] explored the geographical distribution of production activities resulting in carbon dioxide emissions in China, focusing on major cities like Beijing and Shenzhen as case studies. Fang et al. [5] analyzed panel data from 30 provincial capitals in China, quantitatively assessing the influence of urban morphology on CO_2 emissions; their findings indicate a positive relationship between urban area expansion and increased CO_2 emissions. Furthermore, Wang et al. [6] and Shaojian Wang et al. [7] applied an enhanced version of the STIRPAT model to validate their empirical findings in Beijing, concluding that urbanization significantly drives CO_2 emissions, while the share of the tertiary sector serves as a key factor in limiting these emissions.Other scholars have constructed a low carbon city evaluation system, such as [8, 9], who used the entropy weight factor method to construct a low-cost evaluation index framework from the aspects of economy, energy pattern, society and life. And the study was conducted in 10 cities globally, revealing that Stockholm, Vancouver, and Sydney exceeded the benchmark value, suggesting that these locations have attained a significant level of low carbon development. Through an examination of the fundamental correlation framework concerning the spatial arrangement of urban agglomerations during low-carbon development, Qin et al. [10] formulated the RI-AER-FTIE low-carbon planning model tailored for urban agglomerations.

In terms of rural low-carbon development, research by Song et al. [11] developed the STRIPAT extension model and identified the primary sources of carbon emissions in rural settings. Meanwhile, Wang et al. [12] examined the key challenges associated with rural planning and construction management in the Yangtze River Delta region, focusing on a low carbon approach and Yu et al. [13] carried out a study of the carbon balance and the cycling process of the terrestrial ecosystems on a regional scale.

Among these studies, the research on'carbon sinks' mainly focuses on the following three aspects: firstly, the system of urban green spaces primarily emphasizes the evaluation of carbon sink efficiency. Zhao et al. [14] primarily explore the contribution of the urban green space system towards low carbon cities; meanwhile, Wang et al. [15] develops a framework for assessing green carbon sink efficiency in densely populated urban regions; additionally, Xiong [16] creates an evaluation framework for the level of low carbon development utilizing a hierarchical analysis approach. Secondly, in terms of rural ecological space, there is a predominance of research on a particular type of ecological land use. Clark [17] explores the necessity and pitfalls associated with the need for assessing the potential of mountain forest sequestration; Li [18] discusses the role of wetland systems as carbon sinks; Wang [19] assesses the impact of carbon storage and oxygen emission from vegetation in botanical gardens by employing the Li-6400 portable photosynthesis analyzer to measure physiological and ecological indices related to photosynthesis. Li et al. [20] focused on the trend of high carbon sources in mountainous villages and its influencing factors; Ji et al. [21] proposed a quantitative comparison model of carbon sources and sinks in the whole life cycle of rural landscapes. Thirdly, When it comes to the spatial planning of carbon sinks in urban and rural areas, the primary emphasis is on the development of theoretical models. Ye [22] establishes a carbon sink function assessment model based on urban and rural ecological green space; Qiu et al. [23] proposes a carbon emission assessment method for urban design projects at the micro and meso scales from the basic principles of carbon footprint measurement and carbon emission auditing; and gives an overview of the theoretical and practical research on low-carbon oriented urban planning and design at home and abroad.

Summary, although scholars have carried out many studies on low carbon, there are still the following shortcomings: (1) While considering the equilibrium between development and environmental concerns, there is a scarcity of research regarding the planning metrics for rural collective construction land viewed through the lens of carbon sinks. (2) Previous research has predominantly concentrated on carbon sinks within ecosystems from a broad perspective, often utilizing vegetation indices for measurement, which cannot be effectively applied to small-scale rural collective construction areas. Consequently, this study investigates the green metrics involved in planning rural collective construction land from a micro perspective through carbon sink assessment. The aim is to enhance existing green indicators, ensure a balance in the land's carbon sink before and after planning, and provide guidance for low-carbon rural planning and construction practices.

2 Materials and Methods

2.1 Carbon Sink Measurement Indicators

Carbon sink is the sum of the amount of GHGs removed by human activities and the amount of GHGs automatically absorbed by nature over a certain period of time. Currently, the calculation of carbon sinks in the available domestic and international literature predominantly employs the numerical emission factor approach suggested by the 2006 IPCC Guidelines for National Greenhouse Gas Inventories.

$$\text{Emission Factor method (ECM) expresson} : \mathrm{E} = \mathrm{AD} * \mathrm{EF}$$

where E represents GHG emissions; AD is the activity level, i.e., information on the extent to which human activities occur; EF is the emission factor, i.e., a coefficient that quantifies emissions or removals per unit of activity. GHG emissions are obtained by multiplying the activity level of the emission source with the corresponding emission factor.

Based on the literature related to carbon sink measurement, the original carbon sink factors and coefficients of rural collective business construction land were reintegrated (Table 1). At the same time, the carbon sink factors involved in planning and design were sorted out, and appropriate coefficients were selected for reference respectively (Table 2). For the tree species carbon sink, this time, the ficus was selected as the tree species for carbon sink calculation, and the winter carbon sink using unfavourable photosynthesis was 25.842 t/hm^2.a. The canopy area of the tree after maturity was generally 12.56 m^2.

Table 1 Factors and coefficients for former carbon sink sites on rural collective operational construction land (Self-drawn)

	Water t/hm^2.a	Grass t/hm^2.a	Woods t/hm^2.a	Plough t/hm^2.a	Land for growing plants t/hm^2.a
Carbon sink coefficient (Calculated in CO_2)	2.01	0.07	2.36	0.82	2.1
Source (select the latest study parameters)	Qiu et al. [23]	Ye [22]		Qiu et al. [23]	IPCC

Table 2 Factors and coefficients for design carbon sink sites on rural collective operational construction land (Self-drawn)

	Roof garden t/hm^2.a	Green area rate t/hm^2.a	Tree canopy cover area t/hm^2.a
Carbon sink coefficient (calculated in CO_2)	1.65	1.65	1.66
Source (select the latest study parameters)	Ye [22], Qiu et al. etc. [23]		

2.2 *Study Area and Data Analysis*

The subjects were located in Sichuan, China (Fig. 1), The study focuses on assessing the carbon sink potential of rural collective operational construction land through data analysis and model construction. It systematically explores the differences between the original land carbon sink and the design carbon sink.

The research methodology consists of the following main steps:

(1) Data Collection and Preprocessing: Rural collective operational construction land primarily includes areas designated for commercial service facilities, industrial use, mining, logistics and storage, as well as facility agricultural land. The relevant data for this study are derived from the third national land survey, which provides comprehensive baseline information on rural collective operational construction land. Additionally, the status of the original collective operational construction land is obtained through remote sensing imagery, interpreted using ENVI 5.3 software to identify various land use types, such as forests, grasslands, water bodies, and construction land. By processing the remote sensing imagery, accurate land cover data is generated for subsequent carbon sink assessments.

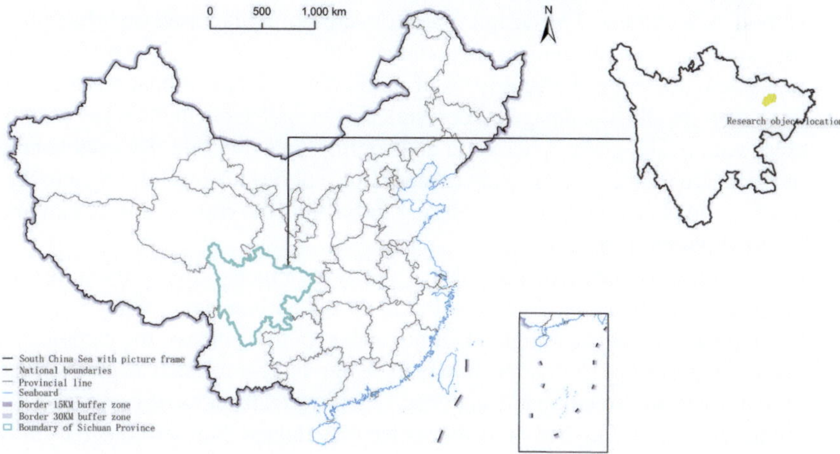

Fig. 1 Location of the subject (Self-drawn)

(2) Data Processing and Spatial Analysis: Following the preprocessing of the collected remote sensing data, spatial analysis is conducted using ARCGIS 10.2 software for data fitting and clipping, ultimately resulting in the carbon sink carrier map of the original land. This spatial data enables the precise identification of land use types and their carbon sink potential, including forests, grasslands, water bodies, and construction land. The core objective of this step is to reveal the distribution of carbon sink carriers in the original land through the integration and analysis of spatial data, providing a foundation for subsequent carbon sink estimation.
(3) Carbon Sink Carrier Area Statistics: After obtaining the carbon sink carrier map, statistical analysis is performed to determine the area of the original carbon sink carriers. By calculating the area of each land use type, we can clearly identify the carbon sink contributions of each land type and provide accurate area data for the next step of carbon sink estimation. This process employs statistical methods to quantitatively analyze the original carbon sink carriers and derive the carbon sink potential of each land type.
(4) Carbon Sink Estimation: Based on literature reviews and previous research, carbon sink coefficients for various land types are utilized to estimate the carbon sink. For different land types (e.g., forests, grasslands, water bodies), the latest carbon sink coefficients are selected for estimating the original carbon sink. Specifically, this study employs the emission factor method to estimate carbon sink volumes. The emission factor method involves multiplying the carbon sink coefficients for different land types by the respective land areas to derive the carbon sink for each land type. The essence of this process is to accurately estimate the original carbon sink by multiplying the carbon sink coefficients by the area, thereby providing a foundation for subsequent analyses of carbon sink restoration and balance.
(5) Design Carbon Sink Carrier Indicators: Building on the carbon sink estimation of the original land, this study also proposes relevant indicators for the design of carbon sink carriers. The design indicators are primarily based on urban planning and design management regulations, with appropriate adjustments made to accommodate the characteristics of rural collective operational construction land. Specifically, the design carbon sink carrier indicators include key parameters such as the green space ratio and rooftop greening rate. By establishing these design indicators, the study explores the carbon sink restoration pathways under various design scenarios, aiming to balance the carbon sink before and after the planning design.
(6) Carbon Sink Restoration and Balance Analysis: Finally, this study compares the original carbon sink volume with the design carbon sink volume to assess the impact of planning design on carbon sinks. In this process, the carbon sink variations resulting from adjustments to design indicators such as green space ratio and rooftop greening are analyzed, and the effectiveness of various design scenarios in achieving carbon sink balance is evaluated. Furthermore, the study considers the differences in carbon sink potential across various land types and employs a regional balance approach to formulate carbon sink restoration

strategies. This comprehensive analysis provides both a theoretical basis and practical guidance for achieving low-carbon rural planning in rural collective operational construction land.

Through these research steps, this study not only offers new methods and frameworks for quantitatively assessing the carbon sink functions of rural collective operational construction land but also presents feasible pathways for low-carbon rural construction through design optimization recommendations.

3 Ideas for the Assessment of Rural Collective Operational Construction Land

3.1 Low-Carbon Countryside is an Inevitable Trend of Rural Development in the Future

Due to the swift economic and social advancements experienced globally in recent years, the ecological environment of the earth has suffered unprecedented damage. According to greenhouse gas emissions data released by Maplecroft of the United Kingdom, the United States and China emit about 6 billion tonnes of carbon dioxide into the atmosphere every year, far more than any other country in the world. In the face of major problems such as frequent extreme weather and melting glaciers due to global warming, the Chinese government is facing unprecedented international pressure to control carbon emissions. In September 2020, China announced its objective to attain carbon peak by the year 2030 and to achieve carbon neutrality by 2060, and regions have launched improvement strategies such as optimising energy structures and developing clean energy projects such as hydro and photovoltaic power generation. Nevertheless, the majority of current policy recommendations and standards emphasize carbon sources, particularly in relation to energy conservation and the reduction of emissions associated with energy use and consumption. Policy guidance and implementation norms are insufficient to maintain and replenish carbon sinks in urban and rural green spaces in economic development, and there is an urgent need to explore how to reasonably balance carbon sinks before and after economic construction.

Developing a low-carbon economy1 is one of the four strategic measures for sustainable development in China. Low carbon means lower (or even lower) greenhouse gas (carbon dioxide-based) emissions. Carbon emissions are generally discussed in terms of 'sources' and 'sinks'. Reducing carbon emissions from sources and increasing carbon absorption from sinks are two common ideas for low-carbon development, which are equally important for improving the environment and climate. Among them, Carbon sinks primarily pertain to China's various green environments, such as its 'mountains, water, fields, gardens, and lakes,' with varying capabilities for carbon sequestration among these distinct green spaces. From the perspective of the process and results of urban construction, most of its rapid development at the expense of the environment, carbon emissions continue to rise, carbon sinks have been losing, urban greenhouse effect, ozone depletion is becoming more and more serious. Mr Wen Tiejun pointed out in his book 'Eight Crises' that several major crises experienced by cities have relied on the countryside for a soft landing. The countryside has always stood at the back of the city, not only in economic terms but also in terms of the environment. The urban-rural dichotomy, despite obvious differences in land use structure and density, is still a whole, interdependent and mirroring each other. It can be said that without the rich and varied carbon sinks in the countryside as support, the urban habitat may have already hit the bottom line of development. Nowadays, the initial road of rural revitalisation, Given the trends of industrial integration and economic growth, it is essential to view urban and rural areas collectively; therefore, creating a low-carbon countryside represents an unavoidable direction for future rural development. In addition to reducing carbon emissions in low-carbon rural development. The reliable upkeep of carbon sink potential holds significant importance for the overall progress of both urban and rural regions.

Rural areas have a huge potential to increase sinks and reduce emissions, which can be effectively stimulated through rural revitalisation (Fig. 2). Considering the unique features of rural regions, the advancement of low-carbon technologies and the implementation of carbon-inclusive guidance frameworks are essential. The proactive guidance of low-carbon initiatives in rural regions is destined to significantly aid China in promptly reaching its carbon peaking and carbon neutrality objectives.

Low-carbon revitalisation of the countryside includes two aspects: (1) maintaining the ecological background, so that the original carbon sink can be continuously consolidated and enhanced; (2) controlling the carbon sources generated by the economic development of the countryside: On one side, implementing clean energy and low-carbon technologies aims to lower carbon emissions; conversely, enhancing the economic efficiency per unit of carbon emissions is also a consideration, among other factors.

Specifically, organisational revitalisation mainly involves raising the level of low-carbon awareness of local residents and reducing carbon emissions from living

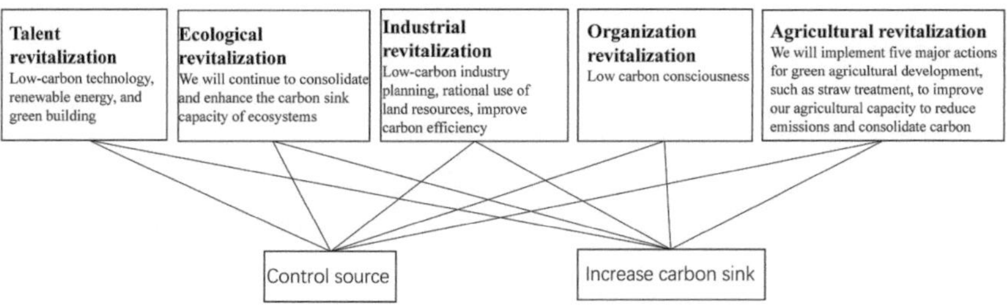

Fig. 2 Distribution of original carbon sink carriers on rural collective operational construction land (Self-drawn)

energy, including the recognition of low-carbon concepts, low-carbon choices in living behaviours, encouragement of low-carbon behaviours in policies and regulations, and low-carbon decision-making for future development, etc.; human resources revitalisation guides the improvement of carbon efficiency and energy production, reduction of construction, transportation and garbage production, and the enhancement of economic efficiency per unit of carbon emissions in rural economic development through the importation of talents related to low-carbon technologies, renewable energies, and green buildings, etc. The revitalisation of agriculture implements green development actions such as straw treatment and enhances the ability of agriculture to reduce emissions and sequester carbon; the revitalisation of ecology focuses on consolidating and enhancing the function of natural carbon sinks; The revitalization of the industrial sector pertains to the planning of low-carbon industries and the efficient utilization of land resources concerning carbon emissions, along with the enhancement of artificial carbon sinks in relation to carbon sequestration.

3.2 The Establishment of Carbon Sink Assessment Ideas

Nature is an important environmental element that determines human health and future well-being. The original natural carbon sinks of rural collective construction land mainly come from blue-green elements, including forests, grass, ploughing and water. It absorbs carbon dioxide through photosynthesis, contributes to the overall climate environment in urban and rural areas, and relies on ecosystems for energy transfer and material exchange, making it a high-quality rural carbon sink. The design of carbon sinks refers to the green space indicators with carbon dioxide absorption functions involved in the planning of rural collective business building land, mainly green space rate, green roofs, trees, etc. (Figs. 3 and 4).

The carbon sink assessment is a comparison of the carbon sequestration capacity of carbon sink carriers before and after planning. It is assumed that the carbon sink of the original collectively operated construction land is the result of the action of the natural environment, which provides the original carbon sequestration, labelled as Y; the design carbon sink is the carbon sequestration after planning and design, labelled as Y'. Rural collectively operated construction land is diverse and complex, with differences in original carbon sinks. How exactly the restoration variables fit with each other to achieve a carbon sink balance often requires a gradual process of hypothesising and validating, starting with a single carrier to multiple carriers.

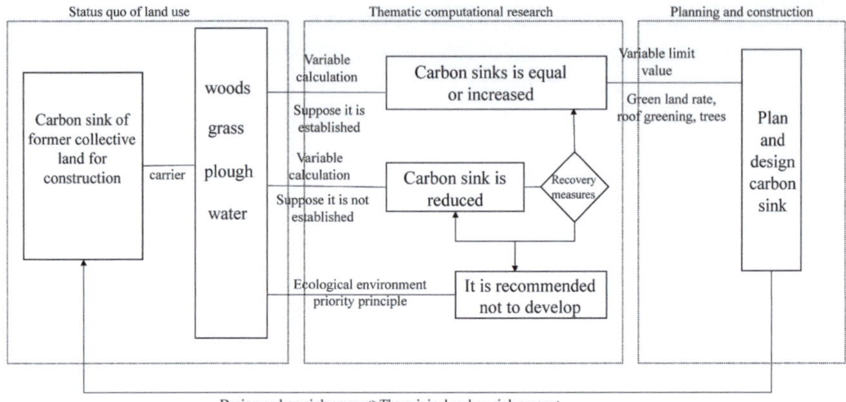

Fig. 3 Ideas for assessing carbon sinks before and after the planning and design of rural collective operational construction land (Self-drawn)

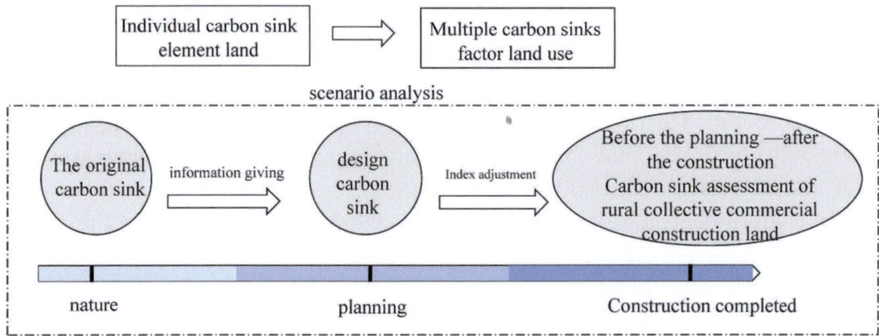

Fig. 4 The process of assessing carbon sinks in rural collectively operated building land (Self-drawn)

4 Scenario Analysis for Carbon Sink Assessment of Rural Collective Management Construction Land

4.1 Single Land Use Carbon Sink Hypotheses and Verification

Four collective operational construction land with an area of 1,000 ha in the rural sample of Chengdu City were selected (Fig. 5). Assuming a 35% green area ratio for the planning layout, the number of floors is not greater than 4, not considering the sloped roof and all the roofs are green, we can get 35% of the green area ratio in the site, and the carbon sink is 577.5 t/hm^2.a; in addition to the green area ratio, road traffic and other facilities, the building density can account for 45%, and the roofs

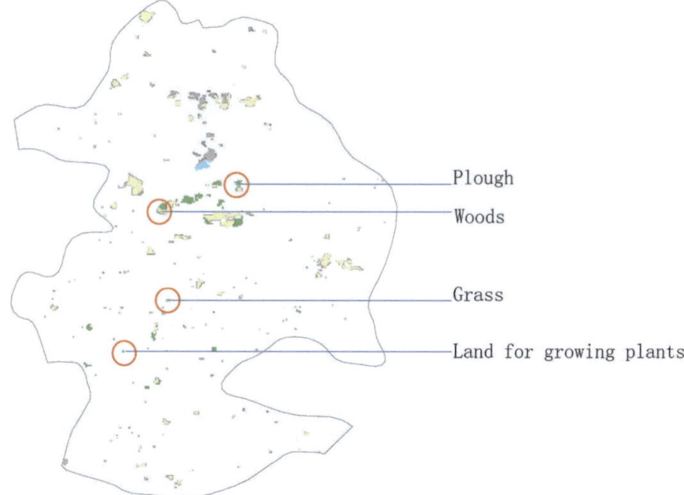

Fig. 5 Schematic representation of the study site (Self-drawn)

need to account for the green area of a 20% discount, so the roofs are 148.5 t/hm^2.a. The total design carbon sink is 726 t/hm^2.a.

The original carbon sink Y_{grass} = 70 t/hm^2.a; Y_{Plough} = 820 t/hm^2.a; Y_{Woods} = 2360 t/hm^2.a; Y_{garden} = 2100 t/hm^2.a is calculated from Tables 1 and 2.

From the calculation results, it can be seen that: (1) when the original site is grass, the carbon sink increase Y'-Y=656 t/hm^2.a; (2)when the original site is plough, Y–Y'=94 t/hm^2.a. The 35% of the green space rate in the planning with the full-roof greening can not balance the carbon sink. Therefore, the original carbon sink can be restored by increasing the planned green space rate, when the green space rate and the discounted green roofs add up to 50%. At the same time, considering the influence of green space rate on building density, the value was finally adjusted to 42% of green space rate, 40% of roof greening, and 8% after discounting; (3) When the original site is woods, the carbon sink reduction Y–Y' = 1634 t/hm^2. a, (4) When the original site is garden, the carbon sink reduction Y–Y'=1374 t/hm^2.a. It can be seen that, when the original rural collective operational construction land is woods and garden, the carbon sink loss changes more if only looking at a single plot. And there is a surplus carbon sink after the recovery of the carbon sink of grass in the original land, which can be used to make up for the carbon sink loss of forest and garden. Therefore, it was considered to extend the study from a single plot to the whole area for further justification.

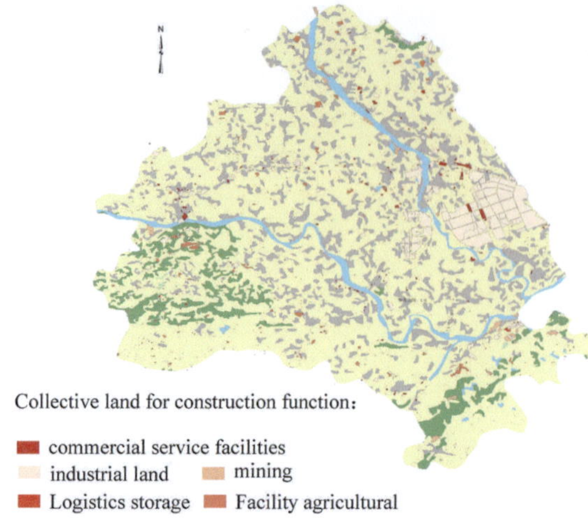

Fig. 6 Overlay of rural collective operational construction land (Self-drawn)

4.2 *Multiple Land Use Carbon Sink Hypotheses and Verification*

The study scope was further expanded by selecting several rural areas for carbon sink assessment. The region encompasses an area of 245.98 km^2, with a measurement of 2.4598 hm (Figs. 6 and 7). The arrangement of the rural collective operational construction land within this area is dispersed and characterized by a fine-grained structure. Additionally, the rural collective operational construction land displays a variety of land use attributes. The original land use attributes in the region are counted separately, which in turn will lead to the amount of the original carbon sink in the region.

Now, after fine calculation of the original carbon sequestration of the whole area, $Y_{\text{original total carbon sink}} = Y_{\text{grass}} + Y_{\text{plough}} + Y_{\text{woods}} + Y_{\text{water}} = 1032.43\text{t}/\text{hm}^2 \cdot \text{a}$; it is found that for the whole collective operational construction land in the area, when the average greening rate reaches 40%, $\text{Y}^{/} = 1566.72 \times 40\% \times 1.65 = 1034.0352 \geq \text{Y}_{\text{original total carbon sink}}$. That is to say, when there are gardens and woods in the collective operational construction land, it is considered to do the balance of occupancy in the whole region in order to ensure the balance of carbon sinks before and after the planning and design.

5 Conclusions

Coordinating economic development with environmental protection remains one of the most significant challenges globally. In China, rapid rural revitalization has created an opportunity to apply innovative strategies in land planning, particularly

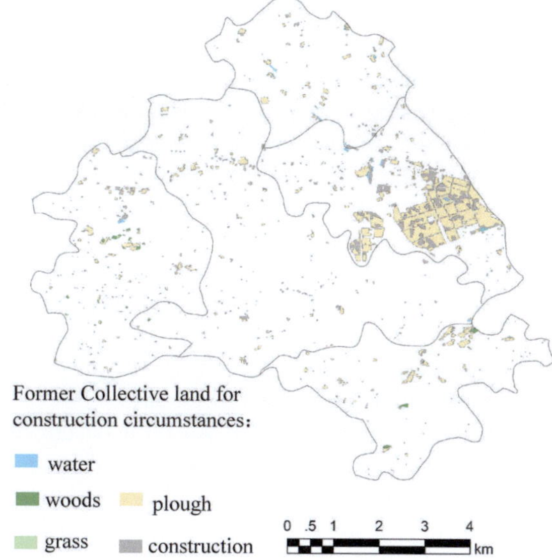

Fig. 7 Distribution of original carbon sink carriers on rural collective operational construction land (Self-drawn)

with regard to carbon sink balance. This study investigates the planning and design of rural collective commercial construction land through a micro perspective, assessing the effectiveness of carbon sink metrics and how adjustments to green space and rooftop greening ratios can achieve carbon sink balance.

The findings from this study reveal that under current urban green space regulations, significant improvements in carbon sink capacity can be made in rural collective operational construction land. Specifically, grassland carbon sinks in original rural collective land could increase by 656 $t/hm^2 \cdot a$. For cultivated land, carbon sink balance requires a combined total of green space and rooftop greening area to reach 50%, while forest land requires an average green space rate of 40% across the entire collective commercial construction land to achieve a carbon sink balance. These insights highlight the potential for adjusting green space parameters to achieve significant improvements in carbon sequestration, contributing to the overall goal of low-carbon rural development.Meanwhlie. The study emphasizes the need for a micro-scale approach when planning rural land, particularly focusing on specific land types such as grasslands, forests, and cultivated land n different scenarios, practical design adjustments for rural collective construction land have been identified, including increasing the green space ratio and adding rooftop greening.

However, this study also reveals several actionable insights for future planning and implementation:

Tailored Green Space Indicators: The research emphasizes the need for site-specific green space standards rather than relying on generalized urban planning regulations. Different land types—such as grasslands, forests, and cultivated lands—require tailored approaches to green space and rooftop greening. Policymakers and urban planners should adjust the green space ratio based on land use types to effectively manage carbon sinks.

Comprehensive Land Use Planning: When planning rural collective construction land, it is essential to adopt a more integrated approach that includes both ecological factors (carbon sink capacity) and practical construction needs (land density, green space availability). For example, increasing the green space ratio for areas such as cultivated and forest land will help mitigate carbon sink losses while supporting overall urban-rural harmony.

Focus on Carbon Sink Restoration: A crucial aspect of the planning process is balancing carbon sinks before and after land development. Ensuring that the original carbon sink potential is not lost in the process of development can be achieved by applying carbon sink restoration measures, such as increasing the green space ratio and implementing green roofing systems.

Dynamic Planning Models: As rural development expands, the carbon sink balance should be continuously reassessed. Future research should incorporate dynamic models that account for changes in climate conditions, land use practices, and technological advancements in green construction techniques. This will help in creating more adaptive strategies for maintaining carbon sinks over time.

Policy and Implementation Frameworks: To ensure that rural revitalization contributes to sustainable development goals, policymakers must create supportive frameworks that integrate carbon sink assessment into land use regulations. These frameworks should provide clear guidelines on green space design and incorporate carbon sink restoration as part of the rural development process.

In conclusion, this study provides important insights into the role of green space and carbon sink management in rural land planning. By focusing on micro-scale planning and utilizing flexible, land-use-specific strategies, this approach can serve as a model for achieving low-carbon, sustainable rural development. As China moves toward its carbon peak and neutrality goals, this research offers practical recommendations for balancing carbon sinks while fostering rural revitalization and ecological sustainability.

6 Limitations and Prospects

While the findings of this study provide valuable insights into the carbon sink balance in rural collective operational construction land, several limitations need to be acknowledged. First, the carbon sink coefficients used are based on existing literature and general models, but actual carbon sink potentials may vary depending on local environmental conditions and specific plant species, which were not fully considered in this study. Second, the research primarily focuses on a micro-scale analysis of land use types, which may not fully represent the complexities of larger rural areas. The generalizability of the results to regions with different climatic, geographic, and economic conditions remains uncertain, and further research with diverse, region-specific data would be necessary to validate these findings. Additionally, the study assumes specific green space ratios and rooftop greening rates based on urban standards, which may not be fully applicable to all rural contexts. Future studies should explore the flexibility of rural land use planning and the specific needs of rural communities to refine these assumptions. Furthermore, the impact of climate change on carbon sequestration capacities was not considered, yet climate change could alter the effectiveness of proposed carbon sink restoration strategies. Lastly, while this study focuses on the ecological aspects of carbon sinks, it does not consider social and economic factors influencing land use decisions, such as land ownership, local policies, and community engagement. Future research should integrate these aspects to provide a more comprehensive understanding of how to achieve sustainable low-carbon rural development. Despite these limitations, this study contributes to the ongoing exploration of sustainable rural development strategies, particularly in balancing carbon sinks in the planning and design of rural collective operational construction land.

Additionally, regional disparities exist in emerging environmental issues such as carbon sink restoration during the future development of rural areas in China. Future efforts should further explore planning indicators related to carbon sink restoration to address environmental challenges in low-carbon development more precisely. For instance, in the planning of collectively-owned commercial construction land in rural areas, parameters like floor area ratio (FAR) and architectural forms may significantly influence local wind-thermal microclimates, thereby affecting carbon sequestration capacity. This suggests a tangible correlation between planning metrics (e.g., FAR) and carbon sink restoration. How to regulate critical elements such as FAR and building mass layouts constitutes an essential research question requiring systematic exploration in low-carbon rural development.

References

1. Wang S, Fang C, Ma H et al (2014) Spatial differences and multi-mechanism of carbon footprint based on GWR model in provincial China. J Geogr Sci 24(4):612–630
2. Junyu Wu, Jinglei Z, Weikang J (2020) Spatiand temporal evolution of ecosystem carbon reserves in the Guangdong-Hong Kong-Macao Greater Bay Area. Landscape Archit 27(10):57–63
3. Feng K, Hubacek K, Sun L et al (2014(Consumption-based CO_2 accounting of China's megacities: the case of Beijing, Tianjin, Shanghai and Chongqing. Ecol Indicators 47(26–31)
4. Fen L, Hongwei M, Pei L (2013) Study and case analysis of urban carbon emission inventory assessment. Urban Dev Stud 20(01):136–139 + 143
5. Fang C, Wang S, Li G (2015) Changing urban forms and carbon dioxide emissions in China: a case study of 30 provincial capital cities. Appl Energy 158(519–531)
6. Wang Z, Yin F, Zhang Y et al (2012) An empirical research on the influencing factors of regional CO_2 emissions: Evidence from Beijing city, China. Appl Energy 100(277–284)
7. Wang S, Liu X, Zhou C et al (2017) Examining the impacts of socioeconomic factors, urban form, and transportation networks on CO_2 emissions in China's megacities. Appl Energy 185(189–200)

8. Tan S, Yang J, Yan J et al (2017) A holistic low carbon city indicator framework for sustainable development. Appl Energy 185(1919–1930)
9. Wang M-X, Zhao H-H, Cui J-X et al (2018) Evaluating green development level of nine cities within the Pearl River Delta, China. J Clean Prod 174(315–323)
10. Menglin Q, Xiaoyuan Li, Qianwen Y, Di D (2019) Evaluation and OPTIMIZATION STRATEGY OF LOW-CARBON SPACE STRUCTURE in Beibu Gulf Urban Agglomeration (Guangxi). Planners 35(13):82–86
11. Limei S, Feng X (2021) Study on carbon emission calculation and influencing factors of rural living environment under the background of rural revitalization. J Human Settlements West China 36(02):36–45
12. Zhu W, Yue X, Yingying Wu (2016) Consensus, dilemma and strategy—exploration of low-carbon rural construction in the Yangtze River Delta region. New Build 04:33–39
13. Guirui Yu, Fang Huajun Fu, Yuling WQ (2011) Progress in carbon expenditure and cycle processes of terrestrial ecosystems at regional scale. Acta Ecol Sin 31(19):5449–5459
14. Caijun Z, Xiaoming L (2010) The role of urban green space system for low-carbon cities. Chinese Landscape Archit 26(06):23–26
15. Min W, Qiaosha S (2016) Evaluation index system and empirical study of green carbon sink efficiency in urban high-density areas—Take Huangpu District of Shanghai as an example. Chinese Landscape Archit 32(08):18–24
16. Qingqing X (2011) Research on the construction of evaluation index system of low-carbon development level in Pearl River Delta cities. Planners 27(06):92–95
17. Clark TA (2021) Carbon capture potential of montane forests: the resilience of loca landscapes from a global perspective. Landscape Archit 28(07):54–68. https://doi.org/10.14085/j.fjyl.2021.07.0054.15
18. Mengying Li (2010) Study on the role of carbon sink in wetland systems in the context of global climate change—Take Tianjin as an example. Chinese Landscape Archit 26(06):27–30
19. Zhongjun W (2010) Study on the benefits of green amount and carbon fixation and oxygen release in Fuzhou Botanical Garden. Chinese Landscape Archit 26(12):1–6
20. Ming LW, Zhen F, Ning W (2015) Take Jingwu Village, Anji County, Zhejiang Province as an example. Huazhong Archit 33(01):174–178
21. Yuanyuan Ji, Wei R, Ting W, Xueyang L (2020) Quantitative exploration of carbon source and carbon sink in the whole life cycle based on the concept of low carbon—Take Tianjin Shilin Linyuan residential area as an example. Chinese Landscape Archit 36(08):68–72
22. Zuda Ye (2011) Establish low-carbon urban planning tool—functional evaluation model of carbon sink in urban and rural ecological green space. City Plann Rev 35(02):32–38
23. Hong Q, Guangjun J, Yaoyu L (2011) Application of the carbon emission assessment method in urban design. Planners 27(05):21–27

Analysis of Rural Evacuation Road Networks Based on Complex Networks Modeling

Wenyin Cheng, Xu Feng, Qinyu Song, and Chuanyang Pan

Abstract Emergency evacuation in rural areas is more vulnerable than that in urban areas, requiring greater inter-village cooperation. The efficiency of inter-village evacuation cooperation is reduced when a road node or section near evacuation shelters with insufficient capacity becomes unusable due to congestion or damage. This paper used the evacuation road network in Xiji Town, Tongzhou district, Beijing, as a case study to supplement the existing theoretical system of rural emergency evacuation and improve rural evacuation efficiency. The topological statistical features of the rural road network were calculated using the complex network modeling method and Gephi. Changes in the size of the maximum connected subgraphs and edge count under deliberate and random attack strategies were analyzed using MATLAB. The results were then analyzed in conjunction with the spatial layout of rural emergency shelters for further analysis. The analysis showed that the rural evacuation road network in Xiji Town was sparse and poorly connected. Intersection numbers 30, 34, 35, 43, 46, 47, 79, 81, 82, 83, 86, 128, 130, 131, 149, 151, and 161 were critical to the road network's resistance to destruction. Therefore, rural shelters near these intersections, such as Laozhua, Fengge, Jinge, Yueshang, Qianzhai, Niumutun, and Xiyi, faced high risks when transferring excess evacuees. Finally, optimization strategies for vulnerable intersections around shelters with insufficient capacity were proposed. These strategies included optimizing layout and evacuation routes, improving transport circulation, enhancing intersection greening and landscaping, and strengthening the emergency functions of public spaces along roads.

Keywords Complex network · Rural road network · Destruction resistance · Shelter

W. Cheng · X. Feng (✉) · Q. Song · C. Pan
College of Horticulture, China Agricultural University, Beijing, China
e-mail: fengxu1204@cau.edu.cn

B. He et al. (eds.), *Climate-Resilient, Low-Carbon, and Sustainable Cities*, Lecture Notes in Civil Engineering 754, https://doi.org/10.1007/978-981-95-2169-2_40

1 Introduction

When disasters occur in rural areas, it is often difficult to match the number of victims with pre-prepared evacuation resources, resulting in resource shortages in some areas and surpluses in others. As inter-village cooperation increases, the timely transfer of rescue workers and victims between neighboring villages will improve the efficiency of relief supplies and significantly enhance emergency evacuation effectiveness. Rural emergency shelters serve as the main hubs for village cooperation, offering relief not only to victims within their own villages but also to those from neighboring villages, if space permits. The rural evacuation road network links these emergency shelters, enabling rescue vehicles to quickly evacuate through the road network. However, when disasters such as earthquakes or mudslides disrupt the road network, or when accidents or congestion occur at junctions, evacuation efficiency is significantly reduced.

The transportation system is key to emergency evacuation, and network topology plays a critical role in the transportation system's disaster response [1]. Many researchers have conducted extensive studies on this topic. Santos et al. [2] represented roads as a network and calculated its topological indices, highlighting the most vulnerable locations on a GIS map. These locations were overlaid with disaster-prone areas to identify critical road network segments and nodes for protection and prioritize reconstruction actions. Cui et al. [3] simulated changes in road network efficiency in Zuoquan County under random, degree centrality, and betweenness centrality attack strategies based on robustness. They found that the robustness of the county road network can be improved by optimizing topology, node connectivity, and increasing the number of nodes to enhance safety and reliability. Shen et al. [4] generated disaster spatial models and investigated their impacts on the complex network indices and connectivity of the road, showing that the resilience of the transport system is highly sensitive to the spatial characteristics of network disruptions. Huang et al. [5] studied the Kangding section of the Dadu River Basin, a disaster-prone area in Southwest China, and established a regional road network model and multi-hazard simulation mechanism, using overall and high-efficiency connectivity as the index system. The study showed that the reliability of urban and rural road networks in mountainous, multi-hazard areas in Southwest China is influenced by the locations of key road sections and geological disaster sites, with key road sections at various levels showing clustered distribution.

Current research mainly focused on large-scale evacuation at the county level and above, emphasizing the disaster response capacity of transport systems in specific regions, such as disaster-prone and mountainous areas, and using complex network models, geographic information systems (GIS), and computer simulations to analyze the robustness and connectivity of road transport systems during disasters [6]. However, few studies have examined the spatial relationship between emergency shelters and evacuation road networks, especially at the Town level. Town-level studies emphasize the spatial utilization of infrastructure, focusing on whether the layout, capacity, and number of emergency facilities, such as shelters and evacuation routes, meet evacuation needs [7]. Although the road network at the Town level experiences less traffic congestion, it faces the issue of road infrastructure distribution, hindering adequate connections between various rural evacuation sites. This problem is exacerbated when the road network is damaged, making it difficult for victims to reach the evacuation sites in a timely manner. In addition, there is insufficient systematic theoretical framework and practical guidance on how to ensure the stability of rural roads and enhance the efficiency of victim transfer between rural evacuation sites.

Therefore, this paper focused on the Town-level evacuation road network, conducted complex network topology analysis, destruction resistance analysis, and explored the geographic relationship with evacuation sites. This research not only helped optimize post-disaster evacuation capacity, enhanced the emergency response capability of rural transport systems, and provided practical guidance for Town-level emergency evacuation management, but also offered theoretical support for future studies on rural security and emergency evacuation. This paper filled existing research gaps and provided both theoretical support and practical methods to improve Town-level emergency evacuation by exploring the following aspects. First, using Xiji Town in Beijing as a case study, the research analyzed the topological features of the rural evacuation road network, evaluated the impact of random and deliberate attacks on its damage resistance, and identified the key nodes of the road network. Second, the study explored the geographical relationship between key road nodes and village evacuation sites, and investigated which evacuation sites and junctions were at higher risk of accidents, considering the number of villages, spatial layout, capacity of evacuation sites, and other relevant factors. Third, the study proposed measures to optimize the layout of vulnerable intersections in Xiji Town, improve evacuation route planning, and enhance intersection greening and landscaping to ensure traffic stability during the transfer between evacuation sites in rural areas. The innovation of this paper lay in its pioneering use of complex network index calculations and destruction resistance analyses for Town-level evacuation road networks, filling the research gap in small-scale rural transport networks. On the other hand, it examined the spatial relationship between rural shelters and evacuation roads, offered a novel research perspective. Finally, it enhanced the post-disaster emergency response capacity and evacuation efficiency of the Town transport network by proposing specific emergency evacuation measures.

2 Methods

2.1 Site Introduction

Xiji Town, in southeastern Tongzhou District, Beijing (see Fig. 1), is highly vulnerable to natural disasters such as floods, earthquakes, ground subsidence, and other hazards. It governs 57 administrative villages, each with at least one shelter. As a result, the Town has 60 rural emergency shelters in total (see Table 1). The space types of theses shelters include green spaces, cultural and recreational areas, open spaces, and basketball courts.

This study identified the permanent population of each village in Xiji Town, along with the locations and capacities of the evacuation shelters. These data were obtained from the Xiji Town government and field research. The analysis revealed that 44 evacuation shelters could accommodate their populations under normal conditions and still have extra space. However, 16 village shelters could not fully accommodate their populations, creating a significant need for transferring victims between villages (see Fig. 2). Therefore, this evacuation situation demanded efficient rural road evacuation.

Transfer between villages is mainly arranged by vehicles provided by the Town government and village committees; therefore, the evacuation road network in Xiji Town mainly consists of carriageways. Rural footpaths within villages, from the evacuation shelter to the boarding location, are not considered. From the spatial distribution of rural emergency evacuation shelters and the evacuation road network in Xiji Town (see Fig. 3), it can be observed that most emergency shelters are located

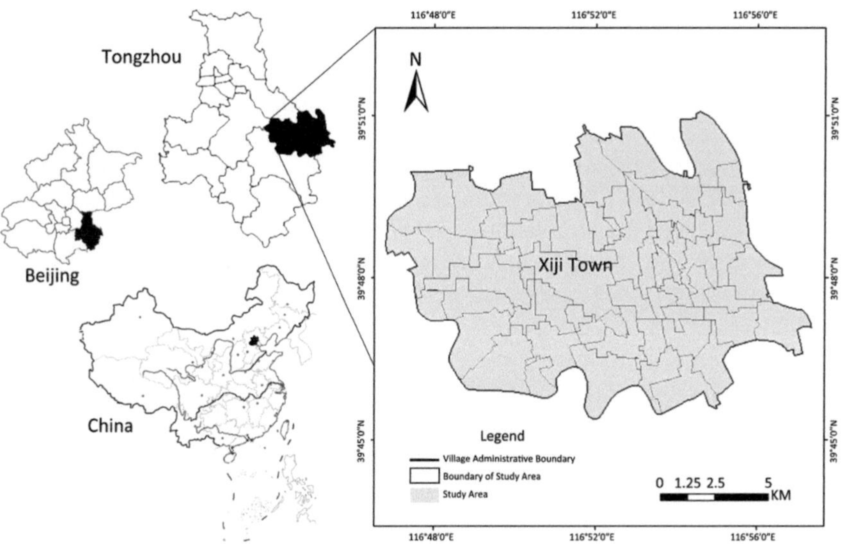

Fig. 1 The location of Xiji town (Self-drawn)

Table 1 IDs and names of rural evacuation shelters in Xiji Town (Self-drawn)

ID	Shelter name	ID	Shelter name	ID	Shelter name
1	Rulin	21	Xiege	41	Niumutun
2	Gongji	22	Yuxin	42	Xinji
3	Xiaoxin	23	Qiandong	43	Yangjiawa
4	Mafang	24	Houdong	44	Lvjiawan
5	Renxin	25	Shidong	45	Anxin
6	Taiping	26	Houge	46	Shangpo
7	Xiaotun	27	Zhaozhua	47	Hehezhan
8	Langxi	28	Yinjiahe	48	Nanxiao
9	Laozhua	29	Xiaoshawu	49	Caoliuge
10	Fengge	30	Dashawu	50	Shagudui
11	Genglou	31	Xiaohui	51	Duliuke
12	Chenheng	32	Shishang	52	Xiji
13	Wangzhua	33	Yueshang	53	Dahuidian
14	Hege	34	Wangshang	54	Huangdong
15	Jintuo	35	Lintun	55	Wenhua
16	Jinge	36	Houfuzhai	56	Xiaojia
17	Wangjun	37	Dongxin	57	Zhangge
18	Wuxin	38	Qianzhai	58	Yunchao
19	Chetun	39	Dudian	59	Xiyi
20	Huzhua	40	Qiaoshang	60	Langdong

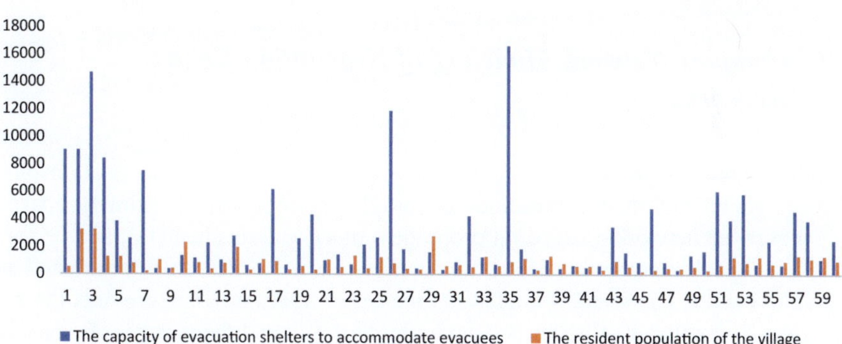

Fig. 2 The capacity of evacuation shelters and the village population (Self-drawn)

within the road network. However, a few shelters are far from the main carriageway, which means they can only be accessed via rural footpaths. Shelters with high evacuation demand are mainly distributed in the southwestern and eastern parts of Xiji Town. Although rural carriageways are available near these shelters, it is unclear

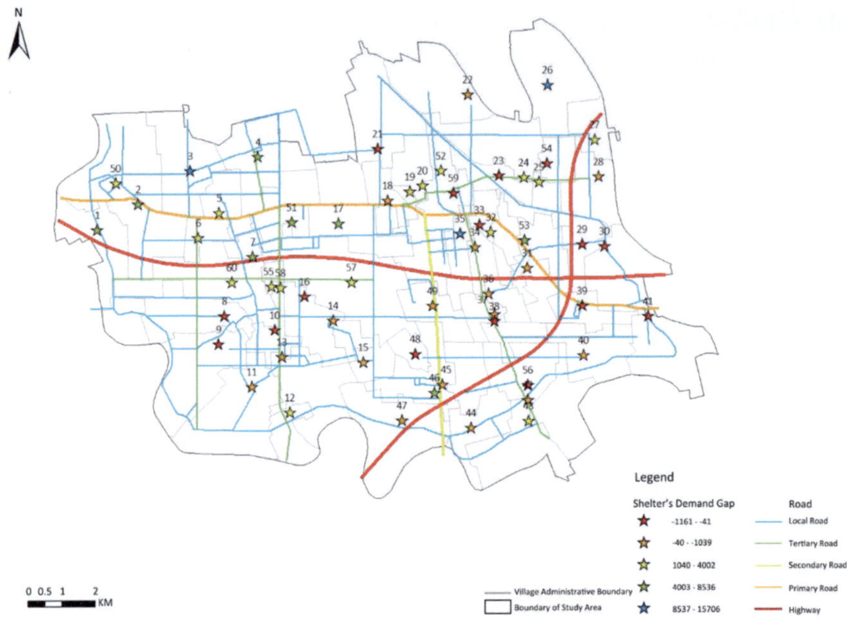

Fig. 3 The location of evacuation shelters and evacuation road network (Self-drawn)

which intersections are prone to congestion or disaster impacts that can significantly reduce the overall efficiency of the evacuation network.

2.2 *Complex Network Modeling of Evacuation Road Networks*

Complex network modeling is a methodology for examining the relationships between nodes and their connections in complex systems [8]. Its primary goal is to uncover the network's structural properties, functional mechanisms, and performance. A specific network, such as a transportation network, can be abstracted as a graph G = (V, E), where V represents the set of nodes and E represents the set of edges. The number of nodes is denoted by N = |V|, and the number of edges is denoted by M = |E|. Each edge in E is associated with a pair of nodes from V. If any pair of nodes (i, j) corresponds to the same edge as (j,i), the network is classified as undirected; otherwise, it is classified as directed.

The study considered road intersections as nodes and assigned numbers to them, representing roads between neighbouring intersections as edges. Two intersection nodes were considered connected if there were direct road segments between them. Since the road network supported bidirectional emergency evacuation flows, the directionality of the edges could be disregarded, resulting in an undirected network.

Table 2 Weights of different road classes (Self-drawn)

Road classes	Weights
Highways	5
Primary roads	4
Secondary roads	3
Tertiary roads	2
Local roads	1

In urban areas, road flows are the main reference for weighting connections between roads in complex network models. However, road flows in rural areas are weak and difficult to measure [9]. On the other hand, road classification significantly affects the quality, width, and management standards of rural roads. Therefore, this study optimized the efficiency and safety of rural road networks by integrating rural road classes into the network topology [10]. Weights were assigned based on different road classes, and a weighted undirected network was constructed (see Table 2).

Road data for Xiji Town were obtained from OSM. This study selected intersections with three or more roads as nodes. To avoid duplicate calculations for structurally similar intersections, only one intersection was selected for every pair of intersections along the same segment that were spaced less than 50 m apart, resulting in a total of 165 intersections. Road intersections were considered to have a side relationship if there was a direct road link between them, and different weights were assigned to the edges based on the classification of the roads. Finally, a complex network model of the evacuation road network in Xiji Town was constructed using Gephi software, resulting in an undirected weighted complex network model containing 165 nodes and 264 edges. The road network topology was then analysed using the following metrics:

Network Density. Network density means the ratio of the number of existing connections to the maximum possible connections in the network. It reflects the closeness of connections between nodes in the network, and a higher density indicates a more interconnected network overall [11]. The formula is as follows:

$$D_e = \frac{2E}{N(N-1)} \tag{1}$$

D_e represents the network density, E denotes the count of existing edges, and N represents the total number of nodes.

Average Path Length. Average path length is defined as the mean of the shortest path length between any two nodes in a network. It indicates the network's overall efficiency and connectivity [12]. A smaller average path length typically suggests that the nodes are more closely interconnected, facilitating faster movement of disaster victims, rescue teams, and emergency supplies throughout the network. The formula is as follows:

$$L = \frac{1}{N(N-1)} \sum_{i \neq j} d_{ij} \tag{2}$$

$$d_{ij} = \min\{size(v_i \leftrightarrow v_j)\} \tag{3}$$

L represents the average path length, N denotes the total number of nodes in the network, and d_{ij} is the shortest path length between node v_i and node v_j.

Network Diameter. Network diameter means the longest of the shortest path lengths between any pair of nodes in a network, representing the network's overall span [13]. It is an important metric for assessing the network's accessibility and compactness. The formula is as follows:

$$D_i = max_{i \neq J} d_{ij} \tag{4}$$

2.3 Destruction Resistance Analysis

In network science, destruction resistance refers to the network's ability to continue to maintain its messaging function after facing an attack or node failure. The destruction resistance analysis complex network is the process of assessing the stability, resilience, and adaptability of a network in the face of disruptive events such as attacks, failures, or disasters, by analyzing the structure, function, and behavior of the network [14]. This study analyzed the changes in network metrics, such as the size of the maximum connected subgraphs and edge count, under random and deliberate attack strategies, identified the most prominent attack strategy, and pinpointed key nodes that could bolster the network's resilience.

The random attack strategy involves randomly failing certain nodes or edges, simulating natural disasters or unexpected events. In contrast, the deliberate attack strategy targets the most vital nodes or edges in the network. This study specifically used weighted betweenness centrality in the context of the deliberate attack strategy. Weighted betweenness centrality refers to the frequency with which a node appears on all weighted shortest paths in the network [15]. The greater the weighted betweenness centrality, the more the network depends on the node's connectivity and the more efficient the network becomes, but the network may also be more fragile. It is calculated as the sum of the ratios of weighted shortest paths passing through node v to the total shortest paths between each pair of nodes (i,j). The formula is as follows:

$$\sigma_{ij} = min\left(\sum_{e \in P} w(e)\right) \tag{5}$$

$$C_B(v) = \sum_{i \neq v \neq j} \frac{\sigma_{ij}(v)}{\sigma_{ij}} \tag{6}$$

σ_{ij} represents the total number of weighted shortest paths between nodes i and j. $C_B(v)$ refers to the weighted betweenness centrality. $\sigma_{ij}(v)$ denotes the number of weighted shortest paths that pass through node v. P is the set of all paths from i to j, and w(e) refers to the weight of edge e.

The maximum connected subgraph is the largest connected component in a network, where nodes in this component can access each other through edges in the network [16]. By analysing the maximum connected subgraph, it is possible to assess whether the network is connected as a whole, or whether there are isolated nodes and subgraphs. If the maximum connected subgraph of the entire network contains most of the nodes, this indicates that the network has good connectivity. The size of the maximum connected subgraph is the ratio of nodes in the maximum connected subgraph to the total number of nodes in the network. The formula is as follows:

$$S = N\prime/N \tag{7}$$

N' denotes the number of nodes in the maximum connected subgraph after the network attack, and N represents the total number of nodes in the network before the attack.

3 Results and Analysis

The road network density was 0.02, which indicated that the evacuation road network in Xiji Town was relatively sparse (see Table 3). The average path length was 7.89, suggesting that travel within this network typically required passing through several intermediate intersections, reflecting the network's complexity and the abundance of detour routes. The network diameter was 18, meaning that 18 intersections had to be traversed to connect the two most distant points, indicating that, in extreme cases, more intermediate points might have been needed to reach the destination. Overall, these indicators revealed that the evacuation road network in Xiji Town was sparse, with long travel distances requiring multiple transit nodes, and limited connectivity, necessitating passage through additional nodes to reach the farthest point in the worst-case scenario.

Table 3 Topological structure index of evacuation road network in Xiji Town (Self-drawn)

Network density	Average path length	Network diameter
0.02	7.89	18

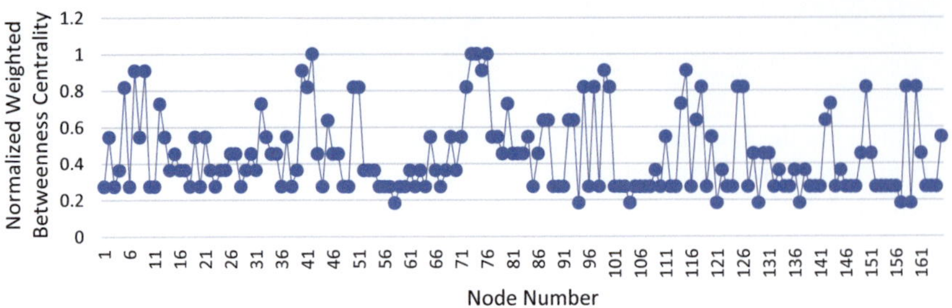

Fig. 4 The weighted betweenness centrality of the road nodes (Self-drawn)

A node with high weighted betweenness centrality indicates that many shortest paths between other nodes pass through it. If this node is removed, communication among other nodes becomes challenging and may even result in network disconnection. Such nodes are typically considered critical within the network, as they serve as key connectors for numerous shortest paths. They play a crucial role in managing evacuation flow, facilitating traffic control during emergencies, and coordinating essential resource distribution. Additionally, road segments with high weighted betweenness centrality should be prioritized for reconstruction to ensure faster network restoration.

The weighted betweenness centrality of nodes in the evacuation road network of Xiji Town ranged between 0 and 1, with most nodes having low weighted betweenness centrality values (see Fig. 4). Nodes with higher values, ranging from 0.8 to 1, accounted for 14.5% of the total and were primarily concentrated in the Xiji Town center. Key roads with these important intersections included primary and secondary roads (see Fig. 5). This distribution indicated frequent traffic exchanges at intersections along these roads during evacuations. From an optimization perspective, these roads should have been prioritized for vehicle flow management.

Displaying the weighted betweenness centrality indicators of intersections in GIS effectively highlighted important intersections within the transportation network (see Fig. 5). An analysis combining the evacuation and transfer needs of rural shelters with large evacuation needs in Xiji Town revealed that several sites on the west side, such as Langxi, Laozhua, Fenggezhua, and JingeZhua, had intersections nearby with relatively low weighted betweenness centrality. Similarly, intersections around shelters on the southeast side, such as Qianzhaifu, Nanxiaozhua, and Xiaojia communities, also displayed low weighted betweenness centrality, therefore, these shelters had a high risk of evacuation. Qiandong, Yueshang, and Dudian, had nearby intersections with high weighted betweenness centrality, and thus, these shelters had a relatively low risk of evacuation.

Overall, it appeared that intersections surrounding shelters with large demand gaps tended to have lower weighted betweenness centrality, while those near shelters with redundant capacity tended to show higher weighted betweenness centrality. This phenomenon suggested that shelters with redundant capacity were better suited for accommodating a larger number of evacuees, particularly from shelters with

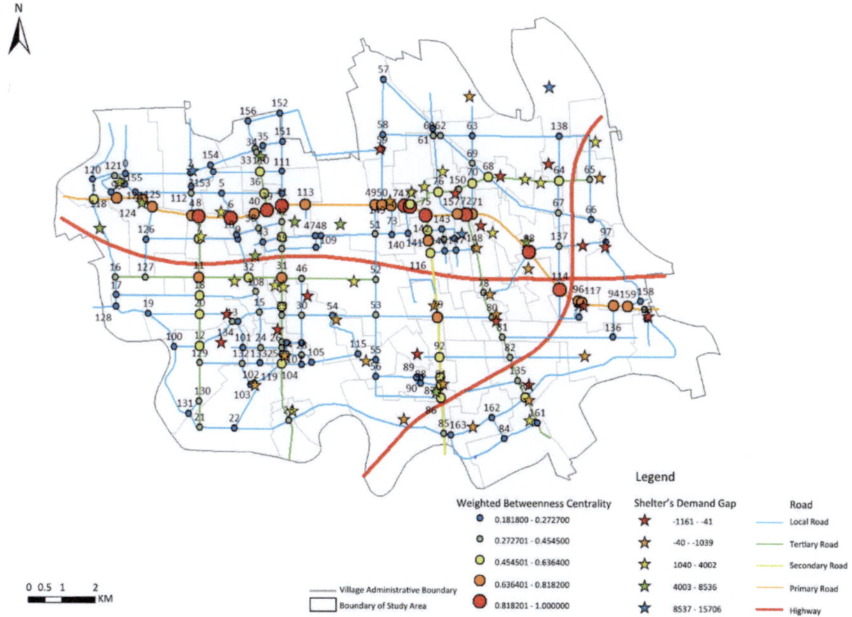

Fig. 5 Spatial distribution of rural shelter's demand gap and weighted betweenness centrality of evacuation road nodes in Xiji Town (Self-drawn)

large demand gaps. However, this raised a critical question: Were the roads around the shelters with redundant capacity sufficiently resistant to destruction? If these intersections were affected by disasters, it might led to a drastic decline in evacuation efficiency. Therefore, it was essential to evaluate the destruction resistance of these intersections during a disaster to ensure the redundancy and smoothness of the evacuation road network.

The destruction resistance of the road network in Xiji Town was investigated using MATLAB. It was found that, when focusing on the network's connectivity after an attack and using the maximum connected subgraph as a metric for evaluation, the network exhibited better resistance under the weighted betweenness centrality attack strategy from step 57 to step 80 (see Fig. 6). This suggested that, although the weighted betweenness centrality attack targeted important nodes, its effect was gradual and did not immediately cause the collapse of the entire network. The weighted betweenness centrality value at step 57 was 0.45, with the corresponding intersection numbers being 30, 34, 35, 43, 46, 47, 79, 81, 82, 83, 86, 128, 130, 131, 149, 151, and 161. These intersections were primarily located along the tertiary roads. When assessing the spatial distribution of evacuation shelters, it was found that the areas surrounding these intersections, which exhibited high evacuation and transfer demand, included the following sites: Laozhua, Fengge, Jinge, Yueshang, Qianzhai, Niumutun, and Xiyi. Furthermore, when analyzing the weighted betweenness centrality of these intersections, it was observed that Laozhua, Fengge, Jinge,

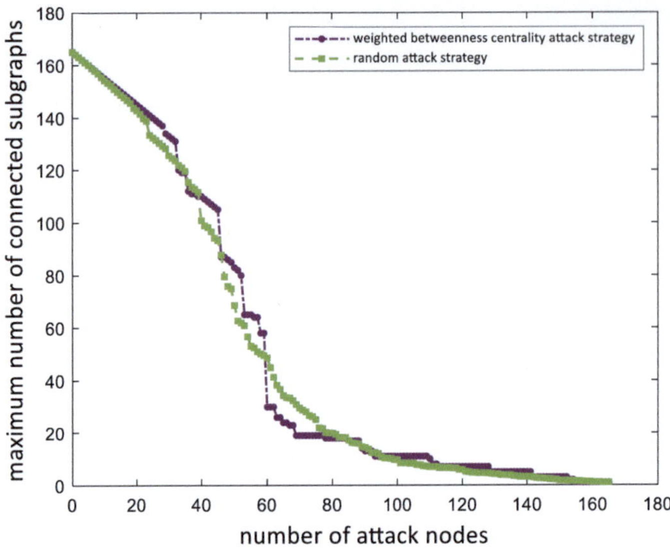

Fig. 6 Changes in the maximum number of connected subgraphs (Self-drawn)

Yueshang, and Xiyi exhibited notably low centrality. This indicated a heightened risk for these five evacuation shelters when transferring excess evacuees.

However, when focusing on network density and edge strength, using edge count as a metric, the random attack strategy resulted in a greater reduction in network resilience than the weighted betweenness centrality attack startegy (see Fig. 7). This indicated that random attack strategy, which did not consider node weights or importance, might have destroyed seemingly unimportant nodes that were actually critical, quickly leading to network collapse. Overall, the network was more significantly affected by the random attack strategy, while some important nodes were more strongly affected by the weighted betweenness centrality attack strategy.

4 Discussion

Based on the analysis of the rural evacuation road network using complex network theory, this study proposed the following strategies for vulnerable intersections:

Optimize the Layout of the Vulnerable Intersections. Redesign the intersection geometry around shelters with insufficient capacity to allow for wider intersections and better sight distances, ensuring more efficient movement during emergencies. Designate adequate assembly points near intersections, considering the needs of pedestrian and vehicular flows during emergency evacuations, and prevent excessive congestion.

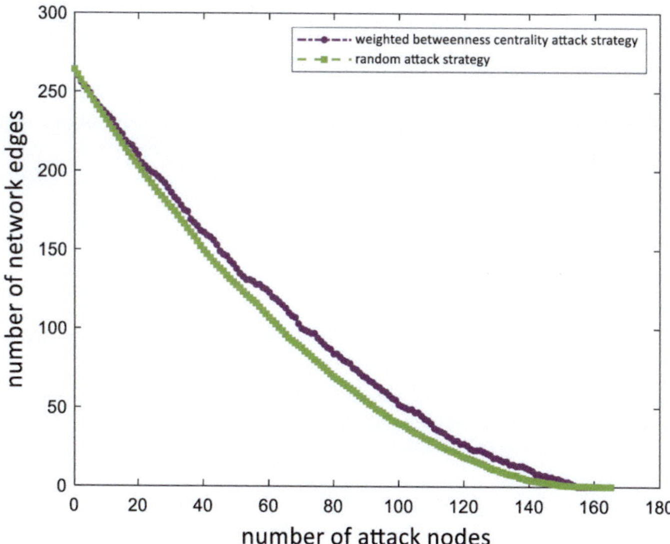

Fig. 7 Changes in the number of network edges (Self-drawn)

Improve Transport Circulation Capacity. Intersections around shelters with insufficient capacity should ensure their ability to handle heavy traffic during emergencies. For example, the durability and load-bearing capacity are improved by strengthening the pavement materials and increasing pavement strength.

Optimize Evacuation Route Planning. Use the results of the analyses to develop efficient evacuation plans that prioritise roads around evacuation sites with insufficient capacity.

Improving road greening and landscaping. Design reasonable plants along the road. They can help relieve psychological pressure, especially during emergency evacuations, and prevent emotional tension caused by an overly closed environment.

Enhance the Emergency Function of Public Space on Roads. Design multifunctional public spaces near roads to serve as temporary shelters or assembly points during disasters, ensuring sufficient capacity and emergency functions. For example, incorporate landscape facilities offering water storage, lighting, seating, and resting, which can be easily converted into emergency facilities post-disaster.

5 Conclusion

This paper presents the workflow and ideas for exploring the rural evacuation road network at the town scale using complex network modeling and destruction resistance analysis. From the perspective of the spatial relationship between refuge sites and the road network, we can gain a clearer understanding of rural critical infrastructures. This understanding can also provide valuable insights for optimizing the urban evacuation road network and improving emergency evacuation efficiency. Nevertheless, the application of specific methods requires further refinement and verification through case studies.

References

1. Zhang XF, Miller-Hooks ES, Denny KT (2015) Assessing the role of network topology in transportation network resilience. J Transp Geogr 46(5):35–45
2. Santos LBF, Jorge AAS, Londe LRT (2023) Vulnerability analysis in complex networks under a flood risk reduction point of view. Front Phys 11(3):1064122
3. Cui Yi F (2017) Robustness of county highway network based on complex network theory. Highway 62(11):151–154
4. Shen ZF, Ji CS, Lu ST (2024) Transportation network resilience response to the spatial feature of Hazards. Transp Res Part D: Transp Environ 128(3):104121
5. Huang YF, Wei MS, Wan DT (2020) Analysis of reliability of road network in mountainous disaster-prone areas in Southwest China. J Tongji Univ (Nat Sci) 48(4):526–535
6. Du JF, Zhang FS, Du ZT (2020) Assessing vulnerability of road networks based on traffic flow betweenness centrality: a case study in Wuxi. J Zhejiang Univ (Sci Ed) 47(2):223–230+243
7. Feng XF, Wang KS, Mao Q (2022) Rural spatial planning methods in the territorial planning: from an integrated perspective of planning and governance. Arch Eng 46(11), 21–31
8. Zhang MF, Huang TS, Guo ZT (2022) Complex-network-based traffic network analysis and Dynamics: a comprehensive review. Phys A: Stat Mech Its Appl 607(23):128063
9. Zhong MF, Hanson BLS (2009) GIS-based travel demand modeling for estimating traffic on low-class roads. Transp Plan Technol 32(5):423–439
10. Chen SF, Saed TUS, Alinizzi MT (2019) Safety sensitivity to roadway characteristics: a comparison across highway classes. Accid Anal; Prev 123(2):39–50
11. Leskovec JF, Kleinberg JS, Faloutsos CT (2005) Graphs over time: densification laws, shrinking diameters and possible explanations. In: Grossman RF, Bayardo RS, Bennett KT (eds) Proceedings of the eleventh ACM SIGKDD international conference on knowledge discovert in data mining 2005, KDD'05. ACM Digital Library, Chicago Illinois USA, pp 177–187
12. Chen FF, Chen ZS, Wang XT (2008) The average path length of scale free networks. Commun Nonlinear Sci Numer Simul 13(7):1405–1410
13. Gu LF, Huang HS, Zhang XT (2012) The clustering coefficient and the diameter of small-world networks. Acta Math Sin, Engl Ser 29(1):199–208
14. Wu JF, Tan YS (2005) Study on measure of complex network invulnerability. J Syst Eng 20(2):128–131
15. Freeman LF (1977) A set of measures of centrality based on Betweenness. Sociometry 40(1):35
16. Wu BF (2019) Networks science and computing. Beijing University of Posts and Telecommunications Press, Beijing

Optimization of Urban Blue-Green Space Layout Based on High-Density Monitoring Sensor Networks: A Case of Xiangyang, China

Wei Xue, Zhihao Sun, Yuli Fan, Qingming Zhan, and Zhonghua Wu

Abstract The current research lacks in-depth examination of the relationship between blue-green space elements and particulate matter at the urban scale, making it difficult to clearly allocate blue-green spaces within urban planning systems. In response, this study focuses on the urban scale and selects Xiangyang, which suffers from severe particulate pollution, as the study area. Collaborating with the Xiangyang Ecological Environment Monitoring Center of Hubei Provincial Department of Ecology and Environment, a high-density near-surface monitoring sensor network for particulate matter and six meteorological elements has been actively established in the central urban area to overcome the issue of insufficient sample points provided by existing discretely distributed monitoring stations. Secondly, following the principles of particulate matter diffusion, we establish a Gaussian kernel density dataset of blue-green space elements that conforms to the spatial process of particulate matter diffusion, and then, a dual-precision nested algorithm is designed to select the best representative elements, allowing for a more refined representation of the interaction between particulate matter and blue-green space elements during the spatial diffusion process. Finally, we input the dataset as independent variables into a geographically weighted regression model to determine the range as well as the intensity of particulate matter reduction by blue-green space elements and propose their precise intervention locations. This study provides a scientific basis for creating a healthy urban living environment.

W. Xue (✉)
State Key Laboratory of Precision Blasting, Jianghan University, Wuhan, China
e-mail: xuewei@jhun.edu.cn

Z. Sun
Wuhan Natural Resources Conservation and Utilization Center, Wuhan, China

Y. Fan
Public Administration School, China University of Geosciences (Wuhan), Wuhan, China

Q. Zhan
School of Urban Design, Wuhan University, Wuhan, China

Z. Wu
Hubei Provincial Department of Ecology and Environment, Xiangyang Ecological Environment Monitoring Center, Xiangyang, China

B. He et al. (eds.), *Climate-Resilient, Low-Carbon, and Sustainable Cities*, Lecture Notes in Civil Engineering 754, https://doi.org/10.1007/978-981-95-2169-2_41

Keywords Spatial planning · Particulate matter pollution · Blue-green space · Xiangyang

1 Introduction

Particulate matter pollution seriously threatens public health [1]. According to statistics from the World Health Organization (WHO) [2], nearly 7 million people die each year worldwide from diseases such as stroke, heart disease, and lung cancer caused by particulate matter pollution. At present, the particulate matter pollution at the urban level remains severe. Globally, about 64% of cities have particulate matter concentrations exceeding the WHO's guidance values [2]. Blue-green spaces have been proven to effectively improve environmental air quality [3], which has received widespread attention in recent years for their role in reducing particulate matter and its negative impacts.

The methods for examining the correlation between particulate matter pollution and blue-green spaces mainly include statistical models based on observational data and numerical simulations. Particulate matter observation data is obtained either by direct sampling at near-surface atmospheric environmental monitoring stations or by observing the aerosol optical depth of the troposphere from satellite-borne sensors and retroactively estimating near-surface particulate matter concentrations. Then, statistical models are constructed using particulate matter observation data to explore the impact of blue-green space scale, shape, and other characteristics on particulate matter concentration and distribution. Numerical simulation involves discretizing the continuous atmospheric fluid motion into a three-dimensional spatial grid. It uses numerical integration to simulate the differential equations of airflow after interactions such as friction and compression with structures, vegetation, and other elements. This process yields the concentration values and diffusion trajectories of particulates in each grid cell along the Earth's longitudinal, latitudinal, and vertical axes. In practical applications, different simulation scenarios are typically constructed by adjusting the characteristics of blue-green spaces as well as the features of particulates and wind. By comparing particulate concentration and distribution across these scenarios, one can determine the reasonable range of characteristics such as the form and scale of blue-green spaces. It is important to note that the computational cost of numerical simulations increases dramatically with the number of grid cells. Even a small urban area may require tens of millions of grid cells, which far exceeds the computational capacity of most laboratories. Therefore, numerical simulation is often applied to neighborhoods and smaller spatial units [4, 5].

Zhao et al. [6] conducted a comprehensive quantitative study on the dust retention effect of lake wetlands in main urban areas of Wuhan, focusing on their morphology, scale, and location. The study indicates that the area and location index of urban lake wetlands contribute the most to the reduction of particulate matter, additionally, the dust retention effect of lake wetlands has a certain range of influence, characterized by locality corresponding to their scale. Specifically, as the area of urban lake wetlands increases, the humidifying effect gradually strengthens, and the reduction effect on PM_{10} (inhalable particles) and $PM_{2.5}$ (fine particles) in the air shows a certain enhancement trend, with a more significant impact on PM_{10} concentration; when the lake wetland area reaches 12.2 hectares, the interaction of cold and warm air flows between the lake wetland and surrounding areas becomes evident, allowing particulate matter in the air to disperse towards areas with lower humidity, thereby improving the air quality around the wetland; furthermore, when the lake wetland area exceeds 12.2 hectares, a multi-patch dispersed layout of lake wetlands has a more significant effect on improving the air quality of the entire urban environment [6, 7].

For green spaces, there is a positive correlation between particulate matter concentration and the scale and density of green spaces, while there is a negative correlation with the fragmentation of green spaces. Based on this, further coupling the scale and morphology of green spaces with particulate matter concentration reveals that increasing the complexity of green space boundaries, aggregation, connectivity, and spatial extensibility significantly enhances their dust retention effect [8, 9]. From the perspective of plant configuration, the dust retention effect of composite green spaces far exceeds that of homogeneous green spaces, with the tree-shrub-herb configuration being optimal, followed by the shrub-herb configuration [10–12]. For the same type of tree species, the more complex the texture and structure of the bark or leaf epidermis, and the higher the leaf area index, the stronger the dust retention effect [10–12]. However, the aforementioned studies lack an exploration of the positioning of blue-green space elements at the urban scale, making it difficult to clearly arrange the locations of blue-green spaces within the urban planning framework.

This study is based on an urban scale, selecting Xiangyang, a city with severe particulate pollution, as the study area. In collaboration with the Xiangyang Ecological Environment Monitoring Center of Hubei Provincial Department of Ecology and Environment, a near-surface high-density monitoring sensor network, with 142 locations, for particulate matter and six meteorological elements has been actively constructed in the central urban area. This approach addresses the issue of existing discretely distributed monitoring stations being unable to provide sufficient sample points. Secondly, following the principles of particulate matter diffusion, we establish a Gaussian kernel density dataset of blue-green space elements that aligns with the spatial process of particulate diffusion, and then, a dual-precision nested algorithm is designed to select the elements that best represent the process, allowing for a more precise characterization of the interaction and scale of influence between particulate matter and blue-green space elements in the spatial diffusion process. Finally, we input the aforementioned dataset as independent variables into a geographically weighted regression (GWR) model to determine the range as well as the intensity of particulate matter reduction by blue-green space elements and to propose their precise intervention locations. This study provides a scientific basis for creating a healthy urban living environment.

2 Research Design

2.1 Research Area

To accurately assess the range and intensity of the impact of blue-green spaces on particulate matter reduction at the urban scale, it is essential first to select a city with severe particulate pollution. Subsequently, a high-density monitoring sensor network for particulate matter and meteorological elements should be actively constructed near the ground level within the city to addresses the issue of sparse sample points caused by the discrete distribution of national and provincial environmental monitoring stations. Based on preliminary field research, this study has chosen Xiangyang (110°45′-113°06′E, 31°13′-32°37′N) in Hubei Province as the study area.

Xiangyang is a significant industrial hub in Hubei Province, with dense industrial activities leading to substantial pollution emissions. Geographically, Xiangyang is located in the Nanyang Basin, where the terrain hinders the outward transmission and diffusion of particulates, resulting in their accumulation. As of now, the annual average concentration of $PM_{2.5}$ in Xiangyang's urban area still does not meet national standards. Through collaboration based on a funded project, our team, in conjunction with the Xiangyang Ecological Environment Monitoring Center of Hubei Provincial Department of Ecology and Environment, has actively established a near-surface high-density monitoring sensor network in the complex and densely populated central urban areas of Xiangyang, including Fancheng District, Xiangzhou District, Xiangcheng District, Yuliangzhou, and Dongjin New District. This network is set up within the main built-up areas, as shown within the boundaries in Fig. 1, to provide data support for accurately assessing the specific impact distance of blue-green space elements on particulate matter at the urban scale. The network consists of 142 integrated monitoring sensors capable of synchronously collecting and storing particulate matter ($PM_{2.5}$ and PM_{10}) and six meteorological elements (temperature, relative humidity, wind speed, and wind direction) every hour. Figure 1 illustrates the specific spatial location of each monitoring sensor.

2.2 Data and Preprocessing

In this study, the particulate matter mass concentrations used are the hourly $PM_{2.5}$ and PM_{10} concentrations ($\mu g/m^3$) collected from the high-density monitoring sensor network. To avoid interference with data collection due to the COVID-19 pandemic, the study focuses on the period when the network was operating stably, before the onset of the pandemic, and during the winter season when particulate pollution is most severe. Specifically, the study period is from December 1, 2018, at 00:00 to February 28, 2019, at 23:00. Subsequently, the data collected during this period are subjected to validity checks according to the requirements of the Ambient Air Quality

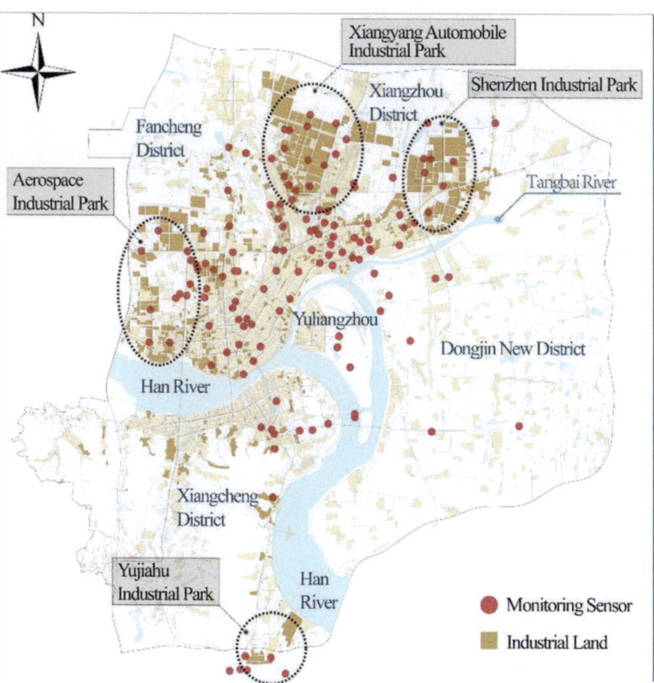

Fig. 1 Location of the study area and layout of high-density monitoring sensor network (Self-drawn)

Standards of the People's Republic of China (GB3095-2012). All stations pass the validity checks.

In this study, bodies of water such as rivers, lakes, and wetlands are considered blue spaces, while green spaces and farmland are collectively regarded as green spaces. Leveraging collaboration from the funded project, the Xiangyang Natural Resources and Planning Bureau provided current land use imagery and vector data for the built-up areas of Xiangyang's central urban area as of 2018. Based on this, the current land use map is converted into vector files, from which green spaces and plaza areas are extracted. We also use the land use vector data to extract the areas of water bodies and farmland. Additionally, the study identifies other main pollution sources related to industry, traffic, and dust within the study area as other explanatory variables.

Through preliminary field research and confirmation with the 2019 Xiangyang Ecological Environment Status Bulletin, pollution sources in the study area are identified as industrial sources dominated by pollution-intensive manufacturing, textiles and apparel, processing, mining and washing, and energy supply industries; traffic sources mainly from urban main roads, highways leading out of the city, and material logistics; and dust sources primarily from construction sites and industrial storage yards.

Therefore, based on the 2017 Xiangyang City Industrial Pollutant Emission Inventory, the Gaode Map API, and the Xiangyang City Government website, we extract the emission source names and their annual emissions (in tons) for key particulate matter precursors, sulfur dioxide (SO_2), nitrogen oxides (NO_x), and volatile organic compounds (VOC_s), respectively, which represents industrial sources. Meanwhile, we also extract freight logistics point of interest (POI) data and bus station POI data representing traffic sources, as well as the names of construction sites in process at that time representing fugitive dust pollution sources. Subsequently, the Baidu Coordinate System is adopted to obtain the spatial information of the pollution sources in above, and then batch-process the data into the CGCS2000 coordinate system.

2.3 Methods

First, the particulate matter concentrations are spatially processed to serve as the dependent variable. To eliminate interference from exogenous particulate pollution, weather conditions are filtered based on wind speed and direction data collected from the high-density monitoring sensor network, focusing on periods with the prevailing wind direction (north wind) and wind speeds not exceeding level 3. Subsequently, the Inverse Distance Weighted interpolation method is used to spatialize the average mass concentrations of $PM_{2.5}$ and PM_{10}. Leave-one-out cross-validation is employed to validate the interpolated predictions against observed values. The validation results shows that the adjusted coefficients of determination (R^2) for the regression equations fitting the predicted and observed concentrations of $PM_{2.5}$ and PM_{10} are 0.81 and 0.85, respectively. The root mean square errors (RMSE) are 5.47 and 9.24, respectively.

Second, a Gaussian kernel density dataset for blue and green space elements is constructed to align with the spatial diffusion process of particulate matter, serving as the initial dataset for independent variables. Unlike previous studies that used characteristic indicators of blue and green space elements, such as the area of blue and green spaces, as independent variables, this study follows the diffusion principles of particulate matter. It utilizes the basic assumptions of the Gaussian diffusion model to transform the areas of blue and green spaces and other variables into a structured dataset of planar Gaussian kernel density values. This approach aims to more accurately characterize the interaction between particulate matter and blue-green space elements during the spatial diffusion process. The planar Gaussian kernel essentially functions as a weight matrix determined by a multiplier and a Gaussian standard deviation (σ). The Gaussian standard deviation can be understood as the rate at which weight values decay with increasing distance from the center point. The multiplier applies a weight to the value at the center point itself, which is reflected in the regression model as a coefficient and does not need to be considered in the independent variable dataset. Specifically, the kernel density values are calculated for the areas of blue and green spaces under a planar Gaussian decay with distance σ. This serves as a hypothetical representation of the impact of these elements on the surrounding environment. Consequently, a dataset is formed with different combinations of independent variables, each representing different Gaussian kernels, or different σ values, to indicate the influence range of each element on the surrounding environment.

Third, a dual-precision nested algorithm is designed to select the most representative elements, thereby refining the characterization of the interaction between particulate matter and blue-green space elements during the spatial diffusion process. The independent variable dataset mentioned above includes all possible hypothetical representations of the impact of blue-green space elements on the surrounding environment. Therefore, it is necessary to select the most representative elements from this initial dataset. Given the extensive computational load involved in the selection process, which severely limits computational efficiency, a dual-precision nested algorithm is developed. Based on previous research findings, the Gaussian standard deviation is set between 100 and 3000 m. Building on this, Gaussian kernel density values of the original feature values are calculated at multiple scales with 400 m intervals. This approach is used to construct Eq. 1, which characterizes the spatial decay features caused by blue space, green space, or other control variables. This method helps in identifying the scales at which the elements most significantly influence particulate matter diffusion, thereby optimizing the model's predictive capabilities.

$$g(d) = \frac{1}{\sqrt{2\pi}\sigma} exp(-\frac{d^2}{2\sigma^2}) \tag{1}$$

In the equation, $g(d)$ represents the value characterizing the impact of blue space, green space, or other control variables at a distance d. The symbol σ denotes the standard deviation of the Gaussian kernel. The results after the selection process are shown in Table 1.

Fourth, a GWR model is constructed to relate particulate matter concentrations to the Gaussian kernel density dataset mentioned above. Considering that a scale of 500 m is sufficient for urban spaces to form their own ventilation systems and represents the minimum scale at which urban space homogeneity might occur [13], a 500-m × 500-m grid is established. The spatializes $PM_{2.5}$ and PM_{10} concentrations are spatially correlated with all factors in the dataset, and then to establish a GWR model. The GWR model is based on the concept of local smoothing, embedding the spatial location of data into the regression parameters. It uses locally weighted least squares to estimate parameters point by point, providing an intuitive description of the spatial variation of the study object at a certain scale and its related driving factors [14]. The expression of the GWR model is shown in Eq. 2, where (u_i, v_i)

Table 1 The filtering results of the dual-precision nested algorithm (Self-drawn)

Adjusted determination coefficient (%)	Selected influencing factors	Gaussian kernel standard deviation (m)
59.82	Blue space	1000
	Green space	400

represents the spatial coordinates of the i-th sampling point, $\beta_k\ (u_i, v_i)$ denotes the k-th regression coefficient at the i-th sampling point, y is the spatialized $PM_{2.5}$ and PM_{10} concentration, and x represents the Gaussian kernel density value of blue space area, green space area, or other control variables representing pollution sources. The specific calculation method for $\beta_k\ (u_i, v_i)$ is shown in Eq. 3, where $W(u_i, v_i)$ is the spatial weight matrix, with its diagonal elements representing the influence weights of surrounding pixels on sampling point i.

$$y_i = \beta_0(u_i, v_i) + \sum_{k=1}^{p} \beta_k(u_i, v_i)x_{ik} + \varepsilon_i \tag{2}$$

$$\widehat{\beta_k}(u_i, v_i) = \left[X^T W(u_i, v_i)X\right]^{-1} X^T W(u_i, v_i)Y \tag{3}$$

To better balance overall bias and standard error, this study chooses the adaptive bi-square kernel function as the spatial weight matrix to determine the spatial extent influencing the prediction of regression points. The calculation method is shown in Eq. 4.

$$W_{ij} = \begin{cases} \left(1 - \left(\frac{d_{ij}}{h_i}\right)^2\right)^2, & d_{ij} \leq h_i \\ 0, & d_{ij} > h_i \end{cases} \tag{4}$$

In this context, d_{ij} represents the Euclidean distance between sampling point ii and sampling point j, and h_i denotes the bandwidth for sampling point i. Then, by combining field survey results, satellite imagery, street view maps, and statistical yearbooks, the regression results of the GWR model are analyzed to determine the specific extent and intensity of the particulate matter reduction effect of blue and green spaces at the urban scale.

3 Results Analysis

3.1 Descriptive Statistics of Geographically Weighted Regression Model

In the GWR model for $PM_{2.5}$, the bandwidth for the Gaussian kernel density values of blue space area and green space area (referred to as $DE_{-bluespace}$ and $DE_{-greenspace}$) is 165, accounting for 11.04% of the total sample size. In the GWR model for PM_{10}, the bandwidth for $DE_{-bluespace}$ and $DE_{-greenspace}$ is 116, accounting for 7.76% of the total sample size. As shown in Table 2, the regression coefficients for each factor in the GWR model vary significantly across different spatial locations. These results indicate a significant spatial non-stationarity in the reduction effects of blue and

Table 2 Statistical description of geographically weighted regression coefficient (Self-drawn)

Particulate matter	Variable coefficients	Mean	Standard deviation	Minimum value	Median	Maximum value
$PM_{2.5}$	$DE_{-bluespace}$	-0.72	6.61	-13.97	-2.72	11.78
	$DE_{-greenspace}$	-1.28	2.78	-8.25	-1.19	5.17
PM_{10}	$DE_{-bluespace}$	-0.83	6.1	-10.32	-3.28	12.36
	$DE_{-greenspace}$	1.51	16.98	-24.98	9.15	25.57

green spaces on both types of particulate matter pollution, and overall, these effects remain as local-scale relationships.

3.2 *Analysis of the Impact of Blue Space on Particulate Matter Pollution*

In the GWR models for $PM_{2.5}$ and PM_{10}, the regression coefficients for $DE_{-bluespace}$ can be both positive and negative, as shown in Fig. 2. On average, blue space has a stronger impact on $PM_{2.5}$. In terms of the areas affected, positive regression coefficients for $DE_{-bluespace}$ are mainly found in regions with relatively high building density and floor area ratio in both pollution models. Specifically, in the $PM_{2.5}$ model, positive coefficients are present in parts of Fancheng District, Xiangcheng District, and the Shenzhen Industrial Park in Xiangzhou District. In the PM_{10} model, positive coefficients appear in Fancheng District and Dongjin New District. Areas with negative regression coefficients for $DE_{-bluespace}$ tend to have relatively low building density and floor area ratio. Specifically, for $PM_{2.5}$, negative coefficients are mostly found in the region from Zhuhai Avenue to Benchi Avenue and westward (see area 1 in Fig. 2), as well as between the Yuliangzhou Ring Road Greenway along both sides of the Tangbai River and Nanshan Road in Dongjin New District (see area 2 in Fig. 2). The area between Zhuhai Avenue and Benchi Avenue appears as a strip approximately 2000 m wide in Fig. 2. For PM_{10}, negative coefficients appear west of Zhuhai Avenue (see area 3 in Fig. 2), between the Yuliangzhou Ring Road Greenway and Nanshan Road in Dongjin New District (see area 4 in Fig. 2), and near the footline of Xianshan Mountain (see area 5 in Fig. 2). However, in the PM_{10} model, the negative coefficients near Zhuhai Avenue no longer appear as a strip and do not extend southward to Dongjin New District but rather expand westward towards the Xiangyang Automobile Industrial Park.

Based on these results, it can be inferred that blue space has a certain reduction effect on both types of pollution, particularly exhibiting a strip approximately 2000 m wide in reducing fine particulate matter pollution between Zhuhai Avenue and Benchi Avenue. According to the Research on Urban Ventilation Corridors in Xiangyang Report, this strip coincides with one of the city's main wind corridors, the Han River-Tangbai River wind corridor. Additionally, the pollution reduction effect of blue

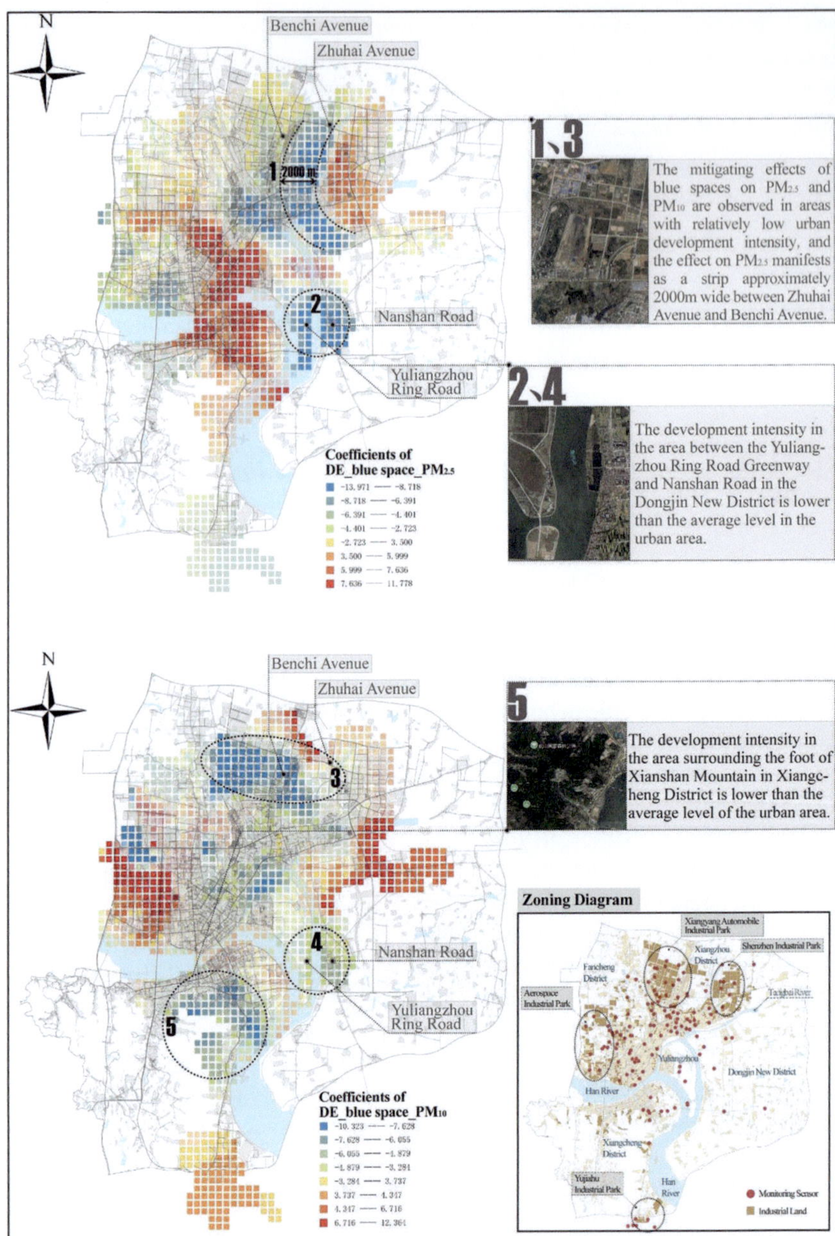

Fig. 2 Analysis of the impact of blue space on $PM_{2.5}$ and PM_{10} (Self-drawn)

space is notably evident in urban areas with lower development intensity, including the region south of Zhuhai Avenue in Xiangzhou District and the area between the Yuliangzhou Ring Road Greenway and Nanshan Road in Dongjin New District. However, this pollution reduction effect can be curtailed by high-intensity urban development, and pollution sources such as SO_2, NO_x, VOC_s, and construction sites may also negate this effect. Therefore, for the two aforementioned areas, attention should be paid to reserving flexible spaces and maintaining the current low-density development status to ensure the smoothness of urban wind corridors. This will promote the coupling and coherence between urban wind corridors and water bodies, thereby enhancing the pollution reduction effect.

3.3 Analysis of the Impact of Green Space on Particulate Matter Pollution

As shown in Fig. 3, the regression coefficients of $DE_{-greenspace}$ in the GWR models of $PM_{2.5}$ and PM_{10} are mainly negative, and at the same time, the average value of the negative regression coefficients of $DE_{-greenspace}$ in the PM_{10} model is smaller, indicating that green space has a particulate matter reduction effect on both pollutants, and the reduction effect on inhalable particulate matter pollution is stronger. In terms of the range of influence, for $PM_{2.5}$, the negative regression coefficients are concentrated between Dengcheng Avenue and Chunyuan Road in Fancheng District (see area 1 in Fig. 3), the area enclosed by the Han River, Jiaoliu Railway Line, Xianshan Mountain, and Wolong Bridge in Xiangcheng District (see area 2 in Fig. 3), and the area enclosed by Airport Road, Xiangyang Avenue, Handan Railway, and Benchi Avenue in Xiangzhou District (see area 3 in Fig. 3). For PM_{10}, negative regression coefficients also appear in roughly the same locations (see areas 4, 5, and 6 in Fig. 3), with a smaller average coefficient value, indicating a stronger mitigating effect of green space on PM_{10}.

Between Dengcheng Avenue and Chunyuan Road in Fancheng District (see areas 1 and 4 in Fig. 3) and in the area enclosed by Jichang Road, Xiangyang Avenue, Handan Railway, and Benchi Avenue in Xiangzhou District (see areas 3 and 5 in Fig. 3), there is a large amount of contiguous farmland but relatively little green space. Therefore, in these two areas, farmland plays a certain role in mitigating both types of pollution. The area enclosed by the Han River, Jiaoliu Railway, Xianshan, and Wolong Bridge has dense green spaces and sparse farmland (see areas 2 and 6 in Fig. 3). Specifically, this area includes seven urban parks and a national AAAA-level scenic area (Xiangyang Ancient City). This indicates that the green spaces in this area play a crucial role in improving local air quality. Furthermore, the results suggest that farmland has a greater contribution to pollution mitigation compared to green spaces, primarily because farmland accounts for about 35.7% of the city's land use, while the greening coverage rate has not yet reached 40%. This also reflects that the construction of green spaces in the study area still needs improvement.

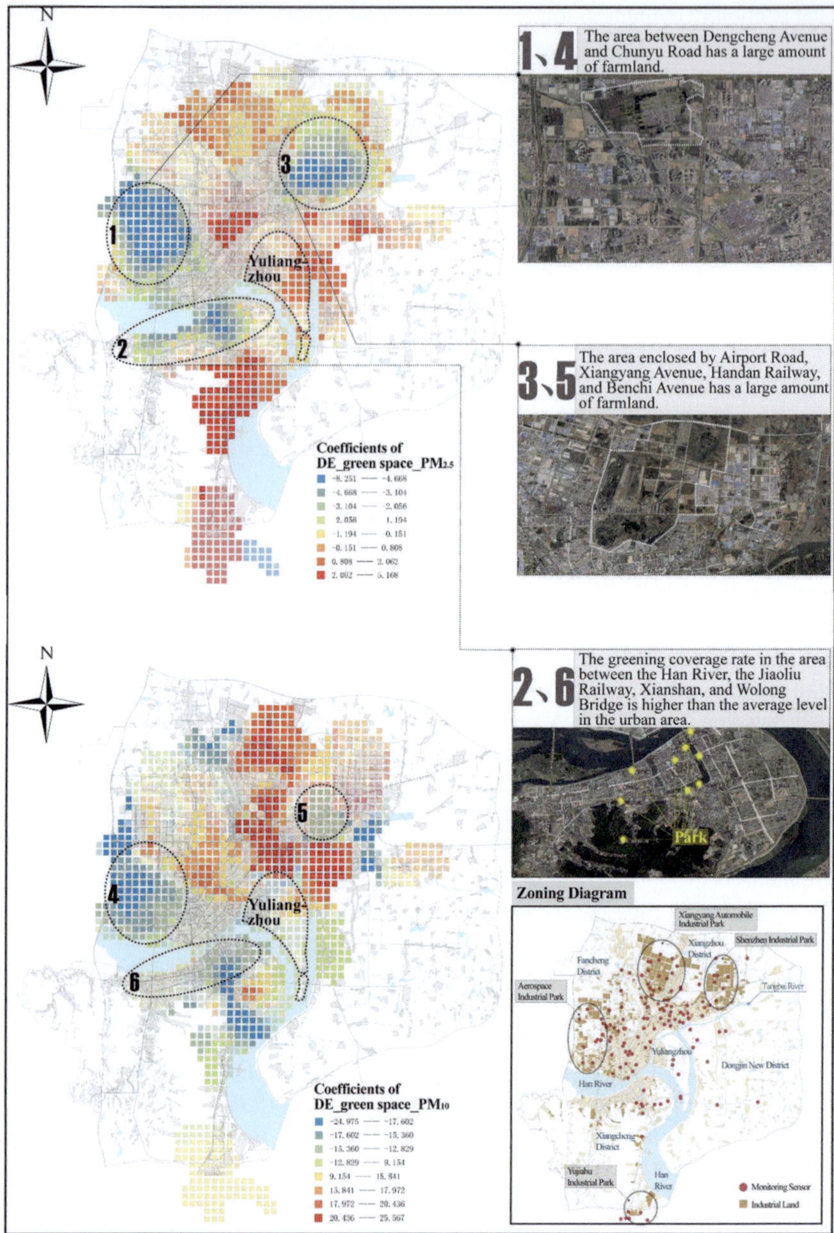

Fig. 3 Analysis of the impact of green space on $PM_{2.5}$ and PM_{10} (Self-drawn)

It is noteworthy that although Yuliangzhou is planned as the urban green heart of Xiangyang, in the models for both types of pollution, $DE_{-greenspace}$ does not show a significant negative correlation in most parts of Yuliangzhou ($|t|$ value ≤ 1.96). This indicates that during the study period (March 1, 2018, to February 28, 2019), the green space in Yuliangzhou did not effectively mitigate particulate matter pollution, which deviates from the expected planning. Further field investigations revealed that Yuliangzhou was undergoing a large-scale greening renovation project at the time, namely the Yuliangzhou Ring Road Landscape Project. This project was criticized for issues such as uncovered bare soil, low frequency of watering for dust suppression, and dust problems from construction trucks. At that time, bare soil and unpaved roads accounted for about 33.77% of the total area of Yuliangzhou.

4 Conclusions

To address the existing research gap in thoroughly examining the relationship between blue-green space elements and particulate matter at the urban scale, which makes it difficult to clearly allocate blue-green spaces within urban planning systems, this study undertook several steps. First, a high-density monitoring sensor network has been established in a heavily polluted city to overcome the limitations of existing discretely distributed monitoring stations that cannot provide enough sample points. Then, following the principles of particulate matter diffusion, we design a dual-precision nested algorithm to create a Gaussian kernel density dataset of blue-green space elements that more accurately reflects the spatial processes of particulate matter diffusion. Finally, we input this dataset as independent variables into a GWR model, and reveal several conclusions as follows:

(1) The mitigating effect of blue space on $PM_{2.5}$ and PM_{10} is evident in areas with relatively low urban development intensity, including the region south of Zhuhai Avenue in Xiangzhou District and the area between the Yuliangzhou Ring Road Greenway along both sides of the Tangbai River and Nanshan Road in Dongjin New District. Between Zhuhai Avenue and Benchi Avenue, the water bodies exhibit a mitigating effect on $PM_{2.5}$ in the form of a strip approximately 2000 m wide. Therefore, to utilize blue space for alleviating particulate matter pollution, development intensity in these areas should be controlled, flexible spaces should be reserved, and the coupling and coherence between urban wind corridors and blue spaces should be strengthened.

(2) Green space has a stronger mitigating effect on PM_{10} pollution compared to $PM_{2.5}$. The green spaces within the area enclosed by the Han River, Jiaoliu Railway, Xianshan, and Wolong Bridge positively impact local particulate matter pollution. Additionally, the extensive contiguous farmland between Dengcheng Avenue and Chunyuan Road in Fancheng District, as well as in the area enclosed by Jichang Road, Xiangyang Avenue, Handan Railway, and Benchi Avenue in Xiangzhou District, also contributes to pollution mitigation.

However, during the study period, the dust retention effect of green space in Yuliangzhou was not significant. Therefore, protecting existing farmland can be a potential path to improve local urban particulate matter pollution. Moreover, there is a need to accelerate the construction of high-quality green spaces, particularly to enhance the role of Yuliangzhou as the urban green heart.

Acknowledgements This study is supported by the Natural Science Foundation Youth Project of Hubei 2023AFB510, Guiding Project of Scientific Research Program of the Hubei Provincial Department of Education B2023254, Natural Science Foundation Youth Project of Wuhan 2024040801020335, and Fundamental Research Support Program of Jianghan University 2023JCYJ09.

References

1. Khamraev K, Cheriyan D, Choi J (2021) A review on health risk assessment of PM in the construction industry–Current situation and future directions. Sci Total Environ 758:143716
2. Southerland VA, Brauer M, Mohegh A et al (2022) Global urban temporal trends in fine particulate matter ($PM_{2.5}$) and attributable health burdens: estimates from global datasets. Lancet Planet Health 6(2):e139–e146
3. Fang Y, Zhao L (2022) Assessing the environmental benefits of urban ventilation corridors: a case study in Hefei, China. Build Environ 212:108810
4. Wu S, Chen C, Wang Y et al (2021) Differential effects of valley city morphology on mesoscale flow field characteristics. Build Environ 205:108283
5. Wu S, Wang Y, Chen C et al (2021) Valley city ventilation under the calm and stable weather conditions: a review. Build Environ 194:107668
6. Zhao L, Li T, Przybysz A et al (2021) Effect of urban lake wetlands and neighboring urban greenery on air PM_{10} and $PM_{2.5}$ mitigation. Build Environ 206:108291
7. Fan Z, Zhan Q, Liu H et al (2022) Investigating the interactive and heterogeneous effects of green and blue space on urban $PM_{2.5}$ concentration, a case study of Wuhan. J Clean Prod 378:134389
8. Diener A, Mudu P (2021) How can vegetation protect us from air pollution? A critical review on green spaces' mitigation abilities for air-borne particles from a public health perspective-with implications for urban planning. Sci Total Environ 796:148605
9. Linden J, Gustafsson M, Uddling J et al (2023) Air pollution removal through deposition on urban vegetation: the importance of vegetation characteristics. Urban For Urban Green 81:127843
10. Zhang L, He J, Gong S et al (2022) Effect of vegetation seasonal cycle alterations to aerosol dry deposition on $PM_{2.5}$ concentrations in China. Sci Total Environ 828:154211
11. Kończak B, Cempa M, Deska M (2021) Assessment of the ability of roadside vegetation to remove particulate matter from the urban air. Environ Pollut 268:115465
12. Wikipedia Homepage. https://en.wikipedia.org/wiki/Xiangyang. Last accessed 30 Sep 2024
13. Oke TR, Mills G, Christen A et al (2017) Urban climates. Cambridge University Press, Cambridge
14. Fotheringham AS, Brunsdon C, Charlton M (2003) Geographically weighted regression: the analysis of spatially varying relationships. Wiley, New Jersey

Rural Revitalization and Transformation from a Catalytic-Sustainable Perspective

Yidi Yao and Jie Huang

Abstract China's fast-paced urbanization has contributed to a significant decrease in rural populations, leading to the phenomenon known as village "hollowing." In response, this study develops an integrated approach that combines the concepts of urban catalyst mechanisms with principles from sustainable tourism to promote rural renewal, while safeguarding both environmental and cultural assets. Urban catalyst theory is particularly effective in such contexts, as it emphasizes minimal intervention strategies capable of producing broad, transformative effects without dismantling existing spatial layouts or community networks. Taking Dongtai Village in Beijing as the focal point, the research explores how targeted actions—such as conserving vernacular architecture, enhancing communal gathering areas, and establishing multifunctional tourism infrastructure—can stimulate long-term development momentum. The findings show that post-renovation, the average courtyard usage rate has improved significantly, with an approximate increase of 30%.

Keywords Urban catalyst · Sustainable · Rural revitalization · Design strategy

1 Introduction

In recent years, the acceleration of urbanization has posed a significant challenge to China's rural areas, resulting in a notable decline in population. This phenomenon, commonly referred to as the "hollowing out" of the countryside, has contributed to the gradual abandonment of many traditional villages, largely due to the loss of young people. Meanwhile, the increase in demand for leisure tourism, especially the rise of weekend and peripheral tourism, has opened new avenues for development in rural areas. With regard to the "hollowing out" of villages, traditional approach

Y. Yao (✉)
School of Design, South China University of Technology, Guangzhou, China
e-mail: yyd@scut.edu.cn

J. Huang
School of Architecture, Guangzhou City University of Technology, Guangzhou, China

B. He et al. (eds.), *Climate-Resilient, Low-Carbon, and Sustainable Cities*, Lecture Notes in Civil Engineering 754, https://doi.org/10.1007/978-981-95-2169-2_42

to building renovation, which often prioritizes the preservation of the appearance of rural dwellings but fails to integrate with emerging industries, is insufficient to provide new attractions that can both attract popularity and alleviate the hollowing out of villages. Therefore, it is imperative to investigate novel approaches to rural revitalization.

Developing new lifestyles in rural bed and breakfasts (B&Bs) with distinctive appeal may offer a solution for revitalizing shrinking rural areas. The B&B model is a new industry in China's rural tourism sector, which is based on the renovation and reuse of rural residences [1, 2]. The new rural B&B industry, focusing on the mixed development of B&Bs and various functions including commerce, hotels, art, workshops, etc. and aiming to build a community meeting diverse needs and attracting a varied demographic to come to travel and consume, has the potential to promote rural revitalization [3]. By strategically selecting renovation sites and using them as catalysts, a "chain reaction" can be triggered, promoting sustainable and gradual development in rural areas with minimal intervention [4]. In this context, specific points can be regarded as urban catalysts. Applying the theory of urban catalysts and introducing new rural catalysts, while leveraging characteristic resources as the foundation, could effectively address rural depopulation. It can be posited that the transformation of a declining rural area into an innovative rural bed and breakfast (B&B) with distinctive attractions may represent a efficacious approach to simultaneously enhance the quality of life in the area and stimulate rural economic growth [5].

2 Literature Review

2.1 Urban Catalyst Theory

Urban catalyst refers to the new elements that can prompt changes in the city and accelerate or change the speed of urban development and construction [4]. The role of urban catalysts emphasizes that through the intervention of specific catalyst elements, it can lead to a "chain reaction", promote the maturity of the objective conditions of urban construction, and thus promote the sustainable and progressive development of the city. The role of urban catalysts in different urban environments varies, and there are four main types, which are preserving, reinforcing, repairing and creating [4] (see Fig. 1). "Preserve" focuses on maintaining core structural identity while adapting existing frameworks. "Strengthen" addresses fragmented urban systems by enhancing connectivity between components. "Repair" revitalizes dysfunctional zones through functional upgrades (e.g., injecting commercial hubs into underperforming areas to reconnect them with the urban fabric). Finally, "Create" establishes new spatial hierarchies and frameworks, typically applied in district-wide regeneration or new developments to redefine urban order through systemic redesign.

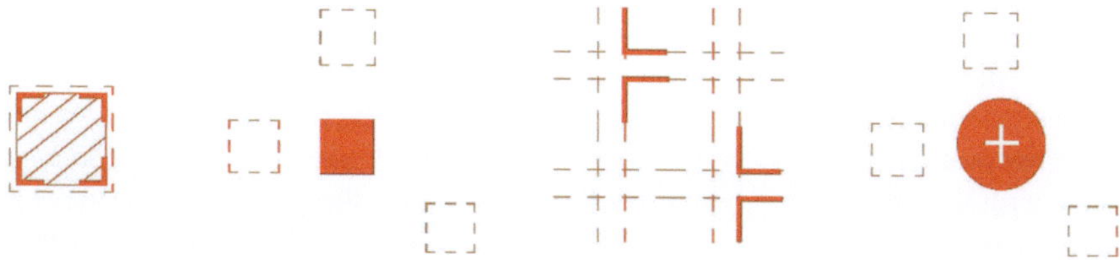

Fig. 1 Analysis diagram of the four catalyst actions: preserve, reinforce, repair and create (self-drawn)

Multiple methods have been used apart from urban catalyst theory. Lu conducted quantitative analysis by utilizing spatial syntax analysis methods, collecting Point of Interest (POI) data, and visualizing relevant data [6]. Mao used space syntax analysis to discuss the transformation of industrial historical stock space [7].

The strength of urban catalyst theory in rural transformation is that designers can influence the overall situation through the smallest possible intervention without destroying the original community structure and social ties: through small-scale, high-impact interventions at key points, the point will lead to a wider community change, innovation and improvement, and promote the optimization of resources and create a demonstration effect. In the 1980s, the catalyst theory gained widespread application in addressing the issues of suburbanization and the decline of city centers in North America. Local government tried to vivify the city center by inserting a series of attraction points such as sports architecture, shopping mall and other large public infrastructure [8, 9].

2.2 Theory of Sustainable Tourism Development

The theory of sustainable tourism development was first put forward in 1980s by Butler [10], who emphasized that tourism resources should be treated as finite and potentially non-renewable assets, and that by setting reasonable capacity limits and implementing protective measures, they can extend the lifecycle of tourist areas and prevent overdevelopments.

The countryside is characterized by a unique set of resources that differ from those found in urban areas. These include geographical resources that are associated with distinctive features and cultural resources associated with folk culture. Integrating local natural landscape and cultural heritage with rural resources for cultural tourism planning, developing tourism projects with rich characteristics, and establishing a strict environmental regulatory mechanism and cultural protection measures can ensure the sustainability of rural ecological and cultural resources. On this basis, villagers and tourists can share in the benefits of tourism and optimize the allocation of resources.

In the study of rural cultural tourism industry, Chen and other scholars have explored the reconstruction of rural communities in the economic, social and spatial dimensions driven by the B&B industry [11]. Ye has taken the Moganshi Valley project as an example to study how to combine the enterprise and rural resources to create a rural cultural tourism industry [2]. Zhai studied the design of homestays in the context of rural construction [12]. Liu discussed how entrepreneurship interacts with rural context on the field of bed and breakfast [13]. In contemporary times, the revitalization and transformation of the countryside under the guidance of cultural and tourism planning will become an important driving force for rural economic development.

3 Methodology

The diagram of the catalytic-sustainable strategies of this study is shown below (see Fig. 2).

The article aims to conclude the revitalization design strategies under the perspective of catalytic-sustainable perspectives for the villages in China. The authors will discuss specific design approaches based on a case study from their own practice.

3.1 Location

The project is situated in Dongtai Village, located northwest of Bohai Town in Huairou District, Beijing, nestled at the base of the Mutianyu section of the Great Wall and along the banks of the Huisha River. The village has a population of nearly 400 households and over 900 people. It is renowned for its abundant chestnut tree

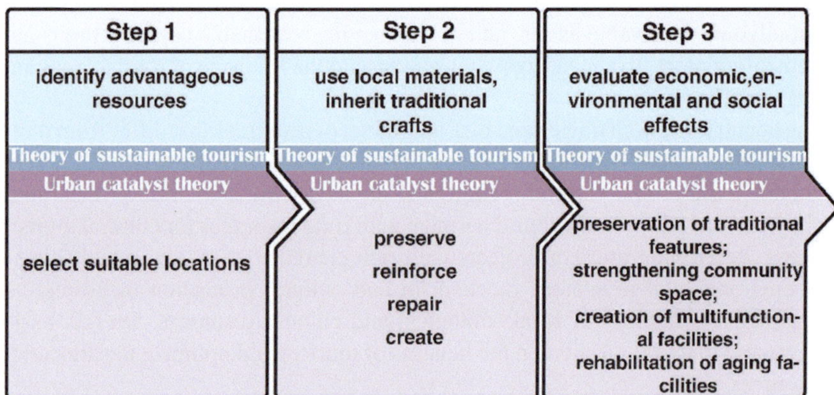

Fig. 2 Diagram of the catalytic-sustainable strategies (self-drawn)

Fig. 3 Characteristic residences in Dongtai village (self-drawn)

resources and its proximity to the Great Wall, making it a notable historical site. The challenge of the case study area is that, despite its proximity to renowned tourist attractions, the surrounding rural hostels have struggled to attract guests due to the lack of multifaceted amenities or diverse functions within the village. Meanwhile, the villages relied on agricultural products, but their sales have been constrained owing to a lack of recognition or familiarity with these products. Therefore, by inserting more functions in the village and creating landmarks to attract people, the village can effectively address the issue of low visitor turnout and stagnant agricultural sales. The project aims to transform Dongtai Village's traditional houses into unique bed and breakfast accommodations, incorporating commercial spaces, hostels, art galleries, workshops, design markets, and workspaces. The goal is to preserve the village's traditional aesthetic, modernize infrastructure, and introduce diverse functions to attract both families and business travelers to the area.

Dongtai Village's architecture is characterized by the traditional "quadrangle" layout with sloped-roof buildings, using local stone and timber (see Fig. 3). The façades feature papier-mâché lattice windows, brick walls, and stone plinths, while modern structures like reinforced concrete are integrated to balance cultural preservation with modernization.

3.2 Field Observation

Our research was primarily conducted through field studies and semi-structured interviews. We conducted a thorough exploration of Dongtai Village, engaging in detailed discussions with the village management to focus on the actual needs of the villagers and the main challenges they face in their living environments. Through in-depth interviews and meticulous research, we identified the following core issues (see Fig. 4).

Problems of architectural style: Due to lack of maintenance, the building facades have deteriorated, showing signs of aging and cracking. The newly added cement plastering on the walls is visually unappealing and clashes with the traditional architectural style. The traditional stone walls are in poor condition. Furthermore, the presence of corrugated iron sheds and unauthorized constructions leads to a disorganized layout, detracting from the overall aesthetics.

Maintenance and space utilization: The courtyard spaces in Dongtai Village have not been fully utilized, with clutter and unauthorized constructions impeding their efficiency. This situation calls for necessary cleanup and space optimization.

Infrastructure and surrounding environment: The infrastructure in Dongtai Village is generally inadequate, with limited road landscaping and recreational facilities, resulting in a poor pedestrian experience. Additionally, there is a noticeable lack of adequate support facilities for homestays, coupled with a limited variety of business types.

Meanwhile, the village has the following advantages.

Excellent geography and rich natural resources: Dongtai Village is strategically located near the Great Wall Cultural Belt and the Ringshui Gorge landscape belt, boasting natural resources like chestnut forests and open fields. It is also close to cultural sites such as the Ringshui Lake Great Wall and Ringshui Camping Camp. Despite its mountainous terrain, the village is easily accessible by car, just 8 km from the Mutianyu section of the Great Wall and 4 km from Ringshui Lake on Beijing's outskirts. Within an hour's drive, it offers access to renowned landmarks like the Yingshan Pagoda Forest and the Ming Dynasty Tombs, enriching its cultural and natural resource base.

Fig. 4 Current status of dwellings in Dongtai village (self-drawn)

Comprehensive facilities: The village has comprehensive facilities including cultural facilities, sports facilities, and a health station, providing a solid foundation for the village's future development.

Rich agricultural resources: The village is blessed with abundant agricultural resources, including mushroom cultivation and chestnut trees. These resources not only support the livelihood of the local residents but also serve as a foundation for developing specialized agriculture and related industrial chains.

Integrated Development Potential: The surrounding area has seen successful homestay projects, notably Jinglin Chestnut Garden Homestay, which has become a regional landmark due to its reputation. These established resources offer insights and potential for integrated development in Dongtai Village.

3.3 Design Strategies

Based on our research findings, we propose that the revitalization and renovation of the Dongtai village houses should prioritize preserving its unique local culture and maintaining traditional residential features, such as sloping roofs, lattice windows and other distinctive elements.

The renovation design should integrate catalyst theory and sustainable tourism development principles, targeting specific houses with tailored strategies to revitalize the village incrementally. Additionally, the project should incorporate diverse functions—art, dining, lodging, and shopping—to establish a restorative, empowering rural hospitality system that enriches experiences for locals and visitors alike.

The first step is to identify strategic locations. We have therefore focused our research on areas near the village's main roads. Through on-site assessments, we carefully selected a series of courtyards to act as "catalyst nodes" for design interventions. These key sites are designed to enhance the village's overall environment. Specifically, we have chosen three courtyards and applied preservation, enhancement, and creation strategies to transform them into boutique B&Bs, community markets, and creative dining spaces. By integrating these "catalyst nodes," the project aims to revitalize and transform Dongtai Village (see Fig. 5).

The second step involves utilizing local materials and embracing traditional craftsmanship during the renovation process. Additionally, it is essential to delineate four key strategies—preservation, reinforcement, repair, and creation—while concurrently integrating diverse functionalities.

Preservation: Craft Distinctive Homestay. For courtyards that are well-preserved, it is it is advisable to prioritize the conservation of original features while using renovations to enhance functionality. This strategy aims to maintain local characteristics and perpetuate the area's cultural heritage. It involves keeping the exterior of traditional dwellings as authentic as possible and implementing essential updates and functional adaptations that respect the original quadrangle layout.

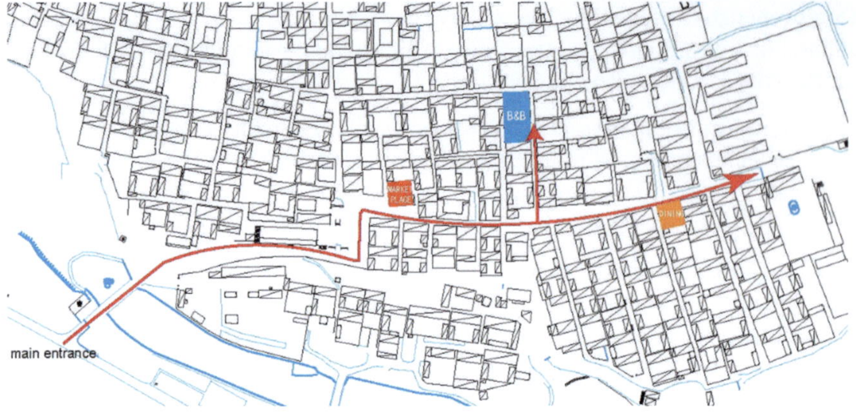

Fig. 5 Schematic diagram of 'Catalyst' activation points in Dongtai village (self-drawn)

We have chosen a relatively well-preserved traditional residential compound in Dongtai Village to transform into a distinctive B&B. The selected courtyard consists of two well-preserved dwellings with eaves and sloping roofs. The northern dwelling, aligned north-to-south, boasts a layout of five bays, one hall, and two rooms. The western dwelling, oriented west-to-east, comprises three bays, one hall, and two rooms. Our design focused on highlighting the regional architectural characteristics while maximizing the preservation of the dwellings' appearances, including the brick façade, tile sloped roofs, and the spatial layout of the courtyard.

The renovation design for the courtyard is divided into three parts: exterior preservation, interior renovation and transformation of outdoor space. For the exterior, our approach is to maintain the architectural integrity. This involves meticulously preserving the external features such as the distinctive sloping roofs, brick walls, and wooden beams. Restoration efforts will focus on repairing damaged elements using traditional materials and techniques to preserve the historical accuracy and aesthetic of the buildings. Special attention will be paid to the façades to ensure their original appearance intact. Any necessary façade repairment will aim to blend seamlessly with the existing structure, ensuring that the refurbishments do not detract from the building's traditional character. For the interior renovation, the internal layout of the buildings will be redesigned to accommodate the needs of a modern homestay while respecting the original structure. This includes creating private guest rooms, bathrooms, and communal areas such as dining rooms and lounges. For the outdoor space transformation, the central courtyard will be cleared of non-essential structures and redesigned as a communal gathering space for guests. This area will be landscaped to enhance its aesthetic appeal and functionality, including seating areas and possibly a small stage for cultural performances. Meanwhile, a new semi-open structure with a sloped roof will be built on the eastern side of the courtyard. This space will be designed for multiple uses, including social interactions, relaxation, and reading,

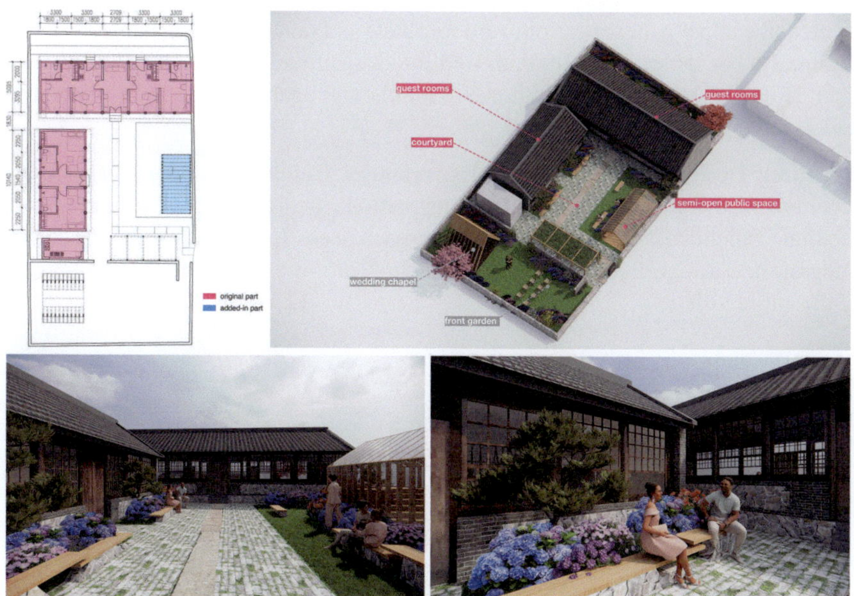

Fig. 6 Plan, axonometric and perspectives of distinctive homestay (self-drawn)

providing a sheltered yet open environment that complements the existing structures (see Fig. 6).

Enhancement: Expand Rural Marketplace. For residential buildings that are in average condition, a reinforcement strategy should be adopted to renovate and repair them, strengthening their structure and functionality. While enhancing the building's efficiency and aesthetics, it is crucial to ensure the continuation of cultural elements. For villages, rural marketplace, which is an important public element of village community and plays an important role in rural life. It is not only an important place for villagers to barter and exchange agricultural products and daily necessities, but also a core area for villagers to communicate with each other. Therefore, in the enhancement strategy, a specific courtyard is selected for transformation, combining local culture to create an innovative rural market: the project in enclosed by three sloped-roofed volumes on the south, west, and north sides, forming an additional layer of "alley" space, between the volumes, which accomodates winter use in the form of a sunroom. The two facades facing the village's public square and main street are left open with commercial windows, creating a lively and cozy rural neighborhood market. Additionally, the existing elevation differences of the site are utilized to from outdoor relaxation space, making it a place for villagers to gather or chat with each other.

The market interior features a flea market showcasing village products and a rural workshop for promotions, meetings, training, handicrafts, and cultural heritage. Inspired by traditional lattice windows, a two-piece operable window design was created: the upper panel opens outward for shade and rain protection, while the lower becomes a bar table. At night, both fold into a secure enclosure (see Fig. 7). This flexible design blends local aesthetics with climate adaptability, ensuring comfort and usability year-round.

Creation: Design Creative Dining Experiences. Buildings in poor condition are recommended for demolition and reconstruction, incorporating new functions to facilitate functional replacement. This approach seeks to modernize the space and integrate new design elements to foster the revitalization and sustainable development of the entire village.

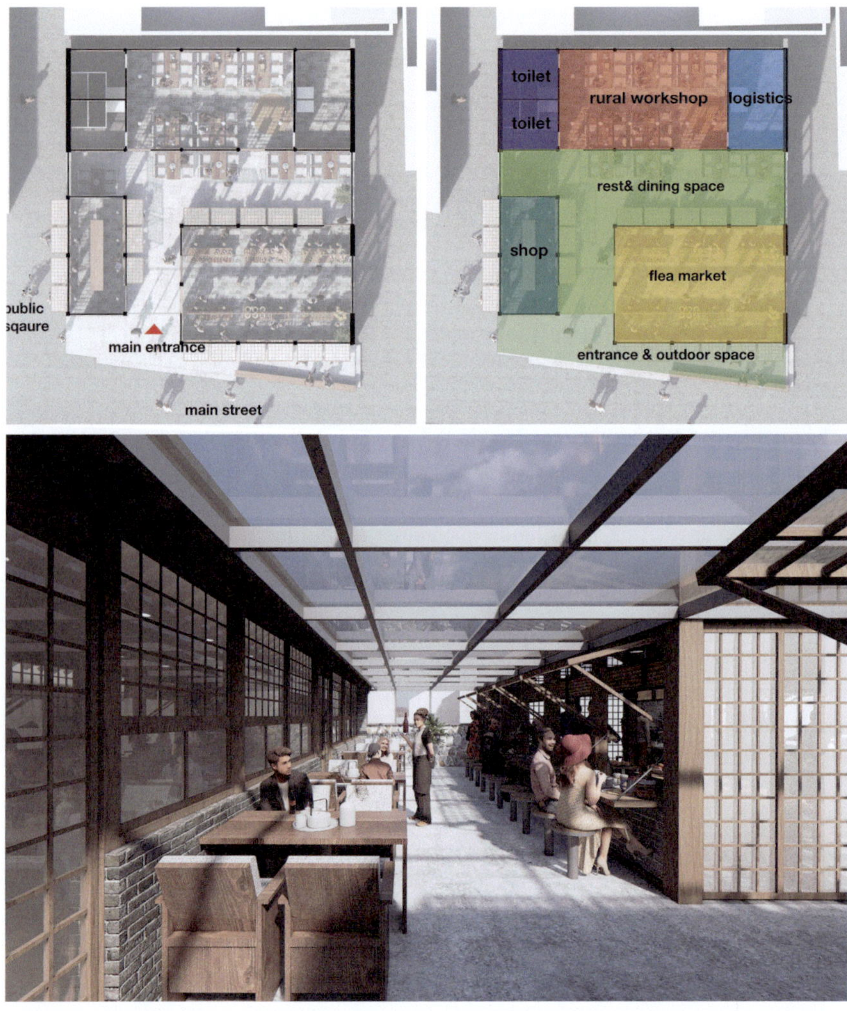

Fig. 7 Plan and perspectives of rural marketplace (self-drawn)

Therefore, the design team transformed the third courtyard into a creative dining space called "Gudong Yard" by incorporating a dining function. n the ren-ovation of the courtyard, the original north and west sloping roofs and the central courtyard of the courtyard were preserved, and the preserved buildings were trans-formed into an indoor restaurant. At the same time, in response to Beijing's climatic characteristics, seats were placed in the central courtyard to create a sheltered semi-outdoor courtyard hot pot dining space, which is used as a sunroom to meet the needs of winter use. Outdoors, the open courtyard space becomes a place for relaxation and enjoyment of food; indoors, the flexible large space and subdivided small rooms provide different dining experiences for different people. The façade elements are also redesigned with latticework inspired by the lattice windows, which open outward, with the upper sash providing shelter from the sun and rain, and the lower sash serving as a tabletop, echoing the local culture and the functional needs of the use (see Fig. 8).

The last step is to evaluate the economic, environmental and social effects of the renovation. As shown in Fig. 9, the use of the courtyard utilization ratio has increased by an average of approximately 30% after the renovation. And an estimated 25% increase in tourists staying is expected. Considering that the renovation is based on minimal intervention, it will have minimal negative impact on the local environment.

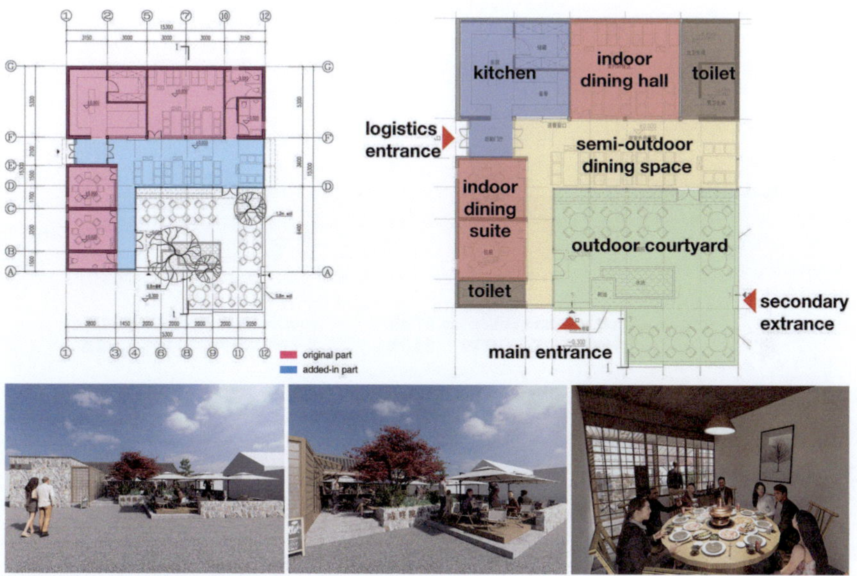

Fig. 8 Plan and perspectives of hot-pot restaurant (self-drawn)

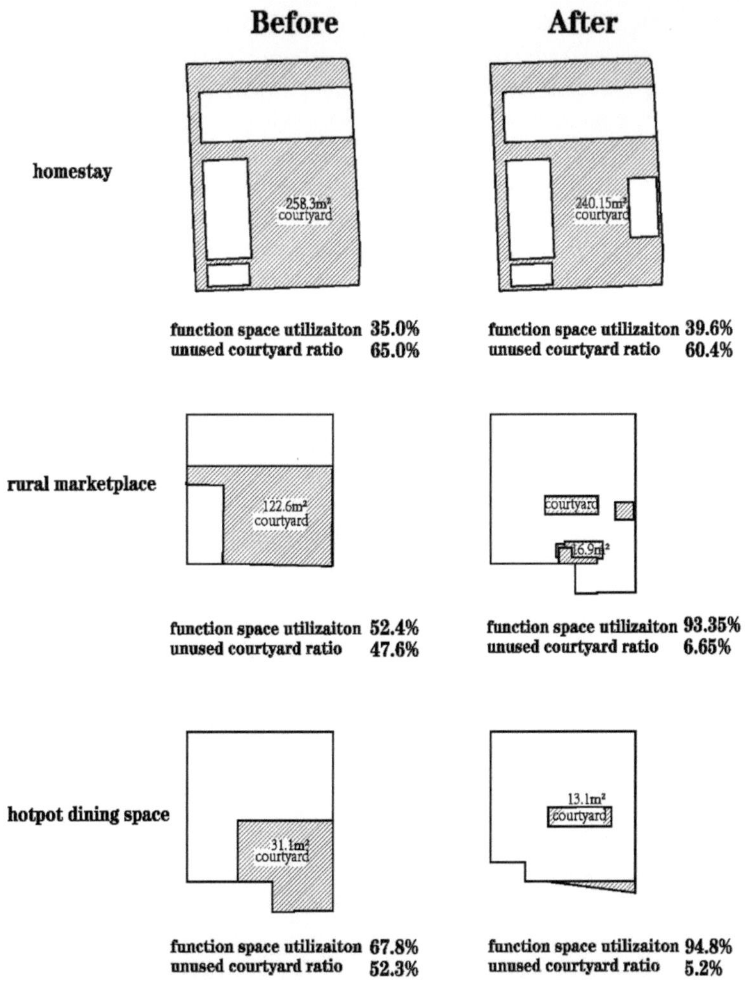

Fig. 9 The before and after utilization of courtyards in three locations (self-drawn)

4 Discussion: Design Principles for Rural Revitalization from a Catalyst Perspective

Under the catalyst perspective for rural revitalization, the approach prioritizes initiating small-scale, strategic interventions that can trigger widespread positive changes throughout a community. This method strives to achieve the maximum social impact while minimizing interference. Summarized from previous research, here are the four key principles.

4.1 Preservation of Traditional Features

In rural revitalization projects, preserving the traditional appearance is a crucial step. For buildings with historical value, it is important to maintain the integrity of their traditional appearance as much as possible. This includes preserving their original external appearance and structural features, while also using traditional materials and techniques for maintenance.

4.2 Strengthening Community Space

Reinforcing community spaces enhances public areas' functionality and appeal. At Qimo Primary School in Huaji County, Guangdong, architects created a community space during the school's construction. A continuous staircase links the basketball court to the rooftop, forming a theater-like plaza. The corridor behind the staircase is illuminated by light filtering through the gaps. This plaza has become the village's social hub, hosting various activities.

4.3 Creation of Multifunctional Facilities

Creating multifunctional facilities aims to enhance the attractiveness and vitality of villages by incorporating various functions. The Songyang Chenjiapu Pioneer Bookstore, transformed from an old villagers' hall, is a successful example. Designers retained the original frame structure of the old building while infusing it with multiple new uses, blending commerce, culture, and public recreation to establish an open and dynamic village library. Eventually the facility serves not only as a landmark for tourists but also as a crucial public space where the village's elderly and children can read and relax.

4.4 Rehabilitation of Aging Facilities

Rehabilitating aging facilities is crucial for the sustainability of village revitalization. This includes the maintenance, refurbishment, and reinforcement of various types of infrastructure such as roads, water systems, sanitation, and environmental facilities to enhance their efficiency and reliability. Such efforts not only improve the quality of life for residents but also lay the groundwork for the long-term development of rural areas. For instance, in Dong villages of Guizhou, China, designers prioritized preserving traditional water systems during restoration, employing ecological tech to modernize circulation while safeguarding cultural heritage.

5 Conclusion

This paper focuses on revitalizing Dongtai Village in Beijing amidst the rural "hollowing out" phenomenon. It applies catalyst theory and sustainable tourism development principles with a three-dimensional design strategy—residential innovation, cultural rejuvenation, and catering innovation—aiming to create a model for rural cultural tourism. This approach enhances infrastructure, boosts the economy, and addresses rural decline, offering villagers improved living standards and tourists a comprehensive cultural experience. After renovation, the utilization rate of courtyards has an average increase around 30%, which largely improves the living quality of the local houses and offers more functions for local residents and tourists.

Furthermore, this paper outlines principles of rural revitalization and transformation from the catalyst perspective, focusing on the preservation of traditional features, strengthening of community spaces, creation of multifunctional facilities, and rehabilitation of aging facilities. The flexible application of these principles not only facilitates the thorough execution of rural revitalization projects but also fosters the development of a new type of rural community, infusing sustainable growth and cultural preservation with renewed vitality.

Acknowledgements This study is supported by the National Natural Science Foundation of China (Youth Program) [Grant No. 52408021]; Basic and Applied Basic Research Foundation of Guangzhou Municipal Science and Technology Bureau (2024A04J3824), and Fundamental Research Funds for the Central Universities (QNMS202315). We thank the anonymous reviewers whose comments and suggestions helped improve and clarify this manuscript.

References

1. Wang W, Ding L, Chan JH, Qi X (2022) Process framework for innovation through tradition and its antecedents in rural heritage B&B. Int J Contemp Hospitality Manag 34(12):4329–4354
2. Ye Y (2020) The brand construction and communication plan of Sunac Mogan Valley—from the perspective of the value co-creation between real estate enterprise and rural revitalization. Master thesis, Zhejiang University. https://link.cnki.net/doi/10.27461/d.cnki.gzjdx.2020.001379
3. Kuo CMM, Tseng CY, Chen LC (2018) Choosing between exiting or innovative solutions for bed and breakfasts. Int J Hospitality Manag 73:12–19
4. Attoe W, Logan D (1989) American urban architecture: catalysts in the design of cities. University of California Press
5. Wang Y, Jiang Y, Geng B, Yan Z, Wang X (2024) Social networks, network interaction and resilience of B&Bs in rural China. Int J Contemp Hospitality Manag 36(2):400–421
6. Lu X (2020) Research on quantification of catalyst elements in the transformation of historic conservation area from the perspective of catalysts theory. Master thesis. Shandong Jianzhu University
7. Mao Y, Ye F (2019) The strategy of spatial renewal, activation and utilization of historical industrial stock in Shapowei, Xiamen. Urban Archit 16(28):71–75
8. Thornley A (2002) Urban regeneration and sports stadia. Eur Plan Stud 10(7):813–818

9. Robertson KA (1995) Downtown redevelopment strategies in the United States: an end-of-the-century assessment. J Am Plann Assoc 61(4):429–437
10. Butler R (1980) The concept of a tourist area cycle of evolution: implications for management of resources. Can Geogr 24:5–12
11. Chen C, Geng J (2023) Research on the evolution of traditional rural communities from the perspective of resilience-guilidng: the case of three typical villages in Moganshan Town. City Plan Rev 47:86–93
12. Zhai J (2016) Research on the design of rural boutique B&B in the context of rural construction. Master thesis, Zhejiang University
13. Liu J, Zhong D, Liu J, Liao Z (2023) B&B accommodation entrepreneurship in rural China: how does embeddedness make a difference? J Hospitality Tour Manag 56:284–294

Characteristics and Spatial Optimization of Primary School Service Performance Under Time and Space Coordination: A Case of Kuiwen District Primary School in Weifang, China

Chang'an Zuo, Junrong Li, Ye Zhang, Wenshuo Zhai, Zhihua Liu, Tianran Wang, and Fenghua Dong

Abstract In response to the problems of uneven allocation of school resources and uneven quality of education in the old urban areas, this paper proposes an evaluation method for the service performance of primary schools in the old urban areas in the urban renewal action, and take Kuiwen District of Weifang City (the central urban area) as an example to conduct an evaluation analysis and put forward an optimization strategy. First of all, through the literature review of the relationship between urban space and educational facilities, the current research frontier and theoretical basis are summarized. Second, the two evaluation indicators of spatial and functional performance under service performance and their standards are determined respectively. Then, based on the information of the Education Bureau and the results of field research, the service performance of each elementary school was evaluated in a refined manner on a school district basis. The evaluation results show that only about 57% of primary schools in the central urban area of Kuiwen District meet the service radius of 500 m, and 54% of the actual demand for primary school degrees is higher than the number of degrees. Some primary schools have problems such as low accessibility, unreasonable school distribution, and mismatch between degree supply and demand. Finally, the evaluation results are classified to propose renewal and optimization strategies for different service performance categories, and three typical elementary schools are selected for the analysis of specific renewal and optimization schemes. Updating and optimizing the service performance of elementary schools in old urban areas is of great significance in improving the public service

C. Zuo (✉) · T. Wang · F. Dong
School of Civil Engineering, Shandong University, Jinan, Shandong, China
e-mail: ca_zuo@sdu.edu.cn

C. Zuo · J. Li · Y. Zhang · W. Zhai · Z. Liu
China Construction Engineering Design and Research Institute Co., Ltd., Jinan, Shandong, China

C. Zuo · T. Wang · F. Dong
Urban and Rural Human Settlement Environment Research Center, Shandong University, Jinan, Shandong, China

B. He et al. (eds.), *Climate-Resilient, Low-Carbon, and Sustainable Cities*, Lecture Notes in Civil Engineering 754, https://doi.org/10.1007/978-981-95-2169-2_43

system and promoting the equalization and fair allocation of educational resources in urban renewal actions.

Keywords Urban renewal · Old town · Primary school · Spatial performance

1 Introduction

Nowadays, urban renewal has become the main means of expanding urban development space in old cities. The 20th Party Congress proposed "accelerating the high-quality and balanced development of compulsory education and urban–rural integration, optimizing the allocation of regional education resources, strengthening preschool education and special education, adhering to the diversified development of the senior high school, and perfecting the student financial aid system covering the whole school cycle" [1, 2]. In old urban areas, the problems of uneven allocation of school resources and uneven quality of education are particularly prominent. Urban renewal brings significant opportunities for the expansion, addition, and integrated layout of education while stimulating the vitality of land in old urban areas and improving urban functions. The performance evaluation of "educational facilities can provide certain data support and scientific optimization suggestions for the renewal planning of educational facilities in urban renewal actions, which has an important impact on the living standard of the residents, and scientific and reasonable planning and allocation of public service facilities is the key to the implementation of the new era of" to The basic requirement of the concept of "people-oriented" is also an important factor in promoting sustainable urban development.

Service performance refers to the comprehensive effectiveness or effect of the services provided by various departments, and only by maximizing the development needs of various functions, rationally allocating various types of resources, and promoting the balanced development of various elements can the comprehensive efficiency be effectively enhanced [3]. The service performance of educational facilities, on the other hand, is a concept used to measure the results and effectiveness of the services provided by schools, to provide measurement and feedback on the degree and efficiency of the realization of school services to students. With the new background of urbanization and the double liberalization of the two-child and household registration systems, the layout and configuration of educational facilities are increasingly facing the inadaptability of the classical theory caused by spatial and temporal heterogeneity, which in turn manifests itself in the uneven spatial distribution of educational facilities [4]. The spatial imbalance between supply and demand, and the uneven spatial quality of the facilities, constrain the process of the development of educational equity. As an important part of the compulsory education stage, elementary school is an important place for children to receive education and enlightenment, thus optimizing the performance evaluation and optimization of elementary school services is of great significance for improving the public service system, promoting equal allocation of basic education resources, and realizing the rapid response and adjustment in response to the demographic changes.

By searching keywords such as 'primary school layout', 'basic education facilities', 'school-age population', and 'school district', it is found that the layout of basic education facilities under the new situation of fertility new policy, stock planning and urban renewal, as well as the supply and demand relationship under education equity are the current research frontiers. Under the new situation, the increasingly complex and dynamic spatial environment problems will inevitably affect the layout and index configuration of educational facilities.

Studies on service performance evaluation of educational facilities in the field of urban and rural planning are mainly based on the perspective of school spatial layout. i.e., the spatial service performance of schools, evaluating the balance, fairness, and accessibility of elementary school layouts using GIS spatial analysis techniques, and proposing optimization suggestions to address the problems of the spatial layout. There are relatively few studies on service performance evaluation from the perspective of urban renewal, and there is a lack of comprehensive research on spatial and functional performance. The rapid development of the city and the drastic changes in population space have made the problems of the stock of spatial environments more and more complicated and dynamic, and the renewal of educational facilities in existing settlements is urgent.

2 Study Area and Data Source

2.1 Overview of the Study Area

Kuiwen District is located in the central urban area of Weifang City, Shandong Province, and now has 8 streets under its jurisdiction, with a total area of 57.6 km^2. According to the seventh. According to the census data, the resident population of Kuiwen District is 475,103. There are 23 elementary schools, 10 nine-year schools, and 10 nine-year schools in the district. Among them, there are 920 classes of elementary school with 37,994 students, high-quality schools, excellent teachers, good basic education resources, and fast development, and most of the elementary schools are concentrated in the old urban areas in the central and northern parts of the city, which is why the old urban areas have been selected as the service performance of the elementary school. The scope of the study of performance evaluation. As an atypical old urban area, the service performance evaluation of its elementary school is representative (Figs. 1 and 2).

Fig.1 Location of Kuiwen district (self-drawn)

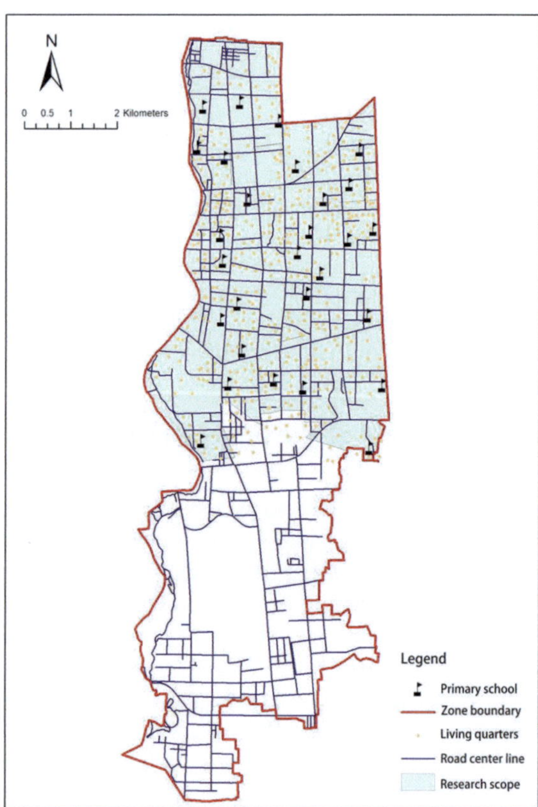

Fig. 2 Spatial distribution of elementary school and settlements in the study area (self-drawn)

2.2 Data Sources

The data used in this paper mainly come from the Weifang Statistical Yearbook and primary and secondary school statistics of the Education Bureau of Kuiwen District, including street population data, enrollment of elementary school, number

Table 1 Sources of data for the study (self-drawn)

Date type		Date content	Year	Date sources
Essential data	Community population data	Community location, community population	2022	Weifang Statistical Yearbook
	School-related data	School location, school boundary, school district scope, number of students, building area	2022	Education Bureau primary and secondary school statistics
	Traffic network data	Road vector data	2022	OSM's data platform

of students, etc., and school data such as school locations, school boundaries, school districts, and floor area, etc., and at the same time, combined with the previous version of the special plan for educational facilities and satellite maps to check the current situation of the school. Traffic network data is downloaded from the OSM data platform. ArcGIS 10.5 software was used to process the data (Table 1).

3 Research Methods

3.1 Evaluation Indicator System

Evaluation principles. School service performance evaluation in urban renewal contains spatial performance evaluation and functional performance evaluation, which comprehensively evaluate the public service capacity of schools in the city oriented to the needs of the population and adjust the schools according to the evaluation results. This paper takes fairness, representation, and effective guarantee as the basic principles, and gives full consideration to the interrelationship between the indicators and the evaluation units. Selects the indicators that can reflect the main characteristics of the evaluation object and are representative and operable, and constructs the performance evaluation system of school services in the old urban areas in urban renewal.

Evaluation indicators. Combining the characteristics of urban spatial systems with the spatial configuration of schools. After collating the relevant literature, it is found that in the practice of educational resource allocation, spatial layout, and functional configuration are two aspects to be concerned with in the current evaluation of service performance. I.e. spatial performance and functional performance. In terms of spatial performance. Policies and adjustments of spatial layout should be adopted to ensure that residents have fair opportunities to enjoy educational resources as much as possible, and that the shortest distance from a residential area to the nearest school can effectively reflect the accessibility of the school and whether the spatial configuration of the school is balanced; and in terms of functional performance, the continuous growth of the birth population and the foreign resident population has made the problem of the shortage of educational facilities more and more prominent, and the demand for schools by urban residents has been growing year by year. The demand for schools by urban residents is growing year by year. Different schools have different scales and can accommodate different numbers of students, and only when the number of students in the school and the surrounding settlements reaches a balance between supply and demand can the efficiency of the school's functional use be maximized. Therefore, this paper uses the supply–demand discrepancy rate as an indicator to characterize the evaluation of school functional performance, and at the same time provides a concrete reference for schools to adjust and respond promptly in the face of complex demographic changes and changes in the demand for schools.

3.2 Evaluation Method

Space performance evaluation. Since China's compulsory education follows the principle of proximity, each elementary school has a relatively fixed enrollment zoning range, and students within the enrollment zoning range are encouraged to choose their corresponding elementary school to attend, the nearest distance method is the most suitable expression for the accessibility of students' schooling, so this paper uses the closest distance between a residential point and an educational facility as a measure of primary and secondary school accessibility, and the distance between a resident and a facility directly affects the cost that residents have to pay to enjoy these services [5]. Taking the center point of the residential area within the elementary school enrollment lot as the residential point, calculating the shortest straight-line distance between it and the corresponding primary and secondary schools, taking the average value of the distance from each residential point to the elementary school as the accessibility indicator, and conducting cluster analysis of the calculation results to get the division standard of elementary school accessibility. In summary, the elementary school accessibility measurement method. The formula for the method is:

$$A_i = \overline{A_j}$$

In the formula: A_i represents the accessibility index of primary school i, and A_j is the shortest straight-line distance from the center point j of the residential grid within the primary school's enrollment area to school i.

Functional performance evaluation. The enrollment capacity of elementary schools and the actual number of school-age students can further clearly reflect the supply and demand of schools. In this study, the supply–demand discrepancy rate of elementary schools is chosen to characterize the functional service efficiency of schools, and the calculation of the supply–demand discrepancy rate of elementary schools requires the estimation of the supply capacity and demand of elementary schools. On the supply side, the supply capacity of elementary schools can be estimated by combining the statistics on education in the Weifang Kuiwen District of the Bureau of Education and the per-student floor area regulations in the Standard for Basic Conditions of Primary and Secondary Schools in Shandong Province. i.e., the total floor area of a school divided by the per-student floor area standard to get the number of students that the school can accommodate. On the demand side, estimates are based on population data from the Annual Statistical Review, current community demographics, and statistics on children who have moved with the family Measured by the difference between supply and demand, i.e. the difference between supply and demand divided by demand, a too small difference between supply and demand reflects the insufficient supply capacity of the school to meet the enrollment demand of school age students in the neighborhood, while a too large difference reflects the insufficient enrollment, which also leads to a waste of resources. A variance value above 10% is defined as a surplus of supply, −10 to 10% as a balance of supply and demand, and below −10% as a shortage of supply.

4 Results of the Performance Evaluation of Educational Facilities Services in Old Urban Areas Under Urban Renewal

4.1 Results of Space Performance Evaluation

ArcGIS software was used to calculate the Euclidean distance from each settlement to the nearest educational facility and to statistically average the distance to all the settlements that reach the school, the results of which represent the relative spatial accessibility of the school. The statistical results cluster analysis method is used to classify the samples, and three categories are obtained, which are used as the basis for assigning values (Tables 2, 3 and Fig. 3).

The differentiation of accessibility among schools is shown in Fig. 2. As a whole, there is a big difference in the accessibility of different schools in the old city of Kuiwen District, and the accessibility of elementary schools is generally characterized by average accessibility. Combined with the analysis of the service radius of the elementary school, 57% of the elementary school' accessibility meets the service radius of 500 m, and 51% of the residents' straight-line distance to the

Table 2 Summary of spatial accessibility clustering for elementary school (self-drawn)

Clustering category	Frequency	Percentage (%)
Spatial accessibility 1	6	21.429
Spatial accessibility 2	12	42.857
Spatial accessibility 3	10	35.714
Add up the total	28	100

Table 3 Results of comparison of clustered differences in elementary school accessibility (self-drawn)

	Cluster category (mean ± standard deviation)			F	P
	Category 1 (n = 6)	Category 2 (n = 12)	Category 3 (n = 10)		
Average distance	304.47 ± 56.017	455.441 ± 47.919	642.412 ± 52.215	86.499	0.000***

Note ***, **, * represent 1, 5, and 10% significance levels, respectively

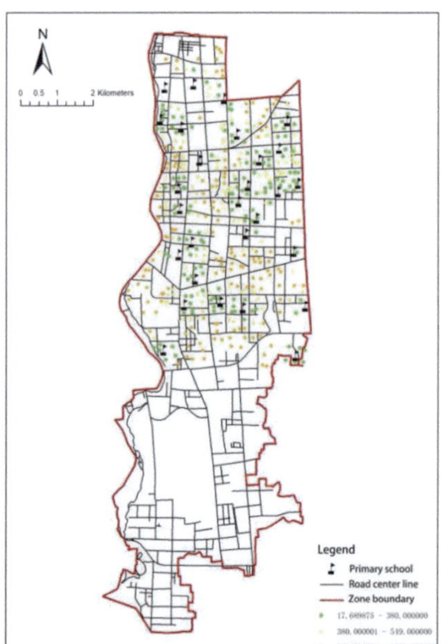

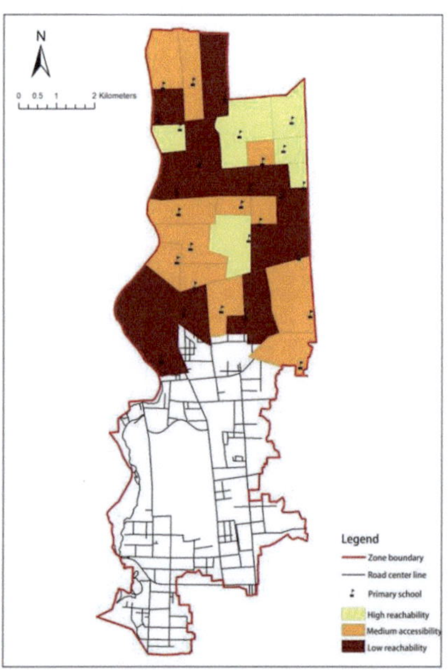

Fig. 3 Spatial differentiation of neighborhood distance (left) and accessibility (right) of elementary schools (self-drawn)

elementary school is less than 500 m. From the spatial distribution point of view, most of the elementary schools with high accessibility are located in the northern part of the Kuiwen District, and the average distance from the settlements within the boundaries of the school districts of the Mingde School, Shengrong Primary School, Chongde School, Yucai school, Wolong Primary school, and Shenglidong Primary School to their closest neighboring schools is relatively small, partly due to the higher level of development of these districts and the better infrastructure of the transportation network, as well as the uniformity in the distribution of the elementary school and the residential districts. The accessibility of Sunjia Primary School, Beigong Street Primary School, Weizhou Road Primary School, Zaochun-yuan Primary School, Rixiang Friendship School and Weizhou Lake Primary School is low, and the average distance from the settlements within the school district to the

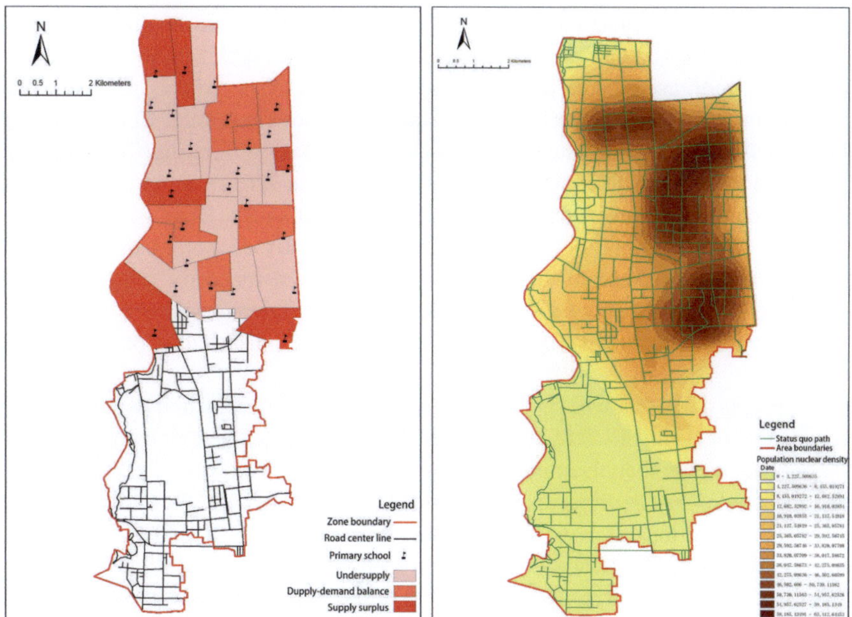

Fig. 4 Spatial differentiation of differential rates of supply and demand for elementary school (left) and population density in the old urban area of Kuiwen district (right) (self-drawn)

nearest school is larger, with some of the distances exceeding the service area of the elementary school by more than 500 m, which is mainly due to the fact that, on the one hand, there is the problem of the scope of the delineated school district being too large, and the settlements included therein being more dispersed or the schools being located in the fringe of the school district, and so on; On the other hand, the complexity of the land use in the area. Such as industrial land, logistics and warehousing, religious land, etc., to a certain extent, impedes the connection between the elementary school and the residents, which affects the radiation ability of the elementary school and leads to the weak accessibility of the elementary school (Fig. 4).

4.2 Difference Between Supply and Demand for Educational Facilities

The number of school-age children in each school district can be estimated based on the analysis of the proximity of schools and the population of each residential neighborhood, and the average floor area per pupil is included in the calculation process, the average student building area is an important index to judge the matching of supply and demand of school material resources, that have a direct impact on the learning environment, teaching conditions and physical and mental health of students. Sufficient floor space can provide teachers and students with bright and spacious classrooms, reorganized activity space, and other types of special classrooms, which to a certain extent can reflect the quality of the school's teaching. Combining the per-pupil floor space stipulated in the Standard for Basic Operating Conditions of Ordinary Primary and Secondary Schools in Shandong Province and calculating it with the actual size of each school in the education statistics, we can get the number of school places that the actual school can provide. By clustering the results of the calculations, the relationship between supply and demand for schools can be further explored, with either a high or a low rate of difference between supply and demand being a concrete manifestation of the unbalanced distribution of educational resources.

The results of the analysis show that the overall supply of elementary schools in the Kuiwen District is insufficient, with 54% of the actual demand for elementary school places being higher than the number of places available. The pressure on elementary school places in the old urban area of Kuiwen District is high, and the current number of schools is overloaded, with the contradiction between supply and demand being more pronounced. The reason for this phenomenon is that the educational resources of Kuiwen District are mainly concentrated in the old city and the living area, the residential area near the school is well-built, the development intensity is larger, and a large number of people are gathered, at the same time, Kuiwen District is the main inflow of foreigners in the city, and the number of children of foreign workers is always higher than that of other municipal districts, and the demand for school enrollment is large. Comparison with the population density analysis map of the old urban area of Kuiwen District reveals that the supply of elementary schools is somewhat consistent with the population density, and most of the under-supplied school districts have a higher population density. Some elementary schools, such as Beiyuan Experimental School, and Lechuan Street Primary School. Chongde School, Kuiwen District Experimental Primary School, Weizhou Lake Primary School, and Baotong Street Primary School, are located in the old urban area, but they are built in better conditions and are larger in scale, and they can provide a large number of school places, with a surplus of supply, which can be used to appropriately expand the scope of radiation or appropriately increase the number of school places for accompanying children to ease the pressure on the number of school places in other schools.

5 Optimization Strategy Discussion

5.1 Overall Problems and Optimization Strategies

Based on evaluating spatial performance and functional performance separately, we comprehensively consider the accessibility of elementary school and the rate of difference between supply and demand and adopt the "indirect quantification" method to divide the above indicators into different levels and rate each level, and then synthesize and visualize the ratings to obtain the final spatial performance evaluation results of educational facilities. The results of the spatial evaluation of the educational facilities are visualized to obtain the final results of the spatial evaluation of the educational facilities, and the problems of the service performance are explored in depth, which can be used as the basis for optimizing the spatial layout and adjusting the functional configuration. The results show that the spatial performance of primary schools in the Kuiwen District is lower than the average level, with a high-level rate of 25% and a low-level rate of 43%, and the quality of elementary schools in the old urban areas varies. Among them, elementary schools in need of optimization and adjustment can be divided into the following five categories (Fig. 5 and Table 4).

The level of accessibility and supply rate of elementary schools in Type 1 is low. There are 8 such schools, accounting for 29% of the total, which is a large proportion, and they generally have the problems of larger school district boundaries, schools

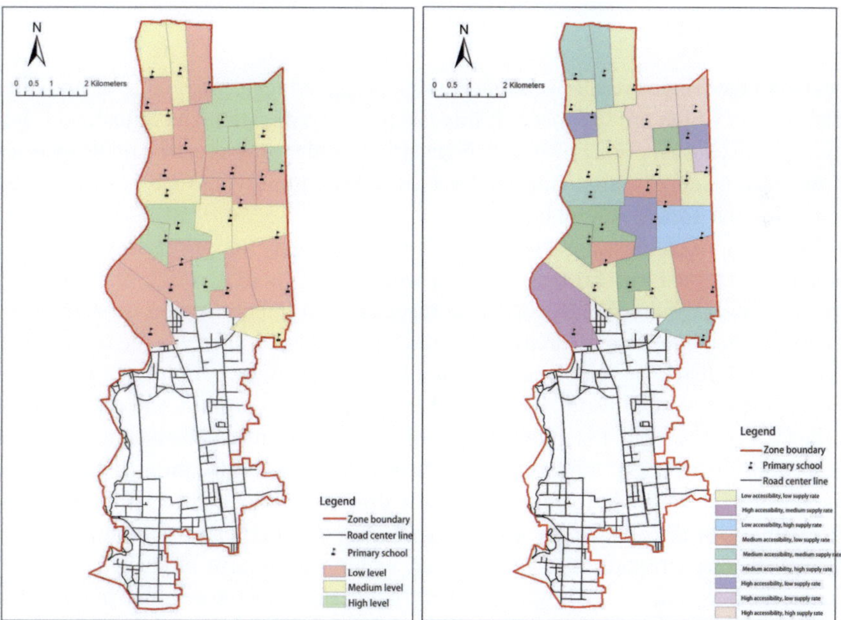

Fig. 5 Spatial distribution of clusters of service performance in elementary school (self-drawn)

Table 4 Problems and optimization strategies for various types of primary schools (self-drawn)

Clustering	Service performance	Features and problems	Optimizing strategy
1	Low accessibility and low supply rate	Larger school district boundaries, schools located in fringe areas of the district, dispersed residential land use, smaller schools	Increase in the number of schools, repositioning of schools, improvement of the road network, increase in the size of school buildings
2	Low accessibility and medium supply rate	Large school district boundaries. Schools located in fringe areas of the district, moderate school size	Adjustment of school districts, addition of schools, and restructuring of school locations
3	Medium accessibility and low supply rate	Moderate school district boundaries, high density road network within the school district, small school size	Expansion of campuses and increase in school space
4	Medium accessibility and Medium supply rate	School district is moderate or small, schools are located near major residential neighborhoods, and schools are moderate in size	–
5	High accessibility and high supply rate	School district is moderate in size, schools are located in the center of the district or near major residential areas, and schools are large in size	Appropriate expansion of enrollment or increase in the number of places for accompanying children

located in the fringe areas of school districts, dispersed residential land, and smaller school sizes which are the focus of this paper on optimization and adjustment. For this type of school, optimization and reconstruction suggestions can be made in terms of increasing schools, adjusting school distribution, improving road networks, and increasing the area of school buildings.

Type 2 is an elementary school with low accessibility and medium supply rate, namely Weizhou Lake Primary School, which is of moderate size and currently provides several school places that can meet the enrolment demand of students of school age in the residential areas within the school district, however, the spatial span of its school district is large, and the school is located in the peripheral areas of the school district, which results in the need for students to enroll in schools at a long distance. For this type of elementary school, consideration can be given to adjusting the school distribution, adjusting the school to the central location of the school district or near the main residential areas within the school district, improving the accessibility of the elementary school; or solving the problem of low accessibility through the construction of a new school, under the premise of sufficient school places, the new school can be optimized in conjunction with the neighboring schools, and the surrounding areas of insufficient school places of the residential area adjusted to the new construction of the nearer and sufficient land schools to improve the service performance of the schools in the area as a whole.

Type 3 is an elementary school with medium accessibility and low provision, with average overall service performance. Its main characteristics include moderate school district boundaries, high density of road network within the district, and high accessibility, but the supply capacity of school places is significantly lower than the demand for school places, and its overall supply capacity is poor due to the high population density and the stereotyped buildings in the vicinity, which make it impossible for the school to be expanded and relocated to a new urban area. For this type of school, there is a need to increase the number of elementary schools or expand the campus as appropriate. Schools that are not suitable for expansion need to consider increasing the size of school premises to accommodate more students.

The accessibility and supply rate of Type 4 elementary schools are at a medium level, with good overall service performance. The school district of this type of elementary school is moderate or small in size, and the time taken by students to reach the schools is relatively short, and the schools are located in the vicinity of major residential districts, which makes it convenient for students from the surrounding areas to attend schools in their proximity. The size of the schools is moderate, and the enrolment of students within the school district can achieve a balance between supply and demand so that school places can be reserved appropriately in the future to cope with demographic trends.

The accessibility and provision of Type 5 elementary schools are at a high level and their comprehensive service performance is good, specifically Matilda School and Yucai School, both of which are nine-year schools. The school districts of this type of elementary school are moderate in scope, with schools located in the central areas of the school districts or near major residential areas, so that most students can go to school within a short distance. Colleagues' schools are large in scale, and the phenomenon of surplus supply has arisen, so that they can appropriately expand the scope of enrolment or increase the number of school places for accompanying children to alleviate the pressure on the number of school places in the surrounding school districts.

5.2 Typical School Optimization Program

Based on the overall problem analysis and the corresponding optimization strategies, a typical school in each of the three categories of low accessibility-low supply rate, low accessibility-medium supply rate, and medium accessibility-low supply rate is selected for the optimization analysis [6].

The typical school with low accessibility and low supply rate is Sunjia Primary School, which has a large area and a long and narrow range. Sunjia Primary School is located in the southeast comer of the school district, so the accessibility of students in the north side of the school district is poor, at the same time, Sunjia Primary School is small in scale, and the number of students can be calculated that the average floor area of Sunjia Primary School is 5.7m^2/person, which is less than the requirement of the provincial standard, and the number of students attending school exceeds the school's capacity, which results in the overall low spatial service performance of the school district. The actual number of students exceeds the carrying capacity of the school, resulting in the overall performance of the school district. Sunjia Primary School District is located on the edge of the northern living area of Kuiwen District, the development intensity of the new residential area is larger, the residential buildings are mainly medium and high rise, and the number of people living in each residential building is larger, to cope with the continuous growth of the demand for school-age students to enroll in schools, it can be considered that there is a new elementary school Yuanfei Road Primary School in the northern part of the district, which can not only alleviate the problem of low accessibility to the north side of the area, but also reduce the pressure on the degree of Sunjia Primary School. This can alleviate the problem of low accessibility of students in the north side of the district, and at the same time reduce the pressure on Sunjia Primary School's degree, it can also reserve a certain amount of degree resources for the new residents after the development and construction of the district in the future, to improve the flexibility of planning (Figs. 6, 7 and Table 5).

The typical school with a low accessibility-medium supply rate is Weizhou Lake Primary School, whose school district is irregularly shaped from northwest to southeast, with a large spatial span. Weizhou Lake Primary School is located in the southern part of the school district, which leads to a longer distance for students in the northern part of the district to go to school, therefore, a new school, Hongwei School, can be built in the northern part of the district, and taking into consideration of the current situation of balancing the number of school places in the district with the number of places in the district, the new Hongwei School can be a nine-year school. The new Hongwei School can be a nine-year school to enhance the flexibility of education services in the district. At the same time, the neighboring school district, Yu Ying School District, is a low accessibility-low supply rate school. Consideration can be given to assigning the west side of Yu Ying School to the new school, to make full use of the service resources of the neighboring schools, and to improve the service performance of the schools in the district in general (Figs. 8, 9 and Table 6).

The typical school with medium accessibility and low supply rate is Xingfu Street Primary School, which has a small school district, and the distance to school is generally shorter, with the majority of students going to school within 500 m, which makes the overall accessibility better. Residents in the school district are concentrated, and the development intensity of the residential area is large, the population is dense, at the same time, the scale of Xingfu Street Primary School is not large, and the number of school places it can provide is limited, showing the phenomenon of oversupply and under-supply. According to the long-term development requirements, it is recommended that Xingfu Street Elementary School be combined with the original site to expand to the south, increase the area of the school building, enhance the number of school places, and at the same time, it is recommended that Xingfu Street Elementary School be adjusted to become a nine-year school, which will enable the school-age population to be stabilized in the school district for a long period while absorbing

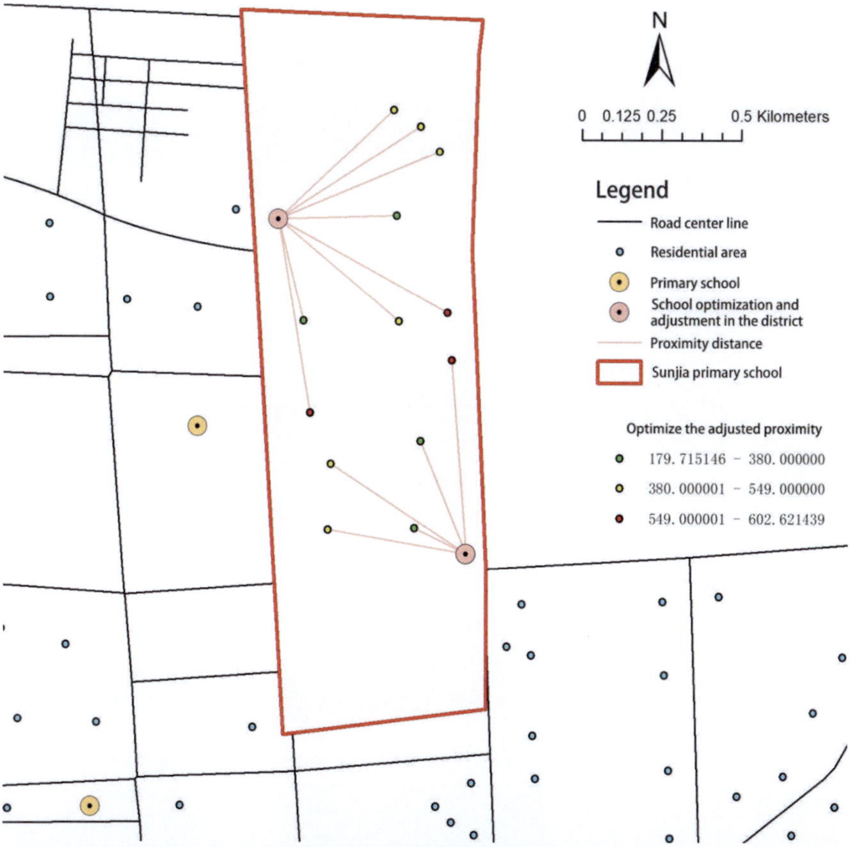

Fig. 6 Sunjia primary school district after optimization (self-drawn)

Fig. 7 Diagram of the newly built Yuanfei Road Primary School (self-drawn)

Table 5 Optimized school scale of Sunjia Primary School district (self-drawn)

School	Number of degrees	Class size	Site area
Sunjia Primary School	1190	34	12,000
Yuanfei Road Primary School	1260	36	31,400

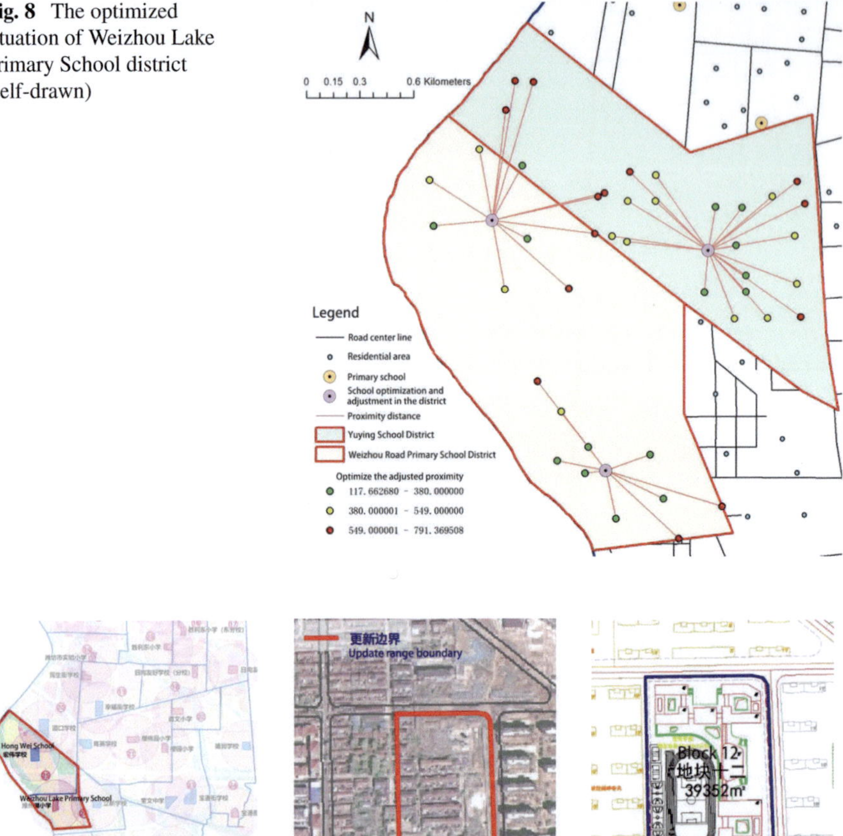

Fig. 8 The optimized situation of Weizhou Lake Primary School district (self-drawn)

Fig. 9 Schematic diagram of the new Hong Wei School (self-drawn)

Table 6 Optimized Weizhou Lake Primary School district school size (self-drawn)

School	Number of degrees	Class size	Site area
Weizhou Lake Primary School	1088	26	21,000
Hong Wei School	1080 (primary)	18 (primary)	40,545

more school-age population, and optimize the overall scope of the district's services (Fig. 10).

Fig. 10 Schematic diagram of the expansion of Xingfu Street Primary School (self-drawn)

6 Conclusions and Implications

The spatial optimization of the old city under the background of renewal can be considered from two aspects: incremental scene and stock scene. On the one hand, in the incremental scenario, it is necessary to focus on strengthening the transformation and upgrading of existing educational facilities, establish a sharing and integration mechanism of high-quality educational resources, and use the complementarity of regional demand changes to balance the scale of each school; on the other hand, in the stock scenario, it can be considered from the aspects of configuring 9-year consistent schools, putting inefficient functions on campus into the community, and transforming the complex functions of school space.

This paper adopts the method of service performance evaluation to give references and suggestions for the optimization and adjustment of elementary schools in old urban areas, which is a useful exploration of guiding urban renewal actions through indicator analysis and performance evaluation. However, there are some limitations, although the study uses data from the Weifang Statistical Yearbook and education statistics of Kuiwen District, population change is a complex process, and this paper does not cover much of the changes in the age structure and social structure of the population in different school districts, and there is an error in the spatial distribution of students between the data used and the actual distribution of students. Subsequent studies need to incorporate more actual research data to verify and correct the number of students in each school district so that the results of service performance evaluation can be more accurate.

References

1. Wang JQ (2023) The three-dimensional direction of cultivating mass supervision awareness in the new era. Res Integr Cult 14(02):76–82
2. Decision of the CPC Dehong Prefecture Committee on further studying and implementing the spirit of the 20th CPC National Congress to comprehensively open up a new situation for the construction of socialist modernization in Dehong in the new era. Dehong Unity News, 4 January 2023

3. Wang YC, Shen J, Xiang WN (2017) Landscape spatial performance evaluation system based on ecosystem services. Landsc Archit (01):35–44
4. Xie SX, Li ZM, Zhou XM (2021) Synthesis and prospect of theoretical research on domestic urban basic education facilities planning based on journal statistics. Central China Archit 39(9):66–71
5. Huang LX, Liu YT (2023) Evaluation study on the spatial layout of basic education facilities under the orientation of equalization: taking public elementary schools in Tianhe district of Guangzhou City as an example [A] people's city. In: Planning empowerment: proceedings of the annual conference on urban planning in China in 2023 (20 Master Plans), China Society of Urban Planning
6. Niu Q, Zhang YX, Wu WX et al (2022) Evaluation methods and optimization strategies of spatial service performance of primary schools based on multi-source data-taking East-West Lake district of Wuhan as an example. Shanghai Urban Plan 3:46–54

Renewal for Space Quality Improvement

A Quantitative Analysis of Curriculum Optimization Strategies for 'Green Building' Education: Insights from Survey Data

Gaochuan Zhang, Qiaoyuan Lin, Xiaolan Li, Tingting Hou, and Bao-Jie He

Abstract As global climate change and environmental pollution increase, the promotion of green buildings has become a consensus. To cultivate quality talents with awareness and skills in green buildings, this study is based on the teaching process of the green building course at Zhejiang University of Science and Technology. A survey questionnaire was designed around four aspects: basic information on students' course learning, students' satisfaction with the course, students' application preferences, and students' learning barriers. A total of 79 valid questionnaires were collected. Following the analysis of the questionnaire, it was revealed that the current teaching model exhibits several issues, including lack of theoretical guidance, analysis focuses only on visual effects, limitations of campus conditions and so on. Therefore, from the four aspects of 'theory module, cognitive module, practice module and practical module', it is proposed to (1) improve the theoretical foundation of students through the integration of simulation technology, (2) perceive the green building environment through the increase of outdoor practice, (3) quantify the energy-saving effect of green building through simulation software, (4) enrich the teaching scenario through the construction of school-enterprise cooperation platform. The study is conducive to improving the quality development of green building talent training courses.

G. Zhang · Q. Lin · T. Hou
School of Civil Engineering and Architecture, Zhejiang University of Science and Technology, Hangzhou, China

G. Zhang
Zhejiang-Singapore Joint Laboratory for Urban Renewal and Future City, Hangzhou, China

X. Li
Zhejiang Southeast Architectural Design Group Co. Ltd., Hangzhou, China

B.-J. He (✉)
Centre for Climate-Resilient and Low-Carbon Cities, School of Architecture and Urban Planning, Chongqing University, Chongqing, China
e-mail: baojie.unsw@gmail.com

Smart City Research Institute in Liyang, Chongqing University, Liyang, Jiangsu, China

B. He et al. (eds.), *Climate-Resilient, Low-Carbon, and Sustainable Cities*, Lecture Notes in Civil Engineering 754, https://doi.org/10.1007/978-981-95-2169-2_44

Keywords Green building · Course reform · Questionnaire survey · Simulation software

1 Introduction

With the increasing severity of global climate change, the governments of various countries and sectors of society have placed growing emphasis on environmental protection. The "2023 Research Report on Carbon Emissions in China's Construction and Urban Infrastructure" indicates that the total carbon emissions from residential buildings in China for the year 2021 were 4.07 billion tons of CO_2, accounting for 38.2% of the national energy-related carbon emissions, highlighting the urgency and importance of energy saving and emissions reduction in the construction sector [1]. The Chinese government has introduced a series of guiding documents, such as the "Green Building Action Plan" and the "14th Five-Year Plan for Building Energy Conservation and Green Building Development," aiming to accelerate the development of green buildings through policy guidance and support [2]. In this context, the development of green buildings has not only become an inevitable trend in the construction industry but has also posed new challenges and requirements for talent cultivation and education systems in the architecture field. Green building courses are an important means of implementing the carbon neutrality goals and methods of China's new urbanization in educational practice, as well as a key component in promoting green building knowledge dissemination and green building design capability cultivation [3]. Course content mainly includes the principles of green building design, energy-efficient green building design, green building construction and maintenance, green building evaluation, and green renovation of existing buildings, focusing on nurturing students' design thinking and capabilities related to green buildings. However, in practical teaching processes, theoretical courses and computer courses remain disconnected, resulting in weak interpretation skills among students, insufficient practical abilities, and a singular evaluation model for teaching.

To ensure that students understand and can proficiently master the knowledge of this course, reforms in teaching have started. In China, Liu [4] and others have conducted teaching reforms from the perspectives of topic settings, teaching methods, and auxiliary means, combining students' participation in design competitions to gain feedback on reform outcomes to enhance green building innovative design capabilities among architecture major students. Gong [5] and others practiced reforms on the teaching content, methods, and assessment modes of green building design courses. They proposed a "Theory-Measurement-Design-Simulation" teaching model to cultivate and enhance the green building design thinking and innovative capacity of students in architecture-related majors. Yanhui [6] and others analyzed the influence of factors related to teachers and students on teaching effectiveness, discussing the challenges posed by teaching materials to both teacher instruction and student learning, and aiming to promote engineering talent cultivation goals from the perspective of applied undergraduate education through reciprocal feedback from research to teaching, balancing theory and technology, cultivating humanistic feelings, and promoting equal interaction between teachers and students. Teaching evaluation results at the end of the semester indicated that the reforms were well-received by students. New teaching modes based on "flipped classrooms" and "blended learning" have also emerged, leveraging contemporary backgrounds [7, 8]. For instance, Maltepe University's School of Architecture and Design has organized a series of off-campus activities, including thematic discussions and workshops, to integrate formal and informal learning experiences [9]. Some studies have linked campus environments with green education, showcasing green features in buildings through green renovations and incorporating interactive touch screens [10, 11]. However, these teaching methods still face the issue of emphasizing theory over application, with skills training remaining centered around the transmission of theoretical knowledge. Students cannot genuinely participate in or comprehend the low-carbon building design process, limiting the effectiveness of design outcomes.

In response, this study conducted a questionnaire survey targeting students enrolled in the "Green Building Course." The survey covered four aspects: basic information on students' course learning, satisfaction with the course, application preferences, and learning barriers. The theoretical main contributions are to (1) find the gaps in student learnings of the perceptions, attitudes and needs; (2) improve course satisfaction level; (3) propose an optimized course plan. The course is currently undergoing a educational reform study, which aims to enhance teaching effectiveness and establish a sustainable framework for the development and cultivation of talent within the green building curriculum, tailored for future needs.

2 Settings of Existing Green Building Course

2.1 Green Building Definition and System

The current green building program emphasizes the instruction of green building design principles, strategies, and methodologies. The course content comprehensively covers foundational knowledge of green buildings, including introductory concepts and evaluation methods, as well as engineering case studies focused on building energy efficiency, ventilation, lighting, thermal engineering, acoustics, and more. Additionally, the course integrates theoretical learning with the practical application of green building software. To complement theoretical learning, practical sessions are also provided, such as site visits to energy-efficient buildings and field surveys. The total duration of the course is 64 semester hours, with 18 semester hours dedicated to theoretical instruction, 16 semester hours for practical teaching, 15 semester hours for classroom seminars, and 15 semester hours for exercises. The program is primarily structured around theoretical education, enabling students to acquire green building knowledge and skills in a systematic and integrated manner.

2.2 Modules of Green Building Course

The existing green building programme is divided into four modules: theory awareness, case awareness, field observation and software practice. The theoretical awareness module focuses on introducing the basic theoretical knowledge of green building, while the case awareness module enhances students' understanding through the analysis of real-world green building cases, considering the physical characteristics of buildings. The field observation module offers students the opportunity to visit green buildings, and the software practice module emphasizes both the theoretical and practical aspects of green building software, aiming to develop students' comprehensive professional expertise in the field.

2.3 Delivery of Green Building Course

The existing programme employs a range of instructional methods, including lectures, post-course assignments, classroom seminars, and practical experiments. Lectures teach theoretical knowledge; after-class assignments such as green building cognitive observation reports are assigned; classroom seminars focus on green building development, low-carbon energy saving and other topics; experimental and practical classes include the use of green software, site visits, field surveys and so on, to exercise students' practical skills.

3 Research Methods

3.1 Questionnaire Design

The survey was conducted using Wenjuanxing and was administered after the conclusion of the "Green Building Course," focusing on four areas of inquiry.

3.1.1 Basic Information on Students' Course Learning (Q1–Q2)

The basic information section included two single-choice questions: "What is your gender?" and "Have you been absent from class for more than two weeks (including two weeks)?".

3.1.2 Students' Satisfaction with the Course (Q3–Q7)

This section included five questions: "Can you clearly distinguish between the environmental factors of wind, light, heat, and sound in building technology?"; "What is your satisfaction level regarding the learning assistance provided by the following four physical environments (wind, light, heat, sound) in your design?"; "How helpful do you find the following three parts of this course (Theory of green building rating evaluation, Green building design and strategy case studies, Methods for field research on green buildings) for your design?"; "How much do you think this course will help you in your future learning and design?"; and "How helpful do you find the field research assignments in this course for your future learning and design?".

3.1.3 Students' Application Preferences (Q8–Q9)

This section included two questions: "After taking this course, would you prioritize the indoor environment or the outdoor environment in architectural technology?"; and "Which of the following aspects do you consider most effective in optimizing/adjusting building design schemes?".

3.1.4 Students' Learning Barriers (Q10–Q13)

This part included four single-choice questions: "Do you feel there are obstacles when learning this course?"; "What are the main reasons for the obstacles you face in this course?"; "Do you think there are barriers to using environmental measurement instruments during field research?"; and "What are the main reasons for the obstacles in using these measurement instruments?".

3.2 *Questionnaire Distribution and Collection*

The subjects of this survey were undergraduate architecture students at Zhejiang University of Science and Technology who enrolled in the "Green Building Course" during the 2021–2022 and 2022–2023 academic years. A total of 87 questionnaires were distributed, and all were returned; 79 were valid, leading to an effective response rate of 90.8%, meeting the statistical sampling standards and requirements for analysis.

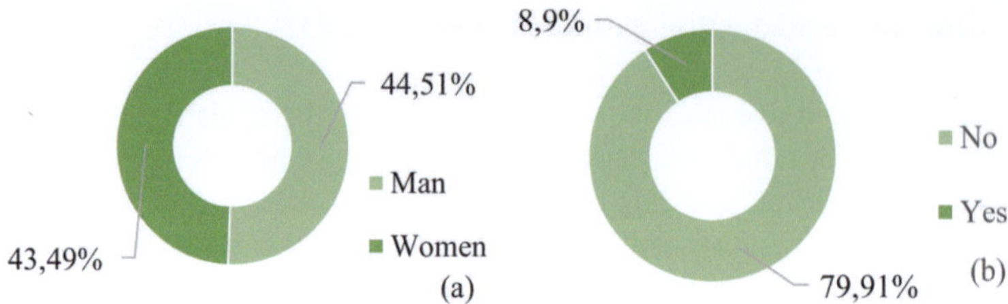

Fig. 1 Basic information on students' course learning (**a** Gender; **b** Absence) (self-drawn)

4 Survey Results

4.1 Basic Information on Students' Course Learning

As shown in Fig. 1a total of 87 students enrolled in the course. Gender distribution was nearly balanced, with 44 male students 50.6 and 43 female students 49.4. In Fig. 1b, it was noted that 8 students were absent for more than two weeks during the course, which may impact the accuracy of the survey results. Thus, these students were excluded from the data analysis to ensure the collected data can genuinely and objectively reflect the course implementation and students' feedback.

4.2 Students' Satisfaction with the Course

Students' satisfaction levels were assessed across the dimensions of their understanding of physical environments (Q3–Q4), course content (Q5) and future assistance (Q6–Q7). As shown in Table 1, students rated their recognition of the four physical environments with an average score of 4.35. The average scores for the wind, light, heat, and sound environments were 4.15, 4.39, 4.13, and 4.04, respectively, indicating students could clearly distinguish between these environments yet had lower satisfaction regarding learning about wind, heat, and sound environments. Regarding satisfaction with course content, feedback on various knowledge modules showed differentiated characteristics. The students' average satisfaction with the theory of green building level evaluation was 3.81, indicating room for improvement in this part of the theoretical course content, suggesting the need for more engaging methods to enhance students' acceptance of theoretical knowledge. In contrast, the modules of green building design and strategy case studies (average satisfaction 4.3) and field research methods (average satisfaction 4.33) received significantly higher ratings, reflecting students' satisfaction with practical and case-rich teaching content. On the future assistance provided by the course, students rated this aspect positively, believing the course content significantly helps future learning, with an average score of 4.03, while practical field research assignments were rated even higher at 4.14, indicating approval for assignments that effectively promote knowledge absorption

Table 1 Descriptive statistics of students' satisfaction with the course (self-drawn)

	Min	Max	Mean	Std. dev	Variance
Q3 Can you clearly distinguish between physical environments?	1.00	5.00	4.35	0.90636	0.821
Q4_1 Satisfaction with the wind environment	2.00	5.00	4.15	0.98825	0.977
Q4_2 Satisfaction with the light environment	2.00	5.00	**4.39**	0.86850	0.754
Q4_3 Satisfaction with the thermal environment	2.00	5.00	4.13	1.05450	1.112
Q4_4 Satisfaction with the sound environment	2.00	5.00	4.04	1.10295	1.216
Q5_1 Theory of green building rating evaluation	2.00	5.00	**3.81**	1.02627	1.053
Q5_2 Design and strategy case studies	2.00	5.00	4.30	0.86757	0.753
Q5_3 Methods for field research	2.00	5.00	4.33	0.94355	0.890
Q6 Future assistance from this course	2.00	5.00	4.03	0.96043	0.922
Q7 Field research assistance for the future	1.00	5.00	4.14	1.00937	1.019

Note 79 samples

and preparation for future careers. Therefore, course design needs to enhance the engaging aspects of theoretical knowledge to facilitate students' comprehension and acceptance. Simultaneously, promoting practical field research methods will provide students with a more intuitive understanding of green buildings.

4.3 *Students' Application Preferences*

As shown in Fig. 2, after taking the course, 11% of the students prioritize the indoor environment in architectural technology, 32% pay attention to the outdoor environment, while most students consider both environments equally.

As reflected in Table 2, regarding the effectiveness of course content in optimizing/adjusting building design schemes, students were surveyed on twelve key focuses, including overall building layout, building form, floor plan, facade, cross-section, spacing, orientation, flowline, entrance, materials, structure, and design explanation. A majority of students indicated that the most effective aspects were building form 59.5, overall building layout 50.6, building facade 41.8, and materials 38.0. For orientation, structure, floor plan, and other factors, only some students found them

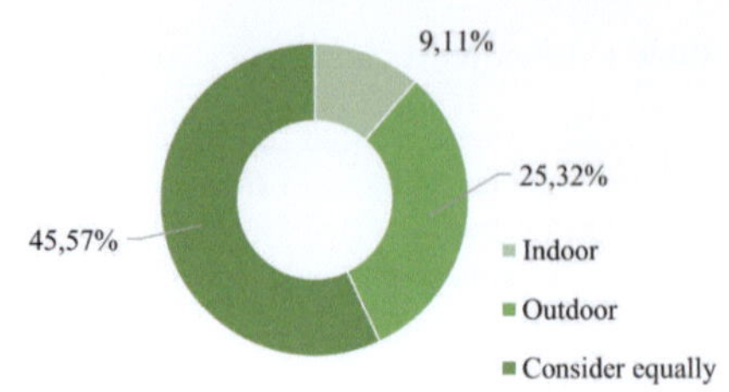

Fig. 2 Consideration preference in building technology application environment (self-drawn)

Table 2 Multiple response analysis on effectiveness of course content for optimizing/adjusting building design schemes (self-drawn)

	Response		Percentage of cases (%)
	Case count	Percentage (%)	
Building form	47	19.80	59.50
Overall layout	40	16.90	50.60
Building facade	33	13.90	41.80
Materials	30	12.70	38.00
Orientation	21	8.90	26.60
Structure	20	8.40	25.30
Floor plan	16	6.80	20.30
Spacing	13	5.50	16.50
Cross-section	8	3.40	10.10
Flow line	5	2.10	6.30
Design explanation	2	0.80	2.50
Entrance	2	0.80	2.50
Total	237	100	300

effective, with respective responses of 26.6, 25.3, 20.3, and 16.5%. A smaller number of students believed that cross-section 10.1, flow line 6.3, design explanation 2.5 could effectively optimize or adjust design schemes.

4.4 *Students' Learning Barriers*

As shown in Fig. 3a, 80% of students felt there were no barriers in learning the course, while 20% indicated encountering obstacles. Further investigation revealed that the main barriers for most students included a lack of introduction to simulation software (44%), and an insufficient introduction to foundational physical knowledge (38%), A smaller proportion pointed to a lack of design case introductions (12%) and a lack of information on the employment market in this field (6%).

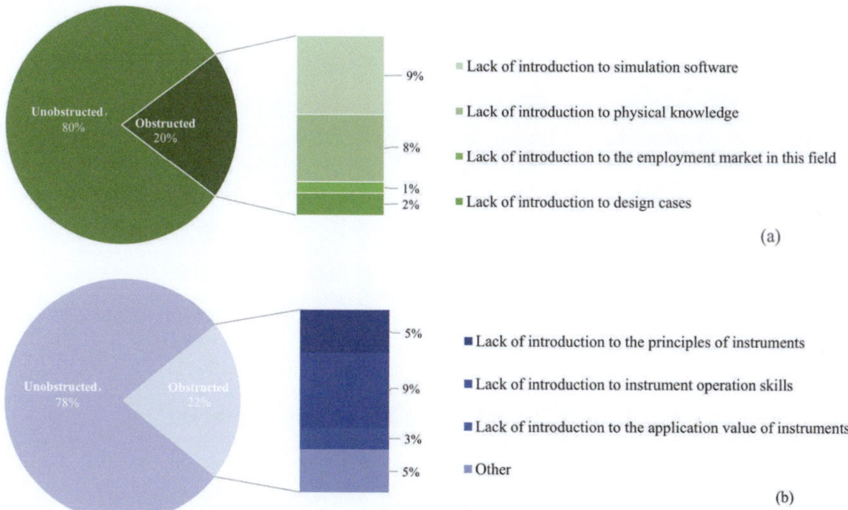

Fig. 3 Learning barriers encountered by students (**a** Obstacles in learning; **b** Obstacles in using instruments during field research) (self-drawn)

In Fig. 3b, 78% of students reported no obstacles in using instruments during field research, while 22% faced obstacles. Among those who experienced barriers in using these instruments, 41% indicated a lack of skills in operating the instruments, 24% cited a lack of understanding of the principles of the instruments, and 12% expressed a lack of appreciation for the value of using these instruments, while 24% selected "other," pointing to issues such as limitations of the research site and convenience of instrument use.

5 Analysis of Course Teaching Optimization

The aim of conducting the questionnaire survey is for instructors to understand teaching effectiveness based on student feedback, thereby exploring ways to improve teaching quality. Analysis of the questionnaire results indicates that the "Green Building Course" has achieved certain educational outcomes while reflecting student needs relating to it. The course teaching needs to adjust four modules: theoretical module, cognitive module, practical module, and operational module. The curriculum optimisation framework is shown in Fig. 4.

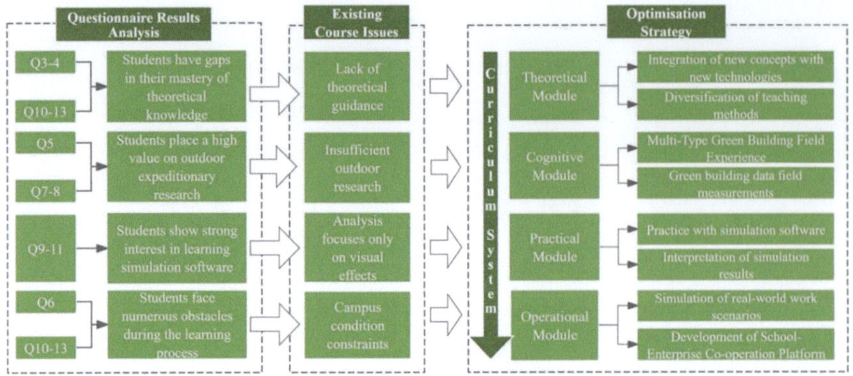

Fig. 4 Framework diagram for green building curriculum reform (self-drawn)

5.1 Theoretical Module: Enhancing Students' Theoretical Foundation and Integrating Simulation Technology

Based on the feedback from questions Q3–Q 4 and Q10–Q13, it's evident that the existing programmes were inadequate in guiding students' theoretical learning. This is particularly evident in the students' deficiencies in understanding the green building evaluation system (average score of 3.18) and foundational knowledge in physics, and students failed to internalise their knowledge effectively, showing a state of forgetting after learning, a status quo that hindered their learning process. Therefore, the teaching plan will incorporate teachings based on the "Green Building Evaluation Standards," using the primary evaluation elements of wind, light, heat, and sound. The course will cover relevant physical principles and delve into practical applications in the context of building physics in green building design. Moreover, a comprehensive interpretation of dual carbon policy goals will be provided, along with an introduction to exemplary green building cases and foundational theories of simulation technology. These will deepen students' understanding of the green building concepts under the dual carbon target framework. Consequently, a deep integration of dual carbon concepts with simulation technology will be executed through diversified instructional methods, including encouraging students to select favorite green building cases and participate in course discussions, enhancing their interest in learning to solidify their theoretical foundation in green building and operational techniques in simulation software for energy efficiency assessment and carbon emissions forecasting.

5.2 Cognitive Module: Increasing Outdoor Practical Activities to Perceive the Green Building Environment

It's essential that theoretical knowledge closely aligns with practical application. The emphasis on outdoor research, as indicated by Q5, Q7, and Q8, suggests that students highly value experiential learning opportunities. However, the current practical courses allocate limited time for hands-on activities, primarily focusing on excursion-type visits, which hinders students' ability to acquire a comprehensive and in-depth understanding of green building concepts To facilitate deeper immersion and flexibility in applying their knowledge, outdoor practical elements should be integrated, divided into two parts: field surveys of green buildings and practical measurements. Regarding field surveys, students will be organized to visit a variety of green building demonstration projects, spanning residential, commercial, and public building types to provide a comprehensive understanding of design approaches and implementation results in different scenarios. Through onsite observation, students will directly perceive the benefits of green buildings in energy efficiency, environmental impact, and comfort while staying updated on advancements in green materials, energy-efficient technologies, and renewable energy applications. In terms of practical measurements, the curriculum will include data collection and analysis in field settings. During this segment, students will be guided to utilize professional measurement tools to collect energy consumption and environmental parameter data from operational green buildings. Once gathered, the data will be analyzed in the practicum module where students will learn to input their measured data into simulation software and conduct processing and analysis to better evaluate the performance of green buildings.

5.3 Practical Module: Introducing Simulation Software to Compare Energy Efficiency of Green Buildings

Determining the energy-saving effects of green building designs and comparing the "green" levels within design proposals is of utmost interest to students. Utilizing physical environment simulation software can provide quantitative answers to these queries. Responses from Q10 to Q11 and observations made during the course indicate students' interest in learning to use simulation software. Although the existing curriculum system also uses green software for design assistance, students' focus is mostly on the presentation of graphical effects and lacks the ability to interpret the energy-saving attributes of design solutions. Currently, prevalent simulation software for green building design includes ENVI-met, Energy Plus, Ecotect, and Design Builder, which can explore sustainable, low-carbon building shapes and optimize desirable qualities like building comfort e.g. ENVI-met, and analyze energy use for heating, cooling, lighting, ventilation, and water e.g. Energy Plus, Design Builder, and assess real-time functionality like sunlight and shading e.g. Ecotect. Such simulation software provides visuals showcasing the effects of energy-saving designs and should be incorporated into the curriculum to enhance students' design capabilities. The theoretical module will introduce various simulation software characteristics, while in the practical module, students can choose software that interests them for post-class analysis, with guidance provided on interpreting the software output results. Presentation of analyses, reports on findings, and visual outputs will help students master the functions of these software tools, enabling energy consumption simulation, environmental performance analysis, and carbon emission calculations. Responses to Q9 indicate students emphasize building form, overall layout, facade, and materials; therefore, highlighting the software's capabilities in analyzing these aspects will excite students to proactively explore software functions and simulation processes, shifting their learning from passive to active engagement and inspiring them towards independent learning and innovative designs.

5.4 Operational Module: Creating School-Enterprise Cooperation Platforms to Enrich Teaching Scenarios

As reflected in Q6, students consider the green building course important for their future studies and designs, yet Q10– Q 13 shed light on the significant barriers students encounter during study and practice. These challenges within the current curriculum include unfamiliarity with instruments, difficulties in instrument usage, and constraints related to the research site-all pointing to a need for more practical platforms. Although universities possess high-level teaching capacities and facilities, they often lack real-world working scenarios. Thus, creating school-enterprise cooperation platforms will allow students, given restricted time and conditions, to engage with more authentic work settings. Simulation technology is highly expandable and open; practical applications of simulation technology in real green scenarios can deepen students' comprehension of green building strategies. Collaborating closely with businesses will help establish more authentic low-carbon building project scenarios, enabling students to learn first-hand practical skills. Schools are encouraged to establish cooperative relations with enterprises to provide training bases, integrating cutting-edge technologies and genuine project cases into the curriculum, thereby enriching the teaching environment. Students will benefit from hands-on experiences within real work contexts and project tasks that stimulate learning and growth. The school-enterprise cooperation platforms can facilitate joint research projects, technology talks, and seminars, advancing academic interactions and technological innovations. Through these collaborative activities, students will not only be exposed to the latest technologies and industry trends but also acquire valuable hands-on experience, laying a solid foundation for future career development. Establishing close partnerships with green building-related companies and research institutions will invigorate the integration of research and practice, fostering the transformation of research outcomes and industry enhancement.

6 Recommendations and Conclusion

Through the questionnaire survey conducted among students enrolled in the "Green Building Course," this study explored students' basic course learning information, satisfaction levels, application preferences, and learning barriers, and proposed the following strategies:

(1) It is found that students have weak theoretical foundations, and the course's theoretical guidance was rigid. In response, the theoretical modules were enhanced, and measures were proposed to strengthen students' theoretical understanding and incorporate simulation, addressing the shortcomings of the existing course in terms of theoretical depth and instructional methods.
(2) It is found that students are highly interested in participating in outdoor investigation, and a cognitive module was added to the course to enhance the outdoor practice environment and help students better perceive the green building context. This change aims to overcome the superficial nature of the original course's practical component and improve the effectiveness of practical learning.
(3) It is found that students attach importance to software simulation and analysis. Consequently, a new practical module was introduced, incorporating simulation software to compare the energy-saving effects of green buildings. This change meets the students' demand for software application learning, and promote the course design from focusing on the traditional point line surface effect to focusing on the energy-saving quantitative effect of the direction of the change.
(4) It was found that students lacked access to a practical platform. To address this, the course includes an increased practical module, with a proposal to establish a collaborative platform between schools and enterprises. This aims to enrich the teaching environment and compensate for the current course limitations, where students are unable to deeply engage with the actual design process due to environmental constraints.

In future, while exploring more scientific and effective teaching models for green building courses, it may be worthwhile to develop and introduce environmental monitoring instruments that can connect with mobile apps or mini-programs, providing more convenient survey methods for students.

Acknowledgements This study is supported by the Zhejiang University of Science and Technology (Teaching Reform Project 2023-jg38).

Appendix

<table>
<tr><td colspan="2">Part one、Basic Information on Students' Course Learning</td></tr>
<tr><td colspan="2">Q1 What is your gender?</td></tr>
<tr><td colspan="2">☐Male ☐Female</td></tr>
<tr><td colspan="2">Q2 Have you been absent from class more than twice?</td></tr>
<tr><td colspan="2">☐Yes ☐No</td></tr>
<tr><td colspan="2">Part two,Students' Satisfaction with the Course(5-pointsatis, faction scale ,maximum scores 5)</td></tr>
<tr><td colspan="2">Q3 Can you clearly distinguish between the physical environments of wind, light, heat, and sound?</td></tr>
<tr><td colspan="2">☐1 ☐2 ☐3 ☐4 ☐5</td></tr>
<tr><td colspan="2">Q4 What is your satisfaction level regarding the learning assistance provided by the following four physical environments?</td></tr>
<tr><td>Wind Environment</td><td>☐1 ☐2 ☐3 ☐4 ☐5</td></tr>
<tr><td>Light Environment</td><td>☐1 ☐2 ☐3 ☐4 ☐5</td></tr>
<tr><td>Heat Environment</td><td>☐1 ☐2 ☐3 ☐4 ☐5</td></tr>
<tr><td>Sound Environment</td><td>☐1 ☐2 ☐3 ☐4 ☐5</td></tr>
<tr><td colspan="2">Q5 How satisfying do you find the following three components of this course for your design?</td></tr>
<tr><td>Green Building Level Evaluation Theory</td><td>☐1 ☐2 ☐3 ☐4 ☐5</td></tr>
<tr><td>Green Building Design and Strategy Cases</td><td>☐1 ☐2 ☐3 ☐4 ☐5</td></tr>
<tr><td>Green Building Field Research Methods</td><td>☐1 ☐2 ☐3 ☐4 ☐5</td></tr>
<tr><td colspan="2">Q6 How much do you believe this course will assist you in your future studies and designs?</td></tr>
<tr><td colspan="2">☐1 ☐2 ☐3 ☐4 ☐5</td></tr>
<tr><td colspan="2">Q7 How helpful do you find the field research assignments in this course for your future studies and designs?</td></tr>
<tr><td colspan="2">☐1 ☐2 ☐3 ☐4 ☐5</td></tr>
<tr><td colspan="2">Part three,Students' Application Preferences</td></tr>
<tr><td colspan="2">Q8 After taking this course, will you prioritize the indoor environment or the outdoor environment in architectural technology?</td></tr>
<tr><td colspan="2">☐Indoor Environment ☐Outdoor Environment ☐Consider Equally</td></tr>
<tr><td colspan="2">Q9 Through this course, which of the following contents do you believe are most effective in optimizing/adjusting building design schemes? (select up to3 options)?</td></tr>
<tr><td colspan="2">☐Overall Building Layout ☐Building Form ☐Floor Plan ☐Facade
☐Cross-section ☐Spacing ☐Orientation ☐Flow Line
☐Entrance ☐Materials ☐Structure ☐Design Explanation</td></tr>
<tr><td colspan="2">Part four,Students' Learning Barriers</td></tr>
<tr><td colspan="2">Q10 Do you feel there were obstacles when learning this course?</td></tr>
<tr><td colspan="2">☐Yes ☐No</td></tr>
<tr><td colspan="2">Q11 What are the main reasons for the obstacles you've encountered in this course?</td></tr>
<tr><td colspan="2">☐Lack of introduction to relevant instruments ☐Lack of introduction to simulation software ☐Lack of introduction to physical knowledge
☐Lack of understanding of the employment market in this field ☐Lack of design case introductions ☐Other</td></tr>
<tr><td colspan="2">Q12 Did you encounter obstacles while using environmental measuring devices during field research?</td></tr>
<tr><td colspan="2">☐Yes ☐No</td></tr>
<tr><td colspan="2">Q13 What were the main obstacles in using these measuring devices?</td></tr>
<tr><td colspan="2">☐Lack of introduction to instrument principles ☐Lack of introduction to instrument operation skills ☐Lack of appreciation for the value of using these instruments ☐Other</td></tr>
</table>

References

1. Cai W et al (2024) Research report on carbon emissions in China's urban infrastructure. Research committee on building energy consumption and carbon emissions data. China Construction Energy Conservation Association, Chongqing University Urban Construction Development and Research Institute. Urban Rural Constr (6):52–61
2. Xu J, Cai Q, Yin C (2024) Exploration into the special planning for green building development in super large cities in the new era—taking Guangzhou as an example. Southern Archit (07):100–108
3. Gao M, Wang S, Wu L et al (2023) Taking 'Practices' as the aim: reflections on green building course reforms. High Educ Archit 2023,32(6):97–104.
4. Liu Y, Liu J, Li J et al (2013) Exploration and practice of teaching reform to enhance innovation in green building design. High Educ Archit 22(06):137–142
5. Gong X, Deng X, Jin Y et al (2020) Teaching practice of green building design based on cultivating students' innovative abilities. High Educ J (08):32–34. https://doi.org/10.19980/j.cn23-1593/g4.2020.08.011
6. Mao Y, Cai W, Zhao Z et al (2022) Teaching reform practices for applied talent cultivation in architectural environment studies. Univ Educ (05):194–196+203
7. Wang X (2022) Research on practical teaching of architectural design theory course based on inquiry-based teaching concept—taking the green building overview course as an example. Archit Cult (09):55–57. https://doi.org/10.19875/j.cnki.jzywh.2022.09.018
8. Jin F (2021) Design and practice of information-based teaching for introductory green building design courses in higher vocational education. Archit Home Abroad (02):170–174. https://doi.org/10.19940/j.cnki.1008-0422.2021.02.028
9. Cantürk Akyildiz E, Özgüven Y (2024) Out-of-school activities in architectural education: MUISCARCH international architecture students congress. Int J Art Des Educ 43(2):190–204
10. Cole LB (2014) The teaching green school building: a framework for linking architecture and environmental education. Environ Educ Res 20(6):836–857
11. Cole LB, Altenburger E (2019) Framing the teaching green building: environmental education through multiple channels in the school environment. Environ Educ Res 25(11):1654–1673

Implementation Pathway of Harmony Village Planning and Case Analysis Based on Tourism Agriculture

Ruiling Liu and Xiang Zhao

Abstract Traditional farming methods and culture in rural areas have been affected by the rapid industrialisation and urbanisation of cities and the modernisation of agriculture. People's pursuit of immediate agricultural production economy has led to the destruction of the rural ecological environment and natural resources. In recent years, rural tourism has become the mainstream of urban suburban tourism, opening up a stable and sustainable development space for tourism agriculture. The report of the Twentieth National Congress of the Communist Party of China (CPC) puts forward the idea of "creating a harmonious and beautiful countryside that magnifies the charm of the original ecological countryside", and tourist agriculture will have a bright future under the strategy of rural revitalisation. This article, through the development history, theoretical studies and case studies of tourism agriculture, builds a theoretical framework for agricultural tourism, identifies six tourism agriculture construction types, and proposes ecological, economic, industrial, cultural and educational aspects of agricultural tourism based on the guidelines of 'comprehensive sightseeing, diversified experience, and economic output'. It also proposes a strategy for the construction of beautiful villages based on the five dimensions of ecology, economy, industry, culture and education in agricultural tourism. It provides a theoretical basis and reference suggestions for the future development of sightseeing agriculture and beautiful countryside construction in Chengdu.

Keywords Sightseeing agriculture · Harmony village · Planning strategies

1 Introduction

As China's urbanization grows faster and faster, the country is paying more and more attention to the issue of "Agriculture, Rural Areas, Farmers". 2023, the No. 1 central document was officially released, emphasizing the need to solidly promote

R. Liu · X. Zhao (✉)
School of Civil Engineering and Architecture, Southwest University of Science and Technology, Mianyang, China
e-mail: rainy@mails.swust.edu.cn

B. He et al. (eds.), *Climate-Resilient, Low-Carbon, and Sustainable Cities*, Lecture Notes in Civil Engineering 754, https://doi.org/10.1007/978-981-95-2169-2_45

the construction of livable, workable and beautiful villages. In recent years, rural tourism has become a new trend in the tourism market. Due to the epidemic, many people have begun to choose suburban tourism. With the implementation of the rural revitalization strategy, rural infrastructure improvement, the rural tourism market gradually opened up, urban residents from the attractions sightseeing gradually to a more experiential tour of the way. After the comprehensive start and beautiful countryside construction, the rural economy is becoming more and more developed, rural tourism agriculture economy has become one of the sources of local GDP. Tourist agriculture is becoming more and more important in the composition of the rural economy. Up to now, a certain amount of research and practical work related to agritourism has begun at home and abroad, but a systematic theoretical framework for agritourism has not yet been formed. Therefore, it is necessary to systematically sort out and summarize domestic and foreign agritourism theories and case studies, sum up excellent experiences, clarify the specific contents of agritourism, and construct a theoretical framework, so as to guide the construction of harmonious and beautiful villages based on agritourism.

2 Method

This paper adopts the literature research method. Through extensive review of relevant literature at home and abroad, including academic journal articles, dissertations, monographs, industry reports, etc., the theoretical basis, research results, and development history of tourism agriculture are comprehensively sorted out and analyses.

This paper adopts the case study method. The 'China TIANFU Agricultural Expo Park', 'Xianghe Village' and 'TIANFU Agricultural Expo Island' in the suburbs of Chengdu City are selected as the research cases, and the construction and development of sightseeing agriculture is explored through in-depth analysis of the cases, and finally, the construction and development of Hemei Agriculture is discussed. The construction and development of tourism agriculture will be discussed through in-depth analysis of the cases, and finally provide a certain theoretical basis for the planning and construction strategy of the beautiful countryside.

3 Tourist Agriculture and Harmony Villages

3.1 Harmony Village

The twentieth CPC National Congress put forward "livable and beautiful countryside construction", livable and beautiful countryside as a new concept of rural construction, is the further expansion of the beautiful countryside, but also livable and livable further expansion of the "livable and beautiful countryside construction" of the new ideas. There's only one word difference between beautiful country and harmony country, the word 'he' represents harmony, good neighborliness and beauty. Compared with the beautiful countryside, it focuses more on the kernel of rural culture, the harmony between people and people, the harmony between people and nature, etc., which are the core of rural culture. The definition of beautiful countryside does not only refer to the beautiful human environment but also includes the pursuit of material consumption and spiritual life, which is the upgraded version of beautiful countryside that includes the integration of ecological, economic, cultural, educational and social development. Different scholars have defined and researched the beautiful countryside from various perspectives, such as landscape gardening, urban and rural planning philosophy and sociology. Domestic scholars Wang Jianing from the philosophical point of view, that the beautiful countryside is the complex of rural society and space, is a harmonious and satisfactory social state of human beings and the harmonious and unified space state of the elements of the place, and that the beautiful countryside is the core of the rural culture [1]. From a planning point of view. From the planning point of view, Cao Chunmei, a domestic scholar, believes that the construction of beautiful villages should be both hard and soft, not only to improve infrastructure and public services, but also to focus on folk culture [2]. From a sociological point of view, domestic scholar Wang Qingzhen believes that the core of the Hemei Rural Strategy is to improve the living environment of villagers and harmonize their neighborhoods, which is a higher stage of the beautiful countryside [3].

3.2 *Tourist Agriculture*

Agriculture in the narrow sense refers to farming. The fusion of "agriculture" and "tourism" has led to the formation of "tourism agriculture". With the development of the times, scholars on "tourism agriculture" research from shallow to deep, respectively, from tourism planning, landscape architecture, urban and rural planning, agronomy and other different perspectives on the definition of tourism. Domestic scholars, such as Guo Huancheng, from the perspective of tourism planning, tourism agriculture is a new type of industry that integrates agriculture, science and technology, aesthetics, culture and visitors' experience with each other, and rationally plans and builds agricultural landscapes and facilities [4]. Domestic scholars, Dong Hong, respectively, from the broad and narrow definition, broadly that tourism agriculture is the space, the carrier of rural land, with rural material, cultural resources development activities. Domestic scholars Dong Hong were defined from the broad and narrow sense, the broad sense that tourism agriculture refers to the space, the rural land as a carrier, by virtue of the rural material, cultural resources development activities. Narrow sense refers to the agricultural resources as the basis for the whole process of processing agricultural products to show as the main content of the form of tourism [5]. Domestic scholars Su Yihua, from the perspective of planning, believe that tourism agriculture is a comprehensive agricultural park that utilizes suburban space, natural resources and culture to build "three lives" through rational planning. Domestic scholar Su Yihua, on the other hand, from the perspective of planning, believes that tourism agriculture is a comprehensive agricultural park that utilizes peri-urban space, natural resources and culture to build "three lives" through rational planning [6]. The domestic scholars Sun Chuan from the agricultural point of view, tourism agriculture is based on agricultural resources and environment, set agricultural production, leisure and tourism, science education, agricultural products sales. Domestic scholars Sun Chuan from the agricultural point of view, tourism agriculture is based on agricultural resources and the environment, agricultural production, leisure and tourism, science education, agricultural products sales of agro-tourism fusion industry [7].

To sum up, the author believes that tourist agriculture is to take the countryside as a carrier, to meet the six functions of landscape viewing, science education, tourism and vacation, thematic fun, economic output, cultural heritage, and is an indispensable part of rural tourism, and the construction of beautiful countryside.

4 History of the Development of Tourist Agriculture

The origins of the development of tourist agriculture can be traced back to the 1860s. With the development of society, especially the rise of the industrial revolution, the progress of urbanization accelerated, people began to gradually get away from the countryside, living in the urban forest built of reinforced concrete. The human desire for nature gave birth to the original form of agricultural tourism—a kind of city residents have the opportunity to get close to nature, experience the farmland wild fun of the new tourism. Italy, Germany, Spain, France, Japan, the United Kingdom and other agriculturally developed countries in accordance with the agricultural fun shaped tourism agriculture is to meet the needs of urban white-collar workers, can be divided into three stages. The development history and characteristics of foreign tourist agriculture are shown in Table 1.

In the 1860s, as a country with developed agriculture in Europe, Italy set up the "National Association of Agriculture and Tourism" and carried out agrotourism activities, and city dwellers began to go to the countryside to experience the rural wilderness and work with villagers in the same way as they ate, lived and labored together. At this stage, the concept of agrotourism was not yet clearly defined, nor were specialized agrotourism parks set up or special tourist facilities constructed.

Development stage: From the 1920s to the 1970s, many European countries saw a boom in rural tourism, with the emergence of agro-tourism parks where tourists could enjoy sightseeing, vacations, crop tastings, adventures and other colorful activities. And there is also a new way of doing business—land rental, the farmer will be a number of land leased to individuals, families, groups, weekdays the owner of the farm to take care of the holidays to the lessee to use.

Table 1 Development history and characteristics of foreign (information from planning and designing tourist agriculture) (self-drawn)

Time	Stage	States	Landmark events	Characteristics
1865	Tourist agriculture in its infancy	Italy	Creation of the National Association for Agriculture and Tourism	Taste the rural interest, eat, live and work with farmers
1930s	The stage of development of tourist agriculture	Europe	–	A form of agritourism that combines sightseeing, vacation and adventure
After World War II	The origins of agritourism parks	Germany	Renting out farm gardens	The owner of the farm rents out portions of the large farm to individuals, families and groups in order to satisfy the need for personal experience and to increase the profitability of the operator
1960s	–	Spain	Paledo State Inn	The farm has a renovated guest house, where you can enjoy recreational programs such as hiking, horseback riding, hang gliding, as well as learning at the Farm School, nature classes, etc.
1970s	–	France	Workers' Orchard	Citizens take advantage of their holidays to grow vegetables in the garden, taste local specialties and enjoy the idyllic scenery
After the 1980s	The period of booming agro-tourism parks	Japan, United Kingdom	Japan Citizen's Park English Rural Parks	A rich agricultural tourism park, where you can take a vacation and have an experience in addition to sightseeing

After the 1980s, the function of tourist agricultural parks is no longer limited to pure sightseeing but also provides and expands the function of vacation and experience facilities, so that people can increase the interaction and participation in agricultural cultivation at the same time.

China's leisure and tourism agriculture first originated in Taiwan, China in the 1970–1980s, in the late 1980s. In the late 1980s, the starting point for the development of tourism agriculture appeared on the mainland—leisure agriculture tours. Relying on local natural resources, large cities organized and held various leisure agricultural activities that attracted countless tourists, such as the "Lychee Festival" held in Shenzhen, which was very popular among Chinese consumers. Sightseeing agriculture to "eat farm food, live in a farmhouse, do farm work, see farm scenery", back to nature as the core concept, in Shenzhen and other metropolitan cities in the daily fast-paced life of urban residents has a great attraction, sightseeing agricultural tourism is also widely noticed, and quickly developed throughout the country. 1998 National Tourism Administration launched the "Huaxia urban and rural tourism". In 1998, the National Tourism Administration launched the "Huaxia urban and rural tourism", and in 2004, the 31 provinces and regions in the country out of the "National Tourism Rural Demonstration Points" of the excellent selection of activities carried out, marking the development of our country's sightseeing agriculture on a new level. Since then, the development of tourist agriculture has been rapid, and such tourism activities are now available in almost every province and district in the country.

5 A Typology Study of Tourist Agriculture

5.1 Categories Based on the Function of the Subject

According to its function of agricultural tourism, the basic form and the business mode of tourism agriculture and other factors, it can be divided into ornamental, vacation, experience, education, industry, research tourism agriculture, as shown in Table 2. 6 types of tourism agriculture in the process of urban–rural integration and development play different functions. The basic forms of tourism agriculture are comprehensive tourism form, multi-distance experience form, economic output form. The business mode of sightseeing agriculture generally includes farmhouse mode, manor mode, natural and humanistic landscape mode, characteristic industrial park mode, park mode and so on. These factors constitute a rich type of tourism agriculture.

5.1.1 Experiential Tourism Agriculture

Experience-type tourism agriculture is similar to vacation-type tourism agriculture, which is the curiosity and desire of most urban residents for nature, farming and agriculture, and turns tourism and vacation destinations to the countryside. It is to experience the farmland as the core purpose, agricultural landscape, farming culture, agricultural production as a carrier, to carry out scenic sightseeing, crop farming, fruit and vegetable picking, processing of agricultural products, leisure fishing, breeding experience and other activities, which can help to improve the understanding of agricultural knowledge and the protection and inheritance of traditional culture. This kind of experiential tourism agriculture not only can get close to nature, but also can actually participate in the experience, feel the charm of agricultural activities, farming culture, non-heritage traditions.

Table 2 Classification of tourist agriculture (self-drawn)

Type of tourist agriculture	Sightseeing agricultural functions	Tourist agricultural business model	Tourist forms of agriculture
Ornamental tourist agriculture	Landscape aesthetics ecological services	Natural and human landscape model	Comprehensive form of sightseeing
Vacation-type tourist agriculture	Recreation recreational health	Natural and human landscapes, park models	
Experiential tourism agriculture	Upgrading agricultural knowledge Study of Farming	Specialized industrial parks and park models	Diversified forms of experience
Educational tourism agriculture	Agricultural cultural heritage natural ecological observations and explorations	The hacienda, agriturismo model	
Industry-based tourist agriculture	Promoting the marketing of agricultural products Industrial restructuring and optimization	Specialized industrial park model	Forms of economic output
Research-based tourist agriculture	Pilot demonstration sites Research showcase	Specialized industrial park model	

5.1.2 Industry-Based Tourist Agriculture

Industry-based tourism agriculture is a more traditional and widely used type. It refers to the integration of characteristic resources, seize its characteristics of industry, combined with modern agricultural technology, production and breeding, agricultural picking, fresh fruit tasting agricultural products display, processing, sale of agricultural products, such as the formation of the main industry chain for the formation of the characteristics of the industrial park model. Such as greenhouses, nurseries, vegetable gardens, orchards, farms in a more transparent way to show visitors; the use of modern agricultural science and technology, cultivate superior varieties of fruits and vegetables, to provide picking and tasting and other experiential activities. According to the local geographic location and climate, summarize the existing farming industry in the village, use its land space, unified planning layout. This type of farm not only attracts tourists to pick and taste fruits and vegetables, but also generates economic benefits from selling agricultural products and by-products.

5.1.3 Educational Tourism Agriculture

Educational tourism agriculture is popular in recent years, is to agricultural resources and the environment as a carrier, to promote agricultural production, natural ecotourism, education and popularization of science and technology to integrate each other to develop a variety of park functions. The main content of the tour viewing, knowledge lectures, through direct observation, personal experience of traditional modern agriculture related to life, agriculture, ecological knowledge and popularization of science education. Vigorously develop such business model as "Agricultural Expo Park", with agricultural production, popularization of science education, tourism and tourism as the main leisure education model.

5.2 Case Studies of Tourist Agriculture

Sightseeing agriculture is a new type of agricultural industry based on traditional agriculture, integrating agriculture, tourism, sightseeing and recreation into one, prompting the target consumers (urban residents) to carry out idyllic sightseeing, leisure and vacation, enjoying the local customs and local conditions, and agricultural practice and education activities. According to the different forms of function, it can be divided into 6 kinds of ornamental, experiential, industrial, educational, vacation, research, etc. In this paper, we selected Chengdu, Sichuan Province, to conduct case study of tourism agriculture as shown in Table 3, each with a different planning and construction paths, among which agriculture, culture and tourism industry integration is the most widely adapted.

5.2.1 Experiential Sightseeing Agriculture—China TIANFU Agricultural Expo Park, the

China Tianfu Agricultural Expo Park is located in Xingyi Town, Xinjin District, with a planning area of 113 km^2 and a core area of 13 km^2, planning to build a "2.5 + X" km^2 high-quality science and technology space on Tianfu Agricultural Expo Island, and to create six major experience scenarios such as exhibition and conference, agriculture, science, culture and creativity, life support and leisure tourism [8]. Inside the expo park, there are Rongjingyuan Family Farm and Yao Tan Wetland. Inside the Expo Park, there are Rongjingyuan Family Farm and Yao Tan Wetland, where visitors can experience the fun of picking and tasting grapes, and in the wetland, they can observe wetland birds and experience traditional activities such as dragon boat rowing. The experience activities allow tourists to enjoy the peace and quiet of the rural landscape while learning about the unique charm of traditional Chinese culture, and allow more young children to gain a deeper understanding of China's farming culture and ecological environment, and to continue their traditional culture.

Table 3 Case study table on tourism agriculture in Chengdu, Sichuan (self-drawn)

The case of tourist agriculture	Business model	Type of tourist agriculture	Case map	Characteristics	Honors
China TIANFU Agricultural Expo Park	Specialty parks	Experiential tourism agriculture		High-quality scientific and creative spaces Family farms Wetland Park	Excellent Agricultural Science and Technology Park in Sichuan Province
The village of Cheung Wo	Specialized industrial parks	Industry-based tourist agriculture		Modern Agricultural Industrial Park	Model Village for Rural Revitalization
TIANFU Agricultural Expo Island	Specialty parks	Educational tourism agriculture		"Grain Silos of Heaven" theme	National Science Popularization Education Base

5.2.2 Industrial-Type Tourism Agriculture—Xianghe Village, the

Xianghe Village is located in Dongchang Town, Dayi County, Chengdu City, Sichuan Province, the village is only a 60-min drive from the center of Chengdu City, occupying a certain advantage in terms of location and transportation. The village is surrounded by Anren Ancient Town, Xiling Snow Mountain and other tourist attractions, and the roads are close to Chengdu Wenqiong Expressway and Chengdu Second Bypass. The village has centralized land improvement and introduced Longyi Group to build a centralized living area for farmers. The village to "rice and fish song" as the theme to create a comprehensive modern agricultural industrial park based on agriculture, integration of tourism, culture, business, education, recreation and other industrial projects, integrated with the Western Sichuan Forest Park. The project builds three systems of ecological operation mode: the aesthetic system of garden life, the ecological system of garden industry, and the production system of garden products.

The Modern Agricultural Industrial Park has successfully increased output from 200 to 800 kg by learning and practicing the new technology of increasing production and efficiency through the symbiosis of fish farming and rice. Through the introduction of good varieties, it has achieved a yield increase of 50–100 kg per mu. We have also developed an intelligent agricultural system to increase the efficiency of production. A series of industries such as research and education, popularization of science, entertainment, recreation and incubation are carried out to form an industrial chain. It is called "the sample of rural revitalization at the foot of snowy mountains".

5.2.3 Educational Tourism Agriculture—TIANFU Agricultural Expo Island, the

Tianfu Agricultural Expo Island is located in Xinjin District, Chengdu City, Sichuan Province, and is the core area of China Tianfu Agricultural Expo Park. Based on the agro-ecological environment, the site is optimized and arranged to create a special activity site with the theme of "Tianfu Granary". The fields, plants and animals of the countryside, as well as farm tools and farmhouses in the countryside are brought into the park so that children can explore the ecology and farming culture. It plans to carry out special activities such as crop picking, popularization of farming and agricultural knowledge, and practical exercises in farming, so as to realize the unity of study and research.

6 Wabi-Sabi Rural Planning Strategy Based on Tourist Agriculture

6.1 Sustainable Development Leads to Rural Revitalization

On the basis of fully utilizing the characteristic natural resources of the countryside, strictly abiding by the red line of arable land, insisting on practicing the development concept of "green water and green mountains are golden silver mountains", and developing leisure and tourism agriculture according to local conditions. Accelerate the pace of integrated development of urban and rural areas, with ecological protection as a prerequisite, rational integration of rural characteristic resources, to create a rural landscape with regional characteristics, rural culture. Following the concept of sustainable development, strengthening the rectification of outstanding environmental problems in the countryside, ensuring the long-term stable development of the countryside, and realizing the "three-effect" integrated development of economic, social and ecological effects.

6.2 Optimizing the Structure of Industries and Empowering Them with Science and Technology to Enhance the Output of the Agricultural Economy

Combined with the National Harmonious Countryside Planning Document, it plans the industrial layout of the countryside, adjusts the structure of the agricultural industry, and promotes the integrated development of one, two and three industries, oriented by the market and supported by science and technology. Focusing on leisure technology agriculture and rural tourism, we make full use of the land production capacity in the countryside, introduce seedlings adapted to the geographic environment and climate, and achieve the characteristic industrial park mode of "four seasons of fruits and four seasons of picking". It promotes the local fruit farmers and farming industry to increase income, drives the employment of the surrounding people, and injects a strong impetus for the revitalization of the countryside.

Against the background of informationization and intelligence, we are actively exploring the path of intelligent agricultural development. The construction of modernized agricultural facilities such as greenhouses effectively improves the yield and quality of crops. Through intelligent breeding technology to improve the level of management, real-time monitoring and precise control of the growing environment of crops, to enhance the gross agricultural product.

6.3 Strengthening the Brand Effect and Emphasizing Publicity and Marketing

Utilizing its local characteristics, it vigorously develops the processing of agricultural products and by-products, and strives to build a special industrial chain with distinctive regional characteristics and brand value. In order to avoid the vicious competition of homogenization, adhere to the differentiated development, the agro-ecological landscape, farming culture as a rural brand promotion point, enhance the cultural value of the brand and commemorative significance, attract tourists to come to travel, and expand the economic benefits of tourism. Strengthen the finishing planning, investment, promotion and operation, through a combination of online and offline ways to enhance the visibility of the countryside, and promote the development of rural agro-tourism.

6.4 Digging Deep Into the History and Traditional Culture, Inheriting and Carrying Forward the Farming Culture

According to the cultural differences of different regions, deeply excavate the farming culture and folk customs of the countryside, and show the charm of traditional culture to tourists. Respecting the traditional culture of the region, extracting regional characteristics, building traditional farm buildings, placing traditional agricultural landmarks, landscape nodes, so that the appearance of villages and villages so that villagers can feel the warmth and resonance. Vigorously constructing the Farming Culture Museum, the Farming Non-legacy Exhibition Hall and other venues, through the collection of information on farming culture, retention, display and other forms of traditional agricultural production methods, lifestyles, rural landscape, etc. Presentation of traditional agricultural production methods. Organize traditional farming activities such as "Farming Culture Festival" and "Farmers' Harvest Festival" to allow visitors to participate in farming activities, such as rice-planting, harvesting, plowing, sowing and so on, to increase the sense of reality of the visitors' experience. The establishment of education and practice bases and the opening of a "second classroom" allow students to learn about the culture and continue to learn about it.

6.5 Strengthening Agricultural Technology Guidance and Training in Cooperation with Universities

Rural construction cannot be separated from the agricultural industrial economy. In addition to training villagers in agricultural technology, it is also necessary to cultivate farmer-entrepreneurs who are skilled in both agricultural technology and business management. Villagers can set up companies or cooperatives, connect with universities to build research and development centers related to the agricultural industry, and form a special industrial park model centered on agricultural production, scientific research, and sightseeing to promote local GDP and economic growth.

Cooperate with universities and research institutes to improve and process agricultural products, and launch industrialized high value-added products in combination with local culture. To create an integrated industry chain of study, education, recreation, recreation and incubation.

7 Conclusion

Under the guidance of the strategy of rural revitalization, how to take tourism agriculture as the basic industry, make it become the core of the construction of beautiful countryside, and integrate it into the industrial structure of the urban peri-urban tourism is one of the new problems in the construction and development of the rural areas nowadays. By summarizing and analyzing the types of tourist agriculture and the better cases of tourist agriculture in Chengdu, it is concluded that only by combining the functional needs of urban residents for suburban tourism, and by proposing strategies based on the five dimensions of tourist agriculture and Harmony Villages with the five dimensions of ecologically sustainable development, optimization of the industrial structure, enhancement of the branding effect, promotion of the traditional culture, and strengthening of the technical training of the universities, it can attract more and further tourists to participate in the construction of the countryside deeply in the countryside. So as to realize the goal of rural revitalization in a sustainable way and achieve the modernization of the countryside in the Chinese style.

References

1. Wang J (2023) Landscape transformation design of idle land in Bikqi township, Tumet left banner under the background of Hemei rural construction. Inner Mongolia Agricultural University. https://doi.org/10.27229/d.cnki.gnmnu.2023.001412
2. Cao C (2024) Research on the construction of harmonious countryside in the new era. Yan'an University. https://doi.org/10.27438/d.cnki.gyadu.2024.000218
3. Wang Q (2024) Research on the status quo and countermeasures of F-village human settlement environment governance in the context of Hemei rural construction. Shandong University of Finance and Economics. https://doi.org/10.27274/d.cnki.gsdjc.2024.000964
4. Guo H, Liu J, Wang Y (2000) Research on the development of tourist agriculture. Econ Geogr (02):119–124. https://doi.org/10.15957/j.cnki.jjdl.2000.02.025
5. Dong H (214) Theoretical analysis of landscape planning in tourist agricultural park. Northwest Agriculture and Forestry University
6. Su Y (2021) Research on the impact of tourist agriculture development on rural economic growth. Changjiang University. https://doi.org/10.26981/d.cnki.gjhsc.2021.000601
7. Sun C (2014) The expression of farming culture in tourism agriculture planning. Southwest University
8. Chengdu X. Tianfu Agricultural Expo Park Management Committee. China Tianfu Agricultural Expo Park Planning Environmental Impact Assessment First Information Public Notice [EB/OL] (27 November 2020) [8 October 2024]. http://www.xinjin.gov.cn/xjxrmzf/c1256263/2020-11/27/content_903802c2fa8c41e0937732da15a12e6a.shtml

Coupling Coordination Between Traditional Village Protection and Tourism Development: Case Study of Laoshan Traditional Villages, Shandong, China

Xiaowen Ma, Enpeng Xu, Lin Wang, Hongsen Li, and Shengyuan Zhang

Abstract The protection and utilization of traditional villages have been a common concern in recent years, and how tourism development and traditional village protection interact in a positive way is a hot research topic. This study introduces the method of statistical coupling collaboration and constructs a collaborative indicator system for traditional village protection and tourism development by drawing on the elements of traditional village protection in cultural landscapes. A collaborative degree model is constructed and analyzed by taking four traditional villages on the eastern foot of Laoshan as an example. Research has proven that the protection and tourism development of traditional villages at the east foot of Laoshan is at a primary level of coordination but has not yet evolved to a well-coupled system coordination level. The coupling coordination situation is closely linked to the development level of the tourism industry, infrastructure construction, material space protection, and the inheritance of intangible cultural elements. The spatial differences in coupled development are related to the case site's resource endowment and industrial foundation. This evaluation system and model can accurately reflect the synergy between traditional village protection and tourism development, thereby effectively guiding the collaborative optimization of various indicators.

Keywords Traditional villages of Laoshan · Traditional village protection · Tourism development · Coupling coordination model

X. Ma · E. Xu · S. Zhang (✉)
College of Arts, Qingdao University of Science and Technology, Qingdao, China
e-mail: 01278@qust.edu.cn

X. Ma · L. Wang
Faculty of Innovation and Design, City University of Macau, Macau, China

H. Li
Institute of Qilu Culture, Shandong Normal University, Jinan, China

B. He et al. (eds.), *Climate-Resilient, Low-Carbon, and Sustainable Cities*, Lecture Notes in Civil Engineering 754, https://doi.org/10.1007/978-981-95-2169-2_46

1 Introduction

Traditional villages [1, 2] play a crucial role in preserving traditional cultural heritage [3], their value and importance in preservation have gradually become widely recognized and developed from the awareness of a few individuals to social consensus [4]. During China's urbanization process, traditional villages are at risk of losing valuable tangible and intangible heritage annually [2, 5]. To address this, the Chinese government has prioritized the protection and development of traditional villages. In December 2012, the Ministry of Housing and Urban–Rural Development, the Ministry of Culture, and the Ministry of Finance jointly issued a notice outlining the list of Chinese traditional villages and the criteria for their classification. In December 2012, the Ministry of Housing and UrbanRural Development of the People's Republic of China, the Ministry of Culture, and the Ministry of Finance issued a notice announcing the list of Chinese traditional villages and the criteria for defining them. As of 2023, China has designated a total of 6,819 traditional villages through six rounds of publication.

In recent years, there has been a increasing number of scholars who have conducted research on the topic of traditional Chinese villages. Compared to previous international studies [6–8] on culture [9], landscape patterns, spatial morphology, and sustainable development of traditional villages [10, 11], research on traditional Chinese villages has placed greater emphasis on their revitalization, development, and construction, particularly focusing on economic and tourism aspects [12–15]. Regarding the research on the protection of traditional villages and the development of tourism, existing literature demonstrates both positive and negative impacts of traditional villages as historical cultural heritage on tourism development [16–18]. The architectural style, landscape resources, and traditional customs of villages are important factors [19, 20] in attracting tourists. Well-preserved traditional villages, with their unique village culture and suitable village environment as tourism resources, can ultimately drive further development of the tourism industry. Moreover, related research shows that tourism development can damage traditional villages [21, 22]. Their interaction is complex and unclear. Therefore, exploring the coordination mechanism between the two and assessing their coordinated state can help improve the level of coordinated development between traditional village protection and tourism development, promoting the sustainable development of traditional villages.

However, there is a scarcity of research exploring the relationship between traditional village protection and tourism development, and few scholars have studied their coupling and coordinated development levels and characteristics. To fill this gap, drawing on the theory of coupling synergy, this study chooses four traditional villages in the eastern foothills of Laoshan in Qingdao as case studies. It researches the coupling and synergy between traditional village cultural landscape protection and tourism development. Constructing an indicator system for traditional village and tourism development explores the degree of integrated development between the two and the mechanisms and causes, providing strategies to facilitate the coordinated development of village protection and tourism, thus promoting rural revitalization and sustainable development in the region.

2 Research Area

Multiple traditional villages are located in the eastern foothills of Mount Lao, adjacent to the Yangkou and Taiqing scenic areas. They are connected through a dedicated tourist road within the Mount Lao scenic area, serving as "rest stations" while developing comprehensive and immersive tourism in Mount Lao (Fig. 1). To preserve the landscape of the Laoshan Scenic Area, the Qingdao Municipal Government has designated this region as a primary protection zone in the "Laoshan Scenic and Historic Spot Masterplan (2021–2035)," The plan positions the area as a relocation and downsizing zone for villages, and prohibits industries such as logging, quarrying, grazing, and artificial breeding. Instead, the tourism industry will become the primary direction for the development of traditional villages in the eastern foothills of Mount Lao.

This study focuses on four typical villages in the eastern foothills of Mount Lao: Diaolongzui Village, Fanling Village, Huangshan Village, and Qingshan Village (Table 1). These villages have a long history and beautiful natural environments. Qingshan Village, an ancient fishing village with a history of more than 600 years, was included in the initial group of traditional villages in China designated for protection in 2012. Diaolongzui Village was ranked among the top 60 most beautiful villages in China during the "Discover 2013 China's Most Beautiful Villages" event 2013. Huangshan Village and Fanling Village are traditional villages with well-preserved environments, rich cultural history, and superior natural conditions. However, with tourism development, issues such as destroying traditional village landscapes and

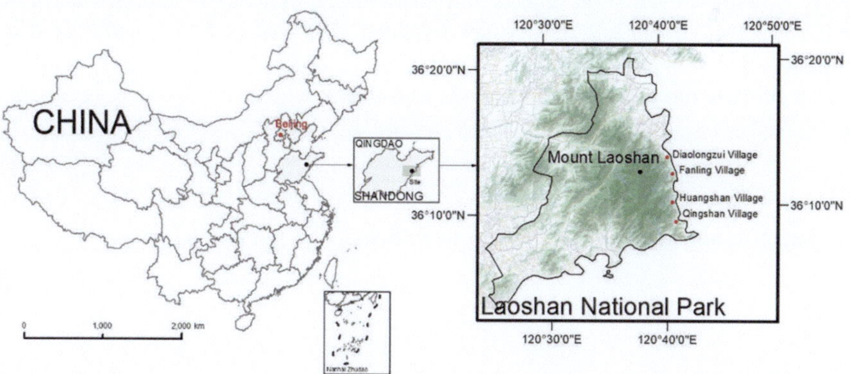

Fig. 1 Map of traditional villages in the east foothills of Laoshan. *Source* Drawn by the authors (self-drawn)

homogenizing tourism products have emerged. Balancing the protection of traditional villages with sustainable tourism development has become an urgent issue to address.

3 Methods and Indices

3.1 Methods

Coupling refers to the dynamic process wherein different systems or elements within a system interact, depend on each other, and mutually enhance one another under specific conditions. The "degree of coupling" indicates the extent of interaction between systems or elements [23]. "Coordination" refers to the harmonious development and positive feedback among systems or components, which is a crucial pathway to achieving sustainable development. The "degree of coordination" quantifies the level of coherence among components within a system, reflecting the system's progression from disorder to order. "Coupled coordinated development" emphasizes the harmonious unity between systems or elements within systems, ultimately achieving comprehensive and orderly overall development [24, 25]. Figure 2 illustrates the coupling coordination mechanism between the protection of traditional villages and tourism development: the protection of traditional villages and tourism development are closely linked. Protecting traditional villages drives rural tourism development, while moderate tourism development can effectively promote the protection and renewal of traditional villages [2].

(1) Data Standardization

To eradicate the impact of different dimensions and magnitudes, the original data of the indicators were standardized. The indicator matrix is represented as $X = \left(x_{ij}\right)_{m\times n}$, with m evaluation objects and n evaluation indicators, among which $i = 1, 2, 3, \ldots, m$; $j = 1, 2, 3, \ldots n$ are included. The matrix X is standardized to obtainv_{ij}.

For positive indicators, the formula is as follows:

$$v_{ij} = \frac{x_{ij} - \min\left(x_j\right)}{\max\left(x_j\right) - \min\left(x_j\right)} \tag{1}$$

For negative indicators, the formula is as follows:

$$v_{ij} = \frac{\max\left(x_j\right) - x_{ij}}{\max\left(x_j\right) - \min\left(x_j\right)} \tag{2}$$

In these formulas, x_{ij} represents the original value, and v_{ij} represents the normalized value.

(2) Determination of Index Weight

Table 1 Village overview. *Source* The authors (self-drawn)

Village	History	Geography	Area (km^2)	Population	Income source
Diaolongzui	380 years old	Nestled amidst three-sided mountains, facing the sea to the east, with Huayan Temple scenic area to the southwest and Yangkou scenic area to the north	3	1,282	Primarily fishing, over 300 households cultivate tea, and more than 120 households run rural tourism businesses
Fanling	Over 600 years old	Surrounded by mountains on three sides, located west of Laoshan Bay, neighboring Diaolongzui village to the north, and separated from Changling village by the Quansxin River	5.4	1,018	Primarily fishing, supplemented by agriculture; catching silverfish in spring and jellyfish in summer; over 40 fishing boats
Huangshan	Over 600 years old	Located west of Laoshan Bay, facing Daguan Island and Shizi Island across the sea, bordered by Huangshangu to the west, Huangshankou village to the south, and Changling village to the north	1.6	975	Primarily fishing, supplemented by agriculture and forestry, a new port of over 6,000 square meters with over 80 fishing boats has been built. The traditional jellyfish processing technique has been passed down through generations and was listed as a district-level intangible cultural heritage in 2010
Qingshan	Over 600 years old	Facing Qingshan Bay to the east, located at the southeaster most tip of Laoshan, within the Taiqing Palace Scenic Area of the Laoshan Scenic Area	3.8	2,326	Primarily fishing, supplemented by agriculture and quarrying

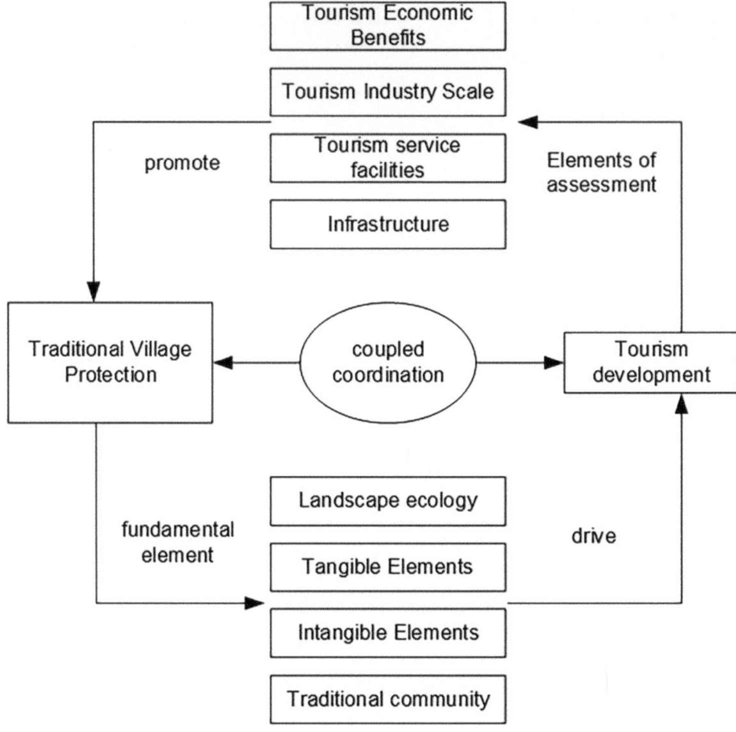

Fig. 2 Mechanism of action of the coupled system of traditional village protection-tourism development. *Source* Drawn by the authors (self-drawn)

The entropy method is employed for determining the weights of the indices, and the calculation process is as follows:

Let p_{ij} represent the characteristic proportion of the i evaluation object for the j indicator. Thus:

$$p_{ij} = \frac{v_{ij}}{\sum_{i=1}^{m} v_{ij}} \tag{3}$$

where $0 \le p_{ij} \le 1$.

Calculate the entropy value e_j for the j indicator:

$$e_j = -\frac{1}{\ln(m)\sum_{i=1}^{m}(p_{ij}\ln p_{ij})} \tag{4}$$

Determine the entropy weight for each indicator:

$$w_j = \frac{1-e_j}{\sum_{j=1}^{n} 1-e_j} \tag{5}$$

(3) Coupling Coordination Status

Coupling coordination can reflect the interaction and influence between systems, measuring the overall synergy effect among multiple systems and revealing the development pattern from uncoordinated to coordinated and from disorder to order. The coupling coordination model for the protection of traditional villages and the development of rural tourism is constructed using the coupling system model as follows:

Comprehensive Evaluation Model:

$$S = \sum_{i=1}^{n} (X_j \times Y_k) \tag{6}$$

where S represents the comprehensive index of traditional village protection or tourism development, X_j represents the weight of each indicator, and Y_k represents the evaluation value of each indicator.

Coupling Degree Model:

$$C = \left(\frac{S_1 \times S_2}{((S_1 + S_2)/2)^2} \right)^c \tag{7}$$

In the equation, C represents the degree of coupling between traditional village preservation and tourism development, with a value range of [0, 1]. A higher value of C indicates a stronger interaction between traditional village preservation and tourism development. When F equals 1, it indicates the optimal coupling state between the two systems.S_1 and S_2 represent the comprehensive indices for traditional village preservation and tourism development, respectively. C is the adjustment coefficient. In practical applications, C should be greater than or equal to 2. In this study, we set $C = 2$.

Coupling Coordination Degree Model:

$$R = \sqrt{(F \times T)} \tag{8}$$

$$T = \alpha \times S_1 + \beta \times S_2 \tag{9}$$

where T is the comprehensive coordination index between systems, α, and β are coefficients such that $\alpha + \beta = 1$. Considering the mutual influence and equal status of traditional village protection and tourism development, α and β are set to 0.5. The coordination degree is divided into ten levels and types, as shown in Table 2, to directly reflect the coupling coordination degree between traditional village protection and tourism development based on the calculated synergy degree (Table 2).

Table 2 Range and grade of the coupling coordination status (self-drawn)

No.	Interval	Grade	Range
1	0.90–1.00	Excellent coordination	Coordination
2	0.80–0.90	Good coordination	
3	0.70–0.80	Intermediate coordination	
4	0.60–0.70	Primary coordination	
5	0.50–0.60	Barely coordination	Transition
6	0.40–0.50	On the verge of disorder	
7	0.30–0.40	Mild disorder	Incoordination
8	0.20–0.30	Moderate disorder	
9	0.10–0.20	Severe disorder	
10	0.00–0.10	Extreme disorder	

3.2 *Index System Construction and Data Sources*

The coupling and coordination evaluation system for the protection of traditional villages and tourism development includes two major systems: village protection and tourism development (see Table 3). This study's traditional village protection system draws on the research findings of other scholars [26, 27] and is based on the theory of cultural landscapes of villages. The traditional village cultural landscape system includes natural, tangible, and intangible elements. Natural elements form the foundation, reflecting the ecological base of the village; tangible elements constitute the structure, supporting the cultural connotation of the village; and intangible elements serve as the outer layer, enriching the cultural characteristics of the village [28]. The traditional village cultural landscape is a product of human activities integrated with specific natural geographic conditions characterized by regionality and cultural specificity. Community culture, guided by human values and consciousness, is an intangible space and an indispensable part of the rural regional system. Yanan [15], after systematically studying the impact of rural tourism on the sustainable development of traditional villages, argues that the development of traditional village tourism should prioritize the protection of community values and emphasize community participation to enhance the authenticity of villages. As a typical rural settlement exemplifying the coupling of human and land relationships, traditional villages, through long term agricultural civilization, have accumulated rich historical information, traditional cultural landscapes, and nostalgic emotions [29]. Compared to typical rural settlements, these villages hold significant cultural continuity value and showcase substantial cultural heritage attributes within the context of modernization. Therefore, this paper argues that the preservation of traditional communities should be emphasized within the category of traditional village protection, and the traditional village protection system should be understood as comprising landscape ecology, tangible elements, intangible elements, and traditional communities.

Table 3 The index system for the coupled and coordinated development of traditional village protection and tourism development (self-drawn)

System	Aspect	Indicator	Definition	Attribute
Traditional village protection	Landscape ecology	Water quality	Measures the natural condition and ecological protection level of traditional villages	+
		Air quality		+
		Ecological pattern protection		+
	Tangible elements	Architectural style	Measures the protection level of the physical environment of traditional villages	+
		Environmental pattern		+
		Street pattern		+
		Water system form		+
	Intangible elements	Dietary habits	Measures the preservation degree of intangible elements and human activities in traditional villages	+
		Frequency of folk activities		+
		Range of folk participation		+
		Frequency of dialect use		+
	Traditional community	Number of immigrants (people)	Measures the vitality and cultural retention of traditional communities in traditional villages	+
		Neighbor relationships		+
		Frequency of public activity participation (times)		+
Tourism development	Tourism economic benefits	Annual tourism operating income (10,000 CNY)	Measures tourism economic income	+
		Annual retail sales of accommodation and catering (10,000 CNY)		+
	Tourism industry scale	Annual tourist reception (people)	Measures the number of tourists and the operational scale of traditional villages	+
		Overseas tourist reception (people)		+
		Number of tourism employees (people)		+
	Tourism service facilities	Number of hotels (units)	Measures the service capacity of operational facilities in traditional villages	+
		Number of parking spaces (units)		+

(continued)

Table 3 (continued)

System	Aspect	Indicator	Definition	Attribute
		Number of restaurants (units)		+
		Number of specialty shops (units)		+
	Infrastructure	Time distance to tourist distribution center (meters)	Measures the accessibility of transportation in traditional villages	+
		Traffic accessibility		+

Current literature on the coupling interaction relationship between the tourism system and other systems is abundant. These studies have comprehensively reviewed the tourism system. For instance, Gao [30] set 20 indicators in four aspects, tourism reception, tourism market, tourism population scale, and tourism economic benefits, to evaluate the level of tourism development. Fang [31] constructed 17 indicators based on the four subsystems of tourism resources, tourism market, tourism industry, and tourism support to measure tourism development. Additionally, research on tourism competitiveness [32] and tourism sustainable development [33–35] also involves tourism system analysis, where these studies, based on tourism resources, tourism market, tourism scale, and tourism benefits, have added the dimension of tourism support environment to the tourism system. In light of this, this paper, drawing on the decomposition of the tourism system by scholars such as Gao Nan and Fang Yelin and combining it with analyses of cultural tourism and sustainable tourism development [36–38], deconstructs the tourism system into four dimensions: tourism economic benefits, tourism industry scale, tourism service facilities, and infrastructure.

The data for this study were collected through multiple survey questionnaires and resident interviews conducted in four traditional villages located on the eastern foothills of Mount Lao from 2022 to 2023. In 2022, resident perception questionnaires were distributed in Qingshan Village, Huangshan Village, Fanling Village, and Diaolongzui Village. The main respondents included local villagers, officials, tourists, and tourism practitioners. Additional data were sourced from the Qingdao Public Data Service Platform, government reports, and other official documents.

4 Results

The coupling coordination degree of traditional village protection and tourism development for Diaolongzui Village, Fanling Village, Huangshan Village, and Qingshan Village has been measured and presented based on the formula in Table 4. Based on the classification of coupling coordination degrees in Table 4, an analysis of the coupling coordination level of village protection and tourism development for Diaolongzui Village, Fanling Village, Huangshan Village, and Qingshan Village will be conducted.

According to the comprehensive evaluation in Fig. 3, the conservation of the four traditional villages at the eastern foot of Mount Laoshan is at the intermediate coordination, primary coordination, barely coordinated, and on the verge of imbalance stages, respectively. They have not reached a level of satisfactory coupling coordination yet. There is a significant disparity in the values of the comprehensive evaluation indicators for the conservation of traditional villages and tourism development. Overall, the value of the comprehensive evaluation indicators for tourism development is relatively low. This indicates that under the background of vigorous tourism development in the Laoshan Scenic Area, the level of tourism development in the traditional villages within the scenic area is relatively low, and the basic conditions for tourism development still need improvement.

Table 4 Indicator system guideline level score (self-drawn)

System (subsystem)	Qingshan village	Huangshan village	Fanling village	Diaolongzui village
Comprehensive index for village preservation S_1	0.62	0.32	0.49	0.40
Landscape ecology (0.43)	0.11	0.03	0.11	0.18
Material elements (0.53)	0.07	0.12	0.14	0.20
Nonmaterial elements (0.61)	0.28	0.12	0.19	0.02
Traditional community (0.29)	0.17	0.05	0.07	0.01
Comprehensive index for tourism development S_2	0.44	0.39	0.29	0.62
Tourism economic benefits (0.28)	0.08	0.03	0.01	0.16
Tourism industry scale (0.49)	0.23	0.11	0.05	0.10
Tourism service facilities (1.02)	0.13	0.26	0.25	0.38
Infrastructure (0.53)	0.15	0.13	0.14	0.11
Coupling coordination degree R	0.72	0.59	0.61	0.70
Coordination index T	0.53	0.36	0.39	0.52

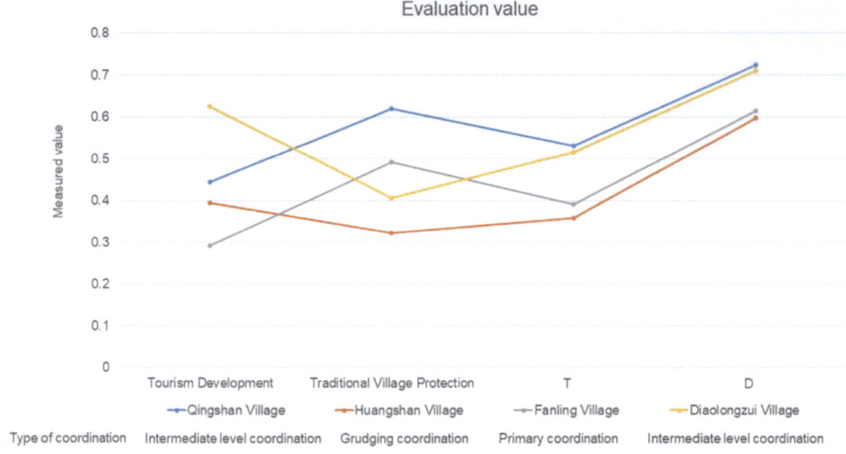

Fig. 3 Coupling coordination measurement results. *Source* Drawn by the authors (self-drawn)

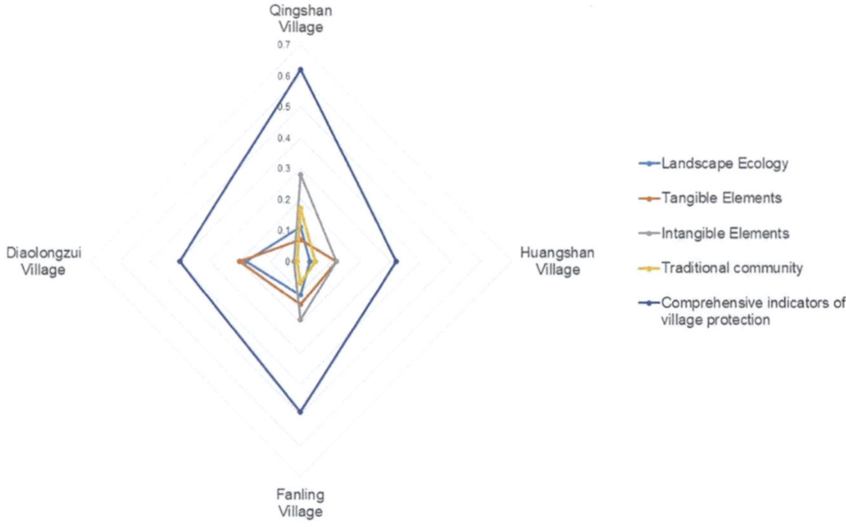

Fig. 4 Traditional village protection subsystem assessment results. *Source* Drawn by the authors (self-drawn)

Based on the analysis, Qingshan Village demonstrates the highest degree of coupled coordination among traditional villages. This is primarily attributed to its highest comprehensive index for village preservation, excellent protection of intangible cultural elements (0.28), the highest community vitality (0.17), and abundant tourism resources. The Qingdao municipal government has also vigorously supported tourism development in this area in recent years. Financial investments have been made in infrastructure, including the construction of archways, tourist access roads, parking lots, a 1400 m^2 entrance hall, and a tourist pier. Furthermore, through investment attraction, the village secured 18 million RMB for renovating homestays, enhancing its material environment significantly. Consequently, Qingshan Village achieves the highest synergy (0.72) between traditional village preservation and tourism development, categorizing it as an intermediate coordination type.

Following Qingshan Village is Diaolongzui Village, which scores notably high in landscape ecology (0.18) and material elements (0.20). Due to its proximity to the early developed Laoshan Yangkou Scenic Area, Diaolongzui Village boasts a comparatively advanced overall tourism development level. The village hosts the highest number of homestays, with a reported 108 establishments in 2023. Its tourism service facilities (0.38) are relatively comprehensive, contributing to a high level of coupled development (0.70). However, rapid growth in tourism and homestays has caused varying degrees of damage to the landscape and traditional village architecture during redevelopment, placing it at an intermediate coordination stage that has yet to reach optimal harmony.

Lastly, Fanling Village and Huangshan Village receive composite coordination scores of 0.61 and 0.59, respectively, categorizing them as primary and low coordination types. These villages maintain relatively harmonious overall appearances and environments (material element scores of 0.14 and 0.12, respectively), preserving a relatively intact village texture and good ecological environment. However, due to historical reasons, only a very small number of historic dwellings have been preserved. Thus, while these villages possess cultural and ecological attractiveness to some extent, their heritage value and overall protection are lower than Qingshan Village, and their tourism industry scale and economic benefits are inferior to Diaolongzui Village. Moreover, community participation in tourism industries and tourism revenue are at their lowest in these villages, highlighting disparities that affect the coordinated development of traditional village preservation and tourism (Figs. 4 and 5).

5 Discussions

The study on the synergetic mechanism between the protection and tourism development of four traditional villages on the eastern slope of Mount Lao indicates that.

Traditional Village Protection as the Foundation of Tourism Development:

(1) The more comprehensive the protection of physical conditions and intangible cultural spaces, the greater the tourism appeal. With the promotion of tourism development, the more complete the infrastructure and tourism facilities, the larger the scale of the tourism industry and the higher the economic benefits of tourism. Consequently, the synergy between traditional village protection and tourism development improves.

Fig. 5 Traditional village tourism development subsystem assessment results. *Source* Drawn by the authors (self-drawn)

(2) Overall, the protection and tourism development of the four traditional villages on the eastern slope of Mount Lao is at the intermediate and primary coordination stages. However, they have not yet developmented to a high level of coupling coordination. Influenced by urbanization and tourism development, traditional villages face disordered village architecture and damaged landscape mechanisms. The tourism industry lacks reasonable planning, rural tourism development is not advanced, tourism features are not distinct, and a cohesive system has yet to form, resulting in a low contribution to traditional village protection. In order to enhance the coupling coordination between traditional village protection and tourism development, it is essential to align policies and financial support, enhance infrastructure and tourism service facilities, and implement specialized plans and regulations. Toplevel design should guide innovative protection and development, maintaining the characteristic landscape features of traditional villages.

(3) The coupling and synergetic characteristics of the protection and tourism development in the four surveyed traditional villages show distinct and regional features. The reasons include differences in resource endowments and tourism development levels, leading to spatial differences in the agglomeration effects of various elements. Therefore, promoting the coupling and synergy between traditional village protection and tourism development requires a comprehensive consideration of the historical conditions, economic foundations, and resource advantages of each region. Policies and measures should be explored according to local conditions to promote integrated development and enhance the degree of coupling coordination between the two systems.

6 Conclusion

This study examines the relationship between traditional village conservation and tourism development through the establishment of a coupling and coordination model. While previous research has discussed the relationship between the two [38–41], few scholars have considered their coupled and coordinated development characteristics. This study reveals the coupled and coordinated development status of different villages, which can contribute to the precise guidance of village conservation and tourism development, as well as the rational allocation of relevant resources. Furthermore, current research on the coordinated development of traditional village conservation and tourism primarily relies on qualitative descriptions, lacking comprehensive quantitative empirical studies on village value, conservation status, and tourism development. The novelty of this study lies in its quantitative empirical research on the coupling and coordination between traditional village conservation and tourism development. It introduces the theory of cultural landscape as the theoretical foundation for evaluating traditional village conservation, expands the scope of traditional village conservation by incorporating the community protection level into the basic indicators of cultural landscape, and reflects the shift in the understanding of traditional villages from historical, scientific, and aesthetic values to social and cultural values. It constructs a more comprehensive, understandable, and applicable evaluation system for the coupling and coordination between traditional village conservation and tourism development. Finally, this study combines resident perception data with socio-economic statistics, integrating qualitative and quantitative analyses, to address the challenge of data retrieval for village-level coupling and coordination analysis. The data for this study primarily came from questionnaire surveys. Due to limited time and resources, it was not possible to incorporate a temporal dimension into the empirical research. However, if future studies can include time points for analysis based on available data, it would provide a better predictive understanding of the development of the case study area.

References

1. Zhang H, Chen J, Zhou C (2017) Review and prospect of research on traditional Chinese villages. Urban Plan 41:74–80
2. Sun J (2017) Traditional villages: theoretical connotations and development paths. J Tour 32(01):1–3
3. UNESCO General Conference (1989) Recommendation on the safeguarding of traditional culture and folklore: adopted by the general conference at its twenty-fifth session, Paris, 15 November 1989. UNESCO, Paris, France
4. Thomas B, Kong D (2020) Cultural landscape as a resource for rural sustainable development: cognition, methods, and practices. Int Urban Plan 35:45–52. https://doi.org/10.19830/j.upi.2020.314
5. Liu X, Wang S (2015) The dilemma and solutions of traditional village protection in China. China Agric Hist 34:99–110

6. Traditional village network—national directory. http://www.chuantongcunluo.com/index.php/Home/Gjml/gjml/id/24.html. Accessed 5 July 2024
7. Yang R, Liu Y, Long H, Qiao L (2015) Spatio-temporal characteristics of rural settlements and land use in the Bohai Rim of China. J Geog Sci 25:559–572. https://doi.org/10.1007/s11442-015-1187-6
8. Zhu J, Xu W, Xiao Y, Shi J, Hu X, Yan B (2023) Temporal and spatial patterns of traditional village distribution evolution in Xiangxi, China: identifying multidimensional influential factors and conservation significance. Herit Sci 11:1–19
9. Qin R, Wei Y (2023) The triple attributes of village cultural heritage landscape reproduction. J Hubei Univ Natl (Philos Soc Sci) 42(11):52–58
10. Xu Q, Wang J (2021) Recognition of values of traditional villages in southwest China for sustainable development: a case study of Liufang village. Sustainability 13:7569
11. Meng Q, Wang C, Xu T, Pi H, Wei Y (2022) Evaluation of the sustainable development of traditional ethnic village tourist destinations: a case study of Jiaju Tibetan village in Danba County, China. Land 11:1008. https://doi.org/10.3390/land11071008
12. Yao G, Xie H (2016) Rural spatial restructuring in ecologically fragile mountainous areas of southern China: a case study of Changgang town, Jiangxi province. J Rural Stud 47:435–448. https://doi.org/10.1016/j.jrurstud.2016.07.014
13. Gao J, Wu B (2017) Revitalizing traditional villages through rural tourism: a case study of Yuanjia village, Shaanxi province, China. Tour Manag 63:223–233
14. Zhou Z, Zheng X (2022) A cultural route perspective on rural revitalization of traditional villages: a case study from Chishui, China. Sustainability 14:2468
15. Yanan L, Ismail MA, Aminuddin A (2024) How has rural tourism influenced the sustainable development of traditional villages? A systematic literature review. Heliyon 10:e25627
16. Wang D (2023) Creating traditional village landscape tourism spaces from the perspective of cultural geography: a case study of Bing'an village, Chishui, Guizhou province. Soc Sci 03:51–56
17. Zhong J, Wei H (2023) The construction of connotations and realization of values in the linguistic landscape of traditional villages: observations from six typical traditional villages in Guangzhou. J Huaqiao Univ (Philos Soc Sci) 06:81–93. https://doi.org/10.16067/j.cnki.35-1049/c.2023.06.005
18. Zadeh Bazargani RH, Kiliç H (2021) Tourism competitiveness and tourism sector performance: empirical insights from new data. J Hosp Tour Manag 46:73–82. https://doi.org/10.1016/j.jhtm.2020.11.011
19. Xie T (2014) Review of rural tourism development in China. Hubei Agric Sci 53:2715–2720. https://doi.org/10.14088/j.cnki.issn0439-8114.2014.11.055
20. Zhou QS (2011) Research on the protection and utilization of historical village and town cultural heritage. Theor Explor 04:86–90
21. Jiang L, Su Q (2013) Creative destruction and local identity transformation in Zhouzhuang Ancient Town. Acta Geogr Sin 68(08):1131–1142
22. Zou J, Zhu Q, Liu P (2018) Study on the vulnerability factors influencing tourism-based traditional villages using interpretive structural modeling. Econ Geogr 38(12):219–225
23. Liao Z (1999) Quantitative evaluation and classification system of coordinated development of environment and economy: a case study of the Pearl River delta urban agglomeration. Trop Geogr 19:7
24. Yang X, Weng G, Zhang X (2013) Review and prospects of domestic research on the application of coupling theory in tourism studies. Reg Res Dev 32:67–73
25. Liang S (1996) Synergetics, and dialectical materialism: reading Haken's "Synergetics: An Introduction." J Shandong Univ Technol (Soc Sci Ed) 04:1–4
26. Zhang Y, Xiao D (2013) Subjective evaluation system of historical and cultural villages and towns. Huazhong Arch 31:4
27. Liu Z (2021) Research on the world cultural heritage value of traditional Chinese villages. J Southwest Univ Natl (Hum Soc Sci Ed) 42:52–58

28. Wu X (2010) The concept, composition, and influence of village cultural landscape heritage. China Townsh Enterp 9:84–93. https://doi.org/10.3969/j.issn
29. Wang S, Sun J (2021) Construction and empirical study of the sustainable development evaluation system for traditional Chinese villages. Acta Geogr Sin 76(04):921–938
30. Gao N, Ma Y, Zhang C (2015) Spatial-temporal coupling and differentiation of tourism industry and regional economy in China's silk road economic belt: based on panel data from nine provinces and cities from 1993 to 2012. Econ Manag 37(09):111–120. https://doi.org/10.19616/j.cnki.bmj.2015.09.014
31. Fang Y, Huang Z, Duan Z, Wang K (2013) Research on the coupling and coordination of tourism development and ecological environment in China. Econ Geogr 33(12):195–201
32. Dwyer L, Kim C (2003) Destination competitiveness: determinants and indicators. Curr Issue Tour 6:369–414
33. Font X, Torres-Delgado A, Crabolu G, Palomo Martinez J, Kantenbacher J, Miller G (2023) The impact of sustainable tourism indicators on destination competitiveness: the European tourism indicator system. J Sustain Tour 31:1608–1630
34. Ko TG (2005) Development of a tourism sustainability assessment procedure: a conceptual approach. Tour Manag 26:431–445. https://doi.org/10.1016/j.tourman.2003.12.003
35. Miller G (2001) The development of indicators for sustainable tourism: results of a Delphi survey of tourism researchers. Tour Manag 22:351–362. https://doi.org/10.1016/S0261-5177(00)00067-4
36. Li Z (2019) Research on the coupling mechanism and coordinated development of rural revitalization and rural tourism system: a case study of Fenghuang County, Hunan Province. Geogr Res 38:643–654
37. Shi P, Li X, Xiong Y (2018) Measurement and prospects of the coupling relationship between regional beautiful China construction and tourism development: a case study of 11 provinces and cities in the Yangtze River economic belt. China Soft Sci 02:86–102
38. Yang L, Long H, Liu P, Liu X (2018) Evaluation system and empirical research on the protection degree of traditional villages: a case study of the first batch of Chinese traditional villages in Hunan province. Hum Geogr 33:121–128+151. https://doi.org/10.13959/j.issn.1003-2398.2018.03.015
39. Yanan L, Ismail MA, Aminuddin A (2024) How has rural tourism influenced the sustainable development of traditional villages? A systematic literature review. Heliyon 10:e25627. https://doi.org/10.1016/j.heliyon.2024.e25627
40. Gao J, Wu B (2017) Revitalizing traditional villages through rural tourism: a case study of Yuanjia village, Shaanxi province, China. Tour Manag 63:223–233. https://doi.org/10.1016/j.tourman.2017.04.003
41. Chen D, Yang Q (2023) Strategies for the preservation and tourism development of traditional villages in the context of rural revitalization. Trans Soc Sci, Educ Hum Res 2:207–212

Optimization Strategies of Street Visual Space Quality Based on Psychological Perception of the Elderly

Yuying Liang and Shuang Liang

Abstract Given the current dearth of street space design that adequately addresses the needs of the elderly, we must understand their needs to cope with an aging society. To make streets more useful and satisfying for the elderly, we propose methods that consider visual continuity, safety, and richness. Introduce psychophysics theory into urban design. Use inductive analysis to sort out elderly-related psychological indicators. Curve-fit psychological perception values with corresponding street space characteristics to explore single-factor influence laws. Findings reveal that the psychological perception of the elderly has a connection with the quality of the street. Among them, the lighting ratio and perceived value of lighting are linearly related; the relationship between the proportion of intersections, the degree of segregation; the proportion of sidewalks, the proportion of sidewalk occupancy, the openness of the sky, and the proportion of seating and the perceived value of street space for the eld is logarithmic, which aligns with the law of Fichtner. This lays a solid foundation for subsequent research into the elderly's psychological perceptions regarding the quality of street space.

Keywords Psychological perception of older people · Spatial quality of streets · Streetscape pictures · Psychophysics · Urban design

1 Introduction

In recent years, the number of older persons has been on the rise. Due to their limited physical strength, short-distance walking is their common mode of travel. Compared with young people, older people are less mobile and prefer to stay in fixed places. The street, as an important place to perceive the urban form, the surrounding environment will have many effects on their body and mind. Finally, Physical limitations and information perception differences expose the elderly to a higher social exclusion

Y. Liang · S. Liang (✉)
College of Civil Engineering and Architecture, Southwest University of Science and Technology, Mianyang, China
e-mail: liiangyuying@mails.swust.edu.cn

B. He et al. (eds.), *Climate-Resilient, Low-Carbon, and Sustainable Cities*, Lecture Notes in Civil Engineering 754, https://doi.org/10.1007/978-981-95-2169-2_47

risk when using streets than other groups. Thus, attention to their psychological perception is crucial.

Existing data on the elderly's perception-related indicators mainly come from field surveys and interviews. Scholars quantify the elderly's perception and evaluation via questionnaires, facilitating the collection of quantitative environmental-perception data. The study of perception originated in environmental psychology. Most literature on the elderly's perception and evaluation has emerged in the past decade. Quantitatively, foreign literature outnumbers domestic, and most are interdisciplinary.

Some scholars [1, 2] use the elderly's attitudes and satisfaction with street and neighborhood facilities and services as perception metrics to analyze facility-satisfaction impacts. Against the backdrop of megacities, Jian Li et al. [3] focus on the link between the elderly's psychological satisfaction and street usage. They found a strong connection between the elderly's street usage rate and psychological satisfaction, especially for living streets. Among street types, the elderly's psychological satisfaction is influenced by multiple factors and not always positively correlated with street satisfaction. Wanlong et al. [4], relying on the subjective perception (expectations and realities) of local elderly in western Hunan and Guizhou, reference domestic and international age-friendly guidelines (e.g., "Healthy Building Evaluation Standard", "Design Code for Elderly Housing Buildings"). Integrating China's mountainous human-settlement characteristics [5–7], they objectively analyze local elderly residents' thematic perception using performance analysis and questionnaires, and assess residential-environment age-friendliness. Zhang et al. [8] use the Perceived Residential Environment Scale and Community Consciousness Scale to survey and evaluate data on the perceived physical environment, subjective well-being, and community consciousness in residential areas, exploring the relationship between neighborhood-environment perception and subjective well-being (life satisfaction, meaning in life, positive/negative impacts) among Chinese elderly, along with the mediating role of neighborhood consciousness.

Nonetheless, most current street-space studies target the general population, neglecting the elderly. This is particularly evident in some developing third-tier small and medium-sized cities, where street-space reality poorly meets age-friendly needs. Moreover, most research focuses on the relationship between a single living-street type and its users. Given these challenges, urban street design should prioritize vulnerable groups like the elderly and their street-space element needs to address aging-society pressures.

1.1 Psychophysics Research Methods

Psychophysics studies and quantifies the relationship between the mind and physical entities. It bridges human sensations with measured data, using simple yet clear mathematical models to depict the link between complex psychological perceptions and external stimuli, thereby exploring the mathematical principles underlying human perception of external stimuli.

Psychophysics has been applied in environmental psychology, urban spatial morphology and other research fields, Yu et al. used both the psychophysical and the spatial perception theory, studied out the parameter quantification method of street elements and psychological perception [9]; Has et al. studied individual cognitive representations, psychological mapping, and perceptual selectivity of environmental elements, Explore the influence of the stimulation of building physical environment on human psychological feelings and behavior [10]; Ribe et al. studied the differentiation of common psychophysical gradients on the perception of photographic beauty and ugliness, And whether people's perception of the visual beauty is objectively shared by the [11]; Chenxue using Weber's law of psychophysics and the description method of waveform graph, Explore the relationship between street and pedestrian space perception [12].

In summary, provided there is corresponding physical and psychological data, psychophysics can be applied to the study of people's perception of street space elements. Given the scarcity of literature in this area, this paper introduces psychophysics, along with research on the scoring of street space elements and related aspects.

2 Selection of Research Indicators

2.1 Study Area

Mianyang is one of China's most representative third-tier cities. It has a large elderly population and is entering a super-aging society. This study will focus on the area within Mianyang city.

This paper draws on Min [13] to the division of street types in Mianyang City, the street types in Mianyang City are divided into: living service streets, landscape and leisure streets, comprehensive streets, commercial streets, and transportation streets, and streets with a high proportion of elderly people are selected as the research sample, and the results of street selection are as follows (Fig. 1 and Table 1).

2.2 Selection of Indicators

Research on street space quality includes objective research, subjective perception research and comprehensive research. Objective research indicators include architecture, green visibility, street height and width, sidewalk width, etc.; subjective research indicators include comfort, safety, attractiveness, visual perception,

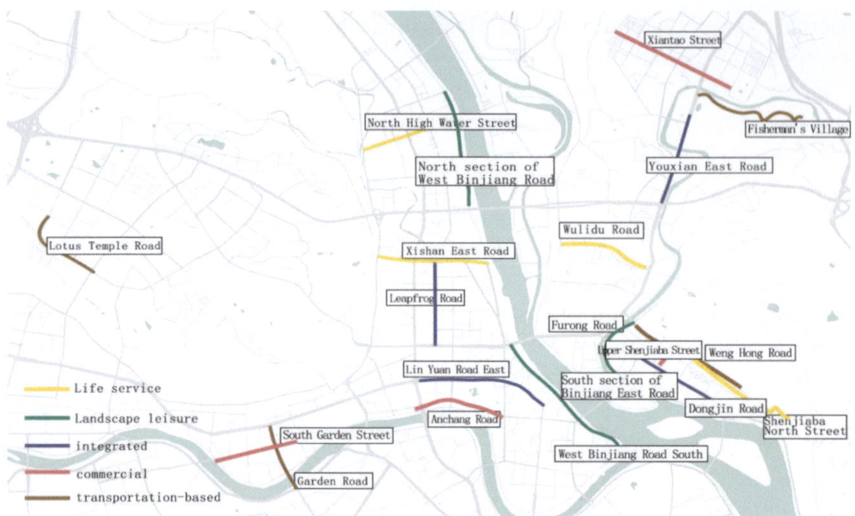

Fig. 1 Street sample distribution (self-drawn)

etc.; comprehensive evaluation research usually studies the subjective–objective correlation.

Street type affects how users use and psychologically perceive the street, and because of the special characteristics of the elderly group itself, this study specializes in combing research methods and street space elements related to the psychology of the elderly, resulting in Table 2.

To gain further insight into the specific street spatial elements that concern older people, the researcher conducted site visits to a number of streets in Mianyang City and administered an open-ended questionnaire to 106 individuals. Of these, 47 were male and 59 were female. The analysis yielded 82 phrases, including interruption of street interface, number of seats, green plantscape, sense of shade, crowded streets, oppressive streets, narrowness and potholes.

Simultaneously, as can be seen from Fig. 2, early classical theories indicate that most researchers initially approached street space perception from the perspective of street aesthetics. Constrained by technology and equipment, their research on street space perception (regarding greenery, health, and facilities) was predominantly subjective and psychological. They also explored ways to enhance urban vitality and space quality from their respective professional viewpoints. Subsequently, with the advancement of the times and technology, researchers adopted a more people-centered approach for comprehensive evaluation. During this period, planning entered the era of modern rationalism, emphasizing street greenery, efficiency, privacy, and aesthetics. In the 1950s, numerous Western scholars started to reexamine modern urban planning. Researchers then truly turned their attention back to human-centered streets, focusing on more user-friendly aspects such as street convenience and comfort. Secondly, the research on spatial quality has evolved from

Table 1 Summary of typical streets (self-drawn)

Type of road	Administrative area	Road name	Road length (km)
Life service	Fucheng	North High Water Street	1.1
		Xishan East Road	1.4
	Youxian	Shenjiaba North Street	1.28
		Wulidu Road	1.29
Landscape leisure	Fucheng	North section of West Binjiang Road	1
		West Binjiang Road South	1.1
	Youxian	South section of Binjiang East Road	1.27
		Furong Road	1.3
Integrated	Fucheng	Leapfrog Road	1.2
		Lin Yuan Road East	1.2
	Youxian	Youxian East Road	1.2
		Dongjin Road	1.2
Commercial	Fucheng	South Garden Street	1.3
		Anchang Road	1.2
	Youxian district	Xiantao Street	1.8
		Upper Shenjiaba Street	0.4
Transportation-based	Fucheng district	Garden Road	1
		Lotus Temple Road	1.15
	Youxian district	Fisherman's Village Road	1.4
		Weng Hong Road	1.1

Table 2 Spatial elements of streets related to older people (self-drawn)

Research methodology	Elements of identity
Subjective study	Street environment, incentives, functionality and safety
Objective study	Greenery, roads, buildings, sky, vehicles, walls, ceilings
Subjective perception	Drama, socialization, authenticity, accessibility
	Pavement smoothness, lighting, sidewalk occupancy, street culture, level of interaction, pleasantness
A comprehensive study	Community infrastructure, street network
	Street connectivity
	Street light levels, sidewalk widths, activity levels, street dwell capacity
	Pavement accessibility, environmental cleanliness, signage visibility, nighttime illumination
	Degree of cleanliness, degree of clustering, number of facilities, degree of illumination

early qualitative analysis to a combination of quantitative and qualitative methods. Research techniques and data sources have also shifted from traditional small-scale data to multi-source big data.

Therefore, these were combined to create a final set of street framework quality measures based on research content previous classical theories and street past studies (Table 3).

(1) Visual continuity: this is about how streets can help the elderly see things clearly. As the population ages, more and more older people are developing mild cognitive impairment. This is a precursor to dementia [14–16]. The environment can affect how well older people think, so this paper looks at how street continuity can help them to navigate safely.
(2) Visual safety: refers to the sense of security that the street can visually provide for the elderly. Security refers here to the sense of safety provided by the street environment. Road unevenness was adopted from Wu's method for obstacles [17].
(3) Visual richness: refers to the visual aesthetics, visual hierarchy and richness of facilities that the street space gives to the elderly. The richness in this paper comes from the green landscape, the hierarchy of architectural spaces, the richness of formal and informal leisure seating facilities, and the socialization of the street.

3 Research and Analysis

3.1 Field Research

First, we noted street features. We also took pictures of the street from the elderly perspective. We sampled at 240 m, the average speed of the elderly over five minutes. We sampled 101 points in 20 streets over five minutes [18]. The camera takes 360° panoramic photos at a height of about 1.6 m, the average height of the human eye [19]. See Table 4 for formulas and methods.

The questionnaire was also designed. The study used a 5-level Likert scale with a well-designed questionnaire on personal attributes, spatial feelings, and other aspects of the elderly. The options ranged from least satisfied to most satisfied. It was conducted from July 23 to July 31, 2023 (8:00 to 19:00). The elderly often have low literacy, so questionnaires and interviews were used to explain the indicators. We selected five types of streets and distributed 10 questionnaires to each. This let us get the elderly's views on the current street space.

Third, DeeplabV3+ extracts street spatial quality indicators from street view images. See Fig. 3.

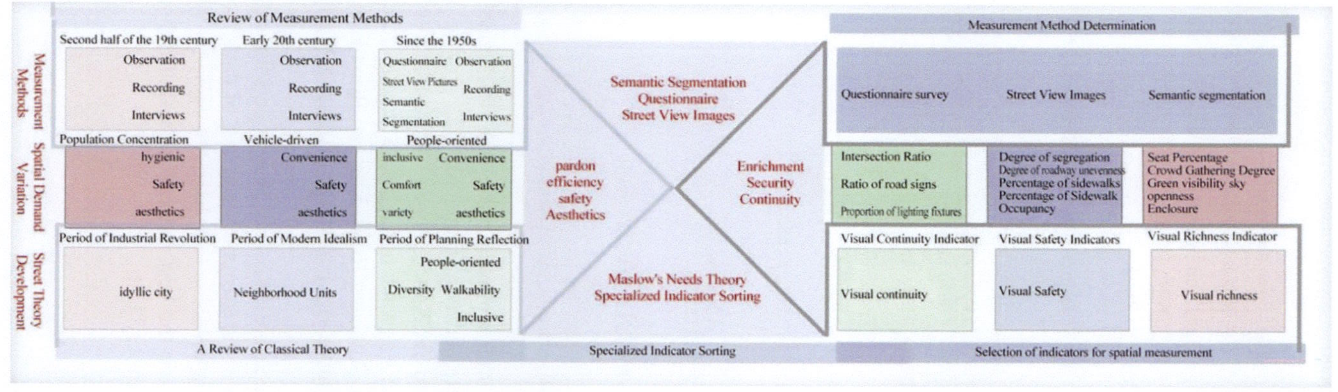

Fig. 2 Selection of measurement indicators and methods (self-drawn)

Table 3 List of indicators for measuring spatial quality of streets (self-drawn)

Target level	Dimension of analysis	Measurement indicators	Acquisition method	Evaluation methodology
Spatial quality of streets	Visual continuity	Intersection ratio	Field measurement	Number of indicators/length of street
		Roadmap ratio		
		Lighting fixtures rate		
	Visual safety	Road unevenness		
		Degree of separation	Street image	Pixels of each index/total image pixels
		Sidewalks rate		
		Pedestrian occupancy rate		
	Visual richness	Seat rate		
		Degree of aggregation of people		
		Green rating		
		Sky openness		
		Enclosure		

3.2 *Data Processing and Analysis*

3.2.1 Questionnaire Analysis

We got 198 questionnaires after removing some that didn't meet our standards. 110 men, 130 women. Most were between 60 and 80.

We imported the data into SPSS for analysis. The results were reliable and valid. Satisfaction data was based on evaluations. Three streets are among the top five. Binjiang West Road South, Anchang Road, Linyuan Road East, Yuejin Road, and Xishan East Road are the most satisfied streets. Binjiang West Road South is the most satisfied street.

Most comprehensive streets are attractive. Most streets for traffic are negative and ranked last. Huayuan Road is the exception. The rest of the streets are negative and at the back. Some landscape streets are the best, but some are negative. Landscape and comprehensive streets are the best in terms of visual richness. Transportation streets are all negative and ranked last.

Table 4 Questionnaire content (self-drawn)

Analysis dimensions	Measurement indicators	Street satisfaction evaluation index system (questionnaire)	Options settings
Visual continuity	Number of cross-connections interrupted (nodes)	Too many intersections on the street to interrupt your walk?	(1–5) cause 1 = least satisfied, least recognized, least consistent to 5 = most satisfied, most recognized, most consistent
	Number of traffic signs (road signs)	There are easily identifiable road signs and text directional signs in the street, so it is not easy to get lost?	
	Number of lighting facilities	Are there enough street lights to make you feel safe?	
Visual safety	Degree of isolation of the facility (boundary)	The isolation facility is normal. Do you feel safe?	
	Road smoothness	The road is flat, there are no potholes, there is no sudden slope?	
	Sidewalk ratio	The sidewalk is comfortable and safe, and not overcrowded?	
	Proportion of sidewalk occupancy	The sidewalk is wide. How do you feel about motor vehicles and other non-pedestrians occupying it?	
Visual richness	Percentage of seating pavilions (including informal)	Are the facilities sufficient and convenient for rest?	
	Crowding	The streets are busy? I'd love to go where there are lots of people?	
	High viewership	Is there enough greenery? Do you feel comfortable or depressed?	
	Openness to the sky	Does the openness of the sky make you feel comfortable or depressed?	
	Enclosure	Does the sense of enclosure of the building in space make you feel comfortable or depressed?	

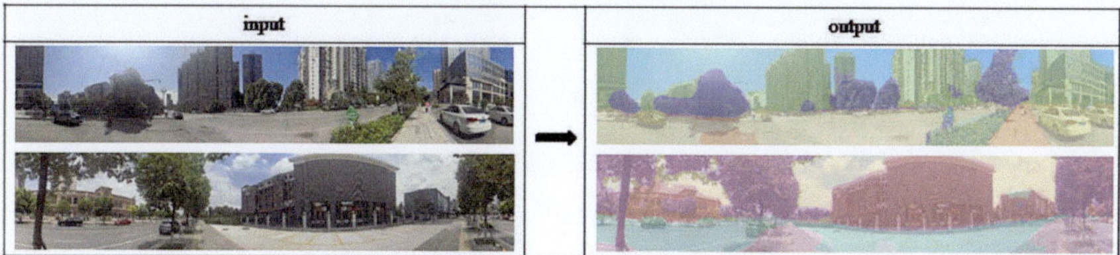

Fig. 3 Example of semantic segmentation extraction (self-drawn)

3.2.2 Spatial Characterization

Integrated streets have the most visual continuity. Landscaped streets are in the middle. Fishers Village Road, a transportation street, has almost no street signs or lighting.

Living-type streets have the most uneven surfaces. Landscaped-type streets are the best in the southern section of West Riverside Road. However, landscaped streets are the safest, integrated streets are the best overall, and the other types of streets are more uneven.

In terms of visual richness, landscaped streets were the best (Fig. 4).

3.3 Analysis of Single-Factor Impact Patterns

To explore the mind–body relationship between how the elderly perceive space and their psychological characteristics, we entered the data into SPSS for curve fitting. For some indicators with non-positive values, we eliminated the 0 s and then performed curve fitting.

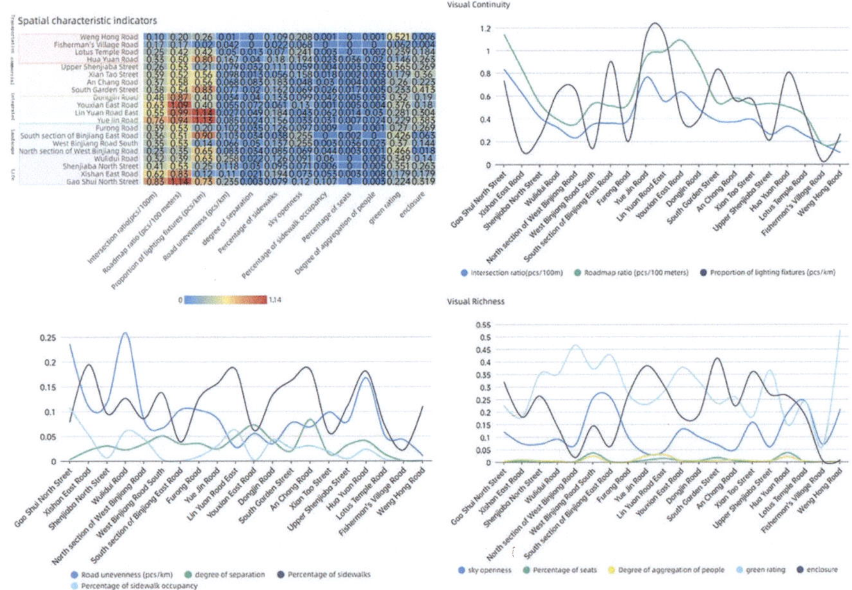

Fig. 4 Intercharacterization (self-drawn)

Table 5 Intersection proportions and their perceptual evaluation models (self-drawn)

Mathematical equation	Model summary					Parameter estimates	
	R-square	F	Degree of freedom 1	Degree of freedom 2	Significance	Constant	b_1
Biometrics	0.908	176.767	1	18	0.000	4.560	−2.126
Logarithms	0.933	250.436	1	18	0.000	2.881	−0.806
Cover with a cloth	0.912	185.561	1	18	0.000	2.948	−0.218

3.3.1 Intersections and Their Perceived Influence

The relationship between the number of intersections and their ratings is stronger than the exponential and power functions. There is a significant advantage, with R^2 equal to 0.933 and the functional expression: $y = 2.881 - 0.806\log x$. An inverse relationship is observed, whereby more intersections lead to less visual continuity and satisfaction (Table 5).

3.3.2 Influence Pattern Between the Proportion of Road Signs and Their Perceptual Evaluation

The three functions between road signs and ratings were lower. Older adults preferred familiar places and rarely used road sign markings. This psychological evaluation of road signs was not significant in any street type (Table 6).

Table 6 Road sign ratio and its perceptual evaluation model (self-drawn)

Mathematical equation	Model summary					Parameter estimates	
	R-square	F	Degree of freedom 1	Degree of freedom 2	Significance	Constant	b_1
Biometrics	0.287	7.232	1	18	0.015	2.918	0.866
Logarithms	0.444	14.384	1	18	0.001	3.804	0.592
Cover with a cloth	0.460	15.334	1	18	0.001	3.816	0.183

3.3.3 Proportion of Lighting Fixtures and Their Influence Patterns

There is a strong relationship between lighting facilities and ratings (Fig. 5). The linear relationship is stronger than the logarithmic and power relationships. R^2 is 0.722, and the functional expression is $y = 2.517x + 0.102$.

However, the fit is not very good. There are cases where there is a high proportion of lighting but a low rating. The survey was repeated based on the responses of the majority of the elderly. Very few would go far from home or travel at night. The psychological evaluation of lighting was not significant on any street type (Table 7).

3.3.4 Road Unevenness and Its Perception Evaluation Influence Laws

There is a link between road unevenness and its perceived ratings (see Fig. 5), but the R^2 is not high and the data distribution is more dispersed. It may be that the unevenness is concentrated in one place, and some of the older adults only moved around in fixed places on the sample street, so although most of the older adults talked about the obstacles affecting them more, the ratings varied widely (Table 8).

3.3.5 Degree of Segregation and Its Perceived Evaluation of the Influence of the Law

Figure 6 shows that the logarithmic and power functions are stronger than the linear function. The logarithmic function is better than the power function. R^2 is 0.963, and the functional expression is $y = 6.852 + 0.967\log x$.

As street segregation increases, older people feel safer and more satisfied with it, regardless of the type of street (Table 9).

3.3.6 Proportion of Sidewalks and Their Perceptual Evaluation of Impact Patterns

The logarithmic relationship between the proportion of sidewalks and their perceptual ratings (see Fig. 6) is stronger than the exponential and power function relationship, with R^2 equal to 0.975 and the functional expression: $y = 5.995 + 1.204\log x$.

Positively contributing to the fact that the higher the proportion of sidewalks, the higher the older adults' visual safety and satisfaction with the street (Table 10).

3.3.7 The Percentage of Sidewalk Occupancy and Its Perceived Evaluation of the Impact Pattern

The fitting results show (see Fig. 6) that the exponential relationship between the percentage of sidewalk occupancy and their perceptual ratings is stronger than the

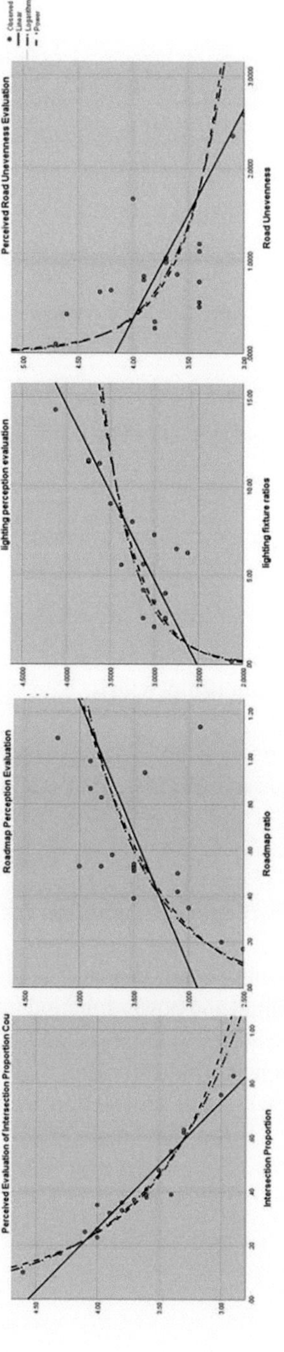

Fig. 5 Curve fitting results (intersections, road signs, lighting, road unevenness) (self-drawn)

Table 7 Proportion of lighting fixtures and their evaluation models (self-drawn)

Mathematical equation	Model summary					Parameter estimates	
	R-square	F	Degree of freedom 1	Degree of freedom 2	Significance	Constant	b_1
Biometrics	0.722	46.709	1	18	0.000	2.517	0.102
Logarithms	0.626	30.144	1	18	0.000	2.583	0.370
Cover with a cloth	0.681	38.347	1	18	0.000	2.572	0.125

Table 8 Evaluation model of road unevenness and its perception (self-drawn)

Mathematical equation	Model summary					Parameter estimates	
	R-square	F	Degree of freedom 1	Degree of freedom 2	Significance	Constant	b_1
Biometrics	0.382	11.126	1	18	0.004	4.160	−0.439
Logarithms	0.430	13.560	1	18	0.002	3.633	−0.398
Cover with a cloth	0.428	13.448	1	18	0.002	3.612	−0.104

logarithmic and power function relationship, with R^2 equal to 0.917 and the functional expression $Y = 4.585x - 24.712$.

There is an inverse inhibitory relationship. The greater the percentage of sidewalk occupancy when using the street, the lower the visual safety and satisfaction of older adults with the street (Table 11).

3.3.8 Sky Openness and Its Perceptual Evaluation Influence Pattern

The relationship between sky openness and its perceptual evaluation (see Fig. 6) is slightly stronger than the exponential and power function relationship, with R^2 equal to 0.887 and the functional expression $y = 2.380 - 0.678\log x$.

This inverse inhibitory relationship indicates that higher sky openness and less shade result in lower visual richness and satisfaction with the street among older adults (Table 12).

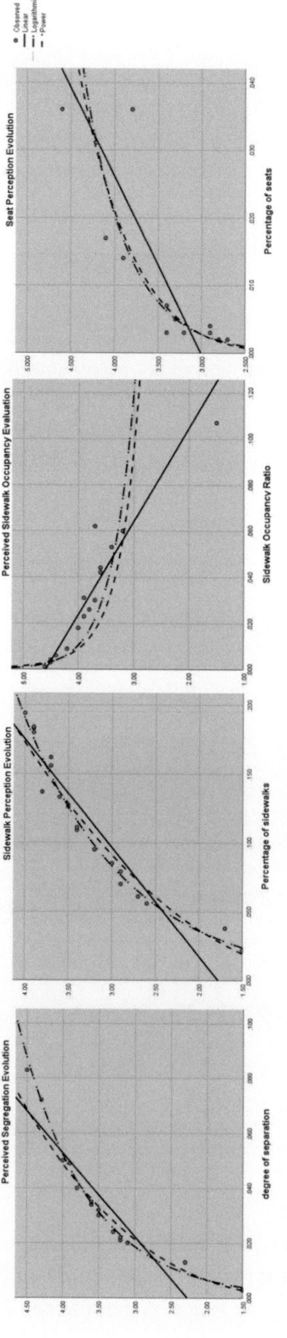

Fig. 6 Curve fitting results (segregation, sidewalk, sidewalk occupancy, sky openness) (self-drawn)

Table 9 Evaluation model of the degree of isolation and its perception (self-drawn)

Mathematical equation	Model summary					Parameter estimates	
	R-square	F	Degree of freedom 1	Degree of freedom 2	Significance	Constant	b_1
Biometrics	0.792	60.933	1	16	0.000	2.270	32.739
Logarithms	0.963	411.572	1	16	0.000	6.852	0.967
Cover with a cloth	0.963	420.186	1	16	0.000	11.363	0.346

Table 10 Evaluation model of pedestrian ratio and its perception (self-drawn)

Mathematical equation	Model summary					Parameter estimates	
	R-square	F	Degree of freedom 1	Degree of freedom 2	Significance	Constant	b_1
Biometrics	0.870	120.017	1	18	0.000	1.777	12.682
Logarithms	0.975	699.938	1	18	0.000	5.995	1.204
Cover with a cloth	0.943	299.577	1	18	0.000	8.792	0.449

Table 11 Sidewalk occupancy ratio and its perception evaluation model (self-drawn)

Mathematical equation	Model summary					Parameter estimates	
	R-square	F	Degree of freedom 1	Degree of freedom 2	Significance	Constant	b_1
Biometrics	0.917	175.928	1	16	0.000	4.585	−24.712
Logarithms	0.646	29.216	1	16	0.000	2.144	−0.407
Cover with a cloth	0.473	14.361	1	16	0.002	2.260	−0.121

3.3.9 Pattern of Influence Between Seat Ratio and Its Perception of Evaluation

The logarithmic relationship between the proportion of seats and their perceptual evaluations (see Fig. 7) is stronger than the exponential and power function relationship, with R^2 equal to 0.824, and the functional expression: $y = 5.949 + 0.501\log x$.

Table 12 Evaluation model of sky openness and its perception (self-drawn)

Mathematical equation	Model summary					Parameter estimates	
	R-square	F	Degree of freedom 1	Degree of freedom 2	Significance	Constant	b_1
Biometrics	0.840	94.244	1	18	0.000	4.605	−5.540
Logarithms	0.887	140.628	1	18	0.000	2.380	−0.678
Cover with a cloth	0.852	103.367	1	18	0.000	2.606	−0.177

There is a positive facilitating relationship, where the higher the proportion of seats, the higher the older people's visual richness and satisfaction with the street (Table 13).

3.3.10 The Influence Law of Crowd Aggregation and Its Perceptual Evaluation

Although there is a significant between the crowd aggregation and its perceived evaluation (see Fig. 7), the R^2 is not high and the fit is not significant. This may be related to the fact that the elderly rarely participate in social activities on the street, because most of them have difficulty in adapting to the various modern developments, so they are unable to integrate into the street, but some of them have talked about how the presence of familiar people on the street increases their evaluation of the visual richness and satisfaction with the street when we conducted our research (Table 14).

3.3.11 The Impact Pattern of Green Visibility and Its Perceptual Evaluation

The impact of green visibility is not readily apparent, and the data are highly dispersed. There are significant discrepancies in the distribution of green visibility on streets. In some instances, areas exhibit markedly low green visibility, a phenomenon observed across all five types of street samples. Consequently, perceptions and fitting effects vary considerably. However, the majority of elderly individuals responded favorably to greening, reporting greater satisfaction with the street when green visibility was optimal (Table 15).

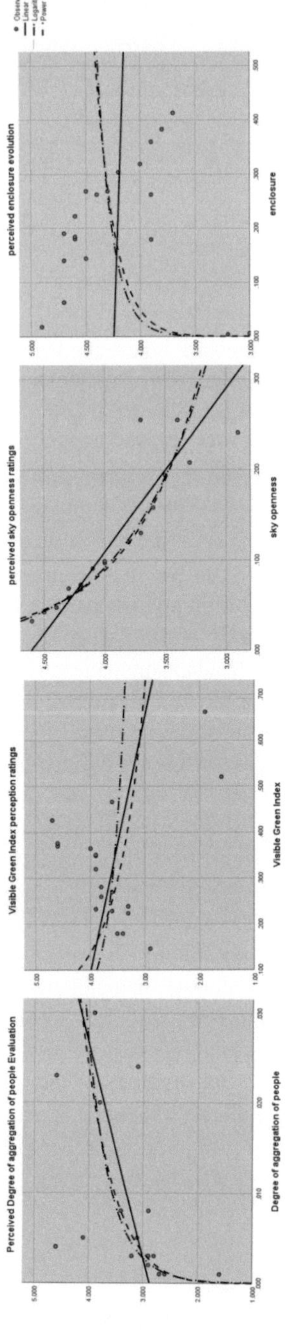

Fig. 7 Curve fitting results (seating, crowd gathering, green visibility, enclosure) (self-drawn)

Table 13 Seat proportions and their perceptual evaluation models (self-drawn)

Mathematical equation	Model summary					Parameter estimates	
	R-square	F	Degree of freedom 1	Degree of freedom 2	Significance	Constant	b_1
Biometrics	0.686	24.084	1	11	0.000	3.007	37.991
Logarithms	0.824	51.635	1	11	0.000	5.949	0.501
Cover with a cloth	0.822	50.868	1	11	0.000	6.931	0.142

Table 14 Crowd aggregation and its perceived evaluation model (self-drawn)

Mathematical equation	Model summary					Parameter estimates	
	R-square	F	Degree of freedom 1	Degree of freedom 2	Significance	Constant	b_1
Biometrics	0.274	6.032	1	16	0.026	2.878	42.146
Logarithms	0.427	11.936	1	16	0.003	5.617	0.446
Cover with a cloth	0.441	12.641	1	16	0.003	6.965	0.148

Table 15 Evaluation model of green visibility and its perception (self-drawn)

Mathematical equation	Model summary					Parameter estimates	
	R-square	F	Degree of freedom 1	Degree of freedom 2	Significance	Constant	b_1
Biometrics	0.087	1.711	1	18	0.207	4.168	−1.792
Logarithms	0.017	0.315	1	18	0.582	3.271	−0.265
Cover with a cloth	0.063	1.218	1	18	0.284	2.827	−0.175

3.3.12 The Influence Law of Enclosure and Enclosure Perception Evaluation

The fit between the degree of enclosure and its perceptual evaluation (see Fig. 7) was not obvious and the data distribution was scattered and did not fit well. Most older people have no perception of enclosure, but in fact they usually choose leisure places with low enclosure for their activities (Table 16).

Table 16 Enclosure and enclosure perception evaluation models (self-drawn)

Mathematical equation	Model summary					Parameter estimates	
	R-square	F	Degree of freedom 1	Degree of freedom 2	Significance	Constant	b_1
Biometrics	0.002	0.030	1	18	0.865	4.241	−0.173
Logarithms	0.133	2.755	1	18	0.114	4.489	0.141
Cover with a cloth	0.180	3.950	1	18	0.062	4.537	0.042

4 Discussion and Conclusion

This paper suggests ways to improve streets in Mianyang City for the elderly. It looks at how people perceive streets and the relationship between people and their surroundings.

At the visual-continuity level, the intersection characteristics, road signs, and lighting facilities, significantly impacts the elderly's behavioral activities. The number of intersection interruptions has a logarithmic correlation with the elderly's sense of visual continuity. In many old-city streets, intersections often have sudden height variations, so it is essential to address intersection inconsistency. Intersection signal timings should be adjusted according to the elderly's walking speed, and audio notifications can be considered. The design and placement of traffic signs need to be standardized. The proportion of illumination is linear with the visual continuity Adequate and gentle lighting is crucial for the elderly's walking continuity.

At the level of visual safety, the degree of isolation, the proportion of sidewalks, and the proportion of sidewalk occupation exhibit a logarithmic-function relationship with visual safety. Road unevenness is significantly correlated with it. Street pavements should be regularly inspected and maintained to avoid inconveniencing the elderly due to height changes at street interfaces. Appropriate measures should be taken to increase isolation facilities or the width of sidewalks, and efforts should be made to minimize sidewalk occupation, so as to make the streets spacious, bright, flat, obstacle-free, and free of potential safety hazards. This will ensure the spatial continuity and smoothness, providing potentially suitable-scaled, aesthetically pleasing community spaces for potential communicative and social activities.

Finally, at the level of visual richness, there is an uneven distribution in the spatial resource allocation of the green-view ratio, resulting in scattered data. Therefore, it is necessary to fill the gaps in areas with less greenery. The openness of the sky and the proportion of seats have a logarithmic relationship with visual richness. High-quality nodes should be created, and abundant green landscapes and shaded, well-ventilated places should be provided. The degree of crowd aggregation is significantly correlated with visual richness, although the functional relationship is not obvious. The data on the degree of enclosure are scattered, and the fitting effect is poor.

Based on the "heart-to-object" relationship, By analyzing the relationship between material spatial elements of different street types and elderly satisfaction, We found that: the proportion of intersection interruption, proportion of lighting facilities, isolation degree, proportion of sidewalk occupation, proportion of sidewalk occupation, sky openness, seat proportion are in line with the "heart-object" relationship of psychophysics, The functions are: $y = 2.881 - 0.806\log x$, $y = 2.517x + 0.102$, $y = 6.852 + 0.967\log x$, $y = 5.995 + 1.204\log x$, $y = 4.585x - 24.712$, $y = 2.380 - 0.678\log x$, $y = 5949 + 0.501\log x$.

This study still has some limitations, Due to the number of street samples, The fitting effect of road sign ratio, road flatness, crowd aggregation degree, green visual rate, enclosure degree is not good; And since there are only four streets of each type, The small sample size, Therefore, the curve fitting of each type of streets is impossible, Then explore the "heart-object" relationship of each type of streets and its relationship with the corresponding streets; in addition, Due to the limitations of the study methodology, The street space scoring data obtained through questionnaire interview may affect the subsequent fitting effect due to insufficient accuracy, All of the above will be optimized and improved in the future studies.

References

1. Jiayao Q, Xiaolan W (2018) Study on the influencing factors of residential satisfaction among urban elderly—take Wuxi and Yantai as an example. J Party School CPC Fujian Prov Comm (04):81–91
2. Zhou, Yao, Chang et al (2019) Factor analysis of the influence of neighborhood satisfaction of the elderly: evidence from Xuzhou, China. J Women Aging
3. Li J, Gao X, Qiu Y, Ling Y (2022) Aging-friendly cities: Investigating the effects of street usage on the psychological satisfaction of older adults in megacities. Front Psychol 20(13):942301
4. Wanlong H, Shiyuan D, Yafeng Z et al (2022) Study on the subject perception of mountain residential environment in western Hunan and Guizhou province. Arch Cult 7:160–162
5. Gen M (2019) Research on the construction and application of aging evaluation system in suburban villages. Huazhong University of Science and Technology
6. Dawei M, Hongbo Z (2017) Evaluation of the pension community environment for the elderly based on hierarchical analysis. Build Energy Effic 45(06):90–94+121
7. Mingzhi C (2020) Study on evaluation and optimization of aging in urban parks based on AHP. Shenzhen University
8. Zhang Z, Zhang J (2017) Perceived residential environment of neighborhood and subjective well-being among the elderly in China: a mediating role of sense of community. J Environ Psychol 51:82–94
9. Yu Z, Yukun Z, Sinan Y (2012) A quantitative study of psychological cognition at the street interface. J Arch S2:126–129
10. Al-Alwan HAS, Al-Bazzaz IA, Ali YHM (2022) The potency of architectural probabilism in shaping cognitive environments: a psychophysical approach. Ain Shams Eng J 13(1):3–7
11. Ribe R (2022) Exploring psychophysical measurement in landscape aesthetics: validity, reliability and signal detection via single-versus opposing-construct rating scales, with or without zeros. J Environ Psychol 83:101862
12. Chenxue S, Jianbo Z (2022) A spatial cognition and quantitative study of street interface rhythm. New Arch 02:112–116

13. Min Z (2023) Research on pedestrian-friendly oriented street design. Southwest University of Science and Technology
14. Luping S, Shui Hong Y, Wei W (2022) Meta-analysis of the prevalence and development trend of mild cognitive impairment in the Chinese elderly population. Chin Fam Med 25(01):109–114
15. Yanxue Y, Ruyu B, Wenlong Y et al (2023) Study on the current status and influencing factors of cognitive dysfunction in people ≥60 years old. Chin Fam Med 26(21):2581–2588
16. Cassarino M, Setti A (2015) Environment as 'brain training': a review of geographical and physical environmental influences on cognitive ageing. Ageing Res Rev 23(Pt B):167–182
17. Yong-Long WU, Guang-Yi YEH (1995) A basic study on trail installation and evaluation of trail width. J Arch 15:33–45
18. Eva B, Hamish F (2013) A study to investigate the walking speed of elderly adults with relation to pedestrian crossings. Physiother Theory Pract 29(2)
19. State Bureau of Technical Supervision. GB 10000-88 (1988) Chinese adult human body size. China Standard Press, Beijing
20. IBM Corp. Released 2019. BM SPSS Statistics for Windows, Version 26.0. IBM Corp, Armonk, NY

Emotional Geography-Driven Renewal of Traditional Villages: A Case of Tielu Village, Sichuan, China

Minghong Yu, Jialiang Zhu, Wen Guo, Qi Zhang, and Congyu Li

Abstract This paper conducts an in-depth study from the perspective of emotional geography on the protection and renewal status of Tielu Village in Weicheng Town, Youxian District, Mianyang City, Sichuan Province, under the background of rural revitalization strategy. Through methods such as field investigation, emotional map drawing, and questionnaire interviews, memories and emotions of villagers and tourists towards the village were collected, and the influence of emotional structure on the reproduction of village space was analyzed. The research found that the spatial structure characteristics of Tielu Village present a 'group axis line' pattern, and based on emotional geography, it proposed renewal concepts, deeply analyzing the existing problems of the village and suggesting protection and reshaping strategies. These strategies aim to protect the original spatial form and lifestyle of the village, ensuring the authenticity of the village layout, and on this basis, to reasonably set up and increase tourism projects and service facilities. The research also emphasizes starting from a holistic and temporal perspective, combining continuity protection with sustainable utilization methods, to protect historical elements from various historical periods and explore a dynamic protection and renewal model. In addition, this paper also explores the emotional perceptions of different groups towards the village, as well as how to enhance the village's attractiveness and sustainability through renewal optimization strategies. The research results not only provide practical guidance for the protection and renewal of Tielu Village, but also offer new perspectives and methods for the study of emotional spaces in traditional villages and village renewal under rural revitalization strategies.

Keywords Traditional villages · Tielu village · Emotional geography · Spatial syntax · Village renewal

M. Yu · J. Zhu (✉) · W. Guo · Q. Zhang · C. Li
School of Civil Engineering and Architecture, Southwest University of Science and Technology, Mianyang, China
e-mail: zhu_jialiang0827@163.com

B. He et al. (eds.), *Climate-Resilient, Low-Carbon, and Sustainable Cities*, Lecture Notes in Civil Engineering 754, https://doi.org/10.1007/978-981-95-2169-2_48

1 Introduction

In the rapid advancement of global modernization, traditional villages, as living representations of cultural heritage, have seen the issues of their protection and revitalization elevated to a national strategic level within the academic and policy-making realms of urban–rural planning [1]. Initiated in 2020, China's nationwide rural revitalization campaign has institutionally embedded the conservation and adaptive reuse of historic settlements within its national development agenda. Through a multi-agency collaboration led by the Ministry of Housing and Urban–Rural Development, 8,155 vernacular communities have been systematically cataloged as protected cultural assets, creating institutional safeguards for the intergenerational transmission of agrarian civilization legacies [2].

Traditional villages not only carry rich historical memories, unique cultural ecologies, exquisite architectural skills, and clear social development contexts, but also face multiple challenges such as hollowing out and degradation of living environments, making the dual pressure of protection and development increasingly prominent [3]. This study, based on the theoretical perspective of emotional geography, comprehensively explores the potential impact of emotional structures on the spatial morphology and functional reproduction of villages through the integrated use of scientific methods such as field investigation and spatial syntax analysis. It reveals the spatial structural characteristics of villages as 'group axis' and innovatively proposes an organic renewal concept based on emotional geography [4–6].

Taking Tielu Village in Weicheng Town, Youxian District, Mianyang City, Sichuan Province as an empirical case, this study provides a comprehensive analysis of villagers' and tourists' emotional perceptions, emotional attachment, and sense of place, thereby offering a solid empirical foundation for formulating protection and renewal strategies [7]. The research emphasizes that it should start from a holistic perspective and a temporal dimension, combining the principles of continuity protection and sustainable utilization, actively exploring dynamic protection and renewal models to achieve the sustainable development of traditional villages [8].

Safeguarding historic settlements constitutes a multidimensional endeavor necessitating the integration of historical preservation, cultural continuity, ecological sustainability, social cohesion, and economic viability [9]. The study advances methodological frameworks for vernacular conservation through place-specific adaptive approaches, while extending emotional geography's analytical applications in heritage contexts. These contributions yield actionable paradigms for regionally differentiated rural–urban transitions, demonstrating synergistic value in bridging academic discourse with sustainable place-making praxis.

2 Spatial Morphological Characteristics of Villages and Research Methods

2.1 *Overview of the Research Area*

Tielu Village, located 27 km from the urban area in Weicheng Town, Youxian District, Mianyang City, Sichuan Province, covers an area of 4 square kilometers and has a population of 1,641, with 12 cooperatives (including 4 specialized ones). The village is renowned for its well-preserved ancient architectural complex, rich intangible cultural heritage, and profound agricultural culture, where nature and ancient architecture coexist harmoniously, forming a unique landscape.

Tielu Village has a long history, with traditional buildings mostly inheriting the styles of the Ming and Qing dynasties, presenting a '回' (hui) and 'U' shaped layout, and is one of the first provincial-level traditional villages in Sichuan Province [6]. The village contains rich folk literature, traditional craftsmanship, and folk culture, forming a unique cultural ecology. As a cultural representative of Mianyang City, Tielu Village has preserved the traditional village layout and architectural features, with Qing Dynasty residences being particularly exemplary. It is not only a cultural heritage but also an important resource for studying Ming and Qing architectural art and rural changes (Fig. 1).

2.2 *Layout of Village Resource Environment and Topography*

Tielu Village is situated in a unique micro-geographical unit, surrounded by mountains on the east, west, and north sides, while facing an open area to the south. The overall terrain exhibits characteristics of descending from east to west and rising from north to south. This geographical layout has resulted in Tielu Village forming a relatively complete natural environment, surrounded by mountains on three sides and facing water to the south. The internal water system of the village flows from north to south in accordance with the terrain, creating a unique water system pattern of 'one dam collecting water and multiple ponds embedded in fields.' At the same time, there are ten wells distributed throughout the village, arranged reasonably according to the different groups within the village.

In terms of architectural layout, the buildings in Tielu Village are arranged according to the mountain, and some buildings show a winding 'sausage-like' layout [6]. On the whole, the architectural layout of the village follows the principle of 'house follows field, field follows water, and field follows mountain', forming a unique style of harmonious coexistence with the natural environment.

In addition, the boundary definition elements of Tielu Village include natural elements and artificial elements. Natural elements mainly include rivers, mountains, forest land, etc., while artificial elements mainly include farmland. These boundary elements jointly define the spatial scope of Tielu Village, and together with the

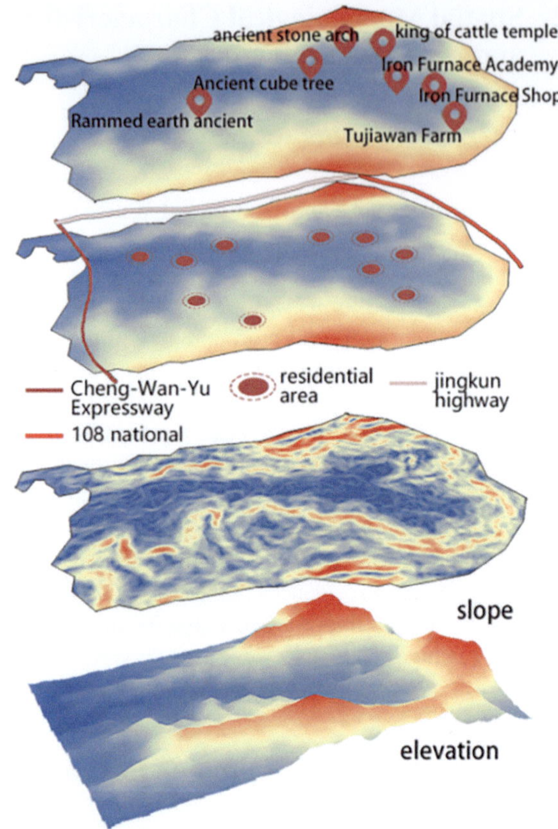

Fig. 1 Resource distribution and environmental layout (self-drawn)

surrounding mountains, water systems, and central farmland constitute a typical 'mountain-water-field-village' settlement style (Fig. 2).

2.3 Research Method

In this study on the renewal of traditional villages in Tielu Village, we adopted a diversified methodological system to ensure the comprehensiveness and depth of the research [8]. First of all, through detailed field investigation, we visited Tielu Village and recorded its natural landscape characteristics, architectural features and residents' daily life status in detail, so as to obtain valuable first-hand information. A structured questionnaire was administered to 120 villagers and 80 tourists in May 2024, covering emotional perceptions of key nodes (e.g., temples, homestays) on a 5-point Likert scale (1 = strongly negative, 5 = strongly positive). Interviews were conducted with 15 village staff to triangulate data. Results indicated that 78% of

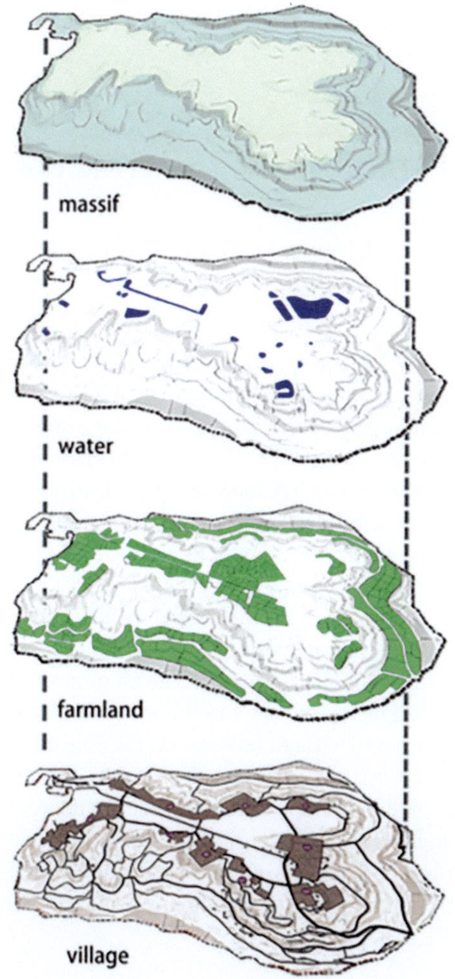

Fig. 2 Terrain characteristics and settlement pattern (self-drawn)

villagers associated ancestral halls with "pride" (mean = 4.2), while tourists rated homestays lower (mean = 2.8) due to inadequate facilities.

In order to deeply understand the emotional cognition of villagers and tourists to the village, we used the theory of emotional geography to draw an emotional map, and combined with questionnaires and in-depth interviews to systematically collect their emotional memory and cognition of the village [9, 10]. After that, we used software such as Photoshop 2021, SketchUp 2021 and ArcGIS 10.8 to draw the emotional map and so on.

In order to further analyze the spatial structure characteristics of the village, we used spatial syntax analysis and used ArcGIS software to quantitatively evaluate the morphological characteristics of the village. This series of research methods and technical means complement each other, which together constitute the methodology system of this study and ensure the scientificity and practicability of the research results [11, 12].

3 Data Collection and Data Processing

3.1 Different Groups of People's Emotional Cognition of Villages

Tielu Village, a traditional village with a history of nearly four hundred years, carries rich folk literature, traditional skills and folk culture. There is a tourist reception center in the village, and there are many functional nodes such as Niuwang Temple, Qing Dynasty Royal Ancestral Hall, Central Square, Iron Furnace Mall, Rammed Earth Characteristic Dwellings, Tujiawan Farm, Ancient Stone Steps, Ancient Stone Arch Bridge, Intangible Cultural Heritage Experience Base, Characteristic Homestay, Meihua Bay, Shilimei Gallery, Tourism Research Base and Iron Furnace Academy (Fig. 3).

In the study of traditional village renewal and optimization strategies, from the perspective of emotional geography, this process involves multi-dimensional and multi-agent participation. The emotional identity, attachment and sense of belonging of villagers, tourists and village staff constitute the basic information and emotional context of village renewal and optimization. As the core subject of cultural inheritance, the villagers have deep emotional identity and attachment to the village, which provides an important reference for the formulation of village renewal strategies. As an external experiencer, tourists' intuitive and short-term emotional experience also plays a key role in the attraction and sustainable development of the village. By optimizing the tourist experience and strengthening its good feeling and sense of

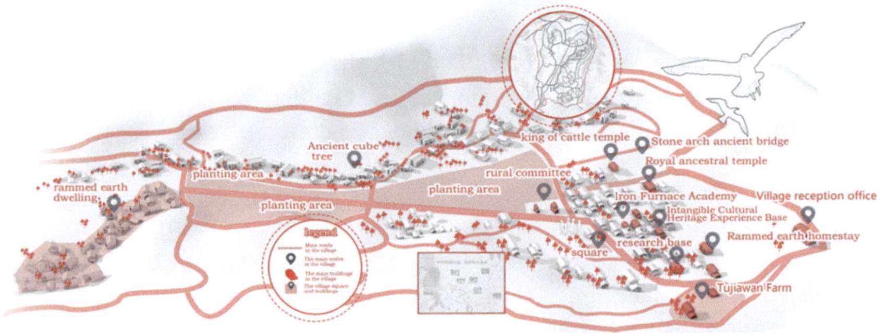

Fig. 3 Aerial view of the current situation and important node map of Tielu village (self-drawn)

belonging to the village, it can enhance the popularity of the village and promote the development of tourism.

Using tribal mapping and in-depth interviews, we found that the main emotional tendencies of three newly built or functionally replaced rammed earth houses in Tielu Village Cultural Service Center, Village Committee and Reception Center are joy and expectation, reflecting the villagers' yearning for a better life in the future. The three nodes of Wang's ancestral temple, Niuwang temple and Xujiawan farm, because they better retain the historical function and cultural atmosphere, the main emotion is the neutral emotion of awe.

Further analysis of emotional cognition and in-depth interview results, new buildings generally show positive emotions such as expectation and happiness. Over time, the node function has an impact on the emotion, and the emotion reacts to the space transformation, making the space more emotional. For example, the high-rise of the temple, the ancient decoration of the ancestral hall, the renewal of the propaganda page of the village committee and the sculptures in the entrance square all reflect the interaction and continuous change of space and emotion (Figs. 4 and 5).

3.2 Space Syntax Analysis

In the in-depth analysis of the spatial structure of Tielu Village, we used several key spatial syntactic indicators to evaluate the organizational characteristics of its village space. Among them, the global integration degree and the local integration degree are the core analysis concepts, which reveal the agglomeration and dispersion of space in the overall range and local range respectively [13–15].

The analysis results show that the integrated core of Tielu Village is mainly concentrated on the axis of Wangjiashangwan residential area and village committee. The Wangjiashangwan axis demonstrating the highest global integration value (1.72) and the outlying clusters showing low local integration (average 0.48). The central axis' topological primacy can be attributed to its strategic connectivity, combining transport efficiency, mixed land uses, and linear spatial organization that facilitate movement between residential, cultural, and commercial nodes. Conversely, the fragmented spatial configuration of peripheral areas—shaped by topographical barriers, mono-functional layouts, and disconnected road networks—creates isolated enclaves prone to demographic decline and cultural erosion. To address this spatial duality, this research advocates a holistic revitalization framework that strengthens core spatial systems while activating marginalized zones through three complementary strategies: implementing hierarchical connectivity improvements including farmland path-waterway boardwalk systems, introducing community-oriented functions such as artisanal workshops and cultural homestays, and establishing adaptive management protocols for continuous spatial performance evaluation. This integrated approach seeks to balance heritage conservation with contemporary livability by fostering systemic spatial resilience in traditional rural settlements. Based on this finding, we propose suggestions to optimize the accessibility and accessibility of the axis, aiming

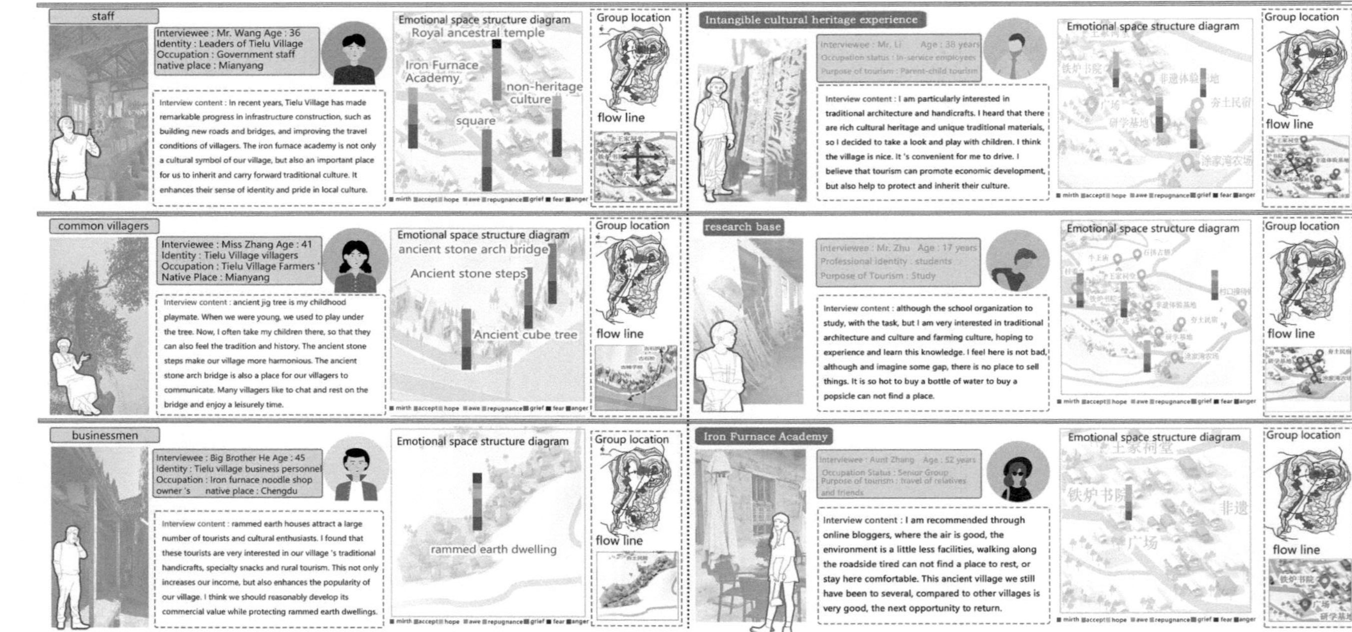

Fig. 4 Emotional cognition questionnaire of different groups of people (self-drawn)

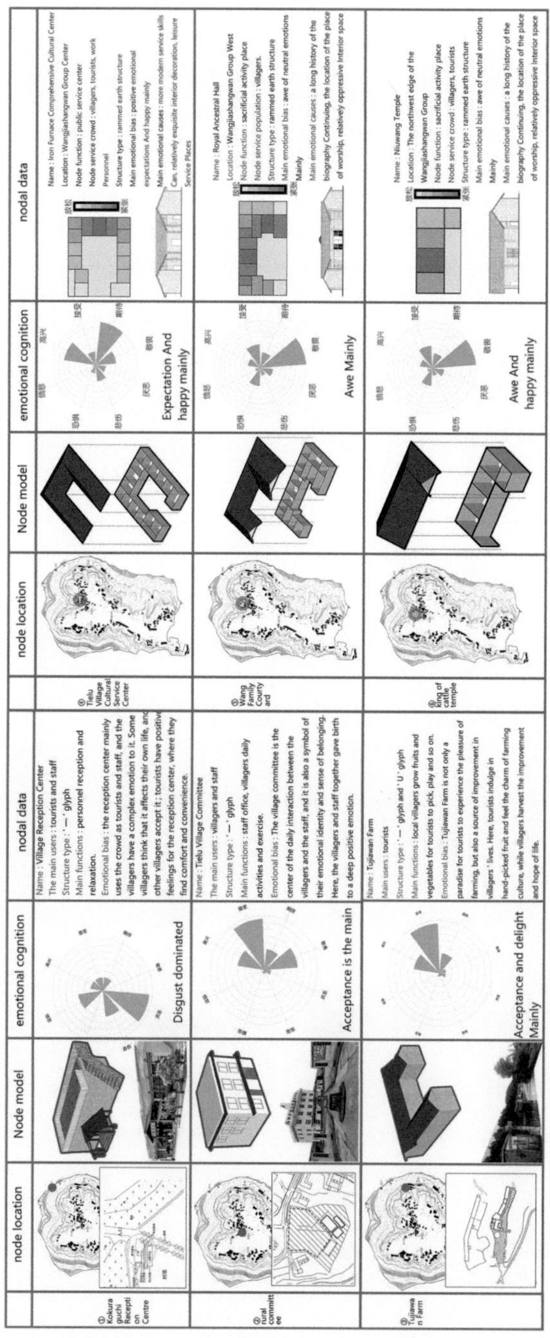

Fig. 5 Emotional cognition of different groups of people on important nodes of villages (self-drawn)

to improve the convenience of residents' transportation and improve the quality of core space. At the same time, we suggest to rationally arrange public service facilities on the axis, and strictly control the style of buildings and streets, so as to ensure the harmonious coexistence of new and old building texture and maintain the consistency of the traditional style of the village (Fig. 6).

In addition, the spatial comprehensibility value of Tielu Village is 0.33 (the moderate value is 0.5, less than 0.5 indicates weak adaptability), indicating that its spatial recognition is relatively weak. This situation is closely related to the scattered layout of the nine residential groups in the village, resulting in a lack of clarity in the spatial structure and difficulty in perceiving the overall space through local space. Therefore, we propose strategies to enhance spatial comprehensibility, including optimizing the identification system and strengthening spatial orientation elements to improve spatial cognition and navigation capabilities (Fig. 6).

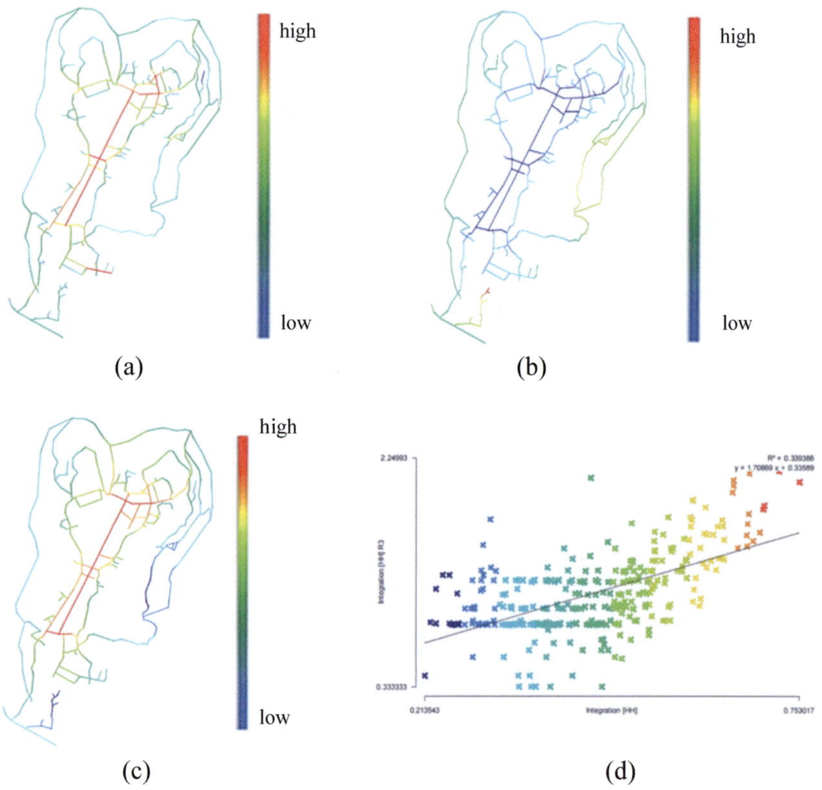

Fig. 6 **a** Global integration map of Tielu village, **b** local integration degree of Tielu village, **c** Tielu village depth value map, **d** comprehensibility of Tielu village (self-drawn)

4 Research Conclusions and Analysis

4.1 *Evaluation of the Status Quo of Tielu Village*

4.1.1 Challenges Facing Villages

The hollowing-out phenomenon not only disrupts demographic structures but also weakens the emotional bonds between villagers and their ancestral spaces. For example, the outflow of young laborers diminishes intergenerational transmission of cultural memories, leading to fragmented emotional identities. Field interviews revealed that elderly villagers expressed nostalgia for communal activities (e.g., festivals at ancestral halls), which are now under threat due to population decline. This underscores the need to integrate emotional geography frameworks into strategies addressing demographic challenges.

(1) Population hollowing and aging: The outflow rate of young and middle-aged labor force is 62% (2023 village committee data), and the proportion of left-behind population over 60 years old is more than 40%, which leads to the inheritance fault of traditional farming skills and folk activities (such as rammed earth construction and Niuwang temple sacrifice). Wang, a 68 year-old villager, confessed in an interview: 'Young people are working in the city, and the ancestral rammed earth crafts have not been learned, and even the temple fair is almost impossible.'

Lag of infrastructure and service facilities: Traffic system: the lack of a unified parking lot in the village, the phenomenon of vehicle occupation is serious in the peak season of tourism; the slow walkway is not a system, and some sections (such as the ancient stone step area) have potential safety hazards due to years of disrepair. Sanitary facilities: the layout of public toilets is uneven, and it takes more than 15 min for peripheral groups to walk to the nearest toilet. Villager Li complained: 'There are many tourists, but the toilets are not enough, and their yards are often mistaken for public toilets.'

Insufficient coordination between architectural style and landscape: The new homestay and reception center adopts modern glass curtain wall, which conflicts with the traditional rammed earth dwelling style and destroys the overall visual continuity of the village.

Infrastructure improvement: planning and construction of reasonable parking lots to meet the parking needs of tourists and villagers. Improve the slow traffic system, add rest facilities, and enhance the travel experience and comfort of tourists and villagers; rational allocation of public toilets in residential areas to improve the level of public health services.

(2) Population loss and the weakening of cultural and emotional ties: based on the perspective of emotional geography, population loss not only leads to the decline of material space, but also leads to deep-seated cultural and emotional fracture.

The dissolution of emotional attachment: the physical isolation between the migrant workers and the village weakens their emotional belonging. According to the questionnaire survey, the emotional recognition of cultural symbols such as 'ancestral temple sacrifice' and 'ancient tree legend' by long-term out-of-town villagers is only 23%, which is significantly lower than that of left-behind villagers (78%). Villager Zhang Mou (35 years old, migrant workers in Chengdu all the year round) said: 'A few days after returning from the Spring Festival every year, I feel that the village is becoming more and more unfamiliar, the old house is demolished, and the pond that I played with as a child is filled.'

The fault of cultural inheritance: intangible cultural heritage skills (such as traditional rammed earth crafts) face the risk of loss due to the aging of inheritors. Chen Mou (72 years old), the only craftsman in the village who mastered the whole set of skills, lamented: 'My son feels tired and does not make money, he would rather send express delivery than learn'. The analysis of the emotional map shows that the emotional markers of the younger generation on the "ancient art experience area" are concentrated on "curiosity" (45%) rather than "identity" (12%), indicating that the intergenerational transmission of cultural memory is blocked.

(3) Planning response framework for challenges and emotions: In response to the above problems, a two-dimensional strategy of 'material repair-emotional remodeling' is proposed.

Material space restoration: infrastructure upgrading: ecological parking lot is added along the main axis, ancient stone steps are repaired and anti-skid handrails are installed to optimize the coherence of slow-moving system. Style control: Formulate 'Guidelines for the Repair of Traditional Buildings', requiring new buildings to adopt traditional elements such as rammed earth facades and sloping roofs.

Emotional bond remodeling: cultural memory activation: set up 'rammed earth skill inheritance base', hire old craftsmen to set up workshops to attract young people to participate; restore the annual temple fair of Niuwang Temple and build it as the core of community emotional cohesion. Strengthen intergenerational interaction: design 'grandparents and grandchildren to build a vegetable garden' project, encourage migrant families to return home during holidays to participate in farming experience, and rebuild the emotional connection between family and land.

4.2 Village Renewal from the Perspective of Emotional Geography

In the study of constructing the emotional spatial structure map of different groups in Tielu Village, we systematically analyzed the emotional response of villagers and tourists to the village space.

First of all, the researchers drew the basic village perception space map of Tielu Village as the basis of the emotional space structure map. Subsequently, we invite the research objects (including villagers and tourists) to mark their emotional responses to different scenic spots (such as temples, ancestral halls, ancient trees, etc.) and scenic spots (such as reception centers, characteristic homestays, squares, etc.) on the map.

The results show that the positive emotions of villagers are generally concentrated in places with profound religious and historical significance, such as temples, ancestral halls and ancient dwellings [3]. These places are not only the spiritual sustenance of villagers, but also the center of community cultural activities, carrying rich historical and cultural values. On the contrary, the negative emotions of villagers mainly point to some construction projects with large investment but less than expected results, especially reception centers and characteristic homestays. These spaces fail to fully meet the villagers' expectations and needs for public space, leading to emotional dissatisfaction.

For tourists, their positive emotions are mainly focused on landscapes rich in historical and cultural elements such as ancient trees and ancient buildings. These landscapes not only attract the attention of tourists, but also deepen their understanding and appreciation of the traditional cultural heritage of Tielu Village. However, the negative emotions of tourists are mainly concentrated on the nodes with poor experience, such as reception centers, characteristic homestays and squares. These nodes may not impress tourists because of imperfect service facilities, insufficient cultural experience or lack of interactivity.

Based on the above findings, we propose the following suggestions: in the future village renewal plan, we should focus on improving places that can stimulate positive emotions, such as increasing cultural activities, improving facilities and service quality to enhance the attractiveness of reception centers and characteristic homestays. At the same time, the nodes that cause negative emotions are optimized and improved, such as increasing interactive devices and expanding rest areas, so as to improve the quality of tourists' experience. In addition, historical sites and ancient buildings should also be protected and restored, which not only helps to enhance the sense of belonging of villagers, but also enhances the satisfaction of tourists, thus jointly promoting the sustainable development of Tielu Village (Fig. 7).

4.2.1 Ideal Spatial Structure Diagram of Different Populations

Adopting an inclusive development approach, Tielu Village's masterplan operationalizes participatory design principles to co-create multifunctional spatial configurations. These calibrated interventions simultaneously conserve indigenous settlement patterns while embedding smart village technologies, thus achieving dialectical synthesis of heritage conservation and sustainable community development through place-based innovation.

The primary measure is to strengthen infrastructure construction, especially the addition of efficient sanitation systems to ensure the health and well-being of villagers

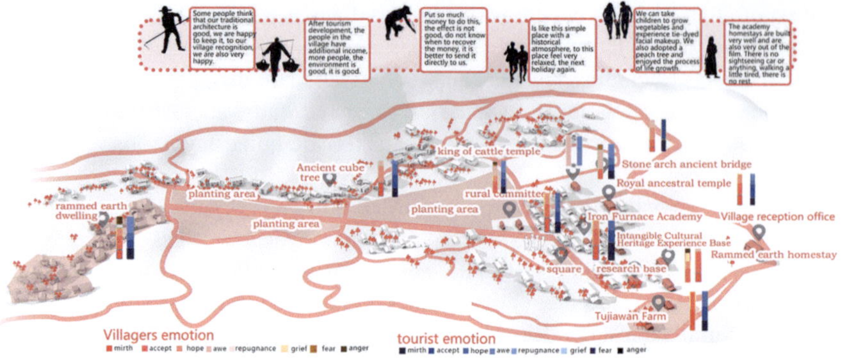

Fig. 7 Emotional spatial structure of different groups (self-drawn)

and tourists, in line with public health and safety standards. Secondly, in order to drive local economic vitality and increase residents' income, we plan to introduce diversified commercial facilities and employment opportunities to promote economic self-sufficiency and sustainable development.

Under the framework of urban and rural planning strategy, the spatial reconstruction of Tielu Village focuses on the strengthening of traffic connectivity. It is proposed to set up bus stops to build an efficient public transport network, enhance the accessibility and openness of the village, and serve the villagers 'commuting and tourists' travel. At the same time, the protection and activation of cultural heritage has become the core of strategic planning, integrating historical relics and cultural tourism resources, and promoting cultural heritage and local characteristics. In view of the village nodes, optimize the functions of the visitor service center and upgrade the facilities, add characteristic homestays, research bases and intangible cultural heritage experience centers, and create a cultural tourism complex integrating living, education and cultural experience, inject new vitality into Tielu Village and promote its transformation to rural tourism destination.

In order to further improve the tourism service system, we conceived the construction of recreational parks and characteristic markets, introduced the internal traffic circulation system, and added leisure and entertainment facilities such as teahouses and farms, aiming to create a diversified and high-quality tourism and leisure environment, enhance tourist satisfaction and loyalty, and lay a solid foundation for the long-term development of the village.

Through this series of spatial structure optimization and function improvement measures, Tielu Village will be reshaped into a dynamic rural model integrating residential suitability, tourism attraction and industrial prosperity, which not only meets the needs of contemporary social development, but also realizes the respect and inheritance of traditional culture (Fig. 8).

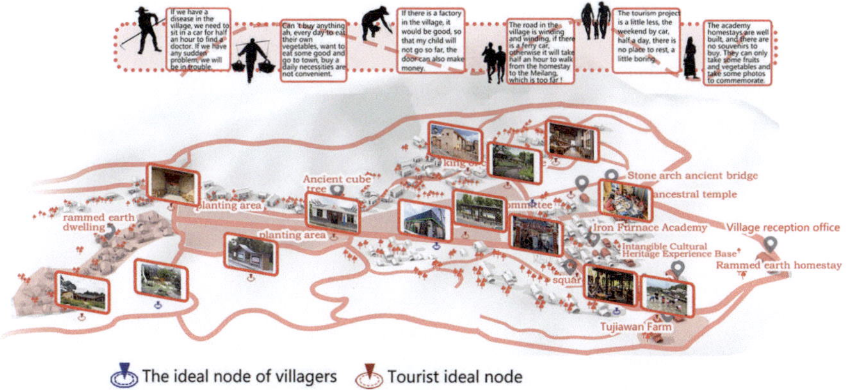

Fig. 8 Ideal spatial structure of different populations

4.2.2 The Layout Plan of Village Planning from the Perspective of Emotional Geography

Under the guidance of the theory of emotional geography, we conducted field research in Tielu Village and conducted in-depth interviews with villagers and tourists to understand their emotional identity, emotional attachment and sense of place belonging to Tielu Village.

In order to transform these emotional identity and spatial conception into actual village renewal strategies, we further combined spatial syntax, established the axis model of the road inside the village, and analyzed the accessibility of the road in depth. This analysis not only helps us to identify the bottlenecks and potential areas in the village space, but also provides a scientific basis for subsequent update and optimization.

Based on the above research, we divide Tielu Village into five functional areas: tourism experience area, historical and cultural protection area, public service area, agricultural experience area and residential living area. In the specific implementation, according to the construction of ideal space for different groups of people, we have added facilities such as hospitals, fishing ponds and agricultural science popularization experience parks to meet the diverse needs of villagers and tourists. At the same time, we have updated or repaired the key nodes such as the entrance square, central square, iron furnace mall, intangible cultural heritage experience base, research base and traditional rammed earth building [16], so as to enhance the overall style and cultural connotation of the village (Fig. 9).

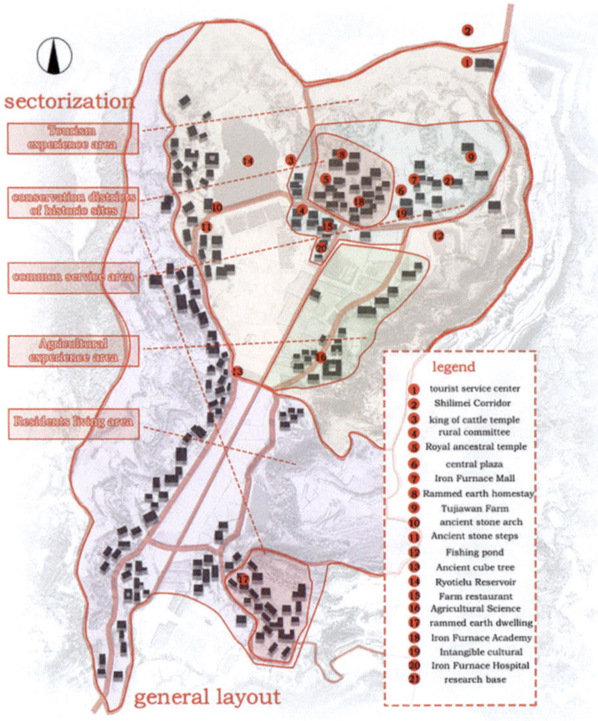

Fig. 9 Village layout planning plan from the perspective of emotional geography

4.3 Village Update Strategy

4.3.1 Protection and Remodel

(1) Firstly, from the perspective of production, life and ecology, the ecological background, industrial foundation and residents' living willingness of Tielu Village are deeply analyzed. On this basis, we remodeled the site, activated the core area of the village by means of points and veins, and gradually restored the mixed state of residential, commercial, cultural and recreational leisure [17].

Building upon an examination of the village's spatial morphology, a targeted industrial framework is proposed to sustain socioeconomic vitality through prioritized sectoral development, while strategically aligning cultural tourism initiatives to foster mutual reinforcement of economic and cultural capital. In addition, we also adopt the strategy of living behavior, emphasizing the priority of living life, the supplement of urban life, and the promotion of leisure culture, so as to form a three-level protection and utilization strategy based on the whole village, public space and single building space.

(2) Grading guidance protection strategy: According to the current function of Tielu Village, we divide it into three regions and formulate corresponding protection measures.

Core Protected Area: This area needs to strictly maintain the traditional architecture and street style to ensure that the historical heritage is properly protected. We plan to demolish the incongruous buildings and repair the damaged traditional buildings to inherit the historical memory.

Excessive control area: In this area, we will focus on the continuation of traditional houses and courtyard features to ensure that the historical and cultural heritage can be inherited. At the same time, we will pay attention to the coordination with the surrounding environment, to avoid excessive development of the traditional style of the village caused damage.

Marginal coordination area: This area needs to pay attention to the protection of construction land and natural environment, and maintain ecological balance and sustainable development. We will strictly control the construction activities to ensure that the expansion of the village will not destroy the surrounding natural environment.

5 Implications for Urban Planning and Design

The research findings of Tielu Village provide valuable insights for urban planning and design, especially in integrating emotional geography into heritage conservation strategies. First, the emotional mapping method can serve as a tool to identify key nodes of cultural significance and community attachment, guiding the prioritization of preservation efforts. Second, spatial syntax analysis offers a quantitative approach to optimize accessibility and spatial coherence, which is critical for balancing tourism development with local livelihoods. For instance, improving the comprehensibility of village layouts (e.g., through signage systems) can enhance both resident navigation and tourist experiences. Additionally, the study highlights the importance of community participation in planning processes, ensuring that emotional attachments and cultural values are embedded in renewal strategies. These approaches can be extended to other traditional villages facing similar challenges, fostering culturally sensitive and sustainable urban–rural integration.

6 Conclusion

In this study, from the perspective of emotional geography, we comprehensively discussed the renewal and protection strategies of Tielu Village, a traditional village, under the background of rural revitalization strategy. We have a deep understanding of the spatial structure characteristics of Tielu Village, the emotional cognition of villagers and tourists, and their expectations for the current situation and future development of the village.

The research reveals the need to be people-oriented, and points out the problems of disunity, imperfect service facilities and insufficient operation and maintenance. In response to these problems, we have proposed a series of renewal and protection strategies, including style protection and promotion, service facilities improvement, cultural inheritance and innovation, emotional space optimization, operation and maintenance strengthening, community participation and supervision, and smart tourism construction.

The conservation and revitalization of Tielu Village must holistically address both the safeguarding of tangible heritage assets and the perpetuation of intangible cultural practices. By optimizing the village style, improving infrastructure and activating cultural experience projects, the cultural characteristics of the village can be strengthened, and the satisfaction and return rate of tourists can be improved. At the same time, by strengthening publicity and marketing, establishing regulatory mechanisms, and encouraging community participation, the self-development capacity and sustainability of villages can be improved.

In addition, we also put forward the strategy of using modern information technology to build a smart tourism platform, in order to enhance the tourist experience and promote the modernization and intelligence of village tourism. By executing these approaches, Tielu Village aims to attain economic growth and communal renewal alongside the preservation and transmission of cultural heritage, thereby offering valuable insights for sustainable growth in comparable rural communities.

References

1. Zhu H, Gao Q (2015) A review of 'Emotional Turn' and emotional geography in western geography. Geogr Res 34(7):1394–1406
2. Chen Y, Liu J (2018) Traditional village conservation in China: policies and practices. Beijing Press
3. Yang X (2022) Research on the change of local emotional space structure of Shuanglang residents under the background of rural gentrification. Unpublished doctoral dissertation, Yunnan Normal University
4. Yang W, Li D, Yan Y et al (2020) Research on memorial space design based on emotional geography. Mod Hortic 43(15):143–145
5. Liu Q (2021) Research on the protection and renewal strategy of Tielu village, a traditional village in Mianyang city. Unpublished master's thesis, Southwest University of Science and Technology
6. Che L, Peng H (2022) Research on the protection and utilization of traditional villages based on space syntax: taking Tielu village (in Weicheng town, Mianyang city) as an example. J Southwest Univ Sci Technol 37(01):59–66 + 100
7. Zhai H (2024) Construction and demonstration of traditional village protection and development evaluation system: taking traditional villages in the Yellow River Basin of Henan Province as an example. Reg Res Dev 43(3):117–122

8. Practice and adjustment of traditional village protection system from the perspective of "System-Life": taking Suzhou as an example. Urban Plan 48(7):90–97, 122 (2024)
9. Du F (2024) The spatial structure of local emotions in ethnic villages: a case study of Shuanglang, Yunnan. J Dali Univ 9(03):8–16
10. Chen C, Li B, Yuan J et al (2018) Spatial morphology cognition of traditional villages based on spatial syntax: a case study of Qinchuan village of Hangzhou. Econ Geogr 38(10):234–240
11. Qi R (2024) Discussion on the public participation model in the protection of traditional villages. J Sichuan Norm Univ (Soc Sci Ed) 51(1):96–103, M0008, M0009
12. Cheng Y, Hu S, Yang R et al (2024) Protection and utilization of traditional Chinese villages for rural revitalization: challenges and prospects. J Nat Resour 39(8):1735–1759
13. Xu H (2015) The study of traditional villages Spaceform-based on space syntax in Nanjing. Unpublished master's thesis, Nanjing Tech University
14. Zhou K, Wang J (2024) From landscape production to local construction: evidence-based theory of traditional village renewal. Urban Plan 48(7):83–89
15. Lai Q, Ma X, Xie X et al (2011) A study of Xuzhou's urban spatial morphology based on space syntax. Planner 27(06):96–100
16. Wan L, Wu E, Chi X et al (2024) New rammed earth construction strategy based on rural endogenous development model: taking the project of 'One Specialty and One Village' in southwest rural area as an example. Arch J 1:105–111
17. 'Research on the authenticity of traditional villages and towns with incomplete historical features': analysis of the protection strategies of traditional villages with incomplete historical features. Arch J (7):I0003–I0004

Tourism Value Assessment of Linear Cultural Heritage of Ancient Shu Road in Cuiyunlang, Sichuan, China

Zhao Tian and Bin Cheng

Abstract The concept of linear cultural heritage is gaining prominence as a novel form of heritage. The intertwining of culture and tourism is placing greater emphasis on heritage tourism. An assessment system, utilizing the analytic hierarchy process and Delphi method, was developed based on resource background (R), environmental conditions (E), and social economy (S). This study delves into the tourism value by scrutinizing resources about the surrounding region across five districts and counties along the ancient Shu Road in Cuiyunlang. Standard values for each evaluation variable were established. The findings suggest that resources are the most pivotal factors, while environmental conditions and social economy are supplementary components. A three-dimensional spatial model was employed to appraise R–E–S coordination, classifying cities into three types: high-value imbalance, comprehensive breakthrough, and low-value limited. The study presents theoretical recommendations for tourism and sustainable development along the Cuiyunlang Ancient Shu Road.

Keywords Cuiyunlang Ancient Shu Road · Linear cultural heritage · Tourism value · Value evaluation · 3D model

1 Introduction

Cultural heritage is receiving increasing attention from society. The approach to cultural heritage protection is evolving. The focus is shifting from specific locations to broader ecosystem continuity. The concept is expanding from "heritage point" to include "heritage line" and "heritage surface" [1, 2]. Linear cultural heritage primarily appears as a new form of heritage as cultural routes and heritage corridors in spatial contexts. It highlights the connections in content, concentrates on linear

Z. Tian · B. Cheng (✉)
School of Civil Engineering and Architecture, Southwest University of Science and Technology, Mianyang, Sichuan, China
e-mail: chengbin1971@126.com

B. He et al. (eds.), *Climate-Resilient, Low-Carbon, and Sustainable Cities*, Lecture Notes in Civil Engineering 754, https://doi.org/10.1007/978-981-95-2169-2_49

spaces, and encompasses a variety of heritage components, showcasing both regional traits and a history of intermingling and accumulation [3].

Under the current background, the concept of heritage activation and utilization and the integration and development of culture and tourism is deeply rooted in the people's hearts. The deep combination of linear cultural heritage and tourism is conducive to protecting the regional culture along the heritage line [4], improving the environment along the heritage line [5], increasing the popularity of the heritage [6], enhancing the sense of regional cultural identity [7], promoting regional economic development and improving personal living standards [8–12]. The theoretical framework for studying the use of linear cultural heritage tourism is still underdeveloped, with limited research conducted. In particular, the exploration of resource value lacks sufficient depth and breadth. A quantitative assessment of the value of linear cultural heritage tourism is essential for effectively advancing this type of tourism [13]. Assessing the tourism value of linear cultural heritage requires various indicators. Commonly used indicators include resource conditions, regional accessibility, social and economic status, natural environment, public amenities, the tourist market, and additional factors [14–28]. Consequently, this paper examines the assessment of tourism value, which is essential for achieving the sustainable development of extensive linear cultural heritage.

This study focuses on the Cuiyunlang Ancient Shu Road as its primary subject, to enhance tourism strategies associated with this historic route. The specific aims of the research are as follows: (1) to develop a novel comprehensive evaluation system for assessing the cultural heritage tourism value; (2) to evaluate and analyze the spatial variations in tourism value along the Cuiyunlang Ancient Shu Road utilizing the established evaluation framework; and (3) to employ a three-dimensional spatial model to assess the developmental coordination along the ancient Shu Road in Cuiyunlang. Through these objectives, this research not only contributes to the theoretical discourse surrounding the reuse and tourism potential of the Cuiyunlang Ancient Shu Road but also serves as a valuable resource for policy development and cross-regional collaboration in the context of tourism along this route.

2 Materials and Methods

2.1 Study Area

The ancient Shu Road is a channel for political, military, economic, cultural, and information exchange that runs through the north-south direction of ancient western China. Its history can be traced back to the Warring States Period, and it is the longest-used road system in China [29]. The Shu Road is characterized by numerous traffic sites and cultural landscapes. In 2015, it was included in the World Heritage Preparatory List by the UNESCO World Heritage Center, and it currently remains on this list [27]. The Cuiyunlang Ancient Shu Road is part of the Jinniu section

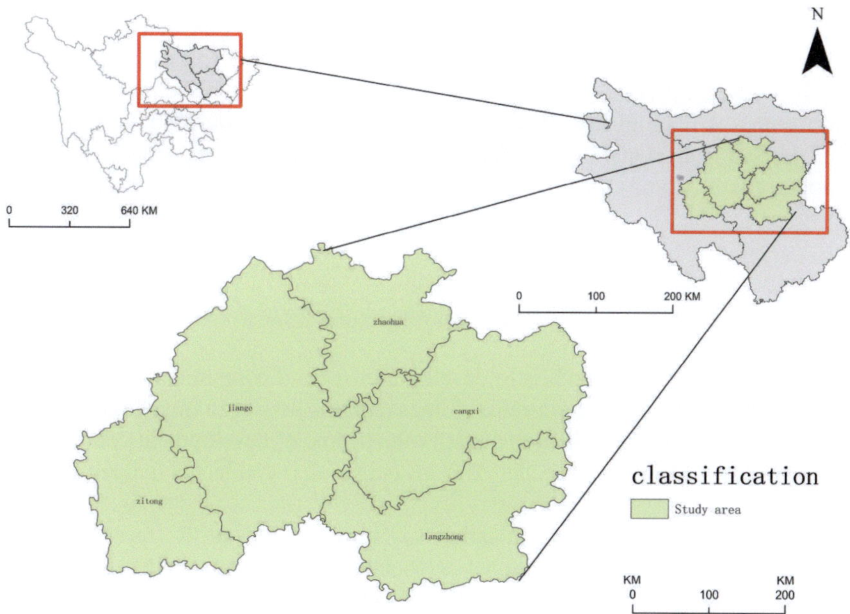

Fig. 1 Overview of the research area (self-drawn)

of the ancient Shu Road, which constitutes a fundamental element of the Jianmen Shu Road site. This section is recognized for its rich resources and is considered the quintessential segment of the Jianmen Shu Road, often referred to as the essence of Shu Road and a marvel of the world. This study adopts a comprehensive approach to the Cuiyunlang Ancient Shu Road, categorizing it into three segments: the western section, the northern end, and the southern section, with Jiange County serving as the focal point. The road extends westward to Zitong, northward to Zhaohua, and southward to Langzhong, traversing five county-level administrative regions, as illustrated in Fig. 1.

2.2 Data Source

The research data primarily encompasses the following components: based on the national basic geographic information data from 2022, the administrative divisions of districts and counties along the ancient Shu Road in Cuiyunlang were delineated. Additionally, a digital elevation model (DEM) with a spatial resolution of 30 m was acquired from the geospatial data cloud platform of the Chinese Academy of Sciences (https://www.gscloud.cn/, accessed on July 3, 2024). A comprehensive inventory of 391 heritage resources located in various districts and counties along the ancient Shu Road in Cuiyunlang was sourced from the respective cultural tourism bureaus of these districts and counties. Information regarding the magnitude and classification of scenic spots was obtained from the Sichuan Provincial Department of Culture and Tourism. Data about nature reserves was provided by the Sichuan Provincial Department of Ecology and Environment, while economic development statistics were derived from statistical yearbooks and government work reports from various districts, counties, and cities.

2.3 Research Methods

The objective of this research is to develop a novel and comprehensive evaluation framework to assess the tourism value of linear cultural heritage along the ancient Shu Road in Cuiyunlang, as well as to quantify the degree of development coordination through a three-dimensional model. This investigation is conducted in three primary phases.

Initially, a comprehensive evaluation system based on the R (resource background)–E (environmental conditions)–S (social economy) framework is established utilizing the analytic hierarchy process (AHP) and the Delphi method, which also facilitates the calculation of the weights assigned to each index. Subsequently, this R–E–S comprehensive evaluation system is employed to assess the value of heritage resources, categorizing them into four distinct grades. Finally, the coordination degree among the R, E, and S factors is quantitatively analyzed using a three-dimensional spatial coordination model, thereby determining the level of coordinated development across various cities.

Evaluation Index System

The index evaluation method involves employing various indices to assess the characteristics of one or more entities. This research utilizes the analytic hierarchy process to develop a novel comprehensive evaluation index aimed at assessing the tourism value of LCH. The criteria for selecting evaluation system indicators must adhere to the principles of hierarchy, guidance, and simplicity. This study employs a methodology that includes a literature review, public surveys, and expert revisions, tailoring the evaluation indicators to the specific context of the study area. Initially, an extensive review of pertinent literature is conducted [14–28], A total of 156 three-level indicators were initially identified, subsequently integrated, and processed, resulting in the selection of 64 three-level indicators. Following this, a substantial number of public questionnaires were disseminated, and through the voting rates of each indicator, 30 third-level indicators were determined. Expert consultations led to further modifications, ultimately yielding a final set of 26 third-level indicators, 8s-level indicators, and 3 first-level indicators. This process culminated in the establishment of a comprehensive evaluation system known as R–E–S, as detailed in Table 1.

This study examines index weight utilizing MATLAB software. The methodology encompasses the following steps: (1) the development of a judgment matrix based on the value evaluation index system, employing the 1–9 scale method as outlined

Table 1 R–E–S system (self-drawn)

Target layer	Feature layer	Indicator layer	Indicator interpretation
R Resource background	R1 Tourism resources	R11 Highest scenic spot level	Grade of the highest-grade tourist attractions in districts and counties Grade of the highest-grade tourist attractions in districts and counties
		R12 Number of scenic spots	Number of scenic spots with grade 3A or above in districts and counties Number of scenic spots with grade 3A or above in districts and counties
	R2 Heritage level	R21 Number of national cultural relics R21 Number of national cultural relics	Number of national protected cultural relics in districts and counties Number of national protected cultural relics in districts and counties
		R22 Number of provincial cultural relics R22 Number of provincial cultural relics	Number of provincial protected cultural relics in districts and counties Number of provincial protected cultural relics in districts and counties
		R23 Number of municipal cultural relics R23 Number of municipal cultural relics	Number of municipal protected cultural relics in districts and counties Number of municipal protected cultural relics in districts and counties
		R24 County and number of cultural relics R24 County and number of cultural relics	Number of protected cultural relics at the county level in districts and counties Number of protected cultural relics at the county level in districts and counties
	R3 Heritage ontology	R31 Legacy features	Does the heritage now have other derivative functions, native functions, etc. Does the heritage now have other derivative functions, native functions, etc.
		R32 Protection status	Cuiyunlang Ancient Shu Road is protected in various districts and counties Cuiyunlang Ancient Shu Road is protected in various districts and counties

(continued)

Table 1 (continued)

Target layer	Feature layer	Indicator layer	Indicator interpretation
		R33 Accessibility	The convenience to each district and county, whether there are main means of transportation, high-speed rail, railway station, bus station, etc. The convenience to each district and county, whether there are main means of transportation, high-speed rail, railway station, bus station, etc.
		R34 Degree of development	The degree to which Cuiyunlang Ancient Shu Road has been developed in various districts and counties The degree to which Cuiyunlang Ancient Shu Road has been developed in various districts and counties
E Environmental conditions	E1 Natural environment	E11 Air quality index	Ambient air quality composite index Ambient air quality composite index
		E12 Days with excellent air quality	The number of days in a year that air quality reaches excellent level The number of days in a year that air quality reaches excellent level
	E2 Artificial environment	E21 Urban greening coverage rate	Urban greening coverage rate
		E22 Nature reserve area	The area occupied by nature reserves in districts and counties The area occupied by nature reserves in districts and counties
		E23 Proportion of environmental	The proportion of urban investment in environmental protection work to the total investment in urban development The proportion of urban investment in environmental protection work to the total investment in urban development
	E3 Social environment	E31 Proportion of tertiary industry	The proportion of urban tertiary industry in urban gross output value The proportion of urban tertiary industry in urban gross output value
		E32 Policy support	Number of relevant tourism plans or policies issued by district and county governments Number of relevant tourism plans or policies issued by district and county governments

(continued)

Table 1 (continued)

Target layer	Feature layer	Indicator layer	Indicator interpretation
		E33 Tourism revenue	Comprehensive tourism income generated by receiving domestic and foreign tourists in one year Comprehensive tourism income generated by receiving domestic and foreign tourists in one year
		E34 Proportion of cultural tourism investment E34 Proportion of cultural tourism investment	The proportion of urban investment in cultural tourism media in the total investment in urban development The proportion of urban investment in cultural tourism media in the total investment in urban development
S Socio-economic	S1 Economic level	S11 Regional GDP (100 million yuan) S11 Regional GDP (100 million yuan)	Gross regional product of districts and counties in one year Gross regional product of districts and counties in one year
		S12 Per capita disposable income	Per capita disposable income of residents in districts and counties within one year Per capita disposable income of residents in districts and counties within one year
		S13 Resident Engel coefficient	The proportion of food consumption expenditure to personal consumption expenditure reflects the living standard of residents in districts and counties The proportion of food consumption expenditure to personal consumption expenditure reflects the living standard of residents in districts and counties
		S14 Unemployment rate	Urban unemployment rate in one year Urban unemployment rate in one year
	S2 Town development	S21 Urbanization rate	Measurement index of urbanization, reflecting the process of urbanization Measurement index of urbanization, reflecting the process of urbanization

(continued)

Table 1 (continued)

Target layer	Feature layer	Indicator layer	Indicator interpretation
		S22 Permanent population size	Number of permanent residents in districts and counties Number of permanent residents in districts and counties
		S23 Highway density	There are dense roads in districts and counties. How many kilometers of road mileage per square kilometer There are dense roads in districts and counties. How many kilometers of road mileage per square kilometer

in Table 2; (2) the engagement of 16 experts to provide scores, all of whom possess specific qualifications, including current employment in universities or research institutions and a minimum of three years of experience in fields such as heritage protection and tourism management, to facilitate pairwise comparisons of hierarchical indicators; (3) the execution of a consistency test for indicators across all levels; and (4) the computation of the index weight.

To ensure the effectiveness of the index weights in the index system, the consistency test of the constructed judgment matrix is carried out.

Consistency indicator CI:

$$CI = \frac{\lambda_{max} - n}{n - 1} \tag{1}$$

(λ_{max} is the largest characteristic quantity, n is the order of the judgment matrix).
Mean RI of random consistency indicators (Table 3).

Calculate the agreement ratio CR:

Table 2 1–9 scale method (self-drawn)

Scale	Meaning
1	The former is as important as the latter
3	The former is slightly more important than the latter
5	The former is obviously more important than the latter
7	The former is much more important than the latter
9	The former is extremely important than the latter
2, 4, 6, 8	Is the corresponding scale between the above two adjacent levels
1/2…1/9	The ratio of importance of the two indexes is the reciprocal of the above grade, and the former is less important than the latter on the surface

Table 3 Self-drawn

n	1	2	3	4	5	6	7	8	9
RI	0.00	0.00	0.58	0.90	1.12	1.24	1.32	1.41	1.45

$$CR = \frac{CI}{RI} \tag{2}$$

When the consistency ratio (CR) is less than 0.1, the judgment matrix is deemed to have passed the consistency test, thereby allowing for the acceptance of the index weights. By the fundamental principle that the minority adheres to the majority, the scoring outcomes provided by experts are processed, followed by the execution of the consistency test and weight calculation utilizing MATLAB software. The findings are presented in Tables 4 and 5. The results indicate that the resource background is the most significant factor influencing tourism value, with the weight values for the three indices in the factor layer being as follows: R (resource background) at 0.6080, E (environmental condition) at 0.1199, and S (social economy) at 0.2721.

Evaluation Criteria

Due to the distinct characteristics of the ancient Shu Road in Cuiyunlang, it is essential to establish standard values for each evaluation index. Utilizing the "natural fracture classification" feature within ArcGIS (ArcGIS 10.8) software, this study categorizes each index variable based on the distribution patterns of the data. Each variable is classified into four grade values: low, moderate, high, and very high. However, certain index variables are not amenable to this classification method due to inherent data limitations, necessitating the use of a direct assignment approach. The final

Table 4 CR inspection result (self-drawn)

Judgment matrix	Maximum eigenvalue	CI	CR	Whether it passed the inspection
R–E–S	3.0735	0.0368	0.0707	Yes
R1–R2–R3	3.0182	0.0091	0.0176	Yes
E1–E2–E3	3.0246	0.0123	0.0236	Yes
S1–S2	2	0	0	Yes
R11–R12	2	0	0	Yes
R21–R22–R23–R24	4.117	0.0390	0.0438	Yes
R31–R32–R33–R34	4.0728	0.0243	0.0273	Yes
E11–E12	2	0	0	Yes
E21–E22–E23	3.0092	0.0046	0.0088	Yes
E31–E32–E33–E34	4.031	0.0103	0.0116	Yes
S11–S12–S13–S14	4.0211	0.0070	0.0079	Yes
S21–S22–S23	3.0037	0.0018	0.0036	Yes

Table 5 Index weight (self-drawn)

Target layer	Weight	Feature layer	Weight	Total weight	Indicator layer	Weight	Total weight
R Resource background	0.6080	R1 Tourism resources	0.1373	0.083478	R11 Highest scenic spot level	0.75	0.062601
					R12 Number of scenic spots	0.25	0.020870
		R2 Heritage level	0.2395	0.145616	R21 Number of national cultural relics	0.5579	0.081239
					R22 Number of provincial cultural relics	0.2633	0.038341
					R23 Number of municipal cultural relics	0.1219	0.017751
					R24 County and number of cultural relics	0.0569	0.008286
		R3 Heritage ontology	0.6232	0.378906	R31 Legacy features	0.1572	0.059564
					R32 Protection status	0.2759	0.104540
					R33 Accessibility	0.0729	0.027622
					R34 Degree of development	0.4941	0.187217
E Environmental conditions	0.1199	E1 Natural environment	0.2014	0.024148	E11 Air quality index	0.5	0.012074
					E12 Days with excellent air quality	0.5	0.012074
		E2 Artificial environment	0.1179	0.014136	E21 Urban greening coverage rate	0.5390	0.007619
					E22 Nature reserve area	0.2973	0.004203
					E23 Proportion of environmental protection investment	0.1638	0.002315
		E3 Social environment	0.6806	0.081604	E31 Proportion of tertiary industry	0.0960	0.007837

(continued)

Table 5 (continued)

Target layer	Weight	Feature layer	Weight	Total weight	Indicator layer	Weight	Total weight
					E32 Policy support	0.1611	0.013152
					E33 Tourism revenue	0.4658	0.038028
					E34 Proportion of cultural tourism investment	0.2771	0.022622
S Socio-economic	0.2721	S1 Economic level	0.25	0.068025	S11 Regional GDP (100 million yuan)	0.1434	0.009755
					S12 Per capita disposable income	0.5056	0.034393
					S13 Resident Engel coefficient	0.2642	0.017972
					S14 Unemployment rate	0.0868	0.005905
		S2 Town development	0.75	0.204075	S21 Urbanization Rate	0.6479	0.132220
					S22 Permanent population size (10,000 people)	0.1222	0.024938
					S23 Highway density	0.2299	0.046917

evaluation criteria for the tourism value of the Cuiyunlang Ancient Shu Road are presented in Table 6.

Three-Dimensional Space Model

In the context of spatial geography, any point D within the three-dimensional spatial framework defined by the X, Y, and Z systems signifies a developmental state of that point within the regional system (refer to Fig. 2). The three subsystems originate from a common point O and are positioned along the diagonal of the three-dimensional spatial structure. These subsystems are considered to be in a synchronous state concerning the regional systems, thereby facilitating coordinated development. Conversely, points that do not lie on the diagonal exhibit a vertical distance from it, which indicates the extent of deviation from this synchronous state. Building upon the findings of previous scholars [30], this study conceptualizes the three element

Table 6 Evaluation criterion (self-drawn)

Indicator layer	Evaluation criteria				Method
	Tall	Higher	Lower	Low	
R11 Highest scenic spot level	5A level	4A level	3A level	Class 3A or below	Direct assignment method
R12 Number of scenic spots	7	6	5	4	Natural fracture method
R21 Number of national cultural relics	8≥ and >4	4≥ and >1	1	0	Natural fracture method
R22 Number of provincial cultural relics	23≥ and >20	20≥ and >14	14≥ and >5	5≥	Natural fracture method
R23 Number of municipal cultural relics	16≥ and >10	10≥ and >8	8≥ and >6	6≥	Natural fracture method
R24 County and number of cultural relics	140≥ and >36	36≥ and >29	29≥ and >11	11≥	Natural fracture method
R31 Legacy features	With derivative functions	Native function	Site function	No function	Direct assignment method
R32 Protection status	Conservation is in good condition	Good protection condition	Poor protection status	Poor protection status	Direct assignment method
R33 Accessibility	Have an airport	With high-speed railway station	With a railway station	With bus terminal	Direct assignment method
R34 Degree of development	High degree of site development	The site has a high degree of development	The degree of site development is low	Poor site development	Direct assignment method
E11 Air quality index	2.9≥ and >2.76	2.75	2.74≥ and >2.30	2.29	Natural fracture method
E12 Number of days with excellent air quality (days)	353≥ and >347	347≥ and > 343	343≥ and 340	340≥	Natural fracture method
E21 Urban greening coverage rate (%)	46.70≥ and >43.89	43.88≥ and >42.61	42.60≥ and >38.41	38.40≥	Natural fracture method

(continued)

Table 6 (continued)

Indicator layer	Evaluation criteria				Method
	Tall	Higher	Lower	Low	
E22 Nature reserve area (ha)	50,572.00≥ and >12,183.60	12,183.59≥ and 7383.01	7383.00≥ and >4866.86	4866.85≥ and >4000.00	Natural fracture method
E23 Proportion of environmental protection investment (%)	2.88≥ and >1.85	1.84≥ and >1.53	1.52≥ and >1.15	1.14≥	Natural fracture method
E31 Proportion of tertiary industry (%)	48.6≥ and >45.81	45.80≥ and >43.11	43.10≥ and >35.51	35.50≥	Natural fracture method
E32 Policy support (number of planned items)	6≥ and >4	4≥ and >3	3≥ and >2	2≥	Direct assignment method
E33 Tourism revenue (100 million yuan)	187.20≥ and 86.57	86.56≥ and 43.88	43.87≥ and 31.29	31.28≥	Natural fracture method
E34 Proportion of cultural tourism investment (%)	2.07≥ and >1.54	1.53≥ and 0.98	0.97≥ and >0.87	0.86≥	Natural fracture method
S11 Regional GDP (100 million yuan)	288.7≥ and >204.05	204.04≥ and >177.06	177.05≥ and >82.66	82.66≥	Natural fracture method
S12 Per capita disposable income (yuan)	31,788≥ and >28,413	28,413≥ and >24,945	24,945≥ and >24,515	24,515≥	Natural fracture method
S13 Resident Engel coefficient (%)	≤34.10	34.11< and ≤34.50	34.51< and ≤34.80	34.81< and ≤35.60	Natural fracture method
S14 Unemployment rate (%)	≤2.60	2.61< and ≤2.76	2.77< and ≤2.95	2.96< and ≤4.00	Natural fracture method
S21 Urbanization rate (%)	39.70≥ and >35.46	35.45≥ and >34.81	34.80≥ and >34.32	34.31≥	Natural fracture method

(continued)

Table 6 (continued)

Indicator layer	Evaluation criteria				Method
	Tall	Higher	Lower	Low	
S22 Permanent population size (10,000 people)	61.60≥ and >50.51	50.50≥ and >27.71	27.70≥ and >13.01	13≥	Natural fracture method
S23 Highway density (km/ km^2)	2.57≥ and >2.09	2.08≥ and >1.67	1.66> and >1.45	1.44≥	Natural fracture method

layers—R, E, and S—as the three subsystems to formulate a three-dimensional spatial model. Additionally, the study integrates the index weights of these element layers to refine the coordination degree formula, thereby capturing the varying degrees of influence that the three element layers exert on tourism value. The resultant calculation formula for the coordination degree, denoted as d_1, is presented. A value of $d_1 = 0$ indicates that the regional system is entirely synchronized and experiences coordinated development; conversely, an increasing value of d_1 signifies greater asynchrony and a lack of coordinated development within the regional system.

$$d_2 = \sqrt{X^2 + Y^2 + Z^2} \tag{3}$$

$$d_1 = \sqrt{\frac{(X - Y)^2 + (X - Z)^2 + (Y - Z)^2}{3}} \tag{4}$$

$$d_1 = \sqrt{\frac{(aR - bE)^2 + (aR - cS)^2 + (bE - cS)^2}{3}} \tag{5}$$

Among, $a = 0.6080$, $b = 0.1199$, $c = 0.2721$.

3 Results

According to the research method, the spatial distribution of tourism value based on the R–E–S system is analyzed.

3.1 Tourism Value of Cuiyunlang Ancient Shu Road

Figure 3 illustrates the spatial distribution of the resource background value (R) along the ancient Shu Road in Cuiyunlang. Notably, the tourism resources (R1) in

Fig. 2 Regional system (self-drawn)

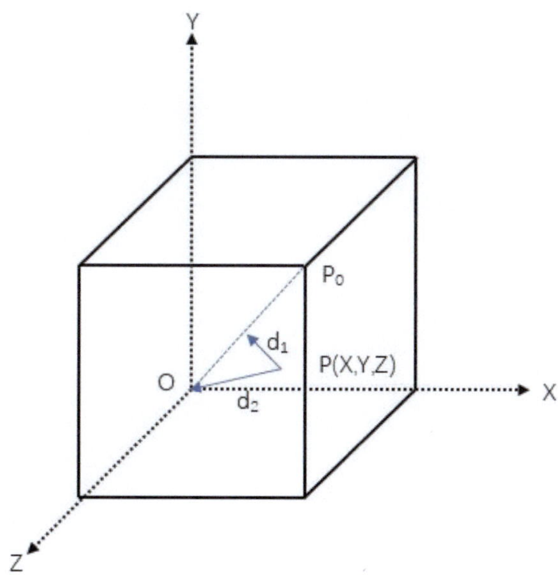

Langzhong City, located to the east of the Cuiyunlang Ancient Shu Road, exhibit significant advantages, particularly at the scenic spot level (R11). In terms of heritage resource classification (R2), both Jiange County and Langzhong City demonstrate a higher heritage resource grade, whereas Zhaohua District is constrained by a limited number of heritage resources, leading to a lower heritage resource grade. The western region of the Cuiyunlang Ancient Shu Road, primarily encompassing Zitong County and Jiange County, displays a high value of heritage ontology (R3), which is predominantly reflected in the values of heritage function (R31), protection status (R32), and development degree (R34). As depicted in Fig. 6, the distribution of resource background (R) values along the Cuiyunlang Ancient Shu Road is generally higher in the western region compared to the eastern region, with Jiange County exhibiting the highest value and Cangxi County showing the least favorable performance. Despite Langzhong City possessing substantial tourism resources and heritage grades, the inadequate protection, development, and correlation associated with the Cuiyunlang Ancient Shu Road contribute to a diminished final value of the heritage background.

Figure 4 illustrates the spatial distribution of environmental condition values (E) along the ancient Shu Road in Cuiyunlang. Based on the comprehensive air quality index (E11) and the number of days with excellent air quality (E12), Zitong County exhibits the highest value for the natural environment (E1), while Jiange County records the lowest. The artificial environment (E2) displays a pattern of elevated values at both ends and diminished values in the central region. In contrast, the social environment (E3) indicates a higher value in the eastern part and a lower value in the western part. As depicted in Fig. 6, the distribution of environmental condition values (E) along the Cuiyunlang Ancient Shu Road is highest in Zhaohua District and lowest in Jiange County.

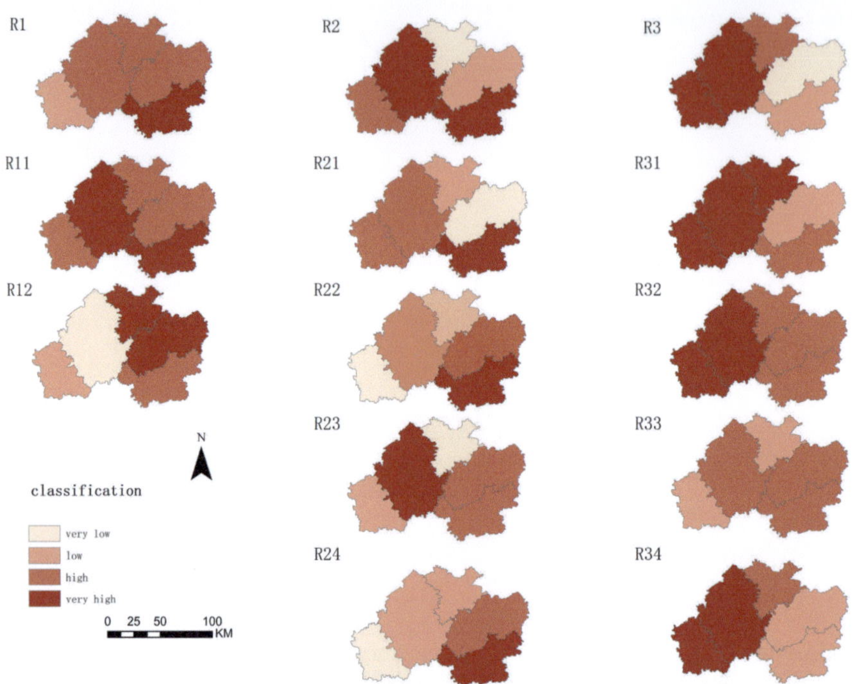

Fig. 3 The spatial distribution of the resource background value (R) (self-drawn)

Figure 5 illustrates the spatial distribution of the socio-economic (S) value associated with the ancient Shu Road in Cuiyunlang. An analysis of the regional gross product (S11), per capita disposable income (S12), Engel's coefficient of residents (S13), and the unemployment rate of residents (S14) reveals a distinctive pattern in the regional economic level (S1). Specifically, the data indicates a high–low–high distribution, with Zitong County and Langzhong City exhibiting the highest economic levels, while Cangxi County demonstrates the lowest. Furthermore, an evaluation of the urbanization rate (S21), permanent population size (S22), and highway density (S23) provides insight into the urban development along the Cuiyunlang Shu Road. The findings indicate that Zitong County and Zhaohua District rank highest in urban development, whereas Jiange County ranks lowest. As depicted in Fig. 6, Zitong County emerges as the leading area in terms of overall social economy (S), while Jiange County is identified as the lowest.

The overall distribution of total value along the ancient Shu Road in Cuiyunlang indicates a higher concentration in the western region compared to the eastern region (see Fig. 6). Zitong County exhibits the highest value, followed by Jiange County, Zhaohua District, and Langzhong City, while Cangxi County records the lowest value.

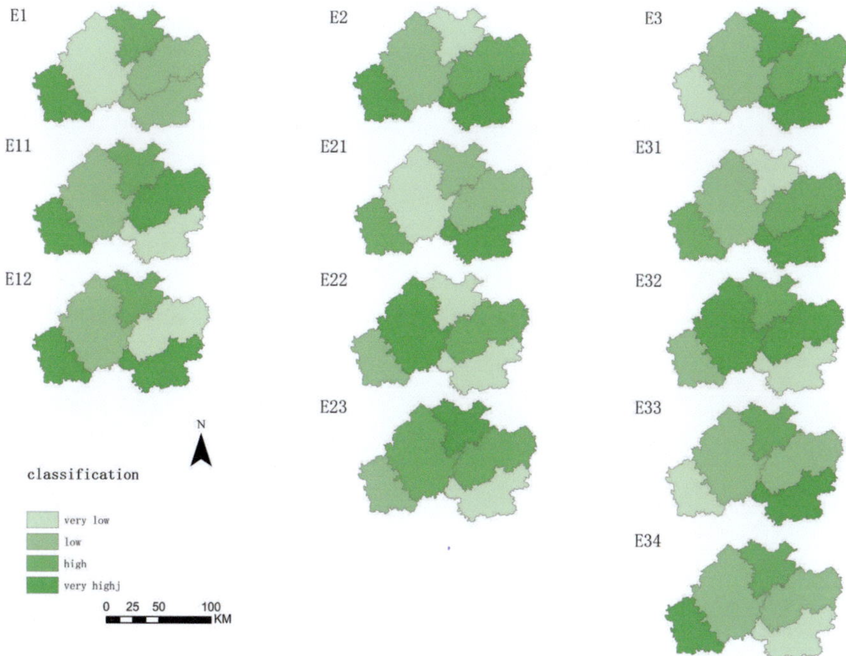

Fig. 4 The spatial distribution of environmental condition values (E) (self-drawn)

3.2 *Tourism Value Coordination of Cuiyunlang Ancient Shu Road*

The calculation of tourism value coordination (d_1) for each study area and county is derived from a specific formula, where a lower value indicates superior coordination. Utilizing the natural breaks method in ArcGIS (ArcGIS 10.8), the coordination levels are categorized into four distinct classifications: uncoordinated, less uncoordinated, more coordinated, and coordinated (refer to Fig. 7). In examining the relationship between tourism value and coordination levels, it is evident that higher coordination does not uniformly benefit cities with varying tourism values. In cities characterized by high tourism value, increased coordination signifies a synchronized development of resources (R), environment (E), and social economy (S), thereby fostering a positive correlation between regional tourism value and coordination, which is advantageous for urban development. Conversely, in cities with lower tourism values, while higher coordination indicates synchronized development of R, E, and S, this synchronization tends to lag, resulting in a lack of distinctive characteristics and presenting challenges to urban development.

As illustrated in Fig. 7, the coordination of tourism value among cities along the ancient Shu Road in Cuiyunlang exhibits an unbalanced pattern. Generally, cities with elevated total tourism values demonstrate lower levels of coordination, whereas

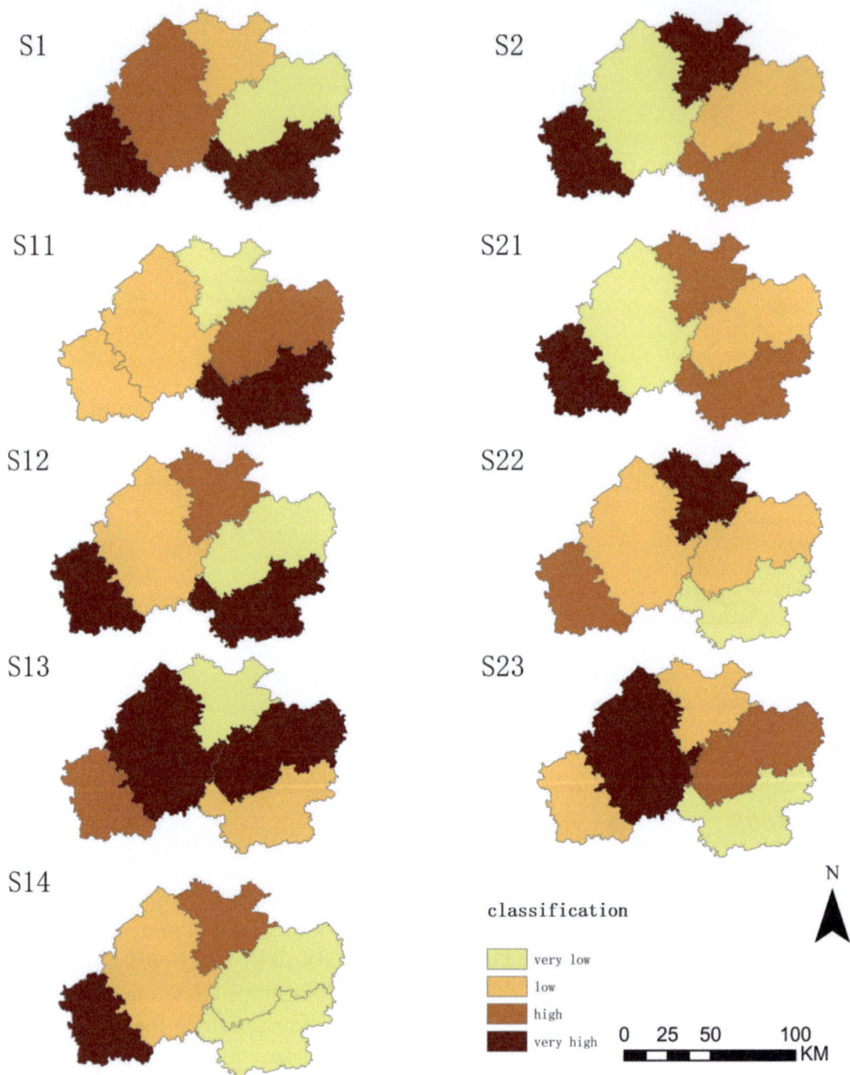

Fig. 5 The spatial distribution of the socio-economic (S) (self-drawn)

those with diminished total tourism values exhibit higher coordination. Based on the distribution of total value and coordination, three categories emerge: "high-value imbalance type," "comprehensive breakthrough type," and "low-value limited type." Jiange County exemplifies the "high-value imbalance" category, characterized by a high total tourism value coupled with the lowest coordination. This is primarily evidenced by Jiange County's substantial heritage background (R), which is undermined by its weak environmental conditions (E) and social economy (S),

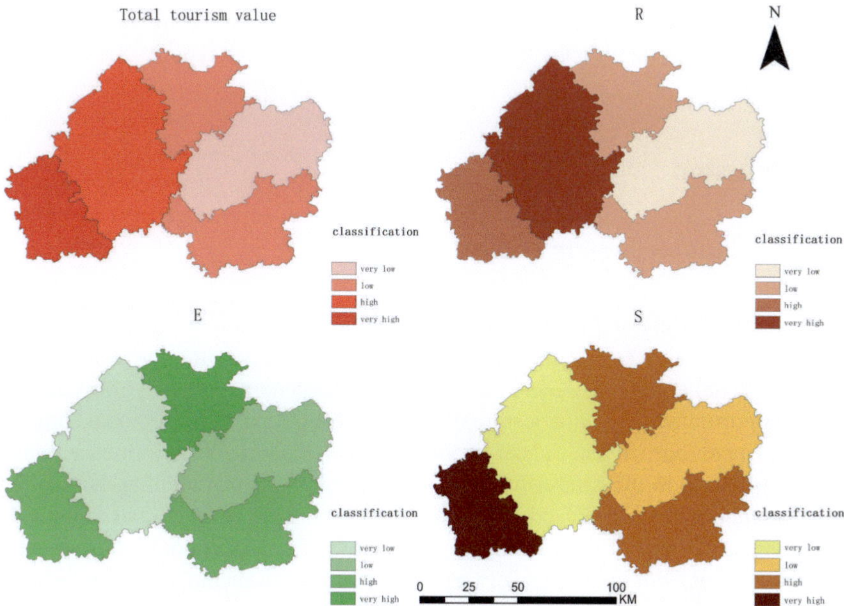

Fig. 6 The spatial distribution of the total value (self-drawn)

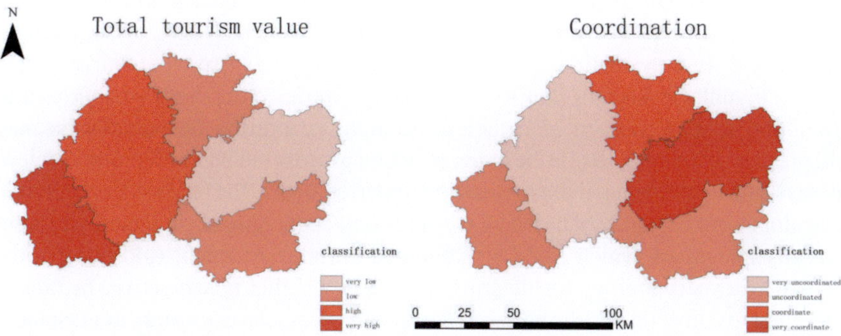

Fig. 7 The spatial distribution of the coordination (self-drawn)

indicating an inadequate utilization of its abundant tourism and heritage resources from the perspective of linear cultural heritage tourism. Similarly, Zitong also reflects a scenario of high total value but poor coordination, categorizing it within the high-value imbalance type. In contrast, Cangxi County, which possesses the lowest tourism value, exhibits the highest coordination among R, E, and S, suggesting the absence of significant restrictive factors; however, this also implies that identifying regional highlights and breakthrough points is challenging, thus categorizing it as a "low-value restricted type." Zhaohua District and Langzhong District, with medium total values, lack uncoordinated elements and therefore require comprehensive strengthening and upgrading, placing them in the "comprehensive breakthrough type."

4 Discussion

4.1 Evaluation System Indicators

For the safeguarding and promotion of heritage, it is essential to integrate tourism with heritage resources. Heritage tourism serves to showcase the intrinsic cultural significance of heritage sites more effectively. For the general public, it offers diverse leisure opportunities and enhances quality of life. Additionally, for regional communities, heritage tourism can bolster economic benefits and cultural influence, while also contributing to the establishment of a distinctive regional identity. Evaluating the comprehensive value of cross-regional linear cultural heritage and effectively protecting, utilizing, or redeveloping it is imperative. The concept of heritage protection has evolved from focusing solely on individual sites to encompassing entire sites and their surrounding environments. To better align heritage protection with tourism, this study has taken into account heritage resources, tourism resources, and regional information when developing the evaluation index system. However, some scholars argue that each cultural heritage site should be evaluated independently, irrespective of its connection to the surrounding area. In this paper, the regional comprehensive index R–E–S evaluation system has been employed to assess the tourism value of the Cuiyunlang Ancient Shu Road, offering a more comprehensive approach to promoting heritage tourism. Nonetheless, accurately capturing the overall value may still present challenges. While the index selection process included a literature review, surveys, and expert input, there may still be imperfections in the index system. Certain indicators were customized based on the study area, but regional characteristics were not fully integrated, which could be addressed in future research. Additionally, the method of expert scoring for determining index weights is subjective, making it important to adjust the results based on the specific characteristics and circumstances of linear cultural heritage in future research.

4.2 Evaluation Criteria Parameters

In the evaluation index of this study, the majority of the data in the data collection stage is sourced from the statistical yearbook and data published by relevant government websites. However, there is currently no effective integration of GIS information data and spatial big data, which could be optimized and refined in future research. For instance, concerning relevant economic indicators and other data [26], night lighting data can be utilized instead of GDP or population data to more accurately reflect the actual intensity of human activities and the economic development of the service industry. This approach could better address data distortions caused by the intensity of certain industries in the region. Additionally, POI data can be employed to illustrate the development of catering services and other economic service sectors. Furthermore, using NDVI covered by vegetation to substitute urban green coverage rate can more precisely reflect regional vegetation growth and changes in the natural environment, offering a more scientific and accurate approach.

4.3 Evaluation Object Unit

When evaluating the tourism potential of the Cuiyunlang Ancient Shu Road, this study centers on a smaller scale, examining five county-level administrative regions along the route. The following factors are taken into consideration: (1) Policy influence: Policies play a pivotal role in promoting the development of heritage tourism. Township-level administrative districts have a lower policy influence compared to county-level districts. (2) Data acquisition: Gathering data at the township level is challenging and may result in missing data, which could hinder the evaluation of Zhao Cheng. (3) Principle of localization: While the Cuiyunlang Ancient Shu Road spans five districts and counties, including all of them in the research scope poses a challenge due to the limited number of samples. However, focusing on smaller administrative units, such as townships, can better highlight regional differences and local characteristics when developing a new heritage tourism corridor route. Therefore, treating the Cuiyunlang area at a more refined level, such as the township level, can enhance the management approach, increase the number of research samples, and highlight small-scale regional differences.

4.4 Evaluation Results

Zitong and Jiange are categorized as "high-value imbalance type", while Zhaohua and Langzhong are classified as "comprehensive breakthrough type". Cangxi falls under the "low-value restricted type" category. For Zitong, which is characterized by a high-value imbalance, there should be an integration of resources, environment, and economy to facilitate coordinated development. In the case of Jiange, there should be a deep exploration of heritage and tourism resources, with a focus on enhancing the existing heritage tourism brand image and promoting environmental optimization and economic development. As for Zhaohua and Langzhong, it is important to explore urban characteristics based on their unique situation and combine the cultural significance of the ancient Shu Road in Cuiyunlang to drive coordinated development. Cangxi, being a low-value and restricted type, faces the most challenging development prospects. Therefore, priority should be given to heritage management and cultural continuity, with economic development being considered following the actual situation.

4.5 Research Significance

Theoretical Contribution: The R–E–S evaluation framework and 3D coordination model proposed in this study provide a novel methodological tool for assessing tourism value in linear cultural heritage. Unlike traditional approaches focusing on single dimensions, the R–E–S system integrates interactions among resource background, environmental conditions, and socioeconomic factors, revealing how multi-dimensional synergy drives sustainable heritage tourism. This framework aligns with the theoretical shift from "point-based" to "linear/surface-based" heritage conservation and offers a replicable model for cross-regional assessments. Additionally, the coordination degree (d_1) quantifies subsystem asynchrony, advancing understanding of dynamic equilibrium in "resource–environment–economy" systems and addressing the lack of coordination mechanism analysis in existing studies.

Practical Relevance: The findings directly inform tourism development and conservation strategies for the Cuiyunlang Ancient Shu Road. For example, "high-value imbalance" areas (Jiange County) require prioritized environmental and economic improvements to unlock resource potential, while "low-value restricted" regions (Cangxi County) need cultural branding to overcome bottlenecks. The classification strategies provide scientific foundations for policymakers to avoid uniform development approaches. Furthermore, the quantified coordination degree (d_1) identifies entry points for cross-regional collaboration: coordinated resource allocation, shared environmental governance, and aligned economic policies can enhance overall tourism value. This model holds reference value for managing similar linear heritage, particularly in balancing conservation-utilization and local–global conflicts.

5 Conclusion

This study developed an R–E–S comprehensive evaluation system to assess the tourism value of the Cuiyunlang Ancient Shu Road, integrating resource background (R), environmental conditions (E), and socioeconomic factors (S). The analytic hierarchy process (AHP) and Delphi method revealed that resource background (R) dominates tourism value with a weight of 0.6080, while environmental conditions (E) and socioeconomic factors (S) serve as supplementary components. Through a three-dimensional spatial coordination model, regions along the route were classified into three distinct types: "high-value imbalance type" (Jiange County and Zitong County, characterized by high resource potential but poor coordination), "comprehensive breakthrough type" (Zhaohua District and Langzhong City, requiring balanced improvements), and "low-value restricted type" (Cangxi County, with limited overall value despite high coordination). The coordination degree (d_1) quantifies subsystem asynchrony, offering insights into mismatches between resource utilization and environmental/economic development.

The R–E–S framework bridges theoretical gaps in multidimensional evaluation of linear cultural heritage and provides actionable strategies for sustainable tourism planning, such as prioritizing environmental optimization in high-resource regions and enhancing cultural branding in low-value areas. Limitations include subjectivity in expert-driven index weighting and constraints from county-level administrative units, suggesting future refinements through township-scale studies and expanded validations across diverse heritage corridors.

References

1. Kongjian Y (2004) World heritage concept challenges China: thoughts on the 28th world heritage conference, Chinese Gardens, vol 11, pp 71–73
2. Xiaoxuan L (2014) Review of the current status of transnational world natural heritage protection. J Nat Resour 29(11):1978–1990
3. Yaolin M, Lan Z, Song Z, Yin W, Tinghai W, Jinfu Y, Xin Y, Xiwen T, Huaiyu S, Jianzhong Y, Chun W (2019) Cross-regional linear cultural heritage protection and utilization. Urban Plan 43(05):40–47
4. Orbaşli DA, Woodward S (2008) A railway 'Route' as a linear heritage attraction: the Hijaz Railway in the Kingdom of Saudi Arabia. J Herit Tour 3(3):159–175
5. Kuiper E, Bryn A (2013) Forest regrowth and cultural heritage sites in Norway and along the Norwegian St Olav pilgrim routes. Int J Biodivers Sci, Ecosyst Serv Manag 9(1):54–64
6. Lois (2003) Cultural tourism and urban management in northwestern Spain: the pilgrimage to Santiago de Compostela, Tourism Geographies 5(4)
7. Hemme D (2005) Landscape, fairies and identity: experience on the backstage of the fairy tale route. J Tour Cult Chang 3(2):71–87
8. Snowball (2010) Cultural heritage routes in South Africa: effective tools for heritage conservation and local economic development? Dev South Afr 27(4)
9. Daniele (2016) Cultural landscape and cultural routes: infrastructure role and indigenous knowledge for a sustainable development of inland areas. Procedia - Soc Behav Sci 223
10. Bras (2010) Network analysis and wine routes: the case of the Bairrada Wine Route. Serv Ind J 30(10)
11. Mell IC, John S (2014) Sustainable urban development in tightly constrained areas: a case study of Darjeeling, India. Int J Urban Sustain Dev 6(1):65–88
12. Rogerson (2007) Tourism routes as vehicles for local economic development in South Africa: the example of the Magaliesberg Meander. Urban Forum 18(2)
13. Timothy DJ, Boyd SW (2006) Heritage tourism in the 21st century: valued traditions and new perspectives. J Herit Tour 1(1):1–16
14. Jian Z (2018) Research on Koguryo world heritage value evaluation and study travel utilization. Resour Dev Mark 34(07):1032–1036
15. Peixiang G, Fuxing L, Yuehao T (2024) Value evaluation of linear cultural heritage tourism resources of Guiliu Canal. J Qiannan Norm Univ Natl 44(01):94–103
16. Jian Z, Weigong J (2013) Research on tourism value evaluation and development of linear cultural heritage of the old Yellow River. Resour Dev Mark 29(05):553–556
17. Bo L, Tianjing G, Shijie H (2018) Research on heritage value evaluation of cultural routes based on analytic hierarchy process-taking the Zijiang section of Wanli tea ceremony as an example. Chin Foreign Arch (12):41–43

18. Zhongchao D, Yinhua L (2011) Comprehensive evaluation of the tourism value of linear heritage corridors based on information entropy—taking the Silk Road in Northwest China as an example. Geogr Arid Areas 34(03):519–524
19. Huanlin R (2017) Value evaluation of cross-regional linear cultural heritage tourism resources-taking the China section of Chang'an-Tianshan corridor road network as an example. Geogr Sci 37(10):1560–1568
20. Juntao C (2017) Origin and evolution: the formation and characteristics of the Hanjiang heritage corridor. Rural Econ Sci Technol 28(05):1–4
21. Juncheng D, Fangyu L (2023) Research on the construction of linear cultural heritage value assessment model—taking China's Grand Canal as an example. China Famous Cities 37(01):90–96
22. Jinwei W, Binna H (2008) Evaluation and empirical research on the development potential of linear cultural heritage tourism. J Yunnan Norm Univ (Philos Soc Sci Ed) (05):120–126
23. Long L, Zhenfang H (2007) Research on the construction and application of tourism value evaluation system of heritage corridor—taking Jiangsu section of ancient canal as an example. China Popul, Resour Environ (06):95–100
24. Zhang (2024) Assessment of Yellow River region cultural heritage value and corridor construction across urban scales: a case study in Shaanxi, China. Sustainability 16(3)
25. Bozic (2016) Developing the cultural route evaluation model (CREM) and its application on the trail of Roman Emperors, Serbia. Tour Manag Perspect 17
26. Briedenhann (2004) Tourism routes as a tool for the economic development of rural areas—vibrant hope or impossible dream? Tour Manag 25(1)
27. Zhang S, Liu J, Pei T, Chan C-S, Wang M, Meng B (2021) Tourism value assessment of linear cultural heritage: the case of the Beijing-Hangzhou Grand Canal in China. Curr Issue Tour 26(1):47–69
28. Božić (2013) Tourist valorization of cultural route 'The Trail of the Roman Emperors'. Eur Res 55(7)
29. Yue (2023) A framework for constructing a heritage corridor system: a case study of the Shu Road in China. Remote Sens 15(19)
30. Hong Z, Chaolin G, Yan Z (2008) Evaluation of coordinated development of tourism in Nanjing: synchronous coordination discrimination method. Econ Geogr (04):682–687

A Compromise-Typed Fuzzy Evaluation of Urban Community Parks' Age-Friendly—A Case of the Urban Area of Mianyang, China

Silin Wu, Dachuan Wang, Liuyin Guo, and Wang Yu

Abstract In the context of deep aging society, urban community parks, as frequent places for daily activities of the elderly, require improvement in their built environment and facilities to be more elderly-friendly. To address the limitations of existing evaluation research that only considers subjective or objective weighting evaluation models, this study proposes a fuzzy comprehensive nonlinear evaluation model for the elderly-friendliness of urban community parks based on a compromise weighting concept that integrates subjectivity and objectivity. The research constructs an evaluation index system for the elderly-friendliness of urban community parks based on the SERVQUAL model and RECITAL tool. The subjective weights are analyzed using the Analytic Hierarchy Process (AHP) based on questionnaire evaluation results of typical urban community parks in Mianyang City. Then, the variable weight fuzzy comprehensive evaluation method is introduced to compromise the subjective and objective weights, resulting in a comprehensive level of elderly-friendliness. The results show that the overall elderly-friendliness evaluation of community parks in Mianyang City is at a moderate level, with the dimensions ranked as follows: Empathy (80) > Reliability (75) > Assurance (74) > Perceptibility (71) > Responsiveness (69). The use of compromise weighting in the fuzzy comprehensive evaluation considers decision-makers' preferences while reducing the subjectivity in weighting, resulting in reliable and realistic evaluation results. This approach can provide methodological references and ideas for evaluation issues in other planning and design domains.

Keywords Age-friendly · Urban community parks · Compromise-typed · Fuzzy evaluation

---

S. Wu · D. Wang (✉) · L. Guo · W. Yu
School of Civil Engineering and Architecture, Southwest University of Science and Technology, Mianyang, Sichuan, China
e-mail: i@alwayswdc.com

B. He et al. (eds.), *Climate-Resilient, Low-Carbon, and Sustainable Cities*, Lecture Notes in Civil Engineering 754, https://doi.org/10.1007/978-981-95-2169-2_50

1 Introduction

Population aging is a significant issue in China, characterized by an "inverted pyramid" demographic structure [1]. The country faces a rapidly growing elderly population, with evolving needs shifting from survival to development, highlighting the urgent need to enhance elderly care services.

The seventh national population census indicates that there are 264 million individuals aged 60 and above in China, comprising 18.7% of the population. In Sichuan Province, this group totals 18.164 million, with Mianyang, the province's second-largest city, having over 1.15 million elderly residents and an aging rate of 23.71%, a notable increase from previous census data (Fig. 1). This trend underscores the rising demand for age-friendly services. Urban community parks, as key venues for elderly activities, should align with national aging strategies to strengthen urban elderly care systems [2].

In the face of the irreversible trend of aging, several scholars internationally have explored the construction of age-friendly societies. Studies have shown that due to constraints in physical functions, lifestyle habits, and economic status, elderly individuals primarily engage in activities within their immediate communities. The built environment of urban community parks offers considerable benefits for enhancing outdoor activities, maintaining emotional well-being, and promoting physical and mental health among older adults [3, 4]. Currently, the literature on the age-friendliness of urban community parks mainly focuses on the overall construction of age-friendly communities [5, 6], summaries of urban or regional construction experiences [7, 8], the renovation of public spaces [9, 10], and the location of public service facilities [11]. Previous research on the evaluation of urban community parks has primarily examined aspects such as park accessibility, recreational environments, and physical facilities [12–14]. Although improvements based on evaluations of these aspects can positively impact the safety and efficiency of elderly individuals using community parks, the evaluation indicators still need to be further refined and expanded to ensure that the results are more comprehensive and accurate. Methodologically, scholars often employ methods such as the SD method [15], AHP, or FAHP [16], which tend to exhibit clear subjective tendencies. In multi-level evaluation systems, the weights at the middle and top levels cannot be directly obtained through the analytic hierarchy process (AHP) or may suffer from insufficient value extraction, leading to suboptimal precision in the evaluation results. To address the shortcomings of existing evaluation research that solely considers either subjective or objective weighting evaluation models, this study introduces the concept of compromise weighting that integrates both subjective and objective perspectives. We propose a fuzzy comprehensive nonlinear evaluation model based on compromise weights to assess the age-friendliness of urban community parks, aiming for a more scientific and effective evaluation.

This study first introduces the SERVQUAL model [17], a widely recognized framework for service quality evaluation, to ensure the comprehensiveness and scientific rigor of the existing indicator system. Building upon this foundation,

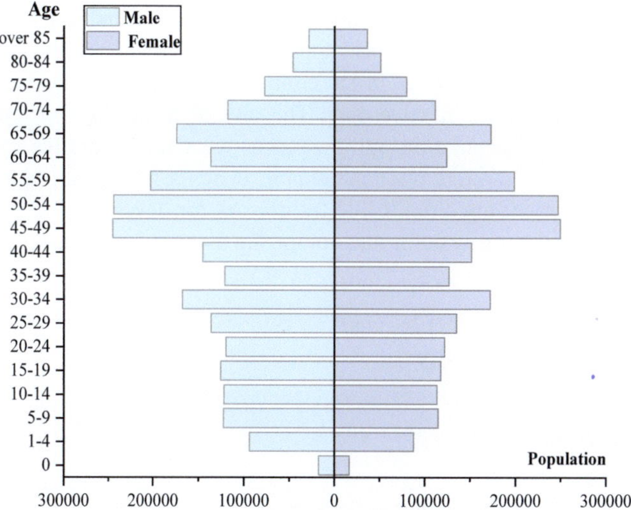

(a) Population Pyramid of the Seventh National Census in Mianyang City

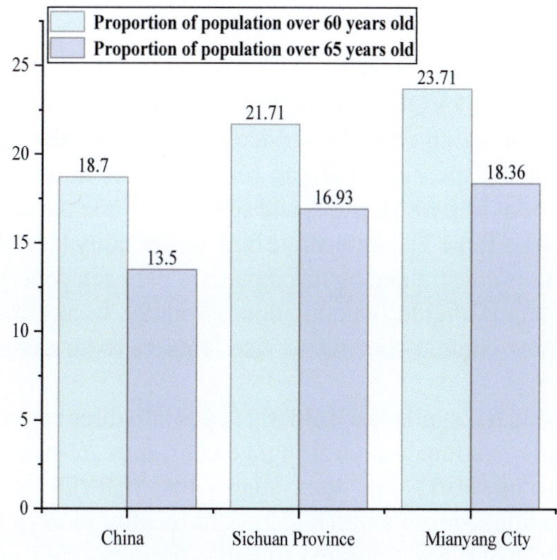

(b) Comparison of Aging in the Seventh National Census

Fig. 1 Results of the national census (Self-drawn)

we integrate the RECITAL tool [18]—a validated instrument for assessing urban green space quality—to supplement aging-friendly evaluation indicators specific to urban community parks. This integration establishes a robust aging-friendly evaluation indicator system. Subsequently, subjective evaluations are quantified, and the Analytic Hierarchy Process (AHP) is applied to derive subjective weights for each factor. To enhance objectivity, a fuzzy comprehensive evaluation based on compromise weighting is introduced, harmonizing subjective and objective weights to generate variable weights. This hybrid approach enables a comprehensive assessment of aging-friendly attributes in urban community parks by incorporating both subjective perceptions and objective data. The city of Mianyang—a region characterized by advanced population aging—serves as a representative case study. The methodology and findings offer actionable insights for cities grappling with aging populations, providing a replicable framework for developing aging-friendly community park evaluation systems and composite assessment methods.

2 Research Methodology

2.1 Research Methods

2.1.1 SERVQUAL Model and RECITA Tool

This study combines the SERVQUAL model and RECITA tool to develop an evaluation index system.

As a highly utilized public service facility, community parks were evaluated in this study through the SERVQUAL model [17]—also known as the service quality gap model—to assess aging-friendly service quality from the user (i.e., elderly) perspective. This model measures the gap between older adults' expectations and their actual perceptions of park facilities and services. While the original five dimensions of SERVQUAL (Table 1) demonstrate high universality [18], their applicability to community parks is constrained by discrepancies between generic service metrics and the unique demands of elderly populations. Notably, weak correlations between certain indicators and context-specific service features reduce practical utility [19, 20].

Recent scholarship has synthesized elderly needs into three core domains: psychological well-being, recreational-cultural engagement, and care-oriented services [21]. Building on this framework, our study adapts the SERVQUAL model by integrating elderly-specific requirements (Fig. 2). This recalibration yields a revised five-dimensional SERVQUAL scale tailored to aging-friendly urban community parks, with reconceptualized definitions for each dimension as detailed in Table 1.

The RECITA tool is used to assess quality aspects related to human health, consisting of 90 items across 11 dimensions [13]. It helps researchers explore how urban green space quality impacts health benefits. Unlike SERVQUAL, which focuses on customer perceptions, RECITA emphasizes facility attributes. By combining these perspectives, a comprehensive evaluation is achieved. Initial assessment factors from RECITA are matched with SERVQUAL's five dimensions, resulting in a framework for evaluating urban community parks' suitability for the elderly, comprising 5 dimensions and 23 indicators (Fig. 3).

Table 1 SERVQUAL scale of community parks' age-friendly interpretations (Self-drawn)

Dimensions	General interpretation	Interpretation of the aging dimension
Tangibility	The actual facilities, equipment, personnel, and informational materials that create an impression of service for customers	The quantity and variety of park facilities and equipment, enhancing their appeal to elderly users
Reliability	The ability to fulfill service promises consistently and accurately	The capacity of service providers to offer reliable services, such as the accessibility and location of parks, ensuring convenience for elderly users
Responsiveness	Willingness to assist customers and provide prompt service	Offering personalized services that cater to their needs, such as the provision of accessible facilities in parks and the organization of engaging activities
Assurance	The trustworthiness of the company and its staff, including their knowledge and courtesy	The ability to instill confidence and safety in the elderly, such as providing secure facilities in parks
Empathy	Caring for customers and providing personalized attention	The ability to offer compassionate services to elderly visitors, such as designing parks with their psychological needs in mind

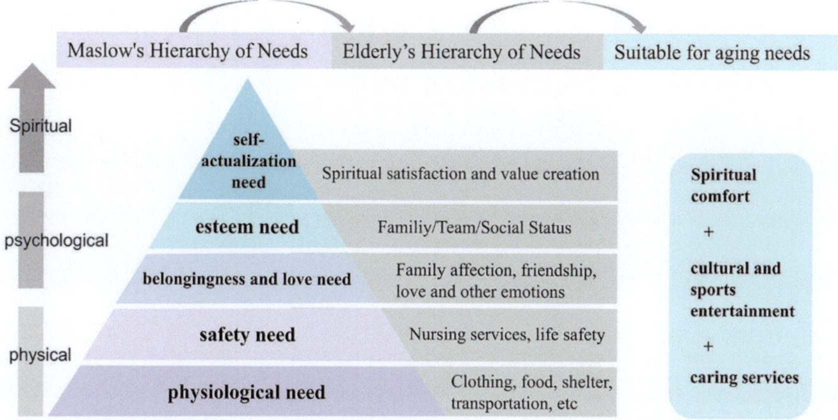

Fig. 2 Hierarchy of needs of the elderly (Self-drawn)

2.1.2 Fuzzy Comprehensive Evaluation Based on Variable Weights

Fuzzy comprehensive evaluation utilizes fuzzy mathematics, applying a weighted average operator to objectively assess the overall situation of the evaluated subject. In evaluating community parks for the elderly, it compares seniors' expectations with their satisfaction, making it a suitable and persuasive method [22].

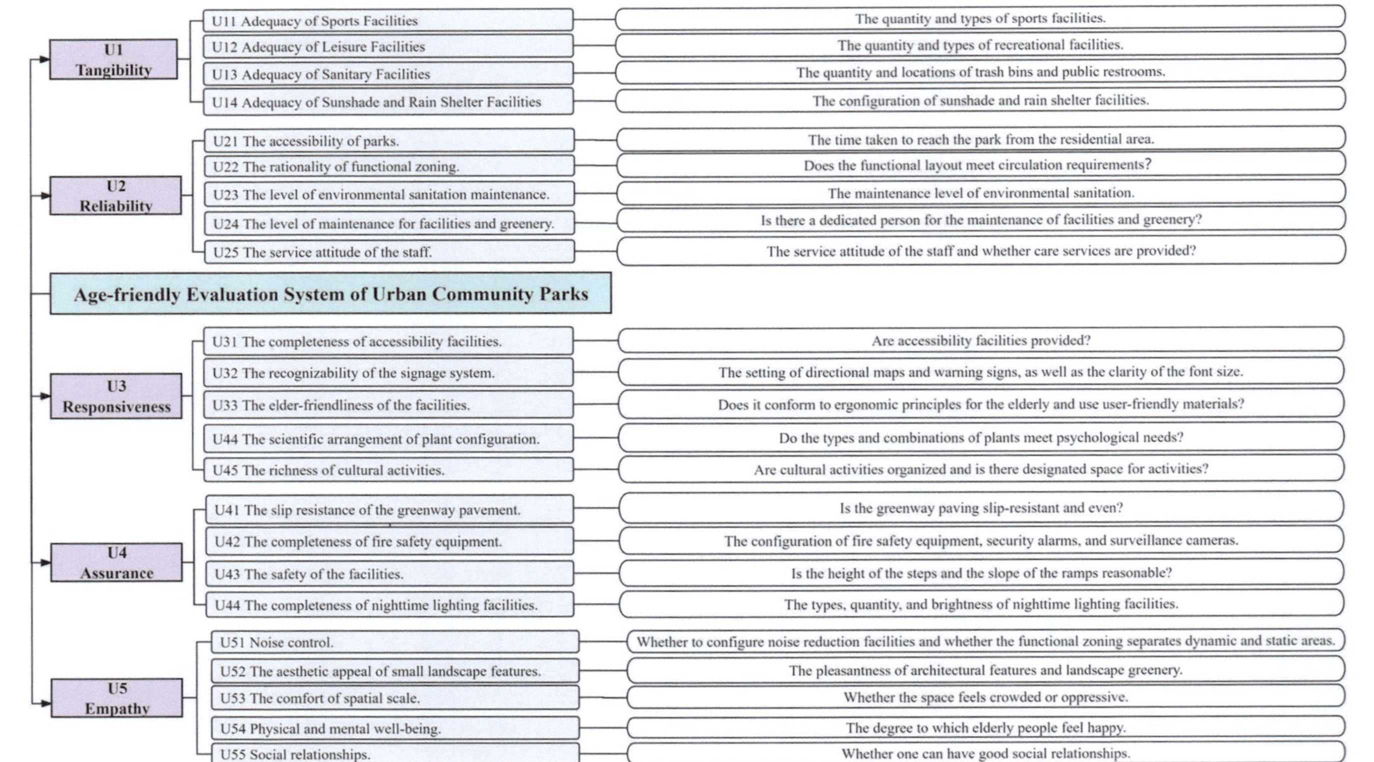

Fig. 3 Evaluation index system of urban community parks' age-friendly (Self-drawn)

AHP-derived weights reflect subjective preferences, but the importance of indicators can vary with their actual status [23]. To address this, we introduce variable weights: indicators positively correlated with their status receive incentives, while negatively correlated ones face penalties. By combining this with fuzzy comprehensive evaluation, we create a membership matrix and determine evaluation levels using the maximum membership principle. Integrating weights and evaluation levels allows for a comprehensive assessment of suitability for the elderly, enhancing the scientific rigor and credibility of the decision-making process.

2.2 Research Area and Data Source

2.2.1 Research Area

The central urban area of Mianyang City, Sichuan Province, specifically Fucheng District and Youxian District, was selected as the research area, covering a total area of 197,140 hectares. The population aged 60 and above is 169,000, accounting for 15.3%. The selection of research subjects follows these principles:

(A) Broad Coverage: Parks in high-density population areas across various administrative districts were chosen to ensure a significant number of elderly individuals are observed. The selected parks should be evenly distributed and primarily surrounded by residential areas, with good accessibility, complete facilities, a stable elderly population, and a strong sense of belonging.

(B) Diverse Element Types: Community parks with a rich variety of internal elements were selected to comprehensively understand their landscapes, facilities, and services. Parks with diverse features tend to attract more visitors and offer better sustainability. Moreover, selected parks should differ in spatial environmental elements to provide a more comprehensive and varied data set.

Ultimately, five exemplary community parks were selected: Science City Cultural Park, Daijiawan Health Park, Times Square, Chinese Scientists Park, and May Day Square (Figs. 4 and 5).

2.2.2 Data Source

Due to Mianyang's hot summers and cold winters, there are fewer elderly individuals engaging in outdoor activities. To ensure the scientific validity and accessibility of the data, random sampling surveys targeting the elderly were conducted on clear, sunny weekends in October and November, between 8:00 AM and 8:00 PM. A total of 200 questionnaires were distributed, yielding 172 valid responses, resulting in an effective response rate of 86.5% (Table 2).

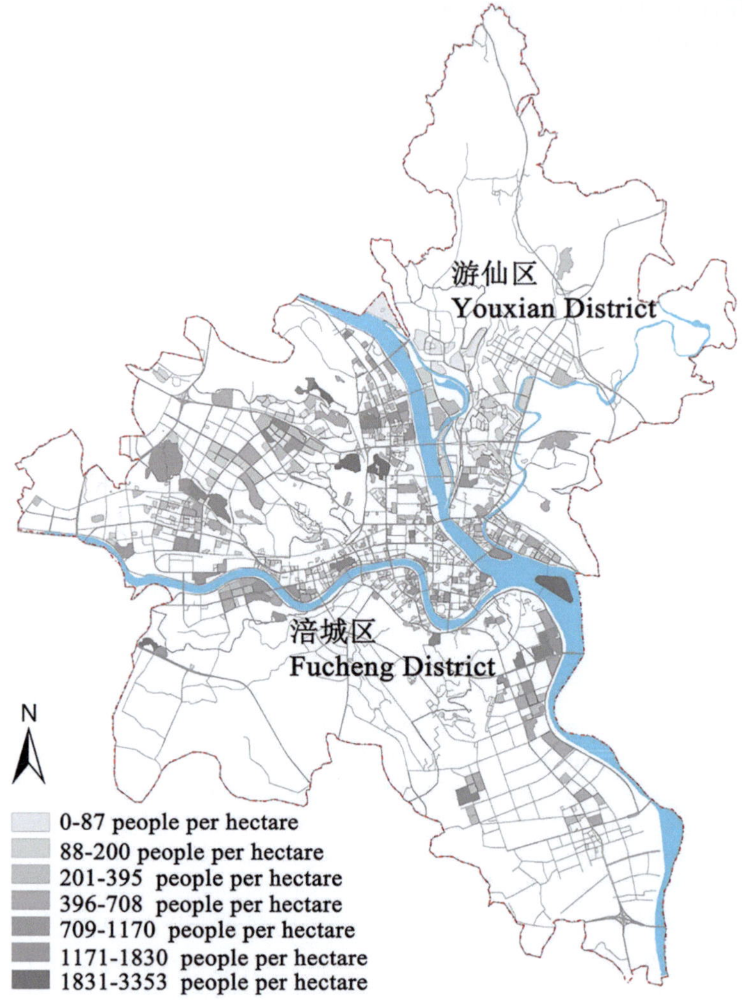

Fig. 4 Distribution of elderly population in Mianyang city center (Self-drawn)

3 Measurement and Evaluation

3.1 Subjective Weighting Using AHP Method

A questionnaire designed for evaluating the age-friendliness of urban community parks was distributed to five experts who have long been engaged in urban and rural planning, management, and education in Mianyang City. Using a 1–5 scale, the weights of five dimensional indicators were determined. Similarly, a survey was

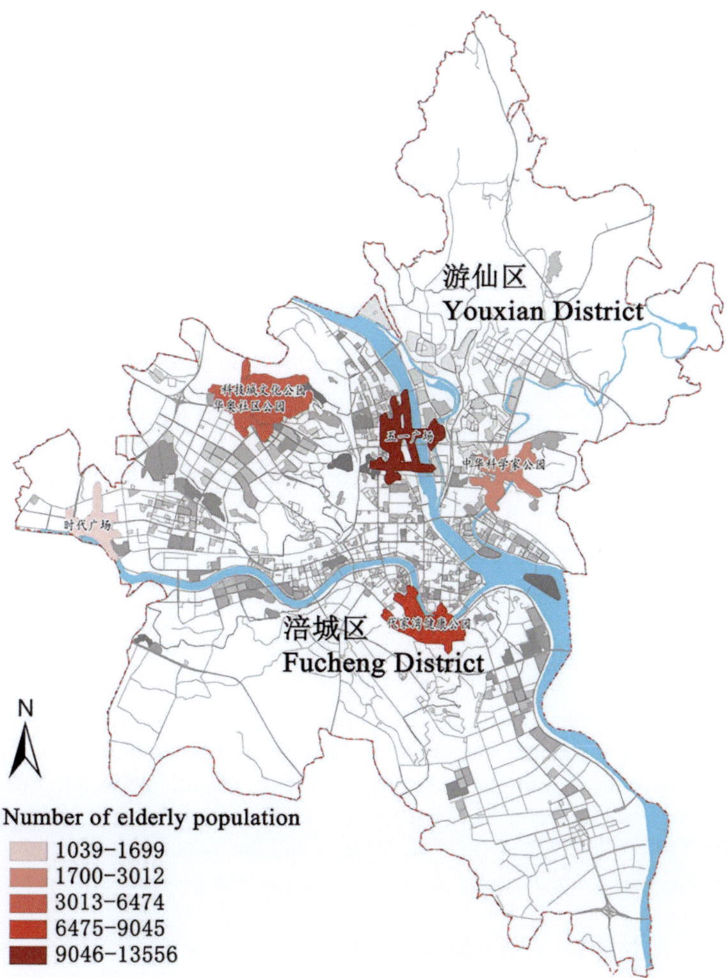

Fig. 5 Number of elderly population in urban community park service area (Self-drawn)

Table 2 Questionnaire survey results for each park (Self-drawn)

Name	Number	Valid	Validity rate (%)
Science and Technology City Cultural Park	40	33	82.50
Dai Jia Wan Health Park	40	33	82.50
Times Square	40	35	87.50
Chinese Scientists Park	40	35	87.50
May Day Square	40	36	90.00
Sum	200	172	86.00

distributed to elderly individuals using community parks to establish the weights of specific indicators for the age-friendliness evaluation.

(A) Establishing the AHP Analysis Model and Constructing the Judgment Matrix

The evaluation criteria consist of five levels: absolutely important, very important, relatively important, slightly important, and equally important, represented by the values 100, 80, 60, 40, and 20, respectively. Evaluators conducted pairwise comparisons of the evaluation indicators and assigned scores. Based on the evaluation results, a judgment matrix was constructed, and the AHP method was used to calculate the subjective weights of each indicator.

(B) Consistency Testing of the Judgment Matrix Using SPSS Software

The judgment matrix was subjected to a consistency test based on the survey results, using the Random Consistency Ratio (CR) for evaluation, resulting in a consistency ratio of less than 0.1. The basic weights of the indicators were calculated, as shown in Table 3.

3.2 *Determine the Variable Weight*

Based on the fundamental weights determined by the Analytic Hierarchy Process (AHP), the variable weight principle is used to adjust these fundamental weights.

According to the definitions and related theorems of variable weight theory, let the subjective weight vector be $B = (b_1, b_2, \ldots, b_t)$, the local state variable weight vector be $Y(X) = \{Y(x_1), Y(x_2), \ldots, Y(x_t)\}$, and the local variable weight vector be $B' = (b'_1, b'_2, \ldots, b'_t)$.

$$b' = \frac{b_i Y(x_i)}{\sum_{i=1}^{t} b_i Y(x_i)}. \tag{1}$$

Which,

$$Y(x_i)\begin{cases} 1, & 0 \le x_i < \mu_1 \\ \frac{c-1}{\mu_2-\mu_1}x_i + \left(1 - \frac{c-1}{\mu_2-\mu_1}\mu_1\right), & \mu_1 \le x_i < \mu_2 \\ C, & \mu_2 \le x_i < \mu_3 \\ \frac{1-c}{1-\mu_3}x_i + \left(1 - \frac{c-1}{1-\mu_3}\right), & \mu_3 \le x_i \le 1 \end{cases} \tag{2}$$

The scoring values are defined by four parameters: μ_1 (negation level), μ_2 (penalty level), μ_3 (incentive level), and C (adjustment level), all of which range within the interval [0, 1]. Based on existing studies, the values of these parameters can be determined through consensus among evaluation experts, taking into account the actual conditions and characteristics of the study area [24]. In this research, five experts with

Table 3 Weighting for age-friendly evaluation system of urban community parks (Self-drawn)

Goal level	Criterion level	Subjective weight	Variable weight	Indicator level	Subjective weight	Variable weight	Secondary indicator evaluation matrix				
							Very satisfied	Satisfied	Neutral	Dissatisfied	Very dissatisfied
U	U1	0.2341	0.2341	U11	0.3684	0.1451	0.1395	0.5640	0.2733	0.0174	0.0058
				U12	0.3684	0.3423	0.2151	0.5349	0.1860	0.0640	0.0000
				U13	0.1928	0.2966	0.2384	0.2733	0.3779	0.1047	0.0058
				U14	0.0704	0.2166	0.0407	0.1744	0.3779	0.3081	0.0988
	U2	0.0928	0.0928	U21	0.1154	0.0742	0.1453	0.5930	0.2326	0.0291	0.0000
				U22	0.1154	0.1775	0.1337	0.5058	0.2674	0.0872	0.0058
				U23	0.4463	0.6866	0.2733	0.3314	0.2500	0.1337	0.0116
				U24	0.2867	0.0106	0.1802	0.4419	0.3372	0.0349	0.0058
				U25	0.0361	0.0555	0.0640	0.4360	0.4302	0.0407	0.0291
	U3	0.3479	0.3479	U31	0.1987	0.4771	0.0407	0.2965	0.5465	0.1163	0.0000
				U32	0.0434	0.1042	0.1221	0.4767	0.3198	0.0814	0.0000
				U33	0.4847	0.1303	0.1628	0.4826	0.3256	0.0291	0.0000
				U34	0.1987	0.1097	0.1860	0.4767	0.2965	0.0407	0.0000
				U35	0.0745	0.1789	0.0756	0.4419	0.3605	0.0930	0.0291
	U4	0.2514	0.2514	U41	0.1630	0.2090	0.1221	0.5233	0.3256	0.0291	0.0000
				U42	0.2551	0.3271	0.0756	0.4302	0.4128	0.0756	0.0058
				U43	0.4747	0.3262	0.1919	0.5291	0.2500	0.0291	0.0000
				U44	0.1072	0.1374	0.1802	0.4651	0.2442	0.0814	0.0291
	U5	0.0739	0.0739	U51	0.0968	0.0866	0.0349	0.3140	0.4360	0.1337	0.0814
				U52	0.1763	0.0514	0.2558	0.3547	0.3547	0.0349	0.0000

(continued)

Table 3 (continued)

Goal level	Criterion level	Subjective weight	Variable weight	Indicator level	Subjective weight	Variable weight	Secondary indicator evaluation matrix				
							Very satisfied	Satisfied	Neutral	Dissatisfied	Very dissatisfied
				U53	0.1868	0.1671	0.2965	0.4419	0.2267	0.0349	0.0000
				U54	0.4850	0.6457	0.3256	0.5058	0.1337	0.0349	0.0000
				U55	0.0551	0.0493	0.0872	0.5581	0.3314	0.0233	0.0000

extensive experience in urban and rural planning design, urban planning management, and urban planning education in Mianyang City collaboratively assigned the following values through deliberation:

Negation level: $\mu_1 = 0.25$.

Penalty level: $\mu_2 = 0.6$

Incentive level: $\mu_3 = 0.75$.

Adjustment level: $C = 0.2$.

Using these in Formula (1) gives:

$$Y(x_i)\begin{cases} 1, & 0 \le x_i < 0.25 \\ \frac{c-1}{\mu_2-\mu_1}x_i + \left(1 - \frac{c-1}{\mu_2-\mu_1}\mu_1\right), & 0.25 \le x_i < 0.6 \\ C, & 0.6 \le x_i < 0.75 \\ \frac{1-c}{1-\mu_3}x_i + \left(1 - \frac{c-1}{1-\mu_3}\right), & 0.75 \le x_i \le 1 \end{cases} \tag{3}$$

The value of x_i is obtained by multiplying the evaluation matrix with the quantization set of evaluation levels, scaled by 1%. The variable weights for each indicator are calculated using Eqs. (2) and (3), as shown in Table 3.

3.3 *Fuzzy Comprehensive Evaluation*

This section uses the fuzzy comprehensive evaluation method to assess elderly satisfaction in urban community parks.

(A) Establishing the Evaluation Scale.

A five-level scale (V1 to V5) is used, corresponding to very satisfied, satisfied, neutral, dissatisfied, and very dissatisfied, $V = \{v_1, v_2, v_3, v_4, v_5\} = \{100, 80, 60, 40, 20\}$. The indicators for elderly satisfaction form set $U = (U_i)$ $(i = 1, 2, 3, 4, 5)$.

(B) Constructing the Fuzzy Evaluation Matrix.

If the membership degree of the i-th element in set U to the first element in set V is r_{i1}, the single-factor evaluation result is represented as $R_i = (r_{i1}, r_{i2}, \ldots\ldots, r_{im})$. These results are combined into a matrix, forming the fuzzy comprehensive evaluation matrix.

$$R = \begin{bmatrix} r_{11}, & r_{12}, & \ldots\ldots, & r_{1n} \\ \ldots & & & \ldots \\ r_{m1}, & r_{m2}, & \ldots\ldots, & r_{mn} \end{bmatrix} \tag{4}$$

(C) Fuzzy Comprehensive Calculation.

$$P = W_i * R_{ij} \tag{5}$$

Using the membership degree matrix R and indicator weights, perform the fuzzy comprehensive operation. P is the evaluation vector, W_i is the weight of the i-th factor, and R_{ij} are the elements of the evaluation matrix.

(D) Final Comprehensive Evaluation.

$$E = P * V \tag{6}$$

Finally, the defuzzification of each criterion layer evaluation set is conducted according to Formula (6) to calculate the elderly accessibility evaluation value and the comprehensive evaluation value for the target layer.

4 Results

4.1 Calculated Results

4.1.1 Construction of Membership Matrix

Utilizing field surveys and percentage statistics to obtain the corresponding membership degree matrix.

$$RU1 = \begin{bmatrix} 0.1395 & 0.564 & 0.2733 & 0.0174 & 0.0058 \\ 0.2151 & 0.5349 & 0.186 & 0.064 & 0 \\ 0.2384 & 0.2733 & 0.3779 & 0.1047 & 0.0058 \\ 0.0407 & 0.1744 & 0.3779 & 0.3081 & 0.0988 \end{bmatrix}$$

$$RU2 = \begin{bmatrix} 0.1453 & 0.593 & 0.2326 & 0.0291 & 0 \\ 0.1337 & 0.5058 & 0.2674 & 0.0872 & 0.0058 \\ 0.2733 & 0.3314 & 0.25 & 0.1337 & 0.0116 \\ 0.1802 & 0.4419 & 0.3372 & 0.0349 & 0.0058 \\ 0.064 & 0.436 & 0.4302 & 0.0407 & 0.0291 \end{bmatrix}$$

$$RU3 = \begin{bmatrix} 0.0407 & 0.2965 & 0.5465 & 0.1163 & 0 \\ 0.1221 & 0.4767 & 0.3198 & 0.0814 & 0 \\ 0.1628 & 0.4826 & 0.3256 & 0.0291 & 0 \\ 0.186 & 0.4767 & 0.2965 & 0.0407 & 0 \\ 0.0756 & 0.4419 & 0.3605 & 0.0930 & 0.0291 \end{bmatrix}$$

$$RU4 = \begin{bmatrix} 0.1221 & 0.5233 & 0.3256 & 0.0291 & 0 \\ 0.0756 & 0.4302 & 0.4128 & 0.0756 & 0.0058 \\ 0.1919 & 0.5291 & 0.25 & 0.0291 & 0 \\ 0.1802 & 0.4651 & 0.2442 & 0.0814 & 0.0291 \end{bmatrix}$$

$$RU5 = \begin{bmatrix} 0.0349 & 0.314 & 0.436 & 0.1337 & 0.0814 \\ 0.2558 & 0.3547 & 0.3547 & 0.0349 & 0 \\ 0.2965 & 0.4419 & 0.2267 & 0.0349 & 0 \\ 0.3256 & 0.5058 & 0.1337 & 0.0349 & 0 \\ 0.0872 & 0.5581 & 0.3314 & 0.0233 & 0 \end{bmatrix}$$

4.1.2 Fuzzy Synthesis Operation

$$PU1 = \left[0.1734\ 0.3838\ 0.2973\ 0.1222\ 0.024\right]$$

$$PU2 = \left[0.2276\ 0.3902\ 0.2638\ 0.1121\ 0.0107\right]$$

$$PU3 = \left[0.0873\ 0.3854\ 0.4335\ 0.0889\ 0.0052\right]$$

$$PU4 = \left[0.1376\ 0.4866\ 0.3182\ 0.0515\ 0.0059\right]$$

$$PU5 = \left[0.2803\ 0.4734\ 0.1965\ 0.0429\ 0.007\right]$$

4.1.3 Comprehensive Evaluation

$$E_{U1} = 100b_{11} + 80b_{12} + 60b_{13} + 40b_{14} + 20b_{15} = 71$$

$$E_{U2} = 75, E_{U3} = 69,\ E_{U4} = 74,\ E_{U5} = 80$$

$$U = A^*R = \left(0.1474\ 0.4175\ 0.3394\ 0.0861\ 0.0104\right)$$

Thus, $E_U = 72$.

The comprehensive evaluation results of elderly accessibility in urban community parks of Mianyang City are shown in Table 4.

Table 4 Evaluation results of age-friendly for urban community parks in Mianyang (Self-drawn)

Goal level	Comprehensive result	Indicator level	Comprehensive result
Evaluation results of age-friendly for urban community parks in Mianyang city	72	Tangibility	71
		Reliability	75
		Responsiveness	69
		Assurance	74
		Empathy	80

4.2 *Analysis of Results*

4.2.1 Overall Level

The fuzzy calculations indicate that the overall satisfaction of urban community parks in Mianyang City is categorized as follows: {Very Satisfied, Satisfied, Neutral, Dissatisfied, Very Dissatisfied} with membership degrees of {0.1774, 0.4175, 0.3394, 0.0861, 0.0104}. The overall fuzzy evaluation value is 72, below the threshold of 75, indicating a neutral level of elderly accessibility recognition. The ranking of evaluation factors is: Empathy > Assurance > Reliability > Responsiveness > Tangibles (Fig. 6).

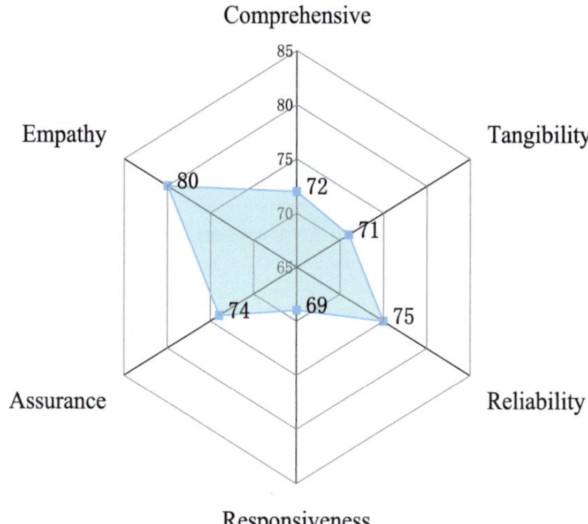

Fig. 6 Evaluation results of age-friendly for community parks in Mianyang City (Self-drawn)

4.2.2 Factor Level

(A) Empathy has the highest evaluation at 80, indicating satisfactory satisfaction among older users. Key indicators like spatial comfort (U53) and overall enjoyment (U54) are rated positively, enhancing elderly accessibility. However, noise control (U51) needs improvement, Fig. 7e.

(B) Reliability, Assurance, and Tangibles are rated neutral (75, 74, and 71). Reliability is slightly above neutral, suggesting the need for staff training to improve service attitudes. The adequacy of sunshade and rain protection (U14) is lacking, impacting satisfaction Fig. 7a–d.

(C) Responsiveness has the lowest ratings across all indicators, indicating significant room for improvement in barrier-free facilities (U31), signage (U32), accessibility (U33), greenery arrangement (U34), and cultural activities (U35). Enhancements in these areas are crucial, as current provisions do not meet the needs of older adults Fig. 7a.

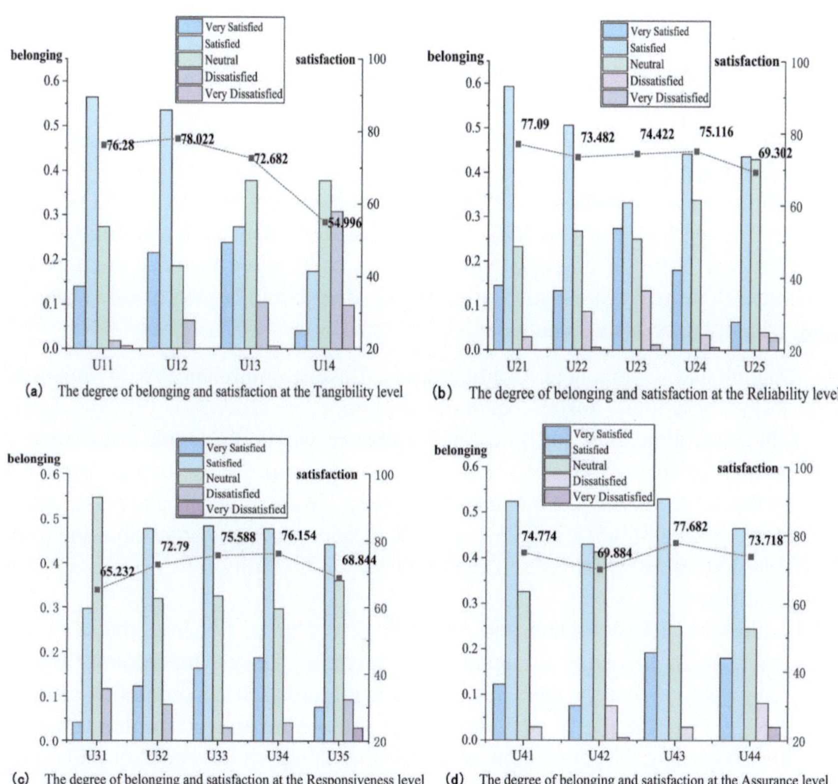

(a) The degree of belonging and satisfaction at the Tangibility level
(b) The degree of belonging and satisfaction at the Reliability level
(c) The degree of belonging and satisfaction at the Responsiveness level
(d) The degree of belonging and satisfaction at the Assurance level

Fig. 7 Membership degree and satisfaction of various indicator factors (Self-drawn)

Fig. 7 (continued)

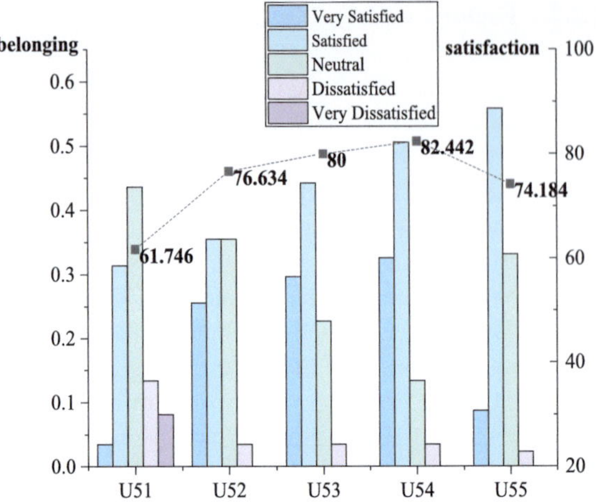

(e) The degree of belonging and satisfaction at the Empathy level

4.2.3 Optimization Recommendations

From the evaluation results, it is necessary to clearly outline some effective recommendations.

Empathy is a fundamental function that urban community parks should possess, serving as a key factor in attracting older adults. Responsiveness can guide future improvements, addressing special needs such as barrier-free facilities and cultural activities. Additionally, sunshade and rain protection should be prioritized to support aging-friendly park development.

(A) Spatial and Component Configuration: Ensure compliance with universal design principles. Install handrails or ramps on stairways, provide accessible restrooms, and add low sinks. Enhance wayfinding with clear signage in appropriate font sizes to boost safety and a sense of belonging for older adults. In vegetation design, create enclosed spaces in plazas and rest areas to block direct sunlight and reduce noise. Consider providing adjustable and foldable umbrellas for tables and chairs, allowing flexibility for different weather conditions.

(B) Addressing Physiological and Psychological Needs: Enhance the scientific arrangement of plants to create inviting resting spaces. Incorporate diverse and layered landscape elements to engage multiple senses, promoting sensory therapy. For example, plant colorful and fragrant flora to attract birds and butterflies, enriching the park's ambiance and stimulating visual, auditory, and olfactory senses. Use varying heights in planting beds to offer tactile experiences, allowing wheelchair users to enjoy scents and textures.

(C) Integrating Intergenerational Communication: Foster a supportive neighborhood environment through intergenerational interactions, reducing feelings of isolation among older adults and encouraging participation in various activities

[25]. Position vibrant children's play areas at the park's visual center or near fitness zones for seniors, enabling them to engage with or observe children's activities. This interaction can uplift their mood and increase physical activity. Additionally, install multi-use equipment suitable for both children and seniors, such as "talking tubes",to facilitate interaction across age groups.

5 Conclusion and Discussion

(A) Conceptual Framework and Methodology
Aging-friendly construction refers to the development of environments that holistically address the physiological, psychological, and behavioral needs of older adults while adapting to their diverse demands. It focuses on optimizing both the physical and socio-cultural environments central to elderly populations' daily lives [26]. As critical connectors within urban public space systems, community parks serve as pivotal spatial platforms for elderly activities. This study establishes a comprehensive evaluation framework for aging-friendly attributes in urban community parks. To mitigate subjectivity and uncertainty in prior research [15, 16], we employ a fuzzy comprehensive nonlinear evaluation method based on compromise weighting, enabling objective, quantitative assessments. A case study of Mianyang City demonstrates the scientific rigor and practical applicability of this approach.

(B) Case Study Findings and Optimization Strategies
The city of Mianyang—a region characterized by advanced population aging—serves as a representative case study. The methodology and findings offer actionable insights for cities grappling with aging populations, providing a replicable framework for developing aging-friendly community park evaluation systems and composite assessment methods. The overall satisfaction level with aging-friendly features in Mianyang's community parks is rated as moderate, reflecting limited approval among elderly users.

- Evaluative scores and ranked dimensions are as follows:
- Empathy (80) > Reliability (75) > Assurance (74) > Perceptibility (71) > Responsiveness (69).

Actionable Recommendations:

- Empathy (e.g., culturally resonant design) should remain a foundational priority to attract elderly engagement.
- Responsiveness (e.g., adaptive maintenance cycles) offers strategic leverage for phased upgrades.
- Perceptibility-related infrastructure, such as shaded seating and weather shelters, demands urgent attention.

- Systemic enhancements—including barrier-free pathways, elderly-oriented cultural programming, and accessible health hubs—are essential to align with national aging strategies and foster aging-friendly park development.

(C) Contributions and Future Directions

This study provides a replicable framework for aging-friendly evaluations in urban contexts. Limitations include:

- Subjectivity in parameterizing fuzzy membership functions, necessitating algorithmic refinements.
- Opportunities to expand the indicator system with dynamic metrics (e.g., seasonal usage intensity, socio-economic inclusivity).

References

1. Yanfeng G, Lijun W, Wenmeng F et al (2020) Challenges and strategic choices for healthy aging in China. Manage World 36(4):86–96
2. Hailong Z, Wenting L, Jinbiao J (2022) Legal safeguards for actively responding to the strategy of population aging. J Hangzhou Norm Univ (Soc Sci Ed) 44(3):11–22
3. Linchuan Y, Xianglong T, Jixiang L (2022) Research on the built environment of aging-friendly travel under the strategy of actively responding to population aging. Shanghai Urban Plan 1:156–162
4. Zhihao Y, Liang Q (2021) Research progress and review on aging-suitable renewal of old residential areas in China. Urban Issues 06:95–102
5. Pauline BDV, Jules S, Stephan M et al (2023) Collective self-build for senior friendly communities. Studying the effects on social cohesion, social satisfaction and loneliness. Hous Stud 38(7):1323–1341
6. Chiaki U, Hiroki I, Mika S et al (2023) A neighbour to consult with is important in dementia-friendly communities: associated factors of self-efficacy allowing older adults to continue living alone in community settings. PsychogeriatrS: Off J Jpn Psychogeriatr Soc
7. Bokolo JA (2023) Inclusive and safe mobility needs of senior citizens: implications for age-friendly cities and communities. Urban Sci 7(4)
8. Wenting Y, Xiaolei Z (2024) Post-occupancy evaluation of age-friendly use in comprehensive community centers: empirical cases from Guangzhou and Shenzhen. South Arch 02:105–114
9. Jusheng S, Yi S, Yamei X (2017) Spatial planning and design strategies based on the characteristics of elderly community activities: a case study of outdoor activity spaces in typical communities in Shenzhen. Urban Plan 41(05):27–36
10. Shaohua T, Yinbao G, Lifeng L et al (2020) Proactive health intervention in community walking environments: a physical activity perspective. Urban Plan 44(12):35–46 + 56
11. Min L, Gangyu H, Jianzhong H (2017) Research on the demand characteristics of community service facilities for the elderly in Shanghai: a perspective based on walking ability differences. Shanghai Urban Plan (01):25–31
12. Linchuan Y (2023) Elderly people's perception of walking environment and optimization of aging-friendly environment. World Arch 7:66–67
13. Wenqing W, Yanhan L, Liang L et al (2023) Study on thermal comfort of elderly in community parks: an exploration from the perspectives of different activities and ages. Build Environ 246
14. Juan L, Hao Y, Kexin C et al (2021) Evaluation and promotion strategies of different types of park services in Shenzhen. J Shenzhen Univ (Sci Technol Ed) 38(04):433–440

15. Tingting K (2023) Research on age-friendly evaluation and optimization of community public spaces based on the SD method: a case study of Nanhu community in Nanjing. In: People's cities, planning empowerment: proceedings of the 2022 China urban planning annual conference (19 Housing and Community Planning). Southeast University School of Architecture, pp 417–426
16. Yuan W, Yanxi X, Ruini W (2023) Design research on shared fitness facilities for the elderly and youth in urban parks based on POE and FAHP. Packag Eng 44(06):352–361 + 374
17. Ming X, Junying Y (2001) Application research of SERVQUAL scale in measuring service quality. Ind Eng Manag (06):6–9
18. Pablo K, Payam D, Lucia A et al (2020) Development of the urban green space quality assessment tool (RECITAL). Urban For Urban Green (Pre-publish) 126895
19. Nguyen PY, Astell-Burt T, Rahimi-Ardabili H, Feng X (2021) Green space quality and health: a systematic review. Int J Environ Res Public Health 18(21):11028
20. Guoling J, Jianwei W (2019) Evaluation of express delivery service quality in Xi'an based on the SERVQUAL-IPA model. Railw Transp Econ 41(10):57–63
21. Jian S (2021) The issue of mismatched supply and demand in elderly care services from the perspective of 'Demand Response' and its solutions. Zhongzhou Acad J 02:28–33
22. Jinchang S, Shuxin D, Yi L et al (2012) Fuzzy comprehensive evaluation method based on cloud model and its applications. Fuzzy Syst Math 26(06):115–123
23. Jijun Z (2000) Fuzzy analytic hierarchy process (FAHP). Fuzzy Syst Math 02:80–88
24. Jin L (2017) Risk evaluation in the decision-making stage of green building based on variable weight principle. J Wuhan Univ Technol (Inf Manag Eng Ed) 39(02):168–171 + 196
25. Jin L (2017) Risk assessment of green building decision-making stage based on variable weight principle. J Wuhan Univ Technol (Inf Manag Eng Ed) 39(02):168–171 + 196
26. Bencheng F, Xue M, Yu Z (2019) International practices of intergenerational mutual assistance in communities and their implications. Arch J 02:50–56

Development Path of Art-Enabled Rural Characteristic Culture: A Case of Wuxi, China

Hui Deng

Abstract The revitalization of rural culture is an important element of the rural revitalization strategy. Under the background of globalization, art, as a unique cultural force, has an irreplaceable role in stimulating the vitality of rural culture, enhancing the confidence of rural culture and promoting the sustainable development of rural society. The purpose of this article is to discuss how art can intervene in the development of rural characteristic culture through various ways, so as to provide new ideas for the revitalization of rural characteristic culture. The article firstly analyzes the interaction between art and rural culture, pointing out that art can inject new vitality into rural culture, while rural culture also provides rich materials for art. Then, the article discusses the coupling relationship between art and rural culture, and the various channels for art to intervene in the development of rural culture. On this basis, the article takes Wuxi's rural cultural development as an example, and details the artistic innovation of local culture, the innovation of cultural creativity, and the integration and introduction of artistic activities, etc. These paths not only enrich the rural culture, but also provide a rich material for art. These paths not only enrich the connotation of rural culture, but also provide new growth points for the development of rural economy. The article emphasizes that the development of rural cultural characteristics should ultimately be implemented in the villagers, and that the villagers should become the main body of rural cultural revitalization by enhancing their cultural consciousness and participation.

Keywords Artistic empowerment · Rural construction · Distinctive culture · Development path

H. Deng (✉)
Wuxi Taihu University, Wuxi, China
e-mail: naussica@qq.com

B. He et al. (eds.), *Climate-Resilient, Low-Carbon, and Sustainable Cities*, Lecture Notes in Civil Engineering 754, https://doi.org/10.1007/978-981-95-2169-2_51

1 Introduction

As the strategy of rural revitalization continues to advance, the protection and development of the countryside is receiving more and more attention. The report of the Nineteenth National Congress clearly puts forward that the overall goal of the strategy of rural revitalization is "industrial prosperity, ecological livability, civilized rural customs, effective governance and affluent living". Civilization is one of the important contents of the implementation of the strategy of rural revitalization, and it is also an important foundation for the construction of civilization in Chinese society. The construction of civilization in the countryside can't be accomplished by the way of quick success and quick profit, nor can it be realized by the formalism of smearing and painting [1]. Rather, it is necessary to integrate good traditional culture with modern culture, infiltrate rural production and social lifestyles in a subtle way, and transform them into people's conscious actions and internalize their beliefs and habits.

2 The Intrinsic Connection Between Art and the Building of a Culture of Rural Identity

2.1 Difficulties Faced by the Development of Rural Special Culture

In the process of development, the rural characteristic culture is influenced by the traditional rural culture, which is still dominated by the traditional social habits and farmers' way of thinking, and lacks the innovative thinking to keep pace with the times, which makes the traditional rural culture unable to adapt to the needs of the contemporary countryside, and seriously affects the development and expansion of the rural culture.

As China's urbanization progresses, the surplus labor force in the countryside is transferred to the cities. Benefiting from the improvement of China's education system, the level of education in the countryside has also been gradually raised; on the other hand, the Government has favoured supporting the development of the cities in terms of technology and capital, while the development of the countryside has lagged behind and been slow. Because of the narrow space for development in the countryside, the talents with higher education tend to develop in the city. This vicious circle has led to a serious loss of cultural talent in the countryside.

The rapid development of urban culture has widened the gap between urban culture and rural culture. The gradual emergence of urban culture as the dominant culture and its prevalence in rural areas has seriously affected the villagers' judgment of advanced culture and their belief that rural culture is a manifestation of "backwardness", resulting in blind imitation of urban culture and the loss of rural distinctive culture.

In contrast to the urban infrastructure of libraries, science and technology centers, museums, etc., which is readily available, the cultural infrastructure in the countryside is very backward and poorly built. This has a particularly strong impact on children, creating a serious lack of cultural self-confidence in the children's community and a longing for urban life. Dramas and variety shows on television and the Internet, which are based on urban culture, have also deepened rural children's longing to escape from the countryside in the future. This subconscious resistance to the rural culture accelerates the decline of the distinctive rural culture [2].

2.2 Interaction Between the Arts and the Distinctive Culture of the Countryside

Culture is a spiritual activity and its product relative to economy and politics. Rural characteristic culture can be understood as a specific rural area, formed through the long-term creation of the people, and recognized by this group, or a more typical ideology different from other areas. It is not the same as material, but is the sublimation of people's living habits and behaviors for a long time. China is a big agricultural country with a long history of farming culture, and a series of colorful and distinctive rural culture has been created and inherited by our ancestors in the vast countryside. Rural culture encompasses all aspects of rural life, productive activities, and ecological resources; according to the theory of cultural structure, it consists of three interdependent dimensions: material, institutional, and spiritual (Table 1).

Art is a wide range of human activities involving creativity and imagination, with aesthetic qualities and aesthetic values, and has a natural advantage in expressing thoughts and conveying ideas. Art itself originates from the cultural and socio-economic environment, and is the crystallization of human wisdom and creativity. It can be said that art works are the carriers reflecting the thoughts and feelings of a nation and its general cognition. The unique language symbols and inherent aesthetic rules of art works can present and convey information in a visual and intuitive way, enabling people to cross the boundaries of language and race, break the language communication barriers, and start a dialogue of heart and mind. Art works often make use of some artistic means and media to shape images, create atmosphere, or reflect reality and emotions, which can eliminate the stereotypes of administrative language and be easily recognized and accepted by the people by transforming the visible into the invisible.

Rural characteristic culture has the essential characteristics of creativity and aesthetics, as well as strong regionality and uniqueness; at the same time, it also has the characteristics of symbolism and easy reproducibility, which can be expressed with the help of images, sounds and other carriers, and break the limitations of time

Table 1 Three dimensions of the rural cultural structure (Self-drawn)

Category	Gene	Factor
Material culture	Natural geography	Landscape, field, forest, lake and grass system
	Production system	Agricultural tools, water, farming practices
	Building form	Settlements, dwellings, shrines, temples, sunbeds
	Household goods	Clothing, eating utensils, transport
	Landscape signage	Pagodas, wells, boundary markers, old trees
Institutional culture	Organisational system	Clan system, village self-government organisations
	Statutory scheme	Village rules and regulations, clan and family traditions
	System of rites and customs	Weddings, funerals, festivals and rituals
	Economy	Agrarian system, mutual aid traditions
	Education system	Private schools, village covenant seminars
Spiritual culture	Value	Filial piety, fraternal duty, loyalty, and trustworthiness; thrift and diligence in household management
	Belief system	Ancestral beliefs, nature spirits, local beliefs
	Knowledge system	Agricultural calendar, folk medicine
	Artistic expression	Theatre, folk music, crafts
	Collective memory	Legends, village histories, and genealogies

and space to spread continuously, and it is mostly constructed and integrated with the history, traditions, customs and so on, and it is similar to the art, which implies the empathy factor of human beings [3]. Rural characteristic culture can be effectively promoted and developed through the flexible integration with art, which can be infiltrated and derived like rain in the spring breeze.

Under the policy guidance of "ecological civilization strategy" and "beautiful countryside construction", in order to avoid the mistake of urban construction, which results in the situation of detaching from the local community, cultural fragmentation and even "one-sidedness", it is necessary to summarize the advanced experience through research and discover the path and method of art-enabled rural cultural development in line with China's national conditions, so as to make useful additions and attempts for the development and research of related fields. In order to avoid repeating the mistakes of urban construction, which may lead to the situation of detachment from the local community, cultural fragmentation and even "one-sidedness of thousands of villages", it is necessary to summarize the advanced experience through research and discover the paths and methods of art-enabled cultural development of rural characteristics in line with China's national conditions, so as to make a useful supplement and attempt to the development and research of the related fields. It is necessary to explore whether we can break the great differences in time and space with artistic diversity, open-mindedness and inclusive cultural attitudes, give positive responses and make up for the construction and development of rural characteristic culture, and at the same time, explore the updating and expansion of the contemporary art system.

3 The Path of Art-Enabled Cultural Development of Rural Specialties

Rural cultural development is an important support surface of the huge systematic project of rural revitalization, and it should start from sorting out the cultural forms of the countryside and giving full play to the advantages of the resources, and incorporate the artistic elements and means according to the local conditions and needs, so as to promote the development of the rural culture. On the basis of comprehensive analysis of domestic and foreign experiences and the results of the recent study on the practice of rural construction, the following is an attempt to summarize the paths and ways of art-enabled cultural development of rural characteristics at various levels.

3.1 In Terms of the Coupling of Art and Rural Culture

Coupling refers to the relationship between systems with different attributes and structures in the natural world, or between different elements of the same system through non-linear interaction to form a relationship of mutual action and influence, including the effect of "cooperation" and "synergy". There is a natural coupling relationship between art and rural culture, and the ways of building the coupling and interaction can be summarized into three categories.

Introductory. Imported art is a form of art or content directly into the cultural construction of the countryside, to a certain extent, it can bring about the improvement of the rural landscape. The most common form is led by the local government, through the organization of large-scale art festivals, gathering artists to hold exhibitions in the countryside, using the artists' "traffic" to enhance the cultural image of the place. The countryside plays the role of "canvas" in this model, providing a stage for artists to create and display their talents. This kind of improvement in the cultural landscape of the countryside is like a flash in the pan, it is short-lived and has a limited period of time, and lacks long-term vitality; once the exhibition is over, its influence reverts to its former self. The negative impact of this model also lies in the waste of resources and even the destruction of the local ecological environment. For example, the Longli International New Media Art Festival was initiated with the intention of building a new platform for the integration of new media art, culture and technology, and injecting vitality into the local rural culture. However, except for the period of the festival, there are few tourists in normal times, and the expected benefits have not been achieved. The utilization of new media technologies involving sound and light has disrupted the lives of local residents, replacing original tranquility with intrusive noise pollution. The bright light interferes with birds' navigation capabilities and sleep patterns, causing circadian rhythm disturbances. This not only results in wasted energy, but also has a negative ecological impact.

Incorporation. Residency is to station art works or art exhibitions in the countryside for a long and stable period of time, continuously influencing the construction of the cultural landscape of the countryside. The biggest advantage is that it can fully respect and explore the characteristics of local customs and cultural resources, and reshape the characteristic culture of the countryside. Specific methods include the improvement of the rural human environment, the protection and renewal of residential buildings, the creation of folklore museums and art galleries, the organization of long-term rural festivals and art exhibitions, and the creation of rural lodgings, etc. The Xucun Project will also help to improve the cultural landscape of the countryside. "The Xucun Project is a typical example of this model. As early as 2007, artist Qu Yan was invited to Xucun to carry out rural renovation and reshaping of rural cultural beliefs. The artist, led by Qu Yan, has been stationed in Xucun for a long time, and has carried out a series of rural construction activities, including repairing and remodeling residential buildings, planning art festivals, and organizing the Xucun Forum, etc., which has formed the brand effect of Xucun culture. The advantage of this model is that it can further improve the rural environment and make the rural cultural landscape last, and to a certain extent, it can also stimulate the initiative and enthusiasm of the villagers in the construction of rural culture, but it still does not have much impact on the production and life of the villagers [4].

Integration. Under the organization and guidance of art creators, the integration mode involves the participation of many subjects, especially local villagers, in art creation, in the creation of rural public space, rural landscape construction and other handicrafts, etc., so as to perfect the rural cultural industry chain. The essence of this model is to weaken the dominant role of art creation organizers and highlight the subjective character of villagers in the construction of rural culture. Emphasizing participation is an important feature of the integrated model, in which designers, artists, scholars, villagers, tourists and other stakeholders take the initiative in the development of the environment, architecture, space, products or services [5]. In the case of rural revitalization in the Tsumari area of Echigo, for example, the government, enterprises, architects, artists, designers, volunteers, operators, and especially local villagers have brought their respective expertise and skills into play. Through a win–win incentive structure for villagers, enterprises, non-profit organizations, farm operators, and individual merchants, the motivation of multiple actors is mobilized, and a mechanism of division of labor and benefit sharing is formed [6]. The Setouchi International Arts Festival addresses the decline of the fishing industry and the island's identity crisis through artistic interventions, inspired by the Echigo-Tsumari model. Deeply integrated with local agriculture and fisheries, the festival promotes the return of young entrepreneurs. Furthermore, cross-island tours will connect dispersed resources, fostering a distinctive branding of the "Art Archipelago". The integrated model is based on the long-term development of the local economy and culture, emphasizes the intrinsic needs of the villagers, stimulates the enthusiasm and subjective role of the villagers, and truly realizes the regeneration of the village's self-breeding capacity.

3.2 From the Point of View of the Channels of Artistic Intervention in the Development of Rural Culture

Single-channel. The single-channel model refers to the intervention of a single art category, such as architecture, sculpture, painting, public art, etc., in rural culture, giving it uniqueness and innovation. This mode usually consists of artists and villagers spontaneously forming the main body of creation, and adopts a bottom-up approach to maximize the respect for the local characteristics of culture. Zhaji Village in Jing County, Anhui Province, is located in the mountainous area and lacks development conditions, but it has preserved more than 200 ancient buildings from Yuan to Qing Dynasty, with high historical and cultural value. In the 1980s and 1990s, Liu Xinsheng, a national-level artist and painter, came to Zhaji Village for sketching, and loved the local conditions so much that he invited famous artists and schools from all over the country to come to Zhaji for sketching classes. Rooted in the ancient village, the painters took the simple life of the countryside as a source of inspiration to create works of art, which drove the local tourism industry. After years of development, the unique artistic temperament of Zhaji has been formed, making Zhaji a well-known "painter village".

Multi-channel. The multi-channel model refers to the involvement of various art forms in the construction of rural culture, forming a comprehensive and diversified rural culture. This mode is usually manifested in the government-led top-down development of rural cultural tourism [7], in which multiple art forms are used in the restoration and renewal of rural landscapes, folklore and folk art performances, and the development of cultural and creative industries, to satisfy people's sense of identity and sense of belonging to the rural culture and at the same time stimulate the vitality of the development of the rural culture. Matera, the best-preserved cave settlement in the Mediterranean, was once labeled "the shame of Italy" in the 1950s due to its extreme poverty and unsanitary conditions. The restoration project, launched after 1986, achieved functional transformation of the space through modernization and revitalization of cultural facilities. It employed digital technology to reconstruct the original appearance of the medieval cave settlement, accompanied by the development of a dedicated website and mobile application. Traditional festivals such as the *Festa della Madonna della Bruna* were organized, while themed tourism routes were developed by leveraging the filming locations of movies like *The Passion of the Christ* and the James Bond film *No Time to Die*. This comprehensive effort transformed Matera from a "symbol of poverty" into a sustainable cultural landmark, offering a global model of "living renewal" for rural heritage preservation. Fiskars, historically renowned for its ironworks, is the birthplace of the iconic Finnish brand Fiskars scissors. Facing decline in the 1980s as the iron industry waned, the village neared abandonment. By the late twentieth century, revitalization efforts began through architectural restoration and adaptive reuse of spaces. Artists, architects, designers, and craftsmen collaborated to forge a harmonious community blending history, art, design, creativity, and nature. Its success has attracted worldwide attention, with its practices shared as a template for rural renewal across the globe.

Under the current environment of China's booming rural construction, art intervenes in rural construction from all aspects in multiple ways, and the integrated and multi-channel mode is the inevitable trend of rural cultural development in the new era.

4 Experience in the Development of Wuxi's Distinctive Rural Culture

4.1 The General Situation of the Development of Rural Special Culture in Wuxi

Let the traditional villages poetic habitat in the land of Jiangsu is the construction of beautiful Jiangsu in the title of the meaning. In recent years, Jiangsu Province attaches great importance to the construction of beautiful villages and cultural villages, traditional villages, the protection of innovative implementation of the characteristics of the idyllic countryside construction action, so far the province announced the provincial traditional villages have 364, the provincial characteristics of the idyllic countryside 407, part of the region to promote the "string of points connected to the line of the piece," initially formed the characteristics of the idyllic countryside. Wuxi is located in the south of Jiangsu Province, has rich and diverse rural resources, and has achieved remarkable results in rural construction. At present, 40 villages such as Yanjiaqiao Village, Yangjian Town, Xishan District, Wuxi City, have been approved as traditional villages in Jiangsu Province. Villages such as Fengxiang in Taoyuan Village, Yangshan Town, Huishan District, and Qiansishe in Taoyuan Village, Yangshan Town, have been named Characteristic Rural Villages of Jiangsu Province. In the construction of new countryside, Wuxi's countryside culture has also achieved vigorous development, and has formed the "Hushan Cultural Tourism" and "Peach Shengjing" beautiful idyllic countryside tour lines, showing unique natural scenery, red culture, farming civilization and rural aesthetics.

4.2 Typical Experience and Cases in the Development of Wuxi's Rural Characteristic Culture

The construction of rural culture in Wuxi is deeply integrated with the industrial development, environmental improvement, ecological construction and other material infrastructure construction in the countryside, and the development and formation of rural distinctive culture is closely related to the integration of the overall concept of rural construction. Good material and ecological environment is the background and strong support for the development of distinctive culture, and it is also the basis for the survival of distinctive culture. Artistic creativity stimulates the life, production and ecological resources of the countryside in a multi-channel and omni-directional way, and forms the characteristic culture integrated with the daily life of the countryside, and maintains the vitality of the development in a benign cycle and interaction.

Qiansishe: Artistic innovations in local culture. Artistic innovation of local culture refers to the use of artistic means to drive the local characteristic cultural resources and form the unique cultural characteristics of the village.

Qiansishe is a village under the jurisdiction of Yangshan Town, Wuxi City. With a total area of over 50,000 square meters, the village comprises 73 households and a population of 296. The villagers are descendants of Zhou Dunyi, a renowned Neo-Confucian philosopher from the Song Dynasty. Peach cultivation serves as the primary local industry. In 2017, the village was shortlisted for Jiangsu Province's first batch of characteristic idyllic village pilot programs. Its development has been supported by provincial pilot funding and municipal special subsidies.

Centering on the peach industry and integrating the "lotus culture," the planning team has crafted distinctive landscapes while providing comprehensive guidance throughout the pilot project. The environmental construction of Qiansishe Village centers on "lotus culture", focusing on the manifestation of lotus cultural symbols in the spatial environment, carrying cultural connotations through the addition of public spaces and landscape vignettes, awakening the villagers' sense of cultural identity, and perpetuating the cultural spirit and local culture of "loving the lotus". At the same time, the Zhou Mingxian Hall is built, which exhibits the genealogy, historical documents and anecdotes of the Zhou family in the village, establishes a good family tradition and provides villagers with a venue for public activities (Fig. 1). Through the shaping of the lotus cultural space, Qiansishe Village has formed a unique rural culture, making it a highlight of Yangshan's rural cultural tourism. Table 2 outlines the cultural framework of the case study, detailing methods of artistic intervention and specific practice, showing that it has deeply tapped into the local characteristics of cultural resources for the integration of art, to achieve the transformation from a traditional agricultural village to a distinctive culture-tourism integrated idyllic destination.

Tianyuan Dongfang: Cultural creativity for innovation. The development of a distinctive rural culture can be based on local traditional culture, or it can take the form of a unique new culture by adopting a creative mode of cultural innovation, i.e.,

Fig. 1 Public activity grounds in Qiansishe Village (Self-drawn)

the creation of a new culture that does not intersect with the original culture of the countryside, thus forming a development mode of a distinctive local culture. This mode is often due to the lack of distinctive culture or natural resources in the villages themselves, or in order to create a new and unique culture in a way that is different from the development mode of villages with the same type of culture and to create a gap between them.

Tianyuan Dongfang, situated in the core zone of Yangshan Town, Wuxi City, is China's pioneering idyllic complex. Developed by Dongfang Garden Industry Group with a total planned area of approximately 6,246 acres, the project commenced construction in 2013. Following the 2017 No.1 Central Document's designation of "Rural Complex" as an innovative rural revitalization model, this case was featured on CCTV's *Economic Half Hour* as a national exemplar.

The founder of Tianyuan Dongfang, with the background of architect and professional perspective of urban planning, initiated the idea of "Neo-Pastoralism" in the field of rural construction, and carried out a large-scale, holistic and comprehensive "Rural Complex" in the town of Yangshan, in the form of cooperation between the enterprise and the local government. At present, it has completed the planning and construction of the modern agricultural industrial park and the new garden community mainly in peach industry, and built a rural tourism resort area in the demonstration area, which includes hotels, academies, bazaars, schools, field paradise and other business forms (Fig. 2). Tianyuan Dongfang is the first to try to form a new brand

Table 2 Analysis of artistic intervention approaches in Qiansishe village (Self-drawn)

Category	Artistic intervention	Specific Practice
Material level	Spatial reproduction of cultural symbols	Around the lotus culture, landmarks were built such as the Lotus Observatory, the Lotus Pavilion and the Zhou Mingxian Hall, and family mural paintings cover 80% of the households
	Artistic interventions in ecosystems	Make use of natural resources such as peach forests and rivers to design peach blossom walkways and ecological barges
	Functional space enhancement	Convert vacant houses into characteristic homestays and agritourism inns
Institutional level	Government guidance and policy support	Relying on policies to obtain financial support to promote the integration of infrastructure development and industry Establishment of a working leadership group to co-ordinate resource allocation and project implementation
	Deep involvement of villagers' self-organisations	Establishment of village self-government councils and supervisory boards to oversee public affairs Operators also serve as cultural interpreters, achieving the role integration of "operators as cultural disseminators"
Spiritual level	Contemporary translation of the family tradition system	Building the Zhou Mingxian Hall to promote the culture of cleanliness and honesty Incorporating the symbolism of "being a clean person" into speciality food production
	Folklore activities and community cohesion	Developing handmade products such as peach blossom brew and rice wine, and disseminating them through the *Sishe* Brand
	Art education and Aesthetic enhancement	Enhance the aesthetic sense of villagers and tourists through public art such as murals and poetry walls

of rural culture and drive the integrated development of urban and rural areas with the new cultural tourism industry. Relying on the ecological resources and policy dividends of Yangshan, the case adopts an innovative cultural and creative model, forming a paragon of the integration of agriculture, culture, and tourism (Table 3).

Yanjiaqiao: Integration and introduction of artistic activities. A variety of art forms and art activities are integrated with the production and life of villages through various channels and in an all-round way, enhancing the interaction between the government's organization, the guidance of professional designers, the participation and collaboration of villagers, and the in-depth experience of tourists, and stimulating the vitality of pluralistic main bodies**.**

Yanjiaqiao, located in eastern Wuxi, covers 8.3 square kilometres and is recognized as the birthplace of both Xiju Opera and modern Chinese industrial enterprises. Since 2005, municipal authorities have implemented a preservation and development plan, prioritizing the restoration of historical architecture and revitalization of the

Fig. 2 Tianyuan Dongfang "Rural Complex" (Self-drawn)

water town's vernacular features (Fig. 3). In 2018, the village became a pilot for the construction of characteristic pastoral villages in Wuxi. Based on the "Yanjiaqiao Village Planning of Yangjian Town, Xishan District, Wuxi City" compiled in 2019, which proposed goals such as "a village for green life experience and for historical and cultural display", it carried out infrastructure construction and industrial cultivation with the support of national financial subsidies and funding from multiple channels, including the Tang Junyuan Educational Foundation.

The government and organizing committee have implemented a scientific preservation strategy for Yanjiaqiao Ancient Village, developing tiered protection measures based on historical buildings' conservation levels and architectural types. Through meticulous restoration of heritage structures and revitalization of cultural assets, they have successfully integrated the village's distinctive commercial traditions with Xiju Opera culture, vividly showcasing its historical legacy. Transformative projects have converted farmland into artistic landscapes and floral displays, while annual festivals including the Kite Festival and Rapeseed Blossom Celebration complement cultural offerings like opera workshops and agricultural education programs. This comprehensive approach has established a premium rural tourism circuit. It was visited and reported by CCTV's *Home in the Mountains* programme in 2024, creating a traffic effect of "the same route as the celebrities" (Table 4).

Table 3 Analysis of artistic intervention approaches in Tianyuan Dongfang (Self-drawn)

Category	Artistic intervention	Specific practice
Material level	Architectural restoration and vernacular landscape creation	Repairing the historical buildings of Shifang Village, preserving the ancient wells, ponds and other original elements, and integrating modern design concepts to transform the cultural bazaar, which includes cafes, themed guesthouses, handmade workshops and other business
	Artisation of agricultural space	Farms are used as landscapes to build agricultural aesthetics in conjunction with harvesting experiences, combining recreational facilities with farmland textures
	Innovative design of cultural and tourism facilities	The world's first "Plants vs Zombies Farm" creates a space for parent–child nature education through game IP scenarios
Institutional level	Innovation in policy mechanisms	As the country's first idyllic garden complex pilot, it has received financial support from the central government and a breakthrough in local land policy A working group was set up to coordinate land transfer, villagers' resettlement and business planning
	Synergistic governance by multiple actors	Enterprise-based, government bridging, farmer participation, multi-party development approach
	Industrial integration mechanisms	In the early stage, the land value is enhanced through cultural tourism, and in the later stage, the operation is fed by real estate, which is the model of "tourism + real estate"
Spiritual level	Native aesthetics education	The Cultural Bazaar conducts cultural experiences and integrates handicrafts into the aesthetics of daily life
	Life philosophy dissemination	Guided by the theory of "Neo-Pastoralism," community design creates a consumer image of "poetic dwelling" through IP events like the Peach Blossom Festival

5 Analysis and Discussion

Through comparative analysis of three cases, these rural cultural development initiatives demonstrate two core mechanisms: (1) Cultural Anchoring through systematic exploration of endemic heritage, guided by artistic reinterpretation frameworks; (2) Pluralistic Governance incorporating multi-stakeholder agency. Although all cases achieved symbiotic economic-cultural outcomes through integrated, multi-channel artistic interventions, marked divergences exist in the distribution of decision-making power and the pathways of cultural dissemination (Table 5).

The artistic intervention in Qiansishe Village primarily operates at the material level, combining cultural symbols with environmental transformation. Through landscape design, it mobilises local characteristic cultural resources and reconstructs

Fig. 3 Riverside promenade restored in Yanjiaqiao Village (Self-drawn)

indigenous cultural elements, thereby shaping distinctive rural cultural features. While the government and cultural elites co-govern the process, emphasising the authority of cultural symbols, their capacity for market-oriented innovation remains underdeveloped. Cultural dissemination relies on educational initiatives and regional tourism networks, yet its reach is largely confined to the local area, demonstrating limited ability to expand beyond the immediate community.

Leveraging institutional policy opportunities, Tianyuan Dongfang exemplifies an urban–rural integration paradigm through market-driven operations. By adopting an innovative cultural-creative model, it cultivates a distinct neo-culture and constructs a symbolic modern idyllic consumption space. Enterprises function as integrators to coordinate resources, while the government assumes a supportive role through policy provision, collectively establishing an efficient market-oriented mechanism. The idyllic lifestyle proliferates rapidly via consumption networks and community linkages, achieving broad outreach. However, commercial pressures risk diluting cultural depth, and the depth of local culture is insufficient.

Yanjiaqiao Village's artistic intervention primarily targets the spiritual dimension, pioneering a niche yet refined localization model through the integration of intangible cultural heritage (ICH) and agrarian traditions. Operating under a tripartite framework of 'government leadership, enterprise execution, and villager participation', this approach prioritises policy guidance and cultural preservation, yet struggles with market-oriented operational efficiency. Key cultural elements—including revolutionary heritage, Xiju Opera, and agrarian aesthetics—are disseminated through educational programs, tourism networks, and media campaigns. While

Table 4 Analysis of artistic intervention approaches in Yanjiaqiao (Self-drawn)

Category	Artistic intervention	Specific practice
Material level	Rehabilitation and functional conversion of historic buildings	Renovate the granary into a themed pavilion to showcase the historical lineage of Yanjiaqiao and become a cultural landmark Restoration of historical buildings to create cultural exhibition spaces such as the Xiju Opera Museum and the Tang Industrial and Commercial Exhibition Hall Renovation of the old street on the riverbank to incorporate modern businesses such as cafes and guesthouse
	Integration of earth art and agricultural landscapes	Combined with the improvement of high-standard farmland, it creates artistic rice paddies and a four-season flower sea, realising the integration of agriculture and art Create landscape installations using old farm tools and objects to enhance the vernacular narrative
	Artistic upgrading of infrastructure	Upgrading of barges, paths, walls and landscaping Modern facilities such as smart public toilets and waste separation kiosks incorporate traditional elements
Institutional level	Policy support and financial security	The government invests in the restoration of ancient villages and seeks funding through provincial pilots Prepare a conservation and development programme in cooperation with universities to form a systematic plan that specifies the norms for building restoration and the introduction of business Establish a working group led by "Party—building guidance + town—village dominance" to coordinate the integrated development of agriculture, culture, and tourism
	Participatory governance mechanisms	Organise villagers to participate in decision-making such as Xiju Opera heritage and agricultural research
	Village-enterprise co-operation and industrial linkage	It cooperates with scientific research units to promote high-quality rice varieties, builds the brand of "Yanjiaqiao Rice", and enhances the added value of agricultural products by combining with the paddy field art festival

(continued)

Table 4 (continued)

Category	Artistic intervention	Specific practice
Spiritual level	Cultural heritage and innovation of Xiju Opera	Establishing a museum of Xiju Opera and a small theatre, holding regular performances, reproducing classic plays and training young inheritors Incorporate elements of Xiju Opera into the tourism experience to enhance cultural interactivity and dissemination
	Clan culture and red gene promotion	Through the theme pavilion to show the Tang family "industrial service" spirit Tapping into the history of the anti-Japanese revolution, designing red education routes and carrying out immersive patriotic education
	Community activities and vernacular identity	Regularly organise activities such as rice culture festivals and farming studies to strengthen the local emotional connection between villagers and tourists Developing Xiju Opera themed cultural and creative products and traditional cuisine to convey cultural values

achieving profound cultural penetration among traditional demographics (middle-aged/elderly groups and educational tourists), it demonstrates limited reach into youth demographics.

In comparison with international cases, in terms of cultural development positioning, the Wuxi cases focus on the localized reconstruction of traditional cultural veins, while international cases pay more attention to the global transformation of cultural resources. Echigo Tsumari, through international artists' local creations, has transformed the local "agricultural culture of the snow country" into the philosophical proposition of "the relationship between humans and nature". Matera has transformed the "memory of poverty" in cave dwellings into a living fossil of Mediterranean civilization, achieving a symbolic reversal from "national shame" to "European Capital of Culture". Fiskars combines cast-iron industrial heritage with Nordic minimalist design genes, interpreting the Scandinavian aesthetics of "sustainable design + eco-craftsmanship" and forming a cultural-creative regeneration of industrial sites. In terms of development implementation subjectivity, the Wuxi cases are predominantly government-led, while international cases focus more on social collaboration. The Echigo Tsumari Art Festival has implemented a collaborative mechanism of "curator + regional revitalization agreement + volunteer network", forming a 20-year-enduring social collaboration network; Fiskars Village has adopted a self-organizing model of "designer cooperative + craftsmen's alliance", forming a closed-loop ecosystem for the art economy. Regarding beneficiary structures, the Wuxi cases tend to follow an "enterprise-led profit—villagers receive rent" framework through housing transfers, while international cases emphasize community sharing. The Fiskars Village Artists' Cooperative allocates profits based on

Table 5 Comparison of the distribution of decision-making power and the path of cultural dissemination among three cases (Self-drawn)

Dimension	Qiansishe	Tianyuan Dongfang	Yanjiaqiao
Distribution of decision-making authority	Co-governance between government and cultural elites + villagers' vectorisation	Enterprise-led + government support + villagers' instrumentalisation	Government-led + enterprise implementation + limited participation of villagers
Subject of communication	Government, School, Cultural elite	Businesses, Artists, Urban consumers	Government, Intangible cultural heritage inheritors, Media
Dissemination of content	Neo-Confucianism + Integrity education	Oriental aesthetics + New rustic lifestyle	Red gene + Xiju Opera culture + Farming aesthetics
Dissemination path	Education chain: Integrity training for government agencies → Dissemination within the vertical system Tourism chain: Check-in at cultural landmarks → Word-of-mouth diffusion among regional tourists	Consumption chain: Art bazaar → Urban young groups → Social media fission Tourism chain: Agri-tourism integration → Visitor experience → Social media diffusion	Education chain: Red study → Party and government organisation, schools → Word of mouth spreading Tourism Chain: Agri-tourism integration → Tourist experience → Social media diffusion Media chain: CCTV programme exposure → National awareness raising
Key benefits	Cultural authority and educational function	Market-based efficiency and cross-border innovation	Cultural depth and historical significance

contribution levels, and Echigo Tsumari has established an "Art Festival Fund" for public affairs like elderly care. These differences reflect divergent pathways for rural revitalization under varying social systems and cultural traditions.

While maintaining cultural subjectivity, the Wuxi cases can draw on the community participation mechanisms and sustainable business models of international cases. In 2024, China's first special local law on harmonious villages was formulated by Wuxi—the Wuxi Harmonious Villages Ordinance—which clarifies the core directions of rural planning, cultural protection, ecological governance, and villager participation from a rule-of-law perspective; it provides strong legal safeguards for village construction. The pathways for art-driven rural revitalization in Wuxi have been endowed with clearer goals and directions. Firstly, enhancing cultural inheritance and revitalization through preservation of traditional village textures and development of intangible cultural heritage resources. Secondly, coordinating village planning with ecological synergy by optimizing industrial structures while balancing environmental benefits. Thirdly, strengthening villagers' agency and benefit distribution—avoiding "artist solos"—through participatory and collaborative rural construction that stimulates villagers' endogenous motivation and cultivates sustainable self-regeneration capabilities.

6 Conclusion

This study explores the typology and pathways of artistic intervention in rural cultural construction. Through a cultural stratification methodology (material/institutional/spiritual), it examines cultural development cases in Wuxi's rural villages. The core propositions of rural cultural development are "whose culture" and "for whom to disseminate". Qiansishe and Yanjiaqiao Village demonstrate that government-led initiatives prioritize preservation and transmission of local traditional culture; the Tianyuan Dongfang project reveals enterprise-driven models focus on achieving economic efficacy. A power-balancing mechanism needs to be established regarding decision-making rights allocation across all three cases to strengthen villagers' agency. Within the same region, rural cultural development should maintain staggered progress and differentiated positioning—preventing homogenization and redundant construction while forming mutually complementary relationships—enabling culture to become a genuine endogenous driver for sustainable community development.

In further research, other aspects such as economic benefits, ecological benefits, and the satisfaction of the subjects should be considered, and quantitative assessment should be made in terms of dimensions such as the degree of cultural inheritance, the growth of villagers' incomes, and the rate of tourists' revisiting the area.

Acknowledgements This study is supported by the "Jiangsu Provincial Qinglan Program Funding for Higher Education Institutions".

References

1. Kong X (2018) Nine dimensions of rural revitalization. Guangdong People's Publishing House, Guangzhou
2. Zeng R (2019) Exploring the development path of rural revitalization from a cultural perspective. Economic Management Press, Beijing
3. Shi W, Wu Y (2021) Characteristic culture + rural revitalization: Models, Methods and Cases. Xiamen University Press, Xiamen
4. Lin Z (2018) Research on the mode of art intervention in rural construction. Public Art 5:99–102
5. Li M, Pi Y (2019) The inspiration of Scandinavian participatory design for Chinese design intervention in rural construction. Contemp Artist 1:61

6. Gu Y (2020) Governance for high-quality development. In: Proceedings of the annual conference on urban planning in China 2020 (16 Rural Planning). China Society of Urban Planning, Chengdu, pp 703–709
7. Zhang F, Li C (2018) Realistic reflection on Art intervention in the construction of rural cultural identity. J Northeast Agric Univ (Soc Sci Ed) 16(06):39

Factors Influencing the Willingness of Elderly People in Urban Communities to Participate in Mutual Aid Elderly Care from the Perspective of Active Aging

Yiwen Xu

Abstract With the advancing process of population aging, the elderly care issues in urban communities are becoming increasingly prominent. Traditional methods of elderly care can no longer meet the diverse needs of the elderly population. Mutual aid elderly care, with its unique advantages, has gradually emerged as a new highlight within the elderly care service system. This paper, based on the theoretical framework of active aging, systematically analyzes the willingness of elderly people in urban communities to participate in mutual aid elderly care and its influencing factors. The study aims to explore how individual characteristics, economic conditions, health status, family structure, and social support collectively influence the elderly's choice of mutual aid elderly care. Through this research, it is hoped to provide scientific support for government policymaking and community optimization of elderly care services, thereby promoting active participation of the elderly and fostering harmonious social development.

Keywords Active aging · Community · Elderly · Mutual aid elderly care

1 Introduction

1.1 Research Background

The aging population has become a major social issue globally, especially in China, where the population is aging due to falling birth rates and increased life expectancy. According to statistical data, by 2023, the population of people aged 60 and above in China has exceeded 260 million, accounting for over 18.9% of the total population. With the rapid increase in the elderly population, traditional forms of care are facing unprecedented challenges, including a shortage of caregivers, a lack of resources, increased economic pressure, and an insufficient social security system, all of which

Y. Xu (✉)
Xuzhou University of Technology, Xuzhou, Jiangsu, China
e-mail: xuyiwen0511@126.com

B. He et al. (eds.), *Climate-Resilient, Low-Carbon, and Sustainable Cities*, Lecture Notes in Civil Engineering 754, https://doi.org/10.1007/978-981-95-2169-2_52

pose serious challenges to the quality of life for the elderly population. Therefore, how to effectively address the challenges of an aging society has become a focus of attention for policymakers, scholars, and all sectors of society.

1.2 Research Progress

In the current research on the willingness of the elderly in urban communities to participate in mutual support for the aged from the perspective of active aging, although there are some theoretical discussions and empirical analysis, there are still some research gaps and deficiencies. First of all, most existing studies focus on the influence of factors such as health status, social support and economic conditions on the participation intention of the elderly, while less attention is paid to the effect of the cultural background, social identity and individual differences of the elderly on the participation intention. In particular, the differences in the concepts and needs of the elderly in different regions and different cultural backgrounds have not received enough attention. China is a vast and culturally diverse country, and the elderly in different regions have significant differences in family structure, social network and economic conditions, which may directly affect the acceptance of mutual support for the elderly [1]. Therefore, future studies should strengthen the investigation of local differences and cultural diversity, and analyze the factors of the elderly's participation in mutual support for the aged from a more comprehensive perspective.

Secondly, although there have been studies on the influence of health status and social support on the elderly's participation in mutual support for the aged, in-depth studies on the psychological factors of the elderly are still scarce. With the growth of age, the psychological pressure and emotional needs faced by the elderly gradually increase, loneliness, anxiety, loss and other psychological difficulties on the quality of life of the elderly can not be ignored. Existing studies pay little attention to the psychological expectations and emotional needs of the elderly. Whether the elderly can accept and actively participate in mutual support for the aged is not only closely related to their physical health, but also closely related to their emotional needs, psychological endurance, social identity and other internal factors. Therefore, future studies should deeply explore the diversity of psychological needs of the elderly, and analyze their willingness to participate from a psychological perspective, especially studies on loneliness, social isolation and other issues, which may provide important theoretical support for policy formulation and optimization of pension models [2].

Moreover, the current research on mutual support for the aged is mostly focused on the individual level, and rarely involves the systematic analysis of the community and policy level. The willingness of the elderly to participate in mutual support for the aged is not only affected by individual factors, but also closely related to external factors such as community environment and policy support. Existing studies often ignore the restricting effect of community environment and social system on elderly people's participation in mutual care for the aged, especially the factors such as community service facilities, volunteer resources, and the soundness of pension policies, which are often not fully discussed. In addition to relying on the willingness of the elderly themselves, the successful implementation of the mutual care model also needs good community infrastructure support, policy guidance and financial investment from the government. In many places, due to the lack of policies or the lag of community construction, it is difficult to carry out mutual support for the elderly effectively. Therefore, future research should pay more attention to the impact of community and policy levels, and explore how to improve community construction, formulate relevant policies, strengthen social mobilization and other means to improve the enthusiasm of the elderly to participate in mutual support for the elderly.

1.3 Research Purpose

Understanding the factors influencing the willingness of the elderly to participate in mutual support for the aged in urban communities is helpful to promote the construction of the elderly service system. With the change of China's social and economic structure, the number of elderly groups has become increasingly large, and the traditional family pension model has been difficult to cope with the increasingly complex pension needs, especially in urban areas, the change of family structure and labor flow make the elderly face a more severe pension dilemma. As a community-based collective pension model, mutual care for the elderly can not only effectively make up for the shortcomings of the traditional pension model, but also give full play to the advantages of internal resources in the community to enhance the sense of social participation and belonging of the elderly. Therefore, the research on the factors affecting the elderly's participation in mutual care for the elderly can provide valuable references for policy makers, promote the implementation of relevant policies, optimize the supply of elderly care services, and improve the quality of life of the elderly.

Secondly, exploring the willingness of the elderly in urban communities to participate in mutual support for the elderly can reveal the needs and preferences of the elderly group. Due to the differences in life experience, cultural background, personality characteristics, health conditions and other aspects, each elderly person has different needs and expectations in old-age care. Understanding the willingness of the elderly to participate in the community pension can not only help us identify the main factors affecting their decision-making, but also reveal the acceptance and adaptability of different groups of elderly people to mutual pension. For example, some older people in better health may be more willing to participate in mutual aged care because they are able to play their part in it, while those in poorer health may be reluctant to participate due to physical inconvenience or psychological stress. Through in-depth analysis of these differences, the research can help design more diversified and personalized elderly care services to ensure that elderly people with different needs can receive appropriate care and support.

Thirdly, the research on the influencing factors of the elderly's participation in mutual support for the aged can provide a scientific basis for optimizing community governance and construction. Urban community, as the main place of the elderly's daily life, its community environment, the distribution of social resources and community service facilities have an important impact on the elderly's participation in mutual support for the elderly. In many places, the initiative of the elderly to participate in mutual support for the aged is often low due to the problems of lagging community construction and insufficient public service resources. Therefore, the study of the factors affecting the participation willingness of the elderly is helpful to find the weak links in the provision of elderly care services in the community, and put forward practical suggestions for improving community construction and perfecting the community service system. For example, by improving the facilities of community activity centers for the elderly, increasing volunteer resources, and promoting the construction of social support networks, the willingness of the elderly to participate in mutual support for the elderly can be effectively improved.

In addition, policy support is one of the important factors affecting the elderly's participation in mutual support for the aged. The government's policy orientation, financial input, social security system and other factors all affect the willingness of the elderly to participate to a certain extent. In some areas, the lack of relevant policy support or policy implementation is not in place, which makes it difficult to effectively promote the mutual care model. Therefore, the study on the willingness of the elderly in urban communities to participate in mutual support for the elderly and its influencing factors can provide an important theoretical basis for the government to formulate policies related to the aging society, so as to better promote the landing and implementation of relevant policies and improve the ability of the elderly to participate in society.

2 The Concept of Elderly People's Willingness to Participate in Mutual Aid Elderly Care

2.1 Mutual Aid Elderly Care

In a broad sense, mutual aid elderly care is a form of social mutual assistance and support, involving volunteers, community organizations, and social resources. This form of care emphasizes collaboration and cooperation between different groups within the community to meet the living needs of the elderly and improve their quality of life [3]. In a narrower sense, mutual aid elderly care focuses more on mutual assistance and support among the elderly themselves. It refers to the relationship of mutual support and cooperation established between elderly individuals or between elderly people and other community members, aimed at fulfilling the various needs of the elderly and enhancing their quality of life. This form of mutual aid emphasizes resource sharing and mutual assistance within the community to address the challenges and issues elderly people may face in their daily lives [4].

2.2 Willingness to Participate

Willingness to participate can be seen as an attitude, reflecting the elderly's attitudes and emotional inclinations toward mutual aid elderly care activities. This includes whether they are willing to participate, the frequency and extent of their participation, and whether they are willing to contribute time and resources to the community's elderly care efforts. Willingness to participate is influenced by individual factors, social factors, and environmental factors [5]. Individual factors include the elderly's health condition, psychological state, and personal interests. Social factors encompass family support, social support, and community culture. Environmental factors refer to the accessibility and convenience of mutual aid elderly care activities.

3 Theoretical Framework of Mutual Elderly Care Based on Active Aging

In 2002, the World Health Organization presented an official report on "Active Aging" at the World Assembly on Ageing. Active aging is defined as the process by which, in later life, individuals enhance their quality of life by maximizing opportunities for health, participation, and security [6].

First, the theory of active aging emphasizes the potential and opportunities of older adults. It posits that older individuals should be viewed as a capable and resourceful part of society, able to make positive contributions. Second, the theory underscores the role and participation of older adults in society, encouraging them to actively engage in community and social activities. In turn, their involvement enriches the community and society by providing valuable resources and experience, promoting diversity and sustainable development. Finally, the theory of active aging has had a positive impact on policy-making. Governments and social organizations can offer training and educational opportunities, develop age-friendly communities, and provide healthcare services [7]. These policy measures help create a more inclusive and supportive social environment.

In this study, the theory of active aging serves as the primary theoretical framework to explore the willingness of elderly individuals in urban communities to participate in mutual care programs and the factors influencing their participation. In addition, hierarchy of needs theory and social exchange theory are employed as supplementary frameworks to gain a more comprehensive understanding of the behavior and decision-making of older adults. hierarchy of needs theory provides insights into the fulfillment of older adults' multiple needs—ranging from physiological and safety needs to social, esteem, and self-actualization needs—helping to identify the impact of these needs on their willingness to engage in mutual care. Social exchange theory, on the other hand, assists in analyzing the interactions and resource exchanges between older adults and their communities, offering insights into their social motivations and benefits for participating in mutual care programs. Together, these supplementary theories contribute to a more in-depth exploration of the research questions, providing a holistic understanding that better supports studies on mutual elderly care in urban communities from the perspective of active aging.

4 The Willingness of Elderly Residents in Urban Communities to Participate in Mutual Care and Its Influencing Factors

This study utilized a stratified sampling method, targeting elderly individuals aged 60 and above in GuLou Districts. Gulou District is situated in the northern half of the urban area of Xuzhou City, with geographical coordinates at approximately 30°16'N and 117°11'E, as depicted in Fig. 1. A total of 500 questionnaires were distributed, and 462 valid responses were collected, resulting in an effective response rate of 92.4%.

4.1 Individual Characteristics

Table 1 presents the data on various individual characteristics and the corresponding proportions of willingness to participate in mutual elderly care.

Firstly, gender plays a significant role in the willingness of elderly individuals to participate in mutual care activities. Women typically show a higher willingness to engage in community-based mutual care programs. They are generally more inclined to express care and offer assistance, making them more likely to integrate into mutual care activities. Additionally, women are often more active in community organizations and volunteer work, further enhancing their willingness to participate.

Secondly, age is another crucial individual characteristic. As people age, they often have more time and interest in participating in mutual care programs. Moreover, age may be associated with accumulated experience and wisdom, which can be valuable in mutual care activities, attracting more elderly individuals to participate.

Educational level and cultural background also play an important role in influencing the willingness of elderly individuals to engage in mutual care [8]. Older adults with higher levels of education and richer cultural backgrounds are more likely to understand the importance of mutual care and be more willing to participate in related activities.

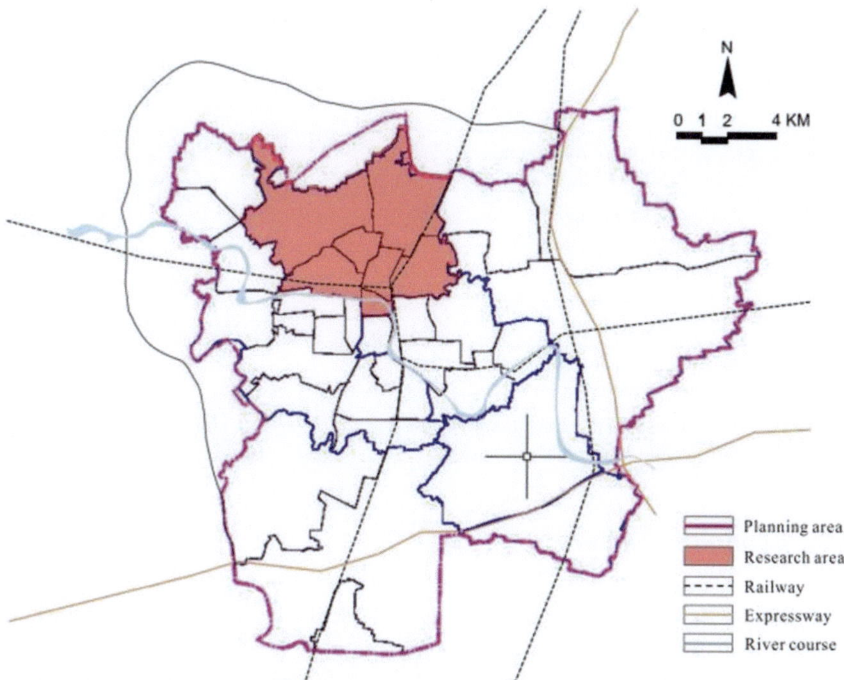

Fig. 1 Geographical location of Gulou Distric (Self-drawn)

Table 1 Statistical summary of individual characteristics (Self-drawn)

Individual characteristic		Sample size (People)	Proportion willing to participate in mutual care (%)
Gender	Male	232	68.1
	Female	230	75.2
Age	60–69 years	168	72.6
	70–79 years	178	68.5
	80 years and above	116	61.2
Education level	Primary school or below	98	59.2
	Middle school	142	64.1
	Middle school	222	71.8
Marital status	Married	334	71.0
	Unmarried or widowed	128	65.6
Number of children	No children	56	53.6
	1–2 children	238	68.9
	3 or more children	238	71.4

Table 2 Statistical summary of economic conditions (Self-drawn)

Economic factor		Sample size (People)	Proportion willing to participate in mutual care (%)
Income level	High monthly income (>5000 RMB)	128	78.9
	Moderate monthly income (3000–5000 RMB)	186	72.0
	Low monthly income (<3000 RMB)	148	60.8
Income stability	Stable income (pension/retirement funds)	238	76.5
	Unstable income (social welfare/personal savings)	224	54.3

4.2 Economic Factors

Table 2 presents the proportions of elderly individuals' willingness to participate in mutual care based on different economic conditions.

On one hand, monthly income is a crucial economic factor influencing elderly individuals' willingness to participate in mutual care activities. Relatively higher monthly income makes it easier for elderly individuals to engage in mutual care, as they can more comfortably cover associated costs, such as membership fees and travel expenses.

On the other hand, income stability is also an important factor affecting the willingness to participate in mutual care. For instance, elderly individuals who rely on pensions or retirement funds may have a more stable economic base and are thus more likely to engage in mutual care activities. In contrast, those dependent on social welfare or other unstable sources of income may be more concerned about economic instability and, as a result, adopt a more cautious attitude towards mutual care involvement.

4.3 Health Status Factors

Table 3 presents the proportions of elderly individuals' willingness to participate in mutual care based on different health conditions.

In terms of physical health, elderly individuals with better physical fitness are generally more likely to actively participate in mutual care activities. This is because good health provides them with greater freedom of movement, enabling them to engage more freely in various community activities.

Table 3 Statistical summary of health status (Self-drawn)

Health status factor		Sample size (People)	Proportion willing to participate in mutual care (%)
Physical health	Good	276	78.3
	Fair	132	67.4
	Poor	54	58.2
Mental health	Positive and optimistic	320	76.8
	Negative or anxious	142	62.7
Self-care ability	Good	198	75.3
	Fair	182	68.1
	Poor	36	54.6
Chronic diseases	Few	268	74.2
	Many	194	64.4

Mental health is equally crucial for the willingness of elderly individuals to participate in mutual care. Those who maintain a positive and optimistic attitude are more likely to seek social opportunities actively. Conversely, elderly individuals with poor mental health, affected by negative emotions such as loneliness and anxiety, may adopt a more passive attitude towards mutual care.

Self-care ability is another important indicator of an elderly person's health status. Elderly individuals who are able to care for themselves are typically more willing to use mutual care methods to maintain their independence and autonomy.

Additionally, elderly individuals with fewer chronic diseases tend to have better overall health and are thus more likely to engage actively in mutual care activities [9]. They can collaborate with others in mutual care programs for health management and disease prevention, thereby improving their quality of life.

4.4 Family Structure Factors

Table 4 presents the proportions of elderly individuals' willingness to participate in mutual care based on different family structures.

Family structure, as an intrinsic organizational form of the basic social unit, significantly impacts elderly individuals' living conditions and their willingness to participate in mutual care. Specifically, elderly individuals with fewer children often face relatively limited internal family support, making them more inclined to seek external assistance and thus more willing to engage in community mutual care. Similarly, those not living with their children are more likely to turn to community mutual care to compensate for the lack of family support.

Moreover, elderly individuals who receive infrequent visits from their children or have poor relationships with them often experience a deficiency in familial

Table 4 Statistical summary of family structure (Self-drawn)

Family structure factor		Sample size (People)	Proportion willing to participate in mutual care (%)
Number of children	Few (<2 children)	86	76.7
	Moderate (2–3 children)	218	71.1
	Many (>3 children)	158	68.9
Living with children	Living with children	288	70.5
	Not living with children	174	74.1
Frequency of visits by children	Rare visits	136	75.7
Relationship with children	Poor relationship	92	77.2

emotional support. This situation drives them to seek belonging and recognition within the community, leading to a more active participation in mutual care programs. Such changes not only respond to shifts in family structure but also reflect elderly individuals' proactive engagement in society as a key aspect of active aging.

4.5 *Social Support Factors*

Table 5 presents the proportions of elderly individuals' willingness to participate in mutual care based on different social support factors.

Social support, as an external force providing assistance and care, has a profound impact on elderly individuals' quality of life and their willingness to participate in mutual care. Firstly, the community, as a crucial environment for elderly individuals, directly influences their willingness to engage in mutual care through the support services it provides. A community that is nurturing and well-equipped with services can stimulate the enthusiasm and initiative of elderly residents, making them more inclined to participate in mutual care activities.

Secondly, government policy support and financial investment are significant factors affecting the willingness of elderly individuals to engage in mutual care. By implementing policies that are favorable to mutual elderly care, the government can create a more supportive environment for elderly individuals. This not only alleviates their financial burden but also enhances their motivation and satisfaction with mutual care participation.

Table 5 Statistical summary of social support (Self-drawn)

Social support factor		Sample size (People)	Proportion willing to participate in mutual care (%)
Community support level	Comprehensive	186	80.6
	Average	174	68.4
	Insufficient	102	59.8
Government policy support level	Adequate	230	82.2
	Average	154	71.4
	Insufficient	78	57.7
Social organization support level	Active	264	79.5
	Average	138	66.7
	Insufficient	80	55.0

Additionally, social organizations contribute by providing various forms of volunteer services, public welfare activities, and cultural exchanges, which create platforms for elderly individuals to express themselves, build relationships, and support one another. This support from social organizations enriches their spiritual and cultural lives and enhances their sense of belonging and happiness, thereby increasing their willingness to engage in mutual care activities.

5 Strategies and Recommendations for Enhancing Elderly Individuals' Willingness to Participate in Mutual Care

5.1 Provide Diverse Mutual Care Activities

Firstly, a variety of mutual care programs should encompass various social activities. Elderly individuals often find satisfaction and happiness through social interactions; therefore, communities can organize different types of social events such as gatherings, seminars, and cultural exchanges. Secondly, skills training is another crucial component of mutual care programs. By offering a range of skill training courses, communities can stimulate elderly individuals' willingness to participate. These courses might include activities such as crafting, cooking techniques, and computer applications. Finally, volunteer services are a significant aspect of mutual care programs [10]. Through volunteering, elderly individuals can give back to the community, which can also provide them with a sense of fulfillment and belonging.

5.2 Provide Economic Support and Incentives

Firstly, governments and community organizations could consider offering free or discounted membership fees to lower the participation threshold for elderly individuals. For instance, communities might establish senior clubs with reduced membership fees to attract more elderly participants. Secondly, providing transportation subsidies is another economic support measure that can enhance elderly individuals' willingness to engage in mutual care activities. Many elderly individuals may face challenges in accessing mutual care locations due to transportation difficulties or financial constraints. Governments and communities could offer transportation subsidies to assist elderly individuals with travel expenses. Additionally, governments and communities could implement reward programs to recognize and honor actively participating elderly individuals, thereby encouraging others to follow their example.

5.3 Strengthen Social Promotion and Education

On one hand, communities and governments can enhance social promotion through various methods, such as creating social media accounts to regularly publish articles, videos, and information about mutual care activities. On the other hand, providing education and training about mutual care is crucial for increasing elderly individuals' willingness to participate. Educational and training programs might cover basic concepts of mutual care, social skills, and volunteer training. Additionally, workshops and training sessions can be used to teach practical skills and knowledge, such as digital technology and social media usage [11]. Promotional and educational materials should be simple and clear, avoiding complex terminology and technical language to ensure ease of understanding.

5.4 Establish Support Systems and Social Networks

Firstly, communities can establish volunteer groups specifically for elderly individuals to provide support and guidance. These groups might consist of younger people, community residents, or experienced elderly individuals who can help newcomers understand community activities and resources. Secondly, encouraging elderly individuals to build social connections and mutual support networks is essential. Communities can regularly organize various social events, such as tea parties, discussion forums, and outdoor activities, to provide opportunities for elderly individuals to interact and engage with one another. Finally, support systems and social networks should be tailored to the needs and characteristics of elderly individuals.

Different elderly individuals face varying challenges, such as health issues or mobility limitations.

In summary, this study explores the willingness of elderly individuals in urban communities to participate in mutual care and the influencing factors, highlighting its importance within the framework of active aging. The research identifies that factors such as individual characteristics, economic conditions, health status, family structure, and social support significantly affect the willingness of elderly individuals to engage in mutual care. Consequently, recommendations are made to create diverse mutual care opportunities, provide economic support and incentives, strengthen social promotion and education, and establish support systems and social networks. These strategies are expected to encourage active participation among elderly individuals in urban communities, advance the development of social care services, achieve the goals of active aging, and build a more inclusive and supportive social environment. We encourage continued collaboration among government, community organizations, and academia in this field to enhance the quality of life and social participation opportunities for elderly individuals.

6 Conclusion

From the perspective of active aging, the willingness of the elderly in urban communities to participate in mutual support for the aged is affected by many factors, including health status, economic conditions, social support network and psychological needs at the individual level, as well as community construction, policy support and social culture in the external environment. Through the systematic study of the factors affecting the willingness of the elderly to participate in mutual support for the aged in urban communities, the following main conclusions can be drawn:

The health status of the elderly is an important factor affecting their willingness to participate in mutual support for the aged. Older people in better health are more likely to actively participate in mutual care, and they can play a role in mutual care, while also receiving psychological and social support. Conversely, older people with poorer health may be less likely to participate in mutual care activities due to physical inconveniences or doubts about their abilities.

The economic conditions of the elderly also have a significant impact on their willingness to participate in mutual support for the elderly. Elderly people with poor economic status may have reservations about mutual care for the elderly because of worries about the financial burden. While sharing resources and helping each other is a core concept, some older people are still concerned about the financial burden it may cause, especially in terms of medical and nursing services.

Social support network is another important factor affecting the participation of elderly people in mutual care. Older people with strong social support, such as good relationships with relatives, friends, and neighbors, are often more willing to participate in mutual care for the elderly because they feel more belonging and security in the community. Conversely, older people who lack social support may refuse to participate in supportive care because of feelings of loneliness and distrust.

The improvement of community construction and service facilities is also an important external factor affecting the willingness of the elderly to participate. In many urban communities, activity centers, public service facilities and volunteer service networks for the elderly are still inadequate, resulting in inadequate resource support and participation opportunities for the elderly. The research shows that the community environment with complete facilities and rich service functions can stimulate the enthusiasm of the elderly to participate in mutual support for the aged. For example, if there is an activity center for the elderly, health consulting services, volunteer assistance, etc., in the community, it can provide convenience and protection for the elderly, so that they feel the care and support of the community, and thus improve the willingness to participate.

Policy support is one of the key factors to promote the development of mutual aged care model. When formulating relevant policies, the government should provide more financial subsidies and social security for the elderly, and strengthen the publicity and guidance of mutual support for the elderly. By setting up special policy funds and providing tax incentives, the burden on communities and families can be reduced and more elderly people can be encouraged to participate. In addition, when formulating policies related to the aging society, the government should also pay attention to the psychological needs of the elderly, promote the social cognitive change of the elderly group, and encourage the social respect and support for the elderly group, so as to form a more inclusive and caring social atmosphere.

Acknowledgements This study is supported by the project grant from Jiangsu Provincial Social Science Fund (Project No. 22SHD003).

References:

1. Wu S, Xiang W (2023) The social construction of urban mutual care services: theoretical perspectives, multiple representations, and practical choices. Acad Exch 01:151–165
2. Liao H (2022) Willingness to participate in mutual care and its influencing factors among elderly in urban communities: an empirical analysis based on survey data from Changsha. Aging Sci Res 10(10):27–40
3. Gan M, Lian J (2024) Exploring long-term mechanisms for mutual care in rural communities under the concept of common prosperity. J Hanshan Norm Univ 45(01):1–9
4. Dou Y (2023) Small-scale renovation: spatial support and governance paths for mutual care in urban communities. J Qiqihar Univ (Philos Soc Sci Ed) 12:82–86
5. Yi Z, Sui Y, Sun J (2023) Gender differences in willingness, preferred forms, and service content preferences for mutual care. Soc Constr 10(06):125–138
6. Fan Y, Zhao L, Zeng H et al (2023) Research progress on elderly mutual care under the context of active aging. J Nurs Sci 38(21):122–125
7. Li Z, Wang J (2023) Research on high-quality development paths for the urban community "Time Bank" mutual care model: a social work analysis perspective. Theor Obs 07:87–91
8. He S, Li B (2022) Historical evolution and advancement paths of community mutual care under urban transformation. Hunan Soc Sci 04:120–126

9. Dou Y (2020) Research on the development path of the "Pre-stage" of neighborhood mutual care for the elderly in urban communities from the perspective of community governance. J Yunnan Univ Natl (Philos Soc Sci Ed) 37(05):83–87
10. Yang H, Zhang G (2020) Study on mutual support for the aged in urban community. J Hubei Polytech Univ (Humlties Soc Sci) 37(03):48–52
11. Zhou RJ, Hong Q, Li XX et al (2020) Factors associated with intention of mutual help in elderly people with home care in urban communities. Chin Gen Pract 23(29):3747–3750, 3762

MIX
Papier aus verantwortungsvollen Quellen
Paper from responsible sources
FSC® C105338

If you have any concerns about our products,
you can contact us on
ProductSafety@springernature.com

In case Publisher is established outside the EU,
the EU authorized representative is:
Springer Nature Customer Service Center GmbH
Europaplatz 3, 69115 Heidelberg, Germany

Printed by Libri Plureos GmbH
in Hamburg, Germany